Fabrication Techniques and Emerging Applications of Hydrogels

Onome Ejeromedoghene
https://orcid.org/0000-0002-5487-3267
Soochow University, China

Guodong Fu
Southeast University, China

Olayinka Kehinde
https://orcid.org/0000-0002-2050-0948
Nile University, Nigeria

IGI Global Scientific Publishing
Publishing Tomorrow's Research Today

Vice President of Editorial	Melissa Wagner
Director of Acquisitions	Mikaela Felty
Director of Book Development	Jocelynn Hessler
Production Manager	Mike Brehm
Cover Design	Jose Rosado

Published in the United States of America by
IGI Global Scientific Publishing
701 East Chocolate Avenue
Hershey, PA, 17033, USA
Tel: 717-533-8845 | Fax: 717-533-7115
Website: https://www.igi-global.com E-mail: cust@igi-global.com

Library of Congress Cataloging-in-Publication Data

LCCN: 2025013490 (CIP Data Pending)
ISBN13: 979-8-3373-2337-4
Isbn13Softcover: 979-8-3373-2338-1
EISBN13: 979-8-3373-2339-8

British Cataloguing in Publication Data
A Cataloguing in Publication record for this book is available from the British Library.

This book, "Fabrication Techniques and Emerging Applications of Hydrogels" is dedicated to the pioneers of polymer science, whose foundational curiosity unveiled the potential of hydrogels. We also dedicate this book to the current community of scientists and engineers, whose collaborative spirit is pushing the boundaries of fabrication and application of smart and functional materials, as well as the next generation of innovators, whose future discoveries will undoubtedly transform these versatile materials into solutions for the world's most pressing challenges.

Table of Contents

Detailed Table of Contents

Hydrogels are versatile polymer networks whose mesh size and responsiveness can be engineered for biomedicine, environmental systems, and soft electronics. This chapter emphasizes crosslinking density as a unifying parameter governing swelling, transport, and mechanical performance. We synthesize the field from early developments to current design strategies, organizing hydrogels by source, network formation, and environmental responsiveness. Particular attention is given to stimuli-responsive systems, including temperature, pH, light, enzymatic cues, and emerging magnetic, electrical, and redox triggers, which enable controlled release, actuation, and self-healing. Advances in nanocomposites, dynamic covalent chemistry, and bioinks are highlighted as drivers of multifunctionality. Remaining challenges include balancing toughness with biocompatibility, achieving predictable degradation, scaling fabrication, and ensuring reliable integration with sensors and biological tissues. Looking forward, opportunities lie in sustainable green synthesis, 3D biofabrication, and next-generation smart hydrogels for theranostics, regenerative medicine, and wearable bioelectronics.

Hydrogels have emerged as an adaptable class of biomaterials because of their significant applications in the biomedical field, environmental remediation, agriculture, cosmetics, and food industry. Over the years, numerous methods have been reported to have control over the properties of hydrogels. This section follows

Masooma Nazar, King Fahd University of Petroleum and Minerals, Dhahran, Saudi Arabia
Ambavaram Vijaya Bhaskar Reddy, BMS Institute of Technology and Management, India
Aqeel Ahmad, King Fahd University of Petroleum and Minerals, Dhahran, Saudi Arabia
Syed Muhammad Shakil Hussain, King Fahd University of Petroleum and Minerals, Dhahran, Saudi Arabia
Othman Charles S. Al-Hamouz, King Fahd University of Petroleum and Minerals, Dhahran, Saudi Arabia
Muhammad Moniruzzaman, New York City College of Technology, City University of New York, USA

Graphene-based hydrogels comprising graphene, graphene oxide or reduced graphene oxide form three-dimensional polymeric networks with high strength, tunable permeability and large surface area. These properties make them suitable for applications in biomedical, soft electronics and water purification. Characterization of GBHs is vital to understand their structure property relationships. FT-IR and Raman spectroscopy reveals the chemical composition, XRD assesses the crystallinity, and SEM/TEM provides morphological insights. On other hand, TGA and DSC evaluate thermal properties, while BET analysis determines surface area, structural connectivity and pore size distribution. This chapter provides a comprehensive overview on GBHs, highlighting their multifunctionality across biomedical, environmental and industrial domains. It emphasizes the structure-property-performance relationship and integrates advanced characterization techniques to align with real-world applications.

Muzammal Hussain, Southeast University, China
Abbas Faheem, University of Wah, Pakistan
Arooba binte Amir, Government College University, Faisalabad, Pakistan
Munir Hussain, Zhejiang Sci-Tech University, China
Aqib Mehmood, University of Wah, Pakistan

Hydrogels have emerged as a focal material in modern science, finding applications across various fields, including medical treatments, industrial processes, and optical devices. Their behavior varies significantly depending on the surrounding medium, making them versatile for a range of applications. In this chapter, we discuss key applications and properties of hydrogels, such as self-crosslinking wound dressings,

The biocompatibility, high water retention, and controlled-release characteristics of hydrogels three-dimensional networks of hydrophilic polymers have made them attractive carriers for sunscreen compositions. The hydrogel-based delivery systems can help to address the instability, skin irritation, and environmental issues that traditional sunscreens frequently face. The physical, chemical, radiation-induced, enzyme-assisted methods, electrospinning, and cryogelation are some of the manufacturing methods for hydrogel-based sunscreens that are examined in this review. Incorporating the hydrogels into sunscreen formulations improves skin hydration, increases the effectiveness of UV protection, and prolongs the stability of active components, such as chemical (oxybenzone, avobenzone) and physical (ZnO, TiO_2, iron oxide) UV filters. Moreover, hydrogels contain organic antioxidants like betalains and anthocyanins, which prevents photoaging and oxidative stress. Hybrid formulations maximize these bioactive chemicals' bioavailability and photostability including hydrogel sunscreens loaded with nanoliposomes. The development of biodegradable polymers, smart hydrogels, and nanotechnology integration promises for the creation of sunscreens that are both efficient and environmentally benign in the future. By outlining their formulation techniques, drawbacks, and possible applications in the cosmeceutical sector, this paper states the potential of hydrogels as multipurpose sunscreen carriers.

Insect hydrogels like other hydrogels are versatile, three-dimensional polymeric networks with significant water-retention capacity, making them valuable in biomedical, agricultural, and environmental applications. This book review explores

The removal of water and wastewater contaminants by adsorptive means has been shown to be very promising. Amongst different adsorbent materials, hydrogels have been found to be a good tool for eliminating both inorganic and organic toxic chemical substances from water and wastewater, because of swelling characteristics, bio-compatibility, non-toxicity, re-usability and high functionalities. From previous studies, it has been shown that hydrogels can be re-used and regenerated for up to 5 times, with only about 10-15% adsorption capacity loss. This chapter focuses on the viability of hydrogels and its modified forms for the treatment of water and wastewater that contains micro-pollutants like pharmaceuticals, pesticides, dyes and heavy metals. Specifically, the mechanisms of interaction/uptake, adsorption isotherm modelling, adsorption capacities and regeneration potentials of (modified) hydrogels for these selected water contaminants shall be discussed.

Catalytic hydrogels represent a groundbreaking class of materials that integrate the unique properties of hydrogels with catalytic functionality, offering transformative solutions across environmental remediation, biotechnology, and industrial processes. This chapter highlights the synthesis and utilization of hydrogel-nanoparticle composites, demonstrating their efficacy in pollutant degradation, energy conversion, and chemical synthesis. Photocatalytic hydrogels, particularly those incorporating TiO_2 and ZnO, exhibit remarkable efficiency in degrading organic contaminants and reducing heavy metals like CrVI. Electrocatalytic hydrogels are explored for their role in oxygen reduction reactions (ORR), hydrogen evolution reactions (HER), and

Recent advancements in printing and additive manufacturing have positioned three-dimensional (3D) and four-dimensional (4D) printing at the forefront of cutting-edge research. These technologies have revolutionized material fabrication, offering significant advantages across various industries, including healthcare, aerospace, and bioengineering. This chapter reviews recent developments in the use of hydrogels for 3D and 4D printing, their role in additive manufacturing, their advantages and limitations in 3D printing, and the latest innovations in 4D printing technologies that leverage hydrogels as stimuli-responsive materials are reviewed.

Hydrogel composites, known for their mechano-tunability, biocompatibility, and stimuli-sensitivity, are revolutionizing biomedical, environmental, and industrial fields. This study explores theoretical modeling (molecular dynamics (MD), coarse-grained simulations, finite element analysis (FEA)) and its synergy with advanced 3D printing (stereolithography, extrusion) to decipher the nuances of hydrogel composites design principles. MD unravels atomic-scale mechanics, designing fracture-resistant networks, whereas FEA engineers' macroscale mechanics towards tissue scaffolds or body sensors. Machine learning speeds up material design, predicting optimal mixtures for drug release or pollutant capture. 3D-printed hydrogels facilitate precision structures: tailored wound healing, enzymes-based biosensors, and aerogels that capture heavy metals. Scalability challenges, as well as fragility, are met with hybrid modeling-experiment approaches, confirmed with

Foreword

Hydrogels, a type of "soft material" composed of hydrophilic polymer networks, have attracted extensive attention from researchers in fields such as materials science, chemistry, biomedical engineering, and even environmental science since their first report in the 1960s. This is due to their unique biocompatibility, tunable physical and chemical properties, and aqueous environment similar to biological tissues. From their initial applications in contact lenses and wound dressings to their current roles as controlled drug release carriers, tissue engineering scaffolds, flexible electronic devices, and even environmental adsorbents, hydrogels have made their mark in every corner of modern technology. We are now at a critical juncture where hydrogel technology is transitioning from the laboratory to industrialization and from single functions to intelligent integration.

Against this backdrop, we deeply feel the necessity to compile a work that comprehensively reflects the latest advancements and future trends in the field of hydrogels. This book, *Fabrication Techniques and Emerging Applications of Hydrogels*, was born in response to this need. It is not merely a summary of traditional knowledge but aims to build a bridge connecting the "basic research" and "frontier applications" of hydrogels, systematically presenting a complete knowledge system—from classical preparation methods to advanced processing techniques, and from fundamental performance characterization to cross-disciplinary innovative applications.

The content of this book has been meticulously organized to ensure a clear structure and logical coherence:

The introductory section of this book lays a solid foundation for readers, systematically introducing the basic concepts, development history, classification, and core functional properties of hydrogels, thereby setting the stage for in-depth discussions that follow.

The book also has chapters that provide an in-depth analysis of the core preparation and processing technologies for hydrogels. The book not only covers traditional synthesis methods but also places significant emphasis on cutting-edge systems such as superporous hydrogels and stimuli-responsive smart hydrogels. A dedicated chapter explores 3D/4D printing and additive manufacturing technologies, which represent the future direction of manufacturing, showcasing the leap of hydrogels from "manufacturing" to "intelligent manufacturing."

Furthermore, other chapters focus on the performance characterization and theoretical simulation of hydrogels. These sections elaborate on characterization strategies for nanocomposite hydrogels containing materials like graphene, analyze their electrochemical properties, and introduce theoretical and three-dimensional simulation methods. These approaches provide powerful tools for understanding and designing hydrogel materials at the microscopic level.

The book culminates in a panoramic view of hydrogels' exciting emerging applications, representing its essential contribution to the field. The exploration begins with biomedical applications—such as drug delivery, tissue engineering, and biosensors—and extends to environmental remediation, including the use of superabsorbent hydrogels for pollution treatment. It further covers applications in catalysis, innovative insect-derived hydrogels, and everyday consumer products like hydrogel-based sunscreens. This broad scope fully demonstrates the immense, interdisciplinary, and multi-dimensional potential of hydrogel technology.

The unique value of this book lies in its broad coverage and forward-looking perspective. The editors have not only invited experts and scholars from renowned institutions to contribute to this masterpiece, ensuring the depth and authority of the content, but have also included cutting-edge topics rarely found in similar publications, such as insect-derived hydrogels, 4D printing, and theoretical simulations. We firmly believe that this systematic yet innovative structure will meet the needs of readers from diverse backgrounds.

This book is suitable for all researchers, engineers, university faculty, and graduate students interested in advanced polymer materials and soft matter science. This book is an essential resource for scholars, industry professionals, and students alike, offering valuable insights to advance fundamental research, drive technological innovation, and inspire future explorations in hydrogel science.

Finally, sincere gratitude is extended to all contributing authors for sharing their valuable knowledge and expertise. The diligent efforts of the editing and publishing teams are also gratefully acknowledged. It is hoped that *Fabrication Techniques and Emerging Applications of Hydrogels* will serve as a valuable guide and a source of inspiration for future contributions to this dynamic field.

Happy Reading!

Yuanfeng Wu
School of Chemistry and Chemical Engineering, Henan Polytechnic University, China

Preface

The evolution of materials science has continually shaped the course of modern technology, medicine, and environmental management. Functional materials possess intrinsic properties that enable them to respond to external stimuli such as heat, light, mechanical stress and specific molecular orientations. These unique properties have led to a number of unique applications in the creation of sustainable technologies. Among the emerging classes of smart and functional materials, hydrogels have received tremendous attention due to their remarkable physicochemical properties, structural versatility, and responsiveness to external stimuli. The writing of this book, *"Fabrication Techniques and Emerging Applications of Hydrogels"* is intended to seek to provide a far-reaching and multidisciplinary understanding of hydrogel science by connecting the gaps between hypothesis, synthesis, and real-world applications.

The structure and composition of hydrogel can be finely tuned to achieve desired properties such as mechanical strength, porosity, biodegradability, and stimuli-responsiveness. These features make them indispensable in numerous domains, including biomedicine, agriculture, environmental remediation, pharmaceuticals, food technology, and soft robotics. The increasing awareness that hydrogels constitute a crucial frontier in sustainable material innovation served as the drive for this book, as it attempts to show how hydrogel technology can make a significant contribution to the current world by examining its chemical, physical, and functional characteristics in an integrated manner.

ORGANIZATION OF THE BOOK

Chapter 1: Introduction to Hydrogels

Hydrogels are 3D polymer networks that absorb large fluid volumes, mimicking soft tissues. Their hydrophilicity comes from -OH, -COOH, or $-NH_2$ groups, enabling tunable swelling. Evolving from simple absorbents to stimuli-responsive systems, hydrogels are classified by origin (natural/synthetic/hybrid), crosslinking, and environmental sensitivity. These properties make them ideal for biomedicine, drug delivery, tissue engineering, and flexible electronics due to their biocompatibility and adaptability. Current research focuses on smart hydrogels and nanotechnology integration, driving innovations in science and technology.

Chapter 2: Hydrogels Processing Techniques

Hydrogels have emerged as an adaptable class of biomaterials because of their significant applications in the biomedical field, environmental remediation, agriculture, cosmetics, and food industry. Over the years, numerous methods have been reported to have control over the properties of hydrogels. This section follows the physical and chemical processes of crosslinking polymers, hybrid strategies, and polymerization techniques that have revolutionized the processing of hydrogels in numerous ways. This chapter also highlights the role of constituent molecules, including monomers, cross-linkers, and nanocomposite materials in synthesizing hydrogels for specific applications, such as stimuli sensitivity, swellability, biodegradability, biocompatibility, and mechanical strength. The biocompatible and stability challenges have also been addressed as a significant hurdle for their scalability.

Chapter 3: Solvent and Solventless Hydrogels

Conventional organic solvents have been widely employed in hydrogel fabrication, but their environmental and health concerns have motivated the search for alternative solvent systems. Ionic liquids have emerged as promising substitutes, offering low volatility, high thermal stability, and improved biocompatibility. Deep eutectic solvents, formed by combining components with unique solvent properties, exhibit low toxicity, biodegradability, and potential for sustainable manufacturing. Surfactants play crucial role in emulsion-based methods, influencing hydrogel structure and properties. Solventless hydrogels, eliminating the need for solvents, provide simplified fabrication processes, reduced environmental impact, and improved biocompatibility. This chapter provides an overview of these solvent systems, discussing their advantages, limitations, and applications in hydrogel synthesis. Understanding the diverse solvent options enables the development of advanced hydrogel materials tailored for various fields, including biomedicine, tissue engineering, and other industrial applications.

Chapter 4: Characterization Strategies for Graphene Containing Hydrogel Systems: Techniques for Evaluating Structural Properties

Graphene-based hydrogels comprising graphene, graphene oxide or reduced graphene oxide form three-dimensional polymeric networks with high strength, tunable permeability and large surface area. These properties make them suitable for applications in biomedical, soft electronics and water purification. Characterization of GBHs is vital to understand their structure property relationships. FT-IR and Raman spectroscopy reveals the chemical composition, XRD assesses the crystallinity, and SEM/TEM provides morphological insights. On other hand, TGA and DSC evaluate thermal properties, while BET analysis determines surface area, structural connectivity and pore size distribution. This chapter provides a comprehensive overview on GBHs, highlighting their multifunctionality across biomedical, environmental and industrial domains. It emphasizes the structure-property-performance relationship and integrates advanced characterization techniques to align with real-world applications.

Chapter 5: Functional Properties of Hydrogels

Hydrogels have emerged as a focal material in modern science, finding applications across various fields, including medical treatments, industrial processes, and optical devices. Their behavior varies significantly depending on the surrounding medium, making them versatile for a range of applications. In this chapter, we discuss key applications and properties of hydrogels, such as self-crosslinking wound dressings, diabetic wound care, and natural polymer-based formulations. Additionally, we explore their self-adaptive wound healing capabilities and their potential in biomedical advancements. Hydrogels are increasingly recognized as promising materials, particularly in the medical field, due to their biocompatibility, tunable properties, and multifunctionality. Given their adaptability and effectiveness, they hold significant potential for future advancements across various disciplines, making them a crucial material for ongoing research and development.

Chapter 6: The Electrochemistry and Electrochemical Properties of Hydrogel-Based Composites

Hydrogel-based composites possess unique electrochemical properties that have substantially triggered the development of efficient, cost-effective, and sustainable energy storage systems. As electrochemical storage technologies like batteries and supercapacitors become central to portable power solutions, there is a growing focus on optimizing performance parameters such as specific energy and power density. Ongoing efforts aim to design advanced materials that outperform existing systems while addressing environmental concerns associated with electronic waste. Key features of hydrogel-based composites—ranging from long-term stability, high proton conductivity, to enhanced energy storage via redox and electrostatic mechanisms—stem from the use of sustainable biomaterials like chitosan and activated carbon. Their flexibility, stretchability, and biocompatibility also make them ideal for wearable electronics and biomedical applications. This chapter explores various hydrogel-based composites, their electrochemical behavior, and the mechanisms driving their performances.

Chapter 7: Super Porous Hydrogels

To overcome the hindrance in the rate of swelling and mechanical strength of hydrogels for different targeted applications that require rapid fluid uptake. Superporous hydrogels have emerged as a new class of materials to hinder these challenges. This chapter will explore the fabrication of SPHs, and various synthetic techniques such as porosigen, phase separation, cross-linking, and gas-blowing methods. The main features of SPHs are high porosity, rapid rate of swelling, higher adsorption capacity, stimuli-responsive behavior to the external environment, biocompatibility, high surface area, and enhanced mass transfer. The SPHs are categorized into three generations: conventional SPHs, SPH composites, and SPH hybrids, with improved mechanical properties and functionalities across various generations. Various applications of SPHs in biotechnology and biomedical engineering are discussed. SPHs have a lot of potential to revolutionize different fields due to their unique properties and tailored functionalities.

Chapter 8: Stimuli-Responsive Intelligent Hydrogels

Recent progress in materials science has led to the emergence of stimuli-responsive intelligent hydrogels, which have gained significant interest due to their unique properties and wide-ranging applications. These hydrogels undergo reversible changes in physical or chemical properties when exposed to external stimuli like temperature, pH, light, electric fields, and biomolecules. Their potential spans biomedical engineering, drug delivery, soft robotics, and environmental sensing. This paper explores their synthesis, properties, and applications. Various stimuli can trigger hydrogel responsiveness, with the choice depending on the application. We discuss methods for incorporating stimuli-responsive elements, including physical entrapment, covalent attachment, and self-assembly. Fabrication techniques such as physical and chemical crosslinking are examined, along with the importance of structure-property relationships in designing precise, responsive hydrogels.

Chapter 9: Innovative Techniques in Hydrogel Based Sunscreen Fabrication and its Application in Healthcare

The biocompatibility, high water retention, and controlled-release characteristics of hydrogels three-dimensional networks of hydrophilic polymers have made attractive carriers for sunscreen compositions.The Physical, chemical, radiation-induced, enzyme-assisted, electrospinning, and cryogelation are some of the manufacturing methods for hydrogel-based sunscreens that are examined in this review. Incorporating the hydrogels into sunscreen formulations improves skin hydration, increases the effectiveness of UV protection, and prolongs the stability of active components, such as chemical (oxybenzone, avobenzone) and physical (ZnO, TiO2, Iron oxide) UV filters. Moreover, hydrogels contains organic antioxidants like betalains and anthocyanins, which prevents from photoaging and oxidative stress. Hybrid formulations maximizes these bioactive chemicals' bioavailability and photostability including hydrogel sunscreens loaded with nanoliposomes.

Chapter 10: Synthesis and Application of Hydrogels of Insect Origin

Insect hydrogels like other hydrogels are versatile, three-dimensional polymeric networks with significant water-retention capacity, making them valuable in biomedical, agricultural, and environmental applications. This book review explores the techniques for synthesis, and multifaceted applications of insect-derived hydrogels, emphasizing their sustainability, biocompatibility, and functionality. Insects such as darkling beetles/super worm (Zophobas morio), black soldier fly (Hermetia illucens), yellow mealworm (Tenebrio molitor), large brown cicada (Graptopsaltria nigrofuscata), wingless flightless cockroach (Eupolyphaga sinensis), silk worm (Antheraea assama), silk worm (Bombyx mori), and many others serve as point source for hydrogel production. Chitosan, silk fibroin, and hemolymph proteins extracted are transformed into hydrogels via physical, ionic, enzymatic, and chemical crosslinking into valid hydrogels for diverse applications in novel food and feed, drivers for drug delivery, pesticide delivery, drinkers for insects, tissue, skin, bone and organ regeneration and more.

Chapter 11: Biomedical Prospects of Hydrogels

Hydrogels have revolutionized biomedical applications, evolving from simple water-absorbing materials to multifunctional drug delivery systems. This section highlights their role in drug delivery, wound care, and therapeutic interventions. Their 3D network enables localized and sustained drug release, enhancing efficacy while reducing side effects. pH-sensitive hydrogels facilitate targeted gastrointestinal delivery, while thermosensitive types are effective for nasal and transdermal administration. In ophthalmology, hydrogels improve drug retention, and their ability to bypass the blood-brain barrier makes them a key research focus. Incorporating nanoparticles like silver, zinc oxide, or copper oxide enhances antimicrobial properties. Despite their potential, challenges such as scalability, stability, and precise drug release hinder clinical adaptation. However, advancements in nanoscale hydrogels, smart materials, and innovative fabrication techniques like 3D printing and microfluidics can improve drug loading and controlled release, driving future biomedical breakthroughs.

Chapter 12: Super-absorbent Hydrogels for Environmental Remediation

The removal of water and wastewater contaminants by adsorptive means has been shown to be very promising. Amongst different adsorbent materials, hydrogels have been found to be a good tool for eliminating both inorganic and organic toxic chemical substances from water and wastewater, because of swelling characteristics, bio-compatibility, non-toxicity, re-usability and high functionalities. From previous studies, it has been shown that hydrogels can be re-used and regenerated for up to 5 times, with only about 10-15% adsorption capacity loss. This chapter focuses on the viability of hydrogels and its modified forms for the treatment of water and wastewater that contains micro-pollutants like pharmaceuticals, pesticides, dyes and heavy metals. Specifically, the mechanisms of interaction/uptake, adsorption isotherm modelling, adsorption capacities and regeneration potentials of (modified) hydrogels for these selected water contaminants shall be discussed.

Chapter 13: Catalytic Hydrogels and Hydrogel-Based Composites: Applications and Innovations

Catalytic hydrogels represent a groundbreaking class of materials that integrate the unique properties of hydrogels with catalytic functionality, offering transformative solutions across environmental remediation, biotechnology, and industrial processes. This chapter highlights the synthesis and utilization of hydrogel-nanoparticle composites, demonstrating their efficacy in pollutant degradation, energy conversion, and chemical synthesis. Photocatalytic hydrogels, particularly those incorporating TiO2 and ZnO, exhibit remarkable efficiency in degrading organic contaminants and reducing heavy metals like CrVI. Electrocatalytic hydrogels are explored for their role in oxygen reduction reactions (ORR), hydrogen evolution reactions (HER), and nitrogen reduction reactions (NRR). Additionally, heterogeneous catalyst hydrogels, including Ag-based and Fe3O4-hydrogel systems, showcase high reusability and stability in treating organic and inorganic pollutants.

Chapter 14: Hydrogel Based Sensor Devices

Sensor devices integrate physical and digital systems for real-time monitoring in healthcare, agriculture, and environmental sectors. Meeting rising demands requires enhanced sensitivity, durability, and compatibility. Hydrogels—hydrophilic, biocompatible, stimuli-responsive, and selective—serve as adaptable sensing platforms. This chapter analyzes structural properties governing hydrogel sensing efficacy. It explores optical, electrochemical, and mechanical mechanisms for transducing stimuli like pH, temperature, and analytes into measurable signals. Fabrication methods, including photopolymerization and 3D printing, enable tailored designs for wearable health monitors, precision agriculture, and pollution detection. Correlating hydrogel properties to performance establishes design principles for optimizing sensitivity and selectivity. Challenges such as spatial inhomogeneity, stability under variable conditions, and biocompatibility validation are critically addressed. This chapter highlights hydrogels' transformative role in next-generation sensors.

Chapter 15: Hydrogel-Based Bioelectronics

Hydrogel-based bioelectronics represent a groundbreaking advancement in biomedical engineering, merging the biocompatibility of hydrogels with the functional capabilities of electronic devices. These materials offer superior flexibility, conductivity, and biocompatibility, making them ideal for applications in biosensors, neural interfaces, cardiac monitoring, and drug delivery systems. This chapter explores the fundamental properties of hydrogels, highlighting their advantages over traditional bioelectronic materials. Innovations in soft robotics and prosthetics also leverage hydrogel properties to improve artificial muscle function and sensory feedback. Despite these advancements, challenges remain, particularly regarding material stability and long-term biocompatibility. Addressing these concerns through scalable manufacturing and interdisciplinary research will be crucial for use in clinical applications. Future perspectives suggest the integration of artificial intelligence and machine learning to enhance bioelectronic performance for neural and cardiac applications.

Chapter 16: The Biotechnology of Hydrogel-based Biocomposites

Hydrogels are a unique type of polymer that are hydrophilic. This chapter explores the applications of hydrogels in tissue engineering, skin regeneration, soft robots and artificial muscle production. For tissue engineering, there is a need to maintain proper hardening and desirable strength of hydrogels at physiological temperatures under biological conditions for proper formation of scaffolds with complex shapes. Hydrogels promote angiogenesis, cell proliferation and wound closure effectively. Hydrogels have been used in the construction of actuators, sensors, communicators, power sources, and computational circuits for soft robots. Hydrogels are also responsive to ambient stimuli, increasing their sensitivity; they are quite flexible, with the potential of enduring large deformations, making them good substrates for innovative research in artificial muscle generation. While hydrogels offer significant advantages in wound healing and tissue engineering, challenges such as optimizing mechanical strength, biocompatibility, and production scalability remain.

Chapter 17: Hydrogels for 3D/4D Printing and Additive Manufacturing

Recent advancements in printing and additive manufacturing have positioned three-dimensional (3D) and four-dimensional (4D) printing at the forefront of cutting-edge research. These technologies have revolutionized material fabrication, offering significant advantages across various industries, including healthcare, aerospace, and bioengineering. This chapter reviews recent developments in the use of hydrogels for 3D and 4D printing, their role in additive manufacturing, their advantages and limitations in 3D printing, and the latest innovations in 4D printing technologies that leverage hydrogels as stimuli-responsive materials are reviewed.

Chapter 18: Theoretical and 3D Simulation of Hydrogel Composites

Hydrogel composites, known for their mechano-tunability, biocompatibility, and stimuli-sensitivity, are revolutionizing biomedical, environmental, and industrial fields. This study explores theoretical modeling (molecular dynamics (MD), coarse-grained simulations, finite element analysis (FEA)) and its synergy with advanced 3D printing (stereolithography, extrusion) to decipher the nuances of hydrogel composites design principles. MD unravels atomic-scale mechanics, designing fracture-resistant networks, whereas FEA engineers' macroscale mechanics towards tissue scaffolds or body sensors. Machine learning speeds up material design, predicting optimal mixtures for drug release or pollutant capture. 3D-printed hydrogels facilitate precision structures: tailored wound healing, enzymes-based biosensors, and aerogels that capture heavy metals. Scalability challenges, as well as fragility, are met with hybrid modeling-experiment approaches, confirmed with AFM and rheometry. AI-based multi-physics models are new trends toward smart hydrogels as well as greener bioinks.

Each thematic area in the book is treated with attention to both underlying principles and practical considerations, emphasizing how hydrogel design can be optimized for specific applications. Moreover, special attention is given to *stimuli-responsive hydrogels*—which can undergo reversible physicochemical changes in response to temperature, pH, ionic strength, light, or biological environment. These dynamic materials embody the transition from passive to intelligent systems, offering promising implications for precision medicine, environmental monitoring, rewritable inks, sensors and soft robotics. Furthermore, sustainability considerations, including the development of biodegradable hydrogels derived from natural polymers and green synthesis techniques that minimize environmental impact, are discussed.

In conclusion, this book aspires to serve as both a foundational text and a reference for ongoing research, while it also aims to impact the theoretical knowledge and hands-on experience to readers in order to fully utilize hydrogels for societal benefit.

The editors and authors have made an effort to convey the material with conceptual clarity, scientific rigor, and relevance to contemporary international objectives.

Acknowledgement

The journey of bringing a book from conceptualization to tangible reality is a collective endeavor, and it is with profound gratitude that we acknowledge all the authors and reviewers who have made our book, "Fabrication Techniques and Emerging Applications of Hydrogels," possible.

Our utmost appreciation goes to all the contributing authors, who are reputable international experts and have generously shared their invaluable knowledge, time, and insights towards completing this book. Their dedication to scientific excellence and their patience throughout the rigorous revision process form the very foundation of this work. We are truly humbled by their commitment.

We are indebted to the numerous peer reviewers who dedicated their time to provide critical, constructive feedback on the manuscripts. Their sharp eyes and scholarly suggestions were instrumental in enhancing the quality, clarity, and rigor of each chapter.

Our sincere thanks go to the entire team at IGI Global Scientific Publishers for their unwavering belief in this project and their expert guidance from inception to completion. We also thank the production and marketing teams for their diligence in shaping this work into its final form.

Finally, we acknowledge the global scientific community dedicated to hydrogel research. It is our sincere hope that this book will serve as a valuable resource, inspiring new innovations and fostering further collaboration in this dynamic and impactful field.

Editorial Advisory Board

Dr. Mengmeng Kang

School of Materials Science and Engineering, Henan Normal University, No. 46, East of Construction Road, Xinxiang, Henan Province 453007, the People's Republic of China

Dr. Bankole I. Oladapo, MSc, PhD, FRSA, SFHEA

University of Dundee, United Kingdom & Dundee International Institute of Central South University, PR China

Professor Michael Olawale Daramola

Department of Chemical Engineering, University of Pretoria, Pretoria, South Africa

Chapter 1
Introduction to Hydrogels

Onome Ejeromedoghene
https://orcid.org/0000-0002-5487-3267
Soochow University, China

Moses Kumi
https://orcid.org/0000-0002-7971-8802
The Hong Kong Polytechnic University, Hong Kong

ABSTRACT

Hydrogels are versatile polymer networks whose mesh size and responsiveness can be engineered for biomedicine, environmental systems, and soft electronics. This chapter emphasizes crosslinking density as a unifying parameter governing swelling, transport, and mechanical performance. We synthesize the field from early developments to current design strategies, organizing hydrogels by source, network formation, and environmental responsiveness. Particular attention is given to stimuli-responsive systems, including temperature, pH, light, enzymatic cues, and emerging magnetic, electrical, and redox triggers, which enable controlled release, actuation, and self-healing. Advances in nanocomposites, dynamic covalent chemistry, and bioinks are highlighted as drivers of multifunctionality. Remaining challenges include balancing toughness with biocompatibility, achieving predictable degradation, scaling fabrication, and ensuring reliable integration with sensors and biological tissues. Looking forward, opportunities lie in sustainable green synthesis, 3D biofabrication, and next-generation smart hydrogels for theranostics, regenerative medicine, and wearable bioelectronics.

1.1. INTRODUCTION

Hydrogels are three-dimensional polymer networks capable of absorbing and retaining substantial amounts of water or biological fluids, giving them a soft and pliable nature reminiscent of living tissues(Kumi, Ejeromedoghene, et al., 2024; Maji & Lee, 2022). The name "hydrogel" is derived from their hydrophilic character ("hydro") and gel-like texture ("gel") (Tripathy et al., 2023). On the molecular level, hydrogels consist of polymer chains joined together by crosslinks, forming a porous network that prevents dissolution while enabling water absorption. Hydrophilic functional groups such as hydroxyl (–OH), carboxyl (–COOH), amine ($–NH_2$), and sulfonic acid

DOI: 10.4018/979-8-3373-2337-4.ch001

groups on the polymer backbone facilitate hydrogen bonding with water molecules, leading to significant swelling when exposed to aqueous environments(Raffe et al., 2024).

A central determinant of hydrogel behavior is crosslinking density, which governs the mesh size of the polymer network. Crosslinking density directly regulates water retention, swelling capacity, and the viscoelastic properties of the material(Rizwan et al., 2017). Loosely crosslinked hydrogels exhibit larger mesh sizes and higher swelling, whereas highly crosslinked networks possess smaller pores, greater stiffness, and reduced water uptake. Thus, the balance between crosslinking density and polymer chemistry defines the dual characteristics of hydrogels: high water content alongside structural stability

In addition to intrinsic structure, environmental conditions—such as pH, temperature, and ionic strength—can dynamically modulate hydrogel swelling and mechanical response(Rumon, 2025). These stimulus-responsiveness features enable the design of "smart" hydrogels for biomedical and industrial applications. Mechanically, hydrogels occupy a unique middle ground between solids and liquids: they are viscoelastic, capable of deforming under stress while maintaining shape recovery, and porous enough to permit diffusion of nutrients, gases, and bioactive molecules(Bashir et al., 2020; L. Li et al., 2024).

In a nutshell, hydrogels are defined by their ability to combine high water content with a stable, three-dimensional polymeric network, resulting in materials that are soft, biocompatible, and functionally versatile. These fundamental properties underline the wide range of scientific and technological uses for hydrogels today.

1.2. HISTORICAL DEVELOPMENT OF HYDROGELS

The concept of hydrophilic polymer networks capable of absorbing water dates back to the early 20th century, though the term "hydrogel" had not yet been coined (Maleki et al., 2025). Early studies focused on naturally occurring polymers such as collagen and gelatin, prized for their inherent water-retention and gelation properties. These natural hydrogels found use in food processing and photographic materials, providing preliminary insights into polymer–water interactions(Karoyo & Wilson, 2021). However, their use was largely empirical and lacked systematic design principles.

A pivotal milestone in hydrogel research occurred in 1960 when Czech scientists, Otto Wichterle and Drahoslav Lim, synthesized the first synthetic hydrogel, poly(2-hydroxyethyl methacrylate) (PHEMA) for biomedical applications (Michálek et al., 2022). PHEMA combined water absorbency, transparency, mechanical stability, and biocompatibility—characteristics that led to the commercialization of soft contact lenses by Bausch & Lomb in 1971, revolutionizing ophthalmology(Zare et al., 2021; Zhu et al., 2022). This success triggered a surge in synthetic hydrogel development, including polyvinyl alcohol (PVA) and polyacrylamide (PAA), which were engineered for applications ranging from wound dressings to artificial cartilage and hygiene products(Khansari et al., 2017; H. Li et al., 2023).

The 1990s introduced the concept of "smart" or stimuli-responsive hydrogels. Materials such as poly(N-isopropylacrylamide) (PNIPAAm) could reversibly swell or deswell in response to environmental cues like temperature, enabling controlled drug release. A notable example is poly(N-isopropylacrylamide) (PNIPAAm), which exhibits a lower critical solution temperature (LCST) near physiological temperature, enabling controlled drug release triggered by temperature changes (Cao et al., 2019). Concurrently, natural polymers such as chitosan and alginate were chemically modified to confer pH-responsiveness, broadening their role in targeted therapies(Idrees et al., 2020).

In the 21st century, advances in nanotechnology and biofabrication expanded hydrogel capabilities. Conductive polymers (e.g., polypyrrole, PEDOT) enabled integration with bioelectronics, while hydrogel-based bioinks such as gelatin methacrylate (GelMA) facilitated 3D printing of living tissues(Rajabi et al., 2021). More recently, dynamic covalent chemistries and nanoparticle incorporation have endowed hydrogels with self-healing, high-strength, and multifunctional properties, paving the way for applications in soft robotics, wearable sensors, and regenerative medicine.

From a classification standpoint, hydrogels can be broadly categorized by source (natural vs. synthetic), crosslinking mechanism (physical vs. chemical), and responsiveness (conventional vs. stimuli-responsive). This dual perspective—historical and structural—underscores the continuous evolution of hydrogels from simple water-retentive materials to sophisticated, multifunctional systems designed for complex biomedical, industrial, and environmental applications (Figure 1).

Figure 1. Brief history of hydrogels

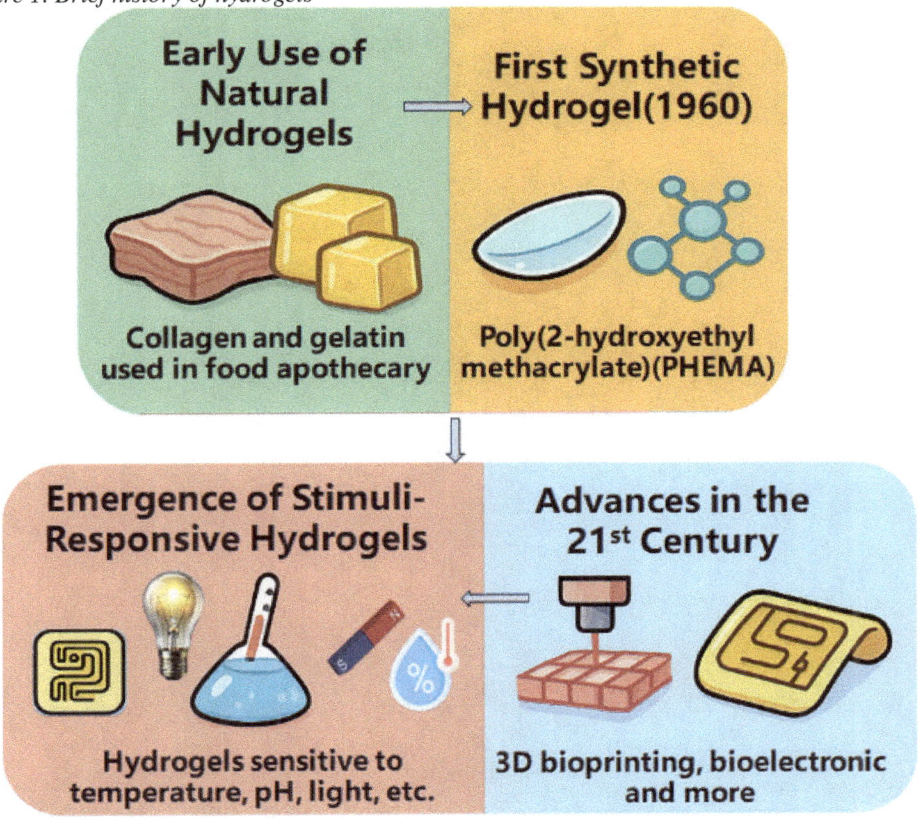

1.3. CLASSIFICATION OF HYDROGELS

Hydrogels are a highly diverse class of materials, and their classification depends on multiple, often interrelated, criteria—including source, crosslinking type, responsiveness to stimuli, particle size, and chain composition(Bustamante-Torres et al., 2021). While these categories are commonly presented separately, in practice they overlap and influence each other. For instance, natural polymers such as alginate or chitosan often undergo ionic or enzymatic crosslinking, whereas synthetic polymers are more commonly stabilized by chemical or radiation crosslinking. Similarly, the responsiveness of a hydrogel is frequently dictated by its chain composition and source material. Understanding these interconnections allows researchers to design hydrogels more systematically for specific applications (Figure 2).

Figure 2. Classifications of hydrogels

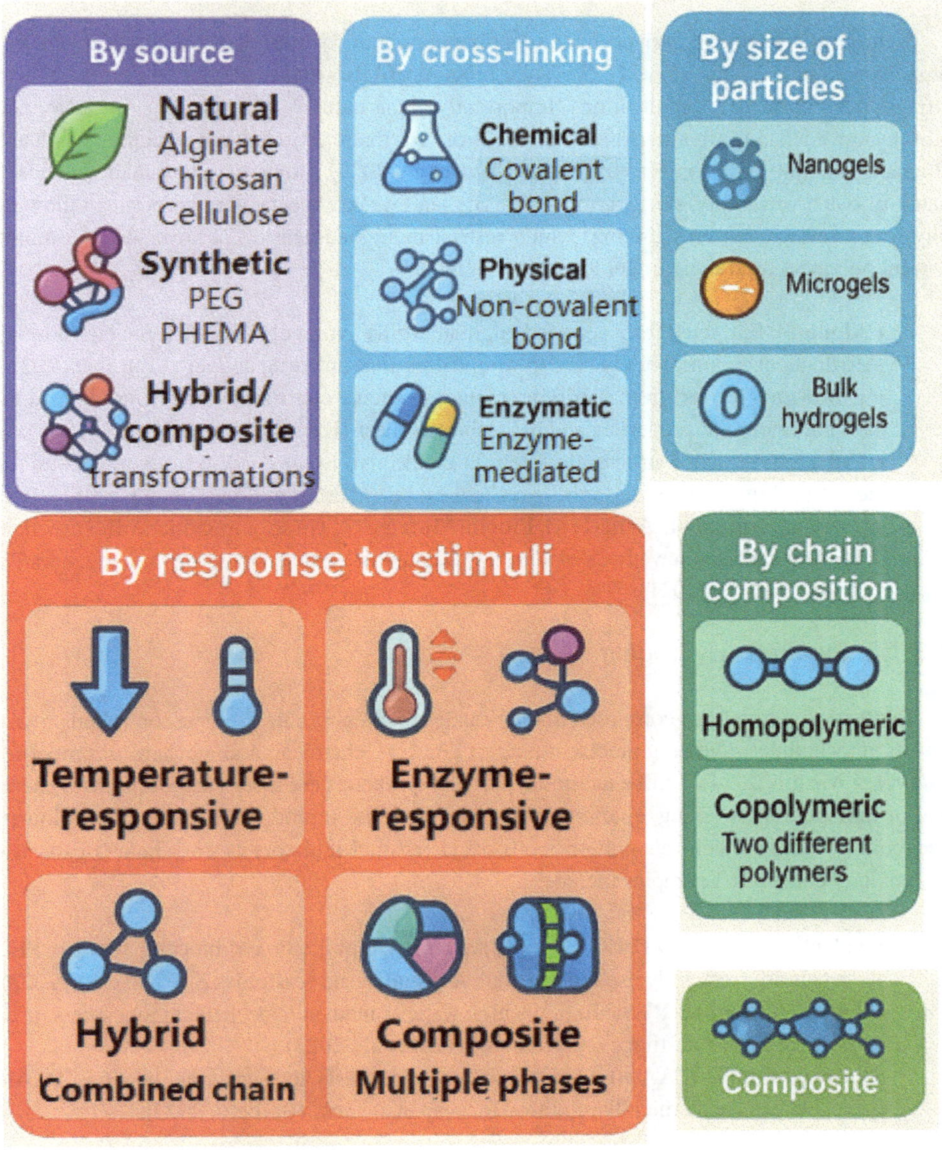

1.3.1. By Source

Hydrogels s can be categorized by origin into natural and synthetic types; more complex structural designs (hybrid/composite) are discussed under chain composition (Section 3.5).

1.3.1.1. Natural Hydrogels

Natural hydrogels are derived from biopolymers obtained from plants, animals, or microorganisms (Kumi, Wang, et al., 2024). Their biocompatibility and biodegradability make them attractive for biomedical applications. Importantly, most natural polymers (e.g., alginate, gelatin, hyaluronic acid) are amenable to physical or enzymatic crosslinking, which avoids harsh chemicals and preserves bioactivity(Kumi, Chen, et al., 2024). However, a key challenge lies in batch-to-batch variability, sterilization sensitivity, and weaker mechanical properties following physical or enzymatic crosslinking, which restrict their load-bearing performance. Common natural polymers used in hydrogel formation include:

- **Alginate**: Extracted from seaweed, alginate forms hydrogels through ionic crosslinking with divalent cations, widely used in wound dressings and drug delivery (Tan et al., 2023).
- **Chitosan**: Derived from chitin in crustacean shells, chitosan hydrogels are valued for their antimicrobial properties and pH responsiveness (Kumi et al., 2025).
- **Collagen**: A major structural protein in connective tissues, collagen hydrogels mimic the extracellular matrix, supporting cell adhesion and proliferation (Dinescu et al., 2018).
- **Gelatin, Hyaluronic Acid (HA), Fibrin, Matrigel®**: These biopolymers offer excellent biological cues and have been extensively applied in tissue engineering and regenerative medicine (Biswal, 2021).

1.3.1.2. Synthetic Hydrogels

Synthetic hydrogels are composed of man-made polymers synthesized via controlled polymerization techniques, allowing precise tuning of physical, chemical, and mechanical properties (Manjit & Mishra, 2024). Unlike natural hydrogels, synthetics are more compatible with chemical or radiation crosslinking, enabling greater stability and reproducibility. Yet, their bioinert backbones often lack cell recognition motifs, and residual initiators or crosslinkers can raise cytotoxicity concerns. Examples include:

- **Polyethylene Glycol (PEG)**: Known for its hydrophilicity and biocompatibility, PEG hydrogels are frequently used in drug delivery and tissue scaffolds (Z. Wang et al., 2023).
- **Polyvinyl Alcohol (PVA)**: Exhibits high water content and mechanical robustness, used in wound care and cartilage replacement (H. Li et al., 2023).
- **Polyacrylamide (PAAm)**: Common in electrophoresis and sensor applications, PAAm hydrogels can be chemically modified for responsiveness (Barbero, 2023).

1.3.2. By Cross-Linking

The method by which polymer chains are interconnected to form the three-dimensional hydrogel network is a critical factor determining the material's stability, swelling, and mechanical properties. A central concept in this context is crosslinking density, which describes the number of crosslinks per unit volume in the polymer network. Crosslinking density directly governs hydrogel mesh size—the average distance between crosslinks—which in turn controls water uptake, solute diffusion, and mechanical stiffness(Grassi et al., 2015). Hydrogels with low crosslinking density exhibit large mesh sizes, high swelling capacity, and softer mechanics, whereas highly cross-linked networks restrict swelling and diffusion but enhance strength and elasticity. Thus, tuning crosslinking density provides a fundamental design parameter for tailoring hydrogel performance across biomedical and engineering applications. In general, a design trade-off emerges: higher crosslink density enhances strength but reduces mesh size and diffusion capacity, affecting drug release and cell viability. Importantly, crosslinking choices are often dictated by source materials: natural hydrogels favor physical/enzymatic methods, whereas synthetic polymers allow chemical or radiation approaches. Hydrogels can be classified based on their crosslinking mechanisms into chemical, physical, enzymatic, and radiation crosslinking types (Maitra & Shukla, 2014).

1.3.2.1. Chemical Crosslinking

Chemical crosslinking involves the formation of covalent bonds between polymer chains, creating a permanent and stable network. This method offers precise control over the hydrogel's mechanical strength, swelling behavior, and degradation rate. Chemical crosslinkers such as glutaraldehyde, genipin, or photo-initiated reactions using UV light are commonly employed (Duquette & Dumont, 2019). Due to the stability of covalent bonds, chemically crosslinked hydrogels are often preferred for applications requiring durability and controlled release, including tissue engineering scaffolds and drug delivery systems (Oryan et al., 2018). However, care must be taken to avoid toxicity from residual crosslinking agents and brittleness at high crosslinking densities.

1.3.2.2. Physical Crosslinking

Physical crosslinking relies on non-covalent interactions such as hydrogen bonding, ionic interactions, hydrophobic associations, or chain entanglements to maintain the hydrogel structure (J. Yang et al., 2023). These interactions are reversible, enabling hydrogels to exhibit dynamic properties like self-healing, stimuli-responsiveness, and injectability. For example, alginate hydrogels formed by ionic crosslinking with calcium ions exemplify physically crosslinked systems. Physical hydrogels typically display lower mechanical strength than chemically crosslinked counterparts but benefit from enhanced biocompatibility and the absence of potentially harmful chemicals (D. Yang, 2022).

1.3.2.3. Enzymatic Crosslinking

Enzymatic crosslinking uses specific enzymes to catalyze bond formation between polymer chains under mild physiological conditions. This approach offers excellent control over gelation kinetics and biocompatibility, making it attractive for *in situ* hydrogel formation and biomedical applications such as cell encapsulation and injectable scaffolds (Z. Li et al., 2023). Enzymes like transglutaminase and horseradish peroxidase are commonly used to crosslink natural polymers like gelatin and hyaluronic acid. However, high enzyme cost, batch variability, and diffusion-limited gelation in thick constructs constrain scalability.

1.3.2.4. Radiation Crosslinking

Radiation crosslinking employs ionizing radiation, such as gamma rays or electron beams, to generate free radicals that induce polymer chain crosslinking (Alcântara et al., 2020). This method enables simultaneous sterilization and hydrogel formation without the need for chemical crosslinkers. Nevertheless, dose control is critical, as excessive irradiation can cause chain scission and yellowing. Radiation crosslinked hydrogels are used in biomedical devices and wound dressings, offering controlled mechanical properties and high purity (Ahmed et al., 2024; Demeter et al., 2023). The process parameters can be fine-tuned to tailor the degree of crosslinking and hydrogel performance.

1.3.3. By Responsiveness

Hydrogels can also be classified based on their ability to respond to external stimuli. This responsiveness enables dynamic changes in their physical or chemical properties, such as swelling, solubility, or mechanical strength, in response to specific environmental signals. Stimuli-responsive hydrogels, often termed "smart hydrogels," have garnered significant interest for controlled drug delivery, biosensing, and actuators (Roy et al., 2022). The challenge for responsive designs lies in narrow operational windows (e.g., pH, temperature, light) and risks of hysteresis or fatigue. These limitations encourage the development of multi-stimuli-responsive and self-healing systems.

1.3.3.1. Stimuli-Responsive Hydrogels

Stimuli-responsiveness often stems from chain composition or functional groups embedded in either natural or synthetic sources. For example, chitosan is inherently pH-responsive, while PNIPAAm exhibits temperature sensitivity. Crosslinking methods (e.g., light-induced bonds in photo-responsive systems) also dictate responsiveness.

- **Thermo-responsive Hydrogels:** These hydrogels exhibit volume phase transitions in response to temperature changes. For example, poly(N-isopropylacrylamide) (PNIPAAm) exhibits a lower critical solution temperature (LCST) around 32°C, shrinking above this temperature and swelling below it (Bischofberger & Trappe, 2015). However, unintended shrinkage at physiological temperatures and slow re-swelling limit applications, prompting strategies for LCST tuning and porous network engineering. Such hydrogels are useful in controlled drug release and tissue engineering, where temperature can modulate material behavior (Sood et al., 2016).
- **pH-Responsive Hydrogels:** These contain ionizable groups that accept or donate protons depending on environmental pH. Changes in pH alter the ionization state, causing swelling or deswelling (Kocak et al., 2017). *In vivo*, however, physiological buffering reduces sensitivity, motivating stronger polyelectrolytes and organ-specific targeting. Natural polymers like chitosan (cationic) and alginate (anionic) exemplify this behavior, enabling targeted delivery in specific regions of the gastrointestinal tract (Chen et al., 2022).
- **Photo-Responsive Hydrogels:** These hydrogels respond to light stimuli, often through photoisomerization or photochemical cleavage of crosslinks, enabling precise spatial and temporal control over hydrogel properties (L. Li et al., 2019). Their major limitations are

shallow light penetration and possible cytotoxic photoproducts, driving interest in NIR or two-photon responsive chemistries. Applications include photo-controlled drug release, photochromic devices, and tissue engineering scaffolds with tunable stiffness (Ejeromedoghene, Zuo, et al., 2023; Jiang et al., 2023).

• **Enzyme-Responsive Hydrogels:** Designed to degrade or alter their structure in response to specific enzymes, these hydrogels enable targeted drug delivery or tissue remodeling (Sobczak, 2022). Variability in patient enzyme expression and risk of off-target cleavage are persistent barriers, motivating protease-selective linkers. For example, hydrogels sensitive to matrix metalloproteinases (MMPs) degrade in disease-affected tissues, releasing therapeutic agents locally.

• **Chemical-Responsive Hydrogels:** These respond to chemical signals such as ionic strength, glucose concentration, or other biomolecules, making them suitable for biosensing and closed-loop drug delivery systems (C. Hu et al., 2022). Glucose-responsive hydrogels have been developed for insulin delivery in diabetes management. Selectivity and signal drift in complex fluids remain challenges, prompting exploration of boronate alternatives and affinity-based scaffolds.

• **Other Stimuli-Responsive Hydrogels:** Beyond the classical categories, hydrogels responsive to magnetic fields, electrical stimuli, and redox conditions are gaining importance in biomedical and wearable applications. Magnetic-responsive hydrogels incorporate magnetic nanoparticles, enabling remote actuation, hyperthermia treatment, and guided drug delivery (Lavrador et al., 2021) . Electrically responsive hydrogels, often based on conductive polymers, undergo rapid swelling, deswelling, or deformation under electrical fields, making them ideal for soft actuators, neural interfaces, and bioelectronic platforms(H. Hu et al., 2024). Redox-responsive hydrogels, containing disulfide or thiol groups, exploit redox gradients in biological environments to achieve controlled degradation and drug release, offering new avenues for tumor-targeted therapy and regenerative medicine (Solanki & Bhatia, 2024). Yet, nanoparticle agglomeration, Joule heating, and nonspecific cleavage are concerns, which motivate surface engineering of fillers and low-power circuitry integration.

1.3.3.2. Non-Responsive Hydrogels

Non-responsive hydrogels maintain their physical and chemical properties regardless of environmental changes. While they lack dynamic behavior, their stable and predictable characteristics make them suitable for structural applications such as implants, contact lenses, and wound dressings, where consistent performance is critical (Štular et al., 2017).

1.3.4. By Size

Hydrogels can be categorized by particle size, dividing them into nanogels, microgels, aerogels, and bulk hydrogels, each possessing distinctive characteristics and applications (Lima et al., 2020). Nanogels, aerogels, microgels, and bulk hydrogels are all specialized forms of hydrogels differentiated primarily by their size and structural characteristics, which define their specific applications and functionalities. Importantly, size interacts with responsiveness and crosslinking: nanogels and microgels often respond rapidly to stimuli due to high surface area, whereas bulk hydrogels emphasize stability for implants. Nanogels are composed of nanoparticles and typically have a size range of less than 100 nanometers. These materials exhibit a very permeable composition and are widely recognized for their capacity to expand and hold substantial quantities of water or biological fluids. Due to their small size, nanogels are excellent carriers for drug delivery, allowing them to circulate in the body and target specific sites such as tumors or inflamed tissues effectively (Sivaram et al., 2015). They are also used in medical imaging and as sensors due to their responsiveness to environmental stimuli. However, they face rapid renal clearance and challenges in large-scale reproducibility, which motivates stealth coatings and continuous-flow production systems. Microgels are larger than nanogels, with particle sizes generally in the micrometer range (from about 1 to 1000 micrometers) (Hamzah et al., 2017). These gels are soft and capable of undergoing significant changes in volume, which makes them useful in applications requiring rapid and reversible reaction to external stimuli, such as temperature or pH. Injector clogging and batch polydispersity remain barriers, which microfluidic fabrication approaches aim to resolve. Microgels are commonly used in encapsulation applications for controlled release of drugs, as well as in regenerative medicine and tissue engineering. Aerogels are a distinct category characterized by their ultra-lightweight and porous structure, but they are not typically classified within the same context as hydrogels like nanogels and microgels, due to their dry nature and the process used to create them (Zhao et al., 2023). Aerogels are formed through the substitution of the liquid component of gels with gas. These materials are renowned for their high thermal insulation capabilities and low density, rendering them highly suitable for a wide range of applications, including sophisticated insulation materials in buildings and spacecraft, as well as filtration systems and sensors. And as the name suggests, bulk hydrogels are larger than 100 μm, and are valued for their structural stability and longevity, making them suitable for more permanent applications such as implants and prostheses. Each type of gel offers unique advantages depending on the application, with nanogels and microgels specifically being pivotal in biomedical fields, while aerogels find extensive use in industrial and environmental applications.

1.3.5. By Chain Composition

The polymeric architecture or chain composition of hydrogels plays a crucial role in determining their physical properties, functionality, and adaptability for various applications (Spicer, 2020). Based on the nature and arrangement of the polymer chains, hydrogels can be categorized into the following structural types:

1.3.5.1. Homopolymeric Hydrogels

These hydrogels are formed from a single type of monomer unit, offering predictable properties and simpler synthesis(Gyles et al., 2017). These hydrogels are often used when a specific, consistent functionality is required, such as in standard drug delivery systems or scaffolds with controlled swelling behavior.

1.3.5.2. Copolymeric Hydrogels

Copolymeric hydrogels are composed of two or more different types of monomers, which may be arranged in random, alternating, block, or graft configurations(Mittal et al., 2016). This structural diversity enables the fine-tuning of mechanical strength, swelling kinetics, degradability, and chemical reactivity. Copolymers offer enhanced design flexibility, making them highly suitable for applications requiring multifunctional performance.

1.3.5.3. Hybrid Hydrogels

Hybrid hydrogels combine distinct types of polymers—typically natural and synthetic—within a single matrix to synergize the favorable properties of each component (Palmese et al., 2019). For instance, a hybrid hydrogel might incorporate the biocompatibility of a natural polymer with the mechanical strength or functional tunability of a synthetic polymer. A classic example is a collagen–PEG hydrogel, which integrates natural cell affinity with synthetic durability. Such designs are especially valuable in tissue engineering and advanced wound healing applications.

1.3.5.4. Composite Hydrogels

Composite hydrogels are structured materials that incorporate additional phases—such as nanoparticles, fibers, or other fillers—within the hydrogel matrix (Utech & Boccaccini, 2016). These components may not participate directly in the network formation but provide enhanced properties such as electrical conductivity, antimicrobial activity, or mechanical reinforcement (Ejeromedoghene et al., 2022; Ejeromedoghene, Oderinde, et al., 2023). Composite hydrogels are widely used in emerging areas like flexible electronics, biosensors, and theragnostics. Table 1 provides an overview of the various classifications of hydrogels based on different parameters. The table can be used by scientists and researchers as a quick reference when working with hydrogels for diverse applications.

Table 1. Classification and Characteristics of Hydrogels

Classification criteria	Examples	Description	Ref.
Source	Natural polymers	Collagen, Gelatin, Fibrin, Hyaluronic Acid (HA), Matrigel, Alginate	(Jose et al., 2020)
	Synthetic polymers	Polyacrylamide (PA), Polyethylene Glycol (PEG)	(Ramazani & Aghahosseini, 2020)
Size of Particles	Nanogels	Particle size < 100 nm, high water absorption, biocompatible	(Pujana et al., 2012)
	Microgels	Particle size in micrometers, soft structures, solvent exchange	(Plamper & Richtering, 2017)
	Bulk Hydrogels	Particle size > 100 μm, structural stability, long-lasting	(Daly et al., 2020)
Crosslinking	Chemical Crosslinking	Covalent bonds between polymer chains, precise control, and high biocompatibility	(Gao et al., 2021)
	Physical Crosslinking	Non-covalent interactions (e.g., hydrogen bonding), reversible properties	(Picchioni & Muljana, 2018)
	Enzymatic Crosslinking	Enzyme-mediated crosslinking, gentle, controlled kinetics	(Nguyen et al., 2019)
	Radiation Crosslinking	crosslinked using ionizing radiation, sterilization, and modification	(Ashfaq et al., 2020)

continued on following page

Table 1. Continued

Classification criteria	Examples	Description	Ref.
Biodegradability	Biodegradable Hydrogels	Gradual breakdown into smaller molecules, suitable for tissue regeneration	(Zhang & King, 2020)
	Non-Biodegradable Hydrogels	Long-lasting, used in medical implants and devices	(Ershad-Langroudi et al., 2024)
Ionic Charge	Anionic Hydrogels	Negatively charged, swell in basic environments	(Kıvanç et al., 2022)
	Cationic Hydrogels	Positively charged, collapse in basic environments	(Ostroha et al., 2004)
	Amphiphilic Hydrogels	Contain both positive and negative charges, versatile	(Chang et al., 2019)
	Neutral Hydrogels	No net charge, applied in a controlled drug release system	(J. Li & Mooney, 2016)
Response to Stimuli	pH-Responsive Hydrogels	pH-dependent volume changes, controlled drug release	(Gupta et al., 2002)
	Temperature-Responsive Hydrogels	Swelling/shrinking with temperature changes, controlled drug release	(Zhu et al., 2020)
	Enzyme-Responsive Hydrogels	Respond to specific enzymes, targeted drug delivery	(Sobczak, 2022)
	Chemical-Responsive Hydrogels	React to chemical cues (e.g., ions, molecules), versatile applications	(Dong et al., 2021)
Chain Composition	Homopolymeric Hydrogels	Single type of polymer chain, controlled properties	(Kuckling et al., 2003)
	Copolymeric Hydrogels	Two or more different polymer chains, with tailored properties	(Singhal & Gupta, 2016)
	Hybrid Hydrogels	A combination of different polymer chains and materials, enhanced performance	(Bao et al., 2019)
	Composite Hydrogels	Multiple phases with diverse properties, versatile applications	(Biondi et al., 2015)
	Crosslinked Hydrogels	Polymer chains crosslinked to form a 3D network, with mechanical strength	(X.-H. Wang et al., 2018)

1.4. CHALLENGES AND FUTURE PERSPECTIVES

Despite tremendous progress in hydrogel science and their wide-ranging applications, several persistent challenges hinder the full realization of their potential in next-generation biomedical devices and smart materials (Figure 3). Mechanical weakness and limited durability, particularly in natural polymer-based or physically crosslinked hydrogels, restrict their use in load-bearing applications such as cartilage replacement or soft robotics, where cyclic stress is frequent. Achieving higher mechanical resilience without compromising biocompatibility remains a major focus of current research. Controlled degradation and stability are equally critical: hydrogels must degrade at rates matched to tissue regeneration or drug delivery timelines, avoiding premature functional loss or long-term accumulation that could trigger toxicity. Additionally, the complexity of synthesizing advanced hydrogels, including multi-responsive or self-healing systems, and the associated high cost of reagents, elaborate procedures, or demanding reaction conditions, limit scalability and commercialization. Biocompatibility and immunogenicity also pose significant challenges; residual crosslinkers, degradation by-products, or embedded nanoparticles may induce immune responses, necessitating rigorous evaluation for clinical translation. Furthermore, integrating hydrogels with electronics or sensor systems while maintaining softness, biocompatibility, and reliable signal transduction presents a frontier in wearable and implantable device design.

Figure 3. Application opportunities of hydrogels as next-generation functional materials

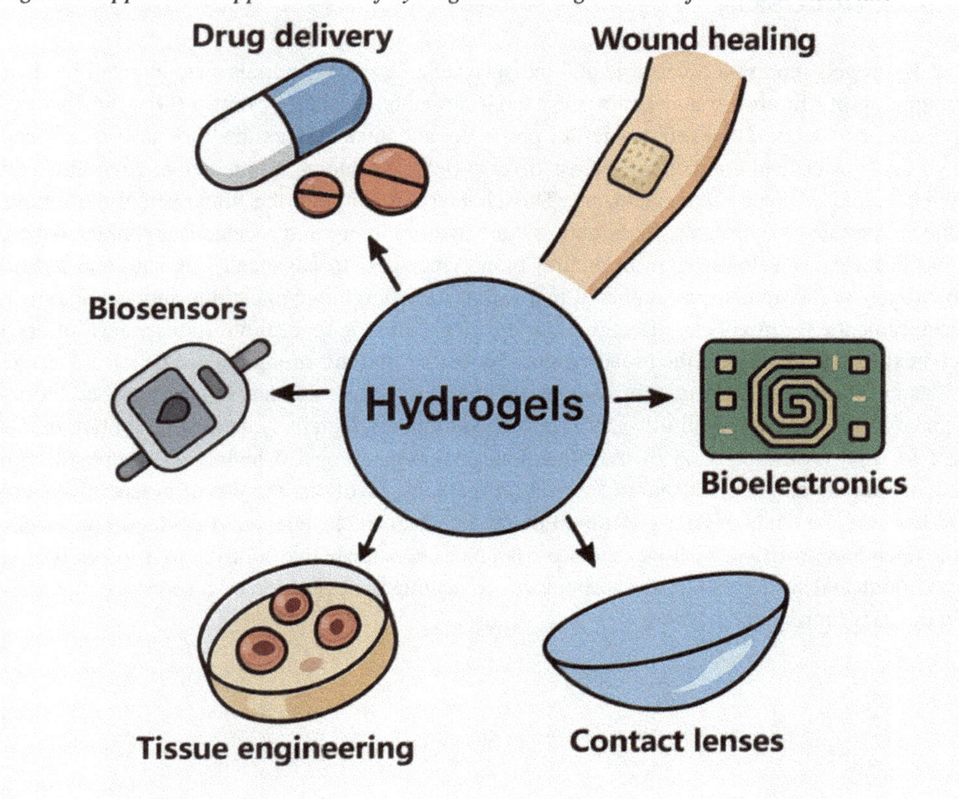

Addressing these challenges has guided the emergence of several promising strategies. The development of multifunctional and smart hydrogels that combine stimuli responsiveness, self-healing, and bioactivity within a single platform offers solutions to both mechanical and functional limitations, with dynamic covalent chemistry and supramolecular interactions enabling adaptable network designs. Nanocomposite and hybrid materials incorporating graphene, silica, or metallic nanoparticles enhance mechanical properties, electrical conductivity, and antimicrobial activity, broadening hydrogel applications in biomedicine, environmental remediation, and flexible electronics. Advances in 3D bioprinting and biofabrication allow the creation of complex, cell-laden hydrogel architectures that closely mimic native tissue structures, paving the way for organ engineering and personalized medicine. Sustainable and green hydrogel synthesis approaches, utilizing renewable resources and cost-effective, environmentally friendly processes, are increasingly prioritized to reduce ecological impact while maintaining functionality. Finally, streamlined clinical translation and regulatory pathways, supported by standardized testing, biocompatibility assessment, and collaboration between academia, industry, and regulatory bodies, are essential to accelerate hydrogel adoption in real-world biomedical and technological applications.

1.5. CONCLUSION

Hydrogels represent a versatile and indispensable class of materials characterized by their unique ability to absorb and retain substantial amounts of water within a three-dimensional polymeric network. Their soft, tissue-like properties and tunable functionalities have established hydrogels as critical components across diverse fields, including biomedicine, environmental science, and advanced manufacturing. This chapter has outlined the fundamental definitions and properties of hydrogels, highlighting their hydrophilicity and mechanical characteristics. The historical development from natural biopolymer gels to advanced synthetic and hybrid hydrogels is a testament to continued innovation for emerging applications and technological requirements. Hydrogel classification by source, crosslinking type, and responsiveness offers a context for understanding the broad-ranging properties and informing material choice. Despite the significant developments, limitations, including mechanical constraints, controlled biodegradability, and biocompatibility, remain, and continued research is encouraged. Future trends are focused on the creation of multifunctional, stimuli-responsive hydrogels combined with nanomaterials and state-of-the-art fabrication methods, involving the use of green solvents to address the demands of next-generation therapies, diagnostics, and smart devices. Ultimately, the relentless growth of hydrogel science guarantees transformative contributions to healthcare, environmental sustainability, and technology, cementing their position as cornerstone materials in the 21st century and beyond.

REFERENCES

Ahmed, M. S., Islam, M., Hasan, M. K., & Nam, K.-W. (2024). A comprehensive review of radiation-induced hydrogels: Synthesis, properties, and multidimensional applications. *Gels (Basel, Switzerland)*, *10*(6), 381. DOI: 10.3390/gels10060381 PMID: 38920928

Alcântara, M. T. S., Lincopan, N., dos Santos, P. M., Ramirez, P. A., Brant, A. J. C., Riella, H. G., & Lugão, A. B. (2020). Simultaneous hydrogel crosslinking and silver nanoparticle formation by using ionizing radiation to obtain antimicrobial hydrogels. *Radiation Physics and Chemistry*, *169*, 108777. DOI: 10.1016/j.radphyschem.2020.108777

Ashfaq, A., Clochard, M.-C., Coqueret, X., Dispenza, C., Driscoll, M. S., Ulański, P., & Al-Sheikhly, M. (2020). Polymerization reactions and modifications of polymers by ionizing radiation. *Polymers*, *12*(12), 2877. DOI: 10.3390/polym12122877 PMID: 33266261

Bao, Z., Xian, C., Yuan, Q., Liu, G., & Wu, J. (2019). Natural polymer-based hydrogels with enhanced mechanical performances: Preparation, structure, and property. *Advanced Healthcare Materials*, *8*(17), 1900670. DOI: 10.1002/adhm.201900670 PMID: 31364824

Barbero, C. A. (2023). Functional materials made by combining hydrogels (cross-linked poly-acrylamides) and conducting polymers (polyanilines)—A critical review. *Polymers*, *15*(10), 2240. DOI: 10.3390/polym15102240 PMID: 37242814

Bashir, S., Hina, M., Iqbal, J., Rajpar, A. H., Mujtaba, M. A., Alghamdi, N. A., Wageh, S., Ramesh, K., & Ramesh, S. (2020). Fundamental concepts of hydrogels: Synthesis, properties, and their applications. *Polymers*, *12*(11), 2702. DOI: 10.3390/polym12112702 PMID: 33207715

Biondi, M., Borzacchiello, A., Mayol, L., & Ambrosio, L. (2015). Nanoparticle-integrated hydrogels as multifunctional composite materials for biomedical applications. *Gels (Basel, Switzerland)*, *1*(2), 162–178. DOI: 10.3390/gels1020162 PMID: 30674171

Bischofberger, I., & Trappe, V. (2015). New aspects in the phase behaviour of poly-N-isopropyl acrylamide: Systematic temperature dependent shrinking of PNiPAM assemblies well beyond the LCST. *Scientific Reports*, *5*(1), 15520. DOI: 10.1038/srep15520 PMID: 26493499

Biswal, T. (2021). Biopolymers for tissue engineering applications: A review. *Materials Today: Proceedings*, *41*, 397–402. DOI: 10.1016/j.matpr.2020.09.628

Bustamante-Torres, M., Romero-Fierro, D., Arcentales-Vera, B., Palomino, K., Magaña, H., & Bucio, E. (2021). Hydrogels classification according to the physical or chemical interactions and as stimuli-sensitive materials. *Gels (Basel, Switzerland)*, *7*(4), 182. DOI: 10.3390/gels7040182 PMID: 34842654

Cao, M., Wang, Y., Hu, X., Gong, H., Li, R., Cox, H., Zhang, J., Waigh, T. A., Xu, H., & Lu, J. R. (2019). Reversible thermoresponsive peptide–PNIPAM hydrogels for controlled drug delivery. *Biomacromolecules*, *20*(9), 3601–3610. DOI: 10.1021/acs.biomac.9b01009 PMID: 31365246

Chang, H., Li, C., Huang, R., Su, R., Qi, W., & He, Z. (2019). Amphiphilic hydrogels for biomedical applications. *Journal of Materials Chemistry. B*, *7*(18), 2899–2910. DOI: 10.1039/C9TB00073A

Chen, S., Zhu, H., & Luo, Y. (2022). Chitosan-based oral colon-specific delivery systems for polyphenols: Recent advances and emerging trends. *Journal of Materials Chemistry. B*, *10*(37), 7328–7348. DOI: 10.1039/D2TB00874B PMID: 35766297

Daly, A. C., Riley, L., Segura, T., & Burdick, J. A. (2020). Hydrogel microparticles for biomedical applications. *Nature Reviews. Materials*, *5*(1), 20–43. DOI: 10.1038/s41578-019-0148-6 PMID: 34123409

Demeter, M., Scări oreanu, A., & Călina, I. (2023). State of the art of hydrogel wound dressings developed by ionizing radiation. *Gels (Basel, Switzerland)*, *9*(1), 55. DOI: 10.3390/gels9010055 PMID: 36661821

Dinescu, S., Albu Kaya, M., Chitoiu, L., Ignat, S., Kaya, D. A., & Costache, M. (2018). Collagen-based hydrogels and their applications for tissue engineering and regenerative medicine. In *Cellulose-Based Superabsorbent Hydrogels* (pp. 1–21). Springer. DOI: 10.1007/978-3-319-76573-0_54-1

Dong, Y., Ramey-Ward, A. N., & Salaita, K. (2021). Programmable Mechanically Active Hydrogel-Based Materials. *Advanced Materials*, *33*(46), 2006600. DOI: 10.1002/adma.202006600 PMID: 34309076

Duquette, D., & Dumont, M.-J. (2019). Comparative studies of chemical crosslinking reactions and applications of bio-based hydrogels. *Polymer Bulletin*, *76*(5), 2683–2710. DOI: 10.1007/s00289-018-2516-6

Ejeromedoghene, O., Hu, Y. P., Oderinde, O., Yao, F., Akinremi, C., Akinyeye, R., Adewuyi, S., & Fu, G. (2022). Transparent and photochromic poly(hydroxyethyl acrylate–acrylamide)/WO3 hydrogel with antibacterial properties against bacterial keratitis in contact lens. *Journal of Applied Polymer Science*, *139*(12), 1–15. DOI: 10.1002/app.51815

Ejeromedoghene, O., Oderinde, O., Adewuyi, S., & Fu, G. (2023). Fabrication of photochromic and electrochromic (PVP-am)/WO3/MoO3 hydrogel composites and glass films with programmable shapes. *Reactive & Functional Polymers*, *186*, 105553. DOI: 10.1016/j.reactfunctpolym.2023.105553

Ejeromedoghene, O., Zuo, X., Oderinde, O., Yao, F., Adewuyi, S., & Fu, G. (2023). Photochromic Behavior of Inorganic Superporous Hydrogels Fabricated from Different Reacting Systems of Polymeric Deep Eutectic Solvents. *Journal of Molecular Structure*, *1271*, 134101. DOI: 10.1016/j.molstruc.2022.134101

Ershad-Langroudi, A., Babazadeh, N., Alizadegan, F., Mousaei, S. M., & Moradi, G. (2024). Polymers for implantable devices. *Journal of Industrial and Engineering Chemistry*, *137*, 61–86. DOI: 10.1016/j.jiec.2024.03.030

Gao, Y., Peng, K., & Mitragotri, S. (2021). Covalently Crosslinked hydrogels via step-growth reactions: Crosslinking chemistries, polymers, and clinical impact. *Advanced Materials*, *33*(25), 2006362. DOI: 10.1002/adma.202006362 PMID: 33988273

Grassi, M., Farra, R., Fiorentino, S. M., Grassi, G., & Dapas, B. (2015). *Hydrogel mesh size evaluation. Polysaccharide Hydrogels: Characterization and Biomedical Applications.* Jenny Stanford Publishing.

Gupta, P., Vermani, K., & Garg, S. (2002). Hydrogels: From controlled release to pH-responsive drug delivery. *Drug Discovery Today*, *7*(10), 569–579. DOI: 10.1016/S1359-6446(02)02255-9 PMID: 12047857

Gyles, D. A., Castro, L. D., Silva, J. O. C.Jr, & Ribeiro-Costa, R. M. (2017). A review of the designs and prominent biomedical advances of natural and synthetic hydrogel formulations. *European Polymer Journal*, *88*, 373–392. DOI: 10.1016/j.eurpolymj.2017.01.027

Hamzah, Y., Hashim, S., & Rahman, W. A. W. A. (2017). Synthesis of polymeric nano/micro-gels: A review. *Journal of Polymer Research*, *24*(9), 1–19. DOI: 10.1007/s10965-017-1281-9

Hu, C., Yang, L., & Wang, Y. (2022). Recent advances in smart-responsive hydrogels for tissue repairing. *MedComm - Biomaterials and Applications*, *1*(2), e23. DOI: 10.1002/mba2.23

Hu, H., Li, D., Salim, T., Li, Y., Cheng, G., Lam, Y. M., & Ding, J. (2024). Electrically driven hydrogel actuators: Working principle, material design and applications. *Journal of Materials Chemistry. C, Materials for Optical and Electronic Devices*, *12*(5), 1565–1582. DOI: 10.1039/D3TC03863G

Idrees, H., Zaidi, S. Z. J., Sabir, A., Khan, R. U., Zhang, X., & Hassan, S. (2020). A review of biodegradable natural polymer-based nanoparticles for drug delivery applications. *Nanomaterials (Basel, Switzerland)*, *10*(10), 1970. DOI: 10.3390/nano10101970 PMID: 33027891

Jiang, J., Xu, S., Ma, H., Li, C., & Huang, Z. (2023). Photoresponsive hydrogel-based soft robot: A review. *Materials Today. Bio*, *20*, 100657. DOI: 10.1016/j.mtbio.2023.100657 PMID: 37229213

Jose, G., Shalumon, K. T., & Chen, J.-P. (2020). Natural polymers based hydrogels for cell culture applications. *Current Medicinal Chemistry*, *27*(16), 2734–2776. DOI: 10.2174/0929867326666190903113004 PMID: 31480996

Karoyo, A. H., & Wilson, L. D. (2021). A review on the design and hydration properties of natural polymer-based hydrogels. *Materials (Basel)*, *14*(5), 1095. DOI: 10.3390/ma14051095 PMID: 33652859

Khansari, M. M., Sorokina, L. V., Mukherjee, P., Mukhtar, F., Shirdar, M. R., Shahidi, M., & Shokuhfar, T. (2017). Classification of hydrogels based on their source: A review and application in stem cell regulation. *Journal of the Minerals Metals & Materials Society*, *69*(8), 1340–1347. DOI: 10.1007/s11837-017-2412-9

Kıvanç, M. R., Ozay, O., Ozay, H., & Ilgin, P. (2022). Removal of anionic dyes from aqueous media by using a novel high positively charged hydrogel with high capacity. *Journal of Dispersion Science and Technology*, *43*(7), 1000–1015. DOI: 10.1080/01932691.2020.1847658

Kocak, G., Tuncer, C., & Bütün, V. (2017). pH-Responsive polymers. *Polymer Chemistry*, *8*(1), 144–176. DOI: 10.1039/C6PY01872F

Kuckling, D., Richter, A., & Arndt, K. (2003). Temperature and pH-dependent swelling behavior of poly (N-isopropylacrylamide) copolymer hydrogels and their use in flow control. *Macromolecular Materials and Engineering*, *288*(2), 144–151. DOI: 10.1002/mame.200390007

Kumi, M., Chen, T., Zhang, Z., Wang, A., Li, G., Hou, Z., Cheng, T., Wang, J., Wang, T., & Li, P. (2024). Integration of Hydrogels and 3D Bioprinting Technologies for Chronic Wound Healing Management. *ACS Biomaterials Science & Engineering*, *10*(10), 5995–6016. Advance online publication. DOI: 10.1021/acsbiomaterials.4c00957 PMID: 39228365

Kumi, M., Ejeromedoghene, O., Sudane, W. D., & Zhang, Z. (2024). Unlocking the biological response of smart Stimuli-Responsive hydrogels and their application in biological systems. *European Polymer Journal*, *209*, 112906. DOI: 10.1016/j.eurpolymj.2024.112906

Kumi, M., Hou, Z., Zhang, Y., Yang, Y., Han, C., Wang, T., & Li, P. (2025). 3D Printed Chitosan-Based Flexible Electrode with Antimicrobial Properties for Electrical Stimulation Therapy in Wound Healing. *Supramolecular Materials*, *100110*, 100110. Advance online publication. DOI: 10.1016/j.supmat.2025.100110

Kumi, M., Wang, T., Ejeromedoghene, O., Wang, J., Li, P., & Huang, W. (2024). Exploring the Potentials of Chitin and Chitosan-Based Bioinks for 3D-Printing of Flexible Electronics: The Future of Sustainable Bioelectronics. *Small Methods*, *8*(9), 2301341. Advance online publication. DOI: 10.1002/smtd.202301341 PMID: 38403854

Lavrador, P., Esteves, M. R., Gaspar, V. M., & Mano, J. F. (2021). Stimuli-Responsive Nanocomposite Hydrogels for Biomedical Applications. *Advanced Functional Materials*, *31*(8), 2005941. Advance online publication. DOI: 10.1002/adfm.202005941

Li, H., Wu, C., Yu, X., & Zhang, W. (2023). Recent advances of PVA-based hydrogels in cartilage repair application. *Journal of Materials Research and Technology*, *24*, 2279–2298. DOI: 10.1016/j.jmrt.2023.03.130

Li, J., & Mooney, D. J. (2016). Designing hydrogels for controlled drug delivery. *Nature Reviews. Materials*, *1*(12), 1–17. DOI: 10.1038/natrevmats.2016.71 PMID: 29657852

Li, L., Scheiger, J. M., & Levkin, P. A. (2019). Design and applications of photoresponsive hydrogels. *Advanced Materials*, *31*(26), 1807333. DOI: 10.1002/adma.201807333 PMID: 30848524

Li, L., Wang, S., Chen, Y., Dong, S., Zhang, C., Liao, L., & Zhang, W. (2024). Hydrogels mimicking the viscoelasticity of extracellular matrix for regenerative medicine: Design, application, and molecular mechanism. *Chemical Engineering Journal*, *498*, 155206. DOI: 10.1016/j.cej.2024.155206

Li, Z., Lu, F., & Liu, Y. (2023). A review of the mechanism, properties, and applications of hydrogels prepared by enzymatic cross-linking. *Journal of Agricultural and Food Chemistry*, *71*(27), 10238–10249. DOI: 10.1021/acs.jafc.3c01162 PMID: 37390351

Lima, C. S. A., de, , Balogh, T. S., Varca, J. P. R. O., Varca, G. H. C., Lugão, A. B., & Camacho-Cruz, A., L., Bucio, E., & Kadlubowski, S. S. (. (2020). An updated review of macro, micro, and nanostructured hydrogels for biomedical and pharmaceutical applications. *Pharmaceutics*, *12*(10), 970. DOI: 10.3390/pharmaceutics12100970 PMID: 33076231

Maitra, J., & Shukla, V. K. (2014). Cross-linking in hydrogels-a review. *American Journal of Political Science*, *4*(2), 25–31.

Maji, S., & Lee, H. (2022). Engineering hydrogels for the development of three-dimensional in vitro models. *International Journal of Molecular Sciences*, *23*(5), 2662. DOI: 10.3390/ijms23052662 PMID: 35269803

Maleki, B., Kargar, P. G., & Ashrafi, S. S. (2025). Perspective Chapter: Introduction to Hydrogels–Definition, Classifications, Applications and. *Ionic Liquids-Recent Advances: Recent Advances*, 149.

Manjit, M., & Mishra, B. (2024). Synthetic Polymer-Based Hydrogels for Tissue Engineering. In *Biomaterial-based Hydrogels: Therapeutics Carrier and Tissue Regeneration* (pp. 357–385). Springer. DOI: 10.1007/978-981-99-8826-6_13

Michálek, J., Podešva, J., & Dušková-Smrčková, M. (2022). True story of poly (2-hydroxyethyl methacrylate)-based contact lenses: How did it really happen. *Substantia*, *6*(2), 79–91.

Mittal, H., Ray, S. S., & Okamoto, M. (2016). Recent progress on the design and applications of polysaccharide-based graft copolymer hydrogels as adsorbents for wastewater purification. *Macromolecular Materials and Engineering*, *301*(5), 496–522. DOI: 10.1002/mame.201500399

Nguyen, H. D., Liu, H.-Y., Hudson, B. N., & Lin, C.-C. (2019). Enzymatic cross-linking of dynamic thiol-norbornene click hydrogels. *ACS Biomaterials Science & Engineering*, *5*(3), 1247–1256. DOI: 10.1021/acsbiomaterials.8b01607 PMID: 33304998

Oryan, A., Kamali, A., Moshiri, A., Baharvand, H., & Daemi, H. (2018). Chemical crosslinking of biopolymeric scaffolds: Current knowledge and future directions of crosslinked engineered bone scaffolds. *International Journal of Biological Macromolecules*, *107*, 678–688. DOI: 10.1016/j.ijbiomac.2017.08.184 PMID: 28919526

Ostroha, J., Pong, M., Lowman, A., & Dan, N. (2004). Controlling the collapse/swelling transition in charged hydrogels. *Biomaterials*, *25*(18), 4345–4353. DOI: 10.1016/j.biomaterials.2003.11.019 PMID: 15046925

Palmese, L. L., Thapa, R. K., Sullivan, M. O., & Kiick, K. L. (2019). Hybrid hydrogels for biomedical applications. *Current Opinion in Chemical Engineering*, *24*, 143–157. DOI: 10.1016/j.coche.2019.02.010 PMID: 31844607

Picchioni, F., & Muljana, H. (2018). Hydrogels based on dynamic covalent and non covalent bonds: A chemistry perspective. *Gels (Basel, Switzerland)*, *4*(1), 21. DOI: 10.3390/gels4010021 PMID: 30674797

Plamper, F. A., & Richtering, W. (2017). Functional microgels and microgel systems. *Accounts of Chemical Research*, *50*(2), 131–140. DOI: 10.1021/acs.accounts.6b00544 PMID: 28186408

Pujana, M. A., Pérez-Álvarez, L., Iturbe, L. C. C., & Katime, I. (2012). Water dispersible pH-responsive chitosan nanogels modified with biocompatible crosslinking-agents. *Polymer*, *53*(15), 3107–3116. DOI: 10.1016/j.polymer.2012.05.027

Raffe, S. N. A. M., Yusof, R., Hanaphi, R. M., & Chik, M. A. W. (2024). Enhanced water absorption in pectin-chitosan hydrogel via modification with deep eutectic solvent: Optimisation and characterisation. *Journal of Molecular Liquids, 413*, 125963. DOI: 10.1016/j.molliq.2024.125963

Rajabi, N., Rezaei, A., Kharaziha, M., Bakhsheshi-Rad, H. R., Luo, H., RamaKrishna, S., & Berto, F. (2021). Recent advances on bioprinted gelatin methacrylate-based hydrogels for tissue repair. *Tissue Engineering. Part A, 27*(11–12), 679–702. DOI: 10.1089/ten.tea.2020.0350 PMID: 33499750

Ramazani, A., & Aghahosseini, H. (2020). The biological properties of hydrogels based on natural polymers. In *Hydrogels based on natural polymers* (pp. 247–269). Elsevier. DOI: 10.1016/B978-0-12-816421-1.00009-4

Rizwan, M., Yahya, R., Hassan, A., Yar, M., Azzahari, A., Selvanathan, V., Sonsudin, F., & Abouloula, C. (2017). pH Sensitive Hydrogels in Drug Delivery: Brief History, Properties, Swelling, and Release Mechanism, Material Selection and Applications. *Polymers, 9*(12), 137. DOI: 10.3390/polym9040137 PMID: 30970818

Roy, A., Manna, K., & Pal, S. (2022). Recent advances in various stimuli-responsive hydrogels: From synthetic designs to emerging healthcare applications. *Materials Chemistry Frontiers, 6*(17), 2338–2385. DOI: 10.1039/D2QM00469K

Rumon, M. M. H. (2025). Advances in cellulose-based hydrogels: Tunable swelling dynamics and their versatile real-time applications. *RSC Advances, 15*(15), 11688–11729. DOI: 10.1039/D5RA00521C PMID: 40236573

Singhal, R., & Gupta, K. (2016). A review: Tailor-made hydrogel structures (classifications and synthesis parameters). *Polymer-Plastics Technology and Engineering, 55*(1), 54–70. DOI: 10.1080/03602559.2015.1050520

Sivaram, A. J., Rajitha, P., Maya, S., Jayakumar, R., & Sabitha, M. (2015). Nanogels for delivery, imaging and therapy. *Wiley Interdisciplinary Reviews. Nanomedicine and Nanobiotechnology, 7*(4), 509–533. DOI: 10.1002/wnan.1328 PMID: 25581024

Sobczak, M. (2022). Enzyme-responsive hydrogels as potential drug delivery systems—State of knowledge and future prospects. *International Journal of Molecular Sciences, 23*(8), 4421. DOI: 10.3390/ijms23084421 PMID: 35457239

Solanki, R., & Bhatia, D. (2024). Stimulus-responsive hydrogels for targeted cancer therapy. *Gels (Basel, Switzerland), 10*(7), 440. DOI: 10.3390/gels10070440 PMID: 39057463

Sood, N., Bhardwaj, A., Mehta, S., & Mehta, A. (2016). Stimuli-responsive hydrogels in drug delivery and tissue engineering. *Drug Delivery, 23*(3), 748–770. DOI: 10.3109/10717544.2014.940091 PMID: 25045782

Spicer, C. D. (2020). Hydrogel scaffolds for tissue engineering: The importance of polymer choice. *Polymer Chemistry, 11*(2), 184–219. DOI: 10.1039/C9PY01021A

Štular, D., Simončič, B., & Tomšič, B. (2017). Stimuli-responsive Hydrogels for Textile Functionalisation: A Review. *Tekstilec, 60*(2).

Tan, J., Luo, Y., Guo, Y., Zhou, Y., Liao, X., Li, D., Lai, X., & Liu, Y. (2023). Development of alginate-based hydrogels: Crosslinking strategies and biomedical applications. *International Journal of Biological Macromolecules, 239*, 124275. DOI: 10.1016/j.ijbiomac.2023.124275 PMID: 37011751

Tripathy, D., Gadtya, A. S., & Moharana, S. (2023). Supramolecular Gel, Its classification, preparation, properties, and applications: A review. *Polymer-Plastics Technology and Materials, 62*(3), 306–326. DOI: 10.1080/25740881.2022.2113892

Utech, S., & Boccaccini, A. R. (2016). A review of hydrogel-based composites for biomedical applications: Enhancement of hydrogel properties by addition of rigid inorganic fillers. *Journal of Materials Science, 51*(1), 271–310. DOI: 10.1007/s10853-015-9382-5

Wang, X.-H., Song, F., Qian, D., He, Y.-D., Nie, W.-C., Wang, X.-L., & Wang, Y.-Z. (2018). Strong and tough fully physically crosslinked double network hydrogels with tunable mechanics and high self-healing performance. *Chemical Engineering Journal, 349*, 588–594. DOI: 10.1016/j.cej.2018.05.081

Wang, Z., Ye, Q., Yu, S., & Akhavan, B. (2023). Poly ethylene glycol (PEG)-based hydrogels for drug delivery in cancer therapy: A comprehensive review. *Advanced Healthcare Materials, 12*(18), 2300105. DOI: 10.1002/adhm.202300105 PMID: 37052256

Yang, D. (2022). Recent advances in hydrogels. []. ACS Publications.]. *Chemistry of Materials, 34*(5), 1987–1989. DOI: 10.1021/acs.chemmater.2c00188

Yang, J., Chen, Y., Zhao, L., Zhang, J., & Luo, H. (2023). Constructions and properties of physically cross-linked hydrogels based on natural polymers. *Polymer Reviews (Philadelphia, Pa.), 63*(3), 574–612. DOI: 10.1080/15583724.2022.2137525

Zare, M., Bigham, A., Zare, M., Luo, H., Rezvani Ghomi, E., & Ramakrishna, S. (2021). pHEMA: An overview for biomedical applications. *International Journal of Molecular Sciences, 22*(12), 6376. DOI: 10.3390/ijms22126376 PMID: 34203608

Zhang, F., & King, M. W. (2020). Biodegradable polymers as the pivotal player in the design of tissue engineering scaffolds. *Advanced Healthcare Materials, 9*(13), 1901358. DOI: 10.1002/adhm.201901358 PMID: 32424996

Zhao, J., Yuan, X., Wu, X., Liu, L., Guo, H., Xu, K., Zhang, L., & Du, G. (2023). Preparation of nanocellulose-based aerogel and its research progress in wastewater treatment. *Molecules (Basel, Switzerland), 28*(8), 3541. DOI: 10.3390/molecules28083541 PMID: 37110772

Zhu, Y., Li, S., Li, J., Falcone, N., Cui, Q., Shah, S., Hartel, M. C., Yu, N., Young, P., de Barros, N. R., Wu, Z., Haghniaz, R., Ermis, M., Wang, C., Kang, H., Lee, J., Karamikamkar, S., Ahadian, S., Jucaud, V., & Khademhosseini, A. (2022). Lab-on-a-Contact lens: Recent advances and future opportunities in diagnostics and therapeutics. *Advanced Materials, 34*(24), 2108389. DOI: 10.1002/adma.202108389 PMID: 35130584

Zhu, Y., Zeng, Q., Zhang, Q., Li, K., Shi, X., Liang, F., & Han, D. (2020). Temperature/near-infrared light-responsive conductive hydrogels for controlled drug release and real-time monitoring. *Nanoscale, 12*(16), 8679–8686. DOI: 10.1039/D0NR01736A PMID: 32253408

Chapter 2
Hydrogels Processing Techniques

Aqib Mehmood
University of Wah, Pakistan

Irum Fatima
https://orcid.org/0009-0008-4202-4170
University of Wah, Pakistan

Muhammad Hamza
East China University of Science and Technology, China

ABSTRACT

Hydrogels have emerged as an adaptable class of biomaterials because of their significant applications in the biomedical field, environmental remediation, agriculture, cosmetics, and food industry. Over the years, numerous methods have been reported to have control over the properties of hydrogels. This section follows the physical and chemical processes of crosslinking polymers, hybrid strategies, and polymerization techniques that have revolutionized the processing of hydrogels in numerous ways. This chapter also highlights the role of constituent molecules, including monomers, cross-linkers, and nanocomposite materials in synthesizing hydrogels for specific applications, such as stimuli sensitivity, swellability, biodegradability, biocompatibility, and mechanical strength. The biocompatible and stability challenges have also been addressed as a significant hurdle for their scalability.

2.1 INTRODUCTION

Hydrogels are networks of polymer chains formed in water that retain a stable three-dimensional structure while absorbing large amounts of water. The network's crosslinks can be physical (chain entanglements or crystalline regions) or chemical (covalent bonds) and arise through covalent bonding, hydrogen bonding, van der Waals forces, or chain intertwining. Their affinity for water can be attributed to hydrophilic units like -OH, -CONH-, $-CONH_2-$, and $-SO_3H$. Gelation can be induced via physical interactions, electrostatic attraction, or chemical crosslinking (Drury

DOI: 10.4018/979-8-3373-2337-4.ch002

& Mooney, 2003; Gupta et al., 2002; Kirschner & Anseth, 2013; Rupali et al., 2010). Physical gelation relies on inherent polymer traits and thus limits fine control of gel properties, whereas chemical crosslinking enables precise and dynamic network formation (M. Bustamante-Torres et al., 2021; Zhao et al., 2018). Conventional hydrogels often have weak structures with poor mechanical durability and limited responsiveness. Engineering the molecular structure and network architecture can improve performance (Naficy et al., 2011; Zhao et al., 2018). Typical hydrogels have low elastic moduli (~10 kPa) and fracture energies (~10 J/m^2), rendering them brittle under even small loads (Li & Gong, 2024). Additionally, with up to 95% water by weight, the concentration of covalent load-bearing strands is very low, so even moderate stresses can exceed the material's capacity, causing mechanical failure (Majcher & Hoare, 2019). Conventional, permanently cross-linked hydrogels often suffer from uncontrolled, bulk (or heterogeneous surface) erosion in physiological environments, especially those based on polyester backbones (e.g., poly(lactic-co-glycolic acid) blends or ester-linked PEG networks). As water infiltrates the network, hydrolysis produces small organic acids (e.g., lactic and glycolic acids) that locally lower pH, catalyze further cleavage, and weaken or embrittle the gel matrix (Garcia-Garcia et al., 2024). Furthermore, because their strands are fixed by irreversible covalent bonds, traditional hydrogels respond slowly to external triggers (pH, temperature, light, ions): the polymer chains cannot reorganize quickly, so swelling–deswelling or shape changes lag behind the stimulus or do not fully develop before damage occurs (Amirthalingam et al., 2023). However, innovative engineering approaches (e.g., double networks, nanocomposites, dynamic bonds) are gradually overcoming these issues barriers. Double networks combine rigid and ductile networks (e.g., alginate/polyacrylamide) for toughness (Dhand et al., 2021). The reinforcement of nanoparticles, such as silica, cellulose nanocrystals, to enhance strength (Lu et al., 2025). Hydrogels incorporating pH-sensitive dynamic bonds (e.g., boronate esters) for controlled degradation (Terriac et al., 2024). Hybrid natural/synthetic networks blend natural polymers, such as alginate, with synthetic polymers to stabilize degradation rates (Ren et al., 2022). Comb-type grafted hydrogels attach mobile side chains to accelerate swelling/deswelling (Yoshida et al., 1995). Likewise, breakthroughs in chemical approaches, for instance, click chemistry, the fusion of various gelation techniques and the inclusion of nanomaterials, have brought superior control over the properties of hydrogels (Zhang & Khademhosseini, 2017).

Crosslinking strategies are employed to fabricate hydrogels. The freeze-thaw crystallization can synthesize reversible hydrogels. As the solvent is evaporated, microcrystals emerge within the framework (Waresindo et al., 2023). Amphiphilic hydrogels contain dual segments, hydrophilic and hydrophobic. Their self-assembling traits are advantageous in forming a diverse range of hydrogels due to their biological origin, and thus, they are valuable in biological fields (Chang et al., 2019). An ionotropic gel is a physical gel formed by crosslinking positively and negatively charged polymers at varying concentrations and pH of the solution. These poly-ionic complexes can also be de-crosslinked by reversible processes (Caló & Khutoryanskiy, 2015; Papadakis & Tsitsilianis, 2017). Natural polymers form hydrogels due to hydrogen bonds, for instance, cellulose (has -OH group) (Chang & Zhang, 2011), alginate (has -OH, -COOH, -R-COO- groups) (Şolpan et al., 2008; Sun & Tan, 2013) and chitosan (has -OH, -NH_2) (Bhattarai et al., 2010). Stereocomplexing signifies intermolecular interactions of chemically identical compounds but with different stereochemistry. Crosslinking by protein interactions can be accomplished through genetically engineered proteins or antigen-antibody interactions (Tirrell & Tirrell, 1996) heating/cooling of a polymer solution (Fernández et al., 2008) and heat-induced aggregation (Bhattacharjee et al., 2016). For example, irradiation and grafting copolymerization (Munoz-Munoz et al., 2012) can be used to establish covalent bonds between the functional groups (Peppas, 2010) (mainly -COOH, $-NH_2$ and -OH) of polymer chains (Gulrez et al., 2011).

2.2 CLASSIFICATIONS OF HYDROGELS

Hydrogels have been categorized on the basis of their method of synthesis, type of crosslinking, the polymeric composition, source, charge, and environment sensitivity.

Figure 1. Classification of Hydrogels

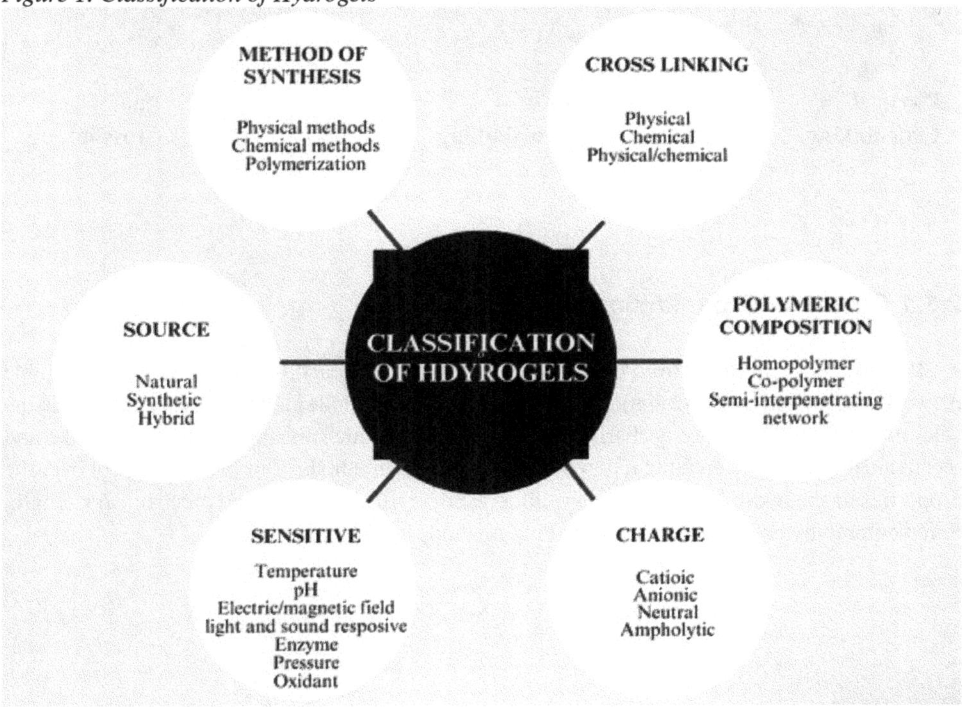

2.3 TYPES OF METHODS EMPLOYED FOR HYDROGELS SYNTHESIS

Hydrogel synthesis methods fall into broad categories: physical (reversible) and chemical (irreversible) crosslinking. Hybrids combining physical and chemical crosslinks are also used to enhance properties.

Figure 2. Methods employed for Intelligent Hydrogel Synthesis

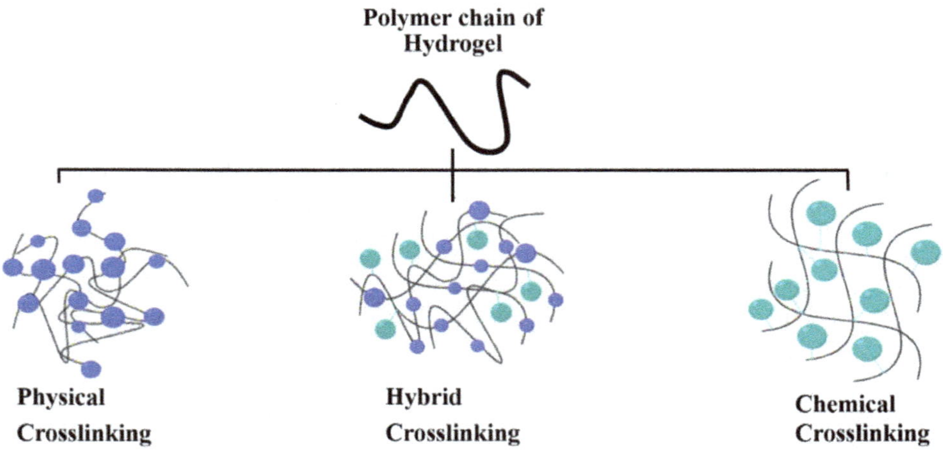

2.3.1 Chemical Crosslinking Methods

Physical gels can suffer from loop defects, whereas chemical crosslinking produces a stable network via covalent bonds. Common chemical crosslinkers include glutaraldehyde, formaldehyde, calcium salts, oxidized polysaccharides and dialdehydes, which enhance both water uptake and mechanical strength by creating a porous network. These methods offer precise control over the properties of the hydrogel, including swelling response, degradability and mechanical strength. Some commonly used chemical crosslinking methods include:

Figure 3. Chemical Crosslinking Methods

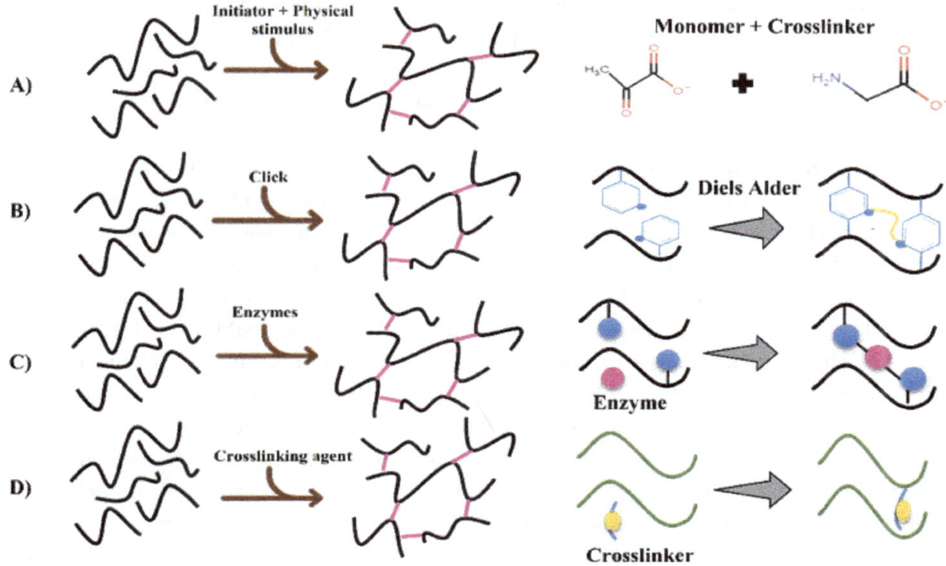

2.3.2 Click-Chemistry Reactions

Click chemistry reactions, such as thiol-ene or azide-alkyne reactions (Yu et al., 2015), offer efficient and selective crosslinking methods for hydrogel synthesis. These reactions allow for the incorporation of stimuli-responsive functional groups through the click chemistry reactions during hydrogel formation (Li et al., 2015). Click chemistry encompasses various high-yield, selective reactions ideal for cross-linking polymers. Common "click" reactions used in hydrogel formation include azide-alkyne cycloadditions (both copper-catalyzed CuAAC and strain-promoted SPAAC), thiol-based couplings (such as radical thiol-ene/yne, nucleophilic thiol-Michael additions, thiol-epoxy, and thiol-maleimide), Diels-Alder cycloadditions (like the classical furan-maleimide and inverse-electron-demand tetrazine-norbornene), oxime/hydrazone (Schiff-base) ligations, and emerging sulfur(VI) fluoride exchange (SuFEx) chemistry (X. Li & Y. Xiong, 2022; Liu et al., 2017; Reddy et al., 2021). These reactions typically proceed under mild, aqueous conditions, forming stable covalent bonds (such as triazoles, thioethers, and cycloadducts) with high selectivity.

A widely used click reaction for hydrogel synthesis is the CuAAC, commonly called the Huisgen reaction. CuAAC requires a copper(I) catalyst, typically 0.1-1 mM $CuSO_4$, along with reducing agents (e.g., sodium ascorbate) and stabilizing ligands (e.g., Tris-hydroxypropyltri-azolylamine, THPTA), to decrease cytotoxicity (J. Gopinathan & I. Noh, 2018; X. Li & Y. Xiong, 2022). The process is conducted at 25-37°C (Finetti et al., 2017), using aqueous buffers or safe organic solvents (e.g., water/DMSO mixtures) at neutral pH (J. Gopinathan & I. Noh, 2018). Radical-mediated thiol-ene reactions require UV light (~365 nm) and photoinitiators, such as lithium phenyl-2,4,6-trimethylbenzoylphosphinate (LAP), to generate thiyl radicals (Laurano et al., 2023). The reaction occurs at room temperature with a near-neutral pH (pH 6-8). Visible-light systems (like eosin Y) reduce cytotoxicity but needed triethanolamine as a co-initiators. Thiol-ene reactions are less inhibited by oxygen than other radical polymerizations (Batibay & Arsu, 2023; P. M. Kharkar et al., 2016).

CuAAC couples azide- and alkyne-functionalized polymers into triazole-crosslinked networks (Lau et al., 2017; Singh et al., 2023). Strain-promoted azide-alkyne cycloaddition (SPAAC) uses cyclooctyne rings (like dibenzocyclooctyne) to enable catalyst-free gelation (He et al., 2023; Mehak et al., 2024). Radical thiol-ene chemistry (often initiated by light) quickly links thiols to alkenes or norbornenes, while base-catalyzed Michael addition of thiols to electron-poor vinyls (such as acrylates, vinyl sulfones, or maleimides) creates covalent thioether networks (Kenari et al., 2021; Devatha P. Nair et al., 2014). Classical Diels-Alder gels form via cycloaddition of furan (a diene) and maleimide (a dienophile) end-groups, allowing reversible cross-links, whereas inverse-electron-demand Diels-Alder (tetrazine ligation) between tetrazines and strained alkenes cures very rapidly (Nagaraj et al., 2025; Wang et al., 2024). Oxime/hydrazone ligations also produce self-healing, reversible hydrogels (Kölmel & Kool, 2017). Click chemistry refers to highly selective and efficient reactions that rapidly form covalent bonds between functional groups.

Organo-hydrogels synthesized via click-chemistry have a broad spectrum of applications in biomedicines. Researchers are advancing pharmaceutical development through hyaluronic acid-based hydrogel designs. In the latest study, researchers presented "thiol-ene" reactions to fabricate chitosan/thiol functionalized-GO hydrogel. The nanocomposite hydrogel possessed biodegradable properties, compressibility, shape retrieval (Salimiyan et al., 2023) and electrical responsiveness (Selvam et al., 2023). While addressing the shortcomings of collagen hydrogels, Rosenquist et al. (2023) reported an injectable collagen hydrogel crosslinked using thiol-maleimide click-chemistry. Crosslinking was due to the reaction between PEG-maleimide and acetyl thiol collagen. Researchers transformed it into a valuable material with enhanced capabilities (Rosenquist et al., 2023).

Multi-network and IPN hydrogels can be formed through sequential click reactions. For example, an IPN hydrogel was created by first conducting a Diels-Alder crosslinking reaction on modified starch and cellulose to establish a primary network, followed by in situ-photopolymerization of acrylamide to generate a secondary network (Bailey et al., 2023). Similarly, triple-network HA-chitosan gels were produced by coupling azide-HA and alkyne-HA via CuAAC while embedding chitosan, which both complexes copper and forms ionic crosslinks (Engkagul et al., 2018).

Many click reactions proceed under gentle conditions (e.g., room temperature, aqueous media) without requiring harsh catalysts. SPAAC needs no metal catalyst and occurs rapidly at ambient temperature (Gopinathan & Noh, 2018), avoiding toxic residues that can be present in conventional initiator-driven methods. The tetrazine–cyclooctyne click (IEDDA) surpasses even CuAAC in kinetics. Rapid gelation is useful for in situ-forming hydrogels. Light-triggered click reactions (thiol–ene or photoinitiated azide–alkyne) provide precise control over when and where the gel forms.

Each click approach has trade-offs. The CuAAC reaction, although reliable, requires copper catalysts that can be toxic. SPAAC bypasses metal but depends on large, synthetically demanding cyclooctyne reagents. Radical thiol–ene reactions require photoinitiators and UV light, which may not be cell-friendly at high doses. Thiol-Michael and thiol-maleimide additions generally proceed under mild bases but are sensitive to pH and can be slow without catalysts. Diels-Alder cross-linking is catalyst-free and reversible, but standard furan–maleimide pairs often need elevated temperature or long periods. Some click bonds (triazoles, thioethers) are extremely stable and non-degradable, which might be undesirable for resorbable biomaterials. Overall, click chemistries typically require careful control of stoichiometry and sometimes an inert atmosphere or deoxygenated conditions (for radical steps). Despite these challenges, click hydrogels generally outperform traditional cross-linking methods (such as glutaraldehyde or EDC/NHS) in yield and selectivity (Ghorpade, 2020; Xu & Bratlie, 2018).

Figure 4. Chemical Crosslinking by Click Chemistry

2.3.3 Enzyme-Catalyzed Crosslinking Reactions

Enzymes such as horseradish peroxidase (HRP, with H_2O_2), transglutaminase, or tyrosinase (with O_2) catalyze bond formation (e.g., coupling phenols, thiols, and glutamine-lysine) under physiological conditions. Reactions proceed at room temperature to body temperature (20-37°C), with a pH of ~6-8, in an aqueous buffer. For instance, HRP/H_2O_2 crosslinks tyramine-functionalized polymers in seconds to minutes (Ziyuan Li et al., 2023).

Sperinde & Griffith (1997) pioneered a new method by employing enzymes to produce PEG-based gels. By the utilization of transglutaminase, they catalyzed a reaction between poly(lysine-co-phenylalanine) and PEG-Qa, thereby forming PEG networks after modifying glutaminyl groups with tetrahydroxy PEG (PEG-Qa) (McHale et al., 2005; Sperinde & Griffith, 1997). Enzymatic crosslinking offers alluring potential for the kinetic manipulation of in situ gel formation, enabled by precise control over enzyme concentrations. This methodology presents the immense advantage of forging strongly covalent bonds, accompanied by swift gelation, often within 10 minutes, all under physiological conditions. One such method entails the collaboration of transglutaminase (TG) and calcium ions as essential cofactors. The union of carboxamide and amine groups through amide linkages flourishes under this catalytic influence, opening new possibilities for hydrogel creation. Alternatively, the horseradish peroxidase (HRP)-catalyzed crosslinking reaction presents an intriguing avenue. In the presence of hydrogen peroxide (H_2O_2), HRP orchestrates the coupling of aniline (Jin et al., 2009), phenol (Kurisawa et al., 2005), and their derivative Tyra mine (Kuo et al., 2015), setting the stage for captivating transformations.

Notably, an injectable collagen-tyramine hydrogel crosslinked by HRP/ H_2O_2 supported in situ cell delivery and stimulated vascularized bone formation in vivo (Chang et al., 2010). These methods offer precise control of gelation kinetics and biocompatibility, and are emerging as powerful tools for tissue engineering.

Figure 5. Chemical Crosslinking of Hydrogels by Enzymatic Reaction

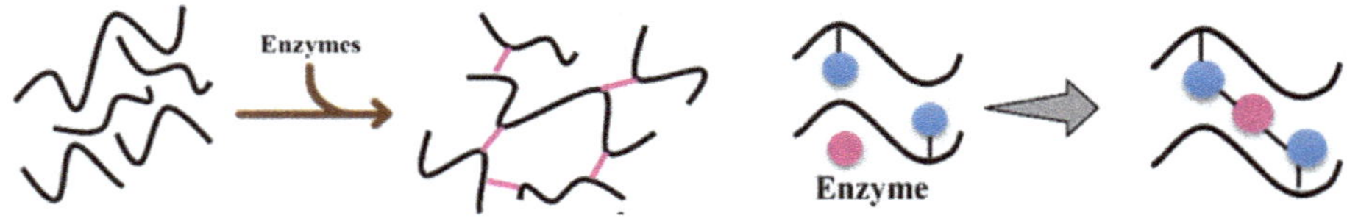

2.3.4 Use of Crosslinking Agents

The amalgamation of a crosslinking agent within hydrogel synthesis culminates in a three-dimensional network facilitated by chemical crosslinking, physical linkage, ionic, and hydrogen bonding (Franklin & Guhanathan, 2014). Chemical crosslinking facilitates the creation of chemical bonds among molecular chains, thus weaving an interconnected three-dimensional network (Winarti et al., 2018). The hydrogel synthesis employs a crosslinking agent that can be derived from either natural or synthetic sources. For the cellulose derivative utilized, several crosslinking agents, namely ECH (Distantina et al., 2012), aldehyde-based reagents (Dalton et al., 2008), urea derivatives (Demitri et al., 2008), and multifunctional carboxylic acids (Sannino et al., 2009), find application in forming the hydrogel structure. However, some crosslinkers can be toxic if not fully reacted. Thus, thorough washing to remove residual reagents is critical to ensure biocompatibility and safety.

Figure 6. Chemical crosslinking via Crosslinking Agent

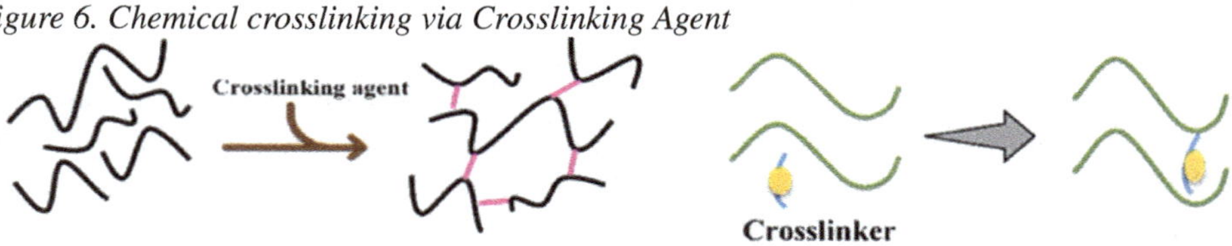

2.3.5 Use of Complementary Groups

Polymers containing hydrophilic groups on their chains can be used to prepare chemical hydrogels. Michael-type addition reactions entail the nucleophilic addition of a nucleophile, often an enolate or enolate equivalent, to an α, β-unsaturated carbonyl compound. Common nucleophilic functional groups include thiols or amines (Tortora et al., 2007). Michael-type additions improved PVA hydrogel's homogeneity by incorporating amine-contained peptides (Guo et al., 2021). In another research, chitosan-based hydrogels were prepared by a thiol-Michael reaction, covalently crosslinked with bismaleimide (Guaresti et al., 2019).

Compounds with aldehyde functional groups can be used to crosslink polymers containing hydroxyl and amino groups, such as forming a Schiff base. Glutaraldehyde is one of the most extensively researched aldehyde crosslinking agents in this field, and it reacts with amino groups.

This reaction produces a myriad of hydrogel formation examples. Albumin, collagen, gelatin, and other proteins undergo crosslinking via this reaction. Moreover, polysaccharides, such as chitosan, guar gum, sodium alginate, and carrageenan, also undergo the process of crosslinking (M. Bustamante-Torres et al., 2021).

The synthesis of polyesters or polyamides involves condensation reactions between hydroxyl/amino groups with carboxyl groups (or derivatives), resulting in additional crosslinking. Researchers employed a facile synthesis of hydrogels by reacting o-phthalaldehyde (OPA) and N-nucleophiles like primary hydrazide, (Moises Bustamante-Torres et al., 2021) amines and aminooxy. A covalent C=C bond is dynamically established in another study through a Knoevenagel condensation reaction between PEGDCA and water-soluble poly(vanillin acrylate) (Jiao et al., 2020).

2.3.6 Chemical Crosslinking by Protein Interactions

The stability of hydrogels in aqueous solutions is essential, emphasizing the preference for covalently crosslinked molecular networks over physical ones. Diverse protein chemistries utilizing residue-specific, sequence-specific, and genetically encoded click chemistries enhance the gel's stability and broaden its mechanical and biochemical properties. By targeting the reactive side of amino acids, genetically engineered hydrogels can be synthesized. Recombinant proteins are chemically crosslinked into hydrogels by using TYR. This linkage is accomplished through either the self-assembly mediated by photosensitizers (ruthenium(II) tris-bipyridyl) or the enzymatic oxidation method (horseradish peroxidase) (Elvin et al., 2005; Jus et al., 2011). The integration of short peptide motifs from Mfp-3 and Mfp-5 modified elastin-like polypeptides. Their enzymatic oxidation yielded hydrogels with hydrophobic and adhesive properties (Park et al., 2019). The same group made some advancements using stem cell delivery/encapsulation (Luo et al., 2018).

Crafting covalent links between proteins using di-tyrosine linkages is accomplished by horseradish peroxidase-hydrogen peroxide system (Bailey, 1991). Nucleophilic cysteine has also been used for the purpose of protein assembly. The structure of cysteine contains a sulfhydryl group, allowing it to react with oxidants (Li et al., 2013) and electrophiles (Devatha P Nair et al., 2014). Therefore, it forms disulphide bonds under oxidized conditions, facilitating the hydrogel synthesis (Salinas & Anseth, 2008). It has been observed that reversible hydrogels formed by this approach are unstable in redox environments. However, Thiol-ene chemistry and Michael addition reactions have improved hydrogels' stability. Therefore, it is an alternative strategy for engineering protein-polymer hybrid hydrogels (Hörner et al., 2019; Kou et al., 2017). The SpyCatcher/SpyTag approach has also synthesized chemically crosslinked hydrogels by protein-peptide interactions (Stanfield & Wilson, 1995). Researchers have adopted this approach to synthesize tandem modular protein-based architectures with multiple spy sequences, which rapidly react to form hydrogels (Gao et al., 2016). Sortase and butelase are used for the enzymatic ligation of several proteins (Schmidt et al., 2017) and peptides (Nguyen et al., 2014; Shadish et al., 2019). Their direct integration into the genetically engineered hydrogel may impede protein ligation efficiency. Therefore, researchers incorporated recombinant protein into the hydrogels post-gelation using sortase (Shadish et al., 2019).

2.3.7 Chemical Crosslinking by Dynamic Covalent Reactions

It involves a set of reversible chemical reactions that can form/break the covalent bond within the hydrogel's network. This is often referred to as a reversible covalent bond. Dynamic covalent chemistry fully governs the physicochemical properties of hydrogels. Environmental changes, such as variations in pH, temperature, or the presence of specific chemicals, lead to cleavage and formation of covalent bonds. The integrity of the hydrogel can also be restored if the material is damaged, thus exhibiting self-healing capabilities.

Figure 7. Crosslinking of Hydrogels by Dynamic Covalent Reactions

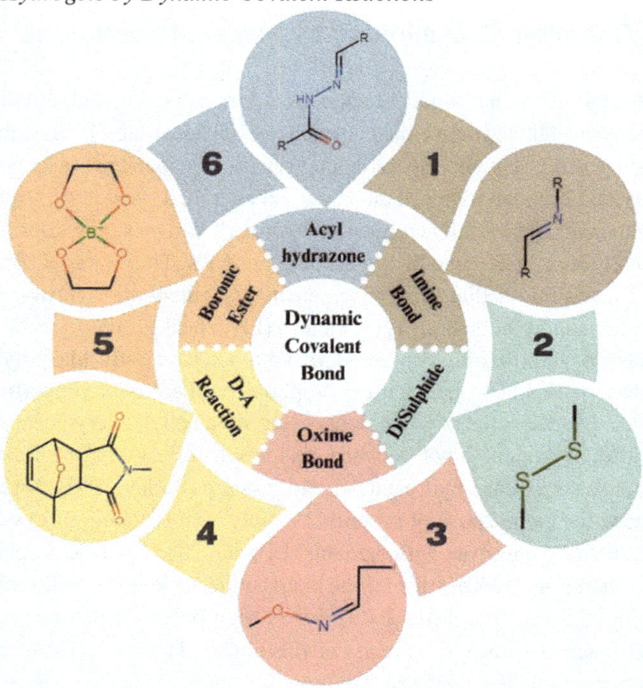

Recently, hydrogels were prepared using recombinant biopolymers. HA and elastic protein were functionalized with benzaldehyde and hydrazine, respectively. The experiments deduced that the higher degree of functionalization offers the faster self-healing potential of hydrogels while lowering the molecular weight of formulations enhances their injectability (de Paiva Narciso et al., 2023).

Boronic ester is a type of reversible covalent bond. It is formed by the condensation reaction of boronic acid derivatives (typically PBA) with a diol. Alkaline pH facilitates these reactions; however, at physiological conditions, the availability of hydroxyl ions is limited. To tackle this problem, researchers used wPBA/glucamine to establish a reversible covalent bond between the HA-based hydrogels at physiological conditions (Lagneau et al., 2023; S. Wang et al., 2019).

Similarly, another group of scientists put forth their research to accelerate the kinetics of the reaction under such conditions. They explored the potential of Na, K, Li, Mg and Ca salts as catalysts for acylhydrazone and oxime crosslinked HA-based hydrogels. Sodium chloride salt has shown a dose-dependent pattern. In contrast, magnesium and calcium chloride salts significantly reduced the gelling time and enhanced the efficiency between acyl hydrazone and oxime dynamic reactions.

Ji et al., (2023) introduced a unique way to connect PVA chains by dynamic crosslinking. Picture this as a dynamic borate-imine-imine-borate bond configuration forming a supportive network among the chains of PVA. Crosslinkers like FPBA and Lys facilitate this reaction. The resulting hydrogel was highly stretchable without losing its structure and exhibited exceptional self-healing ability. It can also track and analyze human body motion in real time and can be employed in voice recognition sensors (Ji et al., 2023).

2.3.8 Physical Crosslinking Methods

Physical crosslinking methods rely on non-covalent interactions to form reversible physical crosslinks within the hydrogel matrix. These interactions can include hydrogen bonding, electrostatic interactions, host-guest interactions, or physical entanglements (Fig. 2.7). Physical crosslinking methods offer advantages such as ease of synthesis, reversibility, and tunable responsiveness.

Figure 8. Methods for Physical crosslinking (A) Hydrogen Bonding, (B) Ionic Interactions, (C) Protein Interactions and (D) Amphiphilic Arafts

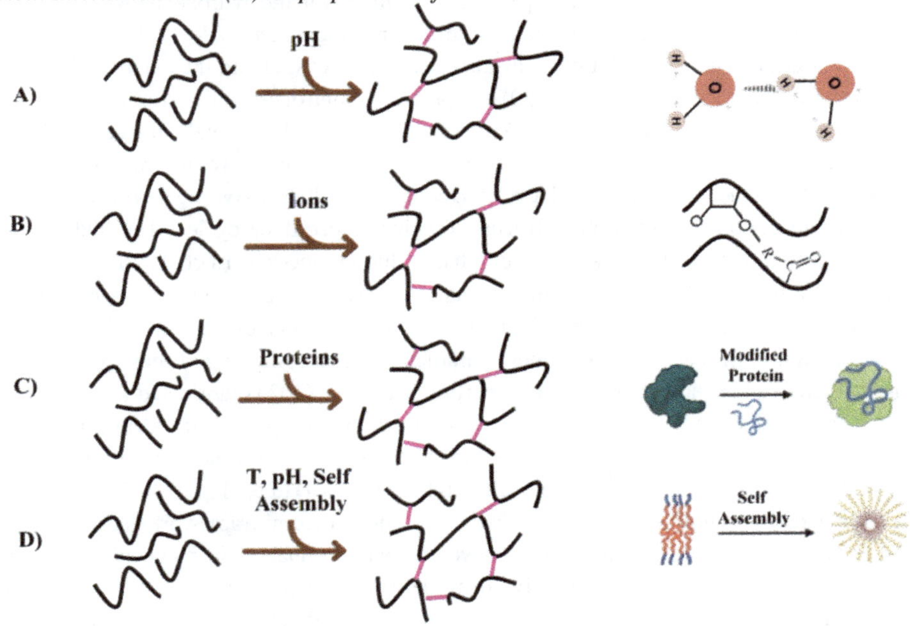

a) Crystallization

Certain polymers can physically crosslink by forming microcrystalline domains. Gelation occurs when polymer chains order into crystalline regions (e.g., during freeze–thaw cycling or through stereocomplexation of enantiomeric polymers), and these ordered domains act as junction points in the network.

b) Freeze-Thaw Gelation

Repeated cycles of freezing and thawing a polymer solution can produce a physical hydrogel. During freezing, ice crystals form in polymer-poor regions, concentrating the polymers into microdomains. Upon thawing, the ice melts and the concentrated polymers reorganize, forming a network of microcrystals, linked by hydrogen bonds. This method is particularly effective for water-soluble, semi-crystalline polymers (e.g., PVA, xanthan) that undergo cryoconcentration (Bernal-Chávez et al., 2023; Holkar et al., 2025). For example, PVA's hydroxyl groups promote crystallization during repeated freeze-thaw cycles (e.g., -20°C for 6-24 hours per cycle) (Sakaguchi et al., 2017; Soto-Bustamante et al., 2023). Key factors include polymer chemistry (e.g., tacticity, molecular weight), concentration, freezing and thawing rates, and the number of cycles (Bernal-Chávez et al., 2023; Holkar et al., 2025). While non-crystallizable polymers (e.g., gelatin) can form cryogels when mixed with crystallizable polymers, nonpolar plastics (such as polyethylene, polypropylene, and polystyrene), which lack hydrophilicity and hydrogen bonding sites and chain mobility, do not form gels at all (Y. Liu et al., 2010; Maitra & Shukla, 2014). The formation of PVA crystallites facilitates physical crosslinking in the hydrogel network. However, hydrogel properties can be tuned by the maximum and minimum thermal level, duration of cycles and concentration. The manipulation of average molecular weight and the extent of hydrolysis (Shoukat et al., 2016; Willcox et al., 1999) can precisely control crystallization.

Zhang et al. (2023) synthesized a PVA/succinoglycan-based hydrogel, adjusted to enhance its swelling rate and drug encapsulation efficiency. The maximum swelling rate was obtained with a 50/50 polymer ratio and 5 F-T cycles (Jeong et al., 2023). Hydrogel with components like PVA and 2,4,6-tri(4-aminophenyl) triazine using freeze-thaw cycles. The hydrogel was synthesized with improved mechanical strength and fluorescence (X. Li et al., 2023; Zhou Li et al., 2023). In the TEMPO-oxidized cellulose nanofiber-cationic guar gum hydrogel synthesis, the components were adjusted with variations in carboxylate, F-T cycles, and solid content. It was observed that with an increase in these parameters, the crosslinking of these polymers was improved, enhancing the structure and properties. Hu et al. (2023) fabricated a flexible PVA network, physically crosslinked through repeated freeze-thaw cycles, with a rigid network of ferric-chelated xanthan gum. This dual-network hydrogel showcased noteworthy in vitro biocompatibility and a remarkable in vivo healing rate of 97.8% (Hu et al., 2023). The integration of keratin and chitosan produces a hybrid hydrogel. The structural framework is established by alkaline freeze-thaw cycles resembling a honeycomb, which enhances its mechanical properties and demonstrates controlled and effective drug delivery of Rhodamine B and bovine serum albumin. In vitro studies showed no harm to normal cells while exhibiting high toxicity against cancer cells due to the release of cisplatin (Yin et al., 2023).

c) Crosslinking by Stereocomplex Formation

Enantiomeric polymers interact to form a more stable crystalline structure by stereoselective interactions. For example, polymers like polylactic acid (PLA) combine (in a racemic mixture) to form stereo-complexes through crystallization. This method can govern the properties of materials, including stability and suitability with biological systems. There are indications that Van der Waal forces are responsible for these physical interactions. Stereocomplexation has been applied to create hybrid hydrogels (e.g., with a gelatin/nano-hydroxyapatite coating) for bone tissue scaffolds (Luo et al., 2020). Another example used both L- and D- forms of a peptide (KYFIL) to form a hydrogel that rapidly self-assembles into plate-like nanostructures in water, controllably tuning its stiffness (Wang et al., 2020). Mechanical properties were less stiff, indicating that this method offers precise control to govern the properties of the resulting biomaterial. The key advantage of crosslinking by stereo complex formation lies in their facile synthesis by dissolving components in water. However, it offers limited scope to the number of polymer compositions.

d) Amphiphilic Block and Graft Copolymers and Thermogelation Techniques

Amphiphilic block copolymers (containing both hydrophilic and hydrophobic segments) can self-assemble into micelles in water; the micellar cores (hydrophobic blocks) aggregate to form a gel network (Varaprasad et al., 2017). Usually, multi-block or graft copolymers are involved in this physical gelation process. For instance, PEG–PLA–PEG triblock polymers form micelles at low concentration and, upon heating (from room temperature to 37°C), undergo a sol-to-gel transition due to increased hydrophobic interactions of PLA blocks, which depend upon the MW and composition of the copolymers (Jeong et al., 1999).

PEG combines with polyester to either form a diblock or triblock copolymer. PEG–PCL–PEG triblocks gel upon heating forming an injectable gel that persisted in vivo for one month. Similarly, diblock copolymers of MPEG/polyesters exhibited temperature-dependent gel-to-sol transition (Kim et al., 2004a). The transition was observed at lower concentrations when the blocks of polyesters were longer. The critical gelation temperature and polymer concentration depend on block lengths and crystallinity. The hydrophilic character of polysaccharides is widely recognized. Synthetically, grafting or modifying their polymer chains by introducing a hydrophobic unit can balance their hydrophilic-hydrophobic nature (Kim et al., 2006). Hydrogels synthesized from thermo-gelation attract applications where temperature can be triggered for the gelation process. It refers to a physical gelation approach for a hydrogel from liquid/sol state to gel state in response to temperature changes, as exemplified by MPEG/polyesters. Since, physical gelation is reversible, thus hydrogel can return to liquid state as temperature is decreased (Michel et al., 2020). The temperature of the system drives the magnitude of hydrophobic interactions. In an aqueous solution, the PEG-PPO hydrogel forms thermally-triggered hydrophobic interactions.

e) Crosslinking by Hydrogen Bonding

It is a weak bond established between the hydrogen atom of one polymer chain and the electronegative element of another chain (Bernhard & Tibbitt, 2021). These intermolecular interactions can be achieved by synthesizing hydrogels using F-T cycles, which induce elasticity and mechanical resilience within the network of hydrogels. These interactions form easily in the polymers bearing –OH, –COOH, –NH_2 groups (e.g., PEG, PVA) and are reversible with stimuli like temperature. PVA and tannic acid were physically crosslinked to form hydrogel-coated porous material as a paint (Bai et al., 2021). However, its industrial applications to separate oil/water were hindered due to the complexity of the synthesis method, extensive hydrogen bonding, although its complex synthesis limited some applications (Bai et al., 2021). More sophisticated designs use multi-step processing: one study chemically crosslinked CMC with ECH, then froze it to induce ice-templated H-bonds, and finally treated it with acid to protonate groups and reinforce H-bonding. The result was a hydrogel that transitioned from a soft to a tough and shape retaining material (Li et al., 2021). CMC-Na/PVA-based hydrogel was synthesized by Wang, et al. (2022).

By using the same strategy, PVA reinforced with cellulose nanofibrils and zinc sulphate ionic hydrogel was designed (Hu et al., 2022). Some of the recently reported Agar-maltodextrin-PVA hydrogel (Nguyen et al., 2021), Na-alginate/PAAM hydrogel (Zhao et al., 2021), dual-crosslinked PVA-AAc-H_2SO_4 hydrogel (Yu et al., 2020), chitosan/PVA hydrogel (Bi et al., 2020) and poly(N-vinylpyrrolidone)/gallic acid composite hydrogel (Zhao et al., 2020) has been synthesized by the action of hydrogen bonding.

f) Crosslinking by Protein Interactions

Protein–protein binding motifs can drive hydrogel assembly. Recombinant proteins are often engineered with specific sequences that self-assemble. Genetic engineers can modify genes to allow the production of peptides containing a predefined sequence of amino acids. Additionally, the genetic code can be engineered to incorporate synthetic amino acids. Since physical and mechanical properties are determined by this sequence, genetically engineered proteins allow us to govern these properties Recognition of biological elements plays a crucial role in the self-assembly processes observed in nature.

An observed recurring pattern, (abcdefg)n is the coiled-coil (CC) motif, which is being extensively studied for engineering protein hydrogel architectures. For example, α-helical coiled-coil peptide motifs (with heptad repeats of hydrophobic/polar residues) can self-assemble into fibrous networks; tuning residues in the motif allows programming of gel mechanics (Lapenta et al., 2018). Any changes in the residues at the hydrophobic interface shape its reactivity to the environment and aid in binding the structures. Out of these two amino acids (a and d) form hydrophobic cores, often these are described as a leucine zipper. Charged amino acids (e and g) form ionic bonds to uphold the structural framework of the complex. Additional research supports the idea that by variations in the sequences of CC motifs, it is possible to program the physical properties of the resulting hydrogels.

g) Hybrid Crosslinking Strategies

Hybrid crosslinking combines permanent (chemical) and reversible (physical or supramolecular) bonds to create networks that can, in principle, harness the strength and stability of covalent crosslinks along with the flexibility and responsiveness of reversible bonds. In theory, such double-network or multi-network hydrogels should exhibit synergistic properties that are not possible with a single crosslink type. For example, covalent bonds serve as a rigid scaffold that resists flow and dissolution, while ionic or supramolecular bonds can dynamically break and reform to dissipate energy and facilitate healing.

Several experimental studies provide evidence of these synergistic effects. For example, hydrogels that combine covalent and noncovalent crosslinks demonstrate better mechanical properties and self-healing ability compared to single-network gels (Yue et al., 2025). Likewise, Liu et al. demonstrate a dual-crosslinked hydrogel in which a permanent poly (HA)-gelatin network is strengthened by hydrophobic interactions. This hybrid gel exhibits high stiffness and toughness (~8 MPa compressive stress at 95% strain), along with rapid recovery and resistance to fatigue (Yu et al., 2021a). According to their words, "hybrid hydrogels can simultaneously achieve high stiffness and toughness by introducing covalent and physical interactions," with the covalent network providing rigidity and the reversible bonds enabling efficient energy dissipation (Yu et al., 2021b). These results show that, when carefully designed, hybrid crosslinks do deliver the combined advantages of both bond types. In fact, comprehensive reviews highlight that hybrid hydrogels have emerged, aiming to improve their strength, drug release efficiency, and therapeutic effectiveness. A sequential procedure is employed to fabricate the initial structure. An intricate lattice is created by chemically bonding polyelectrolytes, precisely poly (2-acrylamido-2-methylpropanesulfonic acid) (PAMPS), forming a rigid and delicate framework. Subsequently, this framework is immersed in a solution comprising a neutral monomer, an initiator, and a minimal quantity of crosslinker. A secondary network, characterized by a loosely connected, supple, and adaptable architecture, takes shape within the confines of the primary structure (Xu et al., 2021).

2.4 FUTURE PERSPECTIVES AND CONCLUSIONS

This review has covered a spectrum of hydrogel formation methods. Chemical crosslinking (free-radical, click, enzymatic) produces strong, stable networks with tunable properties, while physical crosslinking (hydrogen bonds, ionic, crystallization) forms gels under mild conditions (ambient pH, cooling, ionic baths) that are inherently biocompatible and stimuli-responsive, though mechanically weaker. Hybrid networks combining permanent and reversible bonds have emerged to enhance toughness and multifunctionality. Recent advances like double-network hydrogels (record toughness), click and enzymatic bio-gels, and green solvent processes reflect these trends. Looking ahead, priorities include developing greener crosslinking chemistries (biodegradable polymers, non-toxic initiators), engineering dynamic networks (selfhealing, shape-memory gels), and scaling up production (flow processes, ambient cures). Predictive modeling and in-line monitoring will also aid rational design. No single crosslinking method suits all application-sEnzymatic and click methods (e.g., CuAAC, SPAAC, Diels–Alder) offer biocompatibility and precision for medical gels, but can involve catalysts or specialized monomers. Thermoresponsive systems (e.g., Pluronic 407) are valuable for injectable depots. Ultimately, method choice depends on the target: dynamic covalent bonds for self-healing, dual-networks for maximum strength, light-activated systems for cell-encapsulation, etc. Emerging "bio-orthogonal" click reactions expand the toolbox (at the cost of new reagent synthesis). In summary, the processing conditions (freezing, blending, temperature ramps, ionic baths, irradiation, etc.) fundamentally influence hydrogel properties and applications. Physical methods yield mild, reversible gels but limited strength; chemical crosslinkers add robustness and functionality; hybrids merge these advantages. New fabrication technologies (nanocomposite fillers, 3D/4D printing) continue to push the boundaries of hydrogel design for advanced biomaterials.

REFERENCES

Amirthalingam, S., Rajendran, A. K., Moon, Y. G., & Hwang, N. S. (2023). Stimuli-responsive dynamic hydrogels: design, properties and tissue engineering applications []. *Materials Horizons, 10*(9), 3325-3350. DOI: 10.1039/D3MH00399J

Bai, Z., Jia, K., Liu, C., Wang, L., Lin, G., Huang, Y., Liu, S., & Liu, X. (2021). A solvent regulated hydrogen bond crosslinking strategy to prepare robust hydrogel paint for oil/water separation. *Advanced Functional Materials, 31*(49), 2104701. DOI: 10.1002/adfm.202104701

Bailey, A. (1991). The chemistry of natural enzyme-induced cross-links of proteins. *Amino Acids, 1*(3), 293–306. DOI: 10.1007/BF00813999

Bailey, S. J., Hopkins, E., Baxter, N. J., Whitehead, I., de Alaniz, J. R., & Wilson, M. Z. (2023). Diels–Alder photoclick patterning of extracellular matrix for spatially controlled cell behaviors. *Advanced Materials, 35*(46), 2303453. DOI: 10.1002/adma.202303453

Batibay, G. S., & Arsu, N. (2023). Synthesis of photoactive nanocomposite hydrogels via in-situ photoinitiated thiol-ene click polymerization of acrylates containing norbornene-POSS and norbornene-POSS/Ag NPs. *Progress in Organic Coatings, 175*, 107364. https://doi.org/https://doi.org/10.1016/j.porgcoat.2022.107364. DOI: 10.1016/j.porgcoat.2022.107364

Bernal-Chávez, S. A., Romero-Montero, A., Hernández-Parra, H., Peña-Corona, S. I., Del Prado-Audelo, M. L., Alcalá-Alcalá, S., Cortés, H., Kiyekbayeva, L., Sharifi-Rad, J., & Leyva-Gómez, G. (2023). Enhancing chemical and physical stability of pharmaceuticals using freeze-thaw method: Challenges and opportunities for process optimization through quality by design approach. *Journal of Biological Engineering, 17*(1), 35. DOI: 10.1186/s13036-023-00353-9

Bernhard, S., & Tibbitt, M. W. (2021). Supramolecular engineering of hydrogels for drug delivery. *Advanced Drug Delivery Reviews, 171*, 240–256. DOI: 10.1016/j.addr.2021.02.002

Bhattacharjee, S., Maiti, B., & Bhattacharya, S. (2016). First report of charge-transfer induced heat-set hydrogel. Structural insights and remarkable properties. *Nanoscale, 8*(21), 11224–11233. DOI: 10.1039/C6NR01128D

Bhattarai, N., Gunn, J., & Zhang, M. (2010). Chitosan-based hydrogels for controlled, localized drug delivery. *Advanced Drug Delivery Reviews, 62*(1), 83–99. DOI: 10.1016/j.addr.2009.07.019

Bi, S., Pang, J., Huang, L., Sun, M., Cheng, X., & Chen, X. (2020). The toughness chitosan-PVA double network hydrogel based on alkali solution system and hydrogen bonding for tissue engineering applications. *International Journal of Biological Macromolecules, 146*, 99–109. DOI: 10.1016/j.ijbiomac.2019.12.186

Bustamante-Torres, M., Romero-Fierro, D., Arcentales-Vera, B., Palomino, K., Magaña, H., & Bucio, E. (2021). Hydrogels Classification According to the Physical or Chemical Interactions and as Stimuli-Sensitive Materials. *Gels (Basel, Switzerland), 7*(4), 182. Advance online publication. DOI: 10.3390/gels7040182

Bustamante-Torres, M., Romero-Fierro, D., Arcentales-Vera, B., Palomino, K., Magaña, H., & Bucio, E. (2021). Hydrogels classification according to the physical or chemical interactions and as stimuli-sensitive materials. *Gels (Basel, Switzerland)*, *7*(4), 182. DOI: 10.3390/gels7040182

Caló, E., & Khutoryanskiy, V. V. (2015). Biomedical applications of hydrogels: A review of patents and commercial products. *European Polymer Journal*, *65*, 252–267. DOI: 10.1016/j.eurpolymj.2014.11.024

Chang, C., Duan, B., Cai, J., & Zhang, L. (2010). Superabsorbent hydrogels based on cellulose for smart swelling and controllable delivery. *European Polymer Journal*, *46*(1), 92–100. DOI: 10.1016/j.eurpolymj.2009.04.033

Chang, C., & Zhang, L. (2011). Cellulose-based hydrogels: Present status and application prospects. *Carbohydrate Polymers*, *84*(1), 40–53. DOI: 10.1016/j.carbpol.2010.12.023

Chang, H., Li, C., Huang, R., Su, R., Qi, W., & He, Z. (2019). Amphiphilic hydrogels for biomedical applications. *Journal of Materials Chemistry. B*, *7*(18), 2899–2910. DOI: 10.1039/C9TB00073A

Dalton, P. D., Hostert, C., Albrecht, K., Moeller, M., & Groll, J. (2008). Structure and Properties of Urea-Crosslinked Star Poly [(ethylene oxide)-ran-(propylene oxide)] Hydrogels. *Macromolecular Bioscience*, *8*(10), 923–931. DOI: 10.1002/mabi.200800080

de Paiva Narciso, N., Navarro, R. S., Gilchrist, A. E., Trigo, M. L. M., Aviles Rodriguez, G., & Heilshorn, S. C. (2023). Design Parameters for Injectable Biopolymeric Hydrogels with Dynamic Covalent Chemistry Crosslinks. *Advanced Healthcare Materials*, *12*(27), e2301265. DOI: 10.1002/adhm.202301265

Demitri, C., Del Sole, R., Scalera, F., Sannino, A., Vasapollo, G., Maffezzoli, A., Ambrosio, L., & Nicolais, L. (2008). Novel superabsorbent cellulose-based hydrogels crosslinked with citric acid. *Journal of Applied Polymer Science*, *110*(4), 2453–2460. DOI: 10.1002/app.28660

Dhand, A. P., Galarraga, J. H., & Burdick, J. A. (2021). Enhancing Biopolymer Hydrogel Functionality through Interpenetrating Networks. *Trends in Biotechnology*, *39*(5), 519–538. https://doi.org/https://doi.org/10.1016/j.tibtech.2020.08.007. DOI: 10.1016/j.tibtech.2020.08.007

Distantina, S., Rochmadi, S., Fahrurrozi, M., & Wiratni, M. (2012). Preparation of hydrogel based on glutaraldehyde-crosslinked carrageenan. Proceeding of 3rd Int. Conf. on Chemistry and Chemical Engineering IPCBEE.

Drury, J. L., & Mooney, D. J. (2003). Hydrogels for tissue engineering: Scaffold design variables and applications. *Biomaterials*, *24*(24), 4337–4351. DOI: 10.1016/S0142-9612(03)00340-5

Elvin, C. M., Carr, A. G., Huson, M. G., Maxwell, J. M., Pearson, R. D., Vuocolo, T., Liyou, N. E., Wong, D. C., Merritt, D. J., & Dixon, N. E. (2005). Synthesis and properties of crosslinked recombinant pro-resilin. *Nature*, *437*(7061), 999–1002. DOI: 10.1038/nature04085

Engkagul, V., Sereemaspun, A., & Chirachanchai, S. (2018). One pot preparation of chitosan/hyaluronic acid-based triple network hydrogel via in situ click reaction, metal coordination and polyion complexation in water. *Carbohydrate Polymers*, *200*, 616–623. DOI: 10.1016/j.carbpol.2018.07.090

Fernández, E., López, D., Mijangos, C., Duskova-Smrckova, M., Ilavsky, M., & Dusek, K. (2008). Rheological and thermal properties of agarose aqueous solutions and hydrogels. *Journal of Polymer Science. Part B, Polymer Physics*, *46*(3), 322–328. DOI: 10.1002/polb.21370

Finetti, C., Sola, L., Elliott, J., & Chiari, M. (2017). Synthesis of hydrogel via click chemistry for DNA electrophoresis. *Journal of Chromatography. A*, *1513*, 226–234. DOI: 10.1016/j.chroma.2017.07.042

Franklin, D., & Guhanathan, S. (2014). Synthesis and characterization of citric acid-based pH-sensitive biopolymeric hydrogels. *Polymer Bulletin*, *71*(1), 93–110. DOI: 10.1007/s00289-013-1047-4

Gao, X., Fang, J., Xue, B., Fu, L., & Li, H. (2016). Engineering protein hydrogels using SpyCatcher-SpyTag chemistry. *Biomacromolecules*, *17*(9), 2812–2819. DOI: 10.1021/acs.biomac.6b00566

Garcia-Garcia, A., Muñana-González, S., Lanceros-Mendez, S., Ruiz-Rubio, L., Alvarez, L. P., & Vilas-Vilela, J. L. (2024). Biodegradable Natural Hydrogels for Tissue Engineering, Controlled Release, and Soil Remediation. *Polymers*, *16*(18), 2599. https://www.mdpi.com/2073-4360/16/18/2599. DOI: 10.3390/polym16182599

Ghorpade, V. S. (2020). Preparation of hydrogels based on natural polymers via chemical reaction and cross-Linking. In Chen, Y. (Ed.), *Hydrogels Based on Natural Polymers* (pp. 91–118). Elsevier., https://doi.org/https://doi.org/10.1016/B978-0-12-816421-1.00004-5 DOI: 10.1016/B978-0-12-816421-1.00004-5

Gopinathan, J., & Noh, I. (2018). Click Chemistry-Based Injectable Hydrogels and Bioprinting Inks for Tissue Engineering Applications. *Tissue Engineering and Regenerative Medicine*, *15*(5), 531–546. DOI: 10.1007/s13770-018-0152-8

Gopinathan, J., & Noh, I. (2018). Click Chemistry-Based Injectable Hydrogels and Bioprinting Inks for Tissue Engineering Applications. *Tissue Engineering and Regenerative Medicine*, *15*(5), 531–546. DOI: 10.1007/s13770-018-0152-8

Guaresti, O., Basasoro, S., González, K., Eceiza, A., & Gabilondo, N. (2019). In situ cross–linked chitosan hydrogels via Michael addition reaction based on water–soluble thiol–maleimide precursors. *European Polymer Journal*, *119*, 376–384. DOI: 10.1016/j.eurpolymj.2019.08.009

Gulrez, S. K., Al-Assaf, S., & Phillips, G. O. (2011). Hydrogels: methods of preparation, characterisation and applications. *Progress in molecular and environmental bioengineering-from analysis and modeling to technology applications, 117150.*

Guo, Y., Gu, J., Jiang, Y., Zhou, Y., Zhu, Z., Ma, T., Cheng, Y., Ji, Z., Jiao, Y., Xue, B., & Cao, Y. (2021). Regulating the homogeneity of thiol-maleimide Michael-type addition-based hydrogels using amino biomolecules. *Gels (Basel, Switzerland)*, *7*(4), 206. DOI: 10.3390/gels7040206

Gupta, P., Vermani, K., & Garg, S. (2002). Hydrogels: From controlled release to pH-responsive drug delivery. *Drug Discovery Today*, *7*(10), 569–579. DOI: 10.1016/S1359-6446(02)02255-9

He, Y., Liu, L., & Cheng, L. (2023). A Short Review of Research Progress on the Synthesis Approaches of Aza-Dibenzocyclooctyne Derivatives. *Molecules (Basel, Switzerland)*, *28*(9), 3715. https://www.mdpi.com/1420-3049/28/9/3715. DOI: 10.3390/molecules28093715

Holkar, K., Mishra, R., & Jain, E. (2025). Recent Trends in Cryogelation Phenomenon & Factors Affecting Cryotropic Gelation. In *Preprints*: *Preprints*.

Hörner, M., Raute, K., Hummel, B., Madl, J., Creusen, G., Thomas, O. S., Christen, E. H., Hotz, N., Gübeli, R. J., Engesser, R., Rebmann, B., Lauer, J., Rolauffs, B., Timmer, J., Schamel, W. W. A., Pruszak, J., Römer, W., Zurbriggen, M. D., Friedrich, C., & Weber, W. (2019). Phytochrome-based extracellular matrix with reversibly tunable mechanical properties. *Advanced Materials*, *31*(12), 1806727. DOI: 10.1002/adma.201806727

Hu, J., Wu, Y., Yang, Q., Zhou, Q., Hui, L., Liu, Z., Xu, F., & Ding, D. (2022). One-pot freezing-thawing preparation of cellulose nanofibrils reinforced polyvinyl alcohol based ionic hydrogel strain sensor for human motion monitoring. *Carbohydrate Polymers*, *275*, 118697. DOI: 10.1016/j.carbpol.2021.118697

Hu, L., Wang, Y., Liu, Q., Liu, M., Yang, F., Wang, C., Pan, P., Wang, L., Chen, L., & Chen, J. (2023). Real-time monitoring flexible hydrogels based on dual physically cross-linked

Jeong, B., Bae, Y. H., & Kim, S. W. (1999). Thermoreversible gelation of PEG– PLGA– PEG triblock copolymer aqueous solutions. *Macromolecules*, *32*(21), 7064–7069. DOI: 10.1021/ma9908999

Jeong, J., Kim, K., Kim, J., Kim, Y., & Jung, S. (2023). New polyvinyl alcohol/succinoglycan-based hydrogels for pH-responsive drug delivery. *Polymers*, *15*(14), 3009. DOI: 10.3390/polym15143009

Ji, N., Luo, J., Zhang, W., Sun, J., Wang, J., Qin, C., Zhuo, Q., & Dai, L. (2023). A novel polyvinyl alcohol-based hydrogel with ultra-fast self-healing ability and excellent stretchability based on multi dynamic covalent bond cross-linking. *Macromolecular Materials and Engineering*, *308*(3), 2200525. DOI: 10.1002/mame.202200525

Jiao, C., Gao, L., Zhang, H., Yu, B., Cong, H., & Shen, Y. (2020). Dynamic Covalent C= C Bond, Cross-Linked, Injectable, and Self-Healable Hydrogels via Knoevenagel Condensation. *Biomacromolecules*, *21*(3), 1234–1242. DOI: 10.1021/acs.biomac.9b01689

Jin, R., Teixeira, L. M., Dijkstra, P. J., Karperien, M., Van Blitterswijk, C., Zhong, Z., & Feijen, J. (2009). Injectable chitosan-based hydrogels for cartilage tissue engineering. *Biomaterials*, *30*(13), 2544–2551. DOI: 10.1016/j.biomaterials.2009.01.020

Jus, S., Stachel, I., Schloegl, W., Pretzler, M., Friess, W., Meyer, M., Birner-Gruenberger, R., & Guebitz, G. (2011). Cross-linking of collagen with laccases and tyrosinases. *Materials Science and Engineering C*, *31*(5), 1068–1077. DOI: 10.1016/j.msec.2011.03.007

Kenari, F., Molnár, S., & Perjési, P. (2021). Reaction of Chalcones with Cellular Thiols. The Effect of the 4-Substitution of Chalcones and Protonation State of the Thiols on the Addition Process. Diastereoselective Thiol Addition. *Molecules (Basel, Switzerland)*, *26*(14), 4332. Advance online publication. DOI: 10.3390/molecules26144332

Kharkar, P. M., Rehmann, M. S., Skeens, K. M., Maverakis, E., & Kloxin, A. M. (2016). Thiol-ene click hydrogels for therapeutic delivery. *ACS Biomaterials Science & Engineering*, *2*(2), 165–179. DOI: 10.1021/acsbiomaterials.5b00420

Kharkar, P. M., Rehmann, M. S., Skeens, K. M., Maverakis, E., & Kloxin, A. M. (2016). Thiol–ene click hydrogels for therapeutic delivery. *ACS Biomaterials Science & Engineering*, *2*(2), 165–179. DOI: 10.1021/acsbiomaterials.5b00420

Kim, M. S., Hyun, H., Khang, G., & Lee, H. B. (2006). Preparation of thermosensitive diblock copolymers consisting of MPEG and polyesters. *Macromolecules*, *39*(9), 3099–3102. DOI: 10.1021/ma052713d

Kim, M. S., Seo, K. S., Khang, G., Cho, S. H., & Lee, H. B. (2004a). Preparation of methoxy poly (ethylene glycol)/polyester diblock copolymers and examination of the gel-to-sol transition. *Journal of Polymer Science. Part A, Polymer Chemistry*, *42*(22), 5784–5793. DOI: 10.1002/pola.20430

Kirschner, C. M., & Anseth, K. S. (2013). Hydrogels in Healthcare: From Static to Dynamic Material Microenvironments. *Acta Materialia*, *61*(3), 931–944. DOI: 10.1016/j.actamat.2012.10.037

Kölmel, D. K., & Kool, E. T. (2017). Oximes and Hydrazones in Bioconjugation: Mechanism and Catalysis. *Chemical Reviews*, *117*(15), 10358–10376. DOI: 10.1021/acs.chemrev.7b00090

Kou, S., Yang, Z., & Sun, F. (2017). Protein hydrogel microbeads for selective uranium mining from seawater. *ACS Applied Materials & Interfaces*, *9*(3), 2035–2039. DOI: 10.1021/acsami.6b15968

Kuo, K.-C., Lin, R.-Z., Tien, H.-W., Wu, P.-Y., Li, Y.-C., Melero-Martin, J. M., & Chen, Y.-C. (2015). Bioengineering vascularized tissue constructs using an injectable cell-laden enzymatically crosslinked collagen hydrogel derived from dermal extracellular matrix. *Acta Biomaterialia*, *27*, 151–166. DOI: 10.1016/j.actbio.2015.09.002

Kurisawa, M., Chung, J. E., Yang, Y. Y., Gao, S. J., & Uyama, H. (2005). Injectable biodegradable hydrogels composed of hyaluronic acid–tyramine conjugates for drug delivery and tissue engineering. *Chemical Communications (Cambridge)*, (34), 4312–4314. DOI: 10.1039/b506989k

Lagneau, N., Terriac, L., Tournier, P., Helesbeux, J., Viault, G., Séraphin, D., Halgand, B., Loll, F., Garnier, C., Jonchère, C., Rivière, M., Tessier, A., Lebreton, J., Maugars, Y., Guicheux, J., Le Visage, C., & Delplace, V. (2023). A new boronate ester-based crosslinking strategy allows the design of nonswelling and long-term stable dynamic covalent hydrogels. *Biomaterials Science*, *11*(6), 2033–2045. DOI: 10.1039/D2BM01690G

Lapenta, F., Aupič, J., Strmšek, Ž., & Jerala, R. (2018). Coiled coil protein origami: From modular design principles towards biotechnological applications. *Chemical Society Reviews*, *47*(10), 3530–3542. DOI: 10.1039/C7CS00822H

Lau, D., Michael, P., & Binder, W. (2017). CuAAC-Based Click Chemistry in Self-Healing Polymers. *Accounts of Chemical Research*, *50*(10), 2610–2620. Advance online publication. DOI: 10.1021/acs.accounts.7b00371

Laurano, R., Boffito, M., Cassino, C., Midei, L., Pappalardo, R., Chiono, V., & Ciardelli, G. (2023). Thiol-Ene Photo-Click Hydrogels with Tunable Mechanical Properties Resulting from the Exposure of Different -Ene Moieties through a Green Chemistry. *Materials (Basel)*, *16*(5), 2024. https://www.mdpi.com/1996-1944/16/5/2024. DOI: 10.3390/ma16052024

Li, J., Chen, F., Lin, X., & Ding, T. (2021). Hydrogen-bonding-assisted toughening of hierarchical carboxymethyl cellulose hydrogels for biomechanical sensing. *Carbohydrate Polymers*, *269*, 118252. DOI: 10.1016/j.carbpol.2021.118252

Li, L., Tong, Z., Jia, X., & Kiick, K. L. (2013). Resilin-like polypeptide hydrogels engineered for versatile biological function. *Soft Matter*, *9*(3), 665–673. DOI: 10.1039/C2SM26812D

Li, X., & Gong, J. P. (2024). Design principles for strong and tough hydrogels. *Nature Reviews. Materials*, *9*(6), 380–398. DOI: 10.1038/s41578-024-00672-3

Li, X., He, P., Ma, R., Dong, C., Lv, Y., & Dai, L. (2023). Modulation of composite hydrogel consisting of TEMPO-oxidized cellulose nanofibers and cationic guar gum. *International Journal of Biological Macromolecules*, *241*, 124483. DOI: 10.1016/j.ijbiomac.2023.124483

Li, X., & Xiong, Y. (2022). Application of "Click" Chemistry in Biomedical Hydrogels. *ACS Omega*, *7*(42), 36918–36928. DOI: 10.1021/acsomega.2c03931

Li, X., & Xiong, Y. (2022). Application of "Click" Chemistry in Biomedical Hydrogels. *ACS Omega*, *7*(42), 36918–36928. DOI: 10.1021/acsomega.2c03931

Li, Y., Zhou, C., Xu, L., Yao, F., Cen, L., & Fu, G. D. (2015). Stimuli-responsive hydrogels prepared by simultaneous "click chemistry" and metal–ligand coordination []. *RSC advances*, *5*(24), 18242-18251. https://doi.org/DOI: 10.1039/C4RA11946K

Li, Z., Lu, F., & Liu, Y. (2023). A Review of the Mechanism, Properties, and Applications of Hydrogels Prepared by Enzymatic Cross-linking. *Journal of Agricultural and Food Chemistry*, *71*(27), 10238–10249. DOI: 10.1021/acs.jafc.3c01162

Li, Z., Wu, W., Chen, S., Xia, B., Lin, T., Gao, M., Zhao, C., Wu, X., Lin, C., Yang, W., & Wang, Y. (2023). An efficient strategy to design and construct high-strength and fluorescence poly (vinyl alcohol) hydrogels via freeze–Thaw cycles. *Journal of Applied Polymer Science*, *140*(40), 1–9. DOI: 10.1002/app.54492

Liu, M., Zeng, X., Ma, C., Yi, H., Ali, Z., Mou, X., Li, S., Deng, Y., & He, N. (2017). Injectable hydrogels for cartilage and bone tissue engineering. *Bone Research*, *5*(1), 17014. DOI: 10.1038/boneres.2017.14

Liu, Y., Geever, L. M., Kennedy, J. E., Higginbotham, C. L., Cahill, P. A., & McGuinness, G. B. (2010). Thermal behavior and mechanical properties of physically crosslinked PVA/Gelatin hydrogels. *Journal of the Mechanical Behavior of Biomedical Materials*, *3*(2), 203–209. https://doi.org/https://doi.org/10.1016/j.jmbbm.2009.07.001. DOI: 10.1016/j.jmbbm.2009.07.001

Lu, Z., Sheng, R., Zhang, W., & Chen, J. (2025). Strong and tough hydrogels fabricated through molecular and structural engineering and their biomedical applications. *Chemical Engineering Journal, 508*, 160728. https://doi.org/https://doi.org/10.1016/j.cej.2025.160728. DOI: 10.1016/j.cej.2025.160728

Luo, F., Fortenberry, A., Ren, J., & Qiang, Z. (2020). Recent progress in enhancing poly (lactic acid) stereocomplex formation for material property improvement. *Frontiers in Chemistry, 8*, 688. DOI: 10.3389/fchem.2020.00688

Luo, J., Liu, X., Yang, Z., & Sun, F. (2018). Synthesis of entirely protein-based hydrogels by enzymatic oxidation enabling water-resistant bioadhesion and stem cell encapsulation.

Maitra, J., & Shukla, V. K. (2014). Cross-linking in hydrogels-a review. *American Journal of Political Science*, *4*(2), 25–31.

Majcher, M. J., & Hoare, T. (2019). Advanced Hydrogel Structures. In Jafar Mazumder, M. A., Sheardown, H., & Al-Ahmed, A. (Eds.), *Functional Biopolymers* (pp. 279–305). Springer International Publishing., DOI: 10.1007/978-3-319-95990-0_16

McHale, M. K., Setton, L. A., & Chilkoti, A. (2005). Synthesis and in vitro evaluation of enzymatically cross-linked elastin-like polypeptide gels for cartilaginous tissue repair. *Tissue Engineering*, *11*(11-12), 1768–1779. DOI: 10.1089/ten.2005.11.1768

Mehak, Singh, G., Singh, R., Singh, G., Stanzin, J., Singh, H., Kaur, G., & Singh, J. (2024). Clicking in harmony: exploring the bio-orthogonal overlap in click chemistry []. *RSC advances, 14*(11), 7383-7413. DOI: 10.1039/D4RA00494A

Michel, S. S., Kilner, A., Eloi, J.-C., Rogers, S. E., Briscoe, W. H., & Galan, M. C. (2020). Norbornene-functionalized chitosan hydrogels and microgels via unprecedented photoinitiated self-assembly for potential biomedical applications. *ACS Applied Bio Materials*, *3*(8), 5253–5262. DOI: 10.1021/acsabm.0c00629

Munoz-Munoz, F., Ruiz, J.-C., Alvarez-Lorenzo, C., Concheiro, A., & Bucio, E. (2012). Temperature-and pH-sensitive interpenetrating polymer networks grafted on PP: Cross-linking irradiation dose as a critical variable for the performance as vancomycin-eluting systems. *Radiation Physics and Chemistry*, *81*(5), 531–540. DOI: 10.1016/j.radphyschem.2012.01.007

Naficy, S., Brown, H. R., Razal, J. M., Spinks, G. M., & Whitten, P. G. (2011). Progress toward robust polymer hydrogels. *Australian Journal of Chemistry*, *64*(8), 1007–1025. DOI: 10.1071/CH11156

Nagaraj, S., Ezhilarasan, K., & Ramarajan, R. (2025). Click synthesis via 1,3-dipolar addition-A versatile route to tetrazine derivatives and In silico corroboration as anti-muscular dystrophy agent. *Tetrahedron, 175*, 134520. https://doi.org/https://doi.org/10.1016/j.tet.2025.134520. DOI: 10.1016/j.tet.2025.134520

Nair, D. P., Podgorski, M., Chatani, S., Gong, T., Xi, W., Fenoli, C. R., & Bowman, C. N. (2014). The thiol-Michael addition click reaction: A powerful and widely used tool in materials chemistry. *Chemistry of Materials*, *26*(1), 724–744. DOI: 10.1021/cm402180t

Nair, D. P., Podgórski, M., Chatani, S., Gong, T., Xi, W., Fenoli, C. R., & Bowman, C. N. (2014). The Thiol-Michael Addition Click Reaction: A Powerful and Widely Used Tool in Materials Chemistry. *Chemistry of Materials*, *26*(1), 724–744. DOI: 10.1021/cm402180t

Nguyen, G. K., Wang, S., Qiu, Y., Hemu, X., Lian, Y., & Tam, J. P. (2014). Butelase 1 is an Asx-specific ligase enabling peptide macrocyclization and synthesis. *Nature Chemical Biology*, *10*(9), 732–738. DOI: 10.1038/nchembio.1586

Nguyen, T. T., Hoang, B. N., Van Tran, T., Van Nguyen, D., Nguyen, T. D., & Vo, D.-V. N. (2021). Agar/maltodextrin/poly (vinyl alcohol) walled montmorillonite composites for removal of methylene blue from aqueous solutions. *Surfaces and Interfaces*, *26*, 101410. DOI: 10.1016/j.surfin.2021.101410

Papadakis, C. M., & Tsitsilianis, C. (2017). Responsive hydrogels from associative block copolymers: Physical gelling through polyion complexation. *Gels (Basel, Switzerland)*, *3*(1), 3. DOI: 10.3390/gels3010003

Park, B. M., Luo, J., & Sun, F. (2019). Enzymatic assembly of adhesive molecular networks with sequence-dependent mechanical properties inspired by mussel foot proteins. *Polymer Chemistry*, *10*(7), 823–826. DOI: 10.1039/C8PY01581C

Peppas, N. A. (2010). *Biomedical applications of hydrogels handbook*. Springer Science & Business Media.

Reddy, M. M., Boominathan, T., Vijai Anand, A. S., Panicker, R. R., Kaushal, V., Das, A., Jain, N., Vighnesh, I. S., Desikan, R., Vijayakrishna, K., Padala, K., Brahmmananda Rao, C. V. S., Smith, G. S., & Sivaramakrishna, A. (2021). Novel Biocompatible Hydrogels via Click Chemistry. In Jose, J., Thomas, S., & Thakur, V. K. (Eds.), *Nano Hydrogels: Physico-Chemical Properties and Recent Advances in Structural Designing* (pp. 281–304). Springer Singapore., DOI: 10.1007/978-981-15-7138-1_16

Ren, P., Wei, D., Liang, M., Xu, L., Zhang, T., & Zhang, Q. (2022). Alginate/gelatin-based hybrid hydrogels with function of injecting and encapsulating cells in situ. *International Journal of Biological Macromolecules*, *212*, 67–84. Advance online publication. DOI: 10.1016/j.ijbiomac.2022.05.058

Rosenquist, J., Folkesson, M., Höglund, L., Pupkaite, J., Hilborn, J., & Samanta, A. (2023). An Injectable, Shape-Retaining Collagen Hydrogel Cross-linked Using Thiol-Maleimide Click Chemistry for Sealing Corneal Perforations. *ACS Applied Materials & Interfaces*, *15*(29), 34407–34418. DOI: 10.1021/acsami.3c03963

Rupali, K., Mulik, R., Badgujar, L., Paradkar, A., Mahadik, K., Bodhankar, S., & Shekhar, S. (2010). Development and characterization of micro-emulsion formulations for transdermal delivery of aceclofenac: A research. *International Journal of Drug Formulation and Research*, *1*, 359–386.

Sakaguchi, T., Nagano, S., Hara, M., Hyon, S.-H., Patel, M., & Matsumura, K. (2017). Facile preparation of transparent poly(vinyl alcohol) hydrogels with uniform microcrystalline structure by hot-pressing without using organic solvents. *Polymer Journal*, *49*(7), 535–542. DOI: 10.1038/pj.2017.18

Salimiyan, N., Gholami, M., & Sedghi, R. (2023). Preparation of degradable, biocompatible, conductive and multifunctional chitosan/thiol-functionalized graphene nanocomposite hydrogel via click chemistry for human motion sensing. *Chemical Engineering Journal, 471*, 144648. https://doi.org/https://doi.org/10.1016/j.cej.2023.144648. DOI: 10.1016/j.cej.2023.144648

Salinas, C. N., & Anseth, K. S. (2008). Mixed mode thiol– acrylate photopolymerizations for the synthesis of PEG– peptide hydrogels. *Macromolecules, 41*(16), 6019–6026. DOI: 10.1021/ma800621h

Sannino, A., Demitri, C., & Madaghiele, M. (2009). Biodegradable cellulose-based hydrogels: Design and applications. *Materials (Basel), 2*(2), 353–373. DOI: 10.3390/ma2020353

Schmidt, M., Toplak, A., Quaedflieg, P. J., & Nuijens, T. (2017). Enzyme-mediated ligation technologies for peptides and proteins. *Current Opinion in Chemical Biology, 38*, 1–7. DOI: 10.1016/j.cbpa.2017.01.017

Selvam, A., Majood, M., Chaurasia, R., Rupesh, Singh, A., Dey, T., Agrawal, O., Verma, Y. K., & Mukherjee, M. (2023). Injectable organo-hydrogels influenced by click chemistry as a paramount stratagem in the conveyor belt of pharmaceutical revolution []. *Journal of Materials Chemistry B, 11*(45), 10761-10777. DOI: 10.1039/D3TB01674A

Shadish, J. A., Benuska, G. M., & DeForest, C. A. (2019). Bioactive site-specifically modified proteins for 4D patterning of gel biomaterials. *Nature Materials, 18*(9), 1005–1014. DOI: 10.1038/s41563-019-0367-7

Shoukat, H., Pervaiz, F., & Noreen, S. (2016). Novel Crosslinking Methods to Design Hydrogels. *Global Pharmaceutical Sciences Review, 1*(I), 1–5. DOI: 10.31703/gpsr.2016(I-I).01

Singh, G., Majeed, A., Singh, R., George, N., Singh, G., Gupta, S., Singh, H., Kaur, G., & Singh, J. (2023). CuAAC ensembled 1,2,3-triazole linked nanogels for targeted drug delivery: a review []. *RSC advances, 13*(5), 2912-2936. https://doi.org/DOI: 10.1039/D2RA05592A

Şolpan, D., Torun, M., & Güven, O. (2008). The usability of (sodium alginate/acrylamide) semi-interpenetrating polymer networks on removal of some textile dyes. *Journal of Applied Polymer Science, 108*(6), 3787–3795. DOI: 10.1002/app.27945

Soto-Bustamante, F., Bassu, G., Fratini, E., & Laurati, M. (2023). Effect of Composition and Freeze-Thaw on the Network Structure, Porosity and Mechanical Properties of Polyvinyl-Alcohol/Chitosan Hydrogels. *Gels (Basel, Switzerland), 9*(5), 396. https://www.mdpi.com/2310-2861/9/5/396. DOI: 10.3390/gels9050396

Sperinde, J. J., & Griffith, L. G. (1997). Synthesis and characterization of enzymatically-cross-linked poly (ethylene glycol) hydrogels. *Macromolecules, 30*(18), 5255–5264. DOI: 10.1021/ma970345a

Stanfield, R. L., & Wilson, I. A. (1995). Protein-peptide interactions. *Current Opinion in Structural Biology, 5*(1), 103–113. DOI: 10.1016/0959-440X(95)80015-S

Sun, J., & Tan, H. (2013). Alginate-based biomaterials for regenerative medicine applications. *Materials (Basel), 6*(4), 1285–1309. DOI: 10.3390/ma6041285

Terriac, L., Helesbeux, J.-J., Maugars, Y., Guicheux, J., Tibbitt, M. W., & Delplace, V. (2024). Boronate Ester Hydrogels for Biomedical Applications: Challenges and Opportunities. *Chemistry of Materials*, *36*(14), 6674–6695. DOI: 10.1021/acs.chemmater.4c00507

Tirrell, J. G., & Tirrell, D. A. (1996). Synthesis of biopolymers: Proteins, polyesters, polysaccharides and polynucleotides. *Current Opinion in Solid State and Materials Science*, *1*(3), 407–411. DOI: 10.1016/S1359-0286(96)80033-7

Tortora, M., Cavalieri, F., Chiessi, E., & Paradossi, G. (2007). Michael-type addition reactions for the in situ formation of poly (vinyl alcohol)-based hydrogels. *Biomacromolecules*, *8*(1), 209–214. DOI: 10.1021/bm0607269

Varaprasad, K., Raghavendra, G. M., Jayaramudu, T., Yallapu, M. M., & Sadiku, R. (2017). A mini review on hydrogels classification and recent developments in miscellaneous applications. *Materials Science and Engineering C*, *79*, 958–971. https://doi.org/https://doi.org/10.1016/j.msec.2017.05.096. DOI: 10.1016/j.msec.2017.05.096

Wang, H., Li, Z., Zuo, M., Zeng, X., Tang, X., Sun, Y., & Lin, L. (2022). Stretchable, freezing-tolerant conductive hydrogel for wearable electronics reinforced by cellulose nanocrystals toward multiple hydrogen bonding. *Carbohydrate Polymers*, *280*, 119018. DOI: 10.1016/j.carbpol.2021.119018

Wang, H., Wang, E., Huang, Y., & Li, X. (2020). Hybrid hydrogel based on stereocomplex PDLA/PLLA and gelatin for bone regeneration. *Journal of Applied Polymer Science*, *137*(48), 49571. DOI: 10.1002/app.49571

Wang, S., Nawale, G. N., Oommen, O. P., Hilborn, J., & Varghese, O. P. (2019). Influence of ions to modulate hydrazone and oxime reaction kinetics to obtain dynamically cross-linked hyaluronic acid hydrogels. *Polymer Chemistry*, *10*(31), 4322–4327. DOI: 10.1039/C9PY00862D

Wang, T., Gao, D., Yin, H., Zhao, J., Wang, X., & Niu, H. (2024). Kinetic Study of the Diels–Alder Reaction between Maleimide and Furan-Containing Polystyrene Using Infrared Spectroscopy. *Polymers*, *16*(3), 441. https://www.mdpi.com/2073-4360/16/3/441. DOI: 10.3390/polym16030441

Wang, T., Jones, J. D., Niyonshuti, I. I., Agrawal, S., Gundampati, R. K., Kumar, T. K. S., Quinn, K. P., & Chen, J. (2019). Biocompatible, Injectable Anionic Hydrogels Based on Poly (Oligo Ethylene Glycol Monoacrylate-co-Acrylic Acid) for Protein Delivery. *Advanced Therapeutics*, *2*(9), 1900092. DOI: 10.1002/adtp.201900092

Waresindo, W. X., Luthfianti, H. R., Priyanto, A., Hapidin, D. A., Edikresnha, D., Aimon, A. H., Suciati, T., & Khairurrijal, K. (2023). Freeze–thaw hydrogel fabrication method: Basic principles, synthesis parameters, properties, and biomedical applications. *Materials Research Express*, *10*(2), 024003. DOI: 10.1088/2053-1591/acb98e

Willcox, P. J., Howie, D. W.Jr, Schmidt-Rohr, K., Hoagland, D. A., Gido, S. P., Pudjijanto, S., Kleiner, L. W., & Venkatraman, S. (1999). Microstructure of poly (vinyl alcohol) hydrogels produced by freeze/thaw cycling. *Journal of Polymer Science. Part B, Polymer Physics*, *37*(24), 3438–3454. DOI: 10.1002/(SICI)1099-0488(19991215)37:24<3438::AID-POLB6>3.0.CO;2-9

Winarti, C., Kurniati, M., Arif, A., & Sasmitaloka, K. (2018). Cellulose-based nanohydrogel from corncob with chemical crosslinking methods. IOP Conference Series: Earth and Environmental Science.

Xu, X., Jerca, V. V., & Hoogenboom, R. (2021). Bioinspired double network hydrogels: from covalent double network hydrogels via hybrid double network hydrogels to physical double network hydrogels []. *Materials Horizons, 8*(4), 1173-1188. DOI: 10.1088/1755-1315/209/1/012043

Xu, Z., & Bratlie, K. (2018). Click Chemistry and Material Selection for in Situ Fabrication of Hydrogels in Tissue Engineering Applications. *ACS Biomaterials Science & Engineering, 4*(7), 2276–2291. Advance online publication. DOI: 10.1021/acsbiomaterials.8b00230

Yin, X., Du, M., Sun, Z., Zhu, H., Xu, P., & Wang, H. (2023). Solid-waste-based keratin/chitosan hydrogel for controlling drug release in vitro. *European Polymer Journal, 199*, 112451. DOI: 10.1016/j.eurpolymj.2023.112451

Yoshida, R., Uchida, K., Kaneko, Y., Sakai, K., Kikuchi, A., Sakurai, Y., & Okano, T. (1995). Comb-type grafted hydrogels with rapid deswelling response to temperature changes. *Nature, 374*(6519), 240–242. DOI: 10.1038/374240a0

Yu, F., Cao, X., Li, Y., & Chen, X. (2015). Diels-Alder Click-Based Hydrogels for Direct Spatiotemporal Postpatterning via Photoclick Chemistry. *ACS Macro Letters, 4*(3), 289–292. DOI: 10.1021/mz5007427

Yu, H., Rouelle, N., Qiu, A., Oh, J.-A., Kempaiah, D. M., Whittle, J. D., Aakyiir, M., Xing, W., & Ma, J. (2020). Hydrogen bonding-reinforced hydrogel electrolyte for flexible, robust, and all-in-one supercapacitor with excellent low-temperature tolerance. *ACS Applied Materials & Interfaces, 12*(34), 37977–37985. DOI: 10.1021/acsami.0c05454

Yu, S., Ji, Y., Guo, C., Lu, D., Geng, Z., Pei, D., & Liu, Q. (2021a). A dual-cross-linked hydrogel based on hyaluronic acid/gelatin tethered via tannic acid: Mechanical properties' enhancement and stability control. *Iranian Polymer Journal, 30*(3), 307–317. DOI: 10.1007/s13726-020-00891-9

Yu, S., Ji, Y., Guo, C., Lu, D., Geng, Z., Pei, D., & Liu, Q. (2021b). A dual-cross-linked hydrogel based on hyaluronic acid/gelatin tethered via tannic acid: Mechanical properties' enhancement and stability control. *Iranian Polymer Journal, 30*(3), 307–317. Advance online publication. DOI: 10.1007/s13726-020-00891-9

Yue, W., Zhu, X., Wu, M., Qiang, W., Yang, Y., Zhang, Z., Wang, Y., Teng, Y., & Zhou, M. (2025). Research on the application of biomaterial-based responsive hydrogels in the tumor microenvironment. *PeerJ, 13*, e19609. DOI: 10.7717/peerj.19609

Zhang, Y. S., & Khademhosseini, A. (2017). Advances in engineering hydrogels. *Science, 356*(6337), eaaf3627. DOI: 10.1126/science.aaf3627

Zhao, D., Feng, M., Zhang, L., He, B., Chen, X., & Sun, J. (2021). Facile synthesis of self-healing and layered sodium alginate/polyacrylamide hydrogel promoted by dynamic hydrogen bond. *Carbohydrate Polymers, 256*, 117580. DOI: 10.1016/j.carbpol.2020.117580

Zhao, F., Bae, J., Zhou, X., Guo, Y., & Yu, G. (2018). Nanostructured Functional Hydrogels as an Emerging Platform for Advanced Energy Technologies. *Advanced Materials*, *30*(48), 1801796. https://doi.org/https://doi.org/10.1002/adma.201801796. DOI: 10.1002/adma.201801796

Zhao, M., Tang, Z., Zhang, X., Li, Z., Xiao, H., Zhang, M., Liu, K., Ni, Y., Huang, L., Chen, L., & Wu, H. (2020). A self-healing, stretchable, and conductive Poly (N-vinylpyrrolidone)/ gallic acid composite hydrogel formed via hydrogen bonding for wearable electronic sensors. *Composites Science and Technology*, *198*, 108294. DOI: 10.1016/j.compscitech.2020.108294

Chapter 3
Solvent and Solventless Hydrogels

Ahmed Olalekan Omoniyi
https://orcid.org/0000-0002-3841-3383
Changchun University of Science and Technology, China

Adefarati Oloruntoba
University of Calgary, Canada

Omolade Monisayo Adeniji
Wichita State University, USA

Samuel Adesina Ojo
Veritas University, Nigeria

Victor Edet Ebiekpe
Veritas University, Nigeria

Noela Chinyelu Igwemmar
University of Abuja, Nigeria

Sunday Adebayo Kolawole
University of Abuja, Nigeria

ABSTRACT

Conventional organic solvents have been widely employed in hydrogel fabrication, but their environmental and health concerns have motivated the search for alternative solvent systems. Ionic liquids have emerged as promising substitutes, offering low volatility, high thermal stability, and improved biocompatibility. Deep eutectic solvents, formed by combining components with unique solvent properties, exhibit low toxicity, biodegradability, and potential for sustainable manufacturing. Surfactants play crucial role in emulsion-based methods, influencing hydrogel structure and properties. Solventless hydrogels, eliminating the need for solvents, provide simplified fabrication processes, reduced environmental impact, and improved biocompatibility. This chapter provides an overview of these solvent systems, discussing their advantages, limitations, and applications in hydrogel synthesis. Understanding the diverse solvent options enables the development of advanced hydrogel materials tailored for various fields, including biomedicine, tissue engineering, and other industrial applications.

3.1 INTRODUCTION

Hydrogels are three-dimensional crosslinked polymer networks with remarkable capacity to absorb and retain water or biological fluids (Chamkouri & Chamkouri, 2021; El Sayed, 2023). Their high water content gives them tissue-like flexibility, making them valuable in biomedicine

DOI: 10.4018/979-8-3373-2337-4.ch003

and regenerative engineering (Brown & Anseth, 2017; Whitlow et al., 2020; Zhang & Mikos, 2020), drug delivery (Fu et al., 2018; He et al., 2022), and sensing technology (Lim et al., 2020). The hydrophilicity of these networks arises from functional groups such as $-NH_2$, -COOH, -OH, $-CONH_2$, -CONH, and $-SO_3H$ (Halligan et al., 2023).

Hydrogel synthesis often requires solvents to dissolve and disperse polymer precursors, influencing properties such as biocompatibility, biodegradability, swelling, and mechanical strength (Kaur et al., 2022). Conventionally, organic solvents like ethanol, methanol, acetone, and dimethylformamide (DMF) have been used due to their broad solubilizing ability (Jiang et al., 2020; Li et al., 2018, 2019). However, these solvents raise ecological and health concerns. Volatile organic compounds (VOCs) released during processing contribute to smog formation (Ukaogo et al., 2020), while improper disposal contaminates water systems. Many organic solvents are toxic, with prolonged exposure linked to respiratory, neurological, and reproductive disorders (Joshi & Adhikari, 2019).

To mitigate these issues, alternative solvent systems have been explored. Ionic liquids (ILs), composed entirely of ions, offer low vapor pressure, thermal stability, and tunable solvation properties (Liu et al., 2022). By selecting appropriate ILs, hydrogels with tailored porosity, mechanics, and responsiveness can be developed (Wu et al., 2021; Zainal et al., 2021). Deep eutectic solvents (DES), formed by combining hydrogen bond donors and acceptors, are another promising option. They are biodegradable, low-toxicity, and easily synthesized, with demonstrated ability to dissolve diverse polymers (Hooshmand et al., 2020; Perna et al., 2020). Optimizing DES composition can enhance hydrogel performance and biocompatibility (Roda et al., 2019).

Surfactants also play a role in hydrogel design by reducing surface tension and modifying polymer interactions. Their incorporation can tune hydrogel strength, surface properties, and drug-release behavior, broadening applications (Muir et al., 2021). More recently, solventless hydrogels have emerged as a greener alternative, eliminating the need for solvents altogether. Formed through physical crosslinking, supramolecular self-assembly, or covalent reactions, these hydrogels offer simplicity, reduced environmental impact, and cost-effectiveness (El Kadib, 2020).

The solvent choice in hydrogel synthesis strongly influences material performance, safety, and sustainability. Water remains a widely used medium for fabricating hydrogels such as poly(vinyl alcohol) (PVA), poly(ethylene glycol) (PEG), and N,N-dimethylacrylamide (DMA), often polymerized in aqueous systems with radical initiators like ammonium or potassium persulfate. This chapter examines solvent systems in hydrogel fabrication—ranging from conventional organic solvents to ILs, DES, surfactants, and solventless strategies—highlighting their advantages, limitations, and future prospects toward sustainable hydrogel developments.

3.2 CONVENTIONAL ORGANIC SOLVENTS FOR HYDROGEL SYNTHESIS

Conventional organic solvents are widely employed in hydrogel synthesis due to their solvating ability, ease of handling, and compatibility with diverse monomers and crosslinkers (Arabi et al., 2021). Common examples include ethanol, methanol, dimethyl sulfoxide (DMSO), acetone, DMF, and tetrahydrofuran (THF), each enabling hydrogels with tunable properties for applications such as drug delivery and tissue engineering (Xu et al., 2023; Zou et al., 2023). The interactions between solvent, monomer, and crosslinker play a crucial role in defining network architecture and performance.

Polar solvents like ethanol and methanol dissolve both hydrophilic and hydrophobic precursors, producing homogeneous gel networks. Rana et al., (2021) synthesized poly(N-isopropylacrylamide) (PNIPAAm) hydrogels in solvents of varying polarity—dioxane, THF, toluene, and cyclohexane—via free radical polymerization. Hydrogels from polar solvents (THF, dioxane) had lower porosity and swelling (≈275–314%), while those from nonpolar solvents (toluene, cyclohexane) displayed higher swelling (up to ≈900%). Dioxane- and toluene-derived gels exhibited superior mechanical strength, whereas nonpolar-solvent gels combined high porosity with favorable thermal and mechanical properties. Similarly, El-Halah et al., (2020) reported polyacrylamide (PAAm) hydrogels in water/DMSO mixtures displayed greater swelling and higher molar mass than water/ethanol systems, with DMSO-based gels showing enhanced mechanical strength but increased dehydration susceptibility. These studies highlight how solvent choice critically influences hydrogel porosity, swelling, mechanical performance, and stability, enabling tailored material design.

3.2.1 Limitations and Hazards Posed by Conventional Organic Solvents

While conventional organic solvents have been extensively used in hydrogel synthesis, it is important to acknowledge the limitations and potential hazards associated with their use. The widespread adoption of these solvents comes with certain challenges that researchers and practitioners must address to ensure the safety, environmental sustainability, and optimal performance of hydrogels.

3.2.1.1 Environmental Impact

Many conventional organic solvents used in hydrogel synthesis are volatile organic compounds (VOCs) that can contribute to air pollution and have adverse effects on the environment. During the synthesis process, these solvents may evaporate and release harmful compounds into the atmosphere. This do not only pose health risks to researchers working with these solvents but also contributes to air quality degradation. Additionally, the improper disposal of solvent waste can further impact ecosystems and groundwater quality (Khan et al., 2023).

3.2.1.2 Health and Safety Concerns

Several of these solvents are known to have adverse health effects on humans. Prolonged exposure to certain organic solvents can lead to respiratory, skin, and eye irritation (Zolghadri et al., 2023). Some solvents such as DMF, and N-methyl-2-pyrrolidinone (NMP) are also considered carcinogenic or have reproductive toxicity (Bryan et al., 2018). Researchers and laboratory personnel working with these solvents must adhere to stringent safety protocols, including the use of appropriate personal protective equipment and well-ventilated workspaces.

3.2.1.3 Compatibility and Toxicity Issues

Certain hydrogel components, such as specific monomers and crosslinkers, may exhibit poor solubility in conventional organic solvents, leading to incomplete reactions or uneven gel formation. For instance, nonpolar monomers like isoprene and butadiene, when incorporated as comonomers in hydrogel synthesis, face challenges due to their limited solubility in polar solvents. These monomers exhibit high miscibility with nonpolar solvents such as hexane and octane, but are immiscible with polar solvents like water or alcohols, restricting their utility in hydrophilic hydrogel networks. In contrast, polar polymers such as cellulose and PVA are inherently compatible with aqueous systems but exhibit poor miscibility with nonpolar solvents (Senichev & Tereshatov, 2014). Such solubility mismatches can compromise reaction efficiency and hydrogel homogeneity. Furthermore, residual solvent interactions with the hydrogel matrix (e.g., plasticization or incomplete removal) may persist within the gel network (Huang et al., 2022). This residual content can adversely affect hydrogel biocompatibility and safety, particularly in biomedical applications.

3.2.1.4 Regulatory Restrictions

Due to the potential hazards associated with the use of conventional organic solvents, regulatory agencies have imposed restrictions and guidelines on their use and disposal. Researchers and industries must comply with these regulations to ensure the safe handling and disposal of solvent waste.

The production and use of conventional organic solvents often involve the consumption of non-renewable resources, contributing to the depletion of fossil fuels and increasing the carbon footprint (Rajendran et al., 2025). Additionally, the disposal of solvent waste presents challenges in terms of sustainability (Ferrara et al., 2025). As the demand for greener and more sustainable practices increases, there is a growing need to explore alternative solvent systems that minimize environmental impact.

3.3 IONIC LIQUID AS A SOLVENT SYSTEM IN HYDROGEL SYNTHESIS

Ionic liquids (ILs) are a broad scope of solvents (that is, liquid synthetic compounds), composed of non-volatile, organic cations combined with organic or inorganic anions (Lei et al., 2017). ILs have generated deep interest in several research fields due to their highly tunable properties, including a wide variety of low vapor pressures, liquid states at room temperature, and non-flammability (Singh & Savoy, 2020). Due to their unique physicochemical characteristics, ILs are suitable for use in a range of applications, such as nanomaterial forming, biocatalysis, CO_2 capture, treatment of industrial waste stream, electrochemical energy storage, and pharmaceutical substances (Singh & Savoy, 2020).

Utilizing ILs in hydrogel synthesis is gaining traction due to their availability, cost-effective manufacture, and performance characteristics at ambient conditions, such as biocompatibility and high thermal stability (Lei et al., 2017). Additionally, ILs are favoured in hydrogel production owing to their viscosity (typically in the range of 1-100 mPa.s), electrostatic interactions, and solubility

(Shakeel et al., 2019). ILs also have an unparalleled capability of forming liquids or gels when mixed with polymers, proteins, and enzymes without forming precipitates (Castro et al., 2020).

3.3.1 Applications of Ionic Liquids in Hydrogel Synthesis

ILs serve as versatile reaction media for hydrogel synthesis due to their polarity, ionic nature, and ability to dissolve polymers and electrolytes (Ghandi, 2014). ILs such as choline chloride not only facilitate hydrogel formation but also impart ionic conductivity, advantageous for sensors and electrically responsive materials. A range of polymeric hydrogels—including polyurethane, cellulose, hydrophobically modified, block copolymer, and protein-based gels—have been synthesized in ILs (Aminabhavi & Deshmukh, 2016).

For instance, polyurethane (PUR) gels, valued for elasticity and biocompatibility, can form hydrophobic networks in ILs without cosolvents or organic catalysts, as ILs' organic cations and anions stabilize the reaction (Clarke et al., 2018). The intrinsic ionic conductivity of ILs enhances PUR hydrogels for conductive applications. Similarly, cellulose hydrogels benefit from hydrogen bonding and cation–anion interactions in ILs, achieving high swelling, mechanical strength, and ionic pathways that improve conductivity for sensors or energy storage (Zainal et al., 2021) (Figure 3.1).

Figure 1. Potential applications of IL-cellulose-based hydrogels

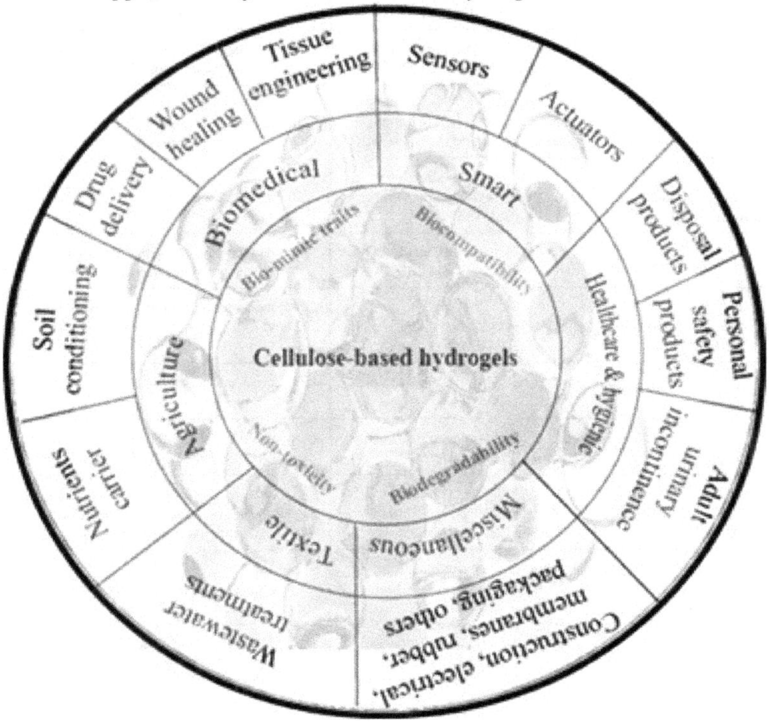

Studies have demonstrated that block copolymer hydrogels such as poly(ethylene glycol)-poly(2-hydroxypropyl methacrylate) can also be synthesized using ILs as reaction media (Tamate et al., 2018). Anionic-cationic interactions in ILs provide an advantage in terms of forming the block copolymer structure, owing to their unique membrane alteration properties and low toxicity profiles comparatively (De Oliveira Silva et al., 2019). A variety of enzyme-based hydrogels are also widely studied in the context of using ILs for formation. Enzymatically cross-linked polysaccharide hydrogels in which glucoamylase, a microbial enzyme, is covalently linked to hydrogels formed using ILs as a reaction medium show enhanced enzymatic stability due to the specific anionic-cationic structures of ILs (Ghorbanizamani et al., 2022). Specific examples of hydrogel synthesized using IL are presented in Table 3.1.

Table 1. Polymeric hydrogel produced using ionic liquids

S/ No	Polymeric hydrogel	Ionic liquids used	Properties of the hydrogel	Ref.
1	Polyacrylamide (PAM)	1-ethyl-3-methylimidazolium acetate (EMIM-Ac)	High water content, good mechanical strength, and biocompatibility.	(Alashkar et al., 2022)
2	Poly(vinyl alcohol) (PVA)	1-butyl-3-methylimidazolium chloride (BMIM-Cl)	Good mechanical strength, biocompatibility, and swelling properties.	(Dotsenko et al., 2018)
3	Poly(ethylene glycol (PEG)	1-ethyl-3-methylimidazolium bis(trifluoromethylsulfonyl)imide (EMIM-Tf2N)	High water content, good mechanical strength, and tunable swelling behavior.	(Uchytil et al., 2011)
4	Poly(lactic acid) (PLA)	1-butyl-3-methylimidazolium hexafluorophosphate (BMIM-PF6)	Good mechanical strength, biocompatibility, and degradability.	(Zhao et al., 2017)
5	Poly(glycolic acid) (PGA)	1-butyl-3-methylimidazolium methylsulfate (BMIM-MeSO4)	Good mechanical strength, biocompatibility, and degradability.	(Mai & Koo, 2017)
6	Poly(lactic-co-glycolic acid) (PLGA)	1-ethyl-3-methylimidazolium bis(trifluoromethylsulfonyl)imide (EMIM-Tf2N)	Tunable swelling behavior and controlled degradation rates.	(Kargupta et al., 2014)
7	Poly(vinylpyrrolidone) (PVP)	1-butyl-3-methylimidazolium tetrafluoroborate (BMIM-BF4)	Good mechanical strength, biocompatibility, and tunable swelling behavior.	(Xie et al., 2012)
8	Poly(N-isopropylacrylamide) (PNIPAm)	1-ethyl-3-methylimidazolium acetate (EMIM-Ac)	Good mechanical strength, biocompatibility, and tunable temperature-responsive properties.	(Chen et al., 2018)
9	Chitosan	1-butyl-3-methylimidazolium chloride (BMIM-Cl)	Good mechanical strength, biocompatibility, and tunable swelling behavior.	(Tian et al., 2015)
10	Alginate	1-ethyl-3-methylimidazolium acetate (EMIM-Ac)	Good mechanical strength, biocompatibility, and tunable swelling behavior.	(Zhang et al., 2014)
11	Poly(methacrylic acid) (PMAA)	1-butyl-3-methylimidazolium ethylsulfate (BMIM-EtSO4)	High water content, good mechanical strength, and biocompatibility.	(Cesarano et al., 1988)
12	Poly(vinylidene fluoride) (PVDF)	1-ethyl-3-methylimidazolium bis(trifluoromethylsulfonyl)imide (EMIM-Tf2N)	Good mechanical strength, biocompatibility, and thermal stability.	(Kim et al., 2011)
13	Poly(ethyleneimine) (PEI)	1-butyl-3-methylimidazolium chloride (BMIM-Cl)	Good biocompatibility and tunable swelling behavior.	(Wu et al., 2010)

continued on following page

Table 1. Continued

S/ No	Polymeric hydrogel	Ionic liquids used	Properties of the hydrogel	Ref.
14	Poly(acrylic acid-co-vinyl pyrrolidone) (PAA-PVP)	1-butyl-3-methylimidazolium bis(trifluoromethylsulfonyl)imide (EMIM-Tf2N)	Tunable swelling behavior and controlled degradation rates.	(Cumplido et al., 2018)
15	Poly(ethylene glycol-co-poly(ethylene oxide)) (PEG-PEO)	1-butyl-3-methylimidazolium chloride (BMIM-Cl)	Good mechanical strength, biocompatibility, and tunable swelling behavior.	(Wu, 2012)

3.3.2 Advantages of Using Ionic Liquids as Solvents for Hydrogel Production

In IL-based hydrogel productions, the anionic-cationic interactions generate an environment for hydrogel formation, which traditional organic solvents cannot facilitate. This is possible due to the unique combination of polar nature of ILs, which provides a specific electrostatic environment that supports intermolecular cooperation (Lei et al., 2017). Additionally, ILs also facilitate catalytic activities and enable superior material transport due to their natural nature of low vapor pressures as well as high thermal stability, which can enable a reduction in energy consumption (Shakeel et al., 2019; Singh & Savoy, 2020). Their combination of polarity and non-polarity also provide the desirable prerequisites for hydrogel formation for a wide range of polymer systems while also eliminating the chance of unwanted by-products (Singh & Savoy, 2020).

ILs also offer environmental advantages due to their safe and economical manufacture and easy disposal. Most IL-based systems are considered safe and operate under milder conditions of hydrogel formulation, generally avoiding harsh conditions, toxic solvents, and hazardous by-products that their traditional organic alternatives possess (Silva et al., 2012). These environmental considerations are becoming increasingly crucial, particularly in biomedical applications where biocompatibility, toxicity, water solubility, and recyclability are of major relevance in addition to physical performance parameters (Clarke et al., 2018).

3.3.3 Challenges and Considerations in Utilizing Ionic Liquids for Hydrogel Formation

While ILs offer unique advantages in hydrogel synthesis, their biocompatibility remains a critical concern. A major challenge lies in residual ILs that are not covalently bonded to the hydrogel network, which may leach out over time and induce cytotoxicity or inflammatory responses in biological systems. This leaching risk, combined with the high cost of IL purification and synthesis, limits their widespread adoption in biomedical applications. Although ILs are often considered milder than traditional organic solvents, their toxicity can persist if uncatalysed by-products remain in the reaction medium (Ghandi, 2014).

Additionally, designing IL-based hydrogels requires careful monomer selection and reaction optimization. Unlike conventional solvents, ILs lack precise, predictable physical properties (e.g., polarity, viscosity) that are crucial for reproducible hydrogel fabrication (Chen et al., 2018). For example, IL viscosity varies significantly with concentration, temperature, and ambient conditions, complicating process scalability (Shen et al., 2016). Furthermore, the chemical structure of ILs dictates their physical behavior, creating a trade-off when tuning hydrogel properties. A specific IL may enhance thermal stability but reduce mechanical flexibility, or improve ionic conductivity while increasing leaching risks (Lei et al., 2017). These interdependencies pose challenges in tailoring hydrogels for biocompatible applications, where mechanical resilience, degradation kinetics, and *in vivo* safety must coexist.

3.4 DEEP EUTECTIC SOLVENTS FOR HYDROGEL SYNTHESIS

The utilization of deep eutectic solvents (DES), a subgroup of ILs, in hydrogel synthesis has gained significant attention in recent years due to their unique properties and potential for improving hydrogel production processes. DES are a class of non-aqueous solvents formed by a combination of a hydrogen bond acceptor and a hydrogen bond donor (Figure 3.2), often involving a quaternary ammonium salt and a hydrogen bond donor such as a carboxylic acid or a polyol (Abbott et al., 2001). These solvents exhibit advantageous characteristics such as low volatility, tunable viscosity, and exceptional solubilization capacity, making them a promising alternative to traditional aqueous solvents.

Figure 2. Common hydrogen bond donors and acceptors used in the preparation of various DES systems

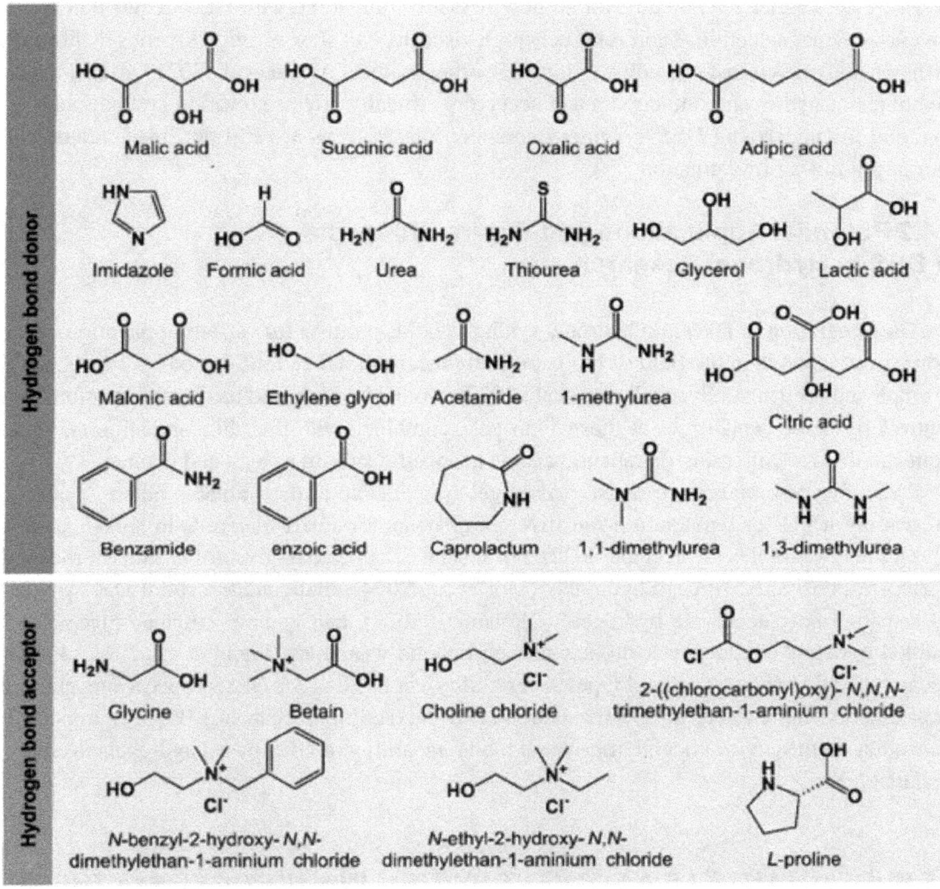

(Satlewal et al., 2018). Copyright 2018 Elsevier Inc.

3.4.1 Benefits and Drawbacks of DES in Hydrogel Production

DES provide a versatile platform for tuning hydrogel properties, allowing optimization of mechanical strength, porosity, and drug-release kinetics (Liao et al., 2022). Many DES are derived from natural compounds, making them biodegradable and environmentally benign (Azzouz & Hayyan, 2023), while some synthetic systems—such as choline chloride–urea—also effectively support hydrogel synthesis. For example, choline chloride–urea DES has been used to prepare cellulose-based hydrogels with enhanced mechanical strength and water absorption compared to conventional solvents (Laitinen et al., 2017). Their low toxicity, biodegradability, and recyclability make DES attractive green alternatives, aligning with sustainable hydrogel manufacturing (Ejeromedoghene et al., 2022).

However, challenges remain. Not all polymers are compatible with DES, requiring careful solvent–polymer selection. Their relatively high viscosity can slow reaction kinetics by limiting diffusion and mass transfer, leading to longer synthesis times (Alsaid et al., 2021). Additionally, the choice of hydrogen bond donors and acceptors critically affects solvation and polymerization, and the underlying DES–polymer interaction mechanisms are still not fully understood, warranting further investigation.

3.4.2 Potential Applications and Future Prospects of DES in Hydrogel Research

The integration of DES into hydrogel synthesis holds promise for various applications. One promising avenue is in the field of biomedical engineering, where DES-based hydrogels could be employed for drug delivery (Pedro et al., 2022), wound healing, and tissue regeneration. The controlled release capabilities of these hydrogels, combined with their biocompatibility, make them suitable for delivering therapeutic agents to specific sites in a sustained manner.

DES offer potentials in responsive hydrogel systems due to their tunable nature, enabling integration of pH- or temperature-sensitive components for smart materials in areas like soft robotics and wearables (Wan et al., 2022). For instance, lactic acid–glycerol DES has produced thermo-responsive PNIPAAm hydrogels (Mor et al., 2024), malic acid–sorbitol DES yielded pH-sensitive polysaccharide hydrogels (Gholamali, 2021), and betaine–ethylene glycol DES enabled ionically conductive hydrogels for sensors and wearables (Wang et al., 2021). Future research should expand hydrogel types accessibility via DES, tailor DES properties to specific requirements, and explore novel formulations and interaction mechanisms. Equally important are studies on the environmental impact and biodegradability of DES-based hydrogels to ensure sustainable use.

3.5 SURFACTANTS AS A SOLVENT SYSTEM IN HYDROGEL SYNTHESIS

Organic compounds consisting of a polar head and a non-polar tail with distinct polarities are known as surfactants, e.g., sodium lauryl sulphate (Figure 3.3) (Zhang et al., 2019). The distinctive structural attributes of surfactants make them valuable for reducing surface and interfacial tension between multiple phases (Wang et al., 2018). The amphiphilic nature of surfactants renders them highly advantageous for diverse industrial applications, including pharmaceuticals, detergents, wetting agents, oil recovery enhancers, and drug delivery (El Mehbad, 2017).

Figure 3. Surfactant molecule

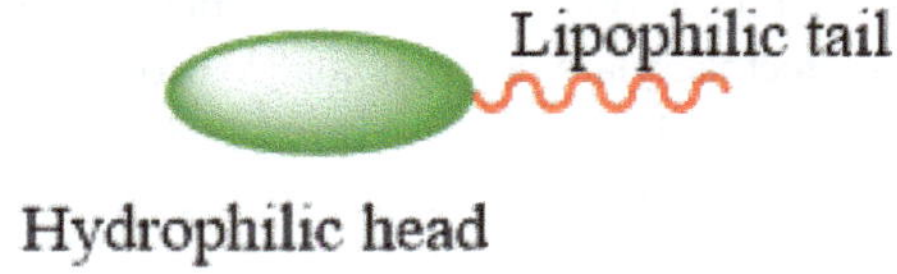

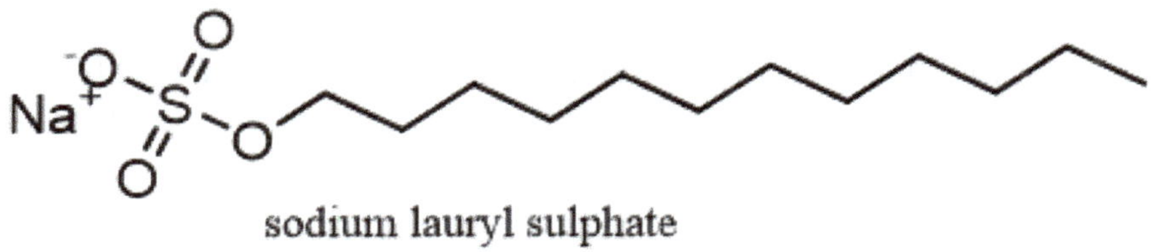

Surfactants play a crucial role in fabricating mechanically responsive hydrogels. These hydrogels, capable of ensnaring water molecules, exhibit diverse structures contingent upon specific conditions such as concentration, molecular geometry, and ionic strength. The gelation ability is primarily facilitated by the feeble Van der Waals forces originating from the hydrophobic moieties of surfactant molecules, functioning as the driving force behind the gel formation. Notably, the micelles generated typically adopt a thread-like morphology, serving as the foundational framework for hydrogel synthesis. Due to this, a rather substantial concentration of surfactant is often requisite for the water gelation process (Chen et al., 2019).

3.5.1 Influence of Surfactants on Hydrogel Properties

The mechanical properties of hydrogels can be enhanced by adjusting monomer or surfactant concentrations. Surfactants are particularly important for imparting self-healing and recovery abilities, as well as other desirable attributes such as reduced surface tension and improved interfacial strength (Zhang et al., 2019). The synergistic use of multiple surfactant types often yields superior properties compared to increasing the concentration of a single surfactant.

Surfactants play a key role in hydrophobic association hydrogels. Li et al., (2018) fabricated aquaphobic hydrogels using polysaccharides and sodium dodecyl sulfate (SDS), achieving stress-breaking strengths above 1.0 MPa through long-lived hydrophobic associations acting as visible cross-links. Similarly, Xu et al., (2018) combined SDS with the amphoteric surfactant dodecyldimethyl betaine (BS-12) to form mixed micelles that stabilized hydrophobic regions as crosslinking points, resulting in tensile strength up to 700 kPa. Zhao et al., (2021) demonstrated that micelles formed by polyacrylamide chains and dodecyl trimethyl ammonium bromide (DTAB) acted as linking points for 3D networks, with mechanical strength and recovery modulated by surfactant concentration (Figure 3.4). Their observations indicated that manipulating the concentration of hydrophobes and surfactants could enhance the mechanical strength and recovery ability of the hydrogel.

Although less explored, hydrophilic association hydrogels can also benefit from simple cationic surfactants. Optimal surfactant formulations improved mechanical strength, notch-insensitivity, time-dependent dynamic moduli, self-healing efficiency, and salt tolerance (Gulyuz & Okay, 2015; Xu et al., 2018). Von Nessen et al., (2013) observed that surfactant type influenced the size responsiveness of poly(N-isopropylmethacrylamide) colloidal hydrogels to temperature changes, supporting sensor applications. In alginate–chitosan hydrogels, anionic surfactants increased swelling, whereas cationic surfactants decreased it due to their relative film stability (Bibi et al., 2019).

Surfactants also facilitate network dynamics in associative polymers. For instance, SDS solubilized hydrophobically modified ethoxylated urethanes by interacting with hydrophobic flower micelles and poly(ethylene glycol) groups, promoting micelle solubilization, polymer chain diffusion, and self-healing behavior (Ilochonwu et al., 2022; Zhang et al., 2019).

Figure 4. The synthesis process, 3D network and recovery mechanism of hydrophobic association hydrogels

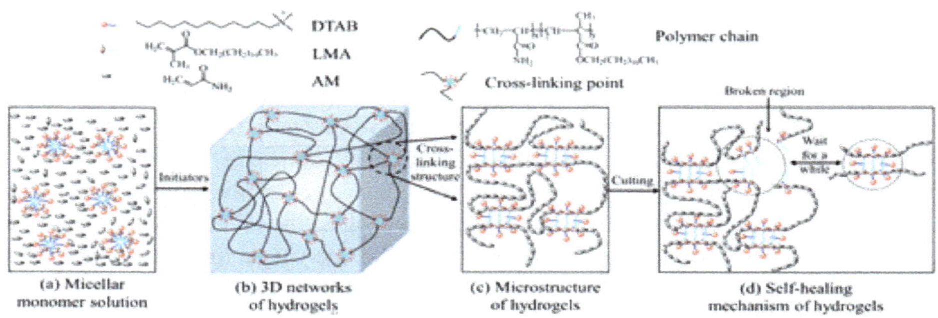

(Zhao et al., 2021). Copyright 2021 Wiley.

3.5.2 Types of Surfactants Used in Hydrogel Fabrication

Surfactants are broadly classified by their dissociation behavior in water into anionic, cationic, non-ionic, and amphoteric (zwitterionic) types (Hussain et al., 2019) (Figure 3.5).

Figure 5. Classification of surfactants according to the ionic charge in its hydrophilic head

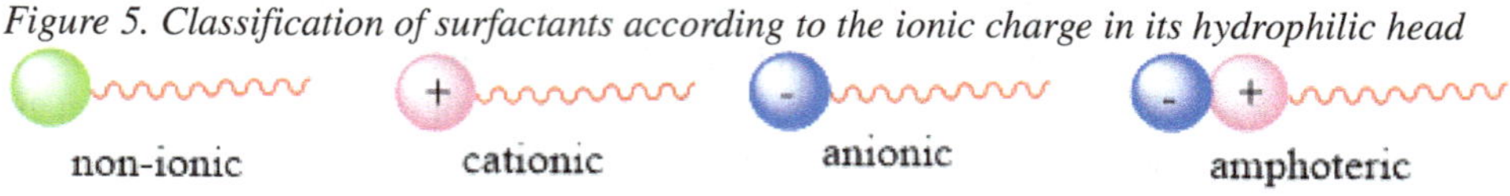

Anionic surfactants dissociate in aqueous media to yield amphiphilic anions paired with counter cations such as Na^+, K^+, or quaternary ammonium ions. Common examples include alkylbenzene sulphonates, lauryl sulphate, dialkyl sulfosuccinate, and lignosulphonates (Camacho Ramirez et al., 2023).

Non-ionic surfactants do not ionize in water; their hydrophilicity arises from groups such as alcohols, phenols, esters, ethers, or amides. Many are derived from the polycondensation of ethylene oxide, producing polyethylene glycol chains with hydrophobic alkyl or alkylbenzene moieties, often sourced from natural fatty acids. Polypropylene oxide derivatives introduce a slightly lipophilic character (Camacho Ramirez et al., 2023).

Cationic surfactants dissociate to form amphiphilic cations with counter anions, typically based on nitrogen compounds such as fatty amine salts and quaternary ammonium compounds. They usually contain long alkyl chains from natural fatty acids. Due to the complexity of high-pressure hydrogenation required for their synthesis, cationic surfactants are generally more expensive than anionic ones (Hussain et al., 2019).

Amphoteric (zwitterionic) surfactants contain both positive and negative charges within the same molecule, reaching charge neutrality at their isoelectric point. Their ability to remain effective across wide pH ranges makes them versatile emulsifiers, foaming agents, and stabilizers (Bashir et al., 2022). Examples include cocamidopropyl betaine, sulfobetaines, amphoacetates, sultaines, phosphobetaines, and amino acid-based surfactants.

3.6 SOLVENTLESS HYDROGELS

Solventless hydrogels, also known as dry hydrogels, represent a distinct class that differs fundamentally from conventional solvent-based hydrogels. Formed without solvents, they align with sustainable and environmentally conscious materials design while offering unique structural and functional advantages (Chai et al., 2017). A key feature of solventless hydrogels is their enhanced mechanical stability, often superior to that of solvent-reliant systems. The absence of solvents during formation promotes higher crosslinking density and network uniformity, yielding compact, interwoven architectures with improved strength and durability (Yang et al., 2014). In addition, the elimination of solvent evaporation prevents void formation, producing structurally consistent hydrogel matrices.

Solventless hydrogels also demonstrate distinct swelling behavior and high intrinsic purity, as no residual solvents or extensive purification steps are required. This minimizes impurities that could impair performance, resulting into enhanced biocompatibility, and strengthens their suitability for biomedical and pharmaceutical applications (Bashir et al., 2020).

3.6.1 Methods for Fabricating Solventless Hydrogels

The processes of fabricating solventless hydrogels include physical, and chemical cross-linking methods (Akhtar et al., 2016), as well as grafting technique (Farrukh et al., 2018).

3.6.1.1 Physical Cross-Linking Method

This technique avoids the utilization of a cross-linking agent, employing various methods including thermal condensation, stereo-complexation, ionic interactions, and self-assembly (Figure 3.6).

Self-assembly involves the spontaneous organization of polymers into ordered three-dimensional networks via weak noncovalent interactions, including hydrogen bonding, van der

Waals forces, hydrophobic interactions, and π–π stacking (Figure 3.6a) (Mann et al., 2018). This approach enables the formation of hydrogels under mild, solvent-free conditions, offering biocompatibility and dynamic responsiveness. Previous studies have demonstrated self-assembled peptide-based hydrogels that mimic the extracellular matrix, providing scaffolds for cell growth and tissue engineering (Guan et al., 2022).

Stereo-complexation occurs when oligomers of opposing chirality, such as *d*- and *l*-lactic acid, interact to form stable cross-linking junctions. For example, poly(L-lactic acid) (PLLA) and poly(D-lactic acid) (PDLA) can form crystalline complexes, generating a robust three-dimensional hydrogel network through stereocomplexation (Figure 3.6b) (Li, 2003). These stereocomplex hydrogels exhibit enhanced thermal and mechanical stability compared to single enantiomer systems, making them suitable for biomedical and drug delivery applications (Im et al., 2021).

Ionic gelation is commonly employed with biopolymers such as alginate, gelatin, or agar, where gelation is triggered by environmental cues like temperature or pH (Figure 3.6c) (Naranjo-Alcazar et al., 2023). For instance, alginate can form hydrogels via ionic cross-linking with divalent cations, producing biocompatible matrices widely used in tissue engineering and controlled drug release (Zhang et al., 2023). Contrarily, ionic-electrostatic interaction involve the use of ions, such as Ca^{2+}, to cross-link polyelectrolyte chains at room temperature and physiological pH (Figure 3.6d) (Heidarian et al., 2020; Li et al., 2021). This mild approach is advantageous for encapsulating bioactive molecules or living cells without compromising their functionality. Previous work has demonstrated the fabrication of calcium-cross-linked alginate–gelatin hydrogels with tunable mechanical properties for wound dressings (Houben et al., 2023). Thermal condensation generates hydrogels through repeated heating and cooling cycles, inducing structural transitions in hydrocolloids. These thermosensitive hydrogels exhibit gelation at specific critical solution temperatures: the lower critical solution temperature (LCST) (Figure 3.6e) or the upper critical solution temperature (UCST) (Figure 3.6f) (Naranjo-Alcazar et al., 2023). For example, PNIPAAm hydrogels undergo reversible volume phase transitions around 32 °C, allowing their use in temperature-responsive drug delivery systems (Cao et al., 2019).

Figure 6. The crosslinking of hydrogels through (a) molecular self-assembly, (b) stereo-complexation, (c) ionic gelation, (d) ionic-electrostatic interaction, (e,f) thermally induced entanglement of polymer chains

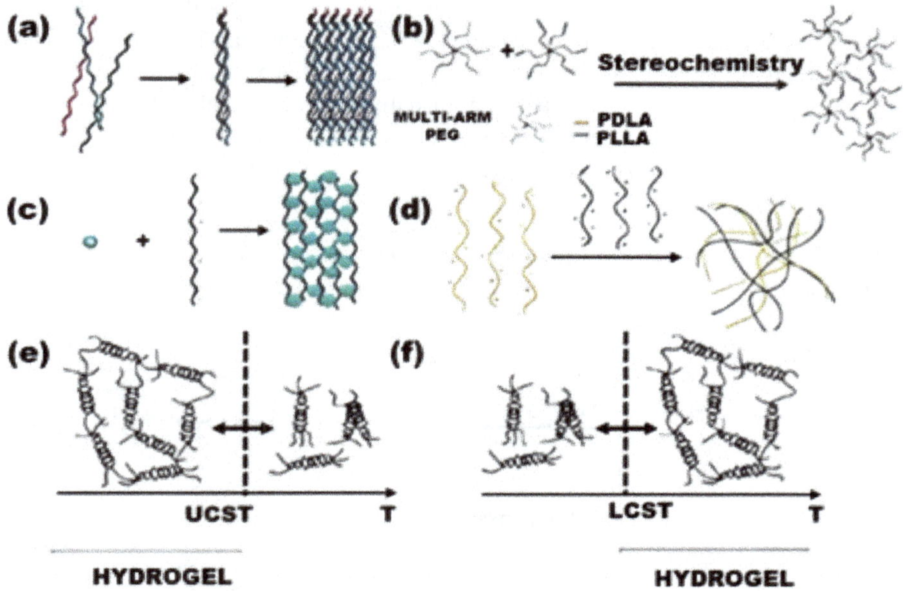

3.6.1.2 Chemical Methods

Chemical methods for solventless hydrogel synthesis rely on cross-linkers and addition reactions to establish covalent connections between polymer chains. Hydrophilic polymers bearing reactive functional groups such as $-NH_2$, –COOH, and –OH can form stable hydrogel networks through covalent bonding with suitable cross-linkers. Enzymatic cross-linking, for instance, utilizes enzymes to catalyze reactions between polymer functional groups under mild, aqueous conditions, enabling the formation of highly hydrated and biocompatible hydrogel networks (Figure 3.7a) (Sánchez-Cid et al., 2022). This approach has been effectively demonstrated in systems such as horseradish peroxidase (HRP)-mediated cross-linking of phenol- or tyramine-modified polymers (Choi et al., 2022), which allow gelation under physiological conditions without the need for toxic reagents.

Aldehyde cross-linking and condensation reactions represent classical chemical routes for hydrogel fabrication, forming stable covalent bonds between polymeric chains (Figure 3.7b) (Liu et al., 2022). Furthermore, cross-linking of hydrogels can be accomplished using UV radiation as well as high-energy radiation sources such as gamma rays (Figure 3.7c) (Godoy-Alvarez et al., 2021; Wu et al., 2020). Such methods enable spatial and temporal control over gelation, making them valuable for biomedical and microfabrication applications (Wu et al., 2020).

Another chemical cross-linking approach is free radical polymerization, which uses free radicals to cross-link hydrophilic polymers with polymerizable groups (Figure 3.7d) (Ahmed, 2015). This method provides flexibility in tuning the hydrogel's mechanical and swelling properties through control of initiator type and concentration (Gao et al., 2020). Lastly, click chemistry has emerged as a modern, efficient cross-linking strategy. Bio-orthogonal click reactions—such as azide–alkyne cycloaddition and thiol–ene coupling—allow rapid, spontaneous, and highly selective covalent bond formation, producing heteroatom-linked networks with minimal byproducts (Figure 3.7e) (Li et al., 2021). These reactions are particularly advantageous for designing solvent-free hydrogels with precise network architectures and functional tunability.

Figure 7. Fabrication of hydrogels by chemical method (a) enzymatic reactions, (b) chemical reaction promoted by a cross-linking agent, (c) high-energy radiation, (d) free-radical polymerization and (e) click reactions

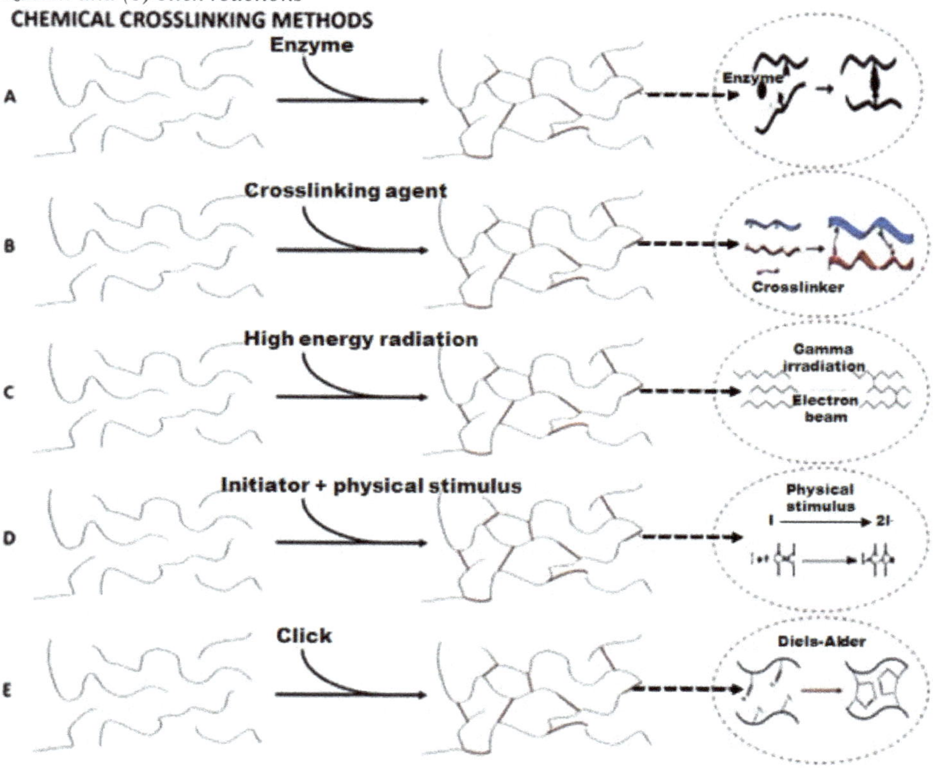

3.6.1.3 Grafting Method

This method can also be applied to the fabrication of solventless hydrogels through the process of grafting them onto a more solid underlying substrate (Figure 3.8). By attaching or grafting polymer chains—referred to as side chains or pendant chains—onto either a substrate or a base polymer, solventless hydrogels can be synthesized (Ibrahim et al., 2020). This involves the incorporation of hydrophilic groups, which exhibit the capability to absorb and retain water, ultimately resulting in the formation of the characteristic three-dimensional network structure found in hydrogels. This technique, illustrated in Figure 3.8, offers the potential to enhance both the mechanical properties and the structural integrity of hydrogels (Fu et al., 2021).

Figure 8. Synthesis of a hydrogel by grafting of acrylamide-co-sodium methacrylate onto chitosan for effective adsorption of Fuchsin basic dye

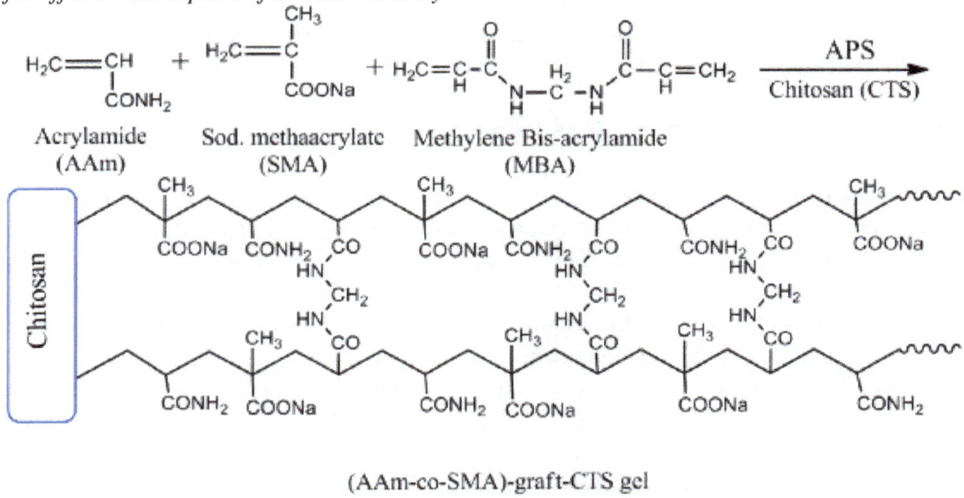

(Ibrahim et al., 2020). Copyright 2020 Elsevier Inc

3.6.2 Advantages of Solventless Hydrogel

Solventless hydrogels exhibit several key advantages:

(a) **Softening necrotic tissue:** These hydrogels hydrate and soften necrotic tissue, supporting debridement by easing the removal of non-viable tissue. This promotes a cleaner wound bed, accelerates healing, and reduces trauma during dressing changes—especially valuable for delicate wounds (Alven & Aderibigbe, 2020).

(b) **Enhanced porosity:** Their porous architecture enables efficient gas and moisture exchange, fostering oxygenation and creating conditions favorable for cell proliferation and tissue regeneration (Jagur-Grodzinski, 2010).

(c) **Environmentally conscious:** By eliminating organic solvents, solventless systems reduce pollution and the risk of solvent residues, offering a greener and safer alternative (Kharissova et al., 2019).

(d) **Higher stability:** The absence of solvents minimizes phase separation, shrinkage, and structural collapse, yielding more stable gels during production and storage (El Kadib, 2020).

(e) **Enhanced mechanical properties:** Solventless hydrogels generally exhibit greater strength and elasticity due to their compact, solvent-free network architecture (Kirillova et al., 2021).
(f) **Tunable properties:** Their formation process allows precise control over gelation time, porosity, mechanics, and drug release, enabling customization for diverse applications (Rhodes & Malavia, 2019).
(g) **Improved biocompatibility:** Without residual solvents, they reduce risks of inflammation or adverse reactions, making them especially suitable for wound dressings, scaffolds, and other biomedical uses (Shang et al., 2019).
(h) **Integration of bioactive motifs:** Solventless hydrogels can incorporate functional motifs that promote specific biological interactions or activities, enhancing their versatility in biomedical applications (Khavas & Alemzadeh, 2021).

3.6.3 Limitations of Solventless Hydrogel

In general, hydrogels prepared without the use of solvents typically exhibit lower swelling capabilities compared to hydrogels that are prepared with solvents. The absence of solvents leads to a reduction in the extent of hydration and water absorption, which can impact applications that require significant water uptake or diffusion (Asadi et al., 2021).

The achievement of uniform dispersion of components within a solventless hydrogel system can present greater challenges compared to solvent-based methods. To ensure the formation of a consistent gel structure, it becomes crucial to thoroughly mix and appropriately distribute monomers, initiators, cross-linkers, and additives (Atia et al., 2023).

Solventless hydrogel systems may come with specific processing limitations, including viscosity control, working duration, and curing techniques. The absence of solvents can complicate the processing procedure, thereby demanding meticulous optimization of formulation parameters and manufacturing protocols (Raees et al., 2023).

3.6.4 Recent Advancements and Applications of Solventless Hydrogels

Solventless hydrogels possess a multitude of attributes that render them highly suitable for various applications, particularly their remarkable healing capabilities. In a notable study, Huang et al. (Huang et al., 2017) effectively investigated the utilization of solventless hydrogels in the realm of supercapacitors. Similarly, Tamesue et al., (2016) investigated free-radical polymerized hydrogels containing cationic substituents, which exhibited strong adhesion to an aqueous micromica dispersion due to the intercalation of the cationic groups within the micromica interlayers.

The advantages and characteristics of solventless hydrogels encompass the ability to soften and hydrate necrotic tissue on wound surfaces. Tamesue et al., (2016) demonstrated that this property contributes to the debridement process by facilitating the removal of non-viable tissue. As a result, it creates a cleaner wound bed, supports enhanced wound healing, and prevents dressing adherence to the wound surface, thus minimizing trauma during dressing changes. This attribute proves particularly valuable for wounds involving delicate or sensitive tissue, as noted by Alven & Aderibigbe, (2020). Furthermore, in the domain of wound healing, Bagher et al., (2020) developed alginate/chitosan hydrogels loaded with hesperidin for rat wound treatment, a finding that aligns with the previous report of Ehterami et al., (2019).

The applications of solventless hydrogels in the biomedical field have witnessed a significant upsurge in recent times. The ensuing instances underscore their growing utilization and the potential benefits they offer (Figure 3.9).

Figure 9. An overview of the various biomedical applications of hydrogels

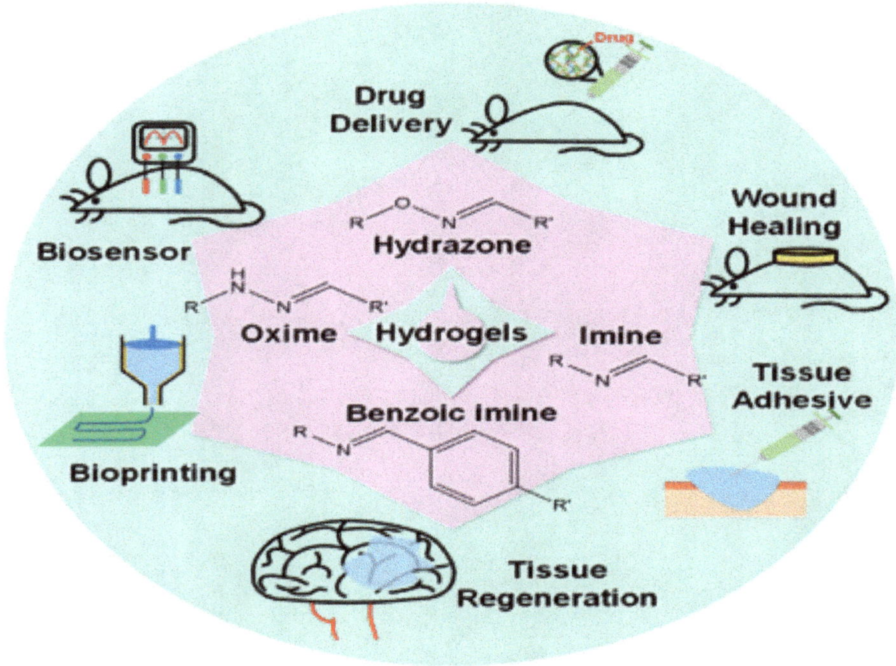

3.7 CONCLUSION

This chapter explores the use of solvents and solventless systems in the synthesis of hydrogels. While traditional organic solvents have been customary in hydrogel fabrication, the imperative need for environmentally-conscious practices has ignited the quest for alternative solvent solutions. The emergence of ILs as viable substitutes holds promise due to their attributes of reduced volatility, heightened thermal stability, and enhanced biocompatibility. Moreover, DES present a novel approach with their inherent low toxicity, biodegradability, and potential for sustainable production. In addition, surfactants, acting as pivotal agents within emulsion-based methods, wield significant influence over hydrogel structure and characteristics. Furthermore, the paradigm of solventless hydrogels, obviating the solvent requirement, ushers in simplified fabrication techniques, diminished environmental complications, and elevated biocompatibility. With the overview of these diverse solvent systems, this chapter serves as a compass to navigate their merits, limitations, and applications within the panorama of hydrogel synthesis. The identification of these solvent alternatives empowers the conception of cutting-edge hydrogel materials, finely tuned to provide various domains of application encompassing biomedicine, pharmaceutical delivery, tissue engineering and functional materials.

REFERENCES

Abbott, A. P., Capper, G., Davies, D. L., Munro, H. L., Rasheed, R. K., & Tambyrajah, V. (2001). Preparation of novel, moisture-stable, lewis-acidic ionic liquids containing quaternary ammonium salts with functional side chains. *Chemical Communications (Cambridge)*, *1*(19), 2010–2011. Advance online publication. DOI: 10.1039/b106357j

Ahmed, E. M. (2015). Hydrogel: Preparation, characterization, and applications: A review. *Journal of Advanced Research*, *6*(2), 105–121. DOI: 10.1016/j.jare.2013.07.006

Akhtar, M. F., Hanif, M., & Ranjha, N. M. (2016). Methods of synthesis of hydrogels. A review. *Saudi Pharmaceutical Journal : SPJ : the Official Publication of the Saudi Pharmaceutical Society*, *24*(5), 554–559. DOI: 10.1016/j.jsps.2015.03.022

Alashkar, A., Al-Othman, A., Tawalbeh, M., & Qasim, M. (2022). A Critical Review on the Use of Ionic Liquids in Proton Exchange Membrane Fuel Cells. In *Membranes* (Vol. 12, Issue 2). DOI: 10.3390/membranes12020178

Alsaid, Y., Wu, S., Wu, D., Du, Y., Shi, L., Khodambashi, R., Rico, R., Hua, M., Yan, Y., Zhao, Y., Aukes, D., & He, X. (2021). Tunable sponge-like hierarchically porous hydrogels with simultaneously enhanced diffusivity and mechanical properties. *Advanced Materials*, *33*(20), 2008235. DOI: 10.1002/adma.202008235

Alven, S., & Aderibigbe, B. A. (2020). Chitosan and cellulose-based hydrogels for wound management. *International Journal of Molecular Sciences*, *21*(24), 9656. DOI: 10.3390/ijms21249656

Aminabhavi, T. M., & Deshmukh, A. S. (2016). Polymeric Hydrogels as Smart Biomaterials. In *Polymer and Composite Materials.*

Arabi, M., Ostovan, A., Li, J., Wang, X., Zhang, Z., Choo, J., & Chen, L. (2021). Molecular imprinting: Green perspectives and strategies. *Advanced Materials*, *33*(30), 2100543. DOI: 10.1002/adma.202100543

Asadi, N., Pazoki-Toroudi, H., Del Bakhshayesh, A. R., Akbarzadeh, A., Davaran, S., & Annabi, N. (2021). Multifunctional hydrogels for wound healing: Special focus on biomacromolecular based hydrogels. *International Journal of Biological Macromolecules*, *170*, 728–750. DOI: 10.1016/j.ijbiomac.2020.12.202

Atia, G. A. N., Shalaby, H. K., Ali, N. G., Morsy, S. M., Ghobashy, M. M., Attia, H. A. N., Barai, P., Nady, N., Kodous, A. S., & Barai, H. R. (2023). New challenges and prospective applications of three-dimensional bioactive polymeric hydrogels in oral and craniofacial tissue engineering: A narrative review. *Pharmaceuticals (Basel, Switzerland)*, *16*(5), 702. DOI: 10.3390/ph16050702

Azzouz, A., & Hayyan, M. (2023). Are deep eutectic solvents biodegradable? *Process Safety and Environmental Protection*, *176*, 1021–1025. DOI: 10.1016/j.psep.2023.06.079

Bagher, Z., Ehterami, A., Safdel, M. H., Khastar, H., Semiari, H., Asefnejad, A., Davachi, S. M., Mirzaii, M., & Salehi, M. (2020). Wound healing with alginate/chitosan hydrogel containing hesperidin in rat model. *Journal of Drug Delivery Science and Technology*, *55*, 101379. DOI: 10.1016/j.jddst.2019.101379

Bashir, S., Hina, M., Iqbal, J., Rajpar, A. H., Mujtaba, M. A., Alghamdi, N. A., Wageh, S., Ramesh, K., & Ramesh, S. (2020). Fundamental concepts of hydrogels: Synthesis, properties, and their applications. *Polymers*, *12*(11), 1–60. DOI: 10.3390/polym12112702

Bashir, S., Houf, W., Villanueva, L., Ren, J., & Liu, J. L. (2022). Nanochemistry: Development of nanomaterials. In *Nanochemistry: From Theory to Application for In-Depth Understanding of Nanomaterials*. DOI: 10.1515/9783110739879-002

Bibi, A., Rehman, S., Faiz, R., Akhtar, T., Nawaz, M., & Bibi, S. (2019). Effect of surfactants on swelling capacity and kinetics of alginate-chitosan/CNTs hydrogel. *Materials Research Express*, *6*(8), 085065. DOI: 10.1088/2053-1591/ab0697

Brown, T. E., & Anseth, K. S. (2017). Spatiotemporal hydrogel biomaterials for regenerative medicine. In *Chemical Society Reviews* (Vol. 46, Issue 21). DOI: 10.1039/C7CS00445A

Bryan, M. C., Dunn, P. J., Entwistle, D., Gallou, F., Koenig, S. G., Hayler, J. D., Hickey, M. R., Hughes, S., Kopach, M. E., Moine, G., Richardson, P., Roschangar, F., Steven, A., & Weiberth, F. J. (2018). Key Green Chemistry research areas from a pharmaceutical manufacturers' perspective revisited. In *Green Chemistry* (Vol. 20, Issue 22). DOI: 10.1039/C8GC01276H

Camacho Ramirez, A., Melendez-Zamudio, M., Guerra Contreras, A., & Brook, M. A. (2023). Lysine-based silicone Surfactants. *Sustainable Chemistry*, *4*(2), 197–208. DOI: 10.3390/suschem4020015

Cao, M., Wang, Y., Hu, X., Gong, H., Li, R., Cox, H., Zhang, J., Waigh, T. A., Xu, H., & Lu, J. R. (2019). Reversible thermoresponsive peptide–PNIPAM hydrogels for controlled drug delivery. *Biomacromolecules*, *20*(9), 3601–3610. DOI: 10.1021/acs.biomac.9b01009

Castro, L. S., Pereira, P., Passarinha, L. A., Freire, M. G., & Pedro, A. Q. (2020). Enhanced performance of polymer-polymer aqueous two-phase systems using ionic liquids as adjuvants towards the purification of recombinant proteins. *Separation and Purification Technology*, *248*, 117051. Advance online publication. DOI: 10.1016/j.seppur.2020.117051

Cesarano, J.III, Aksay, I. A., & Bleier, A. (1988). Stability of Aqueous α-Al2O3 Suspensions with Poly(methacrylic acid) Polyelectrolyte. *Journal of the American Ceramic Society*, *71*(4), 250–255. Advance online publication. DOI: 10.1111/j.1151-2916.1988.tb05855.x

Chai, Q., Jiao, Y., & Yu, X. (2017). Hydrogels for biomedical applications: Their characteristics and the mechanisms behind them. *Gels (Basel, Switzerland)*, *3*(1), 6. Advance online publication. DOI: 10.3390/gels3010006

Chamkouri, H., & Chamkouri, M. (2021). A Review of Hydrogels, Their Properties and Applications in Medicine. *American Journal of Biomedical Science & Research*, *11*(6), 485–493. DOI: 10.34297/AJBSR.2021.11.001682

Chen, J., An, R., Han, L., Wang, X., Zhang, Y., Shi, L., & Ran, R. (2019). Tough hydrophobic association hydrogels with self-healing and reforming capabilities achieved by polymeric core-shell nanoparticles. *Materials Science and Engineering C, 99*, 460–467. DOI: 10.1016/j.msec.2019.02.005

Chen, J., Xie, F., Li, X., & Chen, L. (2018). Ionic liquids for the preparation of biopolymer materials for drug/gene delivery: A review. In *Green Chemistry* (Vol. 20, Issue 18). DOI: 10.1039/C8GC01120F

Choi, S., Ahn, H., & Kim, S. (2022). Tyrosinase-mediated hydrogel crosslinking for tissue engineering. *Journal of Applied Polymer Science, 139*(14), 1–22. DOI: 10.1002/app.51887

Clarke, C. J., Tu, W.-C., Levers, O., Bröhl, A., & Hallett, J. P. (2018). Green and sustainable solvents in chemical processes. *Chemical Reviews, 118*(2), 747–800. DOI: 10.1021/acs.chemrev.7b00571

Cumplido, M. P., Cháfer, A., de la Torre, J., & Lladosa, E. (2018). Capability study of 1-butyl-3- methylimidazolium bis(trifluoromethylsulfonyl)imide and trihexyl(tetradecyl)phosphonium bis(2,4,4-trimethylpentyl)phosphinate as solvents in the separation of 1-propanol from water. *Fluid Phase Equilibria, 469*, 1–8. Advance online publication. DOI: 10.1016/j.fluid.2018.04.011

De Oliveira Silva, G. V., Subramanian, A., Meng, X., Zhang, S., Barbosa, M. S., Baloukas, B., Chartrand, D., Gonzáles, J. C., Orlandi, M. O., Soavi, F., Cicoira, F., & Santato, C. (2019). Tungsten oxide ion-gated phototransistors using ionic liquid and aqueous gating media. *Journal of Physics. D, Applied Physics, 52*(30), 305102. Advance online publication. DOI: 10.1088/1361-6463/ab1dbb

Dotsenko, A. S., Dotsenko, G. S., Senko, O. V., Stepanov, N. A., Lyagin, I. V., Efremenko, E. N., Gusakov, A. V., Zorov, I. N., & Rubtsova, E. A. (2018). Complex effect of lignocellulosic biomass pretreatment with 1-butyl-3-methylimidazolium chloride ionic liquid on various aspects of ethanol and fumaric acid production by immobilized cells within SSF. *Bioresource Technology, 250*, 429–438. Advance online publication. DOI: 10.1016/j.biortech.2017.11.064

Ehterami, A., Salehi, M., Farzamfar, S., Samadian, H., Vaez, A., Ghorbani, S., Ai, J., & Sahrapeyma, H. (2019). Chitosan/alginate hydrogels containing Alpha-tocopherol for wound healing in rat model. *Journal of Drug Delivery Science and Technology, 51*, 204–213. DOI: 10.1016/j.jddst.2019.02.032

Ejeromedoghene, O., Orege, J. I., Oderinde, O., Okoye, C. O., Alowakennu, M., Nnyia, M. O., & Fu, G. (2022). Deep eutectic solvent-assisted stimuli-responsive smart hydrogels – A review. *European Polymer Journal, 181*, 111711. DOI: 10.1016/j.eurpolymj.2022.111711

El-Halah, A., González, N., Contreras, J., & López-Carrasquero, F. (2020). Effect of the synthesis solvent in swelling ability of polyacrylamide hydrogels. *Journal of Polymer Research, 27*(1), 21. DOI: 10.1007/s10965-019-1988-x

El Kadib, A. (2020). Green and functional aerogels by macromolecular and textural engineering of chitosan microspheres. *Chemical Record (New York, N.Y.), 20*(8), 753–772. DOI: 10.1002/tcr.201900089

El Mehbad, N. (2017). Efficiency of N -Decyl- N -benzyl- N -methylglycine and N -Dodecyl- N -benzyl- N -methylglycine surfactants for flow improvers and pour point depressants. *Journal of Molecular Liquids*, *229*, 609–613. DOI: 10.1016/j.molliq.2016.12.075

El Sayed, M. M. (2023). Production of Polymer Hydrogel Composites and Their Applications. *Journal of Polymers and the Environment*, *31*(7), 2855–2879. DOI: 10.1007/s10924-023-02796-z

Farrukh, S., Mustafa, K., & Hussain, A. (2018). *Synthesis and Applications of Carbohydrate-Based Hydrogels Synthesis and Applications of Carbohydrate-Based Hydrogels. August*, 0–24. DOI: 10.1007/978-3-319-76573-0

Ferrara, M., Della Negra, F., & Verzini, M. (2025). Solvent Eco-Impact Metric: A Tool for Chemists to Drive Sustainability in Chemical Processes across Safety, Health, Waste, and Environmental Aspects. *Organic Process Research & Development*, *29*(4), 1110–1124. DOI: 10.1021/acs.oprd.4c00546

Fu, X., Hosta-Rigau, L., Chandrawati, R., & Cui, J. (2018). Multi-Stimuli-Responsive Polymer Particles, Films, and Hydrogels for Drug Delivery. In *Chem* (Vol. 4, Issue 9). DOI: 10.1016/j.chempr.2018.07.002

Fu, Y., Zoetebier, B., Both, S., Dijkstra, P. J., & Karperien, M. (2021). Engineering of optimized hydrogel formulations for cartilage repair. *Polymers*, *13*(9), 1526. DOI: 10.3390/polym13091526

Gao, Y., Zhou, D., Lyu, J. A. S., Xu, Q., Newland, B., Matyjaszewski, K., Tai, H., & Wang, W. (2020). Complex polymer architectures through free-radical polymerization of multivinyl monomers. *Nature Reviews. Chemistry*, *4*(4), 194–212. DOI: 10.1038/s41570-020-0170-7

Ghandi, K. (2014). A Review of Ionic Liquids, Their Limits and Applications. *Green and Sustainable Chemistry*, *04*(01), 44–53. Advance online publication. DOI: 10.4236/gsc.2014.41008

Gholamali, I. (2021). Stimuli-Responsive Polysaccharide Hydrogels for Biomedical Applications: A Review. *Regenerative Engineering and Translational Medicine*, *7*(1), 91–114. DOI: 10.1007/s40883-019-00134-1

Ghorbanizamani, F., Moulahoum, H., Guler Celik, E., & Timur, S. (2022). Ionic liquids enhancement of hydrogels and impact on biosensing applications. In *Journal of Molecular Liquids* (Vol. 357). DOI: 10.1016/j.molliq.2022.119075

Godoy-Alvarez, F. K., González-Torres, M., Giraldo-Gomez, D. M., Sánchez-Sánchez, R., Pérez-Díaz, M. A., González-Del Carmen, M., Figueroa-González, G., Reyes-Hernández, O. D., Sharifi-Rad, J., Cortés, H., Del Prado-Audelo, M. L., & Leyva-Gómez, G. (2021). Synthesis by gamma irradiation of hyaluronic acid-polyvinyl alcohol hydrogel for biomedical applications. *Cellular and Molecular Biology*, *67*(1), 58–63. DOI: 10.14715/cmb/2021.67.1.9

Guan, T., Li, J., Chen, C., & Liu, Y. (2022). Self-Assembling Peptide-Based Hydrogels for Wound Tissue Repair. *Advanced Science (Weinheim, Baden-Wurttemberg, Germany)*, *9*(10), 2104165. DOI: 10.1002/advs.202104165

Gulyuz, U., & Okay, O. (2015). Self-healing Poly(acrylic acid) hydrogels: Effect of surfactant. *Macromolecular Symposia*, *358*(1), 232–238. DOI: 10.1002/masy.201500063

Halligan, E., Tie, B. S. H., Colbert, D. M., Alsaadi, M., Zhuo, S., Keane, G., & Geever, L. M. (2023). Synthesis and Characterisation of Hydrogels Based on Poly (N-Vinylcaprolactam) with Diethylene Glycol Diacrylate. *Gels (Basel, Switzerland)*, *9*(6), 439. Advance online publication. DOI: 10.3390/gels9060439

He, P., Dai, L., Wei, J., Zhu, X., Li, J., Chen, Z., & Ni, Y. (2022). Nanocellulose-based hydrogels as versatile drug delivery vehicles: A review. In *International Journal of Biological Macromolecules* (Vol. 222). DOI: 10.1016/j.ijbiomac.2022.09.214

Heidarian, P., Kouzani, A. Z., Kaynak, A., Paulino, M., Nasri-Nasrabadi, B., Zolfagharian, A., & Varley, R. (2020). Dynamic plant-derived polysaccharide-based hydrogels. *Carbohydrate Polymers*, *231*, 115743. DOI: 10.1016/j.carbpol.2019.115743

Hooshmand, S. E., Afshari, R., Ramón, D. J., & Varma, R. S. (2020). Deep eutectic solvents: Cutting-edge applications in cross-coupling reactions. *Green Chemistry*, *22*(12), 3668–3692. DOI: 10.1039/D0GC01494J

Houben, S., Aldana, A. A., Huysecom, A.-S., Mpinganzima, W., Cardinaels, R., Baker, M. B., & Pitet, L. M. (2023). Hybrid hydrogels with orthogonal transient cross-linking exhibiting highly tunable mechanical properties. *ACS Applied Polymer Materials*, *5*(3), 1819–1827. DOI: 10.1021/acsapm.2c01906

Huang, G., Tang, Z., Peng, S., Zhang, P., Sun, T., Wei, W., Zeng, L., Guo, H., Guo, H., & Meng, G. (2022). Modification of hydrophobic hydrogels into a strongly adhesive and tough hydrogel by electrostatic interaction. *Macromolecules*, *55*(1), 156–165. DOI: 10.1021/acs.macromol.1c01115

Huang, Y., Zhong, M., Shi, F., Liu, X., Tang, Z., Wang, Y., Huang, Y., Hou, H., Xie, X., & Zhi, C. (2017). An intrinsically stretchable and compressible supercapacitor containing a polyacrylamide hydrogel electrolyte. *Angewandte Chemie International Edition*, *56*(31), 9141–9145. DOI: 10.1002/anie.201705212

Hussain, S. M. S., Kamal, M. S., & Fogang, L. T. (2019). Synthesis and physicochemical investigation of betaine type polyoxyethylene zwitterionic surfactants containing different ionic headgroups. *Journal of Molecular Structure*, *1178*, 83–88. DOI: 10.1016/j.molstruc.2018.09.094

Ibrahim, A. G., Sayed, A. Z., Abd El-Wahab, H., & Sayah, M. M. (2020). Synthesis of a hydrogel by grafting of acrylamide-co-sodium methacrylate onto chitosan for effective adsorption of Fuchsin basic dye. *International Journal of Biological Macromolecules*, *159*, 422–432. DOI: 10.1016/j.ijbiomac.2020.05.039

Ilochonwu, B. C., Mihajlovic, M., Maas-Bakker, R. F., Rousou, C., Tang, M., Chen, M., Hennink, W. E., & Vermonden, T. (2022). Hyaluronic acid-PEG-based diels–alder in situ forming hydrogels for sustained intraocular delivery of bevacizumab. *Biomacromolecules*, *23*(7), 2914–2929. DOI: 10.1021/acs.biomac.2c00383

Im, S. H., Im, D. H., Park, S. J., Chung, J. J., Jung, Y., & Kim, S. H. (2021). Stereocomplex polylactide for drug delivery and biomedical applications: A review. *Molecules (Basel, Switzerland)*, *26*(10), 2846. DOI: 10.3390/molecules26102846

Jagur-Grodzinski, J. (2010). Polymeric gels and hydrogels for biomedical and pharmaceutical applications. *Polymers for Advanced Technologies*, *21*(1), 27–47. DOI: 10.1002/pat.1504

Jiang, C., Zhu, T., Liu, H., Yang, G., He, Z., Wang, M., Ji, M., Cong, G., Yu, J., Zhu, C., & Xu, J. (2020). A one-step aqueous route to prepare polyacrylonitrile-based hydrogels with excellent ionic conductivity and extreme low temperature tolerance. *Journal of Materials Chemistry. A, Materials for Energy and Sustainability*, *8*(42), 22090–22099. Advance online publication. DOI: 10.1039/D0TA08177A

Joshi, D. R., & Adhikari, N. (2019). An Overview on Common Organic Solvents and Their Toxicity. *Journal of Pharmaceutical Research International*, (June), 1–18. DOI: 10.9734/jpri/2019/v28i330203

Kabir, S. M. F., Sikdar, P. P., Haque, B., Bhuiyan, M. A. R., Ali, A., & Islam, M. N. (2018). Cellulose-based hydrogel materials: Chemistry, properties and their prospective applications. *Progress in Biomaterials*, *7*(3), 153–174. DOI: 10.1007/s40204-018-0095-0

Kargupta, R., Bok, S., Darr, C. M., Crist, B. D., Gangopadhyay, K., Gangopadhyay, S., & Sengupta, S. (2014). Coatings and surface modifications imparting antimicrobial activity to orthopedic implants. *Wiley Interdisciplinary Reviews. Nanomedicine and Nanobiotechnology*, *6*(5), 475–495. Advance online publication. DOI: 10.1002/wnan.1273

Kaur, J., Kaur Mankoo, R., Dudeja, I., & Kapil, S. (2022). Recent approaches to the synthesis of hydrogels from lignocellulosic biomass: A review. *Cellulose Chemistry and Technology*, *56*(7–8), 891–906. DOI: 10.35812/CelluloseChemTechnol.2022.56.80

Khan, W. U., Ahmed, S., Dhoble, Y., & Madhav, S. (2023). A critical review of hazardous waste generation from textile industries and associated ecological impacts. *Journal of the Indian Chemical Society*, *100*(1), 100829. DOI: 10.1016/j.jics.2022.100829

Kharissova, O. V., Kharisov, B. I., Oliva González, C. M., Méndez, Y. P., & López, I. (2019). Greener synthesis of chemical compounds and materials. *Royal Society Open Science*, *6*(11), 191378. DOI: 10.1098/rsos.191378

Khavas, D. S., & Alemzadeh, I. (2021). *Precise tuning of alginate and gelatin-based bioink for enhanced bioprinting applications. IChEC*, 7–9.

Kim, D. H., Baek, I. H., Hong, S. U., & Lee, H. K. (2011). Study on immobilized liquid membrane using ionic liquid and PVDF hollow fiber as a support for CO2/N2 separation. *Journal of Membrane Science*, *372*(1–2), 346–354. Advance online publication. DOI: 10.1016/j.memsci.2011.02.025

Kirillova, A., Yeazel, T. R., Asheghali, D., Petersen, S. R., Dort, S., Gall, K., & Becker, M. L. (2021). Fabrication of biomedical scaffolds using biodegradable polymers. *Chemical Reviews*, *121*(18), 11238–11304. DOI: 10.1021/acs.chemrev.0c01200

Laitinen, O., Suopajärvi, T., Österberg, M., & Liimatainen, H. (2017). Hydrophobic, superabsorbing aerogels from choline chloride-based deep eutectic solvent pretreated and silylated cellulose nanofibrils for selective oil removal. *ACS Applied Materials & Interfaces*, *9*(29), 25029–25037. DOI: 10.1021/acsami.7b06304

Lei, Z., Chen, B., Koo, Y.-M., & MacFarlane, D. R. (2017). Introduction: Ionic liquids. *Chemical Reviews, 117*(10), 6633–6635. DOI: 10.1021/acs.chemrev.7b00246

Li, D., Wang, S., Meng, Y., Guo, Z., Cheng, M., & Li, J. (2021). Fabrication of self-healing pectin/chitosan hybrid hydrogel via Diels-Alder reactions for drug delivery with high swelling property, pH-responsiveness, and cytocompatibility. *Carbohydrate Polymers, 268*, 118244. DOI: 10.1016/j.carbpol.2021.118244

Li, S. (2003). Bioresorbable Hydrogels Prepared Through Stereocomplexation between Poly(L-lactide) and Poly(D-lactide) Blocks Attached to Poly(ethylene glycol). *Macromolecular Bioscience, 3*(11), 657–661. DOI: 10.1002/mabi.200350032

Li, S., Gao, Y., Jiang, H., Duan, L., & Gao, G. (2018). Tough, sticky and remoldable hydrophobic association hydrogel regulated by polysaccharide and sodium dodecyl sulfate as emulsifiers. *Carbohydrate Polymers, 201*, 591–598. DOI: 10.1016/j.carbpol.2018.08.100

Li, Y., Wang, X., Han, Y., Sun, H.-Y., Hilborn, J., & Shi, L. (2021). Click chemistry-based biopolymeric hydrogels for regenerative medicine. *Biomedical Materials (Bristol, England), 16*(2), 022003. DOI: 10.1088/1748-605X/abc0b3

Li, Y., Wong, E., Volodine, A., Van Haesendonck, C., Zhang, K., & Van Der Bruggen, B. (2019). Nanofibrous hydrogel composite membranes with ultrafast transport performance for molecular separation in organic solvents. *Journal of Materials Chemistry. A, Materials for Energy and Sustainability, 7*(33), 19269–19279. Advance online publication. DOI: 10.1039/C9TA06169J

Li, Y., Yuan, S., Zhou, C., Zhao, Y., & Van Der Bruggen, B. (2018). A high flux organic solvent nanofiltration membrane from Kevlar aramid nanofibers with: In situ incorporation of microspheres. *Journal of Materials Chemistry. A, Materials for Energy and Sustainability, 6*(45), 22987–22997. Advance online publication. DOI: 10.1039/C8TA08518H

Liao, J., Hou, B., & Huang, H. (2022). Preparation, properties and drug controlled release of chitin-based hydrogels: An updated review. *Carbohydrate Polymers, 283*, 119177. DOI: 10.1016/j.carbpol.2022.119177

Lim, J. Y. C., Goh, S. S., & Loh, X. J. (2020). Bottom-Up Engineering of Responsive Hydrogel Materials for Molecular Detection and Biosensing. *ACS Materials Letters, 2*(8), 918–950. Advance online publication. DOI: 10.1021/acsmaterialslett.0c00204

Liu, C., Chen, B., Shi, W., Huang, W., & Qian, H. (2022). Ionic Liquids for Enhanced Drug Delivery: Recent Progress and Prevailing Challenges. In *Molecular Pharmaceutics* (Vol. 19, Issue 4). DOI: 10.1021/acs.molpharmaceut.1c00960

Liu, Y., Wei, H., Li, S., Wang, G., Guo, T., & Han, H. (2022). Facile fabrication of semi-IPN hydrogel adsorbent based on quaternary cellulose via amino-anhydride click reaction in water. *International Journal of Biological Macromolecules, 207*, 622–634. DOI: 10.1016/j.ijbiomac.2022.03.032

Mai, N. L., & Koo, Y. M. (2017). Enzymatic reactions in ionic liquids. In *Emerging Areas in Bioengineering*. DOI: 10.1002/9783527803293.ch3

Mann, J. L., Yu, A. C., Agmon, G., & Appel, E. A. (2018). Supramolecular polymeric biomaterials. *Biomaterials Science*, *6*(1), 10–37. DOI: 10.1039/C7BM00780A

Mor, S., Yadav, R., Bhakuni, K., Rawat, P., Bisht, M., Deenadayalu, N., & Venkatesu, P. (2024). Unraveling the Role of Deep Eutectic Solvents with Varying Hydrogen-Bond Acceptors on the Thermoresponsive Polymer Poly(N -isopropylacrylamide). *The Journal of Physical Chemistry B*, *128*(18), 4554–4565. DOI: 10.1021/acs.jpcb.4c00888

Muir, V. G., Qazi, T. H., Shan, J., Groll, J., & Burdick, J. A. (2021). Influence of Microgel Fabrication Technique on Granular Hydrogel Properties. *ACS Biomaterials Science & Engineering*, *7*(9), 4269–4281. DOI: 10.1021/acsbiomaterials.0c01612

Naranjo-Alcazar, R., Bendix, S., Groth, T., & Gallego Ferrer, G. (2023). Research Progress in Enzymatically Cross-Linked Hydrogels as Injectable Systems for Bioprinting and Tissue Engineering. *Gels (Basel, Switzerland)*, *9*(3), 1–27. DOI: 10.3390/gels9030230

Pedro, S. N., Mendes, M. S. M., Neves, B. M., Almeida, I. F., Costa, P., Correia-Sá, I., Vilela, C., Freire, M. G., Silvestre, A. J. D., & Freire, C. S. R. (2022). Deep eutectic solvent formulations and alginate-based hydrogels as a new partnership for the transdermal administration of anti-inflammatory drugs. *Pharmaceutics*, *14*(4), 827. DOI: 10.3390/pharmaceutics14040827

Perna, F. M., Vitale, P., & Capriati, V. (2020). Deep eutectic solvents and their applications as green solvents. *Current Opinion in Green and Sustainable Chemistry*, *21*, 27–33. DOI: 10.1016/j.cogsc.2019.09.004

Raees, S., Ullah, F., Javed, F., Akil, H. M., Jadoon Khan, M., Safdar, M., Din, I. U., Alotaibi, M. A., Alharthi, A. I., Bakht, M. A., Ahmad, A., & Nassar, A. A. (2023). Classification, processing, and applications of bioink and 3D bioprinting: A detailed review. *International Journal of Biological Macromolecules*, *232*, 123476. DOI: 10.1016/j.ijbiomac.2023.123476

Rajendran, S., Al-Samydai, A., Palani, G., Trilaksana, H., Sathish, T., Giri, J., Saravanan, R., Lalvani, J. I. J., & Nasri, F. (2025). Replacement of Petroleum Based Products With Plant-Based Materials, Green and Sustainable Energy—A Review. *Engineering Reports*, *7*(4), e70108. DOI: 10.1002/eng2.70108

Rana, M. M., Rajeev, A., Natale, G., & De la Hoz Siegler, H. (2021). Effects of synthesis-solvent polarity on the physicochemical and rheological properties of poly(N-isopropylacrylamide) (PNIPAm) hydrogels. *Journal of Materials Research and Technology*, *13*, 769–786. DOI: 10.1016/j.jmrt.2021.05.009

Rhodes, C. A., & Malavia, N. (2019). Formulation of depot delivery systems. In *Parenteral Medications, Fourth Edition*. DOI: 10.1201/9780429201400-21

Roda, A., Matias, A. A., Paiva, A., & Duarte, A. R. C. (2019). Polymer science and engineering using deep eutectic solvents. *Polymers*, *11*(5), 1–22. DOI: 10.3390/polym11050912

Sánchez-Cid, P., Jiménez-Rosado, M., Romero, A., & Pérez-Puyana, V. (2022). Novel Trends in Hydrogel Development for Biomedical Applications: A Review. *Polymers*, *14*(15), 3023. DOI: 10.3390/polym14153023

Satlewal, A., Agrawal, R., Bhagia, S., Sangoro, J., & Ragauskas, A. J. (2018). Natural deep eutectic solvents for lignocellulosic biomass pretreatment: Recent developments, challenges and novel opportunities. *Biotechnology Advances*, *36*(8), 2032–2050. DOI: 10.1016/j.biotechadv.2018.08.009

Senichev, V. Y., & Tereshatov, V. V. (2014). Simple solvent characteristics. In *Handbook of Solvents*. DOI: 10.1016/B978-1-895198-64-5.50006-4

Shakeel, A., Mahmood, H., Farooq, U., Ullah, Z., Yasin, S., Iqbal, T., Chassagne, C., & Moniruzzaman, M. (2019). Rheology of Pure Ionic Liquids and Their Complex Fluids: A Review. In *ACS Sustainable Chemistry and Engineering* (Vol. 7, Issue 16). DOI: 10.1021/acssuschemeng.9b02232

Shang, J., Le, X., Zhang, J., Chen, T., & Theato, P. (2019). Trends in polymeric shape memory hydrogels and hydrogel actuators. *Polymer Chemistry*, *10*(9), 1036–1055. DOI: 10.1039/C8PY01286E

Shen, X., Shamshina, J. L., Berton, P., Bandomir, J., Wang, H., Gurau, G., & Rogers, R. D. (2016). Comparison of Hydrogels Prepared with Ionic-Liquid-Isolated vs Commercial Chitin and Cellulose. *ACS Sustainable Chemistry & Engineering*, *4*(2), 471–480. Advance online publication. DOI: 10.1021/acssuschemeng.5b01400

Silva, S. S., Santos, T. C., Cerqueira, M. T., Marques, A. P., Reys, L. L., Silva, T. H., Caridade, S. G., Mano, J. F., & Reis, R. L. (2012). The use of ionic liquids in the processing of chitosan/silk hydrogels for biomedical applications. *Green Chemistry*, *14*(5), 1463. Advance online publication. DOI: 10.1039/c2gc16535j

Singh, S. K., & Savoy, A. W. (2020). Ionic liquids synthesis and applications: An overview. In *Journal of Molecular Liquids* (Vol. 297). DOI: 10.1016/j.molliq.2019.112038

Tamate, R., Hashimoto, K., Ueki, T., & Watanabe, M. (2018). Block copolymer self-assembly in ionic liquids. *Physical Chemistry Chemical Physics*, *20*(39), 25123–25139. DOI: 10.1039/C8CP04173C

Tamesue, S., Yasuda, K., Noguchi, S., Mitsumata, T., & Yamauchi, T. (2016). Highly tolerant and durable adhesion between hydrogels utilizing intercalation of cationic substituents into layered inorganic compounds. *ACS Macro Letters*, *5*(6), 704–708. DOI: 10.1021/acsmacrolett.6b00337

Tian, Q., Liu, S., Sun, X., Sun, H., Xue, Z., & Mu, T. (2015). Theoretical studies on the dissolution of chitosan in 1-butyl-3-methylimidazolium acetate ionic liquid. *Carbohydrate Research*, *408*, 107–113. Advance online publication. DOI: 10.1016/j.carres.2015.02.015

Uchytil, P., Schauer, J., Petrychkovych, R., Setnickova, K., & Suen, S. Y. (2011). Ionic liquid membranes for carbon dioxide-methane separation. *Journal of Membrane Science*, *383*(1–2), 262–271. Advance online publication. DOI: 10.1016/j.memsci.2011.08.061

Ukaogo, P. O., Ewuzie, U., & Onwuka, C. V. (2020). Environmental pollution: causes, effects, and the remedies. In *Microorganisms for Sustainable Environment and Health* (pp. 419–429). Elsevier., DOI: 10.1016/B978-0-12-819001-2.00021-8 von Nessen, K., Karg, M., & Hellweg, T. (2013). Thermoresponsive poly-(N-isopropylmethacrylamide) microgels: Tailoring particle size by interfacial tension control. *Polymer*, *54*(21), 5499–5510. DOI: 10.1016/j.polymer.2013.08.027

Wan, Y., Huang, S., Sun, Y., Zhu, H., Zheng, Q., Zhang, Q., & Zhu, S. (2022). Superstrong yet water-detachable eutectogel adhesives. *Chemical Engineering Journal*, *442*, 136289. DOI: 10.1016/j.cej.2022.136289

Wang, J., Deng, Y., Ma, Z., Wang, Y., Zhang, S., & Yan, L. (2021). Lignin promoted the fast formation of a robust and highly conductive deep eutectic solvent ionic gel at room temperature for a flexible quasi-solid-state supercapacitor and strain sensors. *Green Chemistry*, *23*(14), 5120–5128. DOI: 10.1039/D1GC01512E

Wang, Y., Yu, Q., Cai, M., Zhou, F., & Liu, W. (2018). Halide-free PN ionic liquids surfactants as additives for enhancing tribological performance of water-based liquid. *Tribology International*, *128*, 190–196. DOI: 10.1016/j.triboint.2018.07.018

Whitlow, J., Pacelli, S., Walston, T., & Paul, A. (2020). Bioactive Hydrogel Platforms for Spatiotemporal Delivery of Baculoviruses in Biomedical Applications. *Advanced Therapeutics*, *3*(1), 1900103. Advance online publication. DOI: 10.1002/adtp.201900103

Wu, C. J. (2012). *Development of elastomeric and adhesive nanocomposite hydrogels for dermatological and wound care applications*. Purdue University.

Wu, N., Yu, H., Sun, M., Li, Z., Zhao, F., Ao, Y., & Chen, H. (2020). Investigation on the structure and mechanical properties of highly tunable elastomeric silk fibroin hydrogels cross-linked by γ-ray radiation. *ACS Applied Bio Materials*, *3*(1), 721–734. DOI: 10.1021/acsabm.9b01062

Wu, T.-Y., Wang, H.-C., Su, S.-G., Gung, S.-T., Lin, M.-W., & Lin, C.-B. (2010). Aggregation Influence of Polyethyleneglycol Organic Solvents with Ionic Liquids BMIBF 4 and BMIPF 6. *Journal of the Chinese Chemical Society (Taipei)*, *57*(1), 44–55. Advance online publication. DOI: 10.1002/jccs.201000008

Wu, X., Zhu, Q., Chen, Z., Wu, W., Lu, Y., & Qi, J. (2021). Ionic liquids as a useful tool for tailoring active pharmaceutical ingredients. *Journal of Controlled Release*, *338*, 268–283. DOI: 10.1016/j.jconrel.2021.08.032

Xie, Y., Wei, Z., Liu, C., Cui, L., & Wang, C. (2012). Morphologic evolution of Au nanocrystals grown in ionic liquid by plasma reduction. *Journal of Colloid and Interface Science*, *374*(1), 40–44. Advance online publication. DOI: 10.1016/j.jcis.2012.01.025

Xu, J., Liu, Y., & Hsu, S. (2019). Hydrogels based on Schiff base linkages for biomedical applications. *Molecules (Basel, Switzerland)*, *24*(16), 3005. DOI: 10.3390/molecules24163005

Xu, J., Ren, X., & Gao, G. (2018). Salt-inactive hydrophobic association hydrogels with fatigue resistant and self-healing properties. *Polymer*, *150*, 194–203. DOI: 10.1016/j.polymer.2018.07.045

Xu, Y., Zhu, H., Zhao, X., Yu, Y., Hu, B., Graff, B., Lalevée, J., Sun, K., & Gao, J. (2023). The solvatochromism and selective discrimination of dopamine substituted polydiacetylene vesicle towards typical water-miscible organic solvents. *European Polymer Journal, 190*, 112007. DOI: 10.1016/j.eurpolymj.2023.112007

Yang, J. A., Yeom, J., Hwang, B. W., Hoffman, A. S., & Hahn, S. K. (2014). In situ-forming injectable hydrogels for regenerative medicine. *Progress in Polymer Science, 39*(12), 1973–1986. DOI: 10.1016/j.progpolymsci.2014.07.006

Zainal, S. H., Mohd, N. H., Suhaili, N., Anuar, F. H., Lazim, A. M., & Othaman, R. (2021). Preparation of cellulose-based hydrogel: A review. *Journal of Materials Research and Technology, 10*, 935–952. DOI: 10.1016/j.jmrt.2020.12.012

Zhang, D., Sha, M., Pan, R., Lin, X., Xing, P., & Jiang, B. (2019). Synthesis and properties study of novel fluorinated surfactants with perfluorinated branched ether chain. *Journal of Fluorine Chemistry, 219*, 62–69. DOI: 10.1016/j.jfluchem.2018.11.001

Zhang, F., Li, S., Zhang, Q., Liu, J., Zeng, S., Liu, M., & Sun, D. (2019). Adsorption of different types of surfactants on graphene oxide. *Journal of Molecular Liquids, 276*, 338–346. DOI: 10.1016/j.molliq.2018.12.009

Zhang, M., Ding, C., Huang, L., Chen, L., & Yang, H. (2014). Interactions of collagen and cellulose in their blends with 1-ethyl-3-methylimidazolium acetate as solvent. *Cellulose (London, England), 21*(5), 3311–3322. Advance online publication. DOI: 10.1007/s10570-014-0372-6

Zhang, S., & Mikos, A. G. (2020). Biomaterials for Regenerative Medicine. *Advanced Healthcare Materials, 9*(23), 1–2. DOI: 10.1002/adhm.202001920

Zhang, Y., Yang, Y., Zhao, X., & Gao, J. (2023). Investigation on ionical cross-linking of alginate by monovalent cations to fabrication alginate gel for biomedical application. *Reactive & Functional Polymers, 183*, 105484. DOI: 10.1016/j.reactfunctpolym.2022.105484

Zhao, H., Nathaniel, G. A., & Merenini, P. C. (2017). Enzymatic ring-opening polymerization (ROP) of lactides and lactone in ionic liquids and organic solvents: Digging the controlling factors. *RSC Advances, 7*(77), 48639–48648. Advance online publication. DOI: 10.1039/C7RA09038B

Zhao, Z., Bai, Y., Sun, J., Lv, K., Lei, S., & Qiu, J. (2021). Tough and self-healing hydrophobic association hydrogels with cationic surfactant. *Journal of Applied Polymer Science, 138*(27), 50645. DOI: 10.1002/app.50645

Zolghadri, S., Rahimpour, E., & Rahimpour, M. R. (2023). Short-term and long-term health problems in exposure to chemicals. In *Crises in Oil, Gas and Petrochemical Industries* (pp. 239–253). Elsevier. DOI: 10.1016/B978-0-323-95154-8.00009-8

Zou, H., Yi, P., Xu, W., Cai, H., He, J., Sun, X., Li, X., Chen, C., Deng, G., Yuan, Y., Li, Z., Fang, M., Shui, J., Liu, X., & Yu, R. (2023). Rapid room-temperature polymerization strategy to prepare organic/inorganic hybrid conductive organohydrogel for terahertz wave responsiveness. *Chemical Engineering Journal, 461*, 141856. DOI: 10.1016/j.cej.2023.141856

Chapter 4

Characterization Strategies for Graphene Containing Hydrogel Systems: Techniques for Evaluating Structural Properties

Masooma Nazar
King Fahd University of Petroleum and Minerals, Dhahran, Saudi Arabia

Ambavaram Vijaya Bhaskar Reddy
https://orcid.org/0000-0003-3178-8817
BMS Institute of Technology and Management, India

Aqeel Ahmad
https://orcid.org/0000-0001-5389-6734
King Fahd University of Petroleum and Minerals, Dhahran, Saudi Arabia

Syed Muhammad Shakil Hussain
King Fahd University of Petroleum and Minerals, Dhahran, Saudi Arabia

Othman Charles S. Al-Hamouz
King Fahd University of Petroleum and Minerals, Dhahran, Saudi Arabia

Muhammad Moniruzzaman
New York City College of Technology, City University of New York, USA

ABSTRACT

Graphene-based hydrogels comprising graphene, graphene oxide or reduced graphene oxide form three-dimensional polymeric networks with high strength, tunable permeability and large surface area. These properties make them suitable for applications in biomedical, soft electronics and water purification. Characterization of GBHs is vital to understand their structure property relationships. FT-IR and Raman spectroscopy reveals the chemical composition, XRD assesses the crystallinity, and SEM/TEM provides morphological insights. On other hand, TGA and DSC evaluate thermal properties, while BET analysis determines surface area, structural connectivity and pore size distribution. This chapter provides a comprehensive overview on GBHs, highlighting their multifunctionality across biomedical, environmental and industrial domains. It emphasizes the structure-property-performance relationship and integrates advanced characterization techniques to align with real-world applications.

DOI: 10.4018/979-8-3373-2337-4.ch004

4.1. INTRODUCTION

Graphene-based hydrogels (GBHs) are three-dimensional (3D) polymeric networks that contain graphene or a derivative of it such as graphene oxide (GO) or reduced graphene oxide (rGO). The remarkable properties of GBHs including mechanical strength, large surface area and adjustable porosity have enabled them to use in various applications from biomedicine and water purification to energy storage and soft electronics (Liao et al., 2018; Lu et al., 2017; Hussain & Maktedar, 2023; Zare et al., 2024). They have been researched for drug delivery, biosensors (Mishra & Jha, 2023), tissue engineering (Zare et al., 2024), oil spill cleanup (Hu et al., 2018; Elhenawy et al., 2021), and supercapacitors (Liu et al., 2017) due to their multifunctionality. Their photothermal and antibacterial characteristics allow tailored photothermal therapies to heal wounds and to cure the cancer effectively (Saeidi et al., 2023; Jiang & Lu, 2021).

GBHs were found promising for environmental applications such as pollution prevention, oil spill cleanup, and water purification. The selective absorption of these particles helps in removing oil spills, while their use in air purification systems detects gases and filters the pollutants (Shen et al., 2014; Sohni et al., 2022). Due to these properties, hydrogels have gained significant potential as sustainable alternatives for environmental applications. GBHs are being investigated extensively in the industrial sector for improved coatings, flexible electronics and energy storage (Tadesse & Lubben, 2023). Their large surface area and superior conductivity boost fuel cells, supercapacitors and battery efficiency (Jiang et al., 2020). The versatility, flexibility, and sustainability of GBHs make them next-generation materials for applications in biomedicine, environmental protection, manufacturing and other fields.

A thorough characterization of GBHs is essential to assess their key features including mechanical strength, conductivity, growth behaviour and porosity, which are crucial for their biological, environmental, and industrial applications. Further, to assess hydrogels suitability for implants, wound dressings, or flexible electronics, mechanical tests such as tensile and compression are required. Research on hydrogel swelling and degradation may provide light on its stability, reactivity, and longevity in various physiological and environmental settings. Before the use of hydrogels in drug delivery or tissue engineering, biocompatibility and cytotoxicity tests are vital to prevent the adverse reactions. Hence, in detail characterization of GBHs enables tuning and optimization to meet the performance demands of advanced biomedical and technological applications.

This chapter provides a comprehensive perspective on GBHs highlighting their multifunctionality across the domains. Unlike the existing literature that focused on single application, this chapter emphasizes the tunability of GBHs properties such as mechanical strength, porosity, and conductivity by manipulating the factors like polymer type, crosslinking density, and graphene concentration. Further, the chapter emphasizes in-depth discussion on the structure–property-performance relationship of GBHs, combining advanced characterization techniques to match with real-world applications. By bridging the gap between material design and application-driven performance, this chapter presents a forward-looking framework for the development and optimization of GBHs as next-generation sustainable materials.

4.2. STRUCTURAL CHARACTERIZATION TECHNIQUES

To analyse the structural characteristics of GBHs, distinct structural characterisation methods including Fourier-transform infrared spectroscopy, X-ray diffraction, and Raman Spectroscopy are essential. Researchers may use the unique information gleaned from each method to tailor the hydrogels' characteristics to their intended uses by gaining the better understanding of materials' molecular interactions, functionality and structure.

4.2.1. Raman Spectroscopy

Raman spectroscopy is a widely used non-destructive technique for the characterisation of graphene-based materials. This technique enables the investigation of vibrational modes at the molecular and material levels, while offering spatial mapping of graphene distribution and uniformity across the hydrogels. Raman spectroscopy provides detailed insights into their molecular structure, chemical composition, bonding, and functional properties through Raman scattering. Schematic representation of a typical Raman spectrometer is presented in Figure 4.1 (Downes & Elfick, 2010).

Figure 1. Schematic overview of a typical Raman spectrometer system. Reprinted with permission from Downes and Elfick, Raman Spectroscopy and related techniques in biomedicine, MDPI, ©2010.

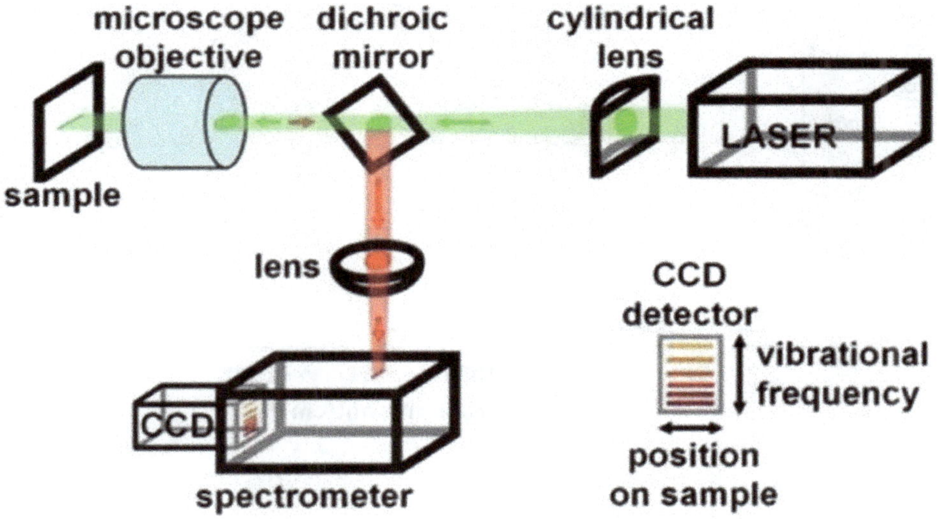

In a study, authors synthesized two GO-based hydrogels viz., benzalkonium bromide/poly-dopamine/reduced graphene oxide hydrogel (BKB/PDA/rGOG) and benzalkonium bromide/g-raphene oxide hydrogels (BKB/GOG) and examined their structural variations using Raman spectroscopy (Wang et al., 2014). The experimental setup consists of an argon ion laser with a 514.5 nm wavelength as the excitation source. The laser beam was focused using a 20x objective lens, and detection was carried out using a charge-coupled device (CCD) array. The transition between GO-to-rGO was explained by Raman spectra. Figure 4.2 indicates that GO exhibits two distinct bands at 1587 cm^{-1} and 1350 cm^{-1}, which are related to the G and D bands, respectively. When comparing the two bands, the G band is much more intense. There is no evidence that BKB alters the GO structure fundamentally based on the intensity ratio of the D/G band. The G and D band locations remain preserved with DA addition; however, the D/G band intensity ratio is much higher than GO. The result demonstrates that GO is partly diminished by DA (Wang et al., 2013). This suggests that rGO can absorb certain aromatic chemicals that GO cannot absorb (Jing & Yong-Fa, 2013).

Figure 2. Results from the Raman analysis of GO, BKB-GOG, and BKB-PDA-GOG. Reprinted with permission from Wang et al., A facile one-pot method to two kinds of graphene oxide-based hydrogels with broad-spectrum antimicrobial properties, Elsevier, 2015.

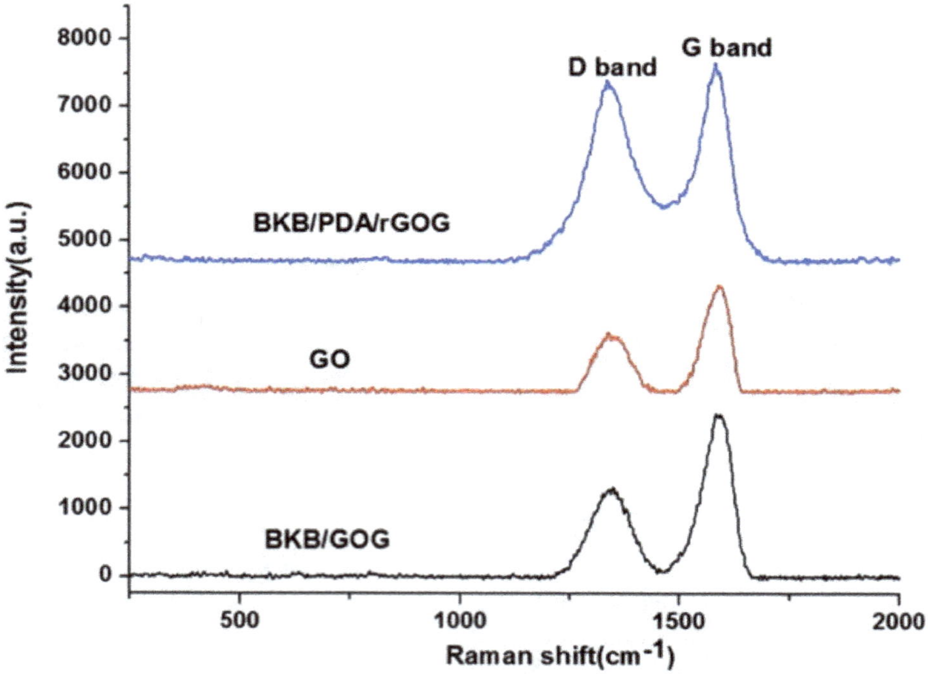

Other study presented the Raman spectroscopy results of the synthesized hydrogels (Figure 4.3). In which, polymerization reduced GO by shifting D and G bands from 1354.1 cm^{-1} and 1612.3 cm^{-1} to 1351.8 and 1602.4 cm^{-1}, respectively. After polymerization, the D and G band intensity ratios were increased, confirming the GO decrease. It also showed that 0.2Fe/rGO/PAM hydrogel has less sp^2 domains and effectively grafts PAM onto rGO surfaces (Dong et al., 2017). It is worth mentioning that hydrogels should be freeze-dried or vacuum-dried prior to Raman measurements to avoid water interference, otherwise water can interfere with Raman signals and suppress the graphene bands. This is particularly critical in composite hydrogels (e.g., biopolymer hydrogels) where peak overlapping may further complicate the spectral interpretation.

Figure 3. Raman spectroscopy outcomes of hydrogels. Reprinted with permission from Dong et al., Developing stretchable and graphene-oxide-based hydrogel for the removal of organic pollutants and metal ions, Elsevier ©2018.

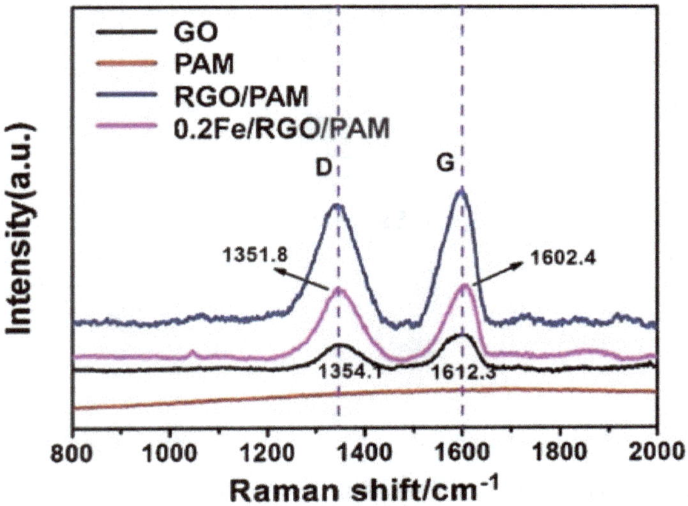

4.2.2. X-ray Diffraction (XRD)

The XRD in GBHs works on the principle of Bragg's Law and typically monitor the crystallinity, interlayer spacing, degree of reduction, and structural organization of graphene sheets within the hydrogel matrix. In XRD analysis, a focused X-ray beam directs at the sample and the interaction of X-rays with the material will be analysed by measuring the angles and intensities of the scattered photons. This characterization provides detailed information about the sample's crystal structure. The diffraction peak pattern that emerges is a valuable source of information regarding the molecular arrangement and crystal structure of the material. The atomic arrangement, crystal orientation, and interlayer spacing will have a role in the diffraction pattern. The illustrative diagram of the X-ray diffraction setup is presented in Figure 4.4 (Hitkari et al., 2018).

Figure 4. Illustrative representation of the X-ray diffraction setup. Reprinted with permission from Hitkari et al., Nanoparticles: An emerging weapon for Mitigation/Removal of various Environmental Pollutants for Environmental safety, Springer ©2018

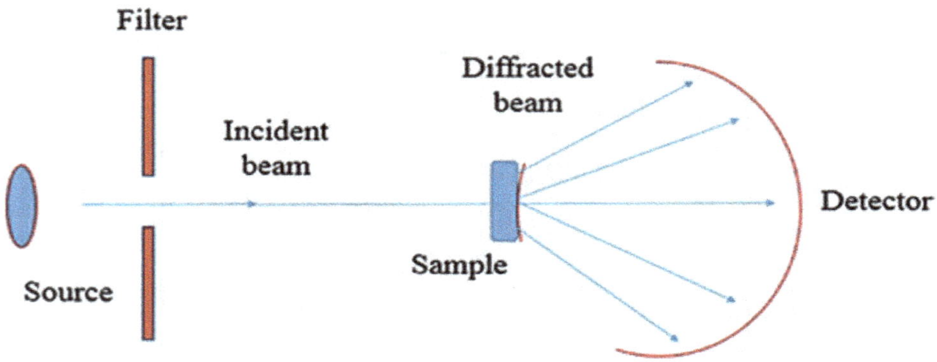

The crystalline structure of several materials including pure graphite, GO, rGO, and composite was studied within a 2θ range of 5° to 60°. Figure 4.5(a) suggests the highly crystalline structure of pure graphite, evidenced by a strong peak at 2θ = 26.46°. In line with the findings given in the literature, this peak vanished, and a new peak situated at 2θ = 11° emerged in the XRD diffractogram of GO (Figure 4.5(b)). One possible reason for this change is the increased interlayer spacing in GO, which results from the presence of oxygen-containing functional groups such as hydroxyl, epoxy, and carboxyl groups attached to the carbon framework (Gul et al., 2016). Figures 4.5(c) and 4.5(d) shows amorphous and crystalline peaks for XRD results of rGO and composite after GO reduction. The broad curve of the 15-30° 2θ region suggests an amorphous phase and the absence of long-range order (Dimiev & Tour, 2014). The effective reduction of GO led to the elimination of oxygen functions and the subsequent repair of C=C bonds in rGO, which may be attributed to this prominent peak centring about 25.0° (De Silva et al., 2018). Comparing the d-value of rGO@LNP to that of rGO, it was observed that there was no substantial change in the d-spacing, with both samples exhibiting a d-value of approximately 3.5 Å.

Figure 5. XRD data of pure graphite, GO, rGO, and rGO@LNP. Reprinted with permission from Sohni et al., Lignin nanoparticles-reduced graphene oxide based hydrogel: A novel strategy for environmental applications, Elsevier ©2023.

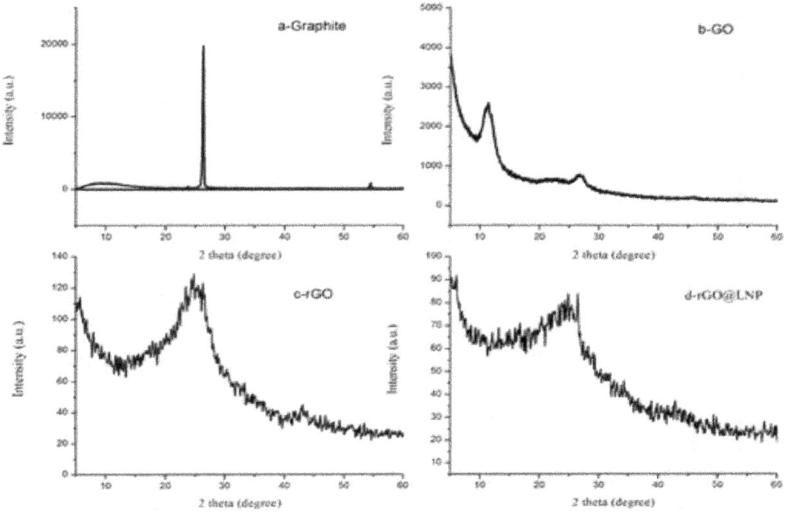

To optimize pore shape, electron and ion transport routes, and conductivity, authors constructed a 3D N-and-S co-doped holey graphene hydrogel (N/S-HGH). The graphene hydrogel was successfully prepared when the GH diffraction peak appeared at 23°, and the XRD curve revealed the typical signal of GO (Figure 4.6) (Zhu et al., 2021). The diffraction angle of holey graphene hydrogels (HGH), N, S co-doped graphene hydrogels (N/S-GH), and N/S-HGHx around 23° slightly shifted to the left, possibly due to H_2O_2 etching and N, S element doping, which can increase graphene's lattice constant (Chen et al., 2019).

Figure 6. XRD spectra of synthesized hydrogels. Reprinted with permission from Zhu et al., Three-dimensional N/S Co-doped holey graphene oxide based hydrogel electrodes for high performance supercapacitors, Elsevier ©2021.

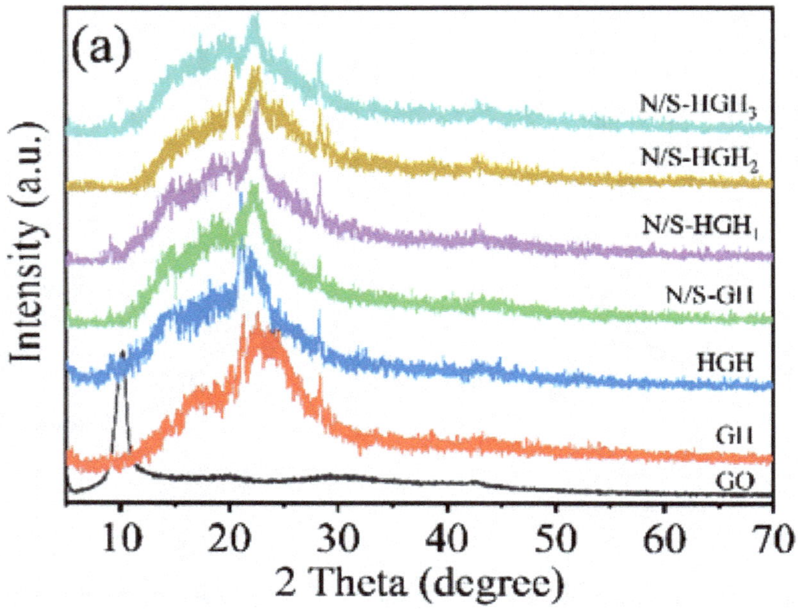

Taken together, XRD has a significant advantage to accurately identify the d-spacing between graphene sheets and to distinguish amorphous and crystalline regions within the hydrogels. It can also detect the phase transitions or incorporation of other crystalline additives into the hydrogels. Like Raman spectroscopy, hydrogels must be freeze-dried prior to XRD analysis, since the presence of water disrupts the diffraction and gives poor signals. Furthermore, XRD needs complementary techniques like XPS to detect the functional groups, oxidation states, and chemical bonding effectively.

4.2.3. Fourier Transform Infrared (FT-IR) Spectroscopy

FT-IR analysis characterizes the chemical structure and interactions within the GBHs and confirms the successful gel formation by recognising the changes in characteristic peaks. It operates based on the absorption of infrared light by chemical bonds in the molecule. When IR radiation passes through a sample, specific frequencies are absorbed by functional groups, leading to vibrational transitions such as stretching and bending. Figure 4.7 shows a schematic representation of the FT-IR analytical setup (Undavalli et al., 2021).

Figure 7. Schematic representation of the FT-IR analytical system. Reprinted with permission from Undavalli et al., Impact of alternative fuels and properties on elastomer compatibility, Elsevier ©2021.

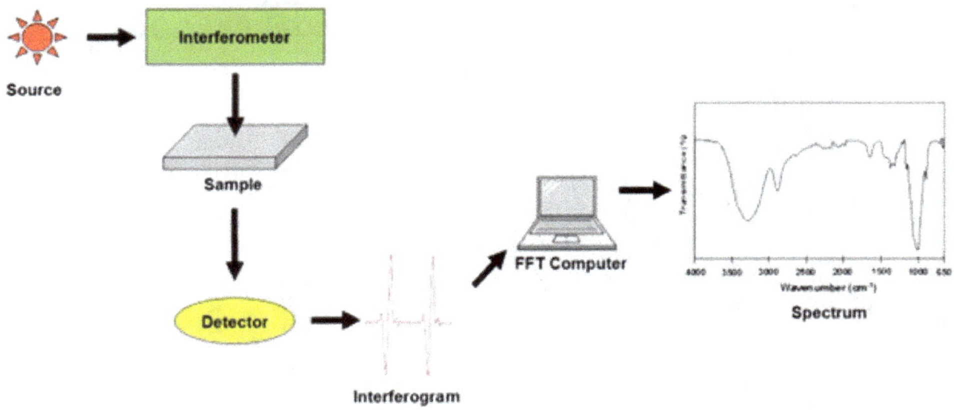

As discussed earlier, a study synthesized two GO based hydrogels namely BKB/GOG and BKB/PDA/rGOG and confirmed their formation by FT-IR analysis (Figure 4.8) (Wang et al., 2013). The copious O-H stretching vibrations are responsible for the large peak at 3300 cm^{-1} seen in GO. In addition, C=O stretching and conjugated C=C peaks appear at 1730 cm^{-1} and 1630 cm^{-1}, respectively. Another innovative approach for the environmental application was revealed by the authors with the fabrication of a 3D rGO hydrogel (rGO@LNP) based on lignin nanoparticles (LNP). In this work, LNPs were created easily via a direct chemical process from discarded walnut shell. By combining this LNP with the rGO network's continuous porous structure, rGO@LNP hydrogel was created. Figure 4.9(a) shows that the FT-IR spectrum of GO includes strong signals at 1622 cm^{-1}, 1407 cm^{-1}, and 1050 cm^{-1}, which reflects the presence of oxygen-containing functional groups, confirming effective oxidation of the graphite. The presence of oxygen-containing functional groups in rGO (Figure 4.9(b)) was confirmed by the high intensities of matching peaks at 1622 cm^{-1}, 1407 cm^{-1}, and 1050 cm^{-1}, which were not present in GO. The presence of oxygen-containing functional groups has been eliminated and the sp^2 plane of C=C has been restored, as seen by the much-decreased matching peaks in the rGO FT-IR spectrum (Figure 4.9(b)) when compared to GO (Jose et al., 2019). As seen in Figure 4.9(b), the addition of LNP to 3D hydrogel caused the composite spectra to exhibit a few peaks. Strong physical interactions between the rGO sheets and the LNP in the composite sample were suggested by the observation that the peaks corresponding to the oxygen functionalities, which were previously located at 1724 cm^{-1}, 1570 cm^{-1}, and 1125 cm^{-1} in the absorption spectra of rGO, moved to 1611 cm^{-1}, 1509 cm^{-1}, and 1119 cm^{-1}, respectively. The production of GO-LNP was suggested by a slightly wider absorption band at 1028 cm^{-1}, which was caused by the presence of an ether functional group (Eivazzadeh-Keihan et al., 2022).

Figure 8. FT-IR profiles of GO, BKB-GOG and BKB-PDA-GOG. Reprinted with permission from Wang et al., A facile one-pot method to two kinds of graphene oxide-based hydrogels with broad-spectrum antimicrobial properties, Elsevier ©2015.

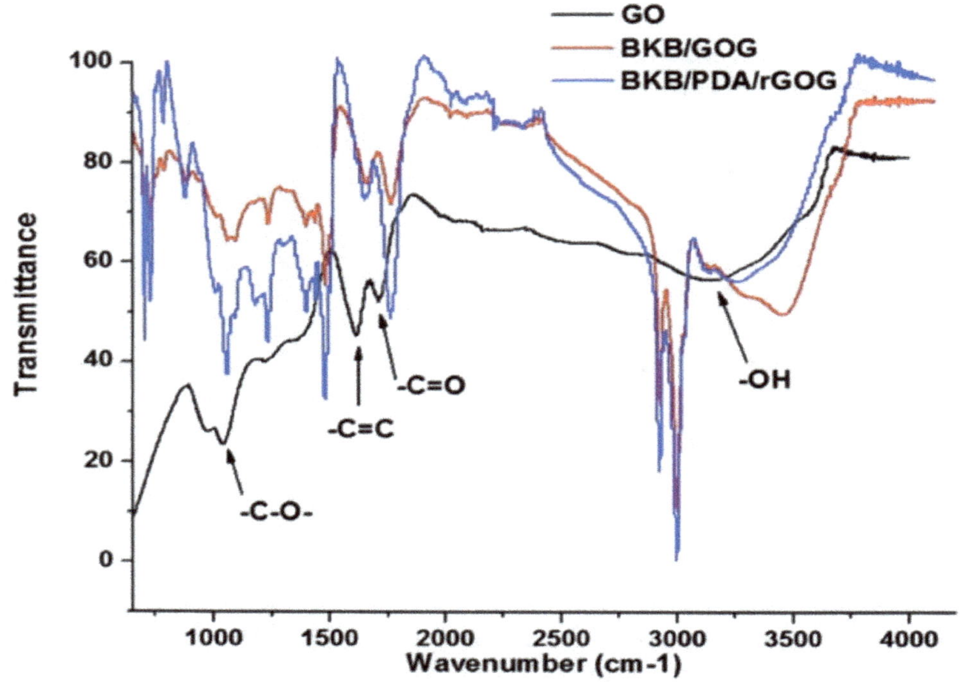

Figure 9. FT-IR Spectra of (a) pure graphite and GO, (b) rGO and rGO@L_{NP}. Reprinted with permission from Sohni et al., Lignin nanoparticles-reduced graphene oxide based hydrogel: A novel strategy for environmental applications, Elsevier ©2023.

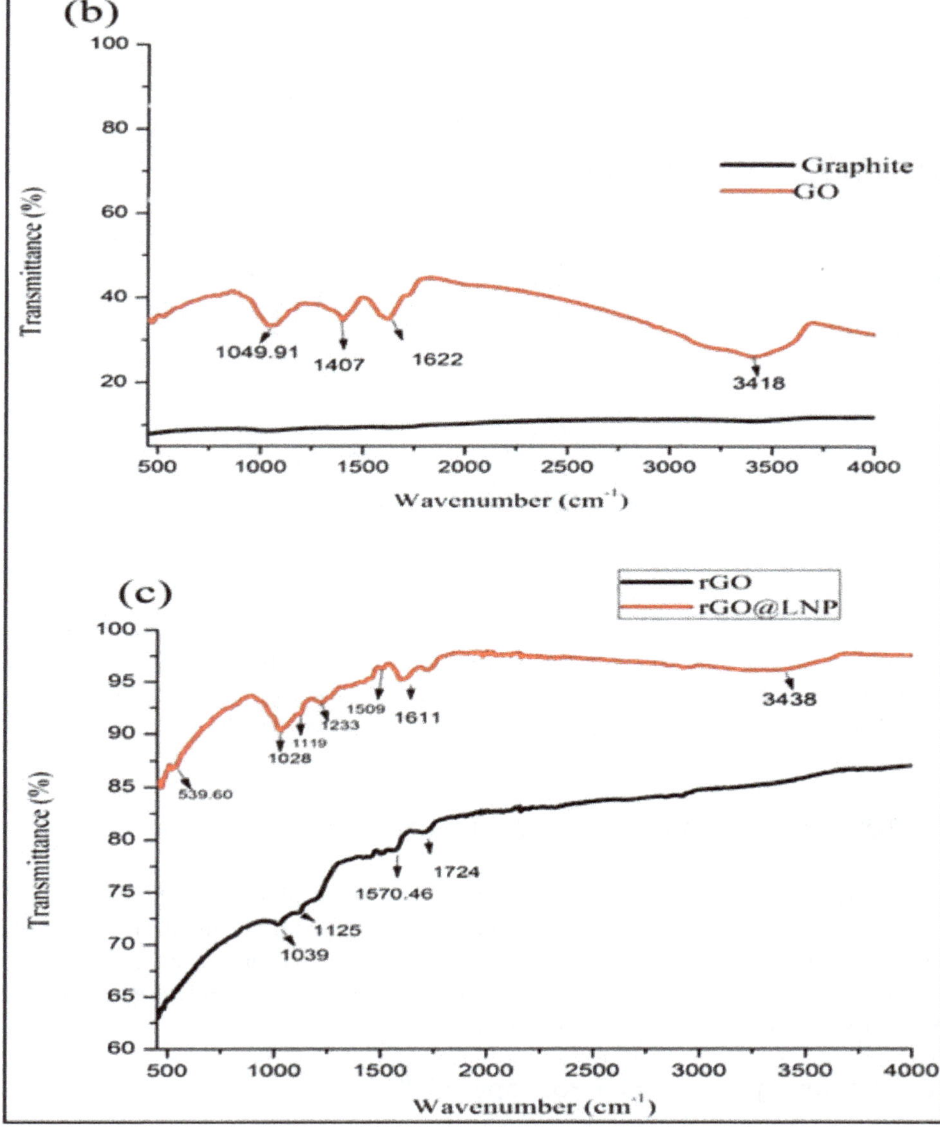

Another study developed 3D rGO-based hydrogel, in which FT-IR analysis revealed a partial reduction of GO within the hydrogel structure (Fig. 4.10(a)) (Tiwari et al., 2013). The band that is located between 3200 cm^{-1} and 3400 cm^{-1} may be ascribed to the stretching vibrations of oxygen and hydrogen that originate from the hydroxyl groups present in GO and the water that is adsorbed on the GO sheets. The absorbance band represents carbonyl and carboxyl groups at 1720 cm^{-1}. Bands at 1627 cm^{-1} may reflect unoxidized graphitic domain skeletal vibrations (C=C) or intercalated water stretching deformation vibrations. The reduction of GO is shown by the absence of the C=O, C-OH, and C-O vibration bands at 1720, 1220, and 1045 cm^{-1}, respectively, in the 3D rGO hydrogels (Figure 4.10(b)).

Figure 10. FT-IR spectra of (a) GO and (b) hydrogels based on 3D rGO. Reprinted with permission from Tiwari et al., Reduced graphene oxide-based hydrogels for the efficient capture of dye pollutants from aqueous solutions, Elsevier ©2013.

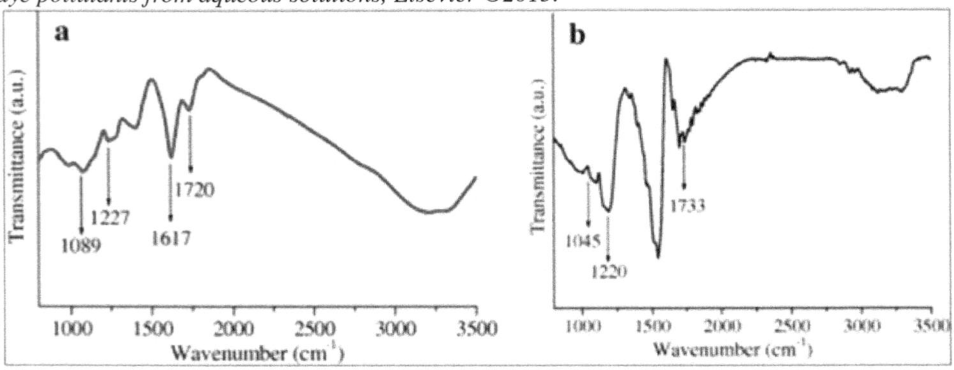

Thus, FT-IR is vital in confirming the hydrogen bonding, covalent bonding, and the incorporation of polymers within the hydrogel matrix. However, strong IR absorption by water can obscure the key peaks when hydrogel retains water. Unlike Raman spectroscopy and XRD, FT-IR has limitations in providing information on layer stacking and crystallinity.

4.3. SURFACE AND MORPHOLOGICAL CHARACTERIZATION

To comprehend the chemical functions, porosity and structural integrity of GBHs, it is essential to characterize their surfaces and morphology. Techniques including SEM, TEM, AFM provide insights into the nanoscale surface roughness, porosity networks, and the organization of graphene sheets. These characterizations enhance the applicability of GBHs in energy storage devices, biological systems and environmental remediation.

4.3.1. Scanning Electron Microscopy (SEM)

SEM plays a key role in the characterization of GBHs due to its ability to provide high-resolution surface imaging. It reveals the surface structure of GBHs and identifies the pore size, shape and connectivity, which are essential for biosensing, catalysis, and supercapacitor performance. In the study, the authors prepared fibrated graphene hydrogel specimens at concentrations of 1.0, 0.5, and 0.1 mg/mL, designated as HG-1, HG-0.5, and HG-0.1, respectively (Lim et al., 2011). The SEM analysis of the samples revealed a porous network composed of interconnected 3D graphene sheets, resembling a loose sponge-like structure as shown in Figure 4.11. These sheets combine to create the graphene hydrogel structure by overlapping, converging, and cross-linking under supercritical circumstances. Since huge GO sheets have limited mobility in solution, they position themselves randomly inside the hydrogel (Bai et al., 2011). The sheets are sturdy and rigid due to the huge conjugated basal planes. Graphene hydrogels have pore sizes that are not very sensitive to the concentration of graphene oxide.

Figure 11. FE-SEM pictures of graphene (a) HG-2 and (b) HGS-2. Reprinted with permission from Lim et al., Fabrication and characterization of graphene hydrogel via hydrothermal approach as a scaffold for preliminary study of cell growth, Elsevier ©2011.

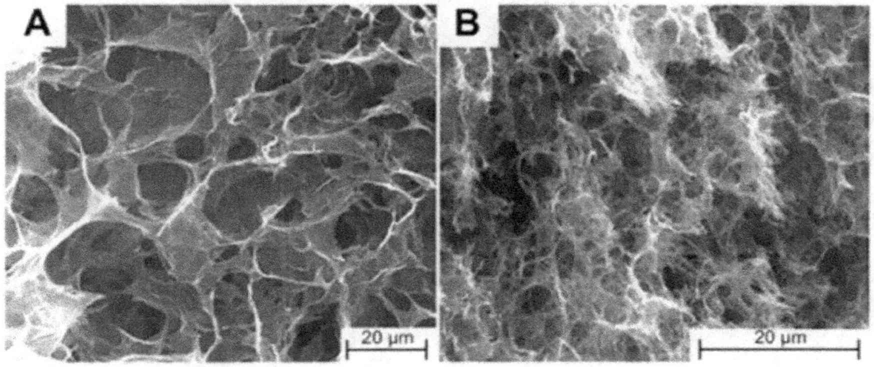

In Figure 4.12, larger SEM images reveal hydrogels surface morphology, while smaller images mounted on them illustrate each sample's cross-section (Lopes et al., 2023). The white dots, which are made up of solid gelatin particles (Adilah et al., 2021) stood out more in the hydrogels that had GO than in the control sample (M). In samples II and III of the hydrogels, the gelatin particles did not aggregate, but in samples IV and VI, they did, suggesting that the GO was evenly distributed throughout the matrix. The smooth, non-crackly surface of the hydrogels showed that GO was well embedded in the polymer matrix. This was possible because GO is hydrophilic, forming strong bonds with whey and gelatin proteins through oxygen-rich functional groups. Because of this, sample IV became much stiffer and a cross-linking effect was produced.

Figure 12. Surface and cross-sectional SEM of whey and gelatin hydrogel samples without GO (M) and with varying concentration of GO (I, II, III and IV). Reprinted with permission from Purcea Lopes et al., Characterization of a Graphene Oxide-Reinforced Whey Hydrogel as an Eco-Friendly Absorbent for Food Packaging, MDPI ©2023.

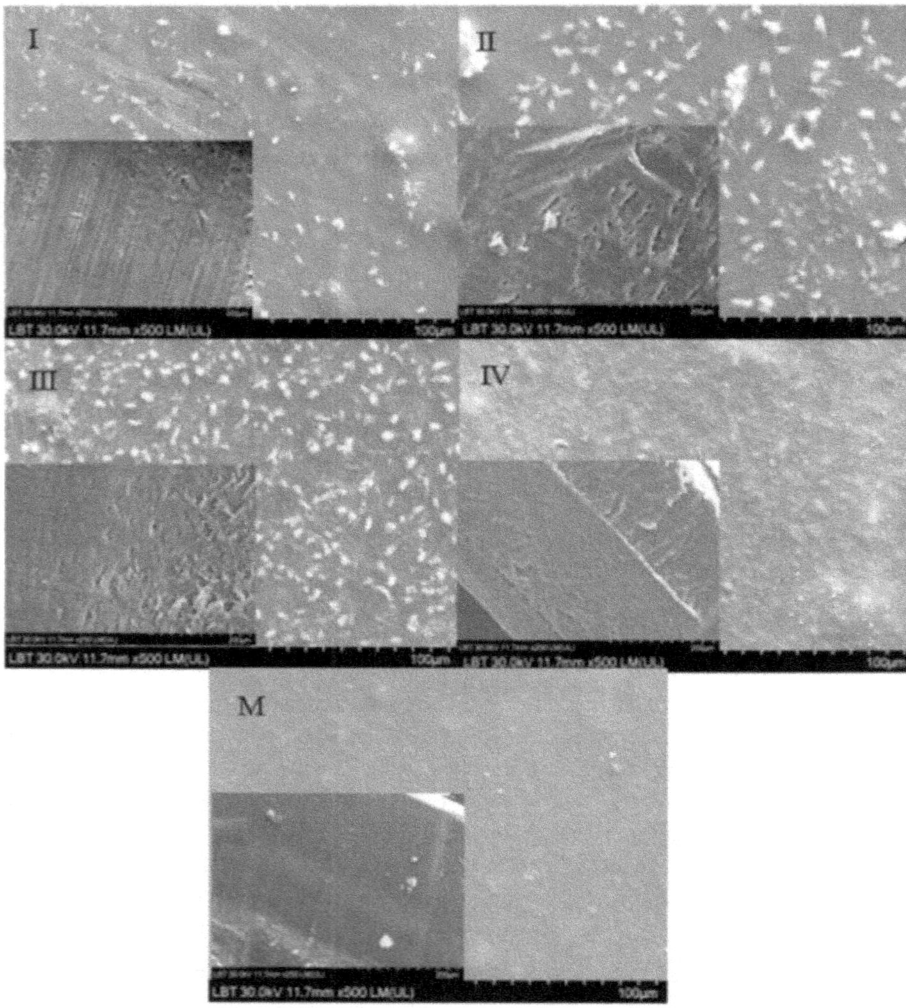

On other hand, Wang et al. synthesized two GO-based hydrogels namely BKB/GOG and BKB/PDA/rGOG using a simple one-pot approach. The SEM analysis of the lyophilized BKB/GOG hydrogel (Figure 4.13(a) and 4.13(b)), revealed that the randomly cross-linked GO nanosheets formed a porous structure with pore diameter ranging from 1 to 20 μm (Wang et al., 2015). The formation of 3D structures through cross-linking of GO nanosheets with BKB may be attributed to partial overlapping of the sheets, facilitated by non-covalent interactions including electrostatic interactions, hydrogen bonding, and π-π stacking. Because of its porous nature, hydrogel allows solutes to diffuse around it more easily. So, these hydrogels may be employed for adsorption processes. In BKB/PDA/rGO hydrogel, the graphene oxide nanosheets stick together better than in BKB/GO hydrogel (Figure 4. 13(c)). This means that the pores of the GO nanosheets in BKB/PDA/rGO hydrogel are smaller than those in BKB/GO hydrogel. More wrinkles and lines could be seen in the hydrogel as presented in Figure 4.13(d). Thus, PDA appears to act as an effective adhesive and promoted stronger connectivity between GO and BKB.

Figure 13. SEM photos of (a), (b) BKB/GOG and (c), (d) BKB/PDA/GOG gels at various magnifications. Reprinted with permission from Wang et al., A facile one-pot method to two kinds of graphene oxide-based hydrogels with broad-spectrum antimicrobial properties, Elsevier ©2015.

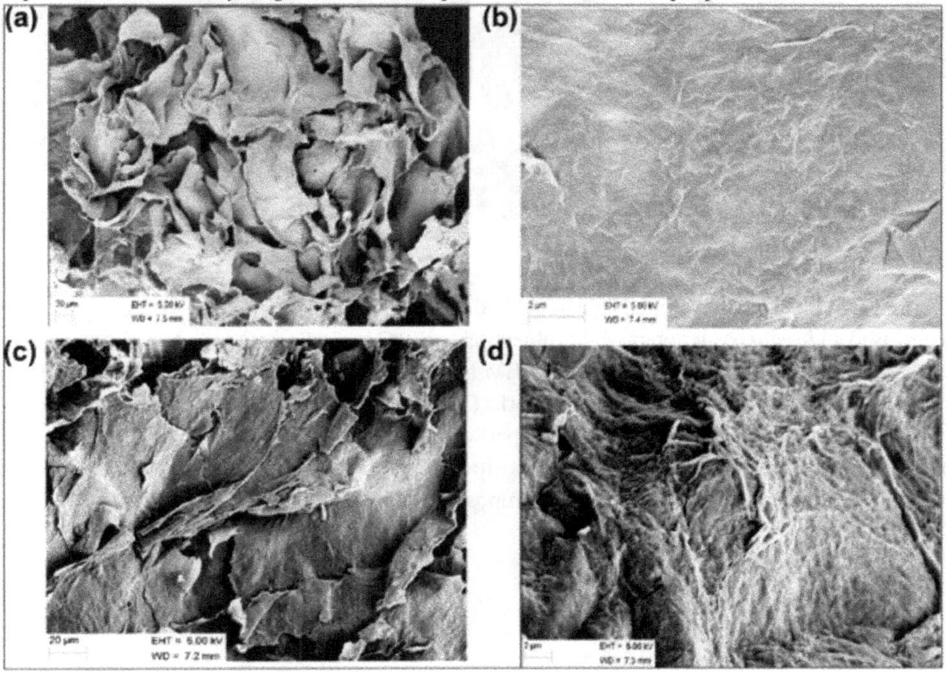

4.3.2. Transmission Electron Microscopy (TEM)

TEM is a powerful tool for the characterization of GBHs considering its high spatial resolution. It provides direct imaging of individual graphene layers and their morphology, along with the determination of layer number, size, and thickness of graphene sheets dispersed within the hydrogel. In a study, Tiwari et al. synthesized hydrogels by reducing GO with sodium ascorbate. The graphene nanosheets with one, two and three layers self-assembled into a defined and linked 3D porous network via strong π-π and van der Waals interactions as the gelation process progresses. Homogeneous pore sizes were observed in the 3D rGO hydrogels as shown in Figures 4.14(a) and 4.14(b). The exfoliation of rGO nanosheet is shown by the presence of many thin films with numerous wrinkles. The creation of single-, double-, and triple-layer graphene was confirmed by high resolution TEM images (Figure 4.14(b)) (Tiwari et al., 2013).

Figure 14. Reduced-GO based 3D hydrogel TEM micrographs: low magnification (a) and high magnification (b). Reprinted with permission from Tiwari et al., Reduced graphene oxide-based hydrogels for the efficient capture of dye pollutants from aqueous solutions, Elsevier ©2013.

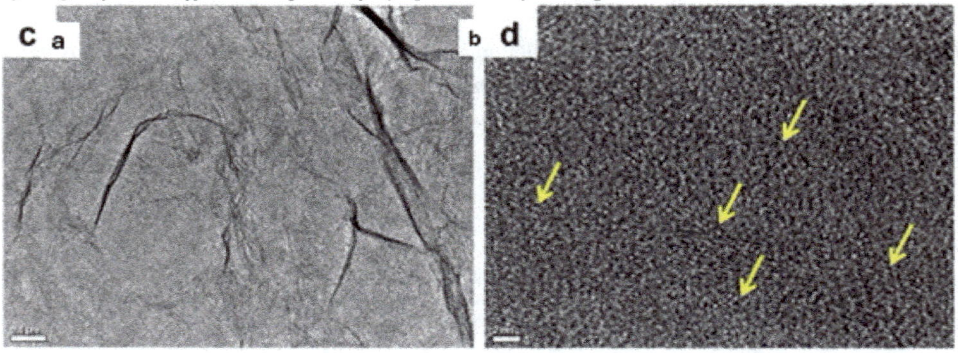

In 2015, Jiao et al. developed GO-based hydrogels with the help of polyethyleneimine (PEI). They used an *in-situ* reduction method subjecting to 90 °C to create a composite hydrogel including rGO-based silver nanoparticles. The SEM and TEM images of rGO hydrogel revealed the cross-linked GO sheets in porous PEI networks (Figure 4.15(a) and 4.15(d)). The micrographs of rGO-based composite gels with silver nanoparticles captured by SEM and TEM (Figure 4.15(b) and 4.15(e)) demonstrated the uniform deposition of nanoparticles onto the rGO nanosheets, a critical step in creating a gel-based nanocomposite system (Jiao et al., 2015).

Figure 15. Comparison of SEM and TEM images of lyophilized rGO/PEI hydrogel (A, D) with rGO/PEI/Ag hydrogel (B, E). Reprinted with permission from Jiao et al., Reduced graphene oxide-based silver nanoparticle-containing composite hydrogel as highly efficient dye catalysts for wastewater treatment, Springer Nature ©2015.

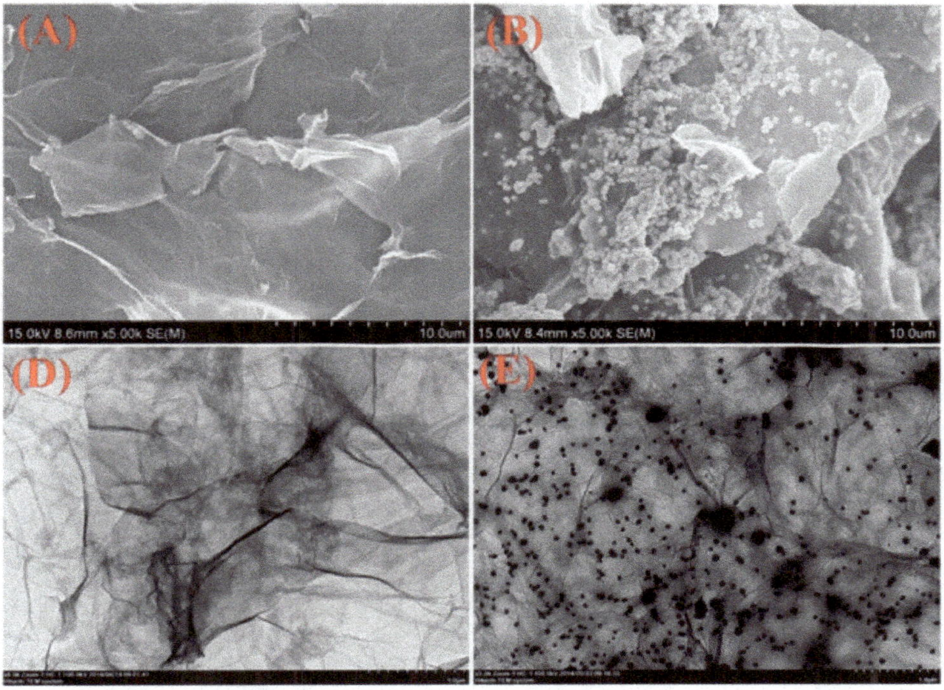

4.3.3. Atomic Force Microscopy (AFM)

AFM is another critical technique for characterizing GBHs due to its ability to provide high-resolution 3D surface information at the nanoscale. In 2021, Khan et al. synthesized GO hybrid hydrogels, and the AFM images shown in Figure 4.16 reveal their intriguing surface roughness behaviour (Khan et al., 2021). The controlled film devoid of cross-linker (CP) exhibits the most severe surface roughness compared to the other samples. Due to its comparatively open and porous nature, the non-crosslinked sample is well understood. When GO is added to cross-linked hydrogel (CP-1) at a concentration of 0.1 to 0.5 µg/mL, the samples turned less rough over time (Table 4.1). This lower level of roughness is because GO has an enormous surface area that covers the holes, and the hydrogel has a more packed structure that makes the surface smooth (Zhu et al., 2010).

Table 1. The average roughness of several hydrogel samples

Sample	Roughness (Center line at Y-Axis) (nm)	Average Roughness (nm)
CP1 (a)	2.245 ± 0.19	2.401 ± 0.17
CP2 (c)	3.186 ± 0.27	3.861 ± 0.23
CP7 (e)	3.376 ± 0.43	4.644 ± 0.36
CP8 (g)	7.161 ± 0.45	7.532 ± 0.49

Figure 16. AFM results of CP-1 (a,b), CP-2 (c,d), CP-7 (e,f), and CP-8 (g,h). Reprinted with permission from Khan et al., Chitosan/poly vinyl alcohol/graphene oxide based pH-responsive composite hydrogel films: Drug release, anti-microbial and cell viability studies, MDPI ©2021.

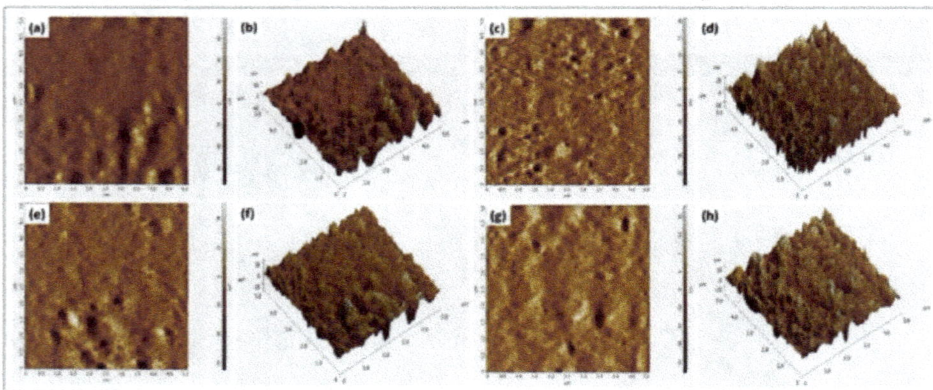

According to the results of the AFM analysis, 3D structured sheets were seen in the rGO/PEI and rGO/PEI/Ag hydrogels (Figure 4.17). Due to the presence of PEI, these sheets are cross-linked, resulting in a massive three-dimensional network-like structure. Low-energy interactions, like hydrogen bonds and π-π stacking, are what make the prominent 3D network-like structure possible. In Figure 4.17(b), an AFM picture of the rGO/PEI/Ag hydrogel revealed that the surface of the rGO nanosheets is evenly and homogeneously changed with the as-formed silver nanoparticles, creating a gel-based nanocomposite system (Jiao et al., 2015).

Figure 17. AFM photos of lyophilized (a) rGO/PEI and B) rGO/PEI/Ag gels with section analysis. Reprinted with permission from Jiao et al., Reduced graphene oxide-based silver nanoparticle-containing composite hydrogel as highly efficient dye catalysts for wastewater treatment, Springer Nature ©2015.

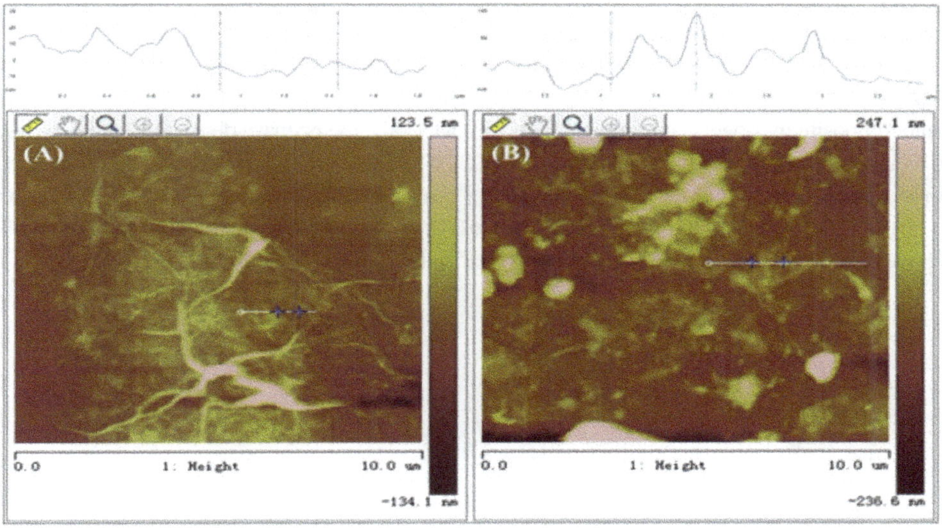

4.4. THERMAL STABILITY AND DEGRADATION ANALYSIS

The superior thermal stability of graphene-based hydrogels may be attributed to their robust π-π interactions and hydrogen bonding within their overall network. Thermal analysis methods like TGA and DSC are used to assess their stability and deterioration. It is imperative to understand these temperature qualities to make the best graphene-based hydrogels for uses like drug delivery, tissue engineering, and cleaning up the environment.

4.4.1. Thermogravimetric Analysis (TGA)

TGA is primarily used to evaluate the thermal stability and composition of GBHs by measuring the weight loss as a function of temperature. It can also estimate the graphene or GO loading within the hydrogel and provides the interaction between graphene and the polymer chains. In a study, authors prepared hybrid hydrogel from poly(acrylic acid) bonded onto amylose (PAA-g-amylose) that had different amounts of GO nanosheets added to it. The thermal performance of PAA-g-amylose and GO/PAA-g-amylose (with 1.0, 3.0, and 5.0 wt% of GO) are shown in Figure 4.18(a) and 4.18(b), respectively, and the results are summarized in Table 4.2. The degradation occurred in two stages. The first stage involves a gradual mass loss between 110 °C and 200 °C, attributed to the removal of water molecules trapped within the GO structure (Abdollahi et al.,

2017). At temperatures more than 200 °C, dehydrated acrylic acid may be formed (Krivorotova et al., 2014). Above 250 °C, decarboxylation of CO_2 is more important. Water and CO_2 are emitted up to 400 °C and chain scission occurs at 350 °C. Between 350 to 516 °C, a second process of breakdown turns functional groups that contain labile oxygen into CO, CO_2 and steam (Abdellaoui-Arous et al., 2012). In contrast to pure PAA-g-amylose hydrogel, the thermal stability of the nanocomposite hydrogel is much higher. This is due to the synergistic impact of the GO and PAA-g-amylose hydrogel matrix. The GO/PAA-g-amylose hybrid hydrogel became thermally stable as the amount of GO added increases (Amirsardari et al., 2015). In the case of nanocomposite hydrogels containing 1.0 and 5.0 wt% GO, the primary degrading step is moved to a higher temperature by about 10 °C and 34 °C. When a polymer breaks down, the first step is for free radicals to develop at weak bonds and/or chain ends. From there, the radicals may move to nearby chains via an interchain reaction. The GO/PAA-g-amylose nanocomposite hydrogel's decreased polymer chain mobility improves heat stability (Wu et al., 2015).

Figure 18. TGA (a) and DTG (b) graphs for GO/PAA-g-amylose hybrid hydrogels and PAA-g-amylose hydrogels. Reprinted with permission from Abdollahi et al., Thermal and mechanical properties of graphene oxide nanocomposite hydrogel based on poly (acrylic acid) grafted onto amylose, Elsevier ©2018.

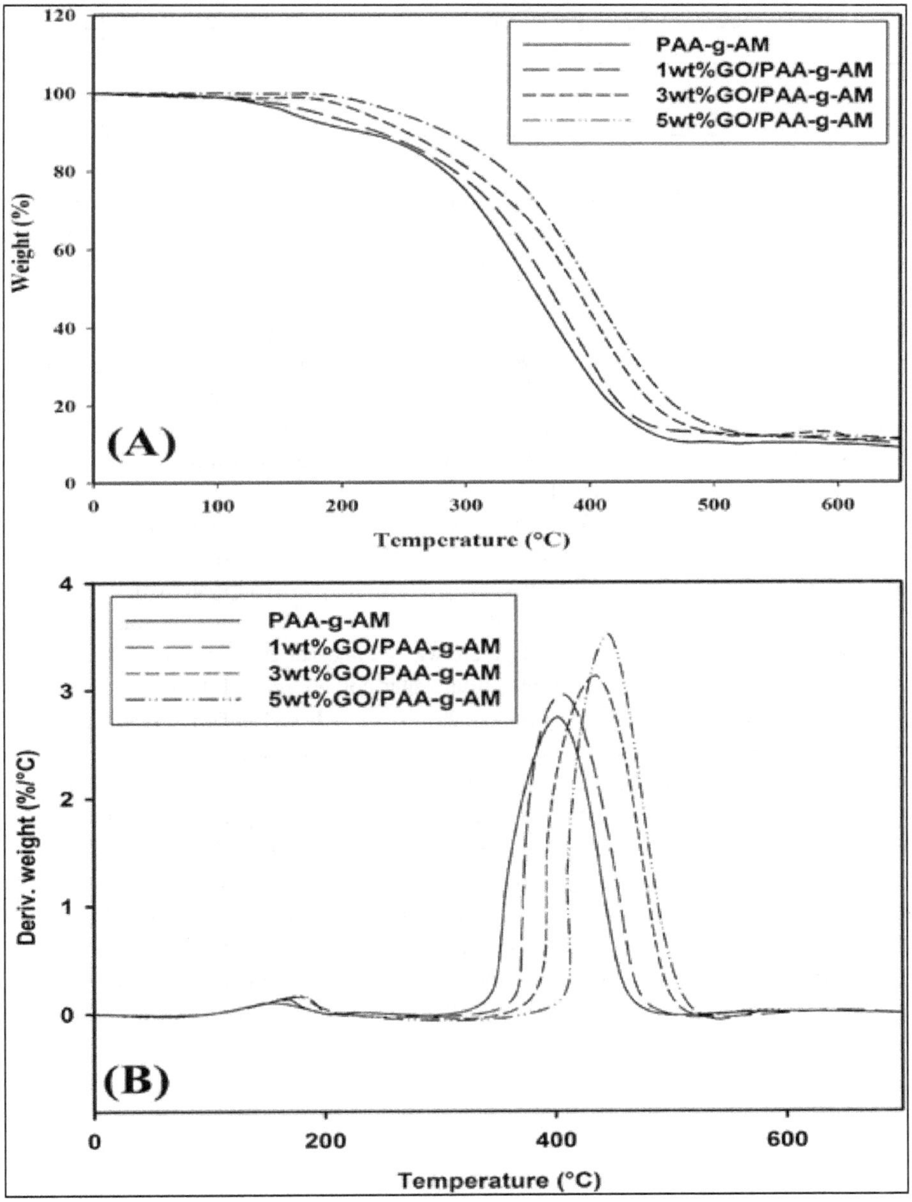

Table 2. TGA measurements of pure PAA-g-amylose with varied GO nanosheet loadings (Abdollahi et al., 2017)

GO content (wt.%)	T_d (°C)	T_{onset}	Weight loss (%) at various temperature (°C)			
			200	300	400	500
0	311	256	9	25	73	89
1	332	281	7	23	68	88
3	359	309	4	20	55	87
5	385	323	2	13	48	85

Reduced chain mobility suppressed chain transfer reaction, lowering the degradation and raising the decomposition temperature (Croitoru et al., 2024). It is probably caused by the fact that the GO nanosheets are evenly distributed and have strong contact with the hydrogel matrix at the interface. Nanocomposite hydrogels show higher thermal stability, and incorporating GO into the polymer matrix reduces volatile byproducts during degradation (Telin et al., 2024). Compared to the pure hydrogel network, the GO-loaded PAA-g-amylose nanocomposite hydrogels exhibited greater T_{onset} and T_{dmax}. Table 4.2 shows that when the GO concentration in the PAA-g-hydrogel matrix increased, the T_{dmax} was moved to higher temperatures. A similar rise to 486 °C in T_{dmax} was seen when 5.0 wt% GO was loaded. This indicates that the nanocomposite hydrogels have enhanced stability, as reference stated (Han et al., 2021).

Sohni et al. created a 3D rGO hydrogel (rGO@LNP) that was manufactured using lignin nanoparticles (LNP). The characterisation investigations compared the rGO and rGO@LNP thermal degradation patterns and stability using TGA analysis. The systematic weight loss in GO is shown by curve in Figure 4.19(a). At first, a 19% reduction in weight was seen between 18 °C and 133 °C. From 116 °C to 251 °C, there was yet another 20% decrease. The weight loss was caused by the evaporation of the interstitial water content up to 100 °C. The further mass loss up to 250 °C was mostly caused by the breakdown of hydroxyl and other reactive functional groups. The decomposition of stable oxygen functions caused the highest weight loss at 49% between 250 °C and 600 °C. This is more explicitly illustrated in the corresponding DTG curve shown in Figure 4.19(b). The constant loss of additional carbon atoms in GO is likely due to its combustion route (Tene et al., 2020). In contrast, the degradation of rGO@LNP is shown by the TGA curve (Figure 4.19(c)), but the weight loss is not visible since the oxygen functions were removed earlier. With a higher thermolytic process initiation temperature in rGO@LNP than in rGO, the composite is more thermo-stable. Lignin, a composite material component, releases volatile phenolic chemicals, which cause the loss to occur between 180 and 300 °C (Follmer et al., 2019). LNP-based hydrogel generated 27% more residue than rGO (12%), indicating a synergistic interaction between the precursors.

Figure 19. TGA and DTG curves for GO (a)&(b), TGA curve for rGO@L_{NP} (c). Reprinted with permission from Sohni et al., Lignin nanoparticles-reduced graphene oxide based hydrogel: A novel strategy for environmental applications, Elsevier ©2023.

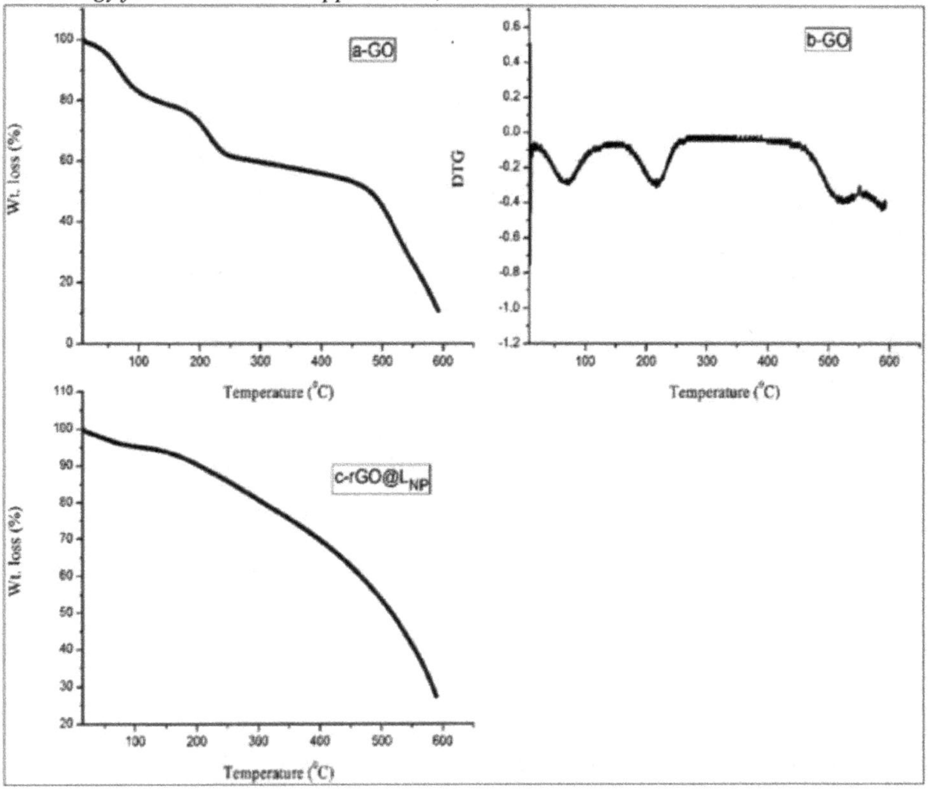

4.4.2. Differential Scanning Calorimetry (DSC)

DSC is a key technique for characterizing GBHs, providing insights into thermal transitions such as Tg, Tm, and Tc. In a study, authors examined the thermo-mechanical stability of the synthesised polyacrylamide (PAM) hydrogels and the synergistic impact of graphene oxide and zirconia as fillers. Graphene percentage-dependent DSC data for polymer gels are shown in Figure 4.20. After graphene was added to PAM hydrogel, the amount of bound water rose, and the free water decreased (Follmer et al., 2019).

Figure 20. DSC analysis of polymer gels containing varying amounts of graphene. Reprinted with permission from Michael et al., Zirconia/graphene nanocomposites effect on the enhancement of thermo-mechanical stability of polymer hydrogels, Elsevier ©2019.

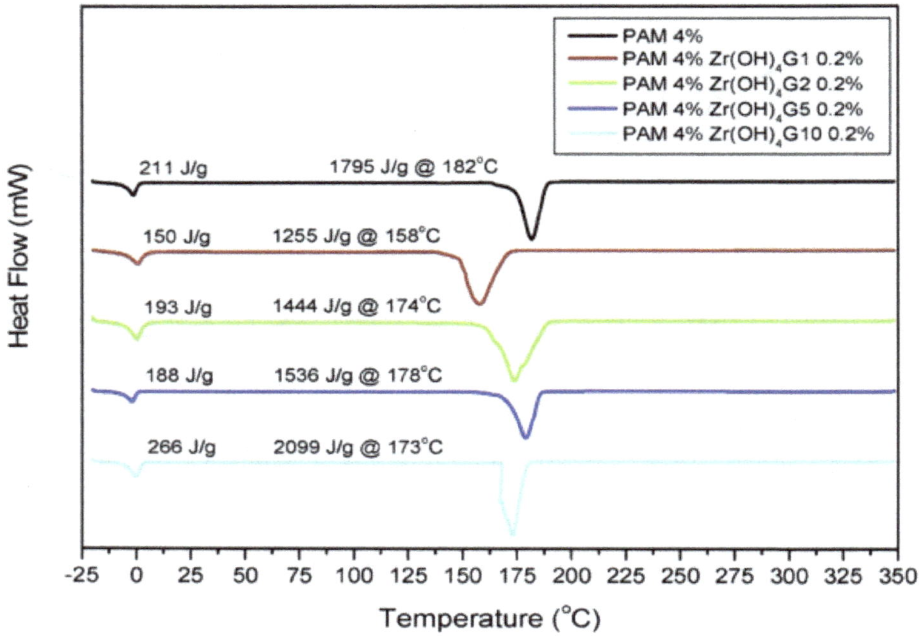

To study possible materials for osteochondral tissue regrowth, Lach et al., made sodium alginate (SA) hydrogels with GO and hydroxyapatite (HAp). These gels were then crosslinked by a calcium chloride solution. DSC was used to assess thermal characteristics and hydrogel water types (Lach et al., 2024). The produced hydrogels showed two primary endothermic peaks in their DSC curves (Figure 4.21). The first one happens around 0 °C and is caused by water that isn't strongly bonded or is only loosely bonded melting. This water then freezes when it cools down at the start of the test. At its highest point, you can see two lesser peaks. The pointed one represents unrestricted water, whereas the broader one represents frozen water.

Figure 21. The detailed DSC curves for produced samples and melting non-structured water: SA/ GO hydrogels. Reprinted with permission from Lach et al., On the influence of graphene oxide and hydroxyapatite modification on alginate-based hydrogel matrix: thermal, physicochemical, and biological considerations, Springer ©2024.

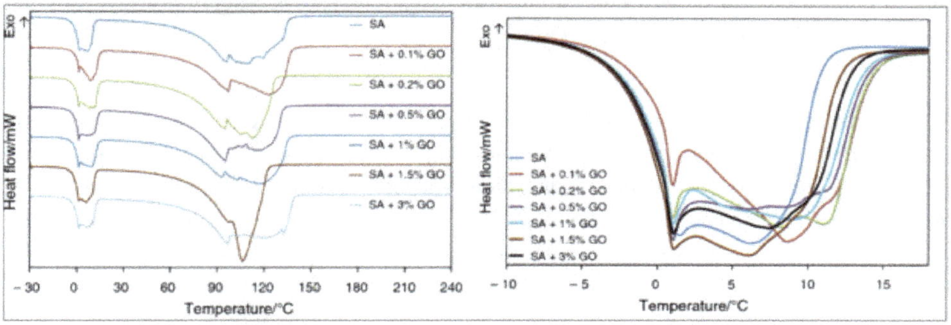

There are typically three kinds of water found in hydrogels i.e., free water, frozen water, and not frozen water. Since free water does not form hydrogen bonds with macromolecules, it behaves like pure water during freezing and melting, while bound water exhibits shifted transition temperatures owing to its limited interaction with macromolecular chains. Non-freezing bound water links to macro chains exclusively via hydrogen bonds and does not experience phase transitions at normal temperatures (Teodorescu et al., 2010). A thorough evaluation of the biocompatibility of hydrogels and their adsorption/desorption processes requires data on many water types present in polymer hydrogels (Gun'ko et al., 2017).

4.5. POROSITY AND SURFACE AREA ANALYSIS

Hydrogels made of graphene have a very porous structure, which increases their surface area and makes mass transfer easier. In hydrogels, variables affect porosity, including graphene concentration, cross-linking density, and hydrogel manufacturing process. The surface area and pore size distribution of these materials are often evaluated using techniques such as nitrogen adsorption-desorption (BET analysis). These hydrogels show great promise for use in biological, adsorption, and catalytic applications due to their controlled porosity and large surface area.

4.5.1. Brunauer-Emmett-Teller (BET) Analysis

A hydrogel adsorbent based on natural compounds was created by Mahmoodi et al. to extract diclofenac from water, which is used as a model pharmaceutical. As a first step, hydrogel adsorbents made of graphene oxide-chitosan (GO-CTS) and amine graphene oxide-chitosan (AGO-CTS) were created using an easy mechanical mixing technique. The BET method was used to examine synthetic materials (Mahmoodi et al., 2021). The pore distribution and isotherms of the N_2 adsorption-desorption AGO2-CTS5 hydrogel are shown in Figure 4.22. A hysteresis loop spanning relative pressures (P/P_0) between 0.4 and 1.0 was seen in a typical type-IV curve, indicating that the AGO2-CTS5 hydrogel is mesoporous. The BET method found that AGO2-CTS5, which is not swollen, has a specific surface area of 7.07 m^2 g^{-1}. In addition, the BJH pore size distribution study confirmed the presence of mesoporous in AGO2-CTS5, with a concentrated pore size of about 1.21 nm, in line with the findings of the N_2 adsorption-desorption investigation. This investigation showed that the produced AGO2–CTS5 hydrogel's porosity structure significantly affects its adsorption capability.

Figure 22. Adsorption-desorption isotherms and AGO_2–CTS5 hydrogel pore distribution. Reprinted with permission from Mahmoodi et al., Graphene oxide-chitosan hydrogel for adsorptive removal of diclofenac from aqueous solution: preparation, characterization, kinetic and thermodynamic modelling, Royal Society of Chemistry ©2021.

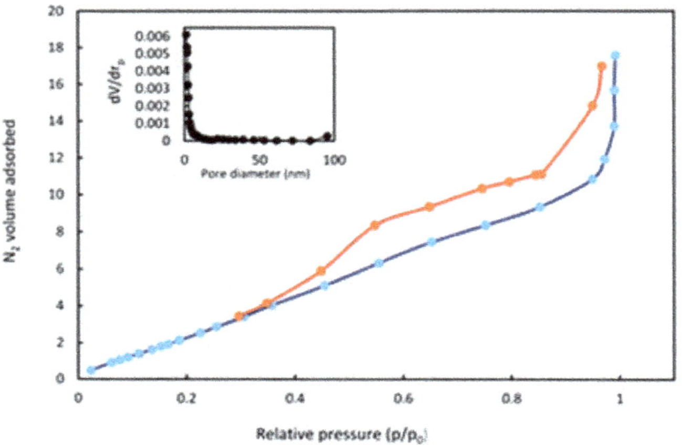

According to Zheng et al. reduced graphene hydrogel (rGH) functionalized with alkylamines were created using a simple two-step process. First, GO was hydrothermally treated with rGH. Then, the rGH was immersed in octylamine (OA), decylamine (DA) and dodecylamine (DDA) (Zheng et al., 2021). To further comprehend the materials' porous properties, N_2 adsorption-desorption was used to analyse their BET surface area and BJH pore size. The isotherm in Figure 4.23(d) has a type IV form and an H3 hysteresis loop, which means that the hydrogels are mesoporous and have a slit-or panel-shaped pore structure with non-rigid clumps. Table 4.3 summarizes the hydrogels' BET surface area, including their pore sizes, volumes, and BJH permeabilities. The specific surface area that was obtained for rGH, rGH-OA, rGH-DA, and rGH-DDA is 27.288, 47.374, 47.791, and 41.996 m^2/g, respectively (Zheng et al., 2021).

Figure 23. The isotherm of N_2 adsorption and desorption on the hydrogel functionalized with alkylamines. Reprinted with permission from Zheng et al., Design of reduced graphene hydrogel with alkylamine surface functionalization through immersion/agitation method and its adsorption mechanism, Elsevier ©2021.

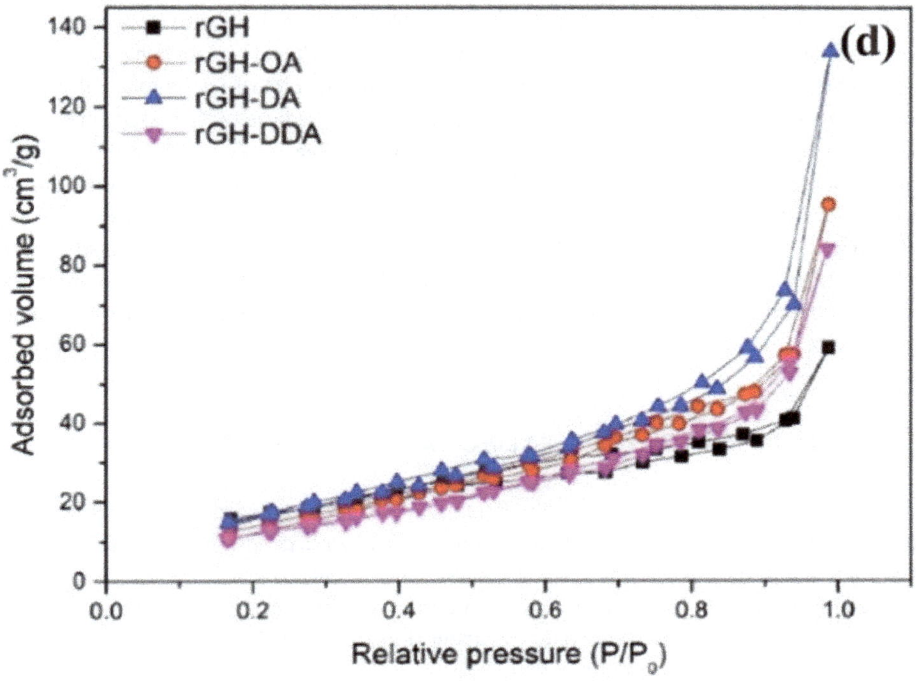

Table 3. BET and BJH studies indicated decreased graphene sample surface area and texture (Zheng et al., 2021).

Sample	Specific surface area (m^2/g)	Pore radius (nm)	Pore volume (cc/g)
rGH	12.125	1.981	0.069
rGH-OA	17.693	1.971	0.140
rGH-DA	20.220	1.717	0.193
rGH-DDA	16.688	1.968	0.125

4.5.2. BET Data Interpretation on GBHs Adsorption

BET surface analysis provides both quantitative and qualitative information about the porous structure of GBHs. By interpreting the BET data, researchers can tailor the hydrogel architecture to meet specific functional applications, ensuring efficient material utilization and performance. Higher BET surface area generally enhances the adsorption capacity of GBHs as more active sites are available for interaction with adsorbate molecules and is beneficial for enhanced electrochemical performance in batteries and supercapacitors, controlled release in biomedical systems and pollutants adsorption in environmental cleanup. If the data indicate a decreased surface are, it may imply aggregation of graphene sheets or partial blockage of pores within the hydrogel matrix, both of which can limit the accessibility of adsorption sites. The shape of the nitrogen adsorption-desorption isotherm (Type I-VI) provides insights into the porosity (Luan et al., 2012). For instance, a Type IV isotherm with a hysteresis loop typically indicates mesoporous structures, which is favourable for adsorption. GBHs generally exhibit surface areas in the range of 50-300 m^2/g, based on synthesis method and graphene loading. Increasing graphene or GO content generally increases surface area, but beyond a threshold, restacking of sheets can reduce porosity.

4.6. SWELLING, STABILITY AND DEGRADATION TESTS

For GBHs, swelling behaviour dictates water retention and responsiveness to environmental changes. Stability tests under varying temperature, pH, and ionic conditions evaluate resistance to degradation and mechanical integrity, while biodegradability and lifetime are assessed via spectroscopic and weight-loss studies. These properties are vital for applications like drug delivery, tissue engineering, and wastewater treatment (Leganes et al., 2020).

4.6.1. Swelling Ratio Measurement

According to a study graphene-based hydrogels perform better than regular ones (Sharma et al., 2022). The addition of high-density hygroscopic salts to the hydrogel matrix solves the problems that come up during the freezing and drying processes (Liu et al., 2018). In addition, a new method for producing hydrogels that can withstand drying and swelling conditions was recently found by Mredha et al. in which they used a quenching procedure to create a double hydrophobic covering comprising a hydrophobic polymer and a viscous oil (Mredha et al., 2020). Due to functional groups on the basal and edge planes, graphite oxide sheets are hydrophilic and may be distributed in a solution to generate stable aqueous colloids for 3D structures (Nassar et al., 2021). Martin et al. chemically linked PAM gels that did not contain graphene and made several mixed graphene-based PAM gels to show that graphene can extend the stability of hydrogels (Martín et al., 2017). Figure 4.24 shows the final features of the materials in their inflated condition, which was found to be entirely homogeneous, and the SEM of the hydrogels, which are representations of their microscopic morphology (Martín et al., 2017). Both in their original inflated form and at greater swelling degrees, the data showed that the hydrogels had excellent dimensional stability. Thus, higher graphene concentration can be expected to enhance form stability with increasing swelling.

Figure 24. Schematic depiction showing completely inflated material homogeneity and SEM illustrations with graphene concentrations (0–2 mg m L^{-1}). Reprinted with permission from Martín et al., Graphene improves the biocompatibility of polyacrylamide hydrogels: 3D polymeric scaffolds for neuronal growth, Springer Nature ©2017.

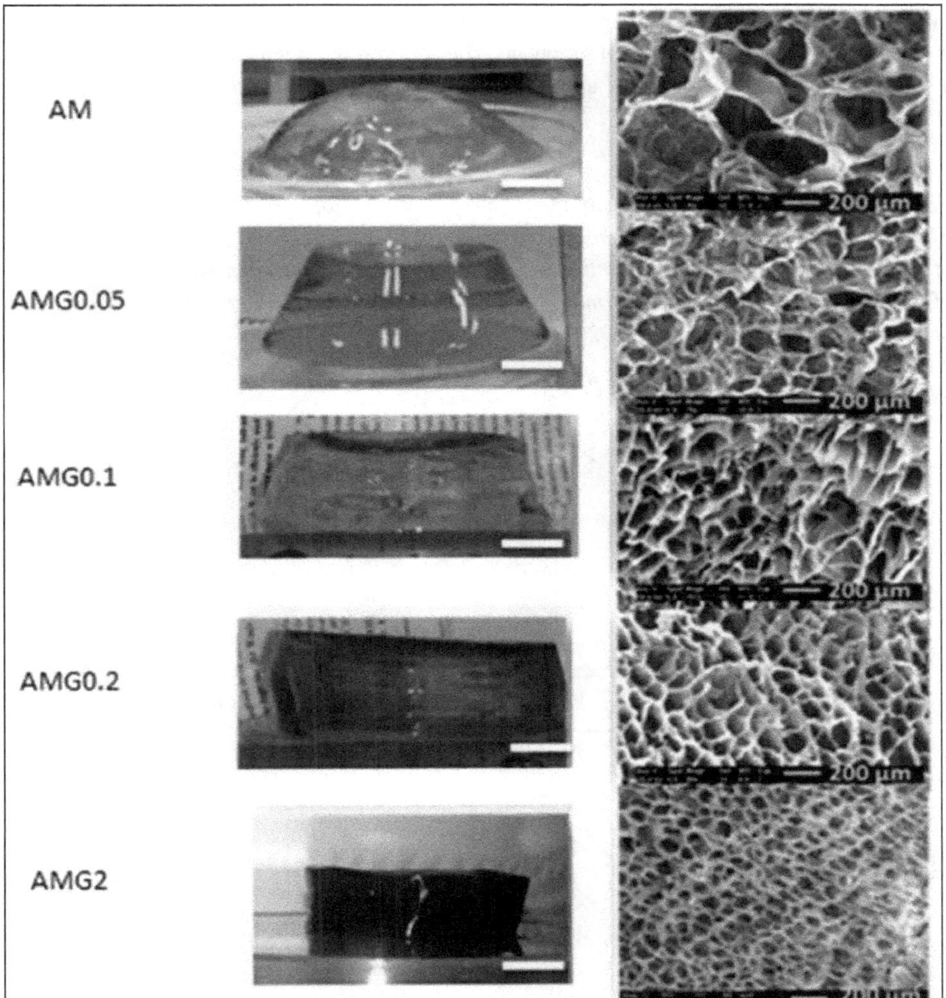

The growth rate and gel fraction number can show how well the hydrogel network is cross-linked. Figures 4.25 demonstrates how graphene dose affects hydrogel swelling and gel fraction. The swelling rate decreases initially and rises with increasing dosages of graphene, as shown in Figure 4.25(a). With the addition of 0.33% graphene, the swelling rate drops to 0.33%. Figure 4.25(b) shows the gel fraction inversely correlated with swelling. For a dose of 0.33% graphene, the gel fraction reaches a maximum of 68.25% and gradually decreases as the dosage rises (Ma et al., 2022). Figures 4.25(c) and 4.25(d) shows how the amount of high thermally conductive graphene in a graphene-based hydrogel (G/PVP-PVA) affects the rate of growth and the gel fraction of the hydrogel. Both the swelling rate and the gel fraction go through phases of reduction and rise, respectively, as the proportion of PVP increases. At a PVP concentration of 4.4%, the swelling rate drops to 219.0%, and the gel fraction rises to 77.1%. Based on the experimental findings, the gel network is regular, and the contact between the PVP and PVA chemical chains is highest at a 4.4% PVP dosage of graphene. The material has a tensile strength of 322.0 kPa, a self-recovery rate of 95.4%, and a maximum thermal conductivity of 1.486 W m^{-1} K^{-1} (Ma et al., 2022).

Figures 25. (a) The rate of swelling as a function of graphene dose. (b) The portion of gel affected by the dose of graphene. (c) The relationship between swelling rate and PVP %. (d) The amount of PVP has an effect on the gel fraction. Reprinted with permission from Ma et al., Preparation of Graphene-Based Hydrogel Thermal Interface Materials with Excellent Heat Dissipation and Mechanical Properties, Elsevier ©2023.

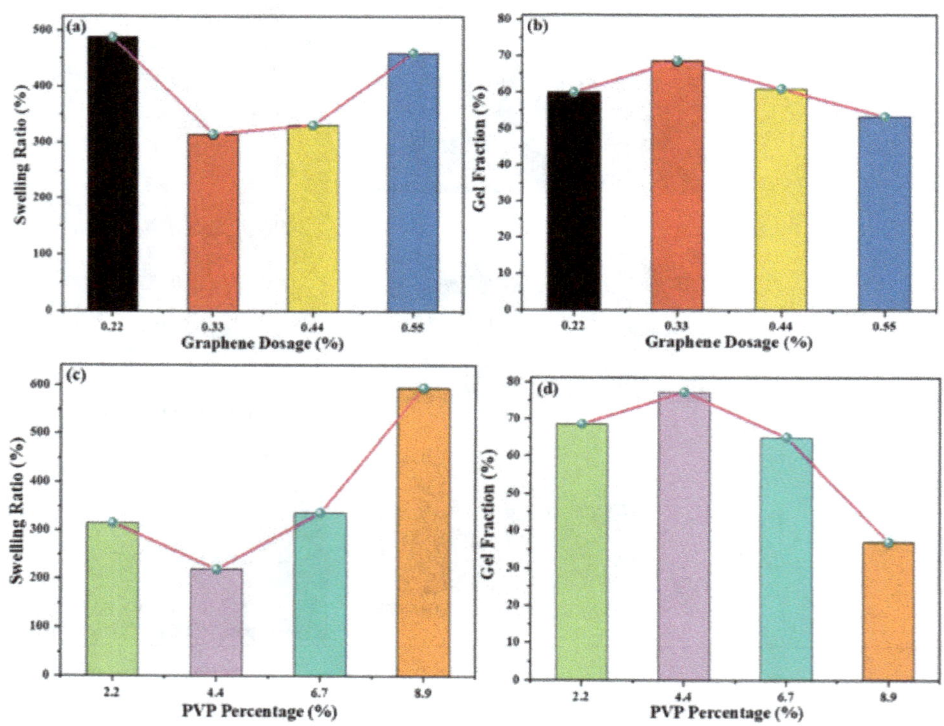

4.7. BIODEGRADABILITY AND STABILITY STUDIES

Recently, studies on GBHs have focused on their biodegradability, highlighting their potential for biomedical applications. In a study, GO was mixed with poly(ethylene glycol) methyl ether methacrylate to make a hydrogel support. The addition of GO to the hydrogel enhanced its tensile strength to 626 kPa. Biocompatibility tests showed that GO/MEO-MA hydrogels were non-toxic to human skin fibroblasts and provided over 95% wound healing within 15 days in rat skin models (Luong et al., 2024). Next study developed a GO-CMC-silk fibroin-Fe_3O_4 nanobiocomposite hydrogel, which proved non-toxic to healthy human cells while showing cytotoxic effects against cancer cells as seen in Figure 4.26. Because of their selectivity, hydrogels have great promise for controlled biodegradation in the medical field (Gorab et al., 2022). Additionally, the authors explored graphene and its derivatives for bone tissue regeneration, emphasizing how factors like oxidation state, functionalization, and environment affect their biocompatibility, biodegradability, and overall safety. To guarantee the safe and successful use of graphene-based hydrogels in biomedical engineering, it is crucial to understand their biodegradation processes (Daneshmandi et al., 2020). Overall, these studies indicate that graphene-based hydrogel biodegradability can be tailored for specific biomedical uses by optimizing component selection and functionalization.

Figure 26. HEK293T and BT549 cell viability percentages after (a) 48 and (b) 72 hours of treatment with the GO-CMC hydrogel/SF/Fe3O4 nanobiocomposite, as well as the HEK293T and BT549 cell survival rates after (c) 2 and (d) 3 days of treatment with cisplatin (as a positive control), along with 96-well plate images taken after 72 hours of treatment. Reprinted with permission from Gorab et al., Decoration of graphene oxide nanosheets with carboxymethylcellulose hydrogel, silk fibroin and magnetic nanoparticles for biomedical and hyperthermia applications, Royal Society of Chemistry ©2023.

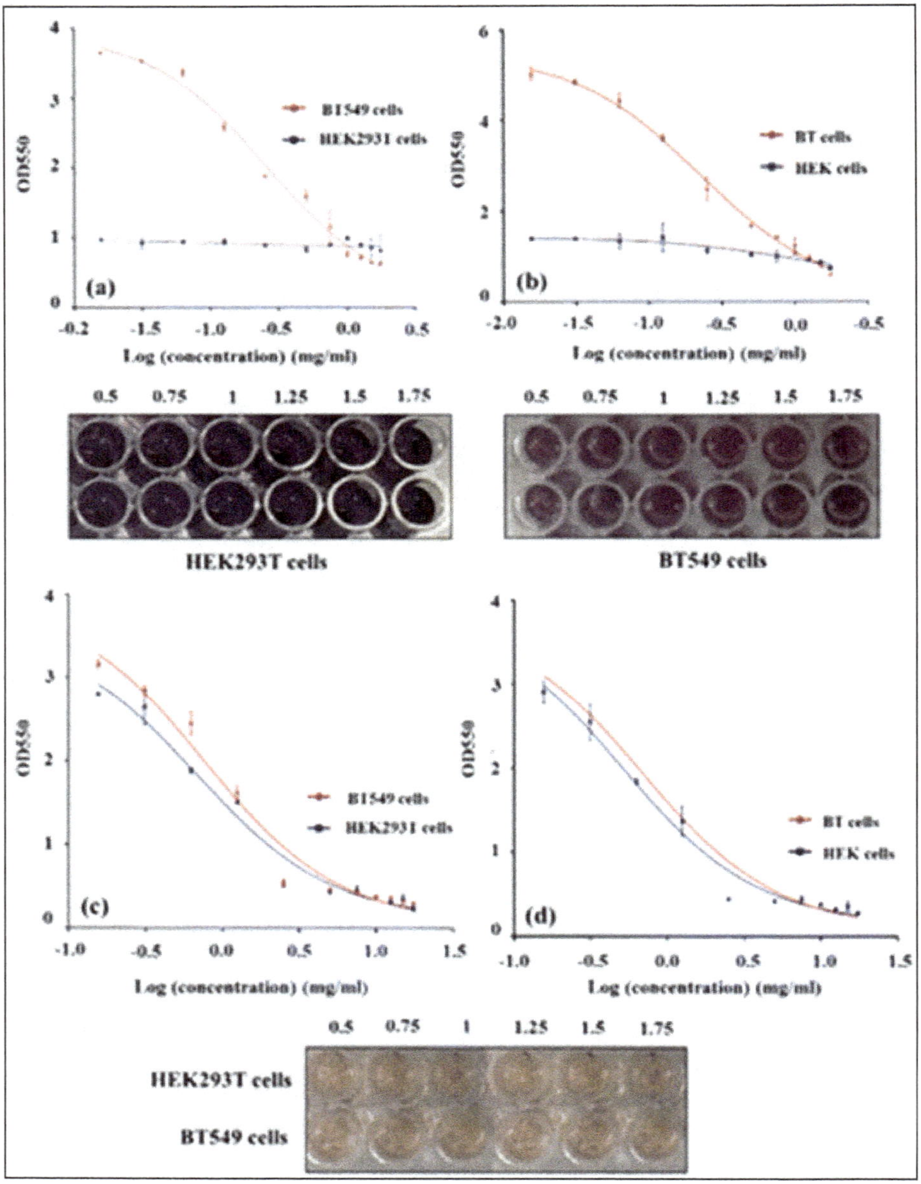

4.7.1. Scalability, Toxicity and Economic Sustainability

4.7.1.1. Scalability

Despite the exceptional multifunctionality of GBHs, their scalable production is being remained as a significant challenge. Additionally, maintaining uniform dispersion of graphene in the hydrogel matrix during bulk synthesis is difficult due to agglomeration and phase separation tendencies. Developing green, one-pot, or *in-situ* synthesis methods using biopolymers or bio-reducing agents could enhance scalability while reducing environmental impact. Furthermore, scalable 3D printing and extrusion techniques are being explored to fabricate customized GBH architectures for industrial and biomedical applications.

4.7.1.2. Toxicity

Although GBHs show excellent promise in biomedical and environmental domains, their long-term safety and toxicity profiles are still under investigation. Graphene derivatives may cause cytotoxic effects depending on their size, surface functionalization, and concentration. Oxidative stress, membrane damage, and inflammatory responses have been reported in some *in-vitro* and *in-vivo* studies. Moreover, residual chemical reagents from synthesis processes (e.g., hydrazine, strong acids) may pose additional biocompatibility concerns. Therefore, comprehensive toxicological evaluations and regulatory compliance are required before their deployment in clinical or environmental settings.

4.7.1.3. Economic Sustainability

The economic feasibility of GBHs is influenced by the cost of raw materials, fabrication processes and reusability. The high cost of graphene synthesis and purification, along with the reliance on advanced characterization tools, currently restricts its widespread adoption. To enhance the economic sustainability, waste-derived graphene sources (e.g., from biomass or waste plastics) are being explored. Low-cost natural polymers (like alginate, chitosan, or gelatin) can be integrated with graphene to reduce overall material cost. Recyclability and reusability of GBHs in environmental applications such as pollutant adsorption or oil recovery can also contribute to cost-effectiveness. By developing eco-friendly synthesis routes, improving material recyclability, and integrating cost-effective components, the economic and environmental sustainability of GBHs can be significantly improved, paving the way for their broader implementation in real-world scenarios.

4.8. COMPARATIVE ANALYSIS OF TECHNIQUES

To compare various characterisation strategies for GBHs, it is necessary to consider various hydrogel compositions and evaluate the advantages and disadvantages of each method. In complicated systems, XRD may yield overlapping peaks, though it helps to determine if graphene in hydrogels is crystalline or amorphous. The surface sensitivity of FT-IR may restrict the resolution, although it is commonly employed to identify functional groups and chemical interactions.

Raman spectroscopy is crucial for differentiating graphene oxide, reduced graphene oxide, and pure graphene; nevertheless, its precision may sometimes be compromised by fluorescence interference. Hydrogel porosity and shape may be seen with excellent resolution using SEM; however, this technique necessitates dehydrating the sample, which might change its structure. Although TEM opens the door to nanoscale graphene dispersion observation, the process of preparing samples for this kind of imaging is laborious and expensive. In TGA, heat stability and degradation patterns are measured, but molecular-level breakdown information is not given. While DSC is valid for assessing crosslinking efficiency and phase transitions, it is essential to prepare the samples accurately before using them. Surface area and porosity are determined via BET analysis; however, drying may change hydrogel structures. Enzymatic degradation results may vary with enzyme activity, while *in-vivo* studies, though insightful, raise ethical issues and require extended durations. Careful method selection enables effective assessment and optimization of GBHs for biological, environmental, and industrial applications.

4.9. CONCLUDING REMARKS, FUTURE PERSPECTIVES

The rapid progress in nanotechnology has shown the importance of 3D GBHs in many fields including electronics, photonics, hybrid materials, energy storage, sensors, healthcare, and the environment. GBHs require advanced characterization techniques to describe their chemical, dynamic and functional properties. AFM is key technique that study the surface roughness and dynamics of GBHs, but it is less effective for large or irregular samples. Next, XRD is good at identifying crystal structures, however it misses the microscopic graphene particles seen in hydrogels. FT-IR can examine graphene-hydrogel matrix chemical interactions, although overlapping peaks are hard to find. The geometry of graphene sheets can be revealed by TEM; however, sample preparation remains a challenge. Despite the widespread use of these technologies, understanding the distribution and dissemination of graphene inside the hydrogel matrix remains a challenge. Improvements in characterization accuracy and resolution are of utmost importance for the development of GBHs, especially when novel techniques are introduced. This chapter provides a comprehensive perspective on GBHs highlighting their application across biomedical, environmental, and industrial domains. The chapter further distinguishes itself by offering an in-depth discussion on the structure-property-performance relationship of GBHs, linking advanced characterization techniques to specific real-world applications.

CONFLICT OF INTEREST

The authors declare that there are no known personal or financial connections or financial interests that could have seemed to affect the work reported in this chapter.

ACKNOWLEDGEMENT

The authors would like to acknowledge the King Fahd University of Petroleum and Minerals, Dhahran, Saudi Arabia.

REFERENCES

Abdellaoui-Arous, N., Hadj-Hamou, A. S., & Djadoun, S. (2012). Thermal analysis and phase behavior of poly(styrene-co-4-vinylpyridine)/poly(styrene-co-acrylic acid) mixtures. *Thermochimica Acta*, *547*, 22–30. DOI: 10.1016/j.tca.2012.08.015

Abdollahi, R., Taghizadeh, M. T., & Savani, S. (2017). Thermal and mechanical properties of graphene oxide nanocomposite hydrogel based on poly (acrylic acid) grafted onto amylose. *Polymer Degradation & Stability*, *147*, 151–158. DOI: 10.1016/j.polymdegradstab.2017.11.022

Adilah, A. N., Hean, C. G., & Hanani, Z. N. (2021). Incorporation of graphene oxide to enhance fish gelatin as bio-packaging material. *Food Packaging and Shelf Life*, *28*, 100679. DOI: 10.1016/j.fpsl.2021.100679

Amirsardari, Z., Aghdam, R. M., Salavati-Niasari, M., & Shakhesi, S. (2015). Enhanced thermal resistance of GO/C/phenolic nanocomposite by introducing ZrB2 nanoparticles. *Composites. Part B, Engineering*, *76*, 174–179. DOI: 10.1016/j.compositesb.2015.01.011

Bai, H., Li, C., Wang, X., & Shi, G. (2011). On the Gelation of Graphene Oxide. *The Journal of Physical Chemistry. C, Nanomaterials and Interfaces*, *115*(13), 5545–5551. DOI: 10.1021/jp1120299

Chen, L., Wen, Z., Chen, L., Wang, W., Ai, Q., Hou, G., Li, Y., Lou, J., & Ci, L. (2019). Nitrogen and sulfur co-doped porous carbon fibers film for flexible symmetric all-solid-state supercapacitors. *Carbon*, *158*, 456–464. DOI: 10.1016/j.carbon.2019.11.012

Croitoru, A., Ficai, D., & Ficai, A. (2024). Novel Photothermal Graphene-Based hydrogels in biomedical applications. *Polymers*, *16*(8), 1098. DOI: 10.3390/polym16081098 PMID: 38675017

Daneshmandi, L., Barajaa, M., Rad, A. T., Sydlik, S. A., & Laurencin, C. T. (2020). Graphene-Based Biomaterials for Bone Regenerative Engineering: A Comprehensive review of the field and considerations regarding biocompatibility and biodegradation. *Advanced Healthcare Materials*, *10*(1), e2001414. DOI: 10.1002/adhm.202001414 PMID: 33103370

De Silva, K. K. H., Huang, H., & Yoshimura, M. (2018). Progress of reduction of graphene oxide by ascorbic acid. *Applied Surface Science*, *447*, 338–346. DOI: 10.1016/j.apsusc.2018.03.243

Dimiev, A. M., & Tour, J. M. (2014). Mechanism of graphene oxide formation. *ACS Nano*, *8*(3), 3060–3068. DOI: 10.1021/nn500606a PMID: 24568241

Dong, C., Lu, J., Qiu, B., Shen, B., Xing, M., & Zhang, J. (2017). Developing stretchable and graphene-oxide-based hydrogel for the removal of organic pollutants and metal ions. *Applied Catalysis B Environment and Energy,* 222, 146-156. DOI:DOI: 10.1016/j.apcatb.2017.10.011

Downes, A., & Elfick, A. (2010). Raman Spectroscopy and related techniques in biomedicine. *Sensors (Basel)*, *10*(3), 1871–1889. DOI: 10.3390/s100301871 PMID: 21151763

Eivazzadeh-Keihan, R., Asgharnasl, S., Aliabadi, H. M., Tahmasebi, B., Radinekiyan, F., Maleki, A., Bahreinizad, H., Mahdavi, M., Alavijeh, M. S., Saber, R., Lanceros-Méndez, S., & Shalan, A. E. (2022). Magnetic graphene oxide–lignin nanobiocomposite: A novel, eco-friendly and stable nanostructure suitable for hyperthermia in cancer therapy. *RSC Advances*, *12*(6), 3593–3601. DOI: 10.1039/D1RA08640E PMID: 35425373

Elhenawy, S., Khraisheh, M., AlMomani, F., Hassan, M. K., Al-Ghouti, M. A., & Selvaraj, R. (2021). Recent developments and advancements in Graphene-Based technologies for oil spill cleanup and Oil-Water separation processes. *Nanomaterials (Basel, Switzerland)*, *12*(1), 87. DOI: 10.3390/nano12010087 PMID: 35010035

Follmer, M., Jestin, S., Neri, W., Derre, A., Bentaleb, A., Mercader, C., & Poulin, P. (2019). Structuration of lignin-graphene oxide based carbon materials through liquid crystallinity. *Carbon*, *149*, 297–306. DOI: 10.1016/j.carbon.2019.04.013

Gorab, M. G., Aliabadi, H. M., Kashtiaray, A., Mahdavi, M., Bani, M. S., Etminan, A., Salehpour, N., Eivazzadeh-Keihan, R., & Maleki, A. (2022). Decoration of graphene oxide nanosheets with carboxymethylcellulose hydrogel, silk fibroin and magnetic nanoparticles for biomedical and hyperthermia applications. *Nanoscale Advances*, *5*(1), 153–159. DOI: 10.1039/D2NA00394E PMID: 36605797

Gul, K., Sohni, S., Waqar, M., Ahmad, F., & Norulaini, N. N., & K, M. O. A. (. (2016). Functionalization of magnetic chitosan with graphene oxide for removal of cationic and anionic dyes from aqueous solution. *Carbohydrate Polymers*, *152*, 520–531. DOI: 10.1016/j.carbpol.2016.06.045 PMID: 27516300

Gun'ko, V., Savina, I., & Mikhalovsky, S. (2017). Properties of water bound in hydrogels. *Gels (Basel, Switzerland)*, *3*(4), 37. DOI: 10.3390/gels3040037 PMID: 30920534

Han, J., Feng, Y., Liu, Z., Chen, Q., Shen, Y., Feng, F., Liu, L., Zhong, M., Zhai, Y., Bockstaller, M., & Zhao, Z. (2021). Degradable GO-Nanocomposite hydrogels with synergistic photothermal and antibacterial response. *Polymer*, *230*, 124018. DOI: 10.1016/j.polymer.2021.124018

Hitkari, G., Singh, S., & Pandey, G. (2018). Nanoparticles: an Emerging weapon for Mitigation/ Removal of various Environmental Pollutants for Environmental safety. *In Springer eBooks* (pp. 359-395). DOI:DOI: 10.1007/978-981-10-8669-4_16

Hu, W., Zhang, P., Liu, X., Yan, B., Xiang, L., Zhang, J., Gong, L., Huang, J., Cui, K., Zhu, L., & Zeng, H. (2018). An amphiphobic graphene-based hydrogel as oil-water separator and oil fence material. *Chemical Engineering Journal*, *353*, 708–716. DOI: 10.1016/j.cej.2018.07.147

Hussain, S., & Maktedar, S. S. (2023). Structural, functional and mechanical performance of advanced Graphene-based composite hydrogels. *Results in Chemistry*, *6*, 101029. DOI: 10.1016/j.rechem.2023.101029

Jiang, H., Zhu, Y., Ye, X., Yue, Z., Wang, L., Xia, J., Xie, J., & Jia, C. (2020). Graphene Hydrogel Film Adsorbed with Redox-Active Molecule Toward Energy Storage Device with Improved Energy Density and Unfading Superior Rate Capability. *ACS Sustainable Chemistry & Engineering*, *8*(26), 9896–9905. DOI: 10.1021/acssuschemeng.0c03129

Jiang, L., & Lu, X. (2021). Functional hydrogel-based supercapacitors for wearable bioelectronic devices. *Materials Chemistry Frontiers*, *5*(20), 7479–7498. DOI: 10.1039/D1QM00657F

Jiao, T., Guo, H., Zhang, Q., Peng, Q., Tang, Y., Yan, X., & Li, B. (2015). Reduced Graphene Oxide-Based Silver Nanoparticle-Containing composite hydrogel as highly efficient dye catalysts for wastewater treatment. *Scientific Reports*, *5*(1), 11873. DOI: 10.1038/srep11873 PMID: 26183266

Jing, X., & Yong-Fa, Z. (2013). Elimination of Bisphenol A from Water via Graphene Oxide Adsorption. *Wuli Huaxue Xuebao*, *29*(04), 829–836. DOI: 10.3866/PKU.WHXB201301211

Jose, J., Rajamani, A. R., Anandaram, S., Jose, S. P., Peter, S. C., & P B, S. (2019). Photophysical and electrochemical studies of anchored chromium (III) complex on reduced graphene oxide via Diazonium chemistry. *Applied Organometallic Chemistry*, *33*(9), e5063. DOI: 10.1002/aoc.5063

Khan, M. U. A., Yaqoob, Z., Ansari, M. N. M., Razak, S. I. A., Raza, M. A., Sajjad, A., Haider, S., & Busra, F. M. (2021). Chitosan/Poly Vinyl Alcohol/Graphene Oxide Based PH-Responsive Composite Hydrogel Films: Drug release, Anti-Microbial and Cell Viability studies. *Polymers*, *13*(18), 3124. DOI: 10.3390/polym13183124 PMID: 34578025

Krivorotova, T., Jonikaite-Svegzdiene, J., Radzevicius, P., & Makuska, R. (2014). Synthesis by RAFT polymerization and properties of anionic cylindrical molecular brushes bearing poly(acrylic acid) side chains. *Reactive & Functional Polymers*, *76*, 32–40. DOI: 10.1016/j.reactfunctpolym.2014.01.004

Lach, A., Skoczeń, M., Ordon, K., Domalik-Pyzik, P., Noga, S., Sekuła-Stryjewska, M., Karnas, E., Byrska, A., Jagiełło, J., Lipińska, L., Zuba-Surma, E., & Pielichowska, K. (2024). On the influence of graphene oxide and hydroxyapatite modification on alginate-based hydrogel matrix: Thermal, physicochemical, and biological considerations. *Journal of Thermal Analysis and Calorimetry*, *149*(12), 6021–6037. DOI: 10.1007/s10973-024-13229-x

Leganés, J., Sánchez-Migallón, A., Merino, S., & Vázquez, E. (2020). Stimuli-responsive graphene-based hydrogel driven by disruption of triazine hydrophobic interactions. *Nanoscale*, *12*(13), 7072–7081. DOI: 10.1039/C9NR10588C PMID: 32188962

Liao, G., Hu, J., Chen, Z., Zhang, R., Wang, G., & Kuang, T. (2018). Preparation, properties, and applications of Graphene-Based Hydrogels. *Frontiers in Chemistry*, *6*, 450. DOI: 10.3389/fchem.2018.00450 PMID: 30327765

Lim, H. N., Huang, N. M., Lim, N., Harrison, N., & Chia, N. (2011). Fabrication and characterization of graphene hydrogel via hydrothermal approach as a scaffold for preliminary study of cell growth. *International Journal of Nanomedicine*, *6*, 1817–1823. DOI: 10.2147/IJN.S23392 PMID: 21931479

Liu, C., Han, G., Chang, Y., Xiao, Y., Zhou, H., & Shi, G. (2017). High-performance supercapacitors based on the reduced graphene oxide hydrogels modified by trace amounts of benzenediols. *Chemical Engineering Journal*, *328*, 25–34. DOI: 10.1016/j.cej.2017.07.032

Liu, T., Liu, M., Dou, S., Sun, J., Cong, Z., Jiang, C., Du, C., Pu, X., Hu, W., & Wang, Z. L. (2018). Triboelectric-Nanogenerator-Based soft Energy-Harvesting skin enabled by toughly bonded Elastomer/Hydrogel hybrids. *ACS Nano*, *12*(3), 2818–2826. DOI: 10.1021/acsnano.8b00108 PMID: 29494127

Lopes, P. M. P., Moldovan, D., Fechete, R., Mare, L., Barbu-Tudoran, L., Sechel, N., & Popescu, V. (2023). Characterization of a Graphene Oxide-Reinforced Whey hydrogel as an Eco-Friendly Absorbent for food packaging. *Gels (Basel, Switzerland)*, *9*(4), 298. DOI: 10.3390/gels9040298 PMID: 37102911

Lu, H., Zhang, S., Guo, L., & Li, W. (2017). Applications of graphene-based composite hydrogels: A review. *RSC Advances*, *7*(80), 51008–51020. DOI: 10.1039/C7RA09634H

Luan, V. H., Tien, H. N., Hoa, L. T., Hien, N. T. M., Oh, E., Chung, J., Kim, E. J., Choi, W. M., Kong, B., & Hur, S. H. (2012). Synthesis of a highly conductive and large surface area graphene oxide hydrogel and its use in a supercapacitor. *Journal of Materials Chemistry. A, Materials for Energy and Sustainability*, *1*(2), 208–211. DOI: 10.1039/C2TA00444E

Luong, A. H., Istiqomah, D., & Lin, W. (2024). Study of mechanical property and biocompatibility of graphene oxide/MEO$_2$MA hydrogel scaffold for wound healing application. *Biomedical Engineering Letters*, *14*(3), 537–548. DOI: 10.1007/s13534-024-00349-4 PMID: 38645584

Ma, J., Du, W., Chen, Z., Wang, W., & Zhang, L. (2022). Preparation of Graphene-Based Hydrogel Thermal Interface Materials with Excellent Heat Dissipation and Mechanical Properties. *Macromolecular Materials and Engineering*, *308*(1), 2200332. DOI: 10.1002/mame.202200332

Mahmoodi, H., Fattahi, M., & Motevassel, M. (2021). Graphene oxide–chitosan hydrogel for adsorptive removal of diclofenac from aqueous solution: Preparation, characterization, kinetic and thermodynamic modelling. *RSC Advances*, *11*(57), 36289–36304. DOI: 10.1039/D1RA06069D PMID: 35492754

Martín, C., Merino, S., González-Domínguez, J. M., Rauti, R., Ballerini, L., Prato, M., & Vázquez, E. (2017). Graphene improves the biocompatibility of polyacrylamide hydrogels: 3D polymeric scaffolds for neuronal growth. *Scientific Reports*, *7*(1), 10942. DOI: 10.1038/s41598-017-11359-x PMID: 28887551

Mishra, S., & Jha, A. (2023). Graphene oxide-based nanocomposite hydrogels for biosensor applications. *In Elsevier eBooks* (pp. 149-180). DOI:DOI: 10.1016/B978-0-323-99638-9.00006-X

Mredha, M. T. I., Le, H. H., Cui, J., & Jeon, I. (2020). Double-Hydrophobic-Coating through Quenching for Hydrogels with Strong Resistance to Both Drying and Swelling. *Advancement of Science*, *7*(6), 1903145. DOI: 10.1002/advs.201903145 PMID: 32195103

Nassar, G., Daou, E., Najjar, R., Bassil, M., & Habchi, R. (2021). A review on the current research on graphene-based aerogels and their applications. *Carbon Trends*, *4*, 100065. DOI: 10.1016/j.cartre.2021.100065

Saeidi, M., Chenani, H., Orouji, M., Rastkhiz, M. A., Bolghanabadi, N., Vakili, S., Mohamadnia, Z., Hatamie, A., & Simchi, A. (2023). Electrochemical wearable biosensors and bioelectronic devices based on hydrogels: Mechanical properties and electrochemical behavior. *Biosensors (Basel), 13*(8), 823. DOI: 10.3390/bios13080823 PMID: 37622909

Sharma, S. S. A., Bashir, S., Kasi, R., & Subramaniam, R. T. (2022). The significance of graphene based composite hydrogels as smart materials: A review on the fabrication, properties, and its applications. *FlatChem, 33*, 100352. DOI: 10.1016/j.flatc.2022.100352

Shen, Y., Fang, Q., & Chen, B. (2014). Environmental applications of Three-Dimensional Graphene-Based macrostructures: Adsorption, transformation, and detection. *Environmental Science & Technology, 49*(1), 67–84. DOI: 10.1021/es504421y PMID: 25510293

Sohni, S., Hassan, T., Khan, S. B., Akhtar, K., Bakhsh, E. M., Hashim, R., Nidaullah, H., Khan, M., & Khan, S. A. (2022). Lignin nanoparticles-reduced graphene oxide based hydrogel: A novel strategy for environmental applications. *International Journal of Biological Macromolecules, 225*, 1426–1436. DOI: 10.1016/j.ijbiomac.2022.11.200 PMID: 36436599

Tadesse, M. G., & Lübben, J. F. (2023). Review on Hydrogel-Based Flexible Supercapacitors for Wearable Applications. *Gels (Basel, Switzerland), 9*(2), 106. DOI: 10.3390/gels9020106 PMID: 36826276

Telin, A., Safarov, F., Yakubov, R., Gusarova, E., Pavlik, A., Lenchenkova, L., & Dokichev, V. (2024). Thermal degradation study of hydrogel nanocomposites based on polyacrylamide and nanosilica used for conformance control and water shutoff. *Gels (Basel, Switzerland), 10*(12), 846. DOI: 10.3390/gels10120846 PMID: 39727603

Tene, T., Usca, G. T., Guevara, M., Molina, R., Veltri, F., Arias, M., Caputi, L. S., & Gomez, C. V. (2020). Toward Large-Scale production of oxidized graphene. *Nanomaterials (Basel, Switzerland), 10*(2), 279. DOI: 10.3390/nano10020279 PMID: 32041339

Teodorescu, M., Cursaru, B., & Stanescu, P. O. (2010). Swelling and Diffusion Characteristics of Hydrogels Synthesized from Diepoxy-terminated Poly(ethylene glycol)s and Aliphatic Polyamines. *Soft Materials, 8*(3), 288–306. DOI: 10.1080/1539445X.2010.495631

Tiwari, J. N., Mahesh, K., Le, N. H., Kemp, K. C., Timilsina, R., Tiwari, R. N., & Kim, K. S. (2013). Reduced graphene oxide-based hydrogels for the efficient capture of dye pollutants from aqueous solutions. *Carbon, 56*, 173–182. DOI: 10.1016/j.carbon.2013.01.001

Undavalli, V. K., Ling, C., & Khandelwal, B. (2021). Impact of alternative fuels and properties on elastomer compatibility. *In Elsevier eBooks* (pp. 113-132). DOI:DOI: 10.1016/B978-0-12-818314-4.00001-7

Wang, X., Liu, Z., Ye, X., Hu, K., Zhong, H., Yuan, X., Xiong, H., & Guo, Z. (2015). A facile one-pot method to two kinds of graphene oxide-based hydrogels with broad-spectrum antimicrobial properties. *Chemical Engineering Journal, 260*, 331–337. DOI: 10.1016/j.cej.2014.08.102

Wang, Y., Zhang, P., Liu, C. F., & Huang, C. Z. (2013). A facile and green method to fabricate graphene-based multifunctional hydrogels for miniature-scale water purification. *RSC Advances, 3*(24), 9240. DOI: 10.1039/c3ra22687e PMID: 23930219

Wu, G., Ma, L., Liu, L., Chen, L., & Huang, Y. (2015). Preparation of SiO_2-GO hybrid nanoparticles and the thermal properties of methylphenylsilicone resins/SiO_2-GO nanocomposites. *Thermochimica Acta*, *613*, 77–86. DOI: 10.1016/j.tca.2015.05.026

Zare, I., Mirshafiei, M., Kheilnezhad, B., Far, B. F., Hassanpour, M., Pishbin, E., Vaghefi, S. S. E., Yazdian, F., Rashedi, H., Hasan, A., Wang, X., Adeli, M., & Makvandi, P. (2024). Hydrogel-integrated graphene superstructures for tissue engineering: From periodontal to neural regeneration. *Carbon*, *223*, 118970. DOI: 10.1016/j.carbon.2024.118970

Zheng, A. L. T., Boonyuen, S., Li, G. Y., Ngee, L. H., & Andou, Y. (2021). Design of reduced graphene hydrogel with alkylamine surface functionalization through immersion/agitation method and its adsorption mechanism. *Journal of Molecular Structure*, *1245*, 131008. DOI: 10.1016/j.molstruc.2021.131008

Zhu, Y., Murali, S., Cai, W., Li, X., Suk, J. W., Potts, J. R., & Ruoff, R. S. (2010). Graphene and graphene oxide: Synthesis, properties, and applications. *Advanced Materials*, *22*(35), 3906–3924. DOI: 10.1002/adma.201001068 PMID: 20706983

Zhu, Z., Wang, Z., Ba, Z., Dong, J., Zhang, Q., & Zhao, X. (2021). Three-dimensional N/S Co-doped holey graphene oxide based hydrogel electrodes for high performance supercapacitors. *Journal of Energy Storage*, *39*, 102658. DOI: 10.1016/j.est.2021.102658

Chapter 5
Functional Properties of Hydrogels

Muzammal Hussain
https://orcid.org/0009-0000-9914-8870
Southeast University, China

Abbas Faheem
https://orcid.org/0009-0002-8465-4492
University of Wah, Pakistan

Arooba binte Amir
Government College University, Faisalabad, Pakistan

Munir Hussain
Zhejiang Sci-Tech University, China

Aqib Mehmood
University of Wah, Pakistan

ABSTRACT

Hydrogels have emerged as a focal material in modern science, finding applications across various fields, including medical treatments, industrial processes, and optical devices. Their behavior varies significantly depending on the surrounding medium, making them versatile for a range of applications. In this chapter, we discuss key applications and properties of hydrogels, such as self-crosslinking wound dressings, diabetic wound care, and natural polymer-based formulations. Additionally, we explore their self-adaptive wound healing capabilities and their potential in biomedical advancements. Hydrogels are increasingly recognized as promising materials, particularly in the medical field, due to their biocompatibility, tunable properties, and multifunctionality. Given their adaptability and effectiveness, they hold significant potential for future advancements across various disciplines, making them a crucial material for ongoing research and development.

DOI: 10.4018/979-8-3373-2337-4.ch005

5.1 INTRODUCTION

Hydrogels are water-rich materials having porous structures. Living systems are also dependent on hydrogel frameworks for their effective functioning. J.D Ferry defined gels as "substantially dilute cross-linked system, which exhibits no flow when in steady- state" (Eelkema & Pich, 2020). Choice and concentration ratio of biopolymers, pH, and pore size allow us to increase functioning of Hydrogels. The electrostatic gels obtained from gelation method have high water density (Le et al., 2017). Water retention in gel is the measure of resistance to any deformation and it relies on gel's microstructures. The Hydrogels prepared by induction methods and enzyme induced crosslinking methods are common in biomaterials and pharmaceutical industries. Hydrogels prepared by organic polymers are less toxic, cost-effective and biocompatible, therefore used for encapsulation of unstable and degraded food ingredients, during, processing, storage or transportation (Liu et al., 2021). Because of their high absorbency, sodium alginate excreted from brown seaweed, forms hydrogels through ionic interactions with Ca^{+2}.The functional properties of hydrogels such as, gaseous exchange and transparency enable simple monitoring of wound healing, cooling effect to burnt area and absorption of extruding liquids from wounds (Straccia et al., 2014). The confined oxygen level to cornea and ocular results various disease, for this purpose bio infinity silicon-hydrogel contact lens is developed which have high oxygen permeability (Guancha-Chalapud et al., 2022). The enhanced swelling property of nanofiber-based hydrogels holds solution to delay drought conditions, by reducing irrigation water demand (up to 40%), increasing water retention capacity in soil, improving plant growth and permits effective usage of fertilizer (X. Li et al., 2018). The crosslinking property of hydrogels allow us to encapsulate cells homogeneously by transacting hydrophilic polymer solution into 3-D network structure of nanoparticles, which are used in tissue engineering (Wu et al., 2017). The hydrogels prepared by microbial polysaccharides had excellent gel strength, transparency and stabilizing capacity with enhanced emulsifying and emulsion properties (Zhao & Sun, 2017).

5.2 SYNTHETIC METHOD AND PROPERTIES OF HYDROGELS

Hydrogels have different physical, chemical and rheological properties, which are responsible for their proper functioning in different circumstances some of them are:

Figure 1. Functional properties of hydrogels

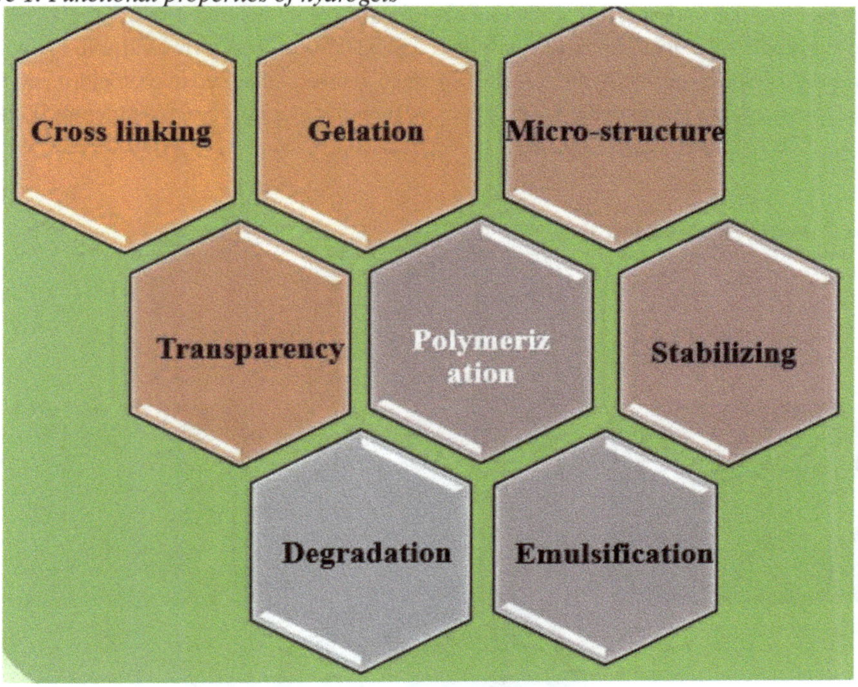

5.2.1 Gelation

The degradation of collagen results gelation, which depends upon inter / intra-molecular interaction, hydrogen bonding, triple-helix content of polyphenols of collagen and molecular weight of gelatin. The hydrogen bonding of polyphenols increases the gel strength but high concentration of polyphenols cause decrease in swelling ratio of hydrogels (Liu & Li, 2016). The Ca^{+2} also enhance the gel strength but causes high temperature during formation and melting of hydrogels, however, the gel strength is independent of Ca+2 ion concentration (Karoyo & Wilson, 2017). The networking at the super-helical level via self-assembly and aggregation leads in an increase in gel strength (Zhao et al., 2016).

5.2.2 Microstructure

On the basis of crosslinking molecules, hydrogels are divided into two kinds: Macromolecular hydrogels (MHGs) and Supramolecular hydrogels (SHGs). MHGs are cross-linked amorphous polymer networks, the radical polymerization of MHGs using acrylamide crosslinkers with water, results transparent hydrogel with 1.33 swelling ratio. SHGs contains low-weight molecular frameworks (oligomers or polymers) having superstructure through non-covalent links, which leads to gelation. SMHGs and MHGs can self-heal after damage by remodeling their networks. The polyampholytes are ionic and covalent crosslinker based amphiphilic polymer chains, forms tough and viscoelastic hydrogels. Hydrogels are further classified into four classes: non-porous, macro-porous, micro-porous and super-porous hydrogel. The size of is micro-porous (100-1000A), macro-porous (0.1-1.1μm), However, non-porous gels (10-100A) had correlated pore size equal to macro-porous. Super- porous hydrogels (SPHs) range to several hundred micrometers, and forms open channel system, responsible for capillary system of water into porous structure (Eelkema & Pich, 2020; Wu et al., 2017).

Figure 2. Properties of hydrogels

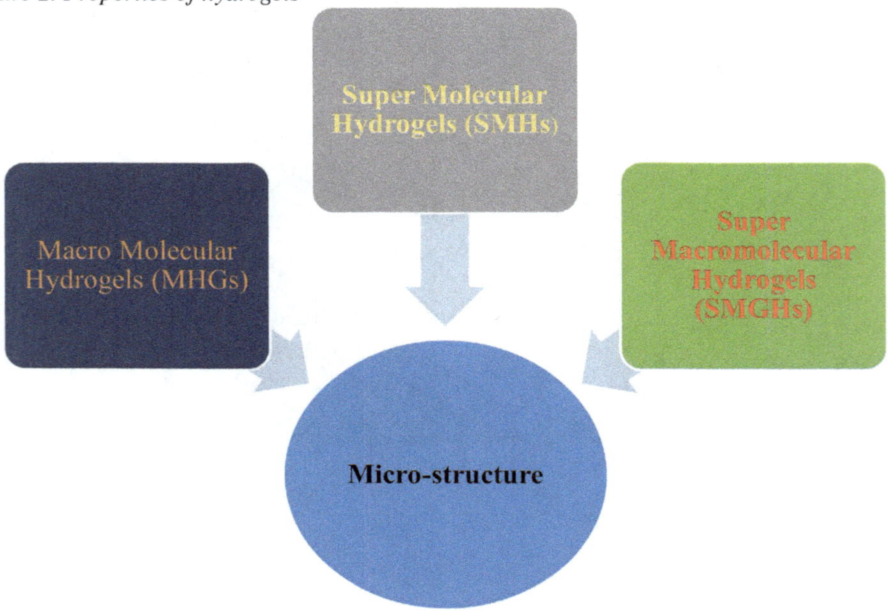

Figure 3. flow chart diagram of micro-structure hydrogel.

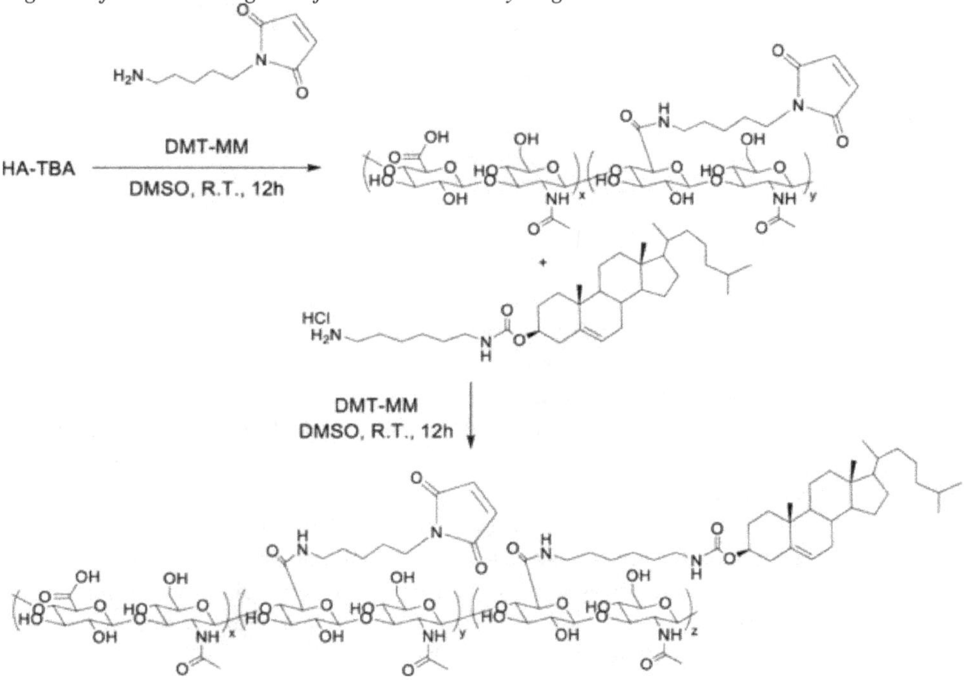

5.2.3 Cross-Linking

The covalent and non-covalent cross-linking of polymeric chains enhances polymeric networking of hydrogels. Therefore, double-networked hydrogels are developed, these are synthesized through two-step sequential free-radical polymerization. These hydrogels are double-networked between brittle and ductile networks, the brittle network has sacrificial bonds, responsible for mechanical strength of hydrogels.

Table 1. Crosslinking Methods using natural polymers

Mechanism	Organic Polymers
Physical cross-linking	
• Hydrophobized polysaccharides • Maturation	Chitosan, dextran, pullulan, and carboxymethyl curdlan Gum Arabic
• Hydrogen bonding • Ionic interaction	Blends of gelatin-agar, starch-carboxymethyl cellulose, hyaluronic acid-methyl cellulose Alginate, chitosan, carrageenan
• Stereo complex formation	Dextran
• Crystallization	Dextran, PVA/chitosan, PVA/starch, PVA/gelatine

continued on following page

Table 1. Continued

Mechanism	Organic Polymers
Chemical Cross-linking	
• High-energy radiation • By functional groups	Polysaccharides Hyaluronic acid-based hydrogel dextran, chitosan alginate
° Schiff-base reaction (aldehydes) ° Michael addition reaction ° Condensation reaction • Enzymatic reaction	Polypeptides, hyaluronic acid, dextran, cellulose and alginate
• Free-radical polymerization	Hyaluronic acid, dextran and chitosan

The intertwined cellulose Nano fibrils transfer mechanical stress throughout the cross-linked domains through H-bonding. The H-bonding and chain entanglement among cellulose hydrates, results 84-96% water content, with excellent flexibility, mechanical strength and elasticity (Chen et al., 2019). The hydrogels in which poly (vinyl alcohol) is cross-linked with tannic acid hydrogels showed remarkable mechanical strength (~16Mpa) and (~1000%) stretchability, with enhanced shape memory performance. The cross-linked PVA-TA thread structures had increased toughness and flexibility of these hydrogels was proved by SEM and WXRD analysis. The cross-linking agents gloxal and glutaraldehyde, cross-linked with poly acrylic acid forms inter / intra molecular acetal formation which increases storage capacity of hydrogels. However, insertion of crosslinkers decreases the elasticity of hydrogels. Here in Table 4.2 there are some cross-linking mechanisms using organic polymers (Cauich-Rodriguez et al., 1996; Ishihara et al., 1990).

5.3 POLYMERIZATIO

To study water content in hydrogel, polymer membranes were formed by solution casting process in which MPC (2-methaceyloyloxyethyl phosphoryl choline) was copolymerized with n-butyl methacrylate. The water retention was increased with temperature and polymer units. In polymer membrane, the hydrophilic groups were bonded with water, causing increase in hydration with temperature, enhanced polymer chain expansion and water storage capacity. A novel method to produce super stretchable and highly elastic polymer hydrogel using low concentration of polymers is redox system technique in which contact change transfer complexes and cyclic transition states are used to initiate the polymerization (Varghese et al., 2020; Zhang et al., 2016). The polymers hydrogels are grouped as:

5.3.1 Homo-Polymeric Hydrogels

These are made by only one kind of monomer. Only one step polymerization method is used, by dissolving monomer in Urea or NaOH, the insertion of epichlorohydrin makes hydrogel transparent.

5.3.2 Co-Polymeric Hydrogels

These are made up of two types of monomers, one should must have hydrophilic nature with different arrangement in structures.

5.3.3 Semi-Interpenetrating Networks

These are made by penetrating one linear polymer in another non-crosslinked polymer.

5.4 INTERPENETRATING NETWORKS

These are made by combination of two polymers, by polymerizing or crosslinking one polymer in the presence of other (Han et al., 2018).

5.4.1 Transparency

Transparent hydrogels are widely used in bio-electrodes, signal detectors, dressing pads, and smart skins. The Nano mesh hydrogel have 70% transmittance across visible region. Due to high transparency UV-shielding these hydrogels are used as protecting dressing. The transparent fiber-reinforced hydrogels showed promising potentials in optical field (Maranchi et al., 2014).

5.4.2 Emulsification

To encapsulate bio-active compounds, emulsion –filled hydrogels are widely used, emulsion under acidic condition make ionic gelation process unfeasible for hydrogel production, so emulsion stability is achieved at 6-PH. The spontaneous emulsion method for producing Translucent-Nano-emulsion hydrogels depends on emulsion of oil and surfactant (Komaiko & McClements, 2015; Silva et al., 2020). This method results high temperature for decreasing transparency and droplet volume. Emulsion filled hydrogels are developed by ionotropic gelation of emulsions in CaCl2 solution and alginate-gelatin as gelatinized starch. Emulsion capacity of hydrogel determines morphological functional characteristics of hydrogels (Silva et al., 2021; Tian et al., 2016).

5.5 STABILITY

The water bonds and cross-links within polymer fibrils are closely associated with stability of hydrogels. The negative effect of hydration on structural stability of polymer hydrogels is prevented by lubricating medium provided by polymer fibrils. The structural change in polymer structures influences the stability of polymer-based hydrogels. To stabilize the 3-dimentional structure of polymer (collagen) gels, crosslinking bonds are developed externally. The studied have shown that the chemically crosslinked polymer gels are more stable than non-crosslinked gels, as during heating treatment the glutaraldehyde (GTA) crosslinked collagen gels shrank, remarkably enhancing the viscoelasticity of gels, where as non-crosslinked hydrogels collapsed due to degradation of collagen fibrils. Another method of creating chemically stable hydrogels is the combination of hydrophilic and crystalline polymers. The hydrophilic polymer preserves the hydrogel's shape, and acts as a powerful catalyst for crystallite reformation, and raises the equilibrium water content, which results electrolytically stable hydrogels (Li et al., 2014; Zhang et al., 2021).

5.6 DEGRADATION

The reduction in molecular density, structural stability or structural deformation leads to degradation of hydrogels. The drug delivery system and tissue regeneration are done by macro-porous hydrogels and degradable hydrogels. The degradation process is done by two processes, hydrolysis of polymer and hydrolysis of crosslinks. The hydrolysis is carried out by by using suitable enzymes or inserting degradable units in polymer chain. Degradation product and swelling behavior are the measure of hydrogel degradation. For cross-link degradation by hydrolysis, Dextran hydroxyethyl methacrylate hydrogels are employed. According to research, all crosslinks binding a single chain to the network must be released before the chain, so, degradation rate is slightly dependent on degree of substitution of methacrylate in dex-HEMA hydrogel. Enzymatic dex-HEMA degradation requires specific polymer chains for active site of dextranase. Enzymatic degradation completely destroys highly substituted hydrogels, and it is also dependent on degree of substitution (Meyvis et al., 2000; Tavakoli et al., 2019).

5.7 SWELLING PROPERTY IN HYDROGELS

Prominent feature of hydrogels is expansion due to water absorption, named as swelling property of hydrogels. The swelling properties depend upon pore size, porous system, cross-linking networks, PH, polymeric interactions. These swelling properties allow us to employ hydrogels for food storage, drug delivery or as wound dressing. Some of the techniques used to enhance or determine swelling properties are:

5.7.1 Aggregation-Induced Swelling Process

Javad Tavakoli et al. used aggregation-induced emission technique to study swelling properties of hydrogels in which swelling process was reflected by fluorescence. The normalization of water uptake to initial weight of hydrogel, to measure swelling property of hydrogels is named as swelling ratio (SR). It is measured by:

$$SR(\%) = \frac{M_t - M_0}{M_0} \times 100$$

The weight analysis is used for measuring SR%, in this process pre-weighed hydrogels are soaked in water, their weight change due to water removal is measured in different interval of time. The highly swellable hydrogels results structural breakdown during weight measurement due poor mechanical strength, causing errors in measurement with the progress of swelling process towards equilibrium. To reduce these error Aggregation Induced Technique is developed, in which tetraphenylethylene (TPE) aggregates with increase in water concentration, and indicate it with strong fluorescent properties. The hydrogels having different initial weight showed increase in florescent intensity until the water fraction (WF) in swelling environment is 90%, but with increase of WF 98-99% there is a gradual decrease in FL. The results show that the diffusion of TPE is closely correlated with water absorption in hydrogel (Xu & Li, 2018).

5.7.2 Hysteresis Free – Swelling Hydrogel

Non-swellable hydrogels are synthesized from polymers having hydrophilic and thermo-responsive characteristics; in which swelling and de-swelling mechanism of hydrogels oppose each other. Investigations have shown that the swelling mechanism of hydrogels is determined by selection of specific ratio of polymer units having no effect on homogeneous structure, and under physiological circumstances, reduction of swelling maintains the functional properties of hydrogels. The swelling is controlled by thermos responsive segment ratio (r): if r=0 the hydrogel consists of hydrophilic units and swelling ratio is more significant than 100% at every temperature, showed that hydrogels swell and change shape in aqueous solution, when r=0.5 and in aqueous solution at 10°C the swelling ratio is 300%, whoever there is an extreme decrease in volume at Tc =25°C, irrespective of r. When r =0.4 the hydrogels swell at 10°C and restore its shape at 37°C with 100% swelling ratio and 90% water absorption, these hydrogels maintained their transparency above Tc. While traditional hydrogels were ruptured (r =0) prior to threefold stretching, hydrogels with r=0.4 exhibited enhanced mechanical characteristics and retained structure despite being stretched more than seven times. Even after repeated mechanical strain, these hydrogels remained unchanged and maintained their structure in aqueous solutions. Furthermore, while in its equilibrium-swollen condition, the "non-swellable" hydrogel resisted a compressive stress of 60 MPa, which is equivalent to so-called "tough hydrogels," but swollen hydrogel ruptured at a 80% strain and endured maximum stress 0.4 MPa (Kim et al., 2014).

5.7.3 Cellulose Nano Fibrils Based Swelling Hydrogels

Reinforcement of porous structures in cellulose nanofiber hydrogels (CNF) and poly vinyl alcohol hydrogels (PVA) had been shown dramatic swelling properties, which has enhanced their wound healing property along with their mechanical properties.

These are prepared by solvent casting / particulate leaching and photo-polymerization techniques, Cheng synthesized porous poly hydrogels (N-isopropylacrylamide) (PNIPAAm) through phase separation in aqueous solution of sodium chloride at different volumes and studied their swelling behavior according to lower critical solution temperature (LCST) and pore size.

Swelling ratio of PVA hydrogels in sodium chloride solution was increased to 770% below their LCST and also showed change in pore size, 935% swelling ratio was enhanced in CNF hydrogels due to larger porous structures and water retention of CNF. However, strong internet-work interactions, decreases intermolecular space when regenerated cellulose chains occupies the PVA region causing low water absorption, hence, exceeded concentration of CNF in hydrogels decreases its swelling ratio (Ganji et al., 2010).

5.7.4 Swelling in Acid Based Hydrogels

Swelling properties of sugar-based hydrogels were investigated at multiple PH and with concentration of methacryl-amidophenyl-boronic acid (MPBA). The hydrogels were analyzed in two sets, one consists of 20mol-% MPBA substituted polyacrylamide hydrogel, second set had 20mol-% MPBA substituted in 20mol-% N-3-dimethylaminopropyl methacrylamide in different concentration of glucose and fructose. Glucose based hydrogels showed shrinking due to crosslinking bridging between internetworks, the fructose hydrogels increased swelling ration by binding and contributing in MPBA ionization, however, these hydrogels showed polyampholytic behavior with increasing PH. The Ph effect on swelling ratio of these polyacids was further studied by soaking hydrogels in different buffers (NaCl, acetic Acid, ethanolamine) at PH ranging 4 to 10. There is higher swelling ratio at PH 6-10, due to acidic shift, in fructose substituted hydrogels to charge stabilization of MPBA after complex formation with fructose. The glucose increases swelling below pKa, and stabilize the charged MPBA, but due to its active crosslinking with two MPBA units at once above pKa, cause shrinking of hydrogel. The shrinking response to physiological pH of glucose hydrogels indicated application of these hydrogels for biomedical sensing and monitoring (Tuncaboylu et al., 2011).

5.7.5 Swelling Kinetics and Porosity

Swelling property of hydrogels correlates with water absorption in porous polymeric networks, before glassy unsolved phase of hydrogels there is a rubbery phase of expansion that allows solvent molecules to penetrate in porous networks. On the basis of porous structure which are used to determine swelling kinetics of hydrogels, are divided in four kinds (Su et al., 2020).

Table 2. Swelling Kinetics of Porous structure.

Type	Morphology	Absorbed Water	Swelling process	Swelling rate	Uses
Non-porous	No porous structure	Bonded	Porous Diffusion	Dependent on sample size, Very-slow	Contact lenses, artificial muscles etc.
Micro-porous	Different size pores having enclosed cellular framework (100-1000A°)	Bonded	convection and diffusion in pores having water	Dependent on sample size, Slow	Biomedical applications and controlled release technology etc.
Macro-porous	Different size pores having enclosed cellular framework (0.1-1μm)	Bonded	Porous Diffusion through water	Dependent on sample size, Fast	Super absorbent in babies' diapers etc.
Super-porous	Lager Pores interconnected open-cell framework	Free	Capillary forces	Independent of sample size, very fast	DDS (in gastrointestinal tract), tissue engineering etc.

5.8 HYDROPHILIC AND HYDROPHOBIC PROPERTY OF HYDROGELS

As the hydrogels mainly consists of water, so water absorption and retention have deep effect on hydrogel function. The hydrogels having balanced Hydrophilic and hydrophobic effect have dramatic increase in mechanical properties. Hydrogel having co-polymerization between hydrophobic component and hydrophilic networks hydrogels have high swelling ratio and does not dissolve in water. The hydrophobic component of hydrogels enhances stiffness and strength, whereas, hydrophilic components make hydrogel stretchable and tough. Some of the hydrophobic monomer crosslinked with hydrophilic networks are:

5.8.1 Acrylate Based Hydrophobic Self-Healing Hydrogels

Non-crosslinking hydrogels are a unique class of hydrogels that do not form permanent chemical crosslinks between polymer chains. Instead, they rely on physical interactions such as hydrophobic associations to maintain their structure. These hydrogels can be synthesized by copolymerizing hydrophobic monomers with acrylamide in a micellar solution containing a surfactant. In this case, the hydrophobic monomers used are Methacrylate (C18) and Dococyl Acrylate (C22), which are polymerized in the presence of acrylamide in a solution of sodium dodecyl sulfate (SDS). This method ensures that the hydrophobic monomers are incorporated into the polymer network while being stabilized by micelles formed by the surfactant.

The polymerization process begins with the formation of micelles in the SDS solution. SDS is an anionic surfactant that forms micelles when its concentration exceeds the critical micelle concentration (CMC). These micelles act as nano-sized reaction sites where the hydrophobic monomers (C_{18} and C_{22}) can be solubilized. Acrylamide, a hydrophilic monomer, is also introduced into the system, allowing for copolymerization to occur in an aqueous medium. The resulting hydrogel network consists of hydrophilic segments (acrylamide) interspersed with hydrophobic domains (C_{18} and C_{22}), which serve as physical crosslinks. The addition of salt, specifically sodium chloride (NaCl), plays a crucial role in the behavior and stability of the hydrogel. When NaCl is introduced into the micellar solution, it influences the solubility and aggregation behavior of SDS micelles. The salt ions shield the electrostatic repulsion between the SDS head groups, allowing the micelles to grow in size. As the micelles expand, they can encapsulate larger amounts of the hydrophobic monomers, enhancing their incorporation into the polymer network. This results in a more robust hydrogel with well-distributed hydrophobic domains.

The mechanical properties of the hydrogel, including its strength and self-healing capabilities, are largely determined by the mobility of the junction zones within the gel network. Junction zones refer to regions where polymer chains are physically entangled or associated through non-covalent interactions, such as hydrophobic interactions. In this hydrogel system, the presence of C_{18} and C_{22} creates hydrophobic interactions that contribute to the gel's structural integrity. Hydrophobic interactions arise due to the tendency of non-polar groups to avoid contact with water. As a result, C_{18} and C_{22} domains cluster together within the hydrogel network, forming nano-sized hydrophobic pockets. These interactions play a dual role: they prevent the hydrogel from excessive water absorption while also enabling self-healing through reversible association and dissociation of hydrophobic domains.

One of the significant effects of the hydrophobic interactions in this hydrogel system is the prevention of water flow or absorption. Unlike conventional hydrogels that absorb large amounts of water due to their hydrophilic nature, the presence of C_{18} and C_{22} reduces the hydrogel's water uptake. This characteristic is beneficial in applications where water retention needs to be controlled, such as in biomedical implants, wound dressings, and drug delivery systems. The elongation capacity of the hydrogel is a critical parameter that determines its flexibility and stretchability. The elongation ratio refers to how much a material can stretch relative to its original length before breaking. In this hydrogel system, the elongation ratios of C_{18} and C_{22} were found to be 3600% and 1700%, respectively. This means that the hydrogel containing C18 can stretch up to 36 times its original length, whereas the hydrogel with C_{22} can stretch up to 17 times its original length. Such high elongation values indicate that the hydrogel possesses excellent flexibility and can withstand significant mechanical deformation without breaking.

5.8.2 Zwitter-Ion Hydrophilic–Hydrophobic Hydrogels

The toughness of hydrogels depends upon the hydrophilic-hydrophobic and ionic interactions between hydrophilic networks. Insertion of zwitter ionic segments in as hydrophilic hetro-network enhanced properties of hydrogels and make them anti-biofouling hydrogels. These hydrogels are synthesized by using 2, Hydroxyethyl methacrylate (HEMA), methyl methacrylate (MMA) and 3-methyl ammonium propane sulfonate (DMAPS) in single-step co-polymerization method. PHEM-5%D hydrogels had strong hydration of zwitter ions and can rehydrate after 7-cycles of dehydration and rehydration. The accumulation of zwitter ions on the hydrogel surface is because of the hydrophobic interaction and ionic repulsion in aqueous solution, which is indicated by increased wettability of the hydrogel surface. As a result, these hydrogels exhibit exceptional anti-biofouling capacity due to strong hydration coating on hydrogel's surface. It is difficult to rehydrate and make these hydrogels non-swellable, however, PHEM-5%D Hydrogels exhibits both properties (Mredha et al., 2019; Wu & Chu, 2008).

5.8.3 Benzene –Substituted Hydrophobic Hydrogels

Phenyl acrylate (PA) (hydrophobic monomer) and Phenyl acryl amide (PAA) (hydrophilic monomer), dimethyl sulfoxide (DMSO) polymerization medium are used in the presences of AIBN initiator (2,2azobis-isobutyronitrile), to make hydrogel containing both hydrophilic and hydrophobic properties. The color shifts of organogel to opaque white from transparent, indicated increase in stiffness and strength of soft hydrogels. By substituting water for DMSO increased the bonding of Hydrophobic components, resulting PA-rich segmented closely packed 3D hard phase (low water density), on the other hand, PAA-segments absorbs water to form soft-phase zone (high water density) that may rupture at a specified strain, aiding release of extra energy, as a result hydrogel have many hard and soft phase 3D- structures having pore size of 300-500nm with 50-60 wt.% water fraction. As alginate and polyampholytes hydrogels distorts ionic interactions and electrostatic interaction were weekend in salt solution, so hydrogels having benzene-benzene and hydrophobic interaction, had higher stability in the absence of ionic interactions, especially in salt water (Jiang et al., 2019).

5.8.4 Maleic Acid Hydrophobic Hydrogel

The studies have shown that multi-dimensional porous structure of hydrophobic-hydrophilic hydrogels induced by swelling; effects drug delivery, especially drugs having high-molecular-weight. To solve this, photo cross-linked biodegradable hydrogel was prepared using derivatives of maleic acid (Dex-Ma) precursors, multiamp poly maleic acid (PGCL-Ma) hydrophobic monomer and hydrophilic polysaccharide dextran. The melic acid enables additional chemical changes with carboxylic acid group along with photo-crosslinking. Hydrophobic-hydrophilic hybrid hydrogels had well-regulated swelling mechanism which correlates with the precursor's feed ratio. The comparative studies showed that higher PGCL-Ma component-based hybrid hydrogel favors diffusion mechanism for swelling, and hybrid hydrogels having higher Dex-Ma component favors diffusion-relaxation swelling mechanism. These hydrogels are bio gradable and efficient for drug delivery with enhanced swelling properties (Mredha et al., 2020).

5.8.5 Hydrophobic Association Hydrogels

During co-polymerization of hydrophilic and hydrophobic monomers, tough hydrophobic micro-domains serve as physical cross-linking points, joined with other polymer chains forming a 3D-microstructure. In case of non-destructive large deformation, hydrophobic group linked with network's short chain is forced out of hydrophobic micro-domain. This results in damage to the material's internal microstructure and modification of elastic modulus upon subsequent deformation. After applied force is removed, free hydrophobic groups that were extracted willingly enter hydrophobic micro-domain and rearrange, permitting quick self-healing to initial condition (Takei et al., 2020).

5.8.6 Double Hydrophobic–Coated Hydrogels

Double-hydrophobic–coated hydrogels were prepared by using hydrophobic monomer of different chain length and silicon oil on the surface of polyacrylamide hydrogels. These hydrogels proved to be highly resistant in both air drying and aqueous medium, as double-hydrophobic-coated PAAm hydrogel had visible variation of size after air drying of 7days, whereas, non-coated shrunk within 2-days. The oil medium in hydrophobic polymer layer provided homogeneous shielding to hydrogel surface in aqueous solution or air drying. The mechanical performance of these gels depends upon polymer density, also showed adhesive strength with high stability and strong resistance to drying or swelling (Qian et al., 2020).

5.8.7 Gelatin Based Hydrophobic Hydrogel

Hydrophobically modified gelatin (HMG) hydrogels are developed by structural crosslinking of hydrophobic segments in 3D micro-networks of gelatin without using cytotoxic chemicals. Gelatin was heated in phosphate-buffered saline solution to produce HMG, which was then cooled, added a mixture of ethanol that contained dodecanal (a hydrophobic monomer), allowed to sit for six hours, and then eliminated using 2-methylpyridine borane for twenty-four hours. The HMG were precipitated after adding a lot of ethanol, collected via centrifugation, and then lyophilized. These hydrogels confirmed to be effective for dual drug release, studied by adsorption and desorption of fluorescein sodium enhanced by hydrophobic segments, as hydrophobic drug to form HMG hydrogels (Pan et al., 2021).

5.9 WOUND HEALING AND TISSUE ENGINEERING PROPERTY OF HYDROGELS

5.9.1 Double-Networked Wound Healing Hydrogel

To synthesize the CS/PVA/SA hydrogels, an initial step involved dissolving 2 g of chitosan in 100 ml of 1% acetic acid at 60 °C. The resulting solution was subsequently filtered, yielding a 2% chitosan solution. Concurrently, 1 g of sodium alginate was dissolved in 100 ml of deionized water, forming a 1% sodium alginate solution. In a separate container, 5 g of polyvinyl alcohol was dissolved in 100 ml of deionized water at 85 °C, resulting in a 5% polyvinyl alcohol solution. Following the preparation of individual solutions, the chitosan, sodium alginate, and polyvinyl alcohol solutions were combined in a 2:2:1 ratio, respectively. Initially, the 2% chitosan solution was added dropwise into the sodium alginate solution over a duration of 2 hours. Subsequently, the 5% polyvinyl alcohol solution was introduced dropwise into the chitosan-sodium alginate solution over a period of 1 hour. The resultant mixture was continuously stirred for 4 hours to ensure homogeneity.

To this solution, an appropriate quantity of Ag_2O/SiO_2 nanoparticles and curcumin was incorporated, following the proportions outlined in Table 1. It was observed that for nanoparticle concentrations exceeding 20%, the polymerization process of the hydrogel was disrupted. Additionally, as the nanoparticle concentration increased, the synthesized hydrogel exhibited unsuitability for biological and mechanical testing. Therefore, the maximum permissible nanoparticle concentration in the hydrogel formulation was determined to be 20%. The total weight of the final solution was standardized at 13 g, comprising 10 g of polymer and 3 g of nanoparticles with varying weight percentages (0, 5, 10, 15, and 20 wt% of Ag_2O/SiO_2). Following a mixing period of 1 hour, the prepared solution was poured into a Petri dish and subjected to drying at 50 °C for 24 hours, as illustrated in **Figure 10.**

5.9.2 Diabetic and Tissue Engineering Hydrogel

For treatment of diabetes and skin ulcer platelet-rich plasma (PRP) hydrogels were used, but as the diabetic wound contains proteases in larger amount, it degrades growth factors derived from PRP, as a result declining the recovery of wound. To solve this, through Schiff base reaction, injectable self-healing chitosan hydrogels (CBPGCTS-SF) were prepared by crosslinking poly ethylene glycol, 4-carboxybenzaldehyde, with glycol chitosan, and silk-fibroin (SF). Because of SF hydrogel's exhibit's stable structure and adjustable degradation rates, bioactive compounds produced from PRP are preserved and compound release on the wound surface is maintained. These hydrogels showed highest efficiency in wound healing and wound closure, when investigated at different time (3-7days) as compared to other traditional hydrogels (Chen et al., 2018; He et al., 2021; Y. Li et al., 2018).

Figure 4 Diabatic and tissue engineering hydrogel

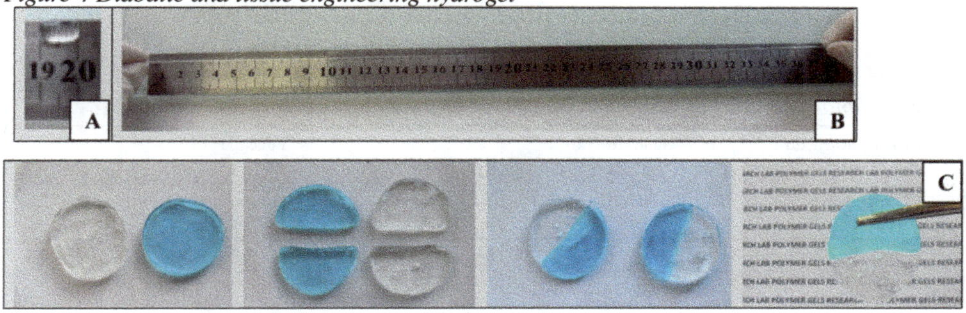

5.9.3 Self-Healable Wound Dressing Hydrogels

Traditional hydrogels fail as wound dressing because of weak mechanical properties, an inability to self-heal under external pressure, or a lack of tissue adhesion. The networking of dopamine-grafted oxidized sodium alginate (OSA-DA) and polyacrylamide (PAM) chains due to hydrogen bonding and dynamic Schiff cross-linking enables self-healable hydrogel. Schiff base reaction results, dynamic covalent cross-linking of aldehyde groups and the amino groups of OSA-DA chains and PAM chains. Between catechol groups, reversible hydrogen bonds are formed. As a result, cross-linked double-network hydrogel obtained. This hydrogel can withstand massive deformations while maintaining its elasticity. Because of reversible interactions such as hydrogen bonding, dynamic covalent cross-linking, hydrogel heal effectively without any external stimulation. Furthermore, the high concentration of catechol groups on the OSA chains, enhanced cell affinity and tissue adhesiveness of hydrogel remarkably, allowing it to be employed in wound dressing (Asadi et al., 2021).

5.9.4 Self-Adaptable Wound Healing Hydrogel

The self-healing Chitosan hydrogels were upgraded by enhancing Schiff-base dynamic networking which allows hydrogel to mobilize like organism or viscous liquid. In chitosan-self-healing hydrogel Schiff base developed difunctional polyethylene networks in hydrogel. This dynamic networking inhibits breakage and regeneration of hydrogel networks in liquid medium, and resulted self-adaptable wound healing hydrogel that can modify its shape under gravity and surface tension by gradually shifting in direction of external stress. Due to rebuilding of cross-linked network via a continuous cleavage and regeneration cycle of the Schiff-based networks it is proved as best drug carries for wound treatment (S. Wang et al., 2019).

5.9.5 Nano Reinforced Multifunctional Hydrogel

The multifunctional hydrogels exhibit enhanced antibacterial, antioxidant, electrical, adhesive, mechanical behavior and wound healing properties. metallic nanoparticles, carbon-based nanoparticles and nano-fibers act as antibacterial agents, and allow hydrogels to multifunction. Some of the components of multifunctional hydrogels are discussed in table (J. Wang et al., 2019).

Table 3. Nano-structures of multifunction wound healing hydrogels.

Hydrogel nanostructure	Nano structure	Function
1. Alginate	Double membrane	Moisture to skin cell for transportation and mobilization, improves neovascularization, covers the wound for long term and provides infection control (Tavakoli et al., 2020).
2. Peptide/Proline	Self-assembled Nano fibrous hydrogel	Contain procollagen components, have pH-switchable antimicrobial properties, release bioactive chemicals in a pH-sensitive manner, and destroys microbial components of biofilms (Liang et al., 2019).
3. Methacrylate	Polydopamine (PDA) modified NPs	Good nanoparticle dispersion in hydrogel networks, enhancing stickiness of tissue, antibacterial efficiency, blood clotting, remarkable structural properties, collagen deposition (Xiang et al., 2019).
4. Chitosan	Polydopamine coated CNTs	Exceptional tissue healing, hemostatic capabilities, antioxidative qualities, anti- bacterial properties (Ming et al., 2021).
5. Zinc cations	PDA functionalized carbon quantum dot (CQD)-decorated ZnO NPs	Injectable, with excellent photothermal and photodynamic antibacterial properties (Wei et al., 2015).

5.9.6 Living Bacterial Hydrogel

Folk treatments kill dangerous microbes, but also inhibit beneficial bacterial growth, altering the wounded skin's microbial environment. It has been stated that a living bacterial hydrogel structure promotes wound healing by secreting antibiotic chemicals from beneficial bacteria. A probiotic Lactobacillus reuteri is enclosed by emulsion polymerization which is subsequently deactivated by covalent cross-linking networks in a hydrogel. These hydrogels shield germs from the immune system, and also stops bacteria from escaping into the surrounding environment, therefore eliminating possible risks. L. reuteri-loaded gelatin hydrogel particles (LRHA gel) can reduce inflammatory cell infiltration, increase collagen deposition, and accelerate wound healing (Phadke et al., 2012).

5.10 SELF-HEALING IN HYDROGEL

Dynamic reaction of N-carboxyethyl chitosan (CEC) and adipic acid dihydrazide (ADH) with oxidized sodium alginate (OSA), resulted polysaccharide based self-healing hydrogels (CEC-I-OSA-I-ADH) were developed. Under physiological circumstances, the CEC-I-OSA-I-ADH hydrogel demonstrates remarkable cytocompatibility, cell encapsulation, and outstanding self-healing ability (up to 95%) without the need for external stimulation. The ability of the hydrogel to self-heal is due to the coexistence of two dynamic bonds: imine linkages created by the Schiff base reaction between OSA and CEC results acylhydrazone links produced by the interaction between OSA and ADH. The concentration of CEC enhanced functional characteristics of the CEC-l-OSA-l-ADH hydrogel. The system's dynamic character resulted self-healable hydrogel, investigated by the rheological recovery test, the macroscopic self-healing test, and the beam-shaped strain compression measurement, making it a feasible option for use as a cell and drug delivery carrier (Lai, 2021).

Rapidly self-healing hydrogels have been identified, in which healing occurs seconds after damage in the material or two hydrogel pieces are parallel to one another. Because of its reversibility and capacity to be switched on and off by pH changes, the healing process may be controlled from outside the body. Furthermore, the hydrogels can withstand several separation and healing cycles without changing mechanical properties or healing kinetics. These polymers based rapid self-healing hydrogels developed by acryloyl-6-aminocaproic acid catalyst exhibits balanced hydrophilic and hydrophobic interactions allows side chains to bond with external metal ions or extracellular proteins. Rapid healing occurs when elastomeric side chains facilitate hydrogen bonding between two hydrogel surfaces, through amide and carboxylic groups (Zhang et al., 2020).

In case of any damage to hydrogel cause release of drugs or bioactive agents. Self-healing hydrogels carrier for bioactive agents were studied on the bases of covalent and non-covalent bonding. Self-healing hydrogels contain a variety of non-covalent cross-links, including reversible ionic and hydrogen connections, hydrophobic or host-guest interactions, and dynamic covalent bonds (including acyl hydrazone, disulfide, and imine groups.) makes them suitable as a self-healing carrier for bioactive or living cells (Xu et al., 2018).

Figure 5. Bonding structure between hydrogel network.

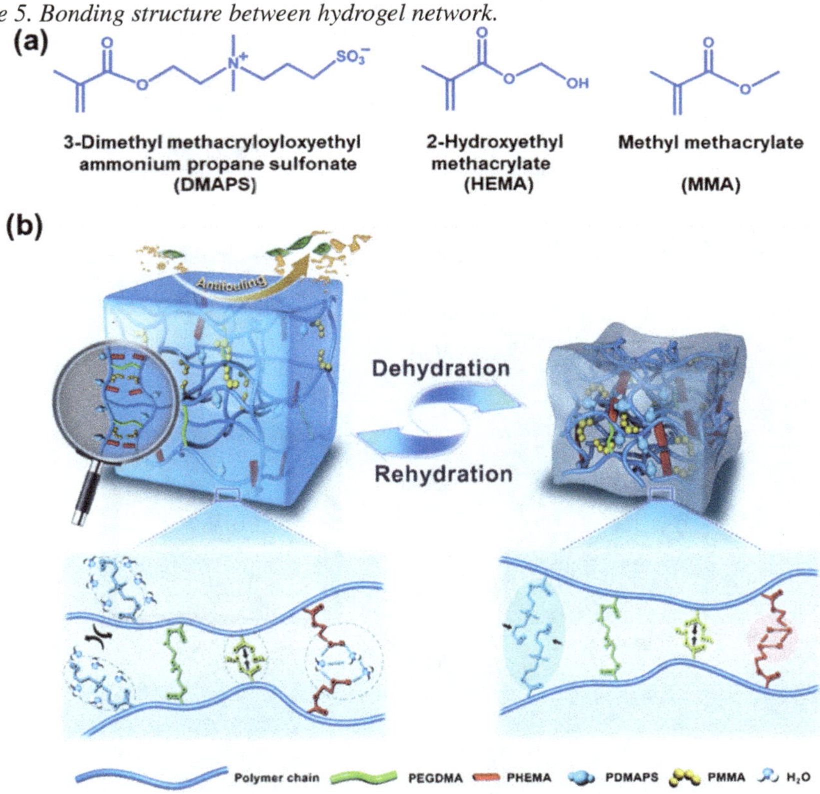

The double networked hydrophobic monomer-based hydrogels showed 88% self-healing property with enhanced mechanical properties and thermal stability. The dual networked hydrogels are prepared by situ micelle copolymerization, in Sodium Alginate (SA) and Acrylamide (Am) monomer in the presence of Lauryl Methacrylate (LMA) and sodium dodecyl sulfate surfactants (SDS). The hydrogen bonds between micelles, given by the SDS-LMA and PAAM chains, and the hydrogen bonds between PAAM and SA are mainly responsible for self-healing. This allows the network chain to repair the broken hydrophobic connection on the sliced surface. The efficiency of these hydrogels depends upon PU units but the internal network structure of the hydrogel becomes more compact as the PU content increases, resulting in a decrease in water loss rate and an extension of the drying time.

Figure 6. Self-healing mechanism of PAAM-PU hydrogel.

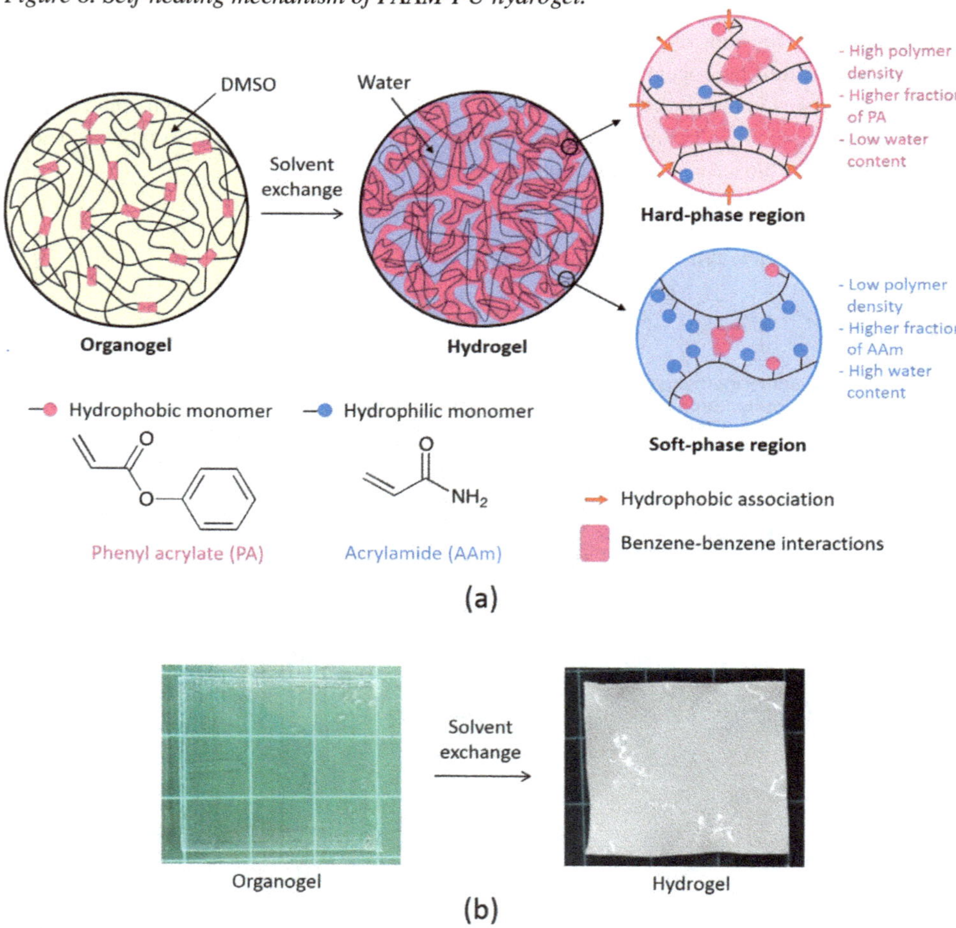

Reproduced from Ref. (Zhang et al., 2020).

Natural vanillin, employed as a crosslinking agent in the production of self-healing chitosan hydrogel. Vanillin's aldehyde group and amino group of one chitosan molecule combines by Schiff-base reaction, while its hydroxyl group connect with another chitosan molecule through H-bond, providing the basis for the formation of a reversible network. The self-healing effect of the hydrogel was mainly caused by the restoration of the Schiff-base link, while the hydrogen bonds remained stable at normal temperatures. The large number of hydrogen bonds kept the chitosan chains together and prevented the chains from moving freely in a tight network. The dynamic exchange of Schiff-base bonds reduced the generation of new hybrid interactions.

5.11 CONCLUSION

In this chapter, we conclude that the degree of crosslinking, gelation, polymerization, and degradation significantly influences the functionality of hydrogels for pharmaceutical and industrial applications. The extent of crosslinking is determined by the nature of the polymer and the specific crosslinking mechanisms employed. To enhance the swelling capacity of hydrogels, we examined various manufacturing techniques and the formation of porous structures under different conditions. The mechanical strength, flexibility, and self-healing properties of hydrogels are primarily governed by the balance between hydrophobic and hydrophilic interactions. Additionally, hydrogen bonding and Schiff base double networking contribute to the self-healing ability of hydrogels, demonstrating exceptional wound healing potential. Given the widespread use of hydrogels in everyday applications, these functional attributes allow for the enhancement of their quality and performance across a diverse range of uses. By optimizing crosslinking strategies and structural modifications, hydrogels can be tailored to meet specific pharmaceutical and industrial requirements, thereby advancing their practical applications in biomedical engineering, drug delivery systems, and tissue regeneration.

REFERENCES

Asadi, N., Pazoki-Toroudi, H., Del Bakhshayesh, A. R., Akbarzadeh, A., Davaran, S., & Annabi, N. (2021). Multifunctional hydrogels for wound healing: Special focus on biomacromolecular based hydrogels. *International Journal of Biological Macromolecules, 170*, 728–750. DOI: 10.1016/j.ijbiomac.2020.12.202 PMID: 33387543

Cauich-Rodriguez, J., Deb, S., & Smith, R. (1996). Effect of cross-linking agents on the dynamic mechanical properties of hydrogel blends of poly (acrylic acid)-poly (vinyl alcohol-vinyl acetate). *Biomaterials, 17*(23), 2259–2264. DOI: 10.1016/0142-9612(96)00058-0 PMID: 8968521

Chen, T., Chen, Y., Rehman, H. U., Chen, Z., Yang, Z., Wang, M., Li, H., & Liu, H. (2018). Ultratough, self-healing, and tissue-adhesive hydrogel for wound dressing. *ACS Applied Materials & Interfaces, 10*(39), 33523–33531. DOI: 10.1021/acsami.8b10064 PMID: 30204399

Chen, W., Li, N., Ma, Y., Minus, M. L., Benson, K., Lu, X., Wang, X., Ling, X., & Zhu, H. (2019). Superstrong and tough hydrogel through physical cross-linking and molecular alignment. *Biomacromolecules, 20*(12), 4476–4484. DOI: 10.1021/acs.biomac.9b01223 PMID: 31644270

Eelkema, R., & Pich, A. (2020). Pros and cons: supramolecular or macromolecular: what is best for functional hydrogels with advanced properties? *Advanced Materials, 32*(20), 1906012. DOI: 10.1002/adma.201906012 PMID: 31919957

Ganji, F., Vasheghani, F. S., & Vasheghani, F. E. (2010). Theoretical description of hydrogel swelling: a review.

Guancha-Chalapud, M. A., Serna-Cock, L., & Tirado, D. F. (2022). Hydrogels are reinforced with Colombian fique nanofibers to improve techno-functional properties for agricultural purposes. *Agriculture, 12*(1), 117. DOI: 10.3390/agriculture12010117

Han, L., Yan, L., Wang, M., Wang, K., Fang, L., Zhou, J., Fang, J., Ren, F., & Lu, X. (2018). Transparent, adhesive, and conductive hydrogel for soft bioelectronics based on light-transmitting polydopamine-doped polypyrrole nanofibrils. *Chemistry of Materials, 30*(16), 5561–5572. DOI: 10.1021/acs.chemmater.8b01446

He, Y., Li, Y., Sun, Y., Zhao, S., Feng, M., Xu, G., Zhu, H., Ji, P., Mao, H., He, Y., & Gu, Z. (2021). A double-network polysaccharide-based composite hydrogel for skin wound healing. *Carbohydrate Polymers, 261*, 117870. DOI: 10.1016/j.carbpol.2021.117870 PMID: 33766357

Ishihara, K., Ueda, T., & Nakabayashi, N. (1990). Preparation of phospholipid polymers and their properties as polymer hydrogel membranes. *Polymer Journal, 22*(5), 355–360. DOI: 10.1295/polymj.22.355

Jiang, H., Duan, L., Ren, X., & Gao, G. (2019). Hydrophobic association hydrogels with excellent mechanical and self-healing properties. *European Polymer Journal, 112*, 660–669. DOI: 10.1016/j.eurpolymj.2018.10.031

Karoyo, A. H., & Wilson, L. D. (2017). Physicochemical properties and the gelation process of supramolecular hydrogels: A review. *Gels (Basel, Switzerland)*, *3*(1), 1. DOI: 10.3390/gels3010001 PMID: 30920498

Kim, A., Mujumdar, S. K., & Siegel, R. A. (2014). Swelling properties of hydrogels containing phenylboronic acids. *Chemosensors (Basel, Switzerland)*, *2*(1), 1–12. DOI: 10.3390/chemosensors2010001

Komaiko, J., & McClements, D. J. (2015). Food-grade nanoemulsion filled hydrogels formed by spontaneous emulsification and gelation: Optical properties, rheology, and stability. *Food Hydrocolloids*, *46*, 67–75. DOI: 10.1016/j.foodhyd.2014.12.031

Lai, W.-F. (2021). Development of hydrogels with self-healing properties for delivery of bioactive agents. *Molecular Pharmaceutics*, *18*(5), 1833–1841. DOI: 10.1021/acs.molpharmaceut.0c00874 PMID: 33818107

Le, X. T., Rioux, L.-E., & Turgeon, S. L. (2017). Formation and functional properties of protein–polysaccharide electrostatic hydrogels in comparison to protein or polysaccharide hydrogels. *Advances in Colloid and Interface Science*, *239*, 127–135. DOI: 10.1016/j.cis.2016.04.006 PMID: 27318757

Li, J., Suo, Z., & Vlassak, J. J. (2014). Stiff, strong, and tough hydrogels with good chemical stability. *Journal of Materials Chemistry. B*, *2*(39), 6708–6713. DOI: 10.1039/C4TB01194E PMID: 32261867

Li, X., Sun, Q., Li, Q., Kawazoe, N., & Chen, G. (2018). Functional hydrogels with tunable structures and properties for tissue engineering applications. *Frontiers in Chemistry*, *6*, 499. DOI: 10.3389/fchem.2018.00499 PMID: 30406081

Li, Y., Wang, X., Fu, Y., Wei, Y., Zhao, L., & Tao, L. (2018). Self-adapting hydrogel to improve the therapeutic effect in wound-healing. *ACS Applied Materials & Interfaces*, *10*(31), 26046–26055. DOI: 10.1021/acsami.8b08874 PMID: 30009601

Liang, Y., Zhao, X., Hu, T., Han, Y., & Guo, B. (2019). Mussel-inspired, antibacterial, conductive, antioxidant, injectable composite hydrogel wound dressing to promote the regeneration of infected skin. *Journal of Colloid and Interface Science*, *556*, 514–528. DOI: 10.1016/j.jcis.2019.08.083 PMID: 31473541

Liu, K., Chen, Y.-Y., Zha, X.-Q., Li, Q.-M., Pan, L.-H., & Luo, J.-P. (2021). Research progress on polysaccharide/protein hydrogels: Preparation method, functional property and application as delivery systems for bioactive ingredients. *Food Research International*, *147*, 110542. DOI: 10.1016/j.foodres.2021.110542 PMID: 34399519

Liu, S., & Li, L. (2016). Thermoreversible gelation and scaling behavior of Ca2+-induced κ-carrageenan hydrogels. *Food Hydrocolloids*, *61*, 793–800. DOI: 10.1016/j.foodhyd.2016.07.003

Maranchi, J. P., Trexler, M. M., Guo, Q., & Elisseeff, J. H. (2014). Fibre-reinforced hydrogels with high optical transparency. *International Materials Reviews*, *59*(5), 264–296. DOI: 10.1179/1743280414Y.0000000032

Meyvis, T., De Smedt, S., Demeester, J., & Hennink, W. (2000). Influence of the degradation mechanism of hydrogels on their elastic and swelling properties during degradation. *Macromolecules, 33*(13), 4717–4725. DOI: 10.1021/ma992131u

Ming, Z., Han, L., Bao, M., Zhu, H., Qiang, S., Xue, S., & Liu, W. (2021). Living bacterial hydrogels for accelerated infected wound healing. *Advancement of Science, 8*(24), 2102545. DOI: 10.1002/advs.202102545 PMID: 34719880

Mredha, M. T. I., Le, H. H., Cui, J., & Jeon, I. (2020). Double-Hydrophobic-Coating through Quenching for Hydrogels with Strong Resistance to Both Drying and Swelling. *Advancement of Science, 7*(6), 1903145. DOI: 10.1002/advs.201903145 PMID: 32195103

Mredha, M. T. I., Pathak, S. K., Cui, J., & Jeon, I. (2019). Hydrogels with superior mechanical properties from the synergistic effect in hydrophobic–hydrophilic copolymers. *Chemical Engineering Journal, 362*, 325–338. DOI: 10.1016/j.cej.2018.12.023

Pan, Z., Ye, H., & Wu, D. (2021). Recent advances on polymeric hydrogels as wound dressings. *APL Bioengineering, 5*(1), 011504. DOI: 10.1063/5.0038364 PMID: 33644627

Phadke, A., Zhang, C., Arman, B., Hsu, C.-C., Mashelkar, R. A., Lele, A. K., Tauber, M. J., Arya, G., & Varghese, S. (2012). Rapid self-healing hydrogels. *Proceedings of the National Academy of Sciences of the United States of America, 109*(12), 4383–4388. DOI: 10.1073/pnas.1201122109 PMID: 22392977

Qian, Z., Wang, H., Bai, Y., Wang, Y., Tao, L., Wei, Y., Fan, Y., Guo, X., & Liu, H. (2020). Improving chronic diabetic wound healing through an injectable and self-healing hydrogel with platelet-rich plasma release. *ACS Applied Materials & Interfaces, 12*(50), 55659–55674. DOI: 10.1021/acsami.0c17142 PMID: 33327053

Silva, K. C. G., Bourbon, A. I., Pastrana, L., & Sato, A. C. K. (2020). Emulsion-filled hydrogels for food applications: Influence of pH on emulsion stability and a coating on microgel protection. *Food & Function, 11*(9), 8331–8341. DOI: 10.1039/D0FO01198C PMID: 32924048

Silva, K. C. G., Bourbon, A. I., Pastrana, L., & Sato, A. C. K. (2021). Biopolymer interactions on emulsion-filled hydrogels: Chemical, mechanical properties and microstructure. *Food Research International, 141*, 110059. DOI: 10.1016/j.foodres.2020.110059 PMID: 33641959

Straccia, M. C., Romano, I., Oliva, A., Santagata, G., & Laurienzo, P. (2014). Crosslinker effects on functional properties of alginate/N-succinylchitosan based hydrogels. *Carbohydrate Polymers, 108*, 321–330. DOI: 10.1016/j.carbpol.2014.02.054 PMID: 24751280

Su, X., Hao, D., Xu, X., Guo, X., Li, Z., & Jiang, L. (2020). Hydrophilic/hydrophobic heterogeneity anti-biofouling hydrogels with well-regulated rehydration. *ACS Applied Materials & Interfaces, 12*(22), 25316–25323. DOI: 10.1021/acsami.0c05406 PMID: 32378403

Takei, T., Yoshihara, R., Danjo, S., Fukuhara, Y., Evans, C., Tomimatsu, R., Ohzuno, Y., & Yoshida, M. (2020). Hydrophobically-modified gelatin hydrogel as a carrier for charged hydrophilic drugs and hydrophobic drugs. *International Journal of Biological Macromolecules, 149*, 140–147. DOI: 10.1016/j.ijbiomac.2020.01.227 PMID: 31982526

Tavakoli, J., Zhang, H., Tang, B. Z., & Tang, Y. (2019). Aggregation-induced emission lights up the swelling process: A new technique for swelling characterisation of hydrogels. *Materials Chemistry Frontiers, 3*(4), 664–667. DOI: 10.1039/C9QM00054B

Tavakoli, S., Mokhtari, H., Kharaziha, M., Kermanpur, A., Talebi, A., & Moshtaghian, J. (2020). A multifunctional nanocomposite spray dressing of Kappa-carrageenan-polydopamine modified ZnO/L-glutamic acid for diabetic wounds. *Materials Science and Engineering C, 111*, 110837. DOI: 10.1016/j.msec.2020.110837 PMID: 32279800

Tian, Z., Liu, W., & Li, G. (2016). The microstructure and stability of collagen hydrogel cross-linked by glutaraldehyde. *Polymer Degradation & Stability, 130*, 264–270. DOI: 10.1016/j.polymdegradstab.2016.06.015

Tuncaboylu, D. C., Sari, M., Oppermann, W., & Okay, O. (2011). Tough and self-healing hydrogels formed via hydrophobic interactions. *Macromolecules, 44*(12), 4997–5005. DOI: 10.1021/ma200579v

Varghese, S. A., Rangappa, S. M., Siengchin, S., & Parameswaranpillai, J. (2020). Natural polymers and the hydrogels prepared from them. In *Hydrogels based on natural polymers* (pp. 17–47). Elsevier. DOI: 10.1016/B978-0-12-816421-1.00002-1

Wang, J., Chen, X.-Y., Zhao, Y., Yang, Y., Wang, W., Wu, C., Yang, B., Zhang, Z., Zhang, L., Liu, Y., Du, X., Li, W., Qiu, L., Jiang, P., Mou, X.-Z., & Li, Y.-Q. (2019). pH-switchable antimicrobial nanofiber networks of hydrogel eradicate biofilm and rescue stalled healing in chronic wounds. *ACS Nano, 13*(10), 11686–11697. DOI: 10.1021/acsnano.9b05608 PMID: 31490650

Wang, S., Xiong, Y., Chen, J., Ghanem, A., Wang, Y., Yang, J., & Sun, B. (2019). Three dimensional printing bilayer membrane scaffold promotes wound healing. *Frontiers in Bioengineering and Biotechnology, 7*, 348. DOI: 10.3389/fbioe.2019.00348 PMID: 31803738

Wei, Z., Yang, J. H., Liu, Z. Q., Xu, F., Zhou, J. X., Zrínyi, M., Osada, Y., & Chen, Y. M. (2015). Novel biocompatible polysaccharide-based self-healing hydrogel. *Advanced Functional Materials, 25*(9), 1352–1359. DOI: 10.1002/adfm.201401502

Wu, D.-Q., & Chu, C.-C. (2008). Biodegradable hydrophobic–hydrophilic hybrid hydrogels: Swelling behavior and controlled drug release. *Journal of Biomaterials Science. Polymer Edition, 19*(4), 411–429. DOI: 10.1163/156856208783719536 PMID: 18318955

Wu, M., Shi, Z., Huang, H., Qu, J., Dai, X., Tian, X., Wei, W., Li, G., & Ma, T. (2017). Network structure and functional properties of transparent hydrogel sanxan produced by Sphingomonas sanxanigenens NX02. *Carbohydrate Polymers, 176*, 65–74. DOI: 10.1016/j.carbpol.2017.08.057 PMID: 28927628

Xiang, Y., Mao, C., Liu, X., Cui, Z., Jing, D., Yang, X., Liang, Y., Li, Z., Zhu, S., Zheng, Y., Yeung, K. W. K., Zheng, D., Wang, X., & Wu, S. (2019). Rapid and superior bacteria killing of carbon quantum dots/ZnO decorated injectable folic acid-conjugated PDA hydrogel through dual-light triggered ROS and membrane permeability. *Small, 15*(22), 1900322. DOI: 10.1002/smll.201900322 PMID: 31021489

Xu, C., Zhan, W., Tang, X., Mo, F., Fu, L., & Lin, B. (2018). Self-healing chitosan/vanillin hydrogels based on Schiff-base bond/hydrogen bond hybrid linkages. *Polymer Testing*, *66*, 155–163. DOI: 10.1016/j.polymertesting.2018.01.016

Xu, Z.-Y., & Li, J.-Y. (2018). Enhanced swelling, mechanical and thermal properties of cellulose nanofibrils (CNF)/poly (vinyl alcohol)(PVA) hydrogels with controlled porous structure. *Journal of Nanoscience and Nanotechnology*, *18*(1), 668–675. DOI: 10.1166/jnn.2018.14185 PMID: 29768893

Zhang, Q., Wu, M., Hu, X., Lu, W., Wang, M., Li, T., & Zhao, Y. (2020). A Novel Double-Network, Self-Healing Hydrogel Based on Hydrogen Bonding and Hydrophobic Effect. *Macromolecular Chemistry and Physics*, *221*(3), 1900320. DOI: 10.1002/macp.201900320

Zhang, S., Shi, Z., Xu, H., Ma, X., Yin, J., & Tian, M. (2016). Revisiting the mechanism of redox-polymerization to build the hydrogel with excellent properties using a novel initiator. *Soft Matter*, *12*(9), 2575–2582. DOI: 10.1039/C5SM02910D PMID: 26906449

Zhang, X., Li, Y., Ma, Z., He, D., & Li, H. (2021). Modulating degradation of sodium alginate/bioglass hydrogel for improving tissue infiltration and promoting wound healing. *Bioactive Materials*, *6*(11), 3692–3704. DOI: 10.1016/j.bioactmat.2021.03.038 PMID: 33898873

Zhao, D., Huang, J., Zhong, Y., Li, K., Zhang, L., & Cai, J. (2016). High-strength and high-toughness double-cross-linked cellulose hydrogels: A new strategy using sequential chemical and physical cross-linking. *Advanced Functional Materials*, *26*(34), 6279–6287. DOI: 10.1002/adfm.201601645

Zhao, Y., & Sun, Z. (2017). Effects of gelatin-polyphenol and gelatin–genipin cross-linking on the structure of gelatin hydrogels. *International Journal of Food Properties*, *20*(sup3), S2822-S2832.

Chapter 6
The Electrochemistry and Electrochemical Properties of Hydrogel-Based Composites

Ibukun Shotonwa
https://orcid.org/0000-0002-5364-717X
Lagos State University, Nigeria

Riliwan A. Kareem
Lagos State University, Nigeria

Molayo G. Agunbiade
Lagos State University, Nigeria

Kenechukwu T. Okonkwo
Lagos State University, Nigeria

Abisola C. Akinwuyi
Lagos State University, Nigeria

David O. Ayeni
Lagos State University, Nigeria

Babatunde J. Dawodu
Lagos State University, Nigeria

Martin O. Agosu
Lagos State University, Nigeria

Abimbola A. Afolabi
Lagos State University, Nigeria

Adebayo Ponle Oduwole
Lagos State University, Nigeria

ABSTRACT

Hydrogel-based composites possess unique electrochemical properties that have substantially triggered the development of efficient, cost-effective, and sustainable energy storage systems. As electrochemical storage technologies like batteries and supercapacitors become central to portable power solutions, there is a growing focus on optimizing performance parameters such as specific energy and power density. Ongoing efforts aim to design advanced materials that outperform existing systems while addressing environmental concerns associated with electronic waste. Key features of hydrogel-based composites—ranging from long-term stability, high proton conductivity, to enhanced energy storage via redox and electrostatic mechanisms—stem from the use of sustainable biomaterials like chitosan and activated carbon. Their flexibility, stretchability,

DOI: 10.4018/979-8-3373-2337-4.ch006

and biocompatibility also make them ideal for wearable electronics and biomedical applications. This chapter explores various hydrogel-based composites, their electrochemical behavior, and the mechanisms driving their performances.

6.1 ELECTROCHEMISTRY AS THE BRIDGE BETWEEN CHEMICAL REACTIONS AND ELECTRICAL ENERGY FOR TECHNOLOGICAL INNOVATION

Electrochemistry is a pivotal branch of chemistry that delves into the interactions between chemical reactions and electrical energy. It encompasses a diverse array of phenomena which include redox reactions in batteries and corrosion processes as well as applications in electroplating, electrolysis, and biosensing technologies (Phogat et al., 2024; Yammine et al., 2024). Electrochemical systems operate by converting chemical energy into electrical energy and vice versa, making them indispensable in the development of sustainable energy storage and conversion technologies. This discipline finds extensive applications across various fields such as materials science, energy systems, environmental remediation, and bioengineering (Gordon, 2023). As society increasingly relies on portable and wearable electronics, biomedical devices, and green technologies, the demand for advanced materials with multifunctional electrochemical properties is also on the rise. Redox (reduction-oxidation) reactions are fundamental to electrochemical processes, involving the transfer of electrons between chemical species. These reactions are central to the operation of batteries, where the movement of electrons through an external circuit provides electrical energy. Recent advancements in battery technologies, such as lithium-ion and sodium-ion batteries, have been driven by the need for higher energy densities, longer cycle lives, and improved safety profiles. Researchers are exploring novel materials and electrolytes to enhance the performance and sustainability of these energy storage systems (Abbas et al., 2020; Arevalo-Cid et al., 2021; Olabi et al., 2023). Electrochemical biosensors leverage on electrochemical principles to detect biological substances, offering advantages like high sensitivity, rapid response times, and portability. These devices are instrumental in medical diagnostics, environmental monitoring and food safety. By transducing biological interactions into measurable electrical signals, electrochemical biosensors facilitate real-time monitoring and analysis of various analytes (Singh et al., 2021). Electrochemical technologies play quite significant roles in environmental protection and remediation. Electrochemical systems are also being explored for air pollution control, offering sustainable solutions for reducing emissions of harmful gases (Chaplin, 2025).

6.2 HYDROGELS: DEFINITION, STRUCTURAL FEATURES, AND DISTINCTIVE PROPERTIES

6.2.1 What Are Hydrogels?

Hydrogels are soft, 3D polymer networks that absorb and retain large amounts of water without dissolving. Their flexibility, high water content, and tissue-like properties make them ideal for biomedical, pharmaceutical, and environmental uses. Composed of cross-linked hydrophilic polymers, their stable matrix holds water many times their dry weight, giving them a gel-like texture (Bashir et al., 2020; Ho et al., 2022; Tran et al., 2020). The presence of hydrophilic functional groups such as hydroxyl (-OH), carboxyl (-COOH), amide (-$CONH_2$), sulfonic (-SO_3H), and amine (-NH_2) contributes to their remarkable water-absorbing ability (Bustamante-Torres et al., 2021; Chafran et al., 2022). Hydrogels swell by interacting with water via hydrogen bonding and osmotic forces while maintaining their structure. Their properties depend on physical or chemical cross-linking, with chemical cross-links forming covalent bonds that create stronger, more durable hydrogels. These are often biodegradable and stimuli-responsive, making them suitable for drug delivery, wound healing, and long-term implants (Bustamante-Torres et al., 2021).

6.2.2 Structural Features of Hydrogels

The internal polymer architecture of hydrogels determines their softness, porosity, and responsiveness, enabling use in drug delivery, wound care, and biosensing. Their tunable mechanics suit tissue scaffolds and flexible devices, while moisture retention and transparency make them ideal for lenses and dressings. Smart hydrogels respond to environmental cues for controlled release and diagnostics (**Table 1**) (Bustamante-Torres et al., 2021; Tran et al., 2020). The cross-link density in hydrogels controls their strength, elasticity, and swelling as higher density improves mechanics but limits swelling. Their porous structure allows diffusion of molecules, essential for drug delivery and biosensing. Pore size affects diffusion rates, crucial for controlled release. Hydrogels can be natural (e.g., alginate, chitosan) or synthetic (e.g., polyacrylamide, PEG); natural types offer better biocompatibility, while synthetics allow more precise property control. Swelling, driven by osmotic pressure, is vital for applications like drug transport, scaffolds, and wound care (Lim et al., 2019; Neumann et al., 2023; Summonte et al., 2021; Xiang et al., 2020).

Table 1. Properties and applications of hydrogels

Properties	Applications	Refs
Stimuli-Responsiveness	Certain hydrogels can alter their swelling behavior in response to environmental triggers like temperature, pH, or ionic strength	(Protsak & Morozov, 2025)
Biocompatibility	Their physical softness and aqueous nature mimic that of natural tissues, minimizing inflammatory responses in medical applications	(Naahidi et al., 2017)
Diffusivity	Their porous nature allows for effective diffusion of therapeutic agents, making them ideal for controlled drug delivery	(Bustamante-Torres et al., 2021)
Mechanical Tunability	Cross-linking and polymer selection allow customization of rigidity, elasticity, and toughness	(Bashir et al., 2020)
Transparency and Moisture Retention	Makes them suitable for contact lenses, wound dressings, and biosensors	(Lee et al., 2024)
Thermo-responsiveness	This enables them to undergo structural or volume changes in reaction to temperature variations. This adaptability is especially beneficial in drug delivery systems that aim for controlled and targeted release within the body	(Matsumoto et al., 2018)
Wound management	Hydrogels help maintain a moist healing environment. This not only reduces patient discomfort but also accelerates tissue regeneration and minimizes scarring	(Ho et al., 2022)

continued on following page

Table 1. Continued

Properties	Applications	Refs
Smart Hydrogels	These are designed to respond to external triggers such as pH, temperature, or light. They are increasingly being used in drug delivery systems, biosensors, and responsive surfaces for diagnostics	(Mantha et al., 2019)
Tissue Engineering	Hydrogels provide a supportive structure that encourages cell attachment, proliferation, and differentiation. They are commonly used as scaffolding materials in the creation of artificial tissues or organs	(Mantha et al., 2019; Naahidi et al., 2017)

6.2.3 The Significance of Hydrogels in Electrochemical Applications

Hydrogels are 3D polymeric networks capable of retaining substantial amounts of water while maintaining their structural integrity. Their high water content, biocompatibility, flexibility, and tunable chemical and mechanical properties have rendered them ideal candidates in various applications, including wound dressings, tissue scaffolds, drug delivery systems, and biosensors. The use of diverse electrochemical techniques (**Table 2**) has revealed hydrogels as ionic conductors, facilitating the transport of ions while maintaining electrical insulation or selective conductivity, depending on the system designs. Their soft, tissue-like properties make hydrogels particularly attractive for flexible and wearable electrochemical devices, where mechanical compatibility with biological tissue is crucial (Wu et al., 2022). Moreover, the ability of hydrogels to be chemically tailored allows for the incorporation of functional groups, enzymes, or conductive particles, enhancing their integration into electrochemical sensing and energy storage platforms (Bashir et al., 2020).

*Table 2. Techniques for Electrochemical Characterization**

Techniques	Measurement modes	Key Applications	Strengths & Insights
Cyclic Voltammetry (CV)	Potentiostatic	Redox reactions, electron transfer kinetics	**Reveals oxidation/reduction peaks, reaction reversibility, and electrochemical surface area (ECSA)**
Electrochemical Impedance Spectroscopy (EIS)	Potentiostatic (AC)	Charge transfer resistance, diffusion, capacitance	**Provides Nyquist and Bode plots, ideal for studying corrosion, battery impedance, and electrode kinetics**
Chronoamperometry (CA)	Potentiostatic	Diffusion-controlled reactions, reaction kinetics	**Measures current over time post potential step; useful for studying diffusion coefficients and reaction rates**
Linear Sweep Voltammetry (LSV)	Potentiostatic	Electrochemical sensors, corrosion studies	**Provides information on oxidation/ reduction potentials, especially in corrosion or sensor applications**
Differential Pulse Voltammetry (DPV)	Potentiostatic (pulsed)	Trace analysis, sensor development	**High sensitivity with sharp peak-shaped responses; detection limits in the micromolar range**
Square Wave Voltammetry (SWV)	Potentiostatic (pulsed)	Environmental monitoring, biosensors	**Faster and more sensitive than CV and DPV; effective for detecting species with close oxidation/reduction potentials**

continued on following page

Table 2. Continued

Techniques	Measurement modes	Key Applications	Strengths & Insights
Tafel Analysis	Potentiostatic	Corrosion studies, electrocatalysis	**Determines corrosion rates and kinetic parameters; often used in electrochemical corrosion testing**
Galvanostatic Charge-Discharge (GCD)	**Galvanostatic**	**Battery and supercapacitor testing**	**Measures capacity, energy density, and stability; commonly used in energy storage device characterization**

* (Choudhary et al., 2017)

Classical hydrogels suffer from poor electronic conductivity, which limits their individual performance in electronics and electrochemical systems. To address this limitation, conductive hydrogels are developed by incorporating conductive polymers, carbon-based materials, or salts to enhance their ionic conductivity (Min et al., 2018; Zhang et al., 2023). For instance, the addition of salts like LiCl to polyacrylamide hydrogels enhances ionic conductivity and slows drying, improving suitability for triboelectric nanogenerators. Incorporating conductive polymers like PEDOT:PSS further combines hydrogel flexibility with high conductivity, enabling applications in wearable sensors and energy devices (Min et al., 2018). In other cases, incorporating conductive polymers such as polypyrrole (PPy), polyaniline (PANI), or poly(3,4-ethylenedioxythiophene) (PEDOT) have made exceeding differences (Min et al., 2018; Zhang et al., 2023). These intrinsically redox-active polymers undergo reversible redox reactions, effectively mediating electron transfer. When embedded in hydrogels, they form networks that merge the hydrogel's ionic conductivity with the polymer's electronic conductivity, enabling mixed conduction essential for fast, reversible redox applications (Li et al., 2020; Min et al., 2018; Shin et al., 2024; Zhang et al., 2023).

Characterizing conductive hydrogel composites is challenging due to their soft, hydrated, and multi-phase nature especially for flexible electronics, biosensors, and biointerfaces. These challenges include:

Deconvoluting Ionic vs. Electronic Conductivity: Conductive hydrogels exhibit mixed ionic-electronic conductivity, making it difficult to accurately separate the two. Standard methods like four-probe measurements often conflate both, especially in hydrated states. This distinction is crucial for applications like bioelectronics and energy storage, where ionic and electronic transport behave differently and impact performance (Tropp et al., 2024).

In Situ Characterization Under Mechanical Stress: Hydrogels' softness and viscoelasticity hinder consistent contact with sensors during mechanical testing, making accurate in situ measurement of electromechanical coupling difficult. Since devices must perform under strain, poor characterization leaves their real-use behavior uncertain (Saeidi et al., 2023; Wang et al., 2020).

Interface and Contact Resistance Artifacts: Contact between soft hydrogels and rigid electrodes is prone to poor adhesion, electrode delamination, and dynamic contact resistance, especially during bending or motion (Yang et al., 2025).

Environmental Sensitivity and Measurement Drift: Conductive hydrogels are sensitive to humidity and temperature, causing changes in conductivity, swelling, and mechanics, and introducing noise or drift in measurements (Sun et al., 2024).

Lack of Standardized Protocols: No universal standard exists for testing conductivity, strength, or swelling in conductive hydrogels, especially under combined conditions (e.g., strain, hydration, voltage). Li and co-workers proposed a unified framework using modular rigs with integrated impedance and tensile testing (Li et al., 2025).

Computational modeling is essential for designing hydrogel composites, offering molecular insights into polymer–filler interactions, charge transport, and structure–property relationships to guide material optimization (Fareed & Shityakov, 2025).

Integrating conductive fillers into hydrogels addresses their poor conductivity, redox activity, fragility, and low charge storage, creating multifunctional materials with tunable properties, enhanced stability, and responsiveness. These composites are ideal for applications in bioelectronics, tissue engineering, soft robotics, energy storage, and wearables.

6.2.4 Redox Reactions in Hydrogels: Mechanisms and Applications

Redox processes, involving electron transfer via coupled oxidation and reduction, drive the functionality of electroactive hydrogels. Incorporating redox-active components gives hydrogels dynamic electrochemical responsiveness to electrical stimuli. This responsiveness (**Table 3**) has paved the way for their application in diverse fields such as biosensors, drug delivery, energy storage, and tissue engineering (Cheng & Liu, 2017). Redox activity in hydrogels relies on electron and ion mobility. An external field triggers oxidation or reduction in the conductive polymer, prompting ion movement to maintain charge balance. This alters the hydrogel's state (swelling, contracting, or releasing embedded agents) making it ideal for controlled, redox-tuned drug delivery (Kilic Boz et al., 2022).

Table 3. Key Features of Redox-Responsive Hydrogels

Properties	Applications	Refs
Reversibility	Many redox hydrogels demonstrate reversible redox cycles, allowing them to repeatedly undergo oxidation and reduction without significant degradation	(Cheng & Liu, 2017)
Electrical Conductivity	Enhanced through the incorporation of conjugated polymers that facilitate long-range electron transport.	(Min et al., 2018)
Ionic-Electronic Coupling	Dual conduction pathways allow for simultaneous ion exchange and electron flow—vital for efficient actuation and sensing	
Stimuli Responsiveness	These hydrogels can be triggered by **electrical potential**, **chemical redox agents**, or **environmental changes**, offering versatile control mechanisms	(Cao et al., 2021)
Functional Versatility	Their redox behavior supports applications in **electrochemical biosensing**, **neuronal interface devices**, **battery systems**, and **responsive membranes**.	(Cheng & Liu, 2017)
Smart Drug Delivery Systems	Redox reactions are exploited to release drugs in a **dose-controlled manner**, either passively or in response to biological signals	(Kilic Boz et al., 2022)
Biosensors	Redox-active hydrogels can be engineered to detect **glucose, lactate, cholesterol**, and other analytes by incorporating redox enzymes or mediator systems.	(Protsak & Morozov, 2025)
Bioelectronics and Neural Interfaces	Owing to their **softness, electrical conductivity, and biocompatibility**, redox hydrogels are used in **implantable electrodes** and **tissue stimulators**	(Shin et al., 2024)
Electrochemical Actuators	Hydrogels that swell or contract in response to redox changes can serve as artificial muscles or soft robotics components	(Protsak & Morozov, 2025)

6.2.5 Electrode Potentials, Electrochemical Window and Electrochemical Interfaces in Conductive Hydrogels

The electrochemical window is the voltage range where a material remains stable without significant oxidation or reduction. A wider window allows for higher operating voltages, leading to greater energy density in capacitive systems (Bahdanchyk et al., 2020). The electrochemical window is key in organic electrosynthesis and battery development, especially with organic materials. Organic electrolytes often decompose above 4.0 V, disrupting redox processes. Thus, ideal electrolytes have a broad electrochemical window exceeding the battery's operating voltage (Méry et al., 2021). Lithium bis(trifluoromethanesulfonyl)imide (LiTFSI) has an electrochemical window of about 3.0 V, operating effectively between 1.9 and 4.9 V. Electrolytes with narrower windows are prone to irreversible breakdown, causing capacity loss and performance decline. The electrochemical window depends on factors like temperature and the energy gap between the electrolyte's frontier orbitals, HOMO and LUMO, which govern its redox behavior (Lv & Suo, 2021).

The electrode potential, or electrochemical potential, is the voltage difference between an electrode and a reference electrode, indicating a species' tendency to gain or lose electrons within the electrochemical window (Boettcher et al., 2021). This parameter is key for analyzing conductive hydrogels' electrochemical behavior, especially in bioelectronics. Their high water content limits the electrochemical window (~1.23 V), but adding ionic liquids, organic solvents, or crosslinked polymers can suppress water electrolysis and extend stability (Bashir et al., 2020). Electrochemical behavior of hydrogels is also influenced by functional groups, pH, and additives. For example, sulfonate- or phosphate-modified hydrogels behave differently from unmodified polyacrylamide or polyvinyl alcohol counterparts (Sandomierski et al., 2024). These chemical modifications can stabilize reactive intermediates or shift the onset potentials of decomposition reactions, thus altering the electrochemical limits.

6.2.6 Electrochemical Behavior at the Hydrogel-Electrode Interface

The interface between a conductive hydrogel and an electrode, known as the electrochemical interface, critically affects charge transfer efficiency, influencing electron and ion movement and thus the performance of bioelectronic devices (Bashir et al., 2020). Advancements in materials engineering have enabled the direct integration of conductive polymers onto hydrogel surfaces, forming seamless, highly conductive interfaces that boost bioelectronic performance through faster electron transport and stable signal output (Wei et al., 2024). Incorporating conductive polymers like PEDOT:PSS into hydrogels enhances both mechanical strength and electrochemical performance. This is especially important in bioelectronics, where the hydrogel-electrode interface is key for neural interfaces, sensors, and implants. Tuning electrode potential and optimizing this interface improves signal fidelity, lowers impedance, and extends device lifespan. Flexible polymer hydrogel electrodes also offer better biocompatibility and signal transmission, making them ideal for long-term use (Wei et al., 2024).

6.3 HYDROGEL-BASED COMPOSITES

Hydrogels, valued for their high water content and biocompatibility, show promise in electrochemical applications. However, their low inherent conductivity limits their standalone performance in electronic and electrochemical systems (Agarwala, 2020). To address their low conductivity, hydrogel-based composites have been developed by incorporating conductive fillers like CNTs, graphene, metal nanoparticles, metal oxides, and conductive polymers. In these composites, the electrochemical window is key to determining their suitability for energy storage applications such as supercapacitors and batteries (Méry et al., 2021; Sadiq et al., 2022).

6.3.1 Conductive Fillers in Hydrogel Composites

Integrating conductive fillers into hydrogel matrices imparts electrical conductivity and improves mechanical strength, while retaining key hydrogel properties such as softness and biocompatibility (Hajalilou, 2025; Li et al., 2025). For example, adding CNTs to polyaniline-based hydrogels enhances supercapacitor performance, while incorporating graphene into conductive polymer hydrogels yields biofibers with improved conductivity and strength, ideal for biomedical use (Liang et al., 2024).

6.3.2 Categories of Conductive Fillers

Conductive additives mainly fall into three groups: carbon-based materials, metallic nanoparticles, and intrinsically conductive polymers. Each offers distinct benefits, with carbon-based materials preferred for their excellent conductivity, chemical stability, and mechanical strength (Entwistle et al., 2022; Yang et al., 2021; Zhang et al., 2010). Metal nanostructures like silver, gold, and copper are prized for their high conductivity. At the nanoscale, they provide enhanced surface reactivity and distinctive electrical properties (Coetzee et al., 2020; Naghdi et al., 2018; Tehrani, 2021). Conductive polymers are organic materials that allow electron delocalization through their conjugated molecular structures, thereby supporting charge transport (Hasan et al., 2025; Namsheer & Rout, 2021).

6.3.3 Strategies for Enhancing Electrochemical Properties

To boost electrochemical performance, hydrogel composites are modified by adding conductive nanomaterials (e.g., graphene, CNTs, MXenes, metal nanowires) to form electron pathways, and doping conductive polymers like polyaniline or PEDOT:PSS to enhance redox activity. Ionic transport is improved with ionic liquids, salts, or ion-conducting channels, while tailored cross-linking strengthens mechanical stability. Advanced designs including dual-network hydrogels, hydrogel-derived carbons, and bio-inspired structures further enhance toughness, conductivity, and durability, focusing on conductivity, ionic transport, stability, and mechanical strength.

6.3.3.1 Surface Modification Techniques

Chemical modification of the polymer backbone or surface can introduce functional groups that enhance electron transfer or ionic mobility. Examples include:

Grafting: Recent advances in hydrogel-based electrode systems have explored the covalent integration of conductive polymers such as polyaniline and PEDOT:PSS onto hydrogel matrices to enhance electrical conductivity and structural stability (Maity & Dawn, 2020; Sheng et al., 2023). $LiFePO_4$ cathodes have low electronic conductivity, which is improved by coating particles with graphitized carbon. The carbon's microstructure, especially sp^2 graphitic networks, boosts electron transport and rate performance. However, integrating such carbon into hydrogel composites is challenging because typical carbon precursors don't reach high graphitization at the moderate sintering temperatures (500–800°C) hydrogels tolerate. Though in situ growth of $LiFePO_4$ on graphene enhances conductivity and capacity, it faces issues with complex synthesis, scalability, and cost (Méry et al., 2021; Olabi et al., 2023; Yammine et al., 2024).

Current research focuses on hydrogel-derived carbons that serve as both structural matrices and conductive networks. These materials aim for tunable porosity, flexibility, and improved graphitization via post-carbonization or doping to boost performance in $LiFePO_4$ and other low-conductivity electrodes. Future advances must enhance conductivity, energy density, rate capability, cycling stability, and low-temperature function to meet next-generation energy storage needs (Yammine et al., 2024).

Crosslinking: Formation of a denser 3D network, enhancing mechanical strength and stability during electrochemical cycling (Sheng et al., 2023).
Doping: Introduction of dopant ions (e.g., Cl^-, SO_4^{2-}) into conductive polymers like PANI to improve conductivity and redox activity (Zeng et al., 2021). Cationic and anionic doping of crystal lattices can further improve material properties, but the exact effects on electrochemical performance remain unclear, requiring better methods to pinpoint substitution sites and measure dopant levels.
Hybridization with Conductive Materials: Adding materials like CNTs, graphene, rGO, or conductive polymers to hydrogel matrices greatly improves electrical performance by forming continuous conductive networks that lower internal resistance and enable efficient charge transport (Zhang et al., 2023).
Nanostructuring and Porosity Control: Designing the hydrogel structure to possess hierarchical porosity can improve ion accessibility and electrolyte diffusion. This is particularly important for high-rate charge/discharge applications (Yom-Tov et al., 2014).
Multi-functionalization: Advanced hydrogels combine stretchability, self-healing, and stimuli-responsiveness, making them ideal for next-generation wearables and implantable biosensors (Li et al., 2025). Combining hydrogels with conductive fillers creates multifunctional composites with enhanced electrical, mechanical, and biological properties. Hydrogels offer softness, high water content, and biocompatibility but lack conductivity. Adding fillers like CNTs, graphene, MXenes, metal nanostructures, or conductive polymers (e.g., polyaniline, PEDOT:PSS) enables hybrid materials with customized functions beyond those of each component alone. (Lin et al., 2025; Niu et al., 2025). Hydrogel–filler composites form percolating conductive networks that enable real-time signal transmission, stimulus responsiveness, and energy storage, while preserving the hydrogel's flexibility, stretchability, and self-healing. This blend of properties makes them ideal for biomedical devices (neural/cardiac interfaces, drug delivery), flexible electronics (e-skins, soft robotics),

and wearables (biosensors, motion trackers) (Han et al., 2025; Li & Wang, 2025). Recent studies show these composites support cell growth, guide neurons, and mimic physiological electrical signals, advancing tissue engineering and bioelectronic medicine. Incorporating nanomaterials like MXenes adds dual ionic-electronic conduction, enabling applications in energy storage and harvesting such as stretchable supercapacitors and biocompatible power sources (Cui et al., 2021).

Overall, combining soft hydrogel matrices with advanced conductive nanomaterials is a fast-growing area in multifunctional materials, poised to transform wearable healthcare, implantable electronics, and smart energy technologies

6.3.4 Effects of the Enhanced Electrochemical Properties in Hydrogel-Based Composites

Hydrogel-based composites are versatile materials for electrochemical applications, combining hydrogels' natural stiffness and biocompatibility with enhanced properties from conductive additives. This section investigates their electrochemical behavior.

6.3.4.1 Conductivity and Ionic Transport

Hydrogel composites exhibit unique ionic transport essential for electrochemical devices. Incorporating conductive fillers forms pathways that greatly enhance conductivity and enable efficient charge transfer. For example, in a study conducted by Li and co-workers (Li et al., 2020), single-walled carbon nanotubes (SWCNTs) are integrated into a hydrogel matrix of polyvinyl alcohol (PVA) and poly(N,N-dimethyl acrylamide) copolymer. Through π-π stacking with pyrene moieties, SWCNTs disperse uniformly, forming conductive pathways that achieve a bulk conductivity of 1.27 S/m at 8 mg concentration. This uniform distribution enhances both ionic and electronic transport within the hydrogel. Similarly, Chyzy and co-workers (Chyzy & Plonska-Brzezinska, 2020) noted that adding conductive fillers like carbon nanotubes and graphene nanosheets to hydrogels improves conductivity by forming interconnected pathways that boost ion mobility. However, they caution that fillers can clump together, reducing electrical performance. Ensuring even dispersion of conductive fillers is crucial to maintain hydrogel conductivity. The study by Li and co-workers (Li et al., 2020) demonstrates that non-covalent functionalization prevents filler clumping while maintaining good conductivity. Additionally, ionic transport mechanisms in these composites are influenced by interactions between the hydrogel's porous structure and the conductive fillers (Li et al., 2020).

6.3.4.2 Redox Properties

Redox properties are vital for sensors and energy devices. Pure hydrogels usually lack this due to high water content and low conductivity, but adding liquid metals like eutectic gallium-indium (EGaIn) can introduce redox functionalities. According to Hajalilou (Hajalilou, 2025), liquid metals form a Ga_2O_3 layer that breaks and reforms under stress, enabling dynamic circuits ideal for flexible sensors. When deformed, they release Ga^{3+} ions that trigger self-healing via cross-linking. They also improve conductivity by reducing graphene oxide and aid energy release.

During hydrogel synthesis, redox reactions expose reactive metal that initiates polymerization with monomers like acrylamide, embedding redox-active particles in the hydrogel without added chemicals.

6.3.4.3 Stability and Durability

The stability of hydrogel composites depends on mechanical strength, water retention, and environmental factors like temperature and pH, supporting their long-term electrochemical use. SWCNTs enhance stiffness, with G' increasing at higher concentrations (Li et al., 2020). Liquid metals like EGaIn improve mechanics and enable self-healing by releasing Ga^{3+} ions upon damage, promoting crosslinking and maintaining functionality under repeated stress (He et al., 2024). Additionally, liquid metal hydrogel composites exhibit improved flexibility, toughness and conductivity making them more suitable for applications such as stretchable circuits and wearable sensors (Hajalilou, 2025; Zhang et al., 2024). Glucose-sensitive hydrogels are promising for biosensing and drug delivery in diabetes care. They undergo reversible structural changes, like swelling or shrinking, in response to glucose levels. This behavior is often driven by functional groups such as phenylboronic acid, which binds glucose and triggers conformational shifts in the hydrogel matrix (Li et al., 2017; Ma & Shi, 2014). These hydrogels can sustain repeated structural transitions without fatigue or loss of responsiveness, making them suitable for long-term use in implantable or wearable glucose monitors and insulin delivery systems. Studies show they maintain glucose sensitivity and mechanical integrity over multiple exposure cycles (Li et al., 2017). Advanced formulations using dynamic covalent cross-linking or self-healing chemistries have further enhanced their durability and performance consistency (Morgan et al., 2022).

6.3.4.4 Charge Storage Capacity

The charge storage capacity of hydrogel-based composites refers to their ability to store electrical energy through electrostatic or faradaic processes. These materials can store charge via two major mechanisms: electric double layer capacitance and pseudocapacitance (Giannakou, 2021; Jain et al., 2023).

Electric Double Layer Capacitance (EDLC): refers to the non-faradaic storage mechanism where ions are physically adsorbed onto the surface of the electrode at the electrode–electrolyte interface, forming an electric double layer without any electron transfer. Hydrogels with high surface area and porous structure can efficiently facilitate EDLC behavior (Khan & Chakraborty, 2024) (Figure 1).

Figure 1. Representative scheme for electric double layer capacitance (EDLC) and pseudocapacitance. The EDLC mechanism is driven by the accumulation of solvated charge ions in the outer Helmholtz layer while the pseudocapacitive storage mechanism emanates from fast and reversible Faradaic charge transfer; a feat instigated by the charge generated from adsorbed ions accumulating in the Stern layer.

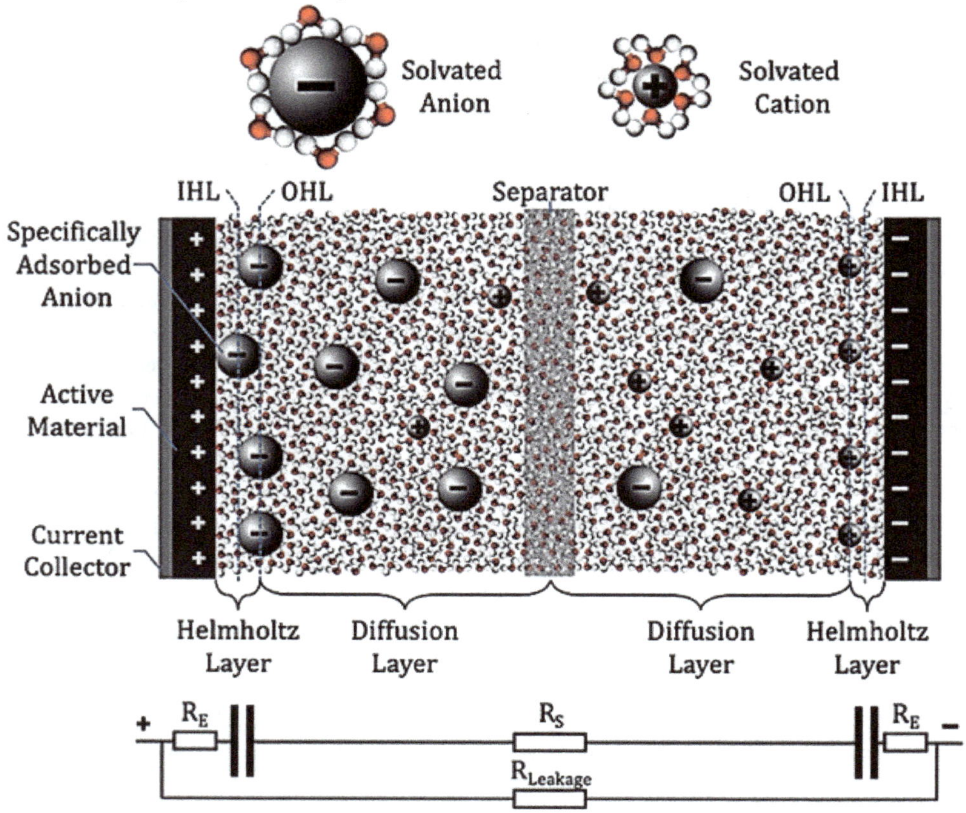

Pseudocapacitance: This process involves fast and reversible redox reactions occurring on the surface or within the bulk of the electroactive materials. Hydrogel composites with redox-active materials like PANI, PPy, or metal oxides (e.g., MnO_2, RuO_2) exhibit enhanced pseudocapacitive behavior due to fast, reversible redox reactions (Fleischmann et al., 2020; George & Balachandran, 2023). MnO_X-based devices, operating in extrinsic pseudocapacitive mode, enable rapid charge/discharge via ultra-short ion intercalation paths (~5 nm), delivering high power densities. Amorphous MnO_2 (a-MnO_2) performs especially well in electrolytes with divalent cations (Ca^{2+}, Mg^{2+}, Ba^{2+}), which boost charge storage. XPS studies show Ca^{2+} ions reversibly insert/extract during Mn^{4+}/Mn^{3+} redox transitions, indicating strong potential for fast-charging energy systems rivaling lithium-ion batteries (Chen et al., 2015; Tang et al., 2021; Yang et al., 2016).

Scaling up advanced hydrogel composites with conductive fillers like carbon nanomaterials, MXenes, and polymers faces key challenges. These include high filler costs, batch variability, complex processing, inconsistent crosslinking scalability, limited shelf life, transport issues, performance retention, and environmental or regulatory constraints—hindering broad commercial adoption (Li et al., 2025; Muthukutty et al., 2024). Addressing these challenges may involve low-cost, green nanofiller synthesis, scalable fabrication via 3D printing or continuous extrusion, use of bio-based or hybrid crosslinkers for eco-friendly production, and advanced automated quality control to ensure uniform dispersion and curing.

6.3.5 Electrochemical Applications

Hydrogel-based composites combine ionic and electronic conductivity, making them ideal for advanced electrochemical applications. In sensors, they support real-time, flexible monitoring. In energy storage, they enhance ion diffusion, electrode–electrolyte contact, and charge storage (Mir et al., 2022; Saini et al., 2009; Sardana et al., 2022). Polyaniline/MWCNT and polyaniline/graphene/Fe_2O_3 hydrogels have shown high capacitance and cycling stability, demonstrating strong potential for flexible supercapacitors (Gupta et al., 2020; Xia et al., 2012).

6.3.5.1 Wearable and Implantable Devices

These multifunctional materials are widely explored for wearable electronics, soft robotics, bioelectronics, and implantable devices. Stretchable, flexible, and self-healing hydrogel-based batteries show great promise for next-gen technologies like smart clothing, biosensors, and fitness wearables (Gao & Yu, 2021; Han et al., 2021).

6.3.5.2 Supercapacitors and Capacitive Devices

Hydrogel-based composites are ideal for flexible supercapacitors in wearable tech due to their high conductivity and flexibility, making them well-suited for smart clothing and portable electronics. (Jiang & Lu, 2021). These materials allow the development of lightweight, stretchable energy storage devices that can be seamlessly integrated into fabrics and wearable systems (Cao et al., 2021). Hydrogel composites offer high specific capacitance, for example, sodium alginate-reduced graphene oxide (rGO) reached 753 F g^{-1} at 1 A g^{-1} (El-Sayed et al., 2022) along with good rate capability. They also deliver high energy density and durability, maintaining stable performance over 5000 cycles, ideal for long-lasting wearable devices (Cao et al., 2021; El-Sayed et al., 2022). However, some challenges persist and these include poor cycling stability and mechanical issues under repeated bending or stretching, which can lead to the peeling of active materials from substrates, limiting long-term reliability (Armelin et al., 2016). Hydrogels face challenges with environmental stability, especially at high temperatures, impacting performance (Lu & Chen, 2020). Additionally, finding safe, nonflammable, and biocompatible electrolytes is a key hurdle for practical use (Ren et al., 2021).

6.3.5.3 Sensors and Biosensors

Hydrogels enhance electrochemical biosensors by providing a supportive structure and high surface area for improved sensitivity. They perform well in vivo with minimal biological disruption, aided by surface treatments that reduce unwanted interactions (Herrmann et al., 2021).

Hydrogel composites effectively detect small metabolites (<1500 Da) like glucose, key biomarkers of metabolic disorders. By sensing metabolites from endogenous and exogenous sources, they help monitor energy production, cell signaling, and epigenetic regulation (Lu & Thompson, 2012; Teunissen et al., 2018). Hydrogels enable pH-related sensing by detecting proteins whose properties shift with pH. For example, hydrogel-coated gold nanoshells can selectively detect high-isoelectric-point tear proteins, aiding dry eye diagnosis without interference from other proteins (Pinelli et al., 2020). A major advantage of hydrogel sensors is their high sensitivity, for instance, detecting just 45 pg/mL of the breast cancer marker HER2 (Wang et al., 2021). These properties make hydrogel composites ideal for electrochemical sensing, especially in glucose and pH-related biomedical monitoring.

6.3.5.4 Batteries and Energy Storage Devices

Hydrogel-based composites are being explored for LIBs and SIBs as cost-effective alternatives, addressing resource limitations of lithium and cobalt (Blomgren, 2017). In SIBs, pairing with hard carbon anodes has achieved reversible capacities around 300 $mAhg^{-1}$, though issues like slow kinetics and high first-cycle irreversibility remain (Phogat et al., 2025). For lithium-ion systems, hydrogels also offer a promising route by enabling alloy formation with Group 13, 14, and 15 metals. For instance, full lithiation of silicon ($Li_{12}Si_4$) delivers a high theoretical capacity of 3578 $mAhg^{-1}$, far exceeding that of conventional electrode materials (Obrovac & Chevrier, 2014). In comparison, sodium-ion systems experience significantly greater volume changes during sodiation than lithiation, making electrode degradation harder to manage. Despite this, sodium-ion batteries have shown superior cycling performance in some cases, achieving capacities close to 600 $mAhg^{-1}$ over more than 150 cycles (Kim et al., 2011; Wu et al., 2014). Titanium-based hydrogels, such as $Na_2Ti_3O_7$, serve as potential anode materials for sodium-ion batteries, capable of storing two sodium ions to form $Na_4Ti_3O_7$ during discharge, offering a theoretical capacity of 178 $mAhg^{-1}$. Nevertheless, issues like poor cycle stability and low initial coulombic efficiency remain significant challenges, mainly due to ongoing side reactions with the electrolyte (Rousse et al., 2013). In terms of electrochemical potential, sodium-ion insertion into $Na_2Ti_3O_7$ occurs at about 0.3 V vs Na/Na^+, which is over 1 volt lower than lithium insertion into $Li_2Ti_3O_7$ (1.6 V vs Li/Li^+). Sodium-ion batteries have a weaker driving force for ion intercalation than lithium systems, leading to lower cell voltages (Rousse et al., 2013). Gel polymer electrolytes (GPEs) offer excellent ionic conductivity and flexibility, making them suitable for supercapacitors, batteries, actuators, and biomedical sensors. Compared to conventional materials, hydrogel-based composites provide unique advantages like biocompatibility, tunable strength, and enhanced ionic transport which are ideal for flexible electronics, biointerfaces, and implantable energy storage systems (Ahmed et al., 2024).

6.3.5.5 Actuators and Artificial Muscles

The demand for complex actuators in robotics and assistive systems has driven interest in stimuli-responsive, self-actuating materials among materials scientists and engineers (Zhang et al., 2019). Artificial muscles are self-actuating materials that mimic real muscle function by producing reversible movement and mechanical work (Greco et al., 2021). Artificial muscles vary by material, control method, and stimulus type. Examples include shape-memory alloys (Bin

Park et al., 2020), dielectric elastomers, and piezoelectric elastomers (Qiu et al., 2019; Peng et al., 2019) and a variety of polymeric materials (Melling et al., 2019). Hydrogel-made artificial muscles offer a range of actuation actions, depending on the type and degree of stimulation, the structure built, and the materials used. Several generic actuation types have been demonstrated, including linear (Park & Kim, 2020), bending (Jiang et al., 2019), rotation (Mirabedini et al., 2017), and the combination of movements (Aziz & Spinks, 2020). Bending is the most common motion in artificial muscles, producing large displacement from small expansion or contraction. Thermo-responsive hydrogels enable fast actuation under heat, as seen in hydrogel-based McKibben muscles, which operate over six times faster than moisture-driven actuators (Bin Park et al., 2020). When properly designed, thermo-responsive hydrogels artificial muscles can be used for a wide range of innovative applications, including drug delivery (Kim et al., 2020), biomedical devices (Selvaraj & Takahata, 2020), smart actuators (Zhang et al., 2020), soft robotics (Wang et al., 2021), smart textiles (Hu et al., 2012), and sensors (Muralter et al., 2020). This is because dynamic macrostructural shape changes in axial, radial, and angular directions are possible. Non-woven cotton fabric loaded with thermos-responsive hydrogel showed regulated bioactive material delivery (Wang et al., 2016). Wearable drug delivery patches use textile fabrics coated with thermo-responsive hydrogels. With added nanoparticles, these hydrogels have broad applications in pharmaceutical and biomedical fields (Rafieian et al., 2019). They are particularly useful for applications involving drug release and delivery (Bellotti et al., 2021).

6.4 CHALLENGES AND FUTURE DIRECTIONS

Hydrogel-based composites have emerged as promising materials for electrochemical and bioelectronic systems owing to their intrinsic biocompatibility, ionic conductivity, and mechanical softness. However, despite accelerating advances in material engineering and functionalization, several core challenges continue to hinder widespread deployment and commercial translation.

6.4.1 Persistent Challenges

A major obstacle lies in the precise deconvolution of ionic and electronic conductivity. Most conductive hydrogels exhibit mixed conduction, and conventional methods such as four-point probe or impedance spectroscopy struggle to isolate these distinct charge carriers under hydrated conditions. This hampers material optimization for applications in energy storage and neural interfaces, where ion versus electron transport plays diverging functional roles.

Interfacial instability at the hydrogel–electrode boundary further limits device reliability. Poor adhesion, dynamic delamination under strain, and fluctuating contact resistance degrade signal fidelity and energy conversion efficiency. These issues are exacerbated by mechanical mismatch between soft hydrogel matrices and rigid conductive substrates, especially in wearables and implantables.

From a materials perspective, inhomogeneous dispersion of conductive fillers such as carbon nanotubes, graphene, or MXenes remains a bottleneck. Aggregation impairs electron percolation networks, weakens mechanical integrity, and reduces reproducibility across batches. In parallel, the lack of standardized testing protocols for electrochemical, mechanical, and swelling properties under biologically relevant conditions complicates inter-study comparisons and regulatory progress.

Environmental sensitivity including dehydration, microbial contamination, and poor thermal stability also limits shelf life and operational robustness. Moreover, the electrochemical window of aqueous hydrogel systems (~1.23 V) restricts high-voltage operation, despite efforts using ionic liquids and water-in-salt electrolytes.

6.4.2 Emerging Strategies and Directions

To address these limitations, future research must converge on multifunctional, adaptive architectures that integrate mechanical durability with high conductivity and long-term stability. Several promising directions are emerging:

- **Interface engineering**: The development of smart, self-healing, or bio-inspired interfaces can mitigate delamination and contact impedance. Incorporation of zwitterionic or amphiphilic linkers may offer dynamic adhesion properties.
- **Computational and AI-driven design**: Molecular modeling, combined with machine learning, can accelerate the prediction of structure–property–performance relationships, enabling rational design of polymer–filler combinations and crosslinking strategies.
- **Scalable, green fabrication**: Sustainable synthesis of nanofillers from bio-derived sources and printable hydrogel inks compatible with 3D/4D fabrication could facilitate industrial-scale production with minimized environmental impact.
- **Hierarchical structuring**: Introducing anisotropy or vascular-like porosity into hydrogel networks can mimic biological tissue properties while enhancing ion mobility, redox activity, and structural resilience.
- **Redox-active hydrogel matrices**: Integrating intrinsic redox moieties such as phenylboronic acid, polyaniline, or metal oxides may reduce reliance on exogenous fillers and support autonomous electron transport, drug release, or actuation.
- **Unified measurement platforms**: Standardized modular rigs that combine tensile, electrochemical, and environmental control could offer reproducible characterization under physiologically relevant conditions.

Looking ahead, hydrogel-based composites are poised to drive innovation in bioelectronics, flexible energy storage, wearable diagnostics, and soft robotics. However, achieving reliable, long-term, and clinically translatable systems will require interdisciplinary integration spanning synthetic chemistry, soft materials physics, bioengineering, and systems-level electronics. The next frontier lies not only in enhancing material properties, but in redefining function through intelligent, integrated design.

REFERENCES

Abbas, Q., Mirzaeian, M., Hunt, M. R. C., Hall, P., & Raza, R. (2020). Current State and Future Prospects for Electrochemical Energy Storage and Conversion Systems. *Energies*, *13*(21), 5847–5887. DOI: 10.3390/en13215847

Agarwala, S. (2020). Electrically conducting hydrogels for health care: Concept, fabrication methods, and applications. *International Journal of Bioprinting*, *6*(2), 1–15.

Ahmed, S., Islam, M., Raut, B., Yun, S., Kim, H. Y., & Nam, K.-W. (2024). A Comprehensive Review of Functional Gel Polymer Electrolytes and Applications in Lithium-Ion Battery. *Gels (Basel, Switzerland)*, *10*(9), 563–600. DOI: 10.3390/gels10090563

Arevalo-Cid, P., Dias, P., Mendes, A., & Azevedo, J. (2021). Redox flow batteries: A new frontier on energy storage. *Sustainable Energy & Fuels*, *5*(21), 5366–5419. DOI: 10.1039/D1SE00839K

Armelin, E., Pérez-Madrigal, M. M., Alemán, C., & Díaz, D. D. (2016). Current status and challenges of biohydrogels for applications as supercapacitors and secondary batteries. *Journal of Materials Chemistry. A, Materials for Energy and Sustainability*, *4*(23), 8952–8968. DOI: 10.1039/C6TA01846G

Aziz, S., & Spinks, G. M. (2020). Torsional artificial muscles. *Materials Horizons*, *7*(3), 667–693. DOI: 10.1039/C9MH01441A

Bahdanchyk, M., Hashempour, M., & Vicenzo, A. (2020). Evaluation of the operating potential window of electrochemical capacitors. *Electrochimica Acta*, *332*, 135503–135542. DOI: 10.1016/j.electacta.2019.135503

Bashir, S., Hina, M., Iqbal, J., Rajpar, A. H., Mujtaba, M. A., Alghamdi, N. A., Wageh, S., Ramesh, K., & Ramesh, S. (2020). Fundamental concepts of hydrogels: Synthesis, properties, and their applications. *Polymers*, *12*(11), 1–60. DOI: 10.3390/polym12112702

Bellotti, E., Schilling, A. L., Little, S. R., & Decuzzi, P. (2021). Injectable thermoresponsive hydrogels as drug delivery system for the treatment of central nervous system disorders: A review. *Journal of Controlled Release*, *329*, 16–35. DOI: 10.1016/j.jconrel.2020.11.049

Blomgren, G. E. (2017). The Development and Future of Lithium Ion Batteries. *Journal of the Electrochemical Society*, *164*(1), A5019–A5025. DOI: 10.1149/2.0251701jes

Boettcher, S. W., Oener, S. Z., Lonergan, M. C., Surendranath, Y., Ardo, S., Brozek, C., & Kempler, P. A. (2021). Potentially Confusing: Potentials in Electrochemistry. *ACS Energy Letters*, *6*(1), 261–266. DOI: 10.1021/acsenergylett.0c02443

Bustamante-Torres, M., Romero-Fierro, D., Arcentales-Vera, B., Palomino, K., Magaña, H., & Bucio, E. (2021). Hydrogels classification according to the physical or chemical interactions and as stimuli-sensitive materials. *Gels (Basel, Switzerland)*, *7*(4), 1–25. DOI: 10.3390/gels7040182

Cao, H., Duan, L., Zhang, Y., Cao, J., & Zhang, K. (2021). Current hydrogel advances in physicochemical and biological response-driven biomedical application diversity. *Signal Transduction and Targeted Therapy*, *6*(1), 1–31. DOI: 10.1038/s41392-021-00830-x

Cao, X., Jiang, C., Sun, N., Tan, D., Li, Q., Bi, S., & Song, J. (2021). Recent progress in multifunctional hydrogel-based supercapacitors. *Journal of Science. Advanced Materials and Devices*, *6*(3), 338–350. DOI: 10.1016/j.jsamd.2021.06.002

Chafran, L., Carfagno, A., Altalhi, A., & Bishop, B. (2022). Green Hydrogel Synthesis: Emphasis on Proteomics and Polymer Particle-Protein Interaction. *Polymers*, *14*(21), 4755–4795. DOI: 10.3390/polym14214755

Chaplin, B. P. (2025). Advanced Electrochemical Technologies for Water Treatment, Resource Recovery, and Sustainable Energy. *ACS ES&T Engineering*, *5*(3), 566–568. DOI: 10.1021/acsestengg.5c00111

Chen, D., Ding, D., Li, X., Waller, G., Xiong, X., El-sayed, M., Liu, M., Chen, D., Ding, D., Li, X., Waller, G., Xiong, X., & El-sayed, M. (2015). Probing the charge storage mechanism of a pseudocapacitive MnO2 electrode using in operando Raman spectroscopy. *Chemistry of Materials*, *27*(19), 6608–6619. DOI: 10.1021/acs.chemmater.5b03118

Cheng, W., & Liu, Y. (2017). Redox-responsive hydrogels. In *Biopolymer-Based Composites: Drug Delivery and Biomedical Applications* (pp. 31–60). Elsevier Ltd. DOI: 10.1016/B978-0-08-101914-6.00002-8

Choudhary, Y. S., Jothi, L., & Nageswaran, G. (2017). Electrochemical Characterization. In *Spectroscopic Methods for Nanomaterials Characterization* (Vol. 2). Elsevier Inc. DOI: 10.1016/B978-0-323-46140-5.00002-9

Chyzy, A., & Plonska-Brzezinska, M. E. (2020). Hydrogel Properties and Their Impact on Regenerative Medicine and Tissue Engineering. *Molecules (Basel, Switzerland)*, *25*(24), 5795–5821. DOI: 10.3390/molecules25245795

Coetzee, D., Venkataraman, M., Militky, J., & Petru, M. (2020). Influence of nanoparticles on thermal and electrical conductivity of composites. *Polymers*, *12*(4), 742–766. DOI: 10.3390/polym12040742

Cui, C., Fu, Q., Meng, L., Hao, S., Dai, R., & Yang, J. (2021). Recent Progress in Natural Biopolymers Conductive Hydrogels for Flexible Wearable Sensors and Energy Devices: Materials, Structures, and Performance. *ACS Applied Bio Materials*, *4*(1), 85–121. DOI: 10.1021/acsabm.0c00807

El-Sayed, N. S., Al Kiey, S. A., Darwish, A., Turky, G., & Kamel, S. (2022). High performance hydrogel electrodes based on sodium alginate-g-poly(AM-c o-ECA-co-AMPS for supercapacitor application. *International Journal of Biological Macromolecules*, *218*, 420–430. DOI: 10.1016/j.ijbiomac.2022.07.117

Entwistle, J., Ge, R., Pardikar, K., Smith, R., & Cumming, D. (2022). Carbon binder domain networks and electrical conductivity in lithium-ion battery electrodes: A critical review. *Renewable & Sustainable Energy Reviews*, *166*, 112624–112645. DOI: 10.1016/j.rser.2022.112624

Fareed, M. M., & Shityakov, S. (2025). Next-Generation Hydrogel Design: Computational Advances in Synthesis, Characterization, and Biomedical Applications. *Polymers*, *17*(10), 1373–1392. DOI: 10.3390/polym17101373

Fleischmann, S., Mitchell, J. B., Wang, R., Zhan, C., Jiang, D. E., Presser, V., & Augustyn, V. (2020). Pseudocapacitance: From Fundamental Understanding to High Power Energy Storage Materials. *Chemical Reviews, 120*(14), 6738–6782. DOI: 10.1021/acs.chemrev.0c00170

Gao, W., & Yu, C. (2021). Wearable and Implantable Devices for Healthcare. *Advanced Healthcare Materials, 10*(17), 1–3. DOI: 10.1002/adhm.202101548

George, J., & Balachandran, M. (2023). Extrinsic pseudocapacitance: Tapering the borderline between pseudocapacitive and battery type electrode materials for energy storage applications. *Journal of Energy Storage, 74*, 109292. DOI: 10.1016/j.est.2023.109292

Giannakou, P. (2021). *Printed Transition Metal Oxide Electrochemical Capacitors for Energy Harvesting Applications* [University of Surrey].

Gordon, A. (2023). Electrochemistry: Advancements and Applications in Modern Science and Engineering. *Research & Reviews: Journal of Chemistry, 12*(1), 5–6.

Greco, C., Kotak, P., Pagnotta, L., & Lamuta, C. (2021). The evolution of mechanical actuation: From conventional actuators to artificial muscles. *International Materials Reviews, 67*(6), 575–619. DOI: 10.1080/09506608.2021.1971428

Gupta, A., Sardana, S., Dalal, J., Lather, S., Maan, A. S., Tripathi, R., Punia, R., Singh, K., & Ohlan, A. (2020). Nanostructured Polyaniline/Graphene/Fe_2O_3 Composites Hydrogel as a High-Performance Flexible Supercapacitor Electrode Material. *ACS Applied Energy Materials, 3*(7), 6434–6446. DOI: 10.1021/acsaem.0c00684

Hajalilou, A. (2025). Liquid Metal–Polymer Hydrogel Composites for Sustainable Electronics: A Review. *Molecules (Basel, Switzerland), 30*(4), 12–15. DOI: 10.3390/molecules30040905

Han, W. B., Ko, G. J., Jang, T. M., & Hwang, S. W. (2021). Materials, Devices, and Applications for Wearable and Implantable Electronics. *ACS Applied Electronic Materials, 3*(2), 485–503. DOI: 10.1021/acsaelm.0c00724

Han, Y. Q., Lei, Z. Y., & Wu, P. Y. (2025). MXene Nanosheet-Enhanced Ionotronic Hydrogels for Wireless Powering and Noncontact Sensing. [English Edition]. *Chinese Journal of Polymer Science, 43*(4), 572–580. DOI: 10.1007/s10118-025-3253-6

Hasan, M. B., Parvez, M. M., Abir, A. Y., & Ahmad, M. F. (2025). A review on conducting organic polymers: Concepts, applications, and potential environmental benefits. *Heliyon, 11*(3), e42375–e42403. DOI: 10.1016/j.heliyon.2025.e42375

He, Y., You, J., Dickey, M. D., & Wang, X. (2024). Controllable Flow and Manipulation of Liquid Metals. *Advanced Functional Materials, 34*(31), 1–25. DOI: 10.1002/adfm.202309614

Herrmann, A., Haag, R., & Schedler, U. (2021). Hydrogels and Their Role in Biosensing Applications. *Advanced Healthcare Materials, 10*(11), 1–25. DOI: 10.1002/adhm.202100062

Ho, T. C., Chang, C. C., Chan, H. P., Chung, T. W., Shu, C. W., Chuang, K. P., Duh, T. H., Yang, M. H., & Tyan, Y. C. (2022). Hydrogels: Properties and Applications in Biomedicine. *Molecules (Basel, Switzerland), 27*(9), 1–29. DOI: 10.3390/molecules27092902

Hu, J., Meng, H., Li, G., & Ibekwe, S. I. (2012). A review of stimuli-responsive polymers for smart textile applications. *Smart Materials and Structures*, *21*(5), 1–24. DOI: 10.1088/0964-1726/21/5/053001

Jain, A., Ziai, Y., Bochenek, K., Manippady, S. R., Pierini, F., & Michalska, M. (2023). Utilization of compressible hydrogels as electrolyte materials for supercapacitor applications. *RSC Advances*, *13*(17), 11503–11512. DOI: 10.1039/D3RA00893B

Jiang, H., Fan, L., Yan, S., Li, F., Li, H., & Tang, J. (2019). Tough and electro-responsive hydrogel actuators with bidirectional bending behavior. *Nanoscale*, *11*(5), 2231–2237. DOI: 10.1039/C8NR07863G

Jiang, L., & Lu, X. (2021). Functional hydrogel-based supercapacitors for wearable bioelectronic devices. *Materials Chemistry Frontiers*, *5*(20), 7479–7498. DOI: 10.1039/D1QM00657F

Kilic Boz, R., Aydin, D., Kocak, S., Golba, B., Sanyal, R., & Sanyal, A. (2022). Redox-Responsive Hydrogels for Tunable and "on-Demand" Release of Biomacromolecules. *Bioconjugate Chemistry*, *33*(5), 839–847. DOI: 10.1021/acs.bioconjchem.2c00094

Kim, S. P., Duin, A. C. T. V., & Shenoy, V. B. (2011). Effect of electrolytes on the structure and evolution of the solid electrolyte interphase (SEI) in Li-ion batteries: A molecular dynamics study. *Journal of Power Sources*, *196*(20), 8590–8597. DOI: 10.1016/j.jpowsour.2011.05.061

Kim, Y. C., Shin, M. D., Hackett, S. F., Hsueh, H. T., Lima e Silva, R., Date, A., Han, H., Kim, B. J., Xiao, A., Kim, Y., Ogunnaike, L., Anders, N. M., Hemingway, A., He, P., Jun, A. S., McDonnell, P. J., Eberhart, C., Pitha, I., Zack, D. J., & Ensign, L. M. (2020). Gelling hypotonic polymer solution for extended topical drug delivery to the eye. *Nature Biomedical Engineering*, *4*(11), 1053–1062. DOI: 10.1038/s41551-020-00606-8

Lee, Y., So, J. H., & Koo, H. J. (2024). A Transparent Hydrogel-Ionic Conductor with High Water Retention and Self-Healing Ability. *Materials (Basel)*, *17*(2), 1–12. DOI: 10.3390/ma17020288

Li, C., & Wang, J. (2025). Highly sensitive and stretchable liquid-free ionic conductive elastomers for soft ionotronics. *Reactive & Functional Polymers*, *215*, 106384–106416. DOI: 10.1016/j.reactfunctpolym.2025.106384

Li, X., Fu, M., Wu, J., Zhang, C., Deng, X., Dhinakar, A., Huang, W., Qian, H., & Ge, L. (2017). pH-sensitive peptide hydrogel for glucose-responsive insulin delivery. *Acta Biomaterialia*, *51*, 294–303. DOI: 10.1016/j.actbio.2017.01.016

Li, X., Huang, X., Mutlu, H., Malik, S., & Theato, P. (2020). Conductive hydrogel composites with autonomous self-healing properties. *Soft Matter*, *16*(48), 10969–10976. DOI: 10.1039/D0SM01234C

Li, Y., Tan, S., Zhang, X., Li, Z., Cai, J., & Liu, Y. (2025). Design Strategies and Emerging Applications of Conductive Hydrogels in Wearable Sensing. *Gels (Basel, Switzerland)*, *11*(4), 1–27. DOI: 10.3390/gels11040258

Li, Z., Ranaweera, C. D., Lin, K., Huang, Y., Molley, T. G., Qin, L., Kruzic, J. J., & Kilian, K. A. (2025). Multi-step functionalization of hydrogels through mechano- and photo-responsive linkages. *Materials Horizons*, *12*(9), 3084–3090. DOI: 10.1039/D4MH00761A

Liang, X., Zhang, M., Chong, C. M., Lin, D., Chen, S., Zhen, Y., Ding, H., & Zhong, H. J. (2024). Recent Advances in the 3D Printing of Conductive Hydrogels for Sensor Applications: A Review. *Polymers*, *16*(15), 2131–2157. DOI: 10.3390/polym16152131

Lim, J. Y. C., Lin, Q., Xue, K., & Loh, X. J. (2019). Recent advances in supramolecular hydrogels for biomedical applications. *Materials Today. Advances*, *3*, 100021–100037. DOI: 10.1016/j.mtadv.2019.100021

Lin, Y., Wu, A., Zhang, Y., Duan, H., Zhu, P., & Mao, Y. (2025). Recent progress of nanomaterials-based composite hydrogel sensors for human–machine interactions. *Discover Nano*, *20*(60), 1–30. DOI: 10.1186/s11671-025-04240-8

Lu, C., & Chen, X. (2020). All-Temperature Flexible Supercapacitors Enabled by Antifreezing and Thermally Stable Hydrogel Electrolyte. *Nano Letters*, *20*(3), 1907–1914. DOI: 10.1021/acs.nanolett.9b05148

Lu, C., & Thompson, C. B. (2012). Perspective Metabolic Regulation of Epigenetics. *Cell Metabolism*, *16*(1), 9–17. DOI: 10.1016/j.cmet.2012.06.001

Lv, T., & Suo, L. (2021). Water-in-salt widens the electrochemical stability window: Thermodynamic and kinetic factors. *Current Opinion in Electrochemistry*, *29*, 100818–100824. DOI: 10.1016/j.coelec.2021.100818

Ma, R., & Shi, L. (2014). Phenylboronic acid-based glucose-responsive polymeric nanoparticles: Synthesis and applications in drug delivery. *Polymer Chemistry*, *5*(5), 1503–1518. DOI: 10.1039/C3PY01202F

Maity, N., & Dawn, A. (2020). Conducting Polymer Grafting : Recent and. *Polymers*, *12*(3), 709–731. DOI: 10.3390/polym12030709

Mantha, S., Pillai, S., Khayambashi, P., Upadhyay, A., Zhang, Y., Tao, O., Pham, H. M., & Tran, S. D. (2019). Smart Hydrogels in Tissue Engineering and Regenerative Medicine. *Materials (Basel)*, *12*(20), 3323–3355. DOI: 10.3390/ma12203323

Matsumoto, K., Sakikawa, N., & Miyata, T. (2018). Thermo-responsive gels that absorb moisture and ooze water. *Nature Communications*, *9*(1), 1–7. DOI: 10.1038/s41467-018-04810-8

Melling, D., Martinez, J. G., & Jager, E. W. H. (2019). Conjugated Polymer Actuators and Devices: Progress and Opportunities. *Advanced Materials*, *31*(22), 1–17. DOI: 10.1002/adma.201808210

Méry, A., Rousselot, S., Lepage, D., & Dollé, M. (2021). A Critical Review for an Accurate Electrochemical Stability Composite Electrolytes. *Materials (Basel)*, *14*(14), 3840–3864. DOI: 10.3390/ma14143840

Min, J. H., Patel, M., & Koh, W. G. (2018). Incorporation of conductive materials into hydrogels for tissue engineering applications. *Polymers*, *10*(10), 1–36. DOI: 10.3390/polym10101078

Mir, A., Kumar, A., & Riaz, U. (2022). A short review on the synthesis and advance applications of polyaniline hydrogels. *RSC Advances, 12*(30), 19122–19132. DOI: 10.1039/D2RA02674K

Mirabedini, A., Aziz, S., Spinks, G. M., & Foroughi, J. (2017). Wet-Spun Biofiber for Torsional Artificial Muscles. *Soft Robotics, 4*(4), 421–430. DOI: 10.1089/soro.2016.0057

Morgan, F. L. C., Fernández-Pérez, J., Moroni, L., & Baker, M. B. (2022). Tuning Hydrogels by Mixing Dynamic Cross-Linkers: Enabling Cell-Instructive Hydrogels and Advanced Bioinks. *Advanced Healthcare Materials, 11*(1), 1–15. DOI: 10.1002/adhm.202101576

Muralter, F., Greco, F., & Coclite, A. M. (2020). Applicability of Vapor-Deposited Thermoresponsive Hydrogel Thin Films in Ultrafast Humidity Sensors/Actuators. *ACS Applied Polymer Materials, 2*(3), 1160–1168. DOI: 10.1021/acsapm.9b00957

Muthukutty, B., Kumar, P. S., Vivekanandan, A. K., Sivakumar, M., Lee, S., & Lee, D. (2024). Progress and Perspective in harnessing MXene–carbon-based composites (0–3D): Synthesis, performance, and applications. *Chemosphere, 355*, 141838–141860. DOI: 10.1016/j.chemosphere.2024.141838

Naahidi, S., Jafari, M., Logan, M., Wang, Y., Yuan, Y., Bae, H., Dixon, B., & Chen, P. (2017). Biocompatibility of hydrogel-based scaffolds for tissue engineering applications. *Biotechnology Advances, 35*(5), 530–544. DOI: 10.1016/j.biotechadv.2017.05.006

Naghdi, S., Rhee, K. Y., Hui, D., & Park, S. J. (2018). A review of conductive metal nanomaterials as conductive, transparent, and flexible coatings, thin films, and conductive fillers: Different deposition methods and applications. *Coatings, 8*(8), 278–304. DOI: 10.3390/coatings8080278

Namsheer, K., & Rout, C. S. (2021). Conducting polymers: A comprehensive review on recent advances in synthesis, properties and applications. *RSC Advances, 11*(10), 5659–5697. DOI: 10.1039/D0RA07800J

Neumann, M., di Marco, G., Iudin, D., Viola, M., van Nostrum, C. F., van Ravensteijn, B. G. P., & Vermonden, T. (2023). Stimuli-Responsive Hydrogels: The Dynamic Smart Biomaterials of Tomorrow. *Macromolecules, 56*(21), 8377–8392. DOI: 10.1021/acs.macromol.3c00967

Niu, L., Bai, B., Zhao, X., Zhang, X., & Su, Z. (2025). From nature to nanotechnology: The synergistic integration of biomimetic nanomaterials and conductive hydrogels for next-generation applications. *Nanoscale, 17*(24), 14558–14586. DOI: 10.1039/D5NR01248A

Obrovac, M. N., & Chevrier, V. L. (2014). Alloy negative electrodes for Li-ion batteries. *Chemical Reviews, 114*(23), 11444–11502. DOI: 10.1021/cr500207g

Olabi, A. G., Allam, M. A., Abdelkareem, M. A., Deepa, T. D., Alami, A. H., Abbas, Q., Alkhalidi, A., & Sayed, E. T. (2023). Redox Flow Batteries: Recent Development in Main Components, Emerging Technologies, Diagnostic Techniques, Large-Scale Applications, and Challenges and Barriers. *Batteries, 9*(8), 1–44. DOI: 10.3390/batteries9080409

Park, H., Kim, D.-R., Kim, H.-J., Wang, W., Han, M.-W., & Ahn, S.-H. (2020). Design and Analysis of Artificial Muscle Robotic Elbow Joint Using Shape Memory Alloy Actuator. *International Journal of Precision Engineering and Manufacturing, 21*(2), 249–256. DOI: 10.1007/s12541-019-00240-8

Park, N., & Kim, J. (2020). Hydrogel-Based Artificial Muscles: Overview and Recent Progress. *Advanced Intelligent Systems, 2*(4), 1900135–1900150. DOI: 10.1002/aisy.201900135

Peng, Y., Liu, Y., Yang, Y., Liu, N., Sun, Y., Liu, Y., Pu, H., Xie, S., & Luo, J. (2019). Development of continuum manipulator actuated by thin McKibben pneumatic artificial muscle. *Mechatronics, 60*, 56–65. DOI: 10.1016/j.mechatronics.2019.05.001

Phogat, P., Rawat, S., Dey, S., & Wan, M. (2025). Advancements and challenges in sodium-ion batteries: A comprehensive review of materials, mechanisms, and future directions for sustainable energy storage. *Journal of Alloys and Compounds, 1020*, 179544. DOI: 10.1016/j.jallcom.2025.179544

Phogat, P., Sharma, S., Jha, R., & Singh, S. (2024). Fundamentals of Electrochemistry. In *Electrochemical Devices: Engineering Materials.* (pp. 1–27). Springer Singapore. DOI: 10.1007/978-981-96-0527-9_1

Pinelli, F., Magagnin, L., & Rossi, F. (2020). Progress in hydrogels for sensing applications : A review. *Materials Today. Chemistry, 17*, 100317–100338. DOI: 10.1016/j.mtchem.2020.100317

Protsak, I. S., & Morozov, Y. M. (2025). Fundamentals and Advances in Stimuli-Responsive Hydrogels and Their Applications: A Review. *Gels (Basel, Switzerland), 11*(1), 1–36. DOI: 10.3390/gels11010030

Qiu, Y., Zhang, E., Plamthottam, R., & Pei, Q. (2019). Dielectric Elastomer Artificial Muscle: Materials Innovations and Device Explorations. *Accounts of Chemical Research, 52*(2), 316–325. DOI: 10.1021/acs.accounts.8b00516

Rafieian, S., Mirzadeh, H., Mahdavi, H., & Masoumi, M. E. (2019). A review on nanocomposite hydrogels and their biomedical applications. *Science and Engineering of Composite Materials, 26*(1), 154–174. DOI: 10.1515/secm-2017-0161

Ren, M., Di, J., & Chen, W. (2021). Recent Progress and Application Challenges of Wearable Supercapacitors. *Batteries & Supercaps, 4*(8), 1279–1290. DOI: 10.1002/batt.202000333

Rousse, G., Arroyo-De Dompablo, M. E., Senguttuvan, P., Ponrouch, A., Tarascon, J. M., & Palacín, M. R. (2013). Rationalization of intercalation potential and redox mechanism for A 2Ti3O7 (A = Li, Na). *Chemistry of Materials, 25*(24), 4946–4956. DOI: 10.1021/cm4032336

Sadiq, M., Khan, M. A., Hasan Raza, M. M., Aalam, S. M., Zulfequar, M., & Ali, J. (2022). Enhancement of Electrochemical Stability Window and Electrical Properties of CNT-Based PVA-PEG Polymer Blend Composites. *ACS Omega, 7*(44), 40116–40131. DOI: 10.1021/acsomega.2c04933

Saeidi, M., Chenani, H., Orouji, M., Adel Rastkhiz, M. S., Bolghanabadi, N., Vakili, S., Mohamadnia, Z., Hatamie, A., & Simchi, A. (2023). Electrochemical Wearable Biosensors and Bioelectronic Devices Based on Hydrogels: Mechanical Properties and Electrochemical Behavior. *Biosensors (Basel)*, *13*(8), 823–883. DOI: 10.3390/bios13080823

Saini, P., Choudhary, V., Singh, B. P., Mathur, R. B., & Dhawan, S. K. (2009). Polyaniline-MWCNT nanocomposites for microwave absorption and EMI shielding. *Materials Chemistry and Physics*, *113*(2–3), 919–926. DOI: 10.1016/j.matchemphys.2008.08.065

Sandomierski, M., Jakubowski, M., Gajewski, P., Szcześniak, K., & Voelkel, A. (2024). Hydrogels with Phosphate Groups as Potential Carriers of Bisphosphonates. *Journal of Inorganic and Organometallic Polymers and Materials*, *34*(3), 1293–1300. DOI: 10.1007/s10904-023-02871-x

Sardana, S., Gupta, A., Maan, A. S., Dahiya, S., Singh, K., & Ohlan, A. (2022). Design and synthesis of polyaniline/MWCNT composite hydrogel as a binder-free flexible supercapacitor electrode. *Indian Journal of Physics and Proceedings of the Indian Association for the Cultivation of Science*, *96*(2), 433–439. DOI: 10.1007/s12648-020-01996-w

Selvaraj, M., & Takahata, K. (2020). Electrothermally driven hydrogel-on-flex-circuit actuator for smart steerable catheters. *Micromachines*, *11*(1), 68–81. DOI: 10.3390/mi11010068

Sheng, K., Gao, Y., Bao, T., & Wang, S. (2023). Covalent coating strategy for enhancing the biocompatibility and hemocompatibility of blood-contacting medical materials. *Pharmaceutical Science Advances*, *1*(1), 100001–100011. DOI: 10.1016/j.pscia.2022.100001

Shin, Y., Lee, H. S., Jeong, H., & Kim, D. (2024). Recent advances in conductive hydrogels for soft biointegrated electronics_ Materials, properties, and device applications. *Wearable Electronics*, *1*(November), 255–280. DOI: 10.1016/j.wees.2024.10.004

Singh, A., Sharma, A., Ahmed, A., Sundramoorthy, A. K., Furukawa, H., Arya, S., & Khosla, A. (2021). Recent advances in electrochemical biosensors: Applications, challenges, and future scope. *Biosensors (Basel)*, *11*(9), 1–31. DOI: 10.3390/bios11090336

Summonte, S., Racaniello, G. F., Lopedota, A., Denora, N., & Bernkop-Schnürch, A. (2021). Thiolated polymeric hydrogels for biomedical application: Cross-linking mechanisms. *Journal of Controlled Release*, *330*, 470–482. DOI: 10.1016/j.jconrel.2020.12.037

Sun, Z., Ou, Q., Dong, C., Zhou, J., Hu, H., Li, C., & Huang, Z. (2024). Conducting polymer hydrogels based on supramolecular strategies for wearable sensors. *Exploration*, *4*(5), 1–42. DOI: 10.1002/EXP.20220167

Tang, X., Zhu, S., Ning, J., Yang, X., Hu, M., & Shao, J. (2021). Charge storage mechanisms of manganese dioxide-based supercapacitors : A review. *New Carbon Materials*, *36*(4), 702–710. DOI: 10.1016/S1872-5805(21)60082-3

Tehrani, M. (2021). Advanced Electrical Conductors: An Overview and Prospects of Metal Nanocomposite and Nanocarbon Based Conductors. *Physica Status Solidi (A). Applications and Materials Science*, *218*, 2000704–2000720.

Teunissen, C. E., Verheul, C., & Willemse, E. A. J. (2018). The use of cerebrospinal fluid in biomarker studies. In *Handbook of Clinical Neurology* (1st ed., Vol. 146, Issue 1998). Elsevier B.V.

Tran, V. T., Mredha, M. T. I., & Jeon, I. (2020). High-water-content hydrogels exhibiting superior stiffness, strength, and toughness. *Extreme Mechanics Letters*, *37*, 100691–100697. DOI: 10.1016/j.eml.2020.100691

Tropp, J., Collins, C. P., Xie, X., Daso, R. E., Mehta, A. S., Patel, S. P., Reddy, M. M., Levin, S. E., Sun, C., & Rivnay, J. (2024). Conducting Polymer Nanoparticles with Intrinsic Aqueous Dispersibility for Conductive Hydrogels. *Advanced Materials*, *36*(1), 2306691–2306700. DOI: 10.1002/adma.202306691

Wang, J., Gao, D., & Lee, P. S. (2021). Recent Progress in Artificial Muscles for Interactive Soft Robotics. *Advanced Materials*, *33*(19), 1–20. DOI: 10.1002/adma.202003088

Wang, W., Han, R., Chen, M., & Luo, X. (2021). Antifouling Peptide Hydrogel Based Electrochemical Biosensors for Highly Sensitive Detection of Cancer Biomarker HER2 in Human Serum. *Analytical Chemistry*, *93*(19), 7355–7361. DOI: 10.1021/acs.analchem.1c01350

Wang, W., Wat, E., Hui, P. C. L., Chan, B., Ng, F. S. F., Kan, C. W., Wang, X., Hu, H., Wong, E. C. W., Lau, C. B. S., & Leung, P. C. (2016). Dual-functional transdermal drug delivery system with controllable drug loading based on thermosensitive poloxamer hydrogel for atopic dermatitis treatment. *Scientific Reports*, *6*(1), 1–10. DOI: 10.1038/srep24112

Wang, Z., Cong, Y., & Fu, J. (2020). Stretchable and tough conductive hydrogels for flexible pressure and strain sensors. *Journal of Materials Chemistry. B*, *8*(16), 3437–3459. DOI: 10.1039/C9TB02570G

Wei, C., Chi, Z., Sidi, C., Yichen, D., Ping, Y., & Ximin, H. (2024). Hydrogel-based soft bioelectronics for personalized healthcare. *Med-X*, *2*(20), 1–22.

Wu, L., Hu, X., Qian, J., Pei, F., Wu, F., Mao, R., Ai, X., Yang, H., & Cao, Y. (2014). Sb-C nanofibers with long cycle life as an anode material for high-performance sodium-ion batteries. *Energy & Environmental Science*, *7*(1), 323–328. DOI: 10.1039/C3EE42944J

Wu, Y., Luo, Y., Cuthbert, T. J., Shokurov, A. V., Chu, P. K., Feng, S. P., & Menon, C. (2022). Hydrogels as Soft Ionic Conductors in Flexible and Wearable Triboelectric Nanogenerators. *Advanced Science (Weinheim, Baden-Wurttemberg, Germany)*, *9*(11), 1–24. DOI: 10.1002/advs.202106008

Xia, X., Hao, Q., Lei, W., Wang, W., Sun, D., & Wang, X. (2012). Nanostructured ternary composites of graphene/Fe_2O_3/polyaniline for high-performance supercapacitors. *Journal of Materials Chemistry*, *22*(33), 16844–16850. DOI: 10.1039/c2jm33064d

Xiang, J., Shen, L., & Hong, Y. (2020). Status and future scope of hydrogels in wound healing: Synthesis, materials and evaluation. *European Polymer Journal*, *130*, 109609–109621. DOI: 10.1016/j.eurpolymj.2020.109609

Yammine, P., El-Nakat, H., Kassab, R., Mansour, A., El Khoury, B., Koumeir, D., Matar, Z., & Chmayssem, A. (2024). Recent Advances in Applied Electrochemistry: A Review. *Chemistry (Basel, Switzerland)*, *6*(3), 407–434. DOI: 10.3390/chemistry6030024

Yang, D., Tian, G., Chen, J., Liu, Y., Fatima, E., Qiu, J., Malek, N. A. N. N., & Qi, D. (2025). Neural electrodes for brain-computer interface system: From rigid to soft. *BMEmat*, *12130*(3), 1–21. DOI: 10.1002/bmm2.12130

Yang, G., Takei, T., Zhu, Y., Iranmanesh, E., Liu, B., Li, Z., Wang, J., Hiralal, P., Amaratunga, G. A. J., Fontaine, O., & Zhou, H. (2021). Constructing an efficient conductive network with carbon-based additives in metal hydroxide electrode for high-performance hybrid supercapacitor. *Electrochimica Acta*, *397*, 1–31. DOI: 10.1016/j.electacta.2021.139242

Yang, L., Cheng, S., Wang, J., Ji, X., Jiang, Y., Yao, M., Wu, P., Wang, M., Zhou, J., & Liu, M. (2016). Investigation into the origin of high stability of δ -MnO2 pseudo-capacitive electrode using operando Raman spectroscopy. *Nano Energy*, *30*, 293–302. DOI: 10.1016/j.nanoen.2016.10.018

Yom-Tov, O., Frisman, I., Seliktar, D., & Bianco-Peled, H. (2014). A novel method for hydrogel nano-structuring. *European Polymer Journal*, *52*(1), 137–145. DOI: 10.1016/j.eurpolymj.2014.01.004

Zeng, Y., Han, G., & Yi, Y. (2021). Impact of n-Doping Mechanisms on the Molecular Packing and Electron Mobilities of Molecular Semiconductors for Organic Thermoelectrics. *Organic Materials*, *4*(1), 1–6.

Zhang, Q., Lu, H., Yun, G., Gong, L., Chen, Z., Jin, S., Du, H., Jiang, Z., & Li, W. (2024). A Laminated Gravity-Driven Liquid Metal-Doped Hydrogel of Unparalleled Toughness and Conductivity. *Advanced Functional Materials*, *34*(31), 1–11. DOI: 10.1002/adfm.202308113

Zhang, Q., Yu, Z., Du, P., & Su, C. (2010). Carbon Nanomaterials Used as Conductive Additives in Lithium Ion Batteries. *Recent Patents on Nanotechnology*, *4*(2), 100–110. DOI: 10.2174/187221010791208803

Zhang, S., Zhao, B., Zhang, D., Yang, M., Huang, X., Han, L., Chen, K., Li, X., Pang, R., Shang, Y., & Cao, A. (2023). Conductive hydrogels incorporating carbon nanoparticles: A review of synthesis, performance and applications. *Particuology*, *83*, 212–231. DOI: 10.1016/j.partic.2023.06.002

Zhang, X., Chen, L., Lim, K. H., Gonuguntla, S., Lim, K. W., Pranantyo, D., Yong, W. P., Yam, W. J. T., Low, Z., Teo, W. J., Nien, H. P., Loh, Q. W., & Soh, S. (2019). The Pathway to Intelligence: Using Stimuli-Responsive Materials as Building Blocks for Constructing Smart and Functional Systems. *Advanced Materials*, *31*(11), 1–48. DOI: 10.1002/adma.201804540

Zhang, X., Xu, X., Chen, L., Zhang, C., & Liao, L. (2020). Multi-responsive hydrogel actuator with photo-switchable color changing behaviors. *Dyes and Pigments*, *174*, 108042–108050. DOI: 10.1016/j.dyepig.2019.108042

Chapter 7
Super Porous Hydrogels

Irum Fatima
https://orcid.org/0009-0008-4202-4170
University of Wah, Pakistan

Humaira Razzaq
University of Wah, Pakistan

Shazia Naheed
University of Wah, Pakistan

Zobiya Asghar
University of Wah, Pakistan

ABSTRACT

To overcome the hindrance in the rate of swelling and mechanical strength of hydrogels for different targeted applications that require rapid fluid uptake. Superporous hydrogels have emerged as a new class of materials to hinder these challenges. This chapter will explore the fabrication of SPHs, and various synthetic techniques such as porosigen, phase separation, cross-linking, and gas-blowing methods. The main features of SPHs are high porosity, rapid rate of swelling, higher adsorption capacity, stimuli-responsive behavior to the external environment, biocompatibility, high surface area, and enhanced mass transfer. The SPHs are categorized into three generations: conventional SPHs, SPH composites, and SPH hybrids, with improved mechanical properties and functionalities across various generations. Various applications of SPHs in biotechnology and biomedical engineering are discussed. SPHs have a lot of potential to revolutionize different fields due to their unique properties and tailored functionalities.

7.1 SUPER POROUS HYDROGEL PRODUCTION

Unraveling the enigmatic nature of hydrogel involves a captivating property where it swells upon immersion in water, showcasing a moderate yet intriguing mechanical robustness. Nonetheless, the current swelling index and mechanical resilience of the materials prove inadequate in enabling rapid and decisive swelling, a quality crucial in numerous scenarios. For certain

DOI: 10.4018/979-8-3373-2337-4.ch007

applications, a hydrogel that can swiftly absorb water and experience an accelerated swelling rate becomes a highly sought-after asset. Consequently,

a groundbreaking generation of hydrogels (Karoyo & Wilson, 2021) has been unveiled, boasting the exceptional ability to imbibe water swiftly, and undergo swelling at an accelerated pace (Karoyo & Wilson, 2021). Super porous hydrogels (SPHs) are three-dimensional hydrophilic structures involving interconnected micropores exhibiting better elastic features, mechanical strength, and water conservation (Patel et al., 2011). Large amounts of liquid can be absorbed fast because of the closeness of interconnected tiny holes (E. M. Ahmed, 2015). The average size of the pores in SPHs is 200 mm (Ahmed et al., 2014), and the maximum swelling often occurs in a fraction of a minute (Alsaid et al., 2021), created by the interaction of non-organic solvents. In contrast to regular hydrogels such as acrylamide and cellulosic by-products, SPH has an exceptionally increased surface area and inter/intra particle voids. This property allows SPH to store large amounts of liquid and swell, thus doubling its surface area. In addition, these hydrogels are evaluated in terms of their compatibility and biodegradability. These characteristics allow them to be used in pharmaceutical and other fields related to material science research. Drugs that need to be delivered to the duodenum or lower gastrointestinal tract have the largest drug absorption window only in that region (Biswal et al., 2020). They swell completely within minutes regardless of their size due to the absorption of water by capillary force rather than by simple absorption (Buwalda et al., 2014). Drugs have good expansion ability and can maximize residence time in the stomach (Chavda et al., 2011).

Figure 1. Superporous Hydrogel with their formula

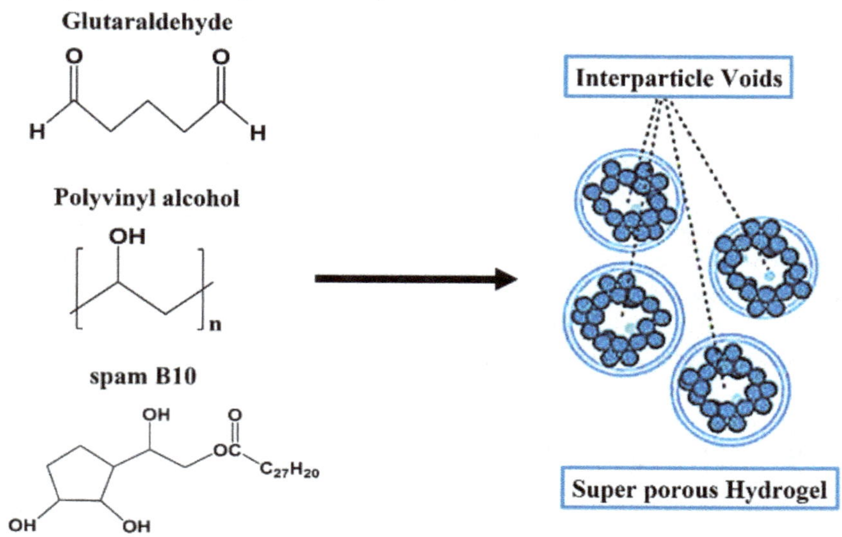

7.1.1 Synthetic Steps

The remarkable permeability of superporous hydrogels offers a multitude of benefits. It facilitates the efficient diffusion of vital nutrients, gases, and waste materials, rendering it an ideal option for tissue engineering frameworks where cells necessitate an ample supply of oxygen and essential elements (Chavda et al., 2012) gently combined with acetic, acrylic, and hydrochloric acids. The production of super porous hydrogel entails employing several distinct methodologies, each contributing to its unique characteristics. These methods encompass the following;

1. Porosigens Technique
2. Phase Separation Technique
3. Cross-Linking Technique
4. Gas Blowing Technique

1. Porosignes Technique

One popular way for creating super porous hydrogels (SPHs), which are three-dimensional networks of polymers with interconnected pores that promote quick swelling

and water absorption is the porosigen technique. Control over pore size and distribution is essential for customizing the SPHs' mechanical characteristics and swelling kinetics for particular uses. In the porosigen technique, hydrogels are prepared in the presence of dispersed water-soluble porosigens. These porosigens can include micronized sucrose (Chavda, 2012), sodium chloride (Chen et al., 1999), and PEG (Chordiya et al., 2013). The crucial aspect of these porosigens is that they may be later removed by washing with water, leaving behind a meshwork that forms the porous hydrogels. Interestingly, a polymer network is formed where water itself can serve as a porosigen in the frozen state (Das, 2013). This unique property expands the options for creating porous hydrogels using this technique. The pore size of the hydrogels, which are prepared using the porosigen technique, is directly influenced by the size of the porosigens utilized during the process (De France et al., 2018).

2. Phase separation Technique

By taking advantage of the thermodynamic incompatibility between various polymer components or between a polymer and a solvent, which produces distinct phases during polymerization, the phase separation approach creates superporous hydrogels (SPHs). The concept of the phase separation technique revolves around the reduction in solvent quality for polymers (Deepthi et al., 2011).

3. Cross-Linking Technique

A key technique for creating super porous hydrogels (SPHs) is cross-linking, in which chains of polymers are joined chemically or physically to create a three-dimensional network with strong water absorption and retention properties. Chemical and physical cross-linking are the two primary categories of cross-linking. Through cross-linking chemicals like glutaraldehyde, divinyl sulfone, or N, N'-methylene bisacrylamide (MBA) covalent connections are created between polymer chains in chemical cross-linking. By creating a permanent and stable network structure, this technique improves the hydrogel's mechanical strength and resistance to swelling. An alternative approach is to produce porous hydrogels by crosslinking individual hydrogel particles, forming crosslinked aggregates of these particles (El Mekawy et al., 2021). In this approach, the pores are situated between these hydrogel particles, with their size significantly smaller than the particles themselves. However, it's essential to note that this technique is limited to absorbent particles having chemically active functional groups on their surface (Feng et al., 2025).

4. Gas Blowing Technique

The gas-blowing technique begins by adding monomers, cross-linking agents, foam stabilizers, and distilled water into a test tube with specific dimensions. The pH is then adjusted to a range of 5 to 6 using 5M NaOH (Feng et al., 2025). The blowing agent, also known as the foaming agent, plays a pivotal role in the foaming process by facilitating the formation of cellular structures within the polymer matrix (Foudazi et al., 2023). After the synthesis, the superporous hydrogels undergo a thorough washing and drying process, preparing them for their intended applications.

The synthetic procedures of superporous hydrogels are elucidated, highlighting the congruity in their synthesis (Ghasemiyeh & Mohammadi-Samani, 2019, Ahmed et al., 2014). These hydrogels possess unparalleled structural properties that render them appealing for applications necessitating swift fluid absorption, controlled discharge of bioactive substances, as well as proficient transportation of nutrients and waste materials (Gupta & Shivakumar, 2012).The figure illustrates the stepwise synthesis process of Superporous Hydrogels (SPH) incorporating Superabsorbent Polymer (SAP) characteristics. The synthesis begins with the addition of a foaming aid to stabilize the foam structure, followed by a crosslinker that forms a three-dimensional polymer network. Neutralization of the monomer (often acrylic acid) is carried out to maintain the pH suitable for polymerization. The monomer dilution step ensures appropriate viscosity and concentration. A redox initiator system is used, where an oxidant (like ammonium persulfate) and a reductant (such as TEMED) initiate rapid polymerization. Simultaneously, bicarbonate is added to release CO_2 gas, creating an interconnected porous structure. This integrated process results in SPHs that possess rapid swelling, high water absorption, and interconnected porosity, making them ideal for applications in drug delivery, tissue engineering, and hygiene products.

Figure 2. Synthetic Steps for SPH

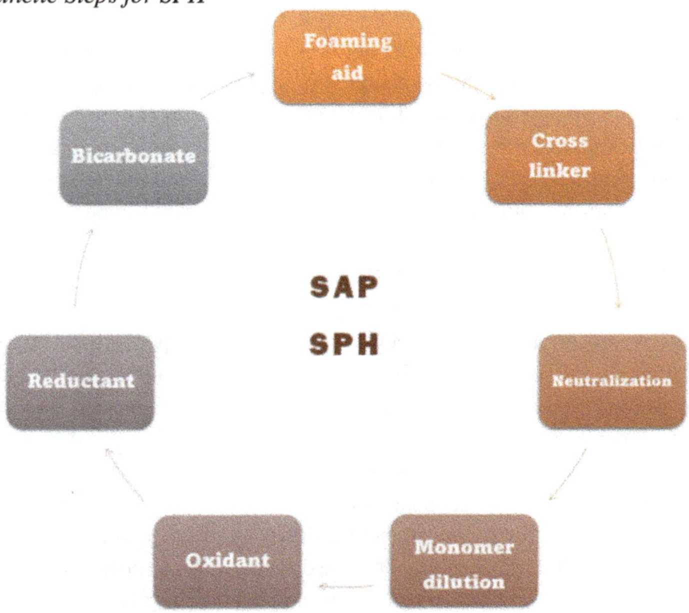

7.2 CHARACTERISTICS OF SPHS

Superporous hydrogels are a class of hydrogel materials with unique properties and characteristics. Here are some of the main features of superporous hydrogels.

Figure 3. Structure of Porous Hydrogels

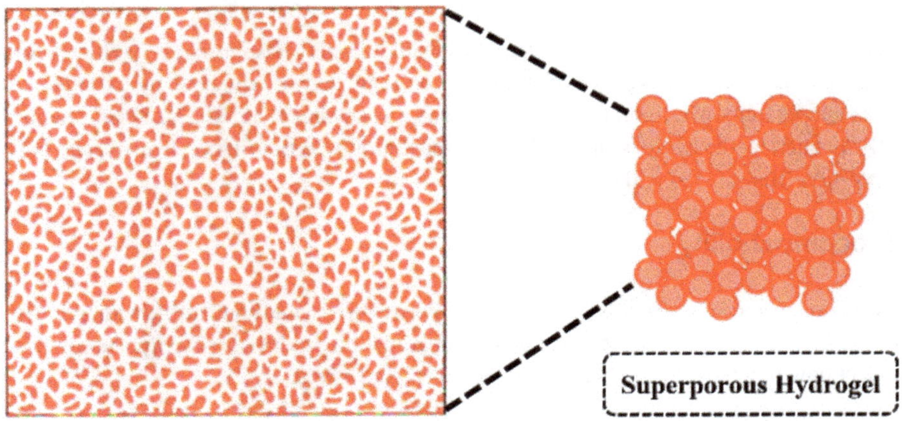

1. High Porosity

High porosity, or the existence of a vast network of interconnected pores inside the hydrogel matrix, is one of the most distinctive features of superporous hydrogels (SPHs). SPHs are extremely effective in applications including medication administration, wound healing, and tissue engineering because of their interconnected pore network, which makes it easier for water molecules and nutrients to travel. The interconnected pores present in hydrogels enable them to absorb and retain substantial quantities of water and other liquid substances (Han et al., 2003).

2. Rapid Swelling and High Absorption Capacity

In situations where immediate water uptake is required, super porous hydrogels (SPHs) outperform ordinary hydrogels due to their high absorption capacity and quick swelling. SPHs' interconnected porous structure enables rapid water penetration through capillary action as opposed to typical hydrogels' principal method of polymer relaxation and diffusion. Because of this, SPHs can absorb water in a matter of seconds to minutes, as opposed to hours for traditional hydrogels to reach equilibrium swelling. Due to their highly porous construction, which increases the surface area available for water interaction, SPHs exhibit a significantly higher swelling ratio. Additionally, cross-linking density, pore size distribution, and polymer composition affect SPHs' high absorption capacity. This advantageous property finds valuable applications in areas like drug delivery, tissue engineering, and wound healing, where a high absorption capacity is essential (Hentze & Antonietti, 2002, Ahmed, 2015).

3. Mechanical Stability

The ability of superporous hydrogels (SPHs) to retain structural integrity in the face of mechanical stress, swelling, and repeated deformation is determined by their mechanical stability. They can maintain structural integrity under external forces; this unique quality makes them appropriate for load-bearing purposes and as scaffolds for tissue regeneration (Ijaz et al., 2025).

4. Stimuli-Responsive Behavior

Superporous hydrogels can be designed to exhibit stimuli-responsive behaviour, allowing their properties and swelling characteristics to be regulated or triggered by external stimuli such as temperature, pH, light, or electrical signals (Issarachot et al., 2023). SPHs' reactivity is mostly dictated by the hydrogel network's functional groups, cross-linking density, and polymer composition (Jhawat et al., 2020).

5. Biocompatibility

Many superporous hydrogels are biocompatible, non-toxic, and compatible with living tissue. Surface chemistry, mechanical characteristics, degradation products, and polymer composition all affect how biocompatible SPHs as individuals (Jyotish et al., 2023).

6. High Surface Area

Superporous hydrogels possess a porous structure and therefore have a large surface area compared to their volume. This large surface area provides ample space for interactions with surrounding materials, making them suitable for applications such as adsorption, separation, and catalysis (Kabiri et al., 2003).

7. Enhanced Mass Transfer

One essential feature that sets superporous hydrogels (SPHs) apart from traditional hydrogels is enhanced mass transfer. Due mainly to the existence of an interconnected porosity network, this feature describes the effective and quick movement of fluids, nutrients, medications, or biomolecules across the hydrogel matrix. The large pores and cross-linked structure of SPH improve mass transfer properties. This property is beneficial in applications such as drug delivery systems where efficient diffusion of therapeutic agents is critical to therapeutic agent efficacy (Kabiri & Zohuriaan-Mehr, 2004).

Superporous hydrogels (SPHs) exhibit rapid swelling due to their interconnected porous structure. They possess high water absorption capacity and excellent fluid transport properties, enabling quick uptake and retention of liquids. Their mechanical strength and responsiveness make them suitable for biomedical and industrial applications.

7.3 GENERATIONS OF SPH

The superporous hydrogels have been classified into three generations and are explained as follows;

1. The first generation SPH named as Conventional SPH
2. Second generation SPH named as SPH Composite
3. Third generation named as super porous hydrogels hybrid

Figure 4. Generations of Superporous hydrogels

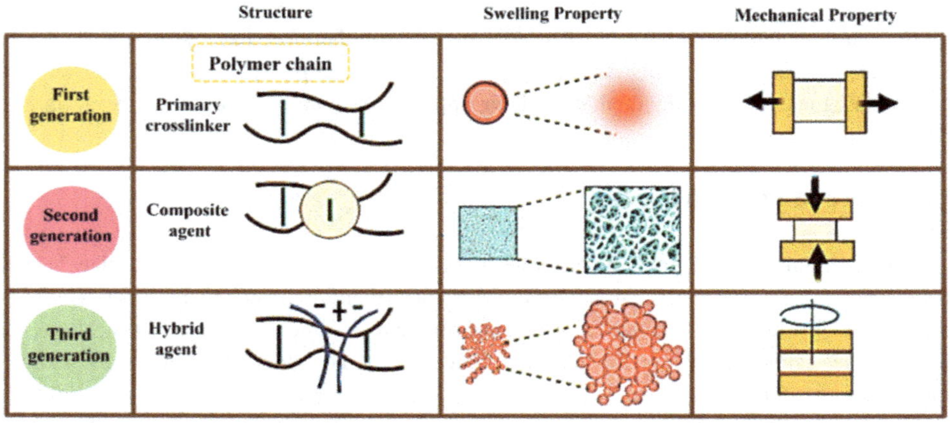

7.3.1 First Generation SPHs

The initial SPHs were developed by Chen et al., who pioneered their production in 1999 (Chen et al., 1999). Their groundbreaking creation boasted rapid swelling abilities and exceptional absorbency, achieved through polymerization and crosslinking with diverse vinyl monomers. To enhance performance, the SPHs necessitated the inclusion of foaming agents, foam stabilisers, and foaming aids, alongside wetting agents to accelerate water absorption to under one minute (Karoyo & Wilson, 2021). However, issues arose during the drying phase where the porous morphology would collapse due to the surface tension of water, causing the constriction of polymer chains (Kashyap et al., 2005). This predicament was efficiently resolved by substituting water with alcohol, counteracting the surface tension, and preserving the porous structure during drying. The first-generation superporous hydrogels demonstrated applications across diverse fields, including tissue-engineering scaffolds, where their high porosity facilitated the transport of nutrients and oxygen, as well as cellular infiltration and proliferation. Furthermore, these hydrogels proved indispensable in drug delivery systems due to their extensive surface area and porosity, which enable the controlled release and encapsulation of therapeutic agents. Presently, researchers are exploring the potential of superporous hydrogels in wound dressings, biosensors, and environmental initiatives (Kashyap et al., 2005).

7.3.2 Second-Generation SPHs (SPH Composite)

The second generation of SPH composites (SPHC) exhibits several key characteristics, including rapid swelling, moderate swelling ratio, and enhanced mechanical properties. A matrix swelling additive, swellable filler, or composite agent is utilized to produce SPH composites. In this composite, a continuous phase acts as the matrix, with a dispersed phase incorporated within it (Kaur et al., 2022). SPH composites exhibit the ability to withstand compression forces of up to 2 N (Keskar et al., 2009).

To fabricate these remarkable composites, the same building blocks—monomers, cross linkers, and initiator systems—are employed as in SPHs, supplemented by the presence of swellable fillers. Filters, in the form of cross-linked water-absorbing hydrophilic polymers, are dispersed into the reaction mixture, subsequently swelling and absorbing a mixture of monomer, cross-linker, initiator, and water-soluble blowing additive. The localized swellings of the fillers contribute to physical cross-linking, thereby facilitating polymer chain formation and an accompanying rise in solution pH during the polymerization process. Each binder or swollen filler can be regarded as an individual reactor throughout polymerization, culminating in a robust connection between composite drug particles via polymer chain links. To the second-generation superporous hydrogels, researchers have devoted efforts to refining control over pore structure, size distribution, and stimulus-response characteristics.

To achieve this, cutting-edge synthetic techniques like templated synthesis, electrospinning, and 3D printing are employed, yielding superior control over the structure and functionality of hydrogels (Khullar et al., 1998). With substantially greater strength, second-generation hydrogels can offer the application necessary for gastric retention.

7.3.3 Third-Generation SPHs (SPH Hybrid)

The next iteration of superporous hydrogels (SPH), known as the third generation, represents a significant improvement over its predecessor. It was developed by combining the properties of SPH hybrids to synthesize SPH with exceptional mechanical and elastic attributes. The second-generation SPHs introduced precross-linked matrix-swelling additives, while the SPH hybrids incorporated hybridizing agents that could be crosslinked post-SPH formation. The hybridizing agent, often a water-soluble or water-dispersible polymer, can create cross-linked structures through chemical or physical cross-linking, resulting in a second network resembling an interpenetrating polymer network. Examples of these hybrid active ingredients include sodium alginate, pectin, chitosan, or synthetic water-soluble hydrophilic polymers like poly(vinyl alcohol) (Kuang et al., 2011).

In 2003 Hossein Omidian conducted a fascinating experiment involving superporous hydrogel hybrids. Researchers utilized acrylamide and methylene bisacrylamide as the monomer and cross-linker, respectively. These hybrids possessed remarkable mechanical and elastic properties. They were later regarded as the third-generation SPH hybrids (SPHH), known for their exceptional mechanical characteristics. Introducing a hybrid agent led to the formation of a cross-linked structure in the SPH. This hybrid agent, a water-soluble or water-dispersible polymer, played a crucial role in creating a crosslinked structure, somewhat similar to an interpenetrating network (IPN) formed through chemical or physical cross-linking (Kumar & Philip, 2007). As a result of this interpenetrating network, these SPH hybrids were also given the name SPH-IPNs.

It is important to note that each hybrid agent may require specific treatment, depending on its type and associated properties. This allowed for the creating various

third-generation SPHs, ranging from those with high modulus to others with high elastic and rubber properties when in their water-swollen states. Some examples of hybrid agents include polysaccharides like sodium alginate, pectin, chitosan, or synthetic water-soluble hydrophilic polymers such as poly(vinyl alcohol) (Kumari et al., 2020). Interestingly, natural hydrocolloids undergo ionotropic gelation when treated with metal ions like calcium or iron. For instance, sodium alginate can form a gel with Ca^{+2} ions, while chitosan does the same with phosphates. This unique property of SPH hybrids is that their gels exhibit remarkable elasticity in their swollen state. In comparison to conventional SPHs and SPH composites, SPH hybrids are far less prone to breaking when stretched due to their exceptional elastic and rubber-like properties (Kumi et al., 2024). These characteristics make SPH hybrids the preferred choice for various applications where resilient gels are necessary. These versatile SPH hybrids can effectively withstand different types of stresses, including tension, compression, bending, and twisting, making them ideal for a wide range of uses. To illustrate, the synthesis of acrylamide-based SPH in the presence of sodium alginate resulted in an SPH hybrid of alginate polyacrylamide capable of enduring compression forces of up to 25 N (Ladola, 2014).

Super-porous hybrid gels (SPHH), or third-generation hydrogels, have been at the forefront of creating superior mechanical stretching capabilities. A distinguishing feature of the third-generation SPHs is the inclusion of nanoparticles or Nano fillers within the hydrogel structure. These nanoparticles serve various purposes, such as drug encapsulation and release, delivery of bioactive molecules, sensing, imaging, and catalysis. For instance, magnetic nanoparticles are incorporated to facilitate targeted drug delivery, while gold nanoparticles enable optical sensing or imaging (Ladola, 2014).

The generations of SPHs evolved to improve structure and functionality.

- First-generation SPHs had fast swelling but weak mechanical strength.
- Second-generation SPHs included composite materials for enhanced strength.
- Third-generation SPHs introduced interpenetrating polymer networks (IPNs) for improved elasticity, durability, and responsiveness to stimuli like pH or temperature.

7.4 APPLICATIONS OF SPHS IN DIFFERENT FIELDS

Superporous hydrogels are a class of materials that possess high porosity and swelling capacity. These hydrogels have numerous applications in various fields due to their exceptional properties. Fig. 5 shows notable applications of superporous hydrogels:

Figure 5. Applications of superporous hydrogels

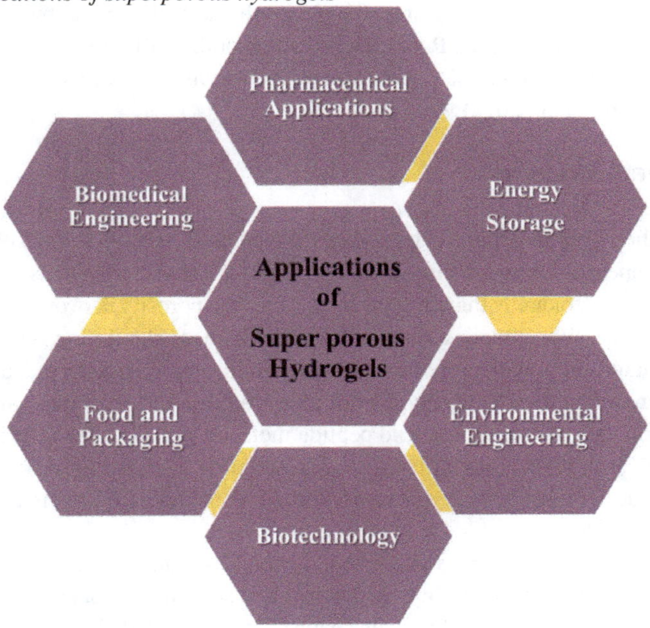

7.4.1 Biomedical Engineering

Superporous hydrogels have extensive applications in biomedical engineering. It can be used as a scaffold for tissue engineering and regenerative medicine. The high porosity of these hydrogels enables efficient cell infiltration, nutrient diffusion, and waste removal, promoting tissue growth and regeneration. In addition, bioactive molecules such as growth factors and drugs for controlled release can be incorporated, making them valuable for drug delivery systems and wound healing applications (Ma et al., 2023).

Superporous hydrogels have several important applications in biomedical engineering. Here are some specific areas where they are used.

1. Tissue Engineering and Regenerative Medicine

Superporous hydrogels serve as scaffolds for tissue engineering and regenerative medicine applications. Their high porosity and interconnected pore structure allow for efficient cell infiltration, nutrient diffusion, and waste removal. This promotes cell attachment, proliferation, and tissue formation. Superporous hydrogels can be seeded with cells and implanted into damaged or diseased tissues to facilitate tissue regeneration and repair. Superporous hydrogels (SPHs) have become very successful biomaterials for tissue engineering and regenerative medicine because of their improved mass transfer, high porosity, quick swelling, and biocompatibility. Because of these characteristics, SPHs are perfect for simulating the extracellular matrix (ECM) and creating an environment that is conducive to cell adhesion, proliferation, and differentiation. SPHs' extremely porous structure ensures tissue growth and cell survival by facilitating effective nutrition and oxygen diffusion. Researchers can customize SPHs for particular tissue types, including bone, cartilage, skin, and neural tissues, by varying the pore size and cross-linking density (Mastropietro et al., 2012).

2. Drug Delivery Systems

Because of their high absorption capacity, variable porosity, stimuli-responsive behavior, and quick swelling, superporous hydrogels (SPHs) are perfect materials for systems that deliver drugs.

Drugs can diffuse gradually thanks to the interconnected porosity network of SPHs, allowing for prolonged release. When treating chronic diseases where regular dosing is impractical this is especially advantageous. To guarantee targeted medication release at the intended location, SPHs can react to environmental stimuli including pH, temperature, strength of ions, and enzymes. For instance, oral drug delivery of protein and peptide molecules is made possible by pH-sensitive SPHs, which release medications in the gastrointestinal tract. Super-absorbent hydrogels are utilized in pharmaceutical conveyance systems for their remarkable capacity to expand and retain copious amounts of fluid.

These hydrogels can encompass and liberate beneficial substances, such as medications, proteins, growth factors, and genes, in a methodical and sustained manner. Figure. 5 shows the SPH-loaded capsule and its drug delivery mechanism at the targeted place. The intricate porous structure of these hydrogels allows for a substantial loading capacity of drugs, while the release rate can be precisely tailored by modifying the gel composition or the cross-linking intensity (Mastropietro et al., 2012).

Figure 6. Mechanism of drug Delivery through Superporous hydrogels

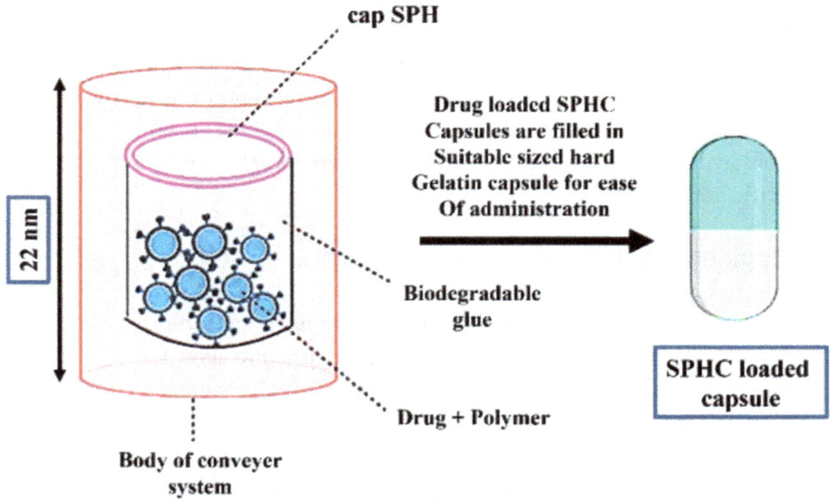

3. Wound Healing

Because of their high retention of water capacity, biological compatibility, mechanical strength, and potential to produce a moist healing environment, superporous hydrogels (SPHs) are essential in wound healing applications. Because of this, they work especially well for burns, diabetic ulcers, and chronic wounds. Antibiotics, growth hormones, antimicrobial peptides, or anti-inflammatory medications can be impregnated into SPHs to provide targeted and prolonged drug release at the wound site. This aids in lowering inflammation, controlling infections, and encouraging tissue regeneration and cell division. Superporous hydrogels find applications in wound healing due to their moisture retaining and biocompatible properties. They can create a moist environment that promotes wound healing by providing a barrier against infection, facilitating the migration of cells involved in the healing process, and accelerating tissue regeneration. Superporous hydrogels can be loaded with bioactive agents to enhance wound healing, such as growth factors, antimicrobial agents, or anti-inflammatory drugs (Mattiasson et al., 2009).

4. Biosensors and Diagnostic Devices

Superporous hydrogels (SPHs) have been extensively investigated in biosensors and diagnostic devices because of their large surface area, rapid swelling behavior, stimuli-responsiveness, and outstanding biocompatibility. These characteristics enable SPHs to function as efficient biorecognition platforms for the high sensitivity and specificity detection of biomolecules, pathogens, and environmental contaminants. The interconnected porous structure of SPHs improves the efficiency of biosensing reactions, allowing for the rapid and accurate detection of glucose, proteins, nucleic acids, and disease biomarkers in biological samples. SPHs can be designed to react to environmental stimuli like pH, temperature, ionic strength, or specific biomolecular interactions; for instance, pH-sensitive SPHs can be used for real-time wound infection monitoring, while glucose-responsive SPHs are used in diabetic patients' continuous glucose monitoring. Superporous hydrogels are used in the development of biosensors and diagnostic devices. The extensive porosity of hydrogels enables the incorporation of sensing elements, such as enzymes, antibodies, or nanoparticles, into their matrix. These sensing elements can selectively bind to target analytes, initiating a measurable response. Superporous hydrogels can be used as matrices for immobilizing biomolecules and facilitating the detection of various analytes, including glucose, cholesterol, DNA, and proteins (Mehta et al., 2018).

5. Organ-on-a-Chip System

In organ-on-a-chip (OoC) systems micro-engineered platforms that mimic the physiological activities of human organs superporous hydrogels (SPHs) are being investigated more. These methods offer a more accurate substitute for conventional 2D cell cultures and animal testing in drug screening, disease modeling, and customized medicine. Because of its high porosity, biomimetic qualities, adjustable mechanical strength, and improved mass transfer capabilities, SPHs are perfect for OoC applications. Cell adhesion, proliferation, and differentiation are facilitated by the linked porosity network of SPHs, which closely matches the natural extracellular matrix of human tissues. Because of this, they can be used to create microenvironments that support the functions of particular organs, liver, kidney, and lung-on-a-chip models. Superporous hydrogels find application in the fabrication of organ-on-a-chip systems, microfluidic devices designed to replicate the structure and function of human organs. These hydrogels serve as scaffolds, mimicking the extracellular matrix (ECM) and offering a three-dimensional environment for cells. The porosity of super porous hydrogels allows for the perfusion of nutrients, oxygen, and waste removal in organ-on-a-chip systems, enabling the study of organ-level functionality, drug testing, and disease modeling (Nagpal et al., 2013).

6. Implantable Devices and Prosthetics

The development of implanted medical devices and prosthetics is paying a lot of attention to superporous hydrogels (SPHs) because of their high porosity, mechanical stability, biocompatibility, and capacity to integrate with biological tissues. Because of these characteristics, SPHs are ideal for applications requiring tissue integration, long-term implantation, and regulated medication release inside the body. SPHs have mechanical tunability, which enables them to closely resemble the flexibility of biological tissues in contrast to rigid synthetic implants. This lowers the possibility of rejection and discomfort. They are therefore perfect for soft-tissue implants like breast implants, vascular grafts, and

artificial cartilage. Superporous hydrogels have potential applications in the development of implantable devices and prosthetics. Their high porosity and swelling capacity enable them to mimic the properties of natural tissues. These hydrogels can be used as coatings or fillers in implants or prosthetic devices to enhance biocompatibility, reduce inflammation, and promote tissue integration (Omidian et al., 2007).

These are just a few examples of the applications of superporous hydrogels in biomedical engineering. Ongoing research and advancements in this field continue to explore new possibilities and optimize the performance of these hydrogels for various biomedical applications.

Superporous hydrogels (SPHs) have diverse applications due to their rapid swelling and high porosity. They are used in drug delivery systems for controlled and targeted release. SPHs are also applied in wound dressings, tissue engineering, biosensors, and agriculture for soil moisture retention and nutrient delivery.

7.4.2 Pharmaceutical Applications

1. Development of Gastric Retention Devices

Gastrointestinal maintenance devices have proven to be highly valuable in efficiently delivering pharmaceuticals. Over the past two decades, significant endeavors have been undertaken to determine the optimal dosage form to withstand the challenging conditions of gastric purging, by ensuring that the medication remains within the stomach and does not pass through the pylorus. The primary objective in developing these devices is to create an oral formulation that swells rapidly upon contact with gastric fluids, effectively preventing its passage through the pyloric sphincter. These devices are particularly advantageous for the localized delivery of drugs targeting stomach-related conditions, such as bacterial ulcers, where medications such as antacids and antibiotics are most effective. They also serve as an effective vehicle for drugs that require prolonged retention within the stomach or are susceptible to degradation in the colon due to alkaline pH. However, it should be acknowledged that extended gastric retention devices are not essential for drugs primarily absorbed in the colon, as they maintain therapeutic blood levels for up to 24 hours. Likewise, these devices are unsuitable for drugs and are unstable in the presence of an acidic pH, such as aspirin and non-steroidal anti-inflammatory drugs (Omidian et al., 2005).

2. Development of Fast-Dissolving Tablets

One significant advantage of quick-dissolving tablet advancements is the ease of administering dosage forms without water or swallowing, making it particularly useful for youngsters and the old. Three different approaches have been developed since the initial success of quick-dissolving tablets: freeze-drying, sublimation or thermal compression, and direct compression. Using freeze-drying, tablets that dissolve within 5 seconds can be produced, while sublimation or thermal compression allows for tablets that dissolve within 15 seconds. However, these two methods are costly, and the resulting tablets are quite fragile. As a result, direct compression was developed, which is less expensive and results in tablets with excellent physical strength.

With this method, the tablets disintegrate within 10 seconds due to rapid water uptake into the tablet's core. The direct compression approach involves incorporating fine particles of superporous hydrogels into the formulation alongside other excipients (Oxley et al., 1993).

3. Development of Diet Aid

SPHs can hypothetically involve a noteworthy parcel of the stomach space because of their quick and expansive enlarging, taking off less space for sustenance, and subsequently smothering cravings. This kind of structure can help to shed pounds on hefty individuals. Keeping up the insight and volume of the enlarged SPH for an extensive timeframe is a significant test of SPHs as a weight misfortune help. Creating unused sedate convey and advances and using them in improvement is basic for drug organizations to outlast. Superporous hydrogels represent a modern class of hydrogel materials that, despite their unique size, exhibit rapid and extensive swelling. Several generations of SPHs have been developed to meet specific drug application needs, including gastric retention. Studies have shown that certain SPH formulations hold promise for challenging applications where exceptional swelling and mechanical properties are required in harsh swelling environments. Additionally, the potential use of these SPHs in solid and semi-solid oral dosage forms has been explored. Preclinical studies have evaluated the safety and effectiveness of specific SPH formulations, paving the way for further development of these materials for drug, food, and biomedical applications. Superporous hydrogels can be made flexible, thereby minimizing their susceptibility to breakage. Various drug and biomedical applications of superporous hydrogels have been developed, and some products are currently in the development stage. The unique properties of superporous hydrogels also hold potential for non-pharmaceutical and non-biomedical applications (Park et al., 2006).

4. Advancements in Occlusion Devices for Chemoembolization Therapy

A combination of chemotherapy and embolization may be called chemoembolization. By reducing the oxygen supply to the growing tumors, embolization has been used to treat

cancer. The systemic toxic quality can be achieved when the embolic technique is coupled with chemotherapeutic agents. In chemoembolization therapy, superporous hydrogels (SPHs) can be loaded with a combination of a chemotherapeutic agent and an anti-angiogenic agent. The rapid and extensive swelling capability of SPHs has been harnessed for this previously unexplored application. Utilizing non-invasive imaging techniques to predict the position and shape of an aneurysm, custom-shaped superporous hydrogels with reduced dimensions can be created. When a highly porous hydrogel is positioned near an aneurysm, it swiftly expands to fill the space, inducing blood clot formation.

Research has demonstrated that superporous hydrogels have effectively prevented aneurysms in up to 95% of cases, without compromising the parent blood vessel or causing adverse reactions. The Hydrocoil is an advanced occlusion device that combines platinum coils with superporous hydrogel, offering a modern approach to aneurysm treatment. Additionally, superporous hydrogels can be engineered to release bioactive substances, enhancing or delaying blood coagulation as needed (Park, 2002).

5. Biomedical Applications

SPHs and SPH composites find diverse applications in the biomedical field, enabling the creation of various biomedical devices. These devices include artificial pancreas, artificial cornea, synthetic skin, articular cartilage, soft tissue substitutes, cell growth substrates for tissue engineering, burn dressings, surgical breast augmentation, and hemoperfusion for uremia and blood detoxification treatment. In cell transplantation and tissue engineering, the ability of superporous hydrogels to facilitate vascular ingrowth is valuable. Furthermore, bone tissue engineering, hydroxyapatite-containing superporous hydrogel composites, and novel poly (2-hydroxyethyl methacrylate) superporous hydrogel frameworks are utilized. These advancements in SPH technology offer a wide range of possibilities in the field of biomedical applications (Park et al., 2019).

7.4.3 Biotechnology

Superporous hydrogels are utilized in different biotechnological applications. They can serve as bolsters for immobilizing proteins or cells in biocatalysis and bioremediation forms. The tall porosity encourages quick mass exchange, empowering proficient substrate transport and item recuperation. In addition, these hydrogels can be functionalized with particular ligands for filtration or chromatography, permitting for specific partition and decontamination of biomolecules. Superporous hydrogels have a few vital applications in biotechnology (Patel et al., 2011). Here are a few particular ranges where they are utilized:

1. Enzyme Immobilization

Because they offer a highly porous and hydrated matrix that improves enzyme stability, reusability, and catalytic effectiveness, superporous hydrogels (SPHs) are widely used for enzyme immobilization. High enzyme loading is made possible by the huge surface area and interconnected pore structure of SPHs, which also preserve ideal substrate and product mass transfer. Because of this, SPH-based enzyme immobilization is extremely useful in applications related to biotechnology, medicine, food processing, and the environment. Enzymes in SPHs that have been immobilized are shielded from denaturation, deterioration, and activity loss brought on by extremes in temperature, pH, and solvents. This greatly lowers process costs in industrial biocatalysis by extending the lifespan of enzymes and allowing for their recurrent usage. Superporous hydrogels can be utilized as underpins for immobilizing proteins in bio-catalysis forms. The tall porosity and expansive surface region of these hydrogels give adequate space for chemical stacking. Immobilization of chemicals on superporous hydrogels improves their steadiness, reusability, and catalytic effectiveness. This empowers the advancement of biocatalysts with moved-forward execution for different mechanical forms, such as the generation of biofuels, pharmaceutical intermediates, and strength chemicals (Patel et al., 2011).

2. Bioremediation

Superporous hydrogels (SPHs) are used more in bioremediation, cleaning the environment with biological agents. Adsorption, breakdown, and detoxification of environmental pollutants in water, soil, and air purification systems, are extremely effective due to their high porosity, quick swelling, and capacity to trap enzymes and microorganisms. Lead (Pb^{2+}), cadmium (Cd^{2+}), mercury (Hg^{2+}), and arsenic (As^{3+}) are among the hazardous heavy metals that can be efficiently removed from industrial effluent by SPHs functionalized with chelating substances, ion-exchange groups, or bioactive compounds. Their quick swelling and large surface area enable effective metal ion absorption, lowering environmental toxicity. Superporous hydrogels discover applications in bioremediation forms for the poison's expulsion from soil, water, or discussion. They can be functionalized with particular ligands or bioactive compounds to specifically adsorb and trap contaminants, such as overwhelming metals, natural compounds, or colors. The tall porosity and swelling capacity of these hydrogels empower effective poison take-up and maintenance, contributing to successful natural cleanup and contamination control (Pati et al., 2016).

3. Bioprocessing and Bioproduction

Acting as very effective scaffolds, biocatalytic supports, and bioreactors for medicinal products, biofuels, and industrially relevant biomolecules. Superporous hydrogels (SPHs)

are transforming bioprocessing and bioproduction. In biotechnological applications, its high porosity, quick swelling, and capacity to encapsulate enzymes, cells, and microbes allow for improved bioconversion efficiency, mass transfer, and scalability.

In the bioprocessing industry, SPHs act as stable enzyme carriers, enabling continuous biocatalysis with enhanced durability of enzymes and reusability. Due to their high catalytic efficiency and extended operating life, immobilized enzymes in SPHs are widely utilized in pharmaceuticals, biofuel, and food. Superporous hydrogels are used in different bioprocessing and bio-production applications. They can be utilized as frameworks for microbial cell immobilization, permitting the generation of profitable metabolites, proteins, or bioactive compounds.

The permeable structure of these hydrogels encourages the dissemination of supplements and gases, guaranteeing ideal cell development and product arrangement. Superporous hydrogels are especially valuable in nonstop bioprocessing frameworks, empowering long-term and steady bio-production (Prinderre et al., 2011).

4. Partiality Decontamination and Chromatography

Superporous hydrogels (SPHs) are used extensively in chromatography and partial decontamination because of their high porosity, quick swelling, and superior adsorption ability. Because of these characteristics, SPHs are very useful in the biochemical, pharmaceutical, and environmental industries for the selective separation, purification, and elimination of pollutants or biomolecules from complicated mixtures. Functionalized SPHs have hydrophobic domains, metal-binding ligands, or charged groups that effectively remove organic contaminants, heavy metals, and hazardous materials from industrial waste and water. Before additional treatment, this enables partial decontamination, lowering contaminant concentrations. Superporous hydrogels can be functionalized with particular ligands, such as antibodies or affinity labels, affinity filtration, and chromatography forms. The high porosity and expansive surface region of these hydrogels give inexhaustible authoritative locales for target atoms. They can specifically capture and filter biomolecules, such as proteins or nucleic acids, from complex blends. Superporous hydrogels offer preferences for official capacity, refinement speed, and versatility compared to conventional chromatographic strategies (Sannino et al., 2006).

5. Biosensing and Biomedical Diagnostics

Superporous hydrogels (SPHs) are used extensively in chromatography and partial decontamination because of their high porosity, quick swelling, and superior adsorption ability. Because of these characteristics, SPHs are very useful in the biochemical, pharmaceutical, and environmental industries for the selective separation, purification, and elimination of pollutants or biomolecules from complicated mixtures.

Proteins, enzymes, and nucleic acids are separated and purified using SPHs as stationary phases in chromatography columns. Enhancing binding efficiency and resolution of their large surface area and adjustable functional groups allows targeted interactions (with target molecules). Superporous hydrogels discover applications in bio-sensing and biomedical diagnostics. can be utilized as frameworks for immobilizing detecting components, such as proteins, antibodies, or nanoparticles, in biosensor gadgets. The tall porosity of these hydrogels permits for proficient mass transport and fast analyte dissemination, empowering delicate and quick discovery of target analytes. Superporous hydrogels are utilized in different bio-sensing stages, counting electrochemical, optical, and microfluidic frameworks, for applications such as illness diagnostics, natural observing, and nourishment security testing. These are a few of the applications of superporous hydrogels in biotechnology. The one-of-a-kind properties of these hydrogels make them important devices for upgrading bioprocessing, bioremediation, biosensing, and decontamination forms, opening up modern conceivable outcomes in biotechnology research and applications (Sikder et al., 2021).

6. Affinity Purification and Chromatography

Specific ligands functionalized on Superporous hydrogels such as antibodies or affinity tags, for affinity purification and chromatography processes. The high porosity and large surface area of these hydrogels provide abundant binding sites for target molecules. They can selectively capture and purify biomolecules, such as proteins or nucleic acids, from complex mixtures.

Superporous hydrogels offer advantages in terms of binding capacity, purification speed, and scalability compared to traditional chromatographic methods (Kumar et al., 2020). These are some of the applications of superporous hydrogels in biotechnology.

The exceptional characteristics of these hydrogels make them valuable tools for enhancing bioprocessing, bioremediation, biosensing, and purification processes, opening up new possibilities in biotechnology research and applications (Spiller et al., 2008).

7.4.4 Environmental Engineering

Superporous hydrogels have applications in natural building and contamination control. They can be used for poison elimination from wastewater, such as overwhelming metals, colors, or natural compounds. The tall porosity and swelling capacity empower compelling adsorption and maintenance of contaminants, giving a potential arrangement for water treatment and remediation forms (Stamatialis et al., 2008). Superporous hydrogels possess applications in natural design especially valuable in water treatment and remediation challenges. Here are a few particular applications of superporous hydrogels in this field:

1. Water Treatment

Superporous hydrogels can be utilized for poison removal from wastewater. The tall porosity and swelling capacity empower proficient adsorption and maintenance of contaminants, such as overwhelming metals, colors, natural compounds, and microorganisms. Superporous hydrogels can be custom-fitted with particular utilitarian bunches or ligands to specifically tie and evacuate target toxins from water, making strides in water quality and decreasing natural contamination (Tambe & Bachhav, 2015).

2. Soil Remediation

Superporous hydrogels discover applications in soil remediation forms. They can be utilized to immobilize or sequester contaminants display in soil, anticipating their filtering into groundwater or take-up by plants. Superporous hydrogels can typify toxins, such as overwhelming metals or natural compounds, and diminish their versatility, permitting for

more viable remediation procedures. Also, these hydrogels can give a controlled discharge of corrections or supplements to improve soil fertility and encourage plant development in polluted ranges (Tang et al., 2005).

3. Ground water Remediation

Superporous hydrogels are utilized in groundwater remediation forms to expel or relieve contaminants shown in underground water sources. They are utilized in penetrable receptive obstructions or groundwater treatment frameworks to trap and adsorb toxins, anticipating their relocation and advanced spread. The tall porosity of superporous hydrogels permits tall contaminant adsorption capacity and productive water stream, making them compelling in tending to groundwater defilement issues (Tellez et al., 1998).

4. Natural Detecting and Observing

Superporous hydrogels discover applications in natural detecting and checking frameworks. They can be utilized as networks for immobilizing detecting components, such as nanoparticles or proteins, in natural sensors. The expansive surface range and tall porosity of these hydrogels permit proficient analyte dissemination and interaction with the detecting components. Superporous hydrogels can be utilized in sensor stages for identifying different natural parameters, such as pH, overwhelming metal concentrations, or toxin levels in soil or water (Thakar et al., 2019).

5. Oil Spill Cleanup

Superporous hydrogels have potential applications in oil spill cleanup operations. They can be utilized to assimilate and contain oil spills in water bodies. The tall porosity and swelling capacity of hydrogels empower them to quickly assimilate and hold expansive amounts of oil. Superporous hydrogels can be sent as coasting absorbents or consolidated into booms or obstructions to anticipate the spread of oil and encourage its expulsion from water surfaces. These are a few of the applications of superporous hydrogels in natural buildings. Hydrogels having one of their properties, make them profitable instruments for water treatment, soil and groundwater remediation, natural detection, and moderation of natural contamination. Continuous investigation proceeds to investigate better approaches to optimize the execution of superporous hydrogels for natural applications (Thakar et al., 2019).

7.4.5 Agriculture

Superporous hydrogels can be used in agriculture to improve water retention and nutrient availability in soil. By incorporating these hydrogels into the soil, they can absorb and store water and slowly release it to plants over time. This saves water and prevents water loss through leaching and evaporation. Additionally, these hydrogels can be loaded with fertilizers and plant growth regulators, allowing for controlled release and enhanced uptake of nutrients by plants (Tolga Demirtaş et al., 2008).

1. Energy Storage

Superporous hydrogels have shown promise in energy storage applications, especially supercapacitors. The hydrogel's high porosity and surface area provide an ideal environment for charge storage. Their use as electrodes or electrolyte matrices enables efficient ion transport and improves the performance of supercapacitor devices. Superporous [Kumari et al., 2020] hydrogels have shown potential applications in various fields, including agriculture. These highly absorbent and highly porous materials can be used in a variety of ways to improve agricultural practices (Venkatesh & Radha, 2021).

2. Soil Moisture Management

Superporous hydrogels are used as water retention agents in arid or drought-prone areas. They absorb and store large amounts of water and slowly release it into plant roots. This feature helps maintain optimal soil moisture levels, reduces water stress on plants, and increases plant growth and yield (Verma et al., 2021).

3. Nutrient Delivery

Hydrogels can be loaded with fertilizers, pesticides, or growth promoters. When hydrogels are introduced into the soil, they gradually release the encapsulated substances, ensuring a controlled and sustainable supply of nutrients to plants. This increases nutrient uptake efficiency, reduces fertilizer leaching, and minimizes environmental impact (Yin et al., 2007).

4. Seed Coating

Superporous hydrogels have the potential to serve as seed coatings, enhancing germination rates and promoting seedling formation. By encapsulating seeds in hydrogels, water, and essential nutrients can be delivered directly to the seeds, promoting germination and early growth. This approach helps increase the success rate of seedling formation, especially under unfavorable soil conditions (Zaman et al., 2015).

5. Erosion Control

Hydrogels can be used for soil erosion in erosion-prone areas preventions, slopes, and bare ground. When mixed with soil or sprayed onto surfaces hydrogels can stabilize soil structure, prevent water runoff, and improve soil retention. This reduces soil erosion from wind and water and encourages vegetation growth. In addition to it, it is highly recommended for soil remediation (Park et al., 2006).

6. Controlled-Release of Agrochemicals

Superporous hydrogels can function as carriers for controlled-release formulations of pesticides, including herbicides and insecticides. The hydrogel matrix encapsulates the active ingredients and releases them gradually over time. This approach ensures sustained and targeted release of pesticides reduces negative environmental impacts and optimizes pesticide effectiveness (Park et al., 2019). These are just a few possible applications of superporous hydrogels in agriculture. Ongoing research and development in this area should reveal further innovative uses of these materials to improve crop production, water management, and sustainable agricultural practices (Bhowmik et al., 2024).

7.4.6 Food and Packaging

Superporous hydrogels find significant applications in food science and the packaging industry. They serve as absorbents, effectively extending the shelf life of food by preventing spoilage caused by moisture sensitive reactions and microbial growth.

Additionally, these hydrogels act as carriers for flavor encapsulation, controlled release of food additives, and edible coatings, preserving food quality and freshness. Importantly, the applications of super porous hydrogels are not limited to the fields mentioned above, and ongoing research and development continuously explore new potential uses for these versatile materials (Zhao et al., 2022). In the realm of food packaging, superporous hydrogels have garnered attention due to their unique properties. Below are some examples of how super porous hydrogels can be applied to food packaging.

1. Moisture Control

Superporous hydrogels can absorb and retain water, thus acting as a barrier against water migration. Incorporation of these hydrogels into food packaging materials can maintain desirable moisture levels and prevent moisture loss or gain, thereby extending the shelf life of fresh produce (Wang et al., 2025).

2. Gas Barrier

Hydrogels can be designed with gas barrier properties such as controlled oxygen and carbon dioxide permeability. Incorporating super porous hydrogels into food packaging films or coatings can modulate gas exchange between the packaged food and the external environment, thereby minimizing oxidation and spoilage of sensitive foods (Kumar et al., 2020).

3. Antimicrobial Properties

Superporous hydrogels can be modified to have antimicrobial properties that can inhibit the growth of bacteria and other microorganisms. Incorporation of these hydrogels into food packaging materials can prevent microbial contamination and extend the shelf life of packaged foods (Barik et al., 2024).

4. Active Ingredient Delivery

Hydrogels can be utilized as carriers and active ingredients such as antioxidants and antimicrobials. These active ingredients are encapsulated in a hydrogel matrix and slowly released over time. Incorporation of such hydrogels into food packaging can provide

positive protection against oxidation and microbial growth, thereby improving the quality and safety of the packaged food (Tian & Liu, 2020).

5. Sensing and Monitoring

Superporous hydrogels can be designed to exhibit responsive behavior to specific environmental conditions, such as variations in temperature, pH, or humidity. By incorporating these hydrogels into food packaging, they act as sensors to monitor the freshness and quality of packaged foods. This helps identify and report deterioration and spoilage, ensuring consumer safety. While super porous hydrogels are promising for food packaging applications, further research and development are needed to address challenges regarding their scalability, cost-effectiveness, and compatibility with various types of food. Undoubtedly, these materials hold great promise in enhancing the performance of food packaging, prolonging shelf life, and advancing food safety (Wang et al., 2023).

CONCLUSION

In conclusion, superporous hydrogels (SPHs) represent a remarkable advancement in hydrogel technology, offering a unique combination of rapid swelling, high water uptake, and interconnected porous architecture that distinguishes them from conventional hydrogels. Through well-defined synthetic strategies involving foaming agents, crosslinkers, and gas-generating components, SPHs are engineered to possess superior responsiveness and mechanical integrity. Their progressive development across generations has further enhanced their structural and functional capabilities, enabling a wide spectrum of advanced applications. From controlled drug delivery and wound healing to tissue engineering, biosensing, and agricultural enhancements, SPHs have demonstrated exceptional versatility and efficiency. This chapter has provided a comprehensive overview of their synthesis, properties, and multifaceted applications, underscoring their immense potential in biomedical, environmental, and industrial domains, and paving the way for future innovations in smart hydrogel systems. Superporous hydrogels, as a novel and distinct class of hydrogels, have gained significant prominence in various fields, irrespective of their initial size, which can expand to a larger volume due to their porous structure. Different generations of superporous hydrogels have been developed to address diverse needs across various fields, including pharmaceuticals, agriculture, and daily usage. Among various applications, superporous hydrogels have found remarkable utility in drug delivery due to their adjustable swelling characteristics. Researchers have reported that these hydrogels can be made elastic to minimize the risk of rupture. With various synthetic approaches and ongoing product development, the unique attributes of super porous hydrogels extend beyond pharmaceutical and biomedical applications.

REFERENCES

Ahmed, E. M. (2015). Hydrogel: Preparation, characterization, and applications: A review. *Journal of Advanced Research, 6*(2), 105–121. DOI: 10.1016/j.jare.2013.07.006 PMID: 25750745

Ahmed, J. A., Chatterjee, A., Chauhan, B. S., Jaimini, M., & Varshney, H. M. (2014). A conceptual overview of superporous hydrogels. *composites, 6*, 9.

Alsaid, Y., Wu, S., Wu, D., Du, Y., Shi, L., Khodambashi, R., & Zhao, Y. (2021). Tunable sponge-like hierarchically porous hydrogels with simultaneously enhanced diffusivity and mechanical properties. *Advanced Materials, 33*(20), 2008235. DOI: 10.1002/adma.202008235 PMID: 33829563

Barik, M., BhagyaRaj, G. V. S., Dash, K. K., & Shams, R. (2024). A thorough evaluation of chitosan-based packaging film and coating for food product shelf-life extension. *Journal of Agriculture and Food Research, 16*, 101164. DOI: 10.1016/j.jafr.2024.101164

Bhowmik, S., Agyei, D., & Ali, A. (2024). Biodegradable chitosan hydrogel film incorporated with polyvinyl alcohol, chitooligosaccharides, and gallic acid for potential application in food packaging. *Cellulose (London, England), 31*(13), 8087–8103. DOI: 10.1007/s10570-024-06080-8

BISWAL, P., Ray, B., & Panda, B. (2020). HYDROGEL: OVERVIEW AND RECENT ASPECTS.

Buwalda, S. J., Boere, K. W., Dijkstra, P. J., Feijen, J., Vermonden, T., & Hennink, W. E. (2014). Hydrogels in a historical perspective: From simple networks to smart materials. *Journal of Controlled Release, 190*, 254–273. DOI: 10.1016/j.jconrel.2014.03.052 PMID: 24746623

Chavda, H., & Patel, C. (2011). Effect of crosslinker concentration on characteristics of superporous hydrogel. *International journal of pharmaceutical investigation, 1*(1), 17.

Chavda, H. V., Patel, R. D., Modhia, I. P., & Patel, C. N. (2012). Preparation and characterization of superporous hydrogel based on different polymers. *International Journal of Pharmaceutical Investigation, 2*(3), 134. DOI: 10.4103/2230-973X.104396 PMID: 23373004

Chen, J., Park, H., & Park, K. (1999). Synthesis of superporous hydrogels: Hydrogels with fast swelling and superabsorbent properties. *Journal of Biomedical Materials Research: An Official Journal of The Society for Biomaterials, The Japanese Society for Biomaterials, and the Australian Society for Biomaterials, 44*(1), 53-62.

Chordiya, M., Senthilkumaran, K., & Gangurde, H. (2013). SUPER POROUS HYDROGELS: A RECENT ADVANCEMENT IN GASTRORETENTIVE DRUG DELIVERY SYSTEM. *Indonesian Journal of Pharmacy/Majalah Farmasi Indonesia, 24*(2).

Das, N. (2013). Preparation methods and properties of hydrogel: A review. *International Journal of Pharmacy and Pharmaceutical Sciences, 5*(3), 112–117.

De France, K. J., Xu, F., & Hoare, T. (2018). Structured macroporous hydrogels: Progress, challenges, and opportunities. *Advanced Healthcare Materials, 7*(1), 1700927. DOI: 10.1002/adhm.201700927 PMID: 29195022

Deepthi, B., Varun, D., Gopal, P., Rao, C., & Sumalatha, G. (2011). Super porous hydrogels–Supreme drug delivery. *Research Journal of Pharmacy and Technology*, *4*(8), 1182–1188.

El-Mekawy, R. E., Elhady, H. A., & Al-Shareef, H. F. (2021). Highly stretchable, smooth, and biodegradable hydrogel films based on chitosan as safety food packaging. *Polymers & Polymer Composites*, *29*(6), 563–573. DOI: 10.1177/0967391120926892

Feng, H., Lv, Y., Liu, Y., Hao, J., Li, F., Chen, N., & Qiu, J. (2025). Synthesis and Characterization of Cationic Modified Starch Grafted Acrylic Acid-Based Absorbent Resin Dust Suppressant. *Environmental Research*, *275*(10), 121147. PMID: 40057107

Foudazi, R., Zowada, R., Manas-Zloczower, I., & Feke, D. L. (2023). Porous hydrogels: Present challenges and future opportunities. *Langmuir*, *39*(6), 2092–2111. DOI: 10.1021/acs.langmuir.2c02253 PMID: 36719086

Ghasemiyeh, P., & Mohammadi-Samani, S. (2019). Hydrogels as drug delivery systems; pros and cons. *Trends in Pharmacological Sciences*, *5*(1), 7–24.

Gupta, N. V., & Shivakumar, H. (2012). Investigation of swelling behavior and mechanical properties of a pH-sensitive superporous hydrogel composite. *Iranian Journal of Pharmaceutical Research : IJPR*, *11*(2), 481. PMID: 24250471

Han, W., Omidian, H., & Rocca, J. (2003). In vivo and in vitro studies on novel gastroretentive superporous hydrogel (SPH) platforms. American Association of Pharmaceutical Scientists (AAPS), Salt Lake City, Utah, USA, October 26–30, 2003, Published in. *AAPS PharmSci*, *5*, W4119.

Hentze, H.-P., & Antonietti, M. (2002). Porous polymers and resins for biotechnological and biomedical applications. *Reviews in Molecular Biotechnology*, *90*(1), 27–53. DOI: 10.1016/S1389-0352(01)00046-0 PMID: 12069045

Ijaz, H., Mahmood, A., Tayyab, M., Sarfraz, R. M., Akram, M. R., Haroon, B., & Albadi, F. O. (2025). Synthesis, characterizations, and evaluation of crosslinked 5-fluorouracil-loaded thiolated chitosan-co-polymethacrylic acid hydrogel for colorectal cancer-targeted drug delivery. *Polymer Bulletin*, 1–39.

Issarachot, O., Bunlung, S., Kaewkroek, K., & Wiwattanapatapee, R. (2023). Superporous hydrogels based on blends of chitosan and polyvinyl alcohol as a carrier for enhanced gastric delivery of resveratrol. *Saudi Pharmaceutical Journal : SPJ : the Official Publication of the Saudi Pharmaceutical Society*, *31*(3), 335–347. DOI: 10.1016/j.jsps.2023.01.001 PMID: 37026050

Jhawat, V., Gulia, M., Maddiboyina, B., Dutt, R., & Gupta, S. (2020). Fate and applications of superporous hydrogel systems: a review. *Current Nanomedicine (Formerly: Recent Patents on Nanomedicine)*, *10*(4), 326-341.

Jyotish, J., Nayak, R., Tripathy, D., Moharana, S., & Mahaling, R. (2023). *Starch-based superabsorbent polymer Bio-based Superabsorbents: Recent Trends, Types, Applications and Recycling*. Springer.

Kabiri, K., Omidian, H., & Zohuriaan-Mehr, M. (2003). Novel approach to highly porous superabsorbent hydrogels: Synergistic effect of porogens on porosity and swelling rate. *Polymer International*, *52*(7), 1158–1164. DOI: 10.1002/pi.1218

Kabiri, K., & Zohuriaan-Mehr, M. J. (2004). Porous superabsorbent hydrogel composites: Synthesis, morphology and swelling rate. *Macromolecular Materials and Engineering*, *289*(7), 653–661. DOI: 10.1002/mame.200400010

Karoyo, A., & Wilson, L. D. (2021). A review on the design and hydration properties of natural polymer-based hydrogels. *Materials (Basel)*, *14*(5), 1095. DOI: 10.3390/ma14051095 PMID: 33652859

Kashyap, N., Kumar, N., & Kumar, M. R. (2005). Hydrogels for pharmaceutical and biomedical applications. *Critical Reviews™ in Therapeutic Drug Carrier Systems, 22*(2).

Kaur, G., Kapoor, S., & Gandhi, N. (2022). *Foods with Edible Coatings: Science, Shelf Life, and Quality Shelf Life and Food Safety*. CRC Press. DOI: 10.1201/9781003091677-17

Keskar, V., Marion, N. W., Mao, J. J., & Gemeinhart, R. A. (2009). In vitro evaluation of macroporous hydrogels to facilitate stem cell infiltration, growth, and mineralization. *Tissue Engineering. Part A*, *15*(7), 1695–1707. DOI: 10.1089/ten.tea.2008.0238 PMID: 19119921

Khullar, P., Khar, R. K., & Agarwal, S. P. (1998). Evaluation of hydrogel-based controlled-release niacin tablets. *Drug Development and Industrial Pharmacy*, *24*(5), 479–483. DOI: 10.3109/03639049809085646 PMID: 9876611

Kuang, J., Yuk, K. Y., & Huh, K. M. (2011). Polysaccharide-based superporous hydrogels with fast swelling and superabsorbent properties. *Carbohydrate Polymers*, *83*(1), 284–290. DOI: 10.1016/j.carbpol.2010.07.052 PMID: 34662962

Kumar, R., & Philip, A. (2007). Gastroretentive dosage forms for prolonging gastric residence time. *International Journal of Pharmaceutical Medicine*, *21*(2), 157–171. DOI: 10.2165/00124363-200721020-00005

Kumar, S., Mukherjee, A., & Dutta, J. (2020). Chitosan based nanocomposite films and coatings: Emerging antimicrobial food packaging alternatives. *Trends in Food Science & Technology*, *97*, 196–209. DOI: 10.1016/j.tifs.2020.01.002

Kumari, P. K., Sharmila, M., & Rao, Y. S. (2020). Super porous hydrogels: A review. *Journal of Pharmaceutical Research International*, *32*, 153–165. DOI: 10.9734/jpri/2020/v32i1330595

Kumi, M., Ejeromedoghene, O., Sudane, W. D., & Zhang, Z. (2024). Unlocking the biological response of smart Stimuli-Responsive hydrogels and their application in biological systems. *European Polymer Journal*, *209*, 112906. DOI: 10.1016/j.eurpolymj.2024.112906

Ladola, M. K. (2014). Superporous hydrogel (SPH): A novel and advanced technique of oral controlled release drug delivery system. *World Journal of Pharmaceutical Research*, *3*(5), 297–319.

Ma, L., Long, T., Yuan, S., Qi, P., Han, L., & Hao, J. (2023). A pH-indicating smart tag based on porous hydrogel as food freshness sensors. *Journal of Colloid and Interface Science*, *647*, 32–42. DOI: 10.1016/j.jcis.2023.05.145 PMID: 37244174

Mastropietro, D. J., Omidian, H., & Park, K. (2012). Drug delivery applications for superporous hydrogels. *Expert Opinion on Drug Delivery*, *9*(1), 71–89. DOI: 10.1517/17425247.2012.641950 PMID: 22145909

Mattiasson, B., Kumar, A., & Galeaev, I. Y. (2009). *Macroporous polymers: production properties and biotechnological/biomedical applications*. CRC Press. DOI: 10.1201/9781420084627

Mehta, M., Pandey, P., Mahajan, S., & Satija, S. (2018). Gastro retentive drug delivery systems: An overview. *Research Journal of Pharmacy and Technology*, *11*(5), 2157–2160. DOI: 10.5958/0974-360X.2018.00398.0

Nagpal, M., Singh, S. K., & Mishra, D. (2013). Superporous hybrid hydrogels based on polyacrylamide and chitosan: Characterization and in vitro drug release. *International Journal of Pharmaceutical Investigation*, *3*(2), 88. DOI: 10.4103/2230-973X.114906 PMID: 24015380

Omidian, H., Park, K., & Rocca, J. G. (2007). Recent developments in superporous hydrogels. *The Journal of Pharmacy and Pharmacology*, *59*(3), 317–327. DOI: 10.1211/jpp.59.3.0001 PMID: 17331335

Omidian, H., Rocca, J. G., & Park, K. (2005). Advances in superporous hydrogels. *Journal of Controlled Release*, *102*(1), 3–12. DOI: 10.1016/j.jconrel.2004.09.028 PMID: 15653129

Oxley, H., Corkhill, P., Fitton, J., & Tighe, B. (1993). Macroporous hydrogels for biomedical applications: Methodology and morphology. *Biomaterials*, *14*(14), 1064–1072. DOI: 10.1016/0142-9612(93)90207-I PMID: 8312461

Park, H., Park, K., & Kim, D. (2006). Preparation and swelling behavior of chitosan-based superporous hydrogels for gastric retention application. *Journal of Biomedical Materials Research. Part A*, *76*(1), 144–150. DOI: 10.1002/jbm.a.30533 PMID: 16258961

Park, K. (2002). Superporous hydrogels for pharmaceutical & other applications. *Drug. Deliver. Technol*, *38*, 40–44.

Park, K., Chen, J., & Park, H. (2019). *Superporous hydrogel composites: a new generation of hydrogels with fast swelling kinetics, high swelling ratio and high mechanical strength Polymeric Drugs and Drug Delivery Systems*. CRC Press.

Patel, Alpesh, Mequanint, & Kibret. (2011). Hydrogel biomaterials *Biomedical engineering-frontiers and challenges*: IntechOpen.

Pati, N. B., Velivela, S., Mayasa, V., & Gupta, V. (2016). Gastroretentive superporous hydrogel tablets of dexlansoprazole. *International Journal of Pharmaceutical Sciences and Research*, *7*(11), 4678.

Prinderre, P., Sauzet, C., & Fuxen, C. (2011). Advances in gastro retentive drug-delivery systems. *Expert Opinion on Drug Delivery*, *8*(9), 1189–1203. DOI: 10.1517/17425247.2011.592828 PMID: 21671821

Sannino, A., Netti, P. A., Madaghiele, M., Coccoli, V., Luciani, A., Maffezzoli, A., & Nicolais, L. (2006). Synthesis and characterization of macroporous poly (ethylene glycol)-based hydrogels for tissue engineering application. *Journal of Biomedical Materials Research. Part A*, *79*(2), 229–236. DOI: 10.1002/jbm.a.30780 PMID: 16752396

Sikder, A., Pearce, A. K., Parkinson, S. J., Napier, R., & O'Reilly, R. K. (2021). Recent trends in advanced polymer materials in agriculture related applications. *ACS Applied Polymer Materials*, *3*(3), 1203–1217. DOI: 10.1021/acsapm.0c00982

Spiller, K. L., Laurencin, S. J., Charlton, D., Maher, S. A., & Lowman, A. M. (2008). Superporous hydrogels for cartilage repair: Evaluation of the morphological and mechanical properties. *Acta Biomaterialia*, *4*(1), 17–25. DOI: 10.1016/j.actbio.2007.09.001 PMID: 18029236

Stamatialis, D. F., Papenburg, B. J., Gironés, M., Saiful, S., Bettahalli, S. N., Schmitmeier, S., & Wessling, M. (2008). Medical applications of membranes: Drug delivery, artificial organs and tissue engineering. *Journal of Membrane Science*, *308*(1-2), 1–34. DOI: 10.1016/j.memsci.2007.09.059

Tambe, S. R., & Bachhav, L. P. D. D. G. (2015). Formulation development and characterization of superporous hydrogel of itopride hydrochloride for gastroretentive drug delivery system. *World Journal of Pharmaceutical Research*, *4*(2), 1552–1573.

Tang, C., Yin, C., Pei, Y., Zhang, M., & Wu, L. (2005). New superporous hydrogels composites based on aqueous Carbopol® solution (SPHCcs): Synthesis, characterization and in vitro bioadhesive force studies. *European Polymer Journal*, *41*(3), 557–562. DOI: 10.1016/j.eurpolymj.2004.10.017

Tellez, C., Benson Iii, A. B., Lyster, M. T., Talamonti, M., Shaw, J., Braun, M. A., & Vogelzang, R. L. (1998). Phase II trial of chemoembolization for the treatment of metastatic colorectal carcinoma to the liver and review of the literature. *Cancer*, *82*(7), 1250–1259. DOI: 10.1002/(SICI)1097-0142(19980401)82:7<1250::AID-CNCR7>3.0.CO;2-J PMID: 9529016

Thakar, H., Sebastian, S. M., Mandal, S., Pople, A., Agarwal, G., & Srivastava, A. (2019). Biomolecule-conjugated macroporous hydrogels for biomedical applications. *ACS Biomaterials Science & Engineering*, *5*(12), 6320–6341. DOI: 10.1021/acsbiomaterials.9b00778 PMID: 33417808

Tian, B., & Liu, Y. (2020). Chitosan-based biomaterials: From discovery to food application. *Polymers for Advanced Technologies*, *31*(11), 2408–2421. DOI: 10.1002/pat.5010

Tolga Demirtaş, T., Karakeçili, A. G., & Gümüşderelioğlu, M. (2008). Hydroxyapatite containing superporous hydrogel composites: Synthesis and in-vitro characterization. *Journal of Materials Science. Materials in Medicine*, *19*(2), 729–735. DOI: 10.1007/s10856-007-3008-7 PMID: 17619984

Venkatesh, C. R., & Radha, G. V. (2021). Formulation and evaluation of venlafaxine hydrochloride superporous hydrogels using factorial design method.

Verma, N. K., Singh, A. K., Yadav, V., Singh, P., Yadav, A., & Jaiswal, S. (2021). Super Porous Hydrogel Based Drug Delivery System: A Review. *South Asian Res. J. Pharm. Sci, 3*(6), 8.

Wang, D., Zhu, C., Yang, Q., Xu, Y., Zhang, D., Wang, D., Liu, F., & Hou, C. (2025). Stretchable, controlled release of active substances, and biodegradable chitosan-polyvinyl alcohol hydrogel film for antibacterial and chilled meat preservation. *Food Chemistry, 477*, 143608. DOI: 10.1016/j.foodchem.2025.143608 PMID: 40023039

Wang, S., Gerlach, G., & Körner, J. (2023). A study of smart hydrogels as sensing elements in gaseous environment for VOC detection. *Polymer, 278*, 126009. DOI: 10.1016/j.polymer.2023.126009

Wang, S., Jiao, C., Gerlach, G., & Körner, J. (2023). Porosity Engineering of Dried Smart Poly (N-isopropylacrylamide) Hydrogels for Gas Sensing. *Biomacromolecules, 25*(5), 2715–2727. DOI: 10.1021/acs.biomac.3c00738 PMID: 38047737

Yin, L., Fei, L., Tang, C., & Yin, C. (2007). Synthesis, characterization, mechanical properties and biocompatibility of interpenetrating polymer network–super-porous hydrogel containing sodium alginate. *Polymer International, 56*(12), 1563–1571. DOI: 10.1002/pi.2306

Zaman, M., Siddique, W., Waheed, S., Sarfraz, R., Mahmood, A., Qureshi, J., & Khalid, U. (2015). Hydrogels, their applications and polymers used for hydrogels: A review. *International Journal of Biology, Pharmacy and Allied Sciences, 4*, 6581–6603.

Zhao, Y., Zhou, S., Xia, X., Tan, M., Lv, Y., Cheng, Y., Tao, Y., Lu, J., Du, J., & Wang, H. (2022). High-performance carboxymethyl cellulose-based hydrogel film for food packaging and preservation system. *International Journal of Biological Macromolecules, 223*, 1126–1137. DOI: 10.1016/j.ijbiomac.2022.11.102 PMID: 36395928

Chapter 8
Stimuli-Responsive Intelligent Hydrogels

Muzammal Hussain
https://orcid.org/0009-0000-9914-8870
Southeast University, China

Aneela Tabasum
University of Wah, Pakistan

Tahira Amir
University of Sharjah, UAE

Khurram Shehzad
https://orcid.org/0009-0004-6486-6310
Jiangsu University, China

ABSTRACT

Recent progress in materials science has led to the emergence of stimuli-responsive intelligent hydrogels, which have gained significant interest due to their unique properties and wide-ranging applications. These hydrogels undergo reversible changes in physical or chemical properties when exposed to external stimuli like temperature, pH, light, electric fields, and biomolecules. Their potential spans biomedical engineering, drug delivery, soft robotics, and environmental sensing. This paper explores their synthesis, properties, and applications. Various stimuli can trigger hydrogel responsiveness, with the choice depending on the application. We discuss methods for incorporating stimuli-responsive elements, including physical entrapment, covalent attachment, and self-assembly. Fabrication techniques such as physical and chemical crosslinking are examined, along with the importance of structure-property relationships in designing precise, responsive hydrogels.

8.1 INTRODUCTION

Hydrogels constitute three-dimensional networks of hydrophilic polymers and have been under scrutiny for a long time since these materials have high water content, softness, and biocompat-

DOI: 10.4018/979-8-3373-2337-4.ch008

ibility, which are their positive attributes. They therefore become suitable candidates for drug delivery, tissue engineering, sensing, and actuation **(Figure 1)**. Such implants are not responsive to exterior stimuli, thereby limiting their application in dynamic settings. The development of stimuli-responsive hydrogels (Moriya et al., 2024) has provided a turnaround for these fields, especially in biologics and robotics. Such smart hydrogels adapt their structure and mechanics to external triggers, facilitating applications in separation processes, target-oriented drug release, advanced sensors(Bordbar-Khiabani & Gasik, 2022), bionic devices, and regenerative medicine. The complexity and responsiveness of these materials hold great promise for scientific advances. These hydrogels are sensitive to stimuli including light(Hussain, Ahmad, Ejeromedoghene, et al., 2025; Hussain et al., 2024), temperature, magnetic fields(Hussain, Ahmad, Khan, et al., 2025), electric fields, pH(Sood et al., 2016), and chemical or biological agents, so show phase change or stiffness variation.

Figure 1. Different uses of Hydrogels

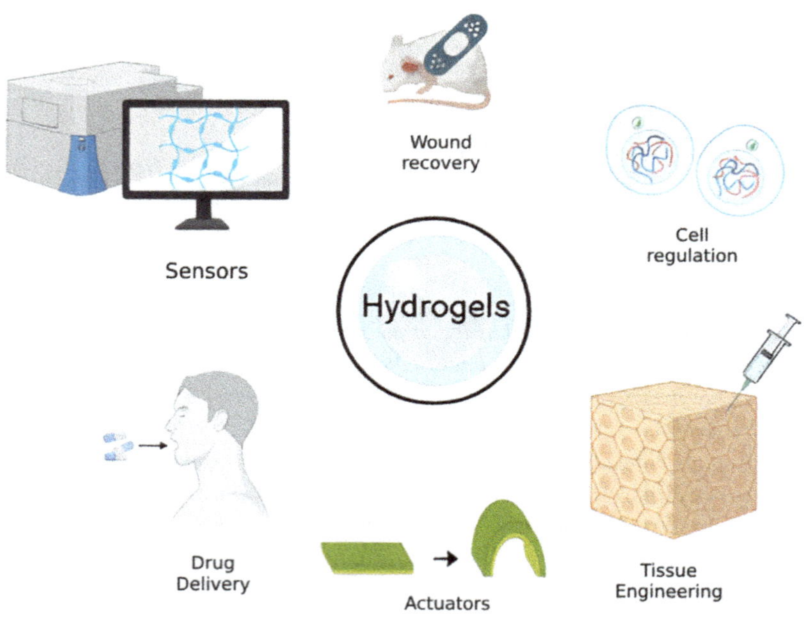

Stimuli-intelligent hydrogels offer solutions across medicine, biotechnology, and environmental science. In drug delivery, they can be tuned to physiological triggers for controlled, site-specific release. In tissue engineering, they replicate dynamic tissue mechanics to support cell growth and regeneration(Bustamante-Torres et al., 2024; Chakrapani et al., 2022). Their analyte-responsive behavior enables highly sensitive biosensors. Their rapid, reversible stiffness and shape changes also make them ideal for soft robotics and actuators, facilitating precise, adaptive motion in human–robot interfaces, prosthetics, and biomedical devices(Nicu & Ciolacu, 2021; Protsak & Morozov, 2025). Finally, environmental and agricultural applications include selective contaminant uptake for water purification and controlled release of fertilizers or pesticides to enhance efficiency and reduce environmental impact(Gautam, 2024).

8.2 SYNTHESIS OF STIMULI-RESPONSIVE INTELLIGENT HYDROGELS

Stimuli-responsive intelligent hydrogels rendering processes commonly define polymeric networks to be cross-linked while undergoing reversible physical or chemical changes as a function of a slight environmental stimulus. These hydrogels are especially made to react to given stimuli operating within ranges such as heat variation, pH changes (Qiu & Park, 2001), light radiation, electromagnetic fields, or certain biomolecular entities, there by incorporating high application potentials in controlled drug delivery systems, tissue engineering, biosensors, and soft robotics (Stuart et al., 2010). Synthesis methods are basically arranged into chemical cross-linking methods, rendering permanent networks, and physical cross-linking methods, relying on reversible secondary forces, the choice of a synthesis method probably impacts the responsiveness, mechanical characteristics, and reversibility of the hydrogels. As a very popular synthetic pathway for temperature-sensitive hydrogels, for instance, those based on poly(N-isopropylacrylamide) (PNIPAM), is free radical polymerization, usually with cross-linking agents such as N, N'-methylenebis(acrylamide) (MBA). With reference to pH-sensitive systems, polymers hold ionizable groups on side chains (e.g, poly(acrylic acid) (PAA) or poly(methacrylic acid) (PMAA)) having swelling capacities that differ depending on whether they exist in the protonated or deprotonated state as decided by the pH in the immediate environment. Among more advanced fabrication routes available to allow fine control over architecture, function, and hence responsivity, are click chemistry, living polymerization methods (ATRP, RAFT) (Schattling et al., 2014), and interpenetrating polymer networks (IPNs). Composite hydrogels can be formed when the hydrogel synthesis incorporates various phases of nanomaterials such as GO, Fe_3O_4 nanoparticles, and carbon nanotubes. More technically speaking, hydrogels are a network of polymers that are swollen in an aqueous medium and hence behave mechanically softer than solids in general. The synthesis is aimed toward achieving suitable kinetics, swelling degree, and trigger specificity with respect to the anticipated use so that the intelligent system can carry out more complex functions such as pulsatile drug release and self-actuation (Manouras & Vamvakaki, 2017). **(Figure 2(a, b))**.

Figure 2. Synthesis of Stimuli-intelligent Hydrogels (a&b)

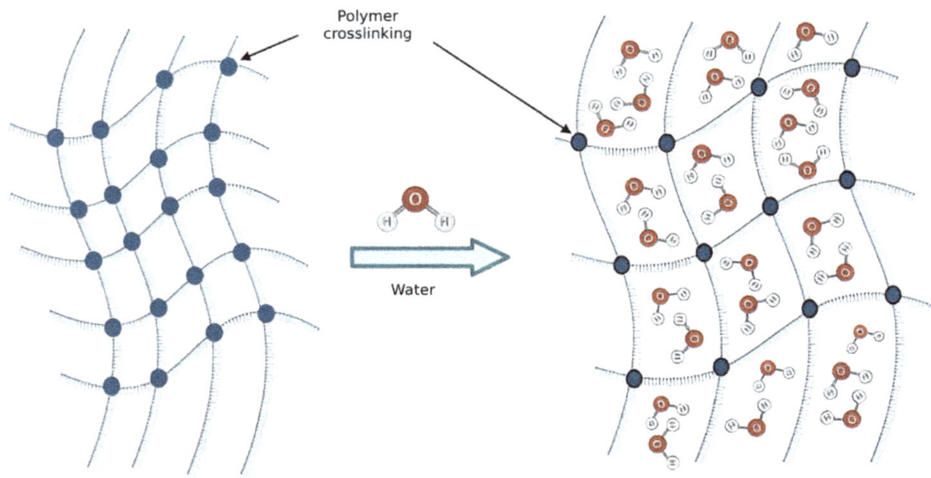

A noteworthy example of a stimuli-responsive hydrogel is the poly(N-isopropylacrylamide) (PNIPAM) hydrogel (Nykänen, 2013; Zhang et al., 2009). This hydrogel exhibits reversible volume changes in response to temperature variations due to the presence of thermo-responsive molecular components such as hydrophilic acylamino groups and isopropyl groups. On the other hand, alginate hydrogels demonstrate the ability to undergo a reversible transition from a gel to a sol state upon the addition or removal of divalent metal ions, owing to the presence of metal-ion responsive crosslinks (Hanyková et al., 2024). These examples vividly illustrate the dependence of gelator-gelator and gelator-solvent interactions on the specific properties of the gelators and crosslinks employed.

8.3 CHARACTERIZATION OF STIMULI RESPONSIVE BEHAVIOR OF INTELLIGENT HYDROGELS

Characterizing the stimuli-responsive behavior of intelligent hydrogels is essential to assess their sensitivity, range of response, and reversibility **(Figure 3).** The following techniques are commonly used for stimuli-responsive behavior evaluation:

Figure 3. Characterization techniques employed for Stimuli reactive intelligent Hydrogels

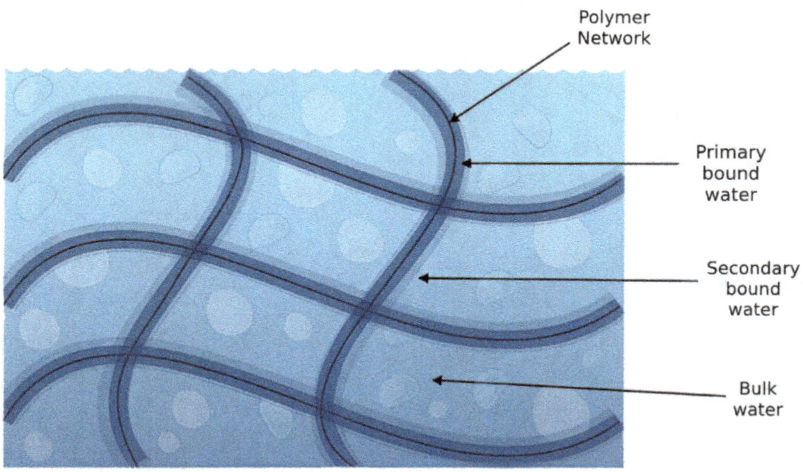

8.3 STIMULI-RESPONSIVE MECHANISMS

Stimuli-responsive intelligent hydrogels exhibit their unique properties and behaviors through specific mechanisms that enable them to respond to external stimuli. These mechanisms can be broadly classified into various categories based on the nature of the stimulus **(Figure. 4).** Here, we discuss some commonly employed stimuli-responsive mechanisms in hydrogels:

Figure 4. Different types of Stimuli responsive Hydrogels

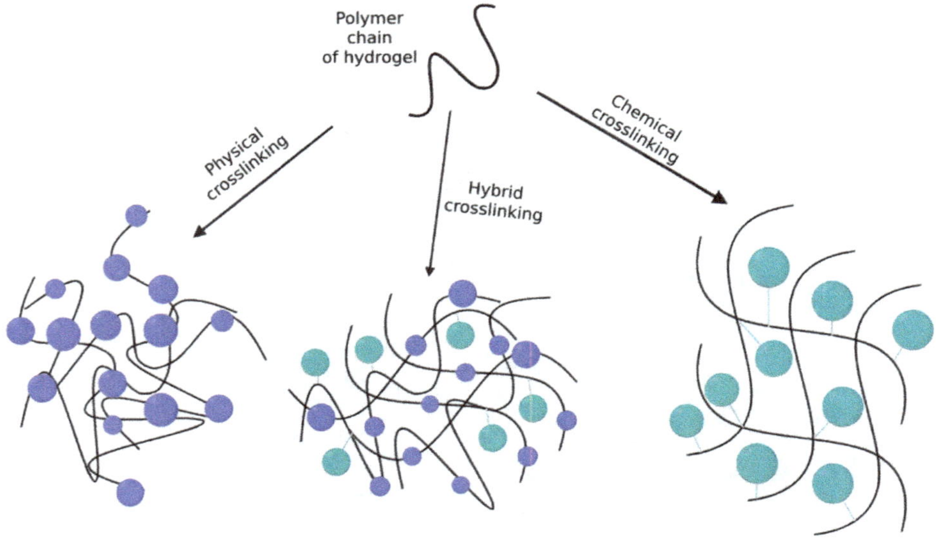

8.3.1 Temperature-Responsive Hydrogels

Thermo-responsive hydrogels contain polymers that phase-separate above their LCST or below their UCST(Zhang et al., 2016). While SPP, EOVE, MOVE, MEO_2MA, and OEGMA are common examples, PNIPAM is the gold standard, possessing hydrophilic amide and hydrophobic isopropyl groups and an LCST of 32–35 °C(Irie, 2005). Below its LCST, PNIPAM swells due to amide–water hydrogen bonding; above it, water is expelled via isopropyl-driven aggregation, and this phase transition reversibly changes the material's electrostatic permittivity. PNIPAM's EP modulation and TiNS-induced anisotropic repulsion in a magnetic field to design an L-shaped actuator that walks unidirectionally with thermal cycling(Ahn et al., 2008). Expanding on this, integrated gold nanoparticles into PNIPAM/TiNS gels to produce worm-like crawling under visible/NIR light, achieving expansion in 0.5 s.

Figure 5. Thermo-responsive behavior of PNIPAM/TiNS hydrogel system

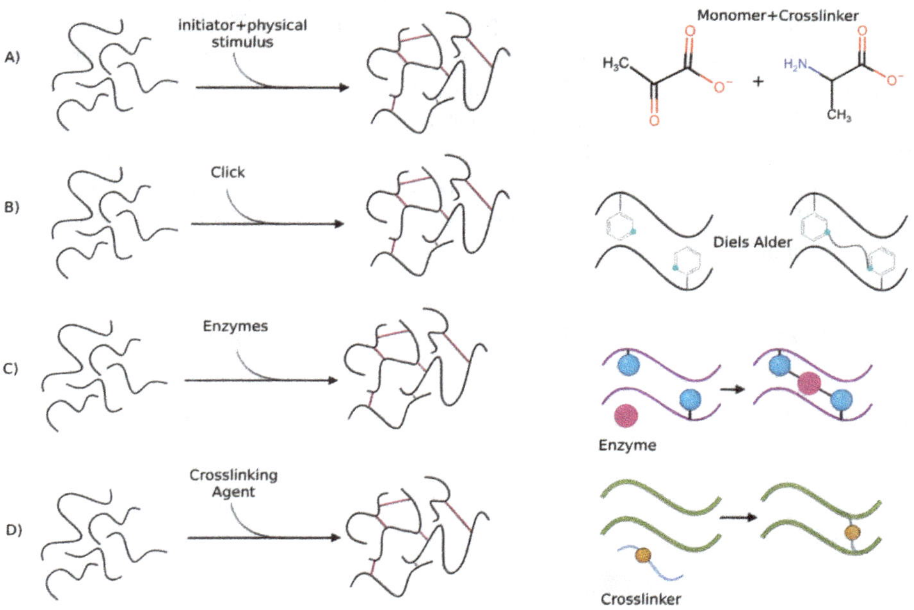

This isovolumetric expansion astonishingly reached up to 80% of the hydrogel's original length. Harnessing this thermo-responsive actuation, a hydrogel rod performed worm-like crawling in the direction opposite to photo scanning. This achievement demonstrates the potential of hydrogel actuators to precisely emulate complex biological motions.

PNIPAM-based thermo-responsive hydrogels are highly effective for soft bionic apparatus. These gels also have substantial biomedical prospects in controlled agent delivery, tissue engineering, and regenerative medicine(Guo, Lu, et al., 2014; Sivokhin et al., 2021). Recently, Ma's group developed thermo-responsive hydrogel microspheres from an N-isopropylacrylamide, ethylene glycol, and L-lactic acid copolymer to mimic the myocardial extracellular matrix (ECM) (Trombino et al., 2021). This work effectively transplanted cardiomyocytes into the damaged region. The microspheres were fluid at 25°C for easy cell injection but gelled at body temperature (37°C) to form an in-situ ECM. This physicochemical property of PNIPAM retained the cells and facilitated heart regeneration, showing promise as a transplantation vector for tissue repair.

8.3.2 Hydrogels with pH-Triggered Responsiveness

pH is another common stimulus for regulating hydrogels. pH-responsive hydrogel networks are made with crosslinks that form or break at specific pH values. These include Schiff base, acylhydrazone, and boronate-catechol crosslinks, as well as ionizable groups. pH-sensitive hydrogels are particularly useful for enhancing drug delivery in cancer treatment and wound healing. This application exploits the pH difference between tumor tissue (pH 6.8–7.0) and wound surroundings (pH 5.5–5.6) versus normal tissues and blood plasma (pH 7.4).

Willner's group developed an intriguing pH-responsive(Hu et al., 2017), two-layer DNA hydrogel (Guo, Qi, et al., 2014). One layer incorporates i-motif structures as pH-sensitive crosslinks. At an acidic pH of 5, cytosine-rich DNA strands assemble into an i-motif structure, markedly increasing stiffness and changing the shape of that layer (Guo, Lu, et al., 2014). Under neutral pH 7, however, this structure disassembles, yielding a less rigid layered hydrogel **(Figure. 6).** This scheme shows the broad potential of pH-responsive DNA hydrogels in actuation, drug delivery, and sensing. Peptides and peptide-based molecules, which contain abundant amphiphilic units, are another common building block for pH-responsive hydrogels.

Figure 6. pH responsive behavior of Hydrogels

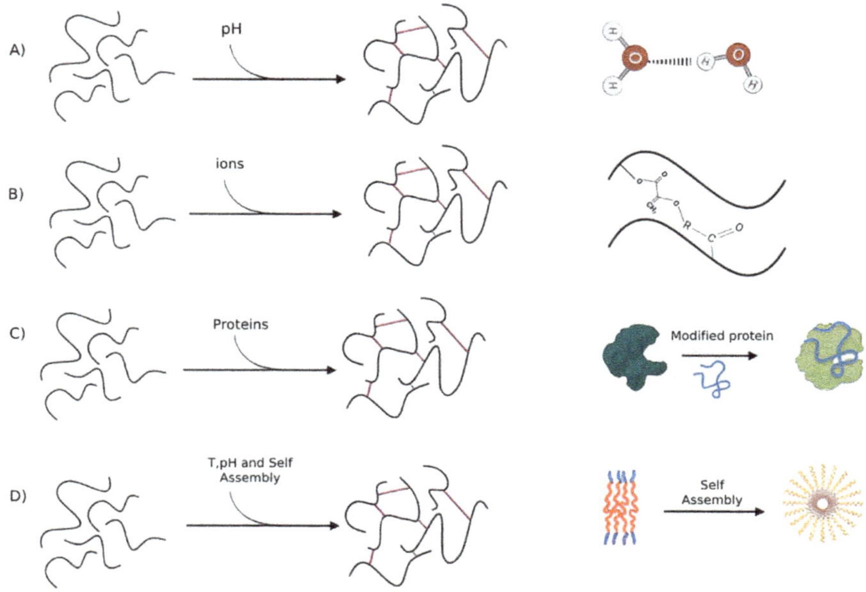

Lampe and his colleagues developed pH-responsive hydrogels from five-residue peptides(Houck et al., 2019; Hu et al., 2019). Through protonation and deprotonation of amine and carboxyl groups, the seven peptides showed diverse morphological and rheological characteristics. In acidic environments, the hydrogels displayed low stiffness with a storage modulus (G') below 80 Pa, but they transitioned into a robust state at pH 7.4 and 10.6. Additionally, each peptide hydrogel's viscosity decreased with increasing shear rate. Notably, the KYFIL hydrogel (G' = 10,000 Pa) maintained its original stiffness after about five strain sweeps from 100% to 0.1%, underscoring its exceptional mechanical properties. Significantly, these peptide hydrogels encapsulated oligodendrocyte progenitor cells and protected them from mechanical membrane disruption during syringe needle injection. The adjustable mechanical properties and cell compatibility make these pentapeptide hydrogels highly suitable for in-cell delivery and tissue engineering.

8.3.3 Light-Activated Hydrogels

Light-activated hydrogels are drawing significant interest for being non-intrusive, offering precise spatiotemporal control, and producing few contaminants. Their promise lies in their ability to convert light signals into a wide range of functions. Making these hydrogels light-controllable involves incorporating light-responsive groups into the gelators, either covalently or non-covalently. Commonly used groups include azobenzene (Ma et al., 2018; Tamesue et al., 2010), diarylethenes (Jia et al., 2021), and spiropyrans (Ter Schiphorst et al., 2016). Others are photo-reversible dimerization groups like coumarins (Kabb et al., 2018), anthracenes (Houck et al., 2019), and pyrimidine derivatives (Luo & Shoichet, 2004), along with photolabile groups, especially nitrobenzyl-based ones.

Azobenzene-containing hydrogels are often made using cyclodextrin (specifically γ- and β-cyclodextrin) to create a host-guest interaction. This interplay forms a photo-responsive gelator or crosslink, producing a hydrogel whose mechanical properties can be adjuste(Zhao & Stoddart, 2009)d with light. Leveraging this host-guest association, (Chen et al., 2014) recently pioneered two different light-responsive hydrogel systems.

In their first system, Tian and Ma used a host polymer, poly-β-cyclodextrin (poly-β-CD), and a guest polymer, poly-α-bromonaphthalene (poly-α-BrNP). Adding poly-azobenzene (poly-Azo) created a crosslinking network in the hydrogel **(Figure. 7).** The gel's strong hydrophobic interactions allowed for rapid self-healing; impressively, two separate pieces could rejoin seamlessly in just one minute with no external trigger. The poly-β-CD also shielded and stabilized the network, dramatically amplifying its room-temperature phosphorescence (RTP) at 552 nm. Introducing poly-Azo then created a light-responsive RTP hydrogel based on the hydrophobic interplay between azobenzene and α-BrNP. In this system, trans-Azo displaced some α-BrNP from the β-CD complex, which suppressed the RTP. When hit with 254 nm light, however, the trans-Azo units isomerized into cis-Azo and detached from the β-CD. This freed the β-CD to interact with α-BrNP again, restoring the RTP intensity.

Figure 7. Construction of photo-controlled RTP supramolecular assembly in aqueous solution through poly-BrNpA/γ-CD/poly-Azo system

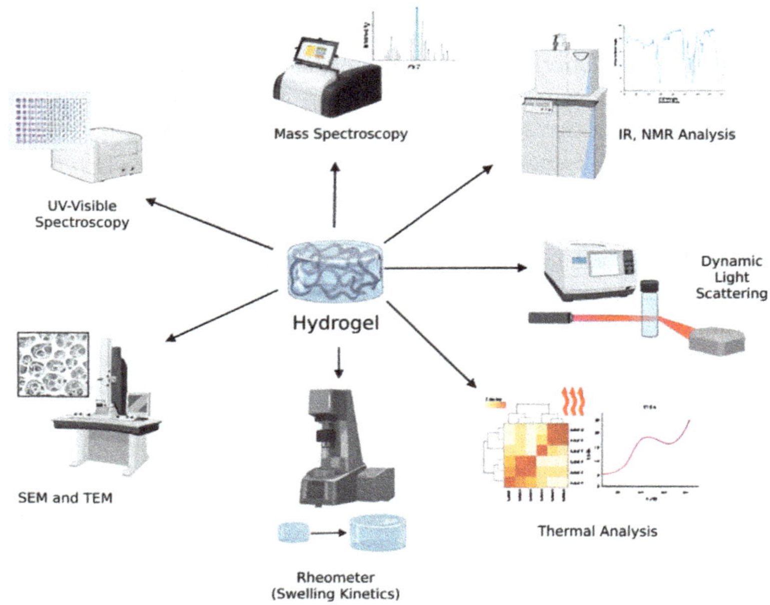

In the second system, a hydrogel was created by combining poly-γ-cyclodextrin (poly-γ-CD), poly-4-bromo-1,8-naphthalic anhydride (poly-BrNpA), and poly-Azo through host-guest interactions. This combination formed a photo-switchable room-temperature phosphorescence (RTP) hydrogel. Building on this, researchers integrated azobenzene derivatives into DNA sequences to create light-responsive nucleotides. (Hu et al., 2019) introduced a light-sensitive hydrogel using azobenzene DNA chains and polyacrylamide. This system had two crosslinker types: covalent glucosamine-boronate esters for memory and self-healing, and azobenzene-modified DNA as a light-sensitive switch.

To achieve light responsiveness, azobenzene was incorporated into the DNA sequence using a wedge-like motif (Kamiya & Asanuma, 2014). This involved placing azobenzene units every three nucleotides opposite an unmodified complementary strand. Under UV irradiation, the cis-state of azobenzene significantly lowered the melting temperature (Tm), causing the duplex to unwind into a quasi-liquid, shapeless state. In contrast, the trans-state of the modified azobenzene stabilized the complementary duplex, reconstructing the hydrogel's structure. Consequently, the hydrogel showed exceptional shape-memory functionality upon reversible irradiation with UV (365 nm) and visible (>420 nm) light. The hydrogel also exhibited light-triggered self-healing and drug release, highlighting its broad application potential.

8.3.4 Electric and Magnetic Field-Responsive Hydrogels

Electric and magnetic field-responsive hydrogels adapt their properties or shape under external fields. Electrically responsive hydrogels, embedded with conductive materials or responsive polymers, modify swelling or conductivity when exposed to electric fields. Magnetic hydrogels, incorporating magnetic nanoparticles or polymers, alter structure or mechanical properties under magnetic fields. These fields enable precise, non-invasive control in confined spaces, unlike light or other stimuli (Ma et al., 2020; Rich et al., 2018; Sitti, 2018; Ze et al., 2020). Such hydrogels are increasingly valued in soft robotics, biomimetics (Jiang et al., 2020; Larson et al., 2016; Li et al., 2016; Yun et al., 2019), drug delivery (Tang et al., 2019), and biotechnology or biomedical engineering (Araujo-Custodio et al., 2019; Noh et al., 2019). These hydrogels convert structural changes into magnetic or electric signals. (Du et al., 2020) developed a milli-robot with neodymium-iron-boron microparticles in its head, enabling crawling, swinging, rolling, and helical propulsion under magnetic fields **(Figure. 8)**, advancing intelligent biomimetic robotics. Magnetic fields also offer remote energy potential. Sitti's team created a biodegradable micro-swimmer using Fe_3O_4 particles and 3D two-photon microprinting, capable of propulsion under rotational magnetic fields and targeted drug delivery (Ceylan et al., 2019).

Figure 8. Magnetization and multimodal locomotion of Magnetic responsive Hydrogels

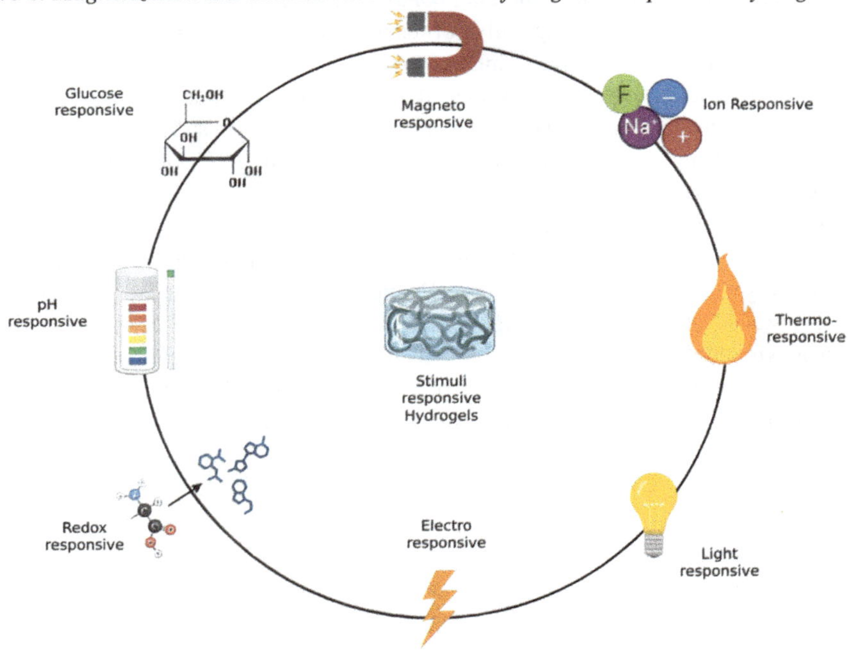

Bian and team engineered a magnetically guided platform for cell adhesion, utilizing methacrylate hyaluronic acid infused with RGD-coated Fe_3O_4 nanoparticles (Wang et al., 2020). By steering the magnetic field, these nanoparticles shift to reveal or hide RGD molecules, prompting cells to attach or release. This innovative approach transforms magnetic signals into precise cellular actions, such as adhesion, growth, movement, and differentiation, demonstrating substantial promise for medical applications (Wu et al., 2016; Xie et al., 2015). Separately, electroactive materials activate various cells-fibroblasts, keratinocytes, muscle, bone, nerve, cardiac, and mesenchymal stem cells-boosting their growth, attachment, differentiation, and mobility. Ma's research team pioneered conductive hydrogels, blending quaternized chitosan derivatives with polyaniline, to enhance skin wound healing (Zhao et al., 2017). These hydrogels excel in fighting bacteria, conducting electrical signals, and neutralizing free radicals, while also promoting growth factor production, collagen buildup, and tissue regeneration, markedly speeding up wound recovery.

8.3.5 Ion-responsive hydrogels

In addition to pH and metal ions, redox reactions, exchange reactions, and enzymatic triggers have been used to precisely control hydrogel crosslinking. Lu employed Ag-lignin to sustain a redox environment that constantly produces catechol groups and radicals, resulting in gelled hydrogels with enduring adhesive properties (Gan et al., 2019). Thordarson's team created transient hydrogels with adjustable lifetimes using the reductive agent tris(2-carboxylethyl) phosphine (TCEP); by varying the initial TCEP concentration, sol-gel transitions induced by disulfide-bonded gelator self-assembly can be actively regulated (Wojciechowski et al., 2018). Grin staff created two degradable systems with thioester bonds that, when in contact with cysteine methyl ester, experience swift bond exchange-driven fragmentation and alteration-providing exciting on-demand applications for wound dressings (Ghobril et al., 2013; Konieczynska et al., 2016).

Willner's constitutional dynamic network (CDN) included AA′, AB′, BA′, and BB′ units with toehold sequences; using E_1 rebalances BA′ in hydrogel X to create a stiffer hydrogel Y rich in AB′/BA′, while E_2 triggers BB′ to transform X into a softer hydrogel Z with reduced AB′/BA′, facilitating self-healing, controlled drug release, and biocatalytic cascades (Yue et al., 2019). Enzymatic recognition is utilized by biochemically responsive hydrogels for theranostics (Culver et al., 2017). Hamachi's non-enzymatic protein-responsive system employs enzyme-activity triggers that attach to target proteins to activate bovine carbonic anhydrase II (bCAII), severing crosslinks to cause a gel-to-sol shift and liberate cargo (Shigemitsu et al., 2020). Li's DNA nanogel breaks down when exposed to flap endonuclease 1, which is overproduced in cancer cells, allowing drugs to be released from cytocompatible matrices (Zhang et al., 2020). These smart hydrogels expand uses in medication delivery, wound repair, and biomaterial development.

Figure 9. The shape memory effect of PAM/SA hydrogel on coordination of SA with metal ions

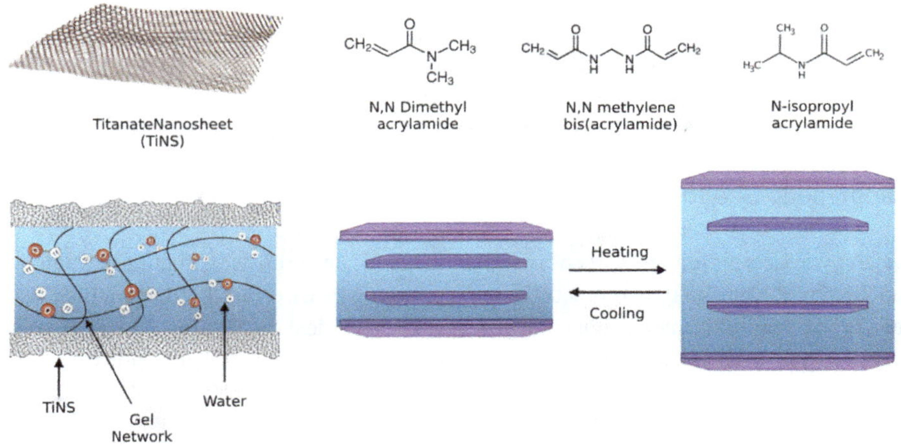

Hydrogel crosslinking is finely controlled by redox reactions, exchange reactions, and enzymatic triggers. Lu employed Ag-lignin to establish a redox atmosphere, producing catechol groups and radicals for persistently adhesive, gelled hydrogels (Gan et al., 2019). Thordarson's group created transient hydrogels with adjustable lifetimes employing tris(2-carboxylethyl) phosphine (TCEP), managing sol-gel transitions through disulfide-bonded gelator self-assembly by varying TCEP concentration (Wojciechowski et al., 2018). Grinstaff's thioester-linked systems, which are degradable and activated by cysteine methyl ester, allow for quick bond exchange and deformation, making them perfect for on-demand wound dressings (Ghobril et al., 2013; Konieczynska et al., 2016). Willner's constitutional dynamic network (CDN) incorporates AA′, AB′, BA′, and BB′ units that utilize toehold sequences; E_1 displaces BA′ in hydrogel X to generate a more rigid hydrogel Y abundant in AB′/BA′, whereas E_2 triggers BB′ to produce a more pliable hydrogel Z, facilitating self-healing, drug delivery, and biocatalytic cascades (Yue et al., 2019). Hydrogels that respond to enzymes facilitate theranostics (Culver et al., 2017). Hamachi's non-enzymatic mechanism employs enzyme-activity triggers to attach proteins, activating bovine carbonic anhydrase II (bCAII) to sever crosslinks, causing a transition from gel to sol and releasing cargo (Shigemitsu et al., 2020). Li's DNA nanogel disintegrates through flap endonuclease 1, which is overexpressed in cancer cells, freeing drugs from cytocompatibility matrices (Zhang et al., 2020). These sophisticated hydrogels improve drug delivery, wound healing, and biomaterial development.

8.4 CASE STUDIES OF STIMULI RESPONSIVE INTELLIGENT HYDROGELS

Stimuli-responsive hydrogels have been deployed in many practical applications. Below are illustrative case studies demonstrating their impact:

8.4.1 Regenerative Medicine and Tissue Engineering

(a) Biomimetic Scaffolds

Smart hydrogels serve as biomimetic scaffolds that replicate the ECM's structure and function, responding to cellular signals to promote tissue regeneration (Raza et al., 2019). Their high-water content, biocompatibility, and tunable properties make them ideal for creating 3D environments that support cell survival and tissue growth. These hydrogels can dynamically adjust their mechanical properties, such as stiffness and porosity, in response to stimuli like pH or enzymes, facilitating cell adhesion, proliferation, and differentiation.

(b) Thermoresponsive Gelatin Scaffolds

Thermoresponsive gelatin scaffolds are particularly promising due to their ability to be 3D-printed into complex shapes at low temperatures and solidify at physiological temperatures (approximately 37°C) to support tissue growth (Zhao & Stoddart, 2009). For instance, gelatin-based hydrogels can be fabricated into intricate structures for tissue engineering, leveraging their lower critical solution temperature (LCST) to transition from a liquid to a gel state, providing a stable scaffold for cells. Recent studies highlight their use in bone and cartilage repair, where they support cell infiltration and matrix deposition.

(c) Enzyme-Responsive Hydrogels

Hydrogels with stiffness or degradation profiles that respond to cell-secreted enzymes, such as matrix metalloproteinases (MMPs), guide stem cell differentiation and tissue regeneration. For example, enzyme-responsive glycosaminoglycan (GAG)-based hydrogels, cross-linked via transglutaminase factor XIII, degrade in response to chondroitinase and MMPs, releasing growth factors like bone morphogenetic protein-2 (BMP-2) to promote osteogenic differentiation of mesenchymal stem cells. These hydrogels mimic the dynamic ECM, allowing precise control over cellular behavior.

(d) Bone and Cartilage Healing

Stimuli-responsive scaffolds have been used to promote bone and cartilage healing by releasing growth factors in situ when triggered by the local environment. For instance, pH-responsive hydrogels can release bioactive molecules in response to the acidic microenvironment of damaged tissues, enhancing repair processes. Recent advancements include 3D-printed hydrogels for bone and joint repair, which exhibit excellent mechanical strength and cell compatibility.

(e) Dynamic Property Adjustment

By adjusting properties like porosity and mechanical modulus in response to pH or enzymatic cues, smart hydrogels facilitate cell adhesion, regulate differentiation, and accelerate tissue regeneration. For example, gelatin/chitosan hydrogel tubes used as nerve guidance conduits demonstrate physiological stability and low inflammatory response, supporting neural tissue regeneration.

8.4.2 Drug Delivery Systems

(a) Temperature-Triggered Drug Release

Hydrogels enable controlled, on-demand release of therapeutics in response to physiological signals. Poly(N-isopropylacrylamide) (PNIPAM) hydrogels are a classic example, where drugs are loaded at low temperatures and released when the gel is heated above its LCST (approximately 32–33°C) due to mild hyperthermia in tumor tissues. A study demonstrated that PNIPAM-based nanogels loaded with anastrozole exhibited temperature- and pH-triggered release, enhancing anticancer drug delivery to tumor sites (Zhang & Wu, 2023).

(b) *pH-Responsive Drug Delivery*

pH-responsive hydrogels offer targeted drug delivery, particularly for cancer therapy. Alginate hydrogels, for instance, remain stable in the neutral pH of the bloodstream but rapidly dissolve in the acidic tumor microenvironment (pH ~6.5), releasing chemotherapy agents like doxorubicin. These hydrogels improve drug efficacy by ensuring site-specific release, reducing systemic side effects.

8.4.3 Commercial Medical Products

Many hydrogel-based devices have achieved commercial success and regulatory approval from the FDA and EMA. Over 30 injectable hydrogel products are approved, including Medtronic's INFUSE® bone graft, a collagen gel delivering BMP-2 for spinal fusion, approved in 2002 (Mandal et al., 2020). Endo's Vantas, a histrelin-acetate hydrogel implant, is used for prostate cancer treatment. Injectable hyaluronic acid fillers like Juvéderm (Allergan) and Restylane (Galderma) are widely used for facial aesthetic corrections. Hydrogel sheets, containing 96% water, are standard in burn dressings worldwide. Space OAR hydrogel protects tissues during cancer radiotherapy, while INFUSE fills bone voids, and hyaluronic acid fillers restore facial volume, demonstrating the real-world impact of smart hydrogels (Mandal et al., 2020).

These examples demonstrate how hydrogel technologies address specific problems: they mimic tissue, release drugs on-demand, or protect vulnerable organs. Importantly, case studies with clinical or commercial success highlight the viability of stimuli-responsive hydrogels when designed and manufactured appropriately.

8.5 OPPORTUNITIES AND LIMITATIONS OF STIMULI-RESPONSIVE HYDROGELS

Stimuli-responsive hydrogels offer unique opportunities in advanced manufacturing and personalized medicine, but they also face inherent limitations. Recent reviews note the following key points:

8.5.1 Opportunities

Stimuli-responsive hydrogels present significant potential in advanced manufacturing and personalized medicine. According to (Alli et al., 2024), their integration with 3D/4D printing enables the creation of dynamic architectures and self-folding structures, ideal for custom implants. The tunability of hydrogels, achieved through adjustments in polymer chemistry and crosslink density, allows customization for applications ranging from soft robotics to smart textiles. In drug delivery, smart hydrogels facilitate precision therapies with reduced side effects by enabling triggered release at pathological sites. Multi-stimuli hydrogels, responsive to factors like pH and temperature, enhance design flexibility. Furthermore, the convergence of hydrogel engineering with nanotechnology and biotechnology expands their potential, such as integrating sensors or combining gels with living cells to create "biohybrids."

8.5.2 Limitations

Despite their promise, stimuli-responsive hydrogels face several challenges. Environmental sensitivity is a key issue; in complex physiological fluids, factors like ionic strength or unintended stimuli can alter their response. (Zhang & Wu, 2023) highlight that the response of ionizable groups in pH-responsive gels can be affected by temperature and ionic strength, leading to unpredictable drug release in vivo. Biocompatibility and toxicity concern also arise, as some crosslinkers or polymerization byproducts, such as those from acid-labile bonds, may be cytotoxic or leave reactive residues. Balancing mechanical strength and rapid response is another challenge natural polymers like collagen or alginate are biocompatible but mechanically weak, while synthetic polymers offer strength but may lack bio friendliness. Additionally, manufacturing scalability remains a hurdle. Mandal et al. note that transitioning lab-scale hydrogel systems to commercial production requires overcoming variability and meeting cGMP standards, with high water content complicating sterilization and storage, and batch-to-batch consistency needing stringent control.

8.6 APPLICATIONS OF STIMULI RESPONSIVE INTELLIGENT HYDROGELS

Stimuli-responsive intelligent hydrogels have garnered significant attention due to their unique properties and versatile responsiveness. These hydrogels find applications in various fields, offering innovative solutions for a wide range of challenges. Here is a detailed overview of Stimuli-responsive hydrogels being used in various fields:

8.6.1 Regenerative Medicine and Tissue Engineering

Stimuli-responsive hydrogels hold a pivotal position in regenerative medicine and tissue engineering, particularly in the development of biomimetic scaffolds that emulate the natural extracellular matrix (ECM) (Zhao et al., 2020). These hydrogels are engineered to respond to specific stimuli, including temperature, pH, or mechanical forces, thereby influencing cell behavior within a favorable microenvironment. They facilitate cell adhesion, regulate cell differentiation, and support tissue regeneration processes, enabling advancements in the field of regenerative medicine.

Here are some specific examples of how stimuli-responsive intelligent hydrogels are utilized in various applications within tissue engineering and regenerative medicine

8.6.2 Drug Delivery Systems

a) Temperature-responsive hydrogels

An example of such hydrogels is poly(N-isopropylacrylamide) (PNIPAAm) hydrogels, which possess a lower critical solution temperature (LCST) near body temperature. These hydrogels have the ability to encapsulate drugs and release them when the temperature rises, such as in the case of localized hyperthermia (Zhang et al., 2020).

b) pH-responsive hydrogels

Alginate-based hydrogels can be modified with pH-sensitive moieties, allowing for drug release in reaction to changes in the pH of the proximate surroundings (Žuržul et al., 2020).

8.6.3 Cell Encapsulation and Transplantation

a) Enzyme-responsive hydrogels

Gelatin-based hydrogels can be engineered to react to specific enzymes, such as matrix metalloproteinases (MMPs), which are often upregulated in inflammatory conditions. These hydrogels can degrade and release encapsulated cells upon exposure to MMPs (Zhu et al., 2019).

b) Light-responsive hydrogels

Hydrogels incorporating light-responsive elements, such as azobenzene derivatives, can release encapsulated cells upon exposure to specific light wavelengths (Zhao & Stoddart, 2009).

8.6.4 Tissue Engineering Scaffolds

a) Photoresponsive hydrogels

Methacrylated hyaluronic acid (MeHA) hydrogels can be crosslinked using photoinitiators and exposed to light to create spatiotemporally controlled patterns of stiffness, allowing for the guided differentiation of stem cells (Niloy et al., 2020).

b) Electroresponsive hydrogels

Electrically conductive hydrogels, such as those based on polypyrrole, can respond to electrical stimulation and modulate the behavior of encapsulated cells (Jia et al., 2021).

8.6.5 Smart Wound Dressings

a) pH-responsive hydrogels

Chitosan-based hydrogels can be designed to alter their properties in reaction to the pH of the wound environment, promoting proper wound healing (Nie et al., 2020).

b) Oxygen-responsive hydrogels

Hydrogels containing oxygen-sensitive molecules, such as porphyrin derivatives, can release oxygen to the wound bed in response to low oxygen levels, thereby promoting healing (Park et al., 2021).

8.6.6 Biosensors

a) Glucose-responsive hydrogels

Hydrogels containing glucose-responsive enzymes, such as glucose oxidase, can swell or change color in the presence of glucose, allowing for non-invasive monitoring of glucose levels in diabetic patients (Kim & Kim, 2020).

b) Ion-responsive hydrogels

Hydrogels modified with ion-selective electrodes can detect specific ions, such as calcium or potassium, which are important for various physiological processes (Wang et al., 2019).

8.7 REGULATORY CONSIDERATIONS FOR STIMULI-RESPONSIVE HYDROGELS

According to U.S. regulations, the majority of hydrogels are classified as medical devices (FD&C Act §201(g)), which means they need a 510(k)-pre-market notification to obtain FDA approval (Mandal et al.). Combination products involving drugs or cells encounter a longer approval process of 7–10 years, hindering their commercialization. Still, certain hydrogels like the injectable collagen/BMP-2 system (INFUSE) for spinal fusion and the PEG-based Space OAR for prostate radiotherapy have obtained FDA approval. Changing guidelines and obstacles in reproducibility of manufacturing and material characterization persist in hindering translation. In Europe, hydrogels that are solely used as devices (such as dressings and dermal fillers) are classified as medical devices, while those containing biologics usually face dual regulation. Timely interaction with regulators and compliance with standards such as ISO 10993 for biocompatibility and the FDA's Quality System Regulation for manufacturing are essential for effective development.

8.8 BEST PRACTICES IN HYDROGEL SYNTHESIS AND SCALE-UP

To ensure quality and facilitate translation of stimuli-responsive hydrogels, the following best practices are recommended:

8.8.1 GMP and Sterility

Implement current Good Manufacturing Practices (cGMP) from the early stages of development to ensure contamination-free, large-scale production. Mandal et al. highlight the importance of addressing batch consistency and GMP compatibility during scale-up. Crosslinking reactions should be conducted in controlled reactors, and sterilization methods, such as gamma irradiation or filtration, must be validated to accommodate the high-water content of hydrogels.

8.8.2 Material Selection

Choose biocompatible and FDA-approved polymers and crosslinkers. Natural polymers like alginate, collagen, and hyaluronic acid are inherently biocompatible, while synthetic polymers like PEG or polyacrylamide derivatives require high purification. Use non-toxic crosslinking agents, such as enzyme-mediated bonding or click reactions, or ensure complete removal of residuals post-gelation. Extensive washing and leachable testing are critical to eliminate residual monomers and catalysts. Thoroughly characterize hydrogel properties, including swelling ratio, mechanical modulus, and degradation rate, under relevant conditions. Conduct stress tests to evaluate sterilization effects and long-term stability. Document precise protocols for temperature, pH, and initiator usage to ensure reproducibility. Apply quality-by-design principles to monitor critical parameters like polymer concentration, crosslinker ratio, and reaction time to minimize variability. Given the sensitivity of hydrogel polymerization, incorporate in-line sensors or automation to enhance process consistency.

(a) Crosslink Density Tuning

Optimize crosslink density to balance responsiveness and mechanical strength. Under-crosslinked gels may degrade prematurely, while over-crosslinked gels may exhibit slower response rates. Conduct pilot studies to map how formulation impacts trigger sensitivity.

By adhering to these practices, researchers and manufacturers can address common challenges, such as batch variability and safety concerns, as noted in the literature, thereby streamlining the path to clinical or industrial applications.

CONCLUSION AND FUTURE PROSPECTS

Stimuli-responsive hydrogels merge polymer science with responsive engineering to advance drug delivery, tissue engineering, and sensing, but fully realizing their promise requires close collaboration among materials chemists, biologists, and regulatory experts, alongside rigorous reporting and best practices. Both hydrogels and their degradation products must undergo thorough safety evaluation for example, polyacrylamide (Amazingel) is safe in breast implants, yet its breakdown product acrylamide is neuro- and nephrotoxic and a suspected carcinogen, necessitating careful assessment. High-throughput synthetic biomimetics such as hyaluronic acid, chitosan, gelatin, and collagen offer cost-effective, tissue-mimicking alternatives that boost biocompatibility. In targeted drug delivery, these materials can release therapeutics with precise spatiotemporal control but face challenges from tumor heterogeneity, high interstitial pressure, and uneven trigger distribution, which can lead to leakage and off-target effects. Biomarker-responsive capsules modified with cell-targeting aptamers (Guo et al., 2014) and multi-trigger systems with controllable on/off switches can enhance selectivity and optimize dosing. Although switchable hydrogels based on noncovalent interactions allow reversible stiffness for actuators and smart devices, they often lack long-term stability and rigidity; integrating dynamic covalent bonds can produce stronger, multifunctional hydrogels suitable for industrial applications (Žuržul et al., 2020). Continuous innovation in sustainable synthesis, manufacturing processes, and regulatory frameworks will be key to unlocking intelligent implants and eco-sensors that enhance health and environmental monitoring.

REFERENCES

Ahn, S., Kasi, R. M., Kim, S.-C., Sharma, N., & Zhou, Y. (2008). Stimuli-responsive polymer gels. *Soft Matter*, *4*(6), 1151–1157. DOI: 10.1039/b714376a

Alli, Y. A., Anuar, H., Bamisaye, A., Manshor, M. R., Etafo, N. O., Bamidele, M. O., Rasheed, M. A., Olatunde, S. K., Akinfenwa, A. S., & Lawal, A. (2024). The appealing prospect of hydrogel in 3D/4D printing technology: Overview and opportunities. *Polymer*, *315*, 127823. DOI: 10.1016/j.polymer.2024.127823

Araujo-Custodio, S., Gomez-Florit, M., Tomas, A. R., Mendes, B. B., Babo, P. S., Mithieux, S. M., Weiss, A., Domingues, R. M., Reis, R. L., & Gomes, M. E. (2019). Injectable and magnetic responsive hydrogels with bioinspired ordered structures. *ACS Biomaterials Science & Engineering*, *5*(3), 1392–1404. DOI: 10.1021/acsbiomaterials.8b01179

Bordbar-Khiabani, A., & Gasik, M. (2022). Smart hydrogels for advanced drug delivery systems. *International Journal of Molecular Sciences*, *23*(7), 3665. DOI: 10.3390/ijms23073665

Bustamante-Torres, M., Rodriguez, P., & Bucio, E. (2024). Smart gels and their applications. In *Engineering of Natural Polymeric Gels and Aerogels for Multifunctional Applications* (pp. 291–312). Elsevier. DOI: 10.1016/B978-0-12-823135-7.00003-6

Ceylan, H., Yasa, I. C., Yasa, O., Tabak, A. F., Giltinan, J., & Sitti, M. (2019). 3D-printed biodegradable microswimmer for theranostic cargo delivery and release. *ACS Nano*, *13*(3), 3353–3362. DOI: 10.1021/acsnano.8b09233

Chakrapani, G., Zare, M., & Ramakrishna, S. (2022). Intelligent hydrogels and their biomedical applications. *Materials Advances*, *3*(21), 7757–7772. DOI: 10.1039/D2MA00527A

Chen, H., Ma, X., Wu, S., & Tian, H. (2014). A rapidly self-healing supramolecular polymer hydrogel with photostimulated room-temperature phosphorescence responsiveness. *Angewandte Chemie International Edition*, *53*(51), 14149–14152. DOI: 10.1002/anie.201407402

Culver, H. R., Clegg, J. R., & Peppas, N. A. (2017). Analyte-responsive hydrogels: Intelligent materials for biosensing and drug delivery. *Accounts of Chemical Research*, *50*(2), 170–178. DOI: 10.1021/acs.accounts.6b00533

Du, X., Cui, H., Xu, T., Huang, C., Wang, Y., Zhao, Q., Xu, Y., & Wu, X. (2020). Reconfiguration, camouflage, and color-shifting for bioinspired adaptive hydrogel-based millirobots. *Advanced Functional Materials*, *30*(10), 1909202. DOI: 10.1002/adfm.201909202

Gan, D., Xing, W., Jiang, L., Fang, J., Zhao, C., Ren, F., Fang, L., Wang, K., & Lu, X. (2019). Plant-inspired adhesive and tough hydrogel based on Ag-Lignin nanoparticles-triggered dynamic redox catechol chemistry. *Nature Communications*, *10*(1), 1487. DOI: 10.1038/s41467-019-09351-2

Gautam, R. K. (2024). Stimulus-responsive hydrogel for tissue engineering. *Hydrogels: Antimicrobial Characteristics, Tissue Engineering, Drug Delivery Vehicle*, 49.

Ghobril, C., Charoen, K., Rodriguez, E. K., Nazarian, A., & Grinstaff, M. W. (2013). A dendritic thioester hydrogel based on thiol-thioester exchange as a dissolvable sealant system for wound closure. *Angewandte Chemie International Edition*, *52*(52), 14070–14074. DOI: 10.1002/anie.201308007

Guo, W., Lu, C. H., Qi, X. J., Orbach, R., Fadeev, M., Yang, H. H., & Willner, I. (2014). Switchable bifunctional stimuli-triggered poly-N-isopropylacrylamide/DNA hydrogels. *Angewandte Chemie International Edition*, *53*(38), 10134–10138. DOI: 10.1002/anie.201405692

Guo, W., Qi, X.-J., Orbach, R., Lu, C.-H., Freage, L., Mironi-Harpaz, I., Seliktar, D., Yang, H.-H., & Willner, I. (2014). Reversible Ag+-crosslinked DNA hydrogels. *Chemical Communications (Cambridge)*, *50*(31), 4065–4068. DOI: 10.1039/c3cc49140d

Hanyková, L., Šťastná, J., & Krakovský, I. (2024). Responsive acrylamide-based hydrogels: Advances in interpenetrating polymer structures. *Gels (Basel, Switzerland)*, *10*(7), 414. DOI: 10.3390/gels10070414

Houck, H. A., Blasco, E., Du Prez, F. E., & Barner-Kowollik, C. (2019). Light-stabilized dynamic materials. *Journal of the American Chemical Society*, *141*(31), 12329–12337. DOI: 10.1021/jacs.9b05092

Hu, J., Zheng, Z., Liu, C., Hu, Q., Cai, X., Xiao, J., & Cheng, Y. (2019). A pH-responsive hydrogel with potent antibacterial activity against both aerobic and anaerobic pathogens. *Biomaterials Science*, *7*(2), 581–584. DOI: 10.1039/C8BM01211C

Hu, Y., Cecconello, A., Idili, A., Ricci, F., & Willner, I. (2017). Triplex DNA nanostructures: From basic properties to applications. *Angewandte Chemie International Edition*, *56*(48), 15210–15233. DOI: 10.1002/anie.201701868

Hussain, M., Ahmad, Z., Ejeromedoghene, O., Shehzad, K., Akhtar, M., & Fu, G. (2025). Hybrid polysaccharide film infused with polyoxometalates for inkless printing and solar ultraviolet sensing. *International Journal of Biological Macromolecules*, *293*, 139308. DOI: 10.1016/j.ijbiomac.2024.139308

Hussain, M., Ahmad, Z., Khan, S., Li, W., Kalulu, M., Ullah, K., Ejeromedoghene, O., & Fu, G. (2025). Innovative Antibacterial Magneto-Conductive Multifunctional Hydrogel From Biocompatible Materials. *Polymers for Advanced Technologies*, *36*(5), e70207. DOI: 10.1002/pat.70207

Hussain, M., Kalulu, M., Ahmad, Z., Ejeromedoghene, O., & Fu, G. (2024). Rapid fabrication of polyoxometalate-enhanced photo responsive films from ethyl cellulose (EC) and polyvinylpyrrolidone (PVP). *International Journal of Biological Macromolecules*, *280*, 136051. DOI: 10.1016/j.ijbiomac.2024.136051

Irie, M. (2005). Stimuli-responsive poly (N-isopropylacrylamide). Photo-and chemical-induced phase transitions. In *Responsive gels: Volume transitions II* (pp. 49-65). Springer.

Jia, Z., Li, G., Wang, J., Su, S., Wen, J., Yuan, J., Pan, M., & Pan, Z. (2021). Polypyrrole/PU hybrid hydrogels: Electrically conductive and fast self-healing for potential applications in body-monitor sensors. *New Journal of Chemistry*, *45*(16), 7321–7331. DOI: 10.1039/D1NJ00616A

Jiang, Z., Tan, M. L., Taheri, M., Yan, Q., Tsuzuki, T., Gardiner, M. G., Diggle, B., & Connal, L. A. (2020). Strong, self-healable, and recyclable visible-light-responsive hydrogel actuators. *Angewandte Chemie, 132*(18), 7115–7122. DOI: 10.1002/ange.201916058

Kabb, C. P., O'Bryan, C. S., Deng, C. C., Angelini, T. E., & Sumerlin, B. S. (2018). Photoreversible covalent hydrogels for soft-matter additive manufacturing. *ACS Applied Materials & Interfaces, 10*(19), 16793–16801. DOI: 10.1021/acsami.8b02441

Kamiya, Y., & Asanuma, H. (2014). Light-driven DNA nanomachine with a photoresponsive molecular engine. *Accounts of Chemical Research, 47*(6), 1663–1672. DOI: 10.1021/ar400308f

Kim, G. J., & Kim, K. O. (2020). Novel glucose-responsive of the transparent nanofiber hydrogel patches as a wearable biosensor via electrospinning. *Scientific Reports, 10*(1), 18858. DOI: 10.1038/s41598-020-75906-9

Konieczynska, M. D., Villa-Camacho, J. C., Ghobril, C., Perez-Viloria, M., Tevis, K. M., Blessing, W. A., Nazarian, A., Rodriguez, E. K., & Grinstaff, M. W. (2016). On-demand dissolution of a dendritic hydrogel-based dressing for second-degree burn wounds through thiol–thioester exchange reaction. *Angewandte Chemie International Edition, 55*(34), 9984–9987. DOI: 10.1002/anie.201604827

Larson, C., Peele, B., Li, S., Robinson, S., Totaro, M., Beccai, L., Mazzolai, B., & Shepherd, R. (2016). Highly stretchable electroluminescent skin for optical signaling and tactile sensing. *science, 351*(6277), 1071-1074.

Li, Y., Sun, Y., Xiao, Y., Gao, G., Liu, S., Zhang, J., & Fu, J. (2016). Electric field actuation of tough electroactive hydrogels cross-linked by functional triblock copolymer micelles. *ACS Applied Materials & Interfaces, 8*(39), 26326–26331. DOI: 10.1021/acsami.6b08841

Luo, Y., & Shoichet, M. S. (2004). A photolabile hydrogel for guided three-dimensional cell growth and migration. *Nature Materials, 3*(4), 249–253. DOI: 10.1038/nmat1092

Ma, C., Lu, W., Yang, X., He, J., Le, X., Wang, L., Zhang, J., Serpe, M. J., Huang, Y., & Chen, T. (2018). Bioinspired anisotropic hydrogel actuators with on–off switchable and color-tunable fluorescence behaviors. *Advanced Functional Materials, 28*(7), 1704568. DOI: 10.1002/adfm.201704568

Ma, C., Wu, S., Ze, Q., Kuang, X., Zhang, R., Qi, H. J., & Zhao, R. (2020). Magnetic multimaterial printing for multimodal shape transformation with tunable properties and shiftable mechanical behaviors. *ACS Applied Materials & Interfaces, 13*(11), 12639–12648. DOI: 10.1021/acsami.0c13863

Mandal, A., Clegg, J., Anselmo, A., & Mitragotri, S. (2020). Hydrogels in the clinic. [In.]. *Bioengineering & Translational Medicine, 5*(2), e10158. DOI: 10.1002/btm2.10158

Manouras, T., & Vamvakaki, M. (2017). Field responsive materials: Photo-, electro-, magnetic- and ultrasound-sensitive polymers. *Polymer Chemistry, 8*(1), 74–96. DOI: 10.1039/C6PY01455K

Moriya, G., Mazumder, R., Padhi, S., & Mishra, R. (2024). Gastrorententive hydrogels responsive to external stimuli for novel drug delivery. *Int J App Pharm, 16*(4), 1–14. DOI: 10.22159/ijap.2024v16i4.51051

Nicu, R., & Ciolacu, D. (2021). Bio-based Stimuli-responsive Hydrogels with Biomedical Applications. In *Sustainability of Biomass through Bio-based Chemistry* (pp. 229–261). CRC Press. DOI: 10.1201/9780429347993-9

Nie, L., Chen, D., Zhong, S., Shi, Q., Sun, Y., Politis, C., & Shavandi, A. (2020). Injectable cell-laden poly (N-isopropylacrylamide)/chitosan hydrogel reinforced via graphene oxide and incorporated with dual-growth factors. *Materials Letters, 280*, 128572. DOI: 10.1016/j.matlet.2020.128572

Niloy, K. K., Gulfam, M., Compton, K. B., Li, D., Huang, G. T.-J., & Lowe, T. L. (2020). Methacrylated hyaluronic acid–based hydrogels maintain stemness in human dental pulp stem cells. *Regenerative Engineering and Translational Medicine, 6*(3), 262–272. DOI: 10.1007/s40883-019-00115-4

Noh, M., Choi, Y. H., An, Y.-H., Tahk, D., Cho, S., Yoon, J. W., Jeon, N. L., Park, T. H., Kim, J., & Hwang, N. S. (2019). Magnetic nanoparticle-embedded hydrogel sheet with a groove pattern for wound healing application. *ACS Biomaterials Science & Engineering, 5*(8), 3909–3921. DOI: 10.1021/acsbiomaterials.8b01307

Nykänen, A. (2013). Thermally sensitive block copolymer hydrogels in bulk and with decreased dimensions.

Park, J. M., Hong, K.-I., Lee, H., & Jang, W.-D. (2021). Bioinspired applications of porphyrin derivatives. *Accounts of Chemical Research, 54*(9), 2249–2260. DOI: 10.1021/acs.accounts.1c00114

Protsak, I. S., & Morozov, Y. M. (2025). Fundamentals and advances in stimuli-responsive hydrogels and their applications: A review. *Gels (Basel, Switzerland), 11*(1), 30. DOI: 10.3390/gels11010030

Qiu, Y., & Park, K. (2001). Environment-sensitive hydrogels for drug delivery. *Advanced Drug Delivery Reviews, 53*(3), 321–339. DOI: 10.1016/S0169-409X(01)00203-4

Raza, F., Zhu, Y., Chen, L., You, X., Zhang, J., Khan, A., Khan, M. W., Hasnat, M., Zafar, H., Wu, J., & Ge, L. (2019). Paclitaxel-loaded pH responsive hydrogel based on self-assembled peptides for tumor targeting. *Biomaterials Science, 7*(5), 2023–2036. DOI: 10.1039/C9BM00139E

Rich, S. I., Wood, R. J., & Majidi, C. (2018). Untethered soft robotics. *Nature Electronics, 1*(2), 102–112. DOI: 10.1038/s41928-018-0024-1

Schattling, P., Jochum, F. D., & Theato, P. (2014). Multi-stimuli responsive polymers–the all-in-one talents. *Polymer Chemistry, 5*(1), 25–36. DOI: 10.1039/C3PY00880K

Shigemitsu, H., Kubota, R., Nakamura, K., Matsuzaki, T., Minami, S., Aoyama, T., Urayama, K., & Hamachi, I. (2020). Protein-responsive protein release of supramolecular/polymer hydrogel composite integrating enzyme activation systems. *Nature Communications, 11*(1), 3859. DOI: 10.1038/s41467-020-17698-0

Sitti, M. (2018). Miniature soft robots—Road to the clinic. *Nature Reviews. Materials*, *3*(6), 74–75. DOI: 10.1038/s41578-018-0001-3

Sivokhin, A. P., Orekhov, D. V., Kazantsev, O. A., Gubanova, O. V., Kamorin, D. M., Zarubina, I. S., Bolshakova, E. A., & Zaitsev, S. D. (2021). Amphiphilic thermoresponsive copolymer bottlebrushes: Synthesis, characterization, and study of their self-assembly into flower-like micelles. *Polymer Journal*, *53*(5), 655–665. DOI: 10.1038/s41428-020-00456-w

Sood, N., Bhardwaj, A., Mehta, S., & Mehta, A. (2016). Stimuli-responsive hydrogels in drug delivery and tissue engineering. *Drug Delivery*, *23*(3), 748–770. DOI: 10.3109/10717544.2014.940091

Stuart, M. A. C., Huck, W. T., Genzer, J., Müller, M., Ober, C., Stamm, M., Sukhorukov, G. B., Szleifer, I., Tsukruk, V. V., Urban, M., Winnik, F., Zauscher, S., Luzinov, I., & Minko, S. (2010). Emerging applications of stimuli-responsive polymer materials. *Nature Materials*, *9*(2), 101–113. DOI: 10.1038/nmat2614

Tamesue, S., Takashima, Y., Yamaguchi, H., Shinkai, S., & Harada, A. (2010). Photoswitchable supramolecular hydrogels formed by cyclodextrins and azobenzene polymers. *Angewandte Chemie*, *122*(41), 7623–7626. DOI: 10.1002/ange.201003567

Tang, J., Qiao, Y., Chu, Y., Tong, Z., Zhou, Y., Zhang, W., Xie, S., Hu, J., & Wang, T. (2019). Magnetic double-network hydrogels for tissue hyperthermia and drug release. *Journal of Materials Chemistry. B*, *7*(8), 1311–1321. DOI: 10.1039/C8TB03301C

Ter Schiphorst, J., Van Den Broek, M., de Koning, T., Murphy, J., Schenning, A., & Esteves, A. (2016). Dual light and temperature responsive cotton fabric functionalized with a surface-grafted spiropyran–NIPAAm-hydrogel. *Journal of Materials Chemistry. A, Materials for Energy and Sustainability*, *4*(22), 8676–8681. DOI: 10.1039/C6TA00161K

Trombino, S., Curcio, F., Cassano, R., Curcio, M., Cirillo, G., & Iemma, F. (2021). Polymeric biomaterials for the treatment of cardiac post-infarction injuries. *Pharmaceutics*, *13*(7), 1038. DOI: 10.3390/pharmaceutics13071038

Wang, M., Chen, M., Niu, W., Winston, D. D., Cheng, W., & Lei, B. (2020). Injectable biodegradation-visual self-healing citrate hydrogel with high tissue penetration for microenvironment-responsive degradation and local tumor therapy. *Biomaterials*, *261*, 120301. DOI: 10.1016/j.biomaterials.2020.120301

Wang, M., Luo, L., Fu, L., & Yang, H. (2019). Ion responsiveness of polyacrylamide/sodium alginate (PAM/SA) shape memory hydrogel. *Soft Materials*, *17*(4), 418–426. DOI: 10.1080/1539445X.2019.1618325

Wojciechowski, J. P., Martin, A. D., & Thordarson, P. (2018). Kinetically controlled lifetimes in redox-responsive transient supramolecular hydrogels. *Journal of the American Chemical Society*, *140*(8), 2869–2874. DOI: 10.1021/jacs.7b12198

Wu, Y., Wang, L., Guo, B., Shao, Y., & Ma, P. X. (2016). Electroactive biodegradable polyurethane significantly enhanced Schwann cells myelin gene expression and neurotrophin secretion for peripheral nerve tissue engineering. *Biomaterials*, *87*, 18–31. DOI: 10.1016/j.biomaterials.2016.02.010

Xie, M., Wang, L., Guo, B., Wang, Z., Chen, Y. E., & Ma, P. X. (2015). Ductile electroactive biodegradable hyperbranched polylactide copolymers enhancing myoblast differentiation. *Biomaterials*, *71*, 158–167. DOI: 10.1016/j.biomaterials.2015.08.042

Yue, L., Wang, S., Wulf, V., & Willner, I. (2019). Stiffness-switchable DNA-based constitutional dynamic network hydrogels for self-healing and matrix-guided controlled chemical processes. *Nature Communications*, *10*(1), 4774. DOI: 10.1038/s41467-019-12697-2

Yun, T. G., Park, M., Kim, D.-H., Kim, D., Cheong, J. Y., Bae, J. G., Han, S. M., & Kim, I.-D. (2019). All-transparent stretchable electrochromic supercapacitor wearable patch device. *ACS Nano*, *13*(3), 3141–3150. DOI: 10.1021/acsnano.8b08560

Ze, Q., Kuang, X., Wu, S., Wong, J., Montgomery, S. M., Zhang, R., Kovitz, J. M., Yang, F., Qi, H. J., & Zhao, R. (2020). Magnetic shape memory polymers with integrated multifunctional shape manipulation. *Advanced Materials*, *32*(4), 1906657. DOI: 10.1002/adma.201906657

Zhang, H., Ba, S., Lee, J. Y., Xie, J., Loh, T.-P., & Li, T. (2020). Cancer biomarker-triggered disintegrable DNA nanogels for intelligent drug delivery. *Nano Letters*, *20*(11), 8399–8407. DOI: 10.1021/acs.nanolett.0c03671

Zhang, H., Zhang, Y., He, L., Yang, B., Zhu, S., & Yao, M. (2016). Thermal-responsive poly (N-isopropyl acrylamide)/sodium alginate hydrogels: Preparation, swelling behaviors, and mechanical properties. *Colloid & Polymer Science*, *294*(12), 1959–1967. DOI: 10.1007/s00396-016-3951-2

Zhang, J. T., Thunga, M., Petersen, S., Bhat, R., Liu, X., Weidisch, R., Fahr, A., & Jandt, K. D. (2009). Temperature-Sensitive Simultaneous Interpenetrating Polymeric Networks With Improved Mechanical Properties and Shrinking Kinetics. *Advanced Engineering Materials*, *11*(3), B12–B19. DOI: 10.1002/adem.200800266

Zhang, Y., & Wu, B. M. (2023). Current advances in stimuli-responsive hydrogels as smart drug delivery carriers. *Gels (Basel, Switzerland)*, *9*(10), 838. DOI: 10.3390/gels9100838

Zhao, C., Tian, S., Liu, Q., Xiu, K., Lei, I., Wang, Z., & Ma, P. X. (2020). Biodegradable nanofibrous temperature-responsive gelling microspheres for heart regeneration. *Advanced Functional Materials*, *30*(21), 2000776. DOI: 10.1002/adfm.202000776

Zhao, X., Wu, H., Guo, B., Dong, R., Qiu, Y., & Ma, P. X. (2017). Antibacterial anti-oxidant electroactive injectable hydrogel as self-healing wound dressing with hemostasis and adhesiveness for cutaneous wound healing. *Biomaterials*, *122*, 34–47. DOI: 10.1016/j.biomaterials.2017.01.011

Zhao, Y.-L., & Stoddart, J. F. (2009). Azobenzene-based light-responsive hydrogel system. *Langmuir*, *25*(15), 8442–8446. DOI: 10.1021/la804316u

Zhu, M., Wang, Y., Ferracci, G., Zheng, J., Cho, N.-J., & Lee, B. H. (2019). Gelatin methacryloyl and its hydrogels with an exceptional degree of controllability and batch-to-batch consistency. *Scientific Reports*, *9*(1), 6863. DOI: 10.1038/s41598-019-42186-x

Žuržul, N., Ilseng, A., Prot, V. E., Sveinsson, H. M., Skallerud, B. H., & Stokke, B. T. (2020). Donnan contribution and specific ion effects in swelling of cationic hydrogels are additive: Combined high-resolution experiments and finite element modeling. *Gels (Basel, Switzerland)*, *6*(3), 31. DOI: 10.3390/gels6030031

Chapter 9
Innovative Techniques in Hydrogel-Based Sunscreen Fabrication and Its Application in Healthcare

Brindha Devi Parthiban
https://orcid.org/0000-0003-2859-4659
Vels Institute of Science, Technology, and Advanced Studies, India

V. Gokul
Vels Institute of Science, Technology, and Advanced Studies, India

Ivo Romauld S.
https://orcid.org/0000-0003-0610-0646
Vels Institute of Science, Technology, and Advanced Studies, India

R. Sudha
Vels Institute of Science, Technology, and Advanced Studies, India

J. Manjunathan
Vels Institute of Science, Technology, and Advanced Studies, India

ABSTRACT

The biocompatibility, high water retention, and controlled-release characteristics of hydrogels three-dimensional networks of hydrophilic polymers have made them attractive carriers for sunscreen compositions. The hydrogel-based delivery systems can help to address the instability, skin irritation, and environmental issues that traditional sunscreens frequently face. The physical, chemical, radiation-induced, enzyme-assisted methods, electrospinning, and cryogelation are some of the manufacturing methods for hydrogel-based sunscreens that are examined in this review. Incorporating the hydrogels into sunscreen formulations improves skin hydration, increases the effectiveness of UV protection, and prolongs the stability of active components, such as chemical (oxybenzone, avobenzone) and physical (ZnO, TiO_2, iron oxide) UV filters. Moreover, hydrogels contain organic antioxidants like betalains and anthocyanins, which prevents photoaging and

DOI: 10.4018/979-8-3373-2337-4.ch009

oxidative stress. Hybrid formulations maximize these bioactive chemicals' bioavailability and photostability including hydrogel sunscreens loaded with nanoliposomes. The development of biodegradable polymers, smart hydrogels, and nanotechnology integration promises for the creation of sunscreens that are both efficient and environmentally benign in the future. By outlining their formulation techniques, drawbacks, and possible applications in the cosmeceutical sector, this paper states the potential of hydrogels as multipurpose sunscreen carriers.

9.1. INTRODUCTION

Hydrogel is one of the new polymers used to create new products in Cosmetic industry. Three-dimensional networks of hydrophilic, cross-linked polymers called hydrogels have the ability to absorb, expand, and retain huge amounts of water or watery liquids. Hydrogel properties are determined by several key factors, including the extent of cross-linking and the hydrophilicity of the polymer chains. Hydroxylic (-OH), carboxylic (-COOH), amidic (-CONH–), primary amidic (-$CONH_2$), and sulphonic (-SO_3H) groups are some of the functional groups responsible for the water-retaining capacity of hydrogels (Urimubenshi et.al. 2022). By varying the synthetic parameters, such as the cross-linker to monomer ratio, the initiator and monomer concentration, the reaction temperature, the vessel, and the duration, hydrogel properties can be altered (Singhal & Gupta 2016). Because of the hydrogel's exceptional capacity to absorb water by filling the porous spaces of the polymer, solutes can diffuse selectively throughout the matrix. These features make it possible to employ hydrogels in membrane separation processes (Kakad et al., 2024). Hydrogels can be made from a range of materials, including synthetic and natural systems, and they can be made for a number of separation technology uses. The term "hydrogel" was initially used in 1894 to describe a colloidal gel. The first discussion of hydrogel applications in the biomedical field was given by Wichterle and Lim in 1960. Because of their high capacity to absorb water, hydrogels swell and take on a soft, rubbery texture (Bashir et. al. 2020).In 1936, DuPont researchers produced the first synthetic hydrogel, which they called poly (2-hydroxyethyl methacrylate) (PHEMA) (Michálek et. al. 2022). The importance of PHEMA hydrogels in contact lens applications was demonstrated by Wichterle and Lim (1960). Some hydrogels are also known as "smart hydrogels" because of their ability to change in response to changes in solvent composition, pH, temperature, and electrical field (Karnakar & Gite 2025).

Hydrogels' have the capacity to absorb water which allows them to expand considerably without disintegrating, absorbing maintaining its strengthening. They are utilized in moisture-retaining applications like wound care because of this unique quality. They are therefore appropriate for use in pharmaceutical and biological applications. Hydrogels frequently resemble the mechanical characteristics of natural tissues in that they are soft and flexible. This characteristic is crucial for applications where the material must adapt to the flexibility and shape of surrounding tissues, such as medication delivery and tissue engineering. Certain hydrogels exhibit "smart" behavior by changing their physical properties in response to outside stimuli such as pH, temperature, light, or ionic strength. The term "stimuli-responsive" refers to these hydrogels, which find applications in tissue engineering, biosensors, and regulated drug delivery. In order to create a network structure, monomers or macromers are usually polymerized during the hydrogel synthesis process. One approach involves employing chemical agents (like free radicals, UV light, or chemical initiators) to crosslink polymer chains, while the other includes crosslinking density, which allows for customization of the hydrogel's mechanical properties and swelling behavior. Physical crosslinking can be caused by non-covalent interactions such as hydrogen bonds, ionic interactions, and Van der Waals forces. Unlike chemical crosslinking, which typically requires a chemical initiator, physical crosslinking typically produces reversible bonds. To make natural hydrogels, proteins or polysaccharides from plants or animals are typically extracted. These hydrogels are then cross-linked to produce the necessary network structure (Zhan et. al. 2022).

Figure 1. Structure of hydrogel

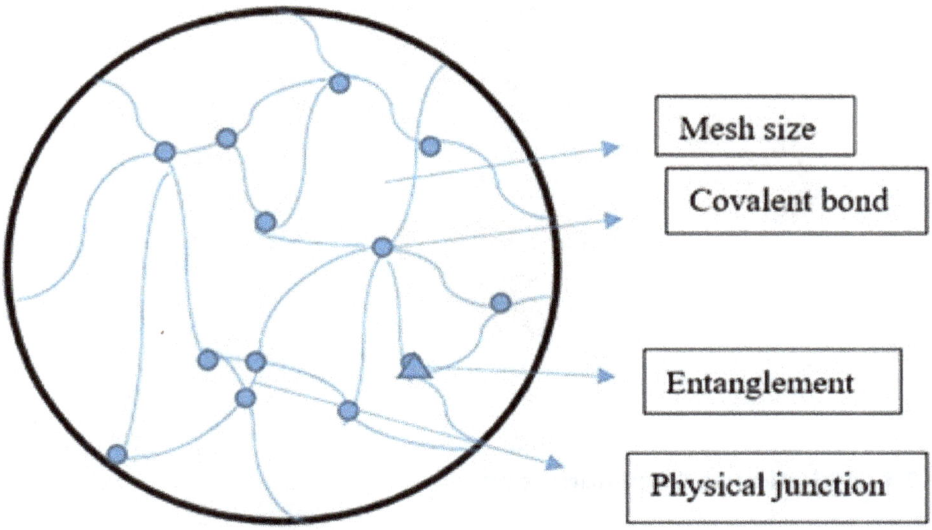

9.2. IMPORTANCE OF HYDROGELS IN SUNSCREEN FORMULATION

Hydrogels are becoming increasingly popular for sunscreen formulation because of their special nature and skin benefits. It is particularly advantageous to incorporate such water gels in cosmetic products such as sunscreens, lotions because of their high water molecule concentration. Among the many benefits of hydrogels are improved hydration and sunscreen application.

Figure 2. Importance of hydrogel in sunscreen formulation.

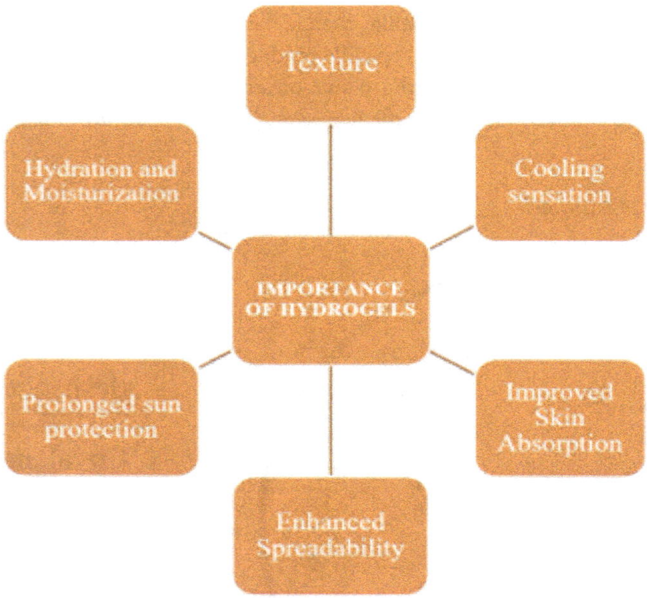

9.2.1. Hydration and Moisturization

Hydrogels' have the ability to withstand skin moisturization and this is one of the most significant advantages when incorporated in sunscreens. Sunscreens with hydrogels can hydrate the skin's outer layer, preventing it from drying out, stiffening. The hydrogel in the sunscreen helps the skin retain moisture throughout the day. Hydrogels strengthen skin hydration and promote general skin health when used with humectants such as glycerine or Aloe-Vera (Sghier et. al. 2024).

9.2.2. Enhanced Spread ability and Texture

Hydrogel-based sunscreens are ideal for oily or acne-prone skin due to their quick absorption and non-greasy, lightweight texture. Hydrogels provide a cool, silky feel without obstructing pores, in contrast to traditional sunscreens that contain thick creams or oils that can make the skin appear oily. This gel-like consistency makes sunscreen more enticing and promotes frequent use by enhancing comfort and preventing pore clogging or breakouts. People who might be put off by sunscreen due to the risk of outbreaks or excessive shininess can use them because they are non-comedogenic. Hydrogels give sunscreen a non-greasy texture that makes it more palatable and comfortable for everyday use, encouraging reliable sun protection (Lu et. al. 2021).

9.2.3. Improved Skin Absorption of Active Ingredients

Aside from hydrating, hydrogels allow for the penetration of the active ingredients of sunscreen. With a hydrogel matrix, active ingredients that provide sun protection, like avobenzone or zinc oxide, are better delivered to the skin. The hydrogel structure enables these ingredients to be more evenly distributed across the skin, ensuring that the sunscreen is applied uniformly and provides effective sun protection. When sunscreens include other skin-improving ingredients such as antioxidants (such as vitamin C) or anti-inflammatory agents such as niacin amide, such absorption efficacy becomes extremely important. Both, Avo-benzone and zinc oxide, two active components in sunscreen, are better absorbed and delivered when hydrogels are used and they guarantee uniform dispersion throughout the skin for efficient sun protection (Aungst, 2012).

9.2.4. Cooling Sensation

One advantage of hydrogels in sunscreens is that they cool the skin. Hydrogels provide a cold, refreshing feeling to the skin when used, especially during hot or sunny conditions. They are naturally absorbed and kept by water. In addition to improving comfort and soothing the skin after exposure, the sunscreen's cooling properties offer instant protection from the sun's heat. When added to sunscreens, hydrogels provide a cool, refreshing feeling. After being exposed to heat or the sun, this promotes relaxation and soothes the skin. Wearing sunscreen is more comfortable due to its cooling qualities, particularly during hot weather.

9.2.5. Prolonged Sun Protection

More water-based hydrogels can provide longer-lasting sun protection than sunscreens. Because of the controlled-release mechanism provided by hydrogels, the ingredients in sunscreen's active ingredients are absorbed and released into the skin gradually over time. This controlled release reduces the need for reapplication throughout the day, resulting in more effective and durable sun protection. Hydrogels enable the controlled release of sunscreen ingredients for prolonged sun protection. This slow absorption reduces the need for frequent reapplication (Van Der Pols et. al. 2006).

9.2.6. On-Greasy Finish for All Skin Types

Sunscreens are therefore perfect for people with oily skin who avoid them since they don't want breakouts or more sheen. The sunscreen's non-greasy texture prevents UV rays from the skin's natural oiliness. The delicate, water-based nature of hydrogels makes them ideal for oily or sensitive skin. They provide protection from the sun without making things shiny or causing breakouts. Sunscreen's non-greasy texture makes it more palatable and comfortable to wear every day (Urimubenshi et. al. 2022).

9.3. FABRICATION TECHNIQUES OF HYDROGELS

In hydrogel-based sunscreens, polymers, UV filters, and antioxidants are combined to create a stable, effective, and skin-friendly formulation. It is vital to build a three-dimensional network of polymers that retains water and permits the slow release of the active chemicals. Physical crosslinking, which relies on non-covalent interactions like hydrogen bonds and ionic contacts, and chemical crosslinking, which forms strong covalent bonds for long-lasting stability, are the two primary crosslinking methods. The film-forming and biocompatible qualities of both natural and synthetic polymers make them widespread (Li et al., 2025). Natural pigments, UV filters, and antioxidants are combined into a three-dimensional polymer network to create hydrogel-based sunscreens, which guarantee even application and enhanced skin compatibility. UV filters like iron oxide, zinc oxide, and titanium dioxide are either evenly distributed or encapsulated within the hydrogel to provide uniform coverage and lessen skin irritation (Li et. al. 2025).

Hydrogels are multifunctional networks of polymers with a large water-holding capacity, making them perfect for use in biological, pharmacological, and environmental applications. They are created utilizing a range of methods tailored to specific structural and functional requirements. Chemical cross-linking, which forms covalent bonds with cross-linkers, yields robust, pliable hydrogels ideal for drug delivery and tissue engineering. Physical cross-linking creates biocompatible and reversible systems suitable for injectable formulations and wound healing, based on non-covalent interactions such as hydrogen bonds and ionic forces (Du & Meng 2024).

Photopolymerization uses light-sensitive precursors to allow spatially controlled gelation, which facilitates applications in biosensing and 3D bioprinting. New technologies like 3D and 4D printing can be used to create hydrogel designs that are responsive to stimuli. For focused medication delivery and diagnostics, hydrogel particles with a large surface area and modifiable release profiles must be produced via emulsion polymerization, microfluidics, nanoprecipitation, and self-assembly. By improving these systems' surface functionalization and stimuli-responsiveness, their therapeutic effectiveness can be raised. Overall, the fabrication method chosen has a significant impact on the physicochemical properties and potential uses of hydrogel-based systems (Yu et. al. 2024).

9.3.1. Physical Selection and Functional Considerations

In order to form stable three-dimensional networks, sunscreen hydrogels need hydrophilic, biocompatible polymers. Both natural and synthetic polymers are employed. Agar, chitosan, alginate, hyaluronic acid, xanthan gum, and gelatin are examples of natural polymers. Polyethylene glycol (PEG), carbopol, polyvinyl alcohol (PVA), hydroxyethyl cellulose (HEC), and polyacrylic acid (PAA) are examples of synthetic polymers. These polymers may contain functional groups (-OH, -COOH, $-NH_2$, $-CONH_2$) that aid in water retention, crosslinking, and UV filter interaction. Natural polymers are usually chosen due to their skin-friendliness and biodegradability, especially when combined with plant-based antioxidants like betalains and anthocyanins (Gholap et. al. 2024).

9.3.2. Physical Cross Linking

The process by which non-covalent interactions like thermally induced gelation, ionic interactions, and hydrogen bonding produce three-dimensional networks in hydrogels is known as physical crosslinking. These interactions can be employed in circumstances where the use of artificial chemicals is not required because they are reversible and do not require artificial crosslinking agents. Due to the reversible nature of these interactions, the gelation process can be controlled and adjusted in response to temperature or pH variations. This adaptability is highly beneficial in biomedical applications where the properties of the hydrogel can be changed to meet specific needs. Moreover, the absence of dangerous crosslinking agents enhances the hydrogel's safety and biocompatibility (Infante et. al. 2021).

9.3.3 Thermal Gelation

Thermally sensitive polymers, such as gelatin and agar, gel reversible when heated and cooled. For instance, gelatin forms a hydrogel by dissolving in warm and cooling water. This technique works well for adding antioxidants, vitamins, and UV filters that are sensitive to heat without degrading the material. As an illustration, agar-gelatin hydrogel loaded with zinc oxide nanoparticles produces a semi-transparent UV-blocking layer suitable for broad-spectrum protection (Sharma et. al. 2024).

9.3.4. Ionic Crosslinking

Polymers such as sodium alginate form hydrogels when multivalent cations such as calcium (Ca_2+) or barium (Ba_2+) are present. This method is particularly effective for encapsulating metal oxide nanoparticles such as ZnO or FeO_3 and natural antioxidants. Crosslinking occurs when sodium ions are replaced by divalent cations in the alginate framework, forming egg-box structures that trap active ingredients (Thongrueng et. al. 2024).

9.3.5. H-Bonding

Functional groups such as amine (-NH_2), carboxyl (-COOH), and hydroxyl (-OH) interact to form hydrogels through hydrogen bonding. Soft, elastic hydrogels that are ideal for topical application can be produced by combining polyethylene glycol (PEG) with polyvinyl alcohol (PVA) or polyacrylic acid (PAA) to form stable networks. Hydrogen-bonded gels are ideal for incorporating betalains and anthocyanins, which require a mild encapsulation environment due to their sensitivity to pH (Rotaru-Zăvăleanu et. al. 2024).

9.3.6. Chemical Cross Linking

Chemically cross-linked hydrogels have recently attracted increased interest because of their high mechanical strength. The literature has employed a variety of methods to produce chemically cross-linked hydrogel, as detailed below (Witono et. al. 2024).

9.3.6.1. Crosslinking with Aldehydes

Chitosan is a naturally occurring polymer with -NH_2 groups that can be chemically crosslinked using aldehydes such as glutaraldehyde. Schiff bases are produced by these reactions, and they create stable hydrogel networks that can contain antioxidants and nanoparticles. For example, chitosan-glutaraldehyde hydrogel with iron oxide nanoparticles and anthocyanins showed improved skin stability, adhesion, and SPF (Li et. al. 2025)

9.3.6.2 Free Radical Polymerization

By using chemical initiators such as ammonium persulfate or ultraviolet light to polymerize hydrophilic vinyl monomers, such as acrylamide or acrylic acid, cross-linked hydrogel networks can be produced. UV filters and antioxidants can be carefully encapsulated using this technique. Photo-initiated polymerization is particularly crucial for the in-situ production of hydrogel sunscreen films (Liu & Chen 2024).

9.3.6.3 Enzymatic Crosslinking

In physiological settings, stable hydrogels are produced through enzyme-mediated crosslinking using natural catalysts such as horseradish peroxidase (HRP) and hydrogen peroxide ($H2O_2$). This technique is eco-friendly and works well with bioactive sunscreen formulations. For example, hyaluronic acid–tyramine conjugates cross-linked by HRP provide a soft hydrogel matrix ideal for delivering antioxidants and UV filters (Zhang & Wang 2023).

Table 1. Chemically cross linked hydrogels.

polymer	Method type	Loaded drug
Chitosan	Cross linking with aldehyde	Indomethacin
chitosan /Pectin	Cross linking by Diels Alder reaction	5-fluorouracil
Gelatin	Cross linking with aldehyde	TGF-β1
Chitosan	Cross linking with aldehyde	Mitoxantrone
Dextran	Addition reaction	Hydrocortisone

Hydrogels are facilitated by hydrophilic groups, particularly NH2, COOH, and OH, which are present in hydrophilic polymers. The formation of an amine–carboxylic acid, an isocyanate–OH/NH2, or a Schiff base in these reactions indicates the presence of covalent bonds between polymer chains

9.3.7. Nanoparticle-Encapsulated Hydrogel Systems

To get around the photo-instability of chemical UV filters and boost SPF efficiency, inorganic UV blockers such as ZnO, TiO_2, and FeO_3 nanoparticles are incorporated into hydrogel matrices.

Benefits:

1. Stop nanoparticles from aggregating.
2. Minimize cytotoxicity and irritation by reducing direct skin contact.
3. Increase the efficiency of UV scattering and reflection because of uniform dispersion.

Direct mixing, in-situ polymerization, or spray-drying methods are used to accomplish encapsulation. Because they can produce translucent, flexible films and are compatible with metal oxides, hydrogels based on chitosan are frequently utilized (Fantini et. al. 2024).

9.3.8. Hydrogel Encapsulation of Natural Pigments

Natural antioxidants that are sensitive to light, heat, and pH include betalains (found in beetroot and amaranth) and anthocyanins (found in red cabbage and berries). Hydrogels must be encapsulated in order to maintain their functionality.

Encapsulation methods:

1. Emulsion crosslinking is the process of trapping pigment droplets inside a polymer shell using water-in-oil emulsions.
2. Ionic gelation is the process of adding pigment-loaded polymers, like chitosan, dropwise into cross-linker solutions, like TPP.
3. Integration of Nano-liposomes: For dual protection, natural pigments are first embedded in hydrogel matrices after being encapsulated in liposomes.

These techniques enable the hydrogels to progressively release pigments, guaranteeing the skin's ongoing photo-protection and antioxidant activity (Urimubenshi et. al. 2022).

Figure 3. Encapsulation and release process in hydrogel based systems.

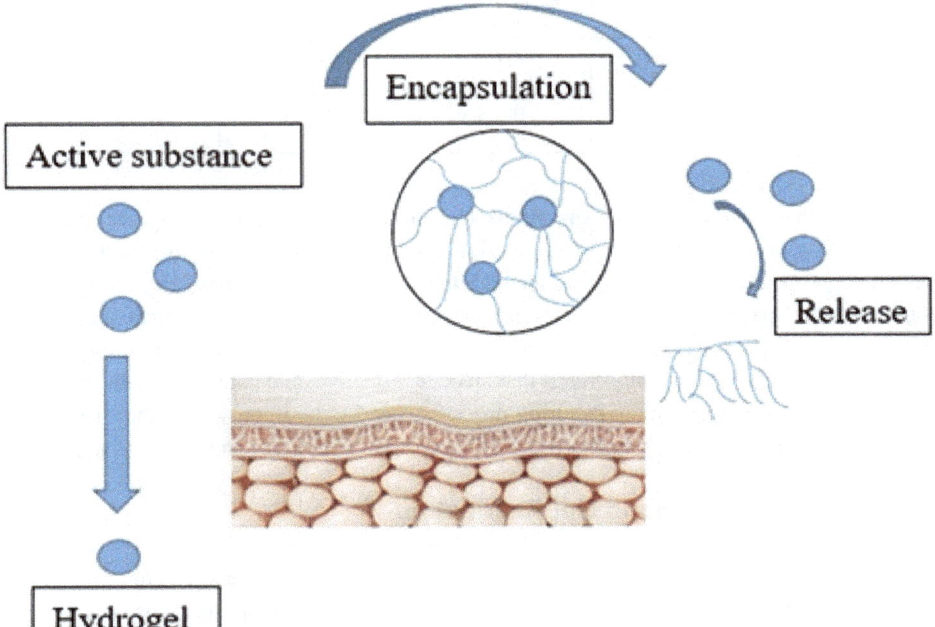

9.3.9. Controlled Release Mechanism

One of the main benefits of hydrogel-based sunscreens is controlled release, which enables the slow distribution of antioxidants and UV filters in response to moisture, pH, and temperature.

1. Diffusion-controlled release: Active compounds diffuse passively through the matrix of hydrated gel.
2. Swelling-controlled release: When hydrogel comes into touch with skin, it swells, causing the components to expand and release.
3. Stimulus-responsive release: Smart hydrogels adjust release kinetics in response to environmental cues, such as sweat, skin pH, or UV intensity.

This characteristic ensures consistent sun protection during exposure periods, improves skin tolerance, and reduces the need for frequent sunscreen reapplications (Cassayre et. al. 2024).

Table 2. Comparative analysis of fabrication techniques

Technique	Crosslinking type	Advantages	Drawbacks	Suitability for sunscreen
Thermal gelation	physical	Easy to process	Poor mechanical strength	Suitable for sensitive ingredients
Ionic crosslinking	physical	Biocompatible	Low water resistance	Good for alginate and chitosan based gels
Aldehyde crosslinking	chemical	Stable and strong gels	Cause toxicity	Effective for UV-filters
UV-initiated polymerization	chemical	Precise control over structure	Requires UV-safe setup	In-situ curable sunscreen sheets
Enzymatic crosslinking	Chemical (biogenic)	Eco-friendly	Higher cost	Ideal for antioxidant loaded gels
Nanoparticle embedding	Hybrid	High SPF	Requires stabilization	Essential for ZnO,Fe_2O_3

Manufacturing hydrogel-based sunscreens requires an interdisciplinary approach that combines materials, polymer science, and dermatological safety. The techniques used, which include selecting appropriate polymer matrices and adding photoactive and antioxidant chemicals, must ensure that the final product offers accurate UV protection, long term environmental stability, and visual appeal. The advancements in nanoparticle encapsulation, and enzymatic crosslinking hold great promise for future sunscreens that are safe, effective, and consistent with sustainable cosmetic innovation. These manufacturing processes need to be improved further if hydrogel-based sunscreen products are to be more widely accepted by customers and adhere to regulations.

9.3.10. Microgel and Nano Gel Fabrication

Microgels and nanogels, which are produced using hydrogel fabrication techniques that use particles at the micro- and nanoscale, are widely used in biological applications such as tissue engineering, drug delivery, and diagnostics. Microgels are typically made using methods like emulsion polymerization, which involves dispersing monomers in an oil phase and polymerizing them into spherical particles, or microfluidics, which offers precise control over droplet size and homogeneity. Extrusion fragmentation, precipitation polymerization, and photopolymerization are additional techniques with varying degrees of control and scalability. Because they are more surface-active and smaller, nanogels are frequently created by ionic gelation, nanoprecipitation, or inverse emulsion polymerization. Nanogel architecture and functionalization can be fine-tuned thanks to sophisticated methods like RAFT polymerization and click chemistry. Both microgels and nanogels are perfect for regulated and targeted therapeutic administration since they can be made to react to stimuli like pH, temperature, or enzymes. To maximize effectiveness for certain biomedical tasks, their manufacture necessitates careful consideration of crosslinking density, biocompatibility, and release kinetics.

9.4. CONVENTIONAL vs HYDROGEL BASED SUNSCREEN

Sunscreen lotion is a well-known example of UV-protective technology. Both organic and inorganic ingredients make up the majority of sunscreen formulations. Organic sunscreens, like oxybenzone, often work by absorbing ultraviolet light. The inorganic component, such as titanium dioxide, frequently acts as a physical sunscreen by absorbing, reflecting, and dispersing UV rays. When sunscreen is applied topically, it protects the stratum corneum and the skin's various layers. Even while the exact cause of the majority of skin cancers, although there is ongoing discussion over melanoma and non-melanoma, the primary goal of sunscreen chemicals is undoubtedly to shield the skin from UV radiation as much as possible (Jiang et. al. 2024). Multiple components of an ideal sunscreen include UV absorption and reflection, stabilising strategies such as encapsulation or complexation for sustained efficacy, and the ability to trap free radicals, such as with antioxidants, to reduce photochemical damage. One of the most important ingredients in contemporary skincare products is sunscreen, which protects the skin from UV radiation's damaging effects, such as sunburn, early aging, DNA damage, and even skin cancer. Sunscreens have historically been made as lotions, creams, sprays, and gels with a blend of inorganic (physical) and organic (chemical) UV filters (Miranda et. al. 2024). These traditional formulas have been effective for decades, but as consumer concerns about their sensory feel, compatibility with skin, and environmental impact have grown, new creative solutions have emerged, most notably hydrogel-based sunscreens (Martinović et. al. 2024). The issue with some chemical filters is their photo-stability; many of them deteriorate when exposed to sunlight, losing their efficacy over time and necessitating regular reapplication. Additionally, some of these chemical ingredients have sparked worries about possible hormone disturbance, allergic reactions, and skin sensitivity. Because of their toxicity to the environment, chemicals like oxybenzone and octinoxate are currently prohibited in a number of nations after being connected to coral reef bleaching.

In order to address the drawbacks of traditional sunscreens, hydrogel-based sunscreens have become a viable substitute. Hydrogels are networks of three-dimensional, hydrophilic polymers that can retain a lot of water while still being structurally sound, and their special physicochemical characteristics make them perfect for topical skincare applications. Hydrogel sunscreens are made with a combination of natural (like chitosan, xanthan gum, and alginate) and synthetic (like PVA, PEG, and HEC) polymers that offer a cooling, non-greasy, and breathable matrix for UV filters and bioactive agents (de Araújo Andrade et. al. 2022).

Table 3. *Difference between conventional and hydrogel based sunscreens*

Parameter	Conventional-sunscreen	Hydrogel based sunscreen
Base material	Oil/cream based	Water based hydrogel
Skin feel	Greasy, heavy	Lightweight, refreshing, non-greasy.
Hydration	Limited, may cause dryness	Excellent skin hydration
UV filter stability	Prone to photo-degradation	Improved with encapsulation and slow release
Reapplication frequency	Frequent due to quick degradation	Reduced due to controlled release
Prone clogging	Possible especially in oil skin	Rare, non-comedogenic
Cooling sensation	Minimal	High, soothes overheated skin
Environmental impact	High with chemical filters	Lower, eco-friendly option available
Antioxidant incorporation	limited	Easy to incorporate plant based antioxidants

9.4.1. ORGANIC SUNSCREEN COMPONENTS

Figure 4. organic components of sunscreen.

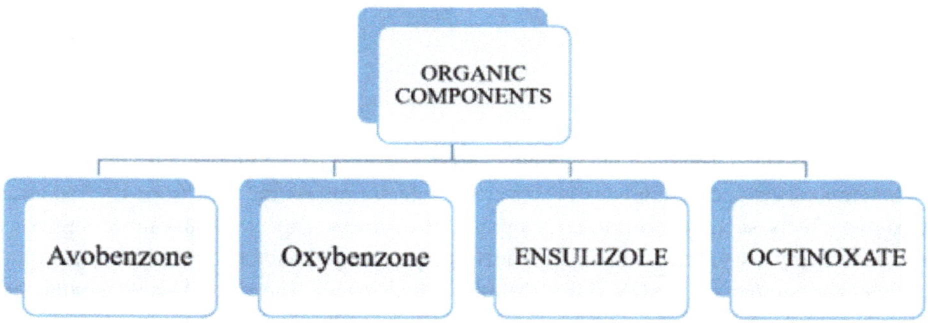

AVOBENZONE

Sunscreens contain avobenzone, an oil-soluble chemical molecule that absorbs UVA rays from 310 to 400 nm. It works well as a broad-spectrum UV filter since it is a derivative of dibenzoyl-methane. Avobenzone functions by absorbing UV rays and transforming them into less hazardous heat energy, in contrast to physical sunscreens like zinc oxide. However, because avobenzone is photo-unstable, its effectiveness is diminished when it is exposed to sunshine (Doan et. al. 2024).

OXYBENZONE

Oxybenzone is an organic substance that is frequently used as a UV filter in sunscreens and personal hygiene products. It helps shield the skin from sun damage by absorbing and dissipating ultraviolet (UV) rays. It is a member of the benzophenone family. Because oxybenzone blocks (Zhang et. al. 2025).

ENSULIZOLE

Phenylbenzimidazole Sulfonic Acid, or ensulizole, is an organic molecule that dissolves in water and is utilised as a UVB sunscreen in skincare and cosmetic goods. It offers some UVA protection and efficiently absorbs UVB rays (280–320 nm), although for broad-spectrum coverage, it is typically used in conjunction with other filters like titanium dioxide or avobenzone. For oily or acne-prone skin, Ensulizole is the perfect sunscreen because it is non-greasy and lightweight, unlike many other chemical sunscreens (Arnanz et. al. 2025).

OCTINOXATE

A common organic UV filter included in sunscreens and personal care products is octinoxate, sometimes referred to as octinocate or ethylhexyl methoxycinnamate. Absorption of UVB rays, which range in wavelength from 280 to 320 nm, is its primary purpose. This reduces the harm that prolonged exposure to the sun does to the skin and helps prevent sunburn. Due to its lipophilic nature, which facilitates easy skin absorption, it is commonly used as an ingredient in lotions, creams, and lip balms. It is well known for being stable and working well with other sunscreen ingredients like avobenzone and zinc oxide that provide more thorough UV protection (Schalka et. al. 2014).

9.5. ADVANTAGES OF HYDROGEL IN SUNSCREEN

Figure 5. Applications of hydrogel in sunscreen.

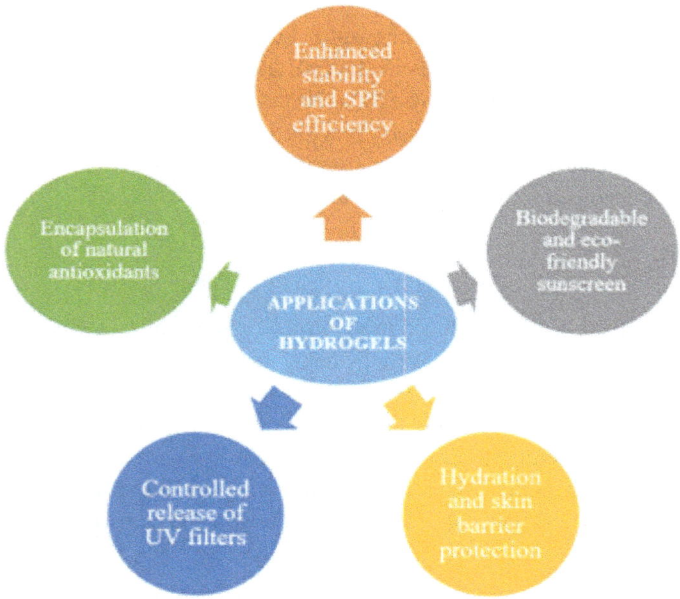

9.5.1. Controlled Release of UV Filters

The controlled and prolonged release of UV filters in hydrogel-based sunscreens enhances their stability, effectiveness, and suitability for all skin types. UV filters can be incorporated into hydrogel matrices to regulate their release, postponing photo-degradation and gradually increasing protection. Hydrogels can be used to encase physical and chemical UV filters. The hydrogel network gradually swells in response to temperature and skin moisture, allowing the sunscreen's active ingredients to be released in a controlled manner. This guarantees long-term UV protection while lowering skin irritation and bloodstream penetration—common problems with some chemical sunscreens. Hydrogel-formulated sunscreen generally offers higher SPF efficacy, lower toxicity, and extended release, making it a promising alternative to conventional sunblocks (Heydari et. al. 2025).

9.5.2. Encapsulation of Natural Antioxidants

Anthocyanins, betalains, and polyphenols are examples of natural antioxidants that can be added to hydrogel-based sunscreens for improved UV protection and skin healthThese bioactive compounds have potent antioxidant and photoprotective properties that help scavenge free radicals, lower oxidative stress, and prevent UV-induced photoaging. Encapsulating such natural pigments in hydrogels ensures stability, controlled release, and improved bioavailability.Due to the fact that betalains and anthocyanins are very sensitive to light, pH, and temperature, hydrogel encapsulation is very important to their activity. Upon application to the skin, the hydrogel matrix facilitates a gradual release of these antioxidants while protecting them from degradation. A biodegradable, eco-friendly, and efficient substitute for chemical sunscreens can be formulated by using natural antioxidants in hydrogel-based sunscreens, which provide both broad-spectrum UV protection and skin-caring benefits (Korkmaz & El 2025).

Figure 6. Hydrogel- based sunscreen system with natural antioxidants.

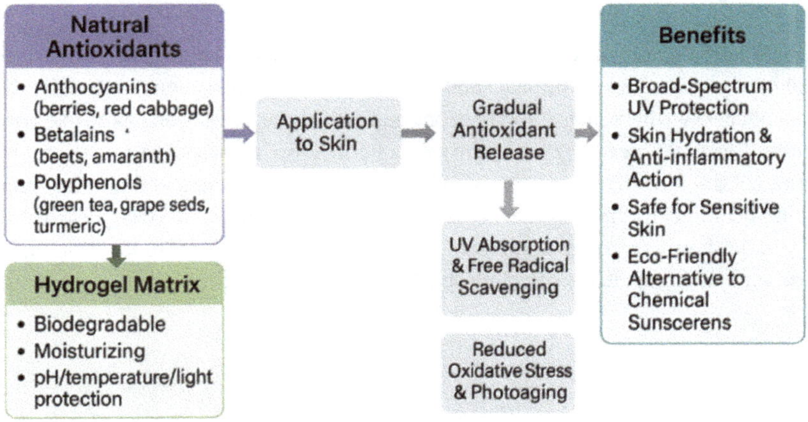

9.5.3. Hydration and Skin Barrier Protection

Hydrating the skin while offering UV protection is one of the main benefits of hydrogel-based sunscreens. Chemical filters found in traditional sunscreens, especially those with high SPF, can cause skin irritation, dryness, and disturbance of the skin's natural barrier. Because of their high water content and biocompatibility, hydrogels have a moisturizing effect that supports the integrity and hydration of skin. Networks of hydrophilic polymers that can hold a lot of water make up hydrogels. They help seal in moisture and stop trans-epidermal water loss (TEWL) by covering the skin with a thin, permeable layer. This moisturizing effect increases the skin's suppleness and suppleness, which is particularly beneficial for those with dry, sensitive, damaged skin (Del Rosso & Kircik 2025).

9.5.4. Enhanced Stability and SPF Efficiency

Hydrogel-based sunscreens offer significant advantages over conventional formulations in terms of longevity and SPF (Sun Protection Factor) efficacy, reducing a number of their negative aspects. Traditional sunscreens, particularly those containing chemical UV filters, often exhibit photo-instability and require frequent reapplications to maintain their efficacy. Hydrogel-based delivery methods make UV filters more stable, ensuring long-lasting protection and reducing deterioration from prolonged sun exposure. Hydrogels act as protective carriers by encasing UV filters, antioxidants, and bioactive compounds in their polymeric matrix. This keeps the active ingredients from oxidizing, photo-degrading, or evaporating, improving formulation stability and prolonging shelf life. For example, zinc oxide (ZnO) and titanium dioxide (TiO_2), which are also frequently used in physical sunscreens, have a propensity to aggregate over time, decreasing their ability to block UV rays. Their encapsulation in a hydrogel matrix gives them uniform dispersion, preventing agglomeration and boosting UV protection (Cefali et. al. 2019).

9.5.5. Biodegradable and Eco-Friendly Sunscreen

Hydrogel-based sunscreens have drawn a lot of attention as an environmentally friendly and biodegradable substitute for traditional sunscreens. Traditional sunscreens, especially those that include chemical UV filters such as oxybenzone and octinoxate, have been linked to environmental pollution, coral reef bleaching, and marine contamination. Hydrogel formulations, however, incorporate natural bioactive ingredients and biodegradable polymers and are thus more skin-friendly and ecologically friendly. Some biodegradable and renewable polymers that naturally degrade without harming aquatic habitats and constitute hydrogels include chitosan, alginate, hyaluronic acid, and derivatives of cellulose (de Souza et. al. 2025).

Figure 7. Biodegradable and eco-friendly sunscreen.

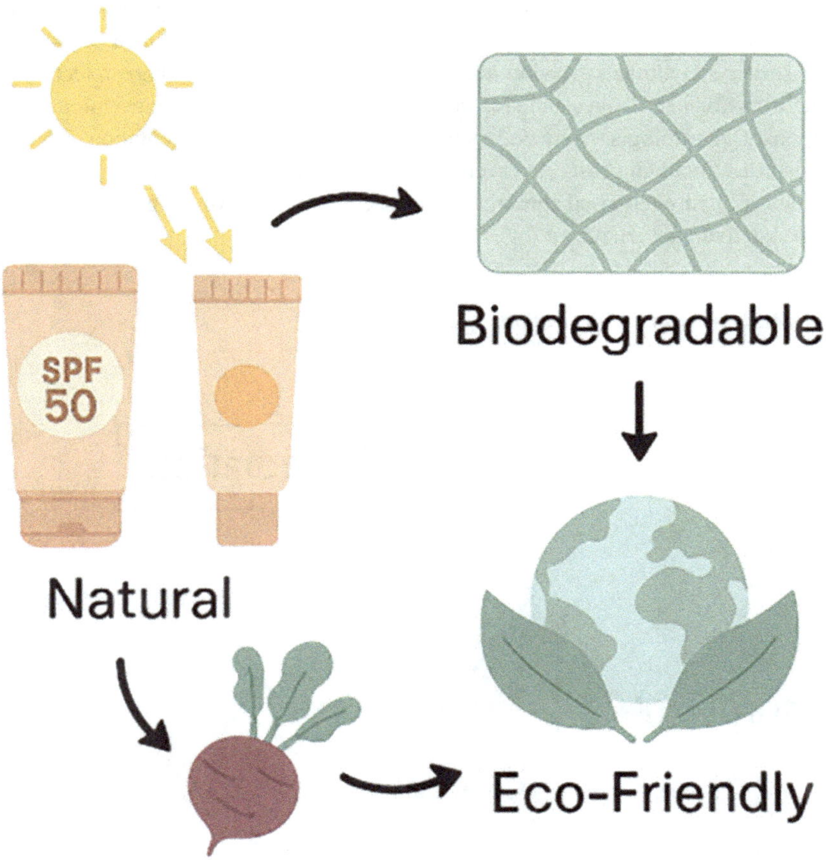

9.5.6. Wound Healing and UV Protection

Hydrogel-based sunscreens are very helpful in healthcare applications because of their special dual-functionality in UV protection and wound healing. While the major goal of conventional sunscreens is to stop UV-induced skin damage, hydrogel formulations offer extra therapeutic advantages by preserving moisture, encouraging cellular renewal, and supplying bioactive substances for improved skin healing. Because of the high water content of hydrogels, the damaged region is kept wet and excessive drying is avoided, creating the ideal conditions for wound healing. This fluidity promotes cell migration, lowers inflammation, and speeds up tissue healing. Hydrogels can also lower the incidence of secondary problems by serving as a barrier against microbial infections. To speed up the healing process, some hydrogel formulations contain natural extracts, growth hormones, or antimicrobial agents (Rennekampff et. al. 2010).

Since hydrogel-based sunscreens can offer improved UV protection, skin hydration, and post-treatment recovery benefits, they are becoming increasingly popular in dermatology and cosmetic surgery. Hydrogels provide a calming, non-greasy texture that protects skin from damaging ultraviolet (UV) rays and aids in skin healing, in contrast to conventional sunscreen formulations. Dermatologists frequently advise people with hyperpigmentation, sensitive skin, or skin disorders including rosacea, eczema, or psoriasis to use broad-spectrum sunscreen. In these situations, hydrogel-based sunscreens because hydrogels have a high water content, they help keep skin hydrated (Haney, 2024).

Figure 8. Mechanism of UV protection.

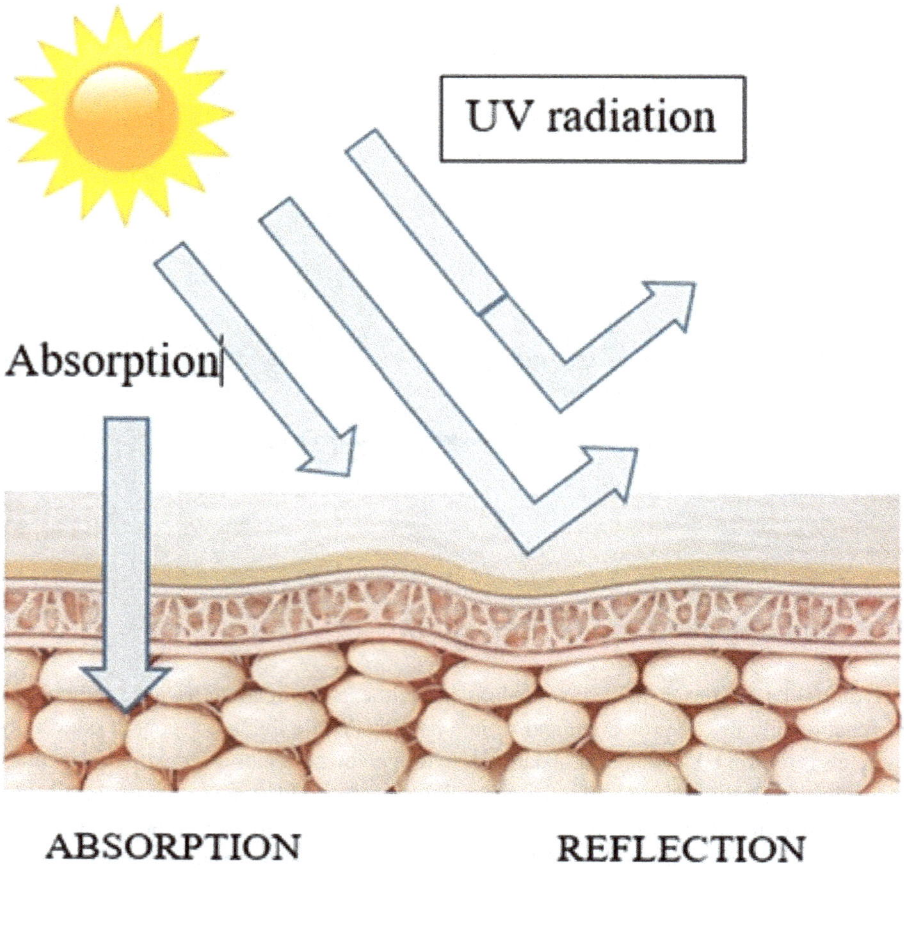

9.5.7. Sports and Outdoor Use

Sunscreens with a hydrogel basis are especially helpful for outdoor workers and athletes who need consistent, long-lasting protection from the sun while exercising. The pain that comes from prolonged sun exposure and perspiration is lessened by their non-greasy, lightweight nature. Water resistance is increased and precise control over the release rates of active substances is made possible by advanced production procedures such enzyme-mediated crosslinking and nanoparticle inclusion. With longer UV protection, a cooling effect, and eco-friendly substitutes for conventional formulas, these advancements make hydrogel sunscreens perfect for outdoor enthusiasts. Examples of products that are appropriate for high-sweat environments include Heliocare 360° Water Gel SPF 50+, which provides protection that is both sweat-proof and water-resistant (Sahoo, 2023).

9.5.8. Anti-Aging and Cosmeceutical Property

The potential of hydrogel-based sunscreens to moisturize the skin, provide antioxidants, and stop UV-induced oxidative damage has led to their increased inclusion in anti-aging skincare regimens. These formulas enhance skin suppleness and lessen pigmentation, fine wrinkles, and other signs of photoaging when combined with active ingredients like retinol, vitamin C, or plant polyphenols. The translucent and breathable nature of hydrogel sunscreens makes them ideal for daily use in cities under makeup because they provide a pleasant and smooth base. The skin's natural barrier and moisture levels are supported by these products in addition to providing protection from UV damage (Alves et. al. 2024).

9.6. CHALLENGES AND LIMITATIONS OF HYDROGEL BASED SUNSCREEN

9.6.1. Stability and Degradation Issues

The type of polymer, crosslinking density, temperature, pH of the surrounding environment, and UV exposure all affect how stable hydrogel is. More mechanical and thermal stability is generally offered by synthetic polymers like PEG and PVA, while natural polymers like chitosan and alginate are more vulnerable to microbial and hydrolytic breakdown. Since covalent bonds provide greater resistance to deformation and breakage over extended exposure, chemically cross-linked hydrogels exhibit greater stability than physically cross-linked ones. For sunscreens, photo stability is especially important. By serving as a physical barrier, encasing UV filters and antioxidants (such as avobenzone and anthocyanins) in hydrogel matrix lessens their deterioration in the presence of sunlight (Gulati et. al. 2024). Hydrogel networks can be made even more durable by adding photo stabilizers or dispersions of nanoparticles (ZnO, TiO_2). Enzymatic activity, oxidation, hydrolysis, and exposure to the environment all contribute to the breakdown of hydrogel. The kind of polymer and degree of crosslinking can be used to customize the pace of deterioration. Schiff base crosslinked chitosan hydrogels, for instance, have stronger connections and break down more slowly, providing longer stability. Research has indicated that when hydrogels containing betalains and anthocyanins are stored properly (low light, cool temperature, neutral pH), they can retain their antioxidant activity for as long as 4-6 months (Batista et. al. 2022)

9.6.2. Limited Water Resistance

Hydrogel-based sunscreens are primarily water-soluble, in contrast to conventional oil-based sunscreens, which create a hydrophobic barrier on the skin. This characteristic lessens their resistance to perspiration and exposure to water, which lessens their usefulness for outdoor sports and swimming activities. Water resistance can be increased by creating hydrogels with hydrophobic changes or adding film-forming chemicals, although doing so may change the product's skin feel and biodegradability (de Araújo et. al. 2024). Finding a balance between water resistances and preserving the hydrogel's inherent hydrophilic properties is the difficult part. Hydrogels are naturally hydrophilic, which makes them less water-resistant than oil-based sunscreens. Films made with hydrogels coated with nanoparticles (such as ZnO in agar-gelatin) can withstand being removed by light perspiration or water splashes. Adhesion and resistance to rinse-off can be increased by applying water-insoluble coatings or by adding film-forming substances like methylcellulose or polyacrylic acid (Heydari et. al. 2025).

9.6.3. Controlled Release Mechanism Challenges

For long-lasting UV protection, hydrogel-based sunscreens use controlled release mechanisms. A major obstacle, though, is guaranteeing constant release rates for antioxidants and UV filters. Numerous factors influence the rate of spread, such as: density of crosslinking in the hydrogel matrix, Temperature and humidity levels in the environment, the integrated active chemicals' hydrophilicity. A sunscreen that releases too slowly might not offer protection right away, and one that releases too quickly might cause UV filters to deplete too quickly and need to be reapplied frequently. To solve this problem, advanced Nano encapsulation technologies are being investigated; nevertheless, they increase formulation complexity and expense (Karisma et. al. 2021).

9.6.4. Skin Penetration and Irritation Risks

The great permeability of hydrogels can improve the skin's absorption of active substances. In some dermatological and cosmetic applications, this is advantageous, but when chemical UV filters or other active ingredients pass through the epidermal layer, there are hazards. Over-penetration might cause systemic absorption, allergic responses, or skin irritation, which raises safety issues. (Heydari et. al. 2025).

9.6.5. Formulation Complexity and Cost

In comparison to traditional sunscreens, hydrogel-based sunscreen formulations are more technically sophisticated because to the complex polymer networks, controlled release technologies, and biodegradable changes. Advanced processing techniques are needed to incorporate biocompatible cross-linkers, natural antioxidants, and UV filters based on nanoparticles, which raises the cost of manufacture. Additionally, differences in raw material quality, processing conditions, and polymer crosslinking might make it difficult to achieve batch-to-batch consistency.

This limits the commercial viability of mass-market sun care products by making large-scale production economically difficult (Almawash et. al. 2022).

9.6.6. Aesthetic and Sensory Limitations

Customers anticipate sunscreens to be pleasant on the skin, lightweight, and non-greasy. Although hydrogels are renowned for their ability to hydrate, some formulations may leave behind an unpleasant, tacky, or sticky residue. Additionally, the sunscreen may take on an unwanted color due to the natural pigments utilized for UV protection, such as betalains and anthocyanins. Consumer approval may be impacted if the skin becomes visibly stained as a result. To get around this restriction, researchers are looking into colorless or translucent formulations, but it's still difficult to preserve the stability and potency of natural colors in these formulations.

9.6.7. Environmental Challenges and Biodegradability Limitations

Despite the eco-friendly nature of hydrogel-based sunscreens, certain of the synthetic polymers employed in hydrogel formulations might not break down well in the environment. Even hydrogels based on biopolymers need particular circumstances (such as pH levels and microbial activity) to break down efficiently. Furthermore, many UV filtering nanoparticles, such iron oxide and zinc oxide, have the potential to build up in aquatic environments, which raises questions regarding their toxicity to marine life. Comprehensive environmental impact studies are necessary to ensure that hydrogel-based sunscreens are safe for marine life and biodegradable (Upadhyay & Bhagat 2025).

9.6.8. Safety Testing Requirements

Regulators require thorough safety analyses prior to approving new sunscreen formulas for release onto the market. For goods that incorporate natural extracts or nanoparticles, these assessments usually involve in vitro cytotoxicity testing, phototoxicity and photoallergenicity experiments, and skin irritation and sensitization investigations (OECD TG 404, 406). The integrity of the product must be guaranteed by evaluating stability and shelf life in a variety of environmental settings. It's also necessary to confirm SPF and UVA/UVB protection levels using defined standards, such ISO 24444 for SPF and ISO 24443 for UVA protection. Further testing is necessary to establish the microbiological stability of hydrogel-based sunscreens, which frequently contain new polymers or natural antioxidants. This is especially crucial because of their high water content, which can promote microbial growth if stored incorrectly (Zou et. al. 2022).

9.6.9. Labelling and Claims

Regulatory agencies carefully examine whether product claims are true, backed up by data, and verifiable Hydrogel-based sunscreens should prominently display important information such as the SPF rating, broad-spectrum protection, and the concentrations of the active ingredients. It is necessary to provide comprehensive usage and reapplication instructions along with accurate information on the water resistance duration, which is usually 40 or 80 minutes. Labels should include the product's expiration date and recommended storage conditions to ensure its efficacy and safety. In countries such as the EU, formulations containing nanomaterials must include the word "(nano)" in the ingredient list, for example, "Titanium Dioxide (nano)." Additionally, trustworthy testing and third-party certification are required to back up marketing claims like "natural," "biodegradable," and "reef-safe." These actions promote consumer confidence, legal compliance, and sunscreen label transparency (Staton, 2021).

9.6.10. Environmental and Marine Safety Regulation

Government efforts to reduce the environmental impact of sunscreen chemicals are growing, especially in environmentally sensitive areas. Several nations, including Hawaii, Palau, and Thailand, have outlawed sunscreens containing oxybenzone and octinoxate due to their harmful effects on coral reefs. Therefore, hydrogel-based sunscreens that contain mineral UV filters and plant-based antioxidants are becoming safer alternatives. These formulations require a detailed evaluation of their polymer matrices' biodegradability, the potential for nanoparticle leakage into aquatic environments, and the overall ecotoxicity to marine life.

Regulatory bodies and environmentally conscious consumers are placing an increasing amount of importance on products with proven environmental safety. Sustainability certifications such as COSMOS, ECOCERT, and Blue Label are increasingly important markers of trust and regulatory compliance (Grant et. al. 2025).

9.7. CONSUMER INSIGHTS ON HYDROGEL-BASED SUNSCREENS

Product preferences for sunscreens are rapidly shifting in favor of products that combine superior UV protection, enhanced skin compatibility, sensory appeal, and environmental responsibility. Hydragel-based sunscreens have grown in popularity due to their non-greasy, lightweight texture. Skin types that are sensitive, oily, or prone to acne are particularly well-suited for them. Their quick absorption, lack of sticky residue or white cast, and soothing, cooling properties make them highly desirable, especially in humid environments (Hussein et. al. 2025).

Surveys show that consumers between the ages of 18 and 35 prioritize texture and ease of application when choosing sunscreens. Many consumers are avoiding chemical filters such as oxybenzone and octinoxate because of their negative effects on marine habitats, illustrating how an environmental consciousness is influencing consumer choices. The market is searching more and more for formulations that include mineral UV filters or natural pigments like betalains and anthocyanins. Hydrogel sunscreens with biodegradable polymers and plant-based antioxidants help to achieve these goals by striking a balance between skin health and environmental sustainability. The demand for sunscreens designed for specific skin types and climates has surged. Hydrogel systems are in a good position to enable targeted delivery and AI-driven customization because of their compatibility with smart-release technologies. Since consumers favor non-comedogenic, dermatologist-tested products free of parabens, sulfates, and artificial fragrances, trust and transparency are still essential. Regional differences also play a role; Asian markets value multifunctional benefits, while Western consumers place more weight on claims that a product is cruelty-free, reef-safe, and organic. The market is ready for novel sunscreen products based on hydrogel, as these trends collectively demonstrate (Yahya et. al. 2025).

9.8. FUTURE PERSEPTIVES AND INNOVATIONS

9.8.1. Advancements in Hydrogel Formulations

Sunscreens made of hydrogel are becoming more popular because of their lightweight texture, biocompatibility, and moisturizing qualities. Future advancements, however, will use a variety of creative strategies to improve their effectiveness, stability, and market appeal.

9.8.1.1. Smart Hydrogels with UV-Responsive Behavior

The creation of hydrogels that react to UV light by altering their characteristics is a potential advance. Smart hydrogels like these can: Reducing needless skin absorption, UV-blocking compounds only release when exposed to sunshine, Alter the color or texture to show that it has to be reapplied, by dynamically raising their SPF in response to UV intensity, they offer adaptive protection (Heydari et. al. 2025).

9.8.1.2. Hybrid Hydrogels for Multifunctional Protection

The development of hydrogels that change their properties in response to UV light is a promising development. These intelligent hydrogels can: UV-blocking chemicals reduce unnecessary skin absorption and only release when exposed to sunlight, To indicate that it needs to be reapplied, change the texture or color, Adaptive protection is provided by their dynamic SPF increase in response to UV intensity (Battistin et. al. 2024).

9.8.2. Sustainable and Eco-Friendly Hydrogel Sunscreens

9.8.2.1. Biodegradable and Reef Safe Ingredients

Chemical UV filters like oxybenzone and octinoxate are being phased out by regulatory agencies because of their effects on marine life. Future sunscreens made of hydrogel will use Biodegradable biopolymers for safer formulations, like alginate and chitosan, Mycosporine-like amino acids (MAAs) are naturally occurring UV protectors derived from marine creatures, Flavonoids and carotenoids, two plant-derived UV blockers, are environmentally beneficial substitutes for sunscreen (Bhardwaj et. al. 2025).

9.8.2.2. Zero Waste and Sustainable Packaging

In an effort to lessen their influence on the environment, Hydrogel sunscreens will also use zero-waste packaging. This comprises: Dissolvable or edible film coatings that do away with conventional plastic containers, Biodegradable sachets composed of elements derived from seaweed, Hydrogel dispensers that can be refilled to save plastic waste (Singh et al., 2024).

9.8.3. Enhancing Consumer Experience and Market Adoption

9.8.3.1. Transport and Non-Sticky Formulation

The sticky residue left by hydrogel sunscreens is one of their drawbacks. Future studies will concentrate on: hydrogels that are colorless and sheer, with no discernible white cast, enhanced texture thanks to changes in polymer crosslinking, Formulations that absorb quickly and have breathable, light effects (Schaldemose et. al. 2023).

9.8.3.2. Personalization and AI- Driven Formulation

Personalized formulas are becoming more and more popular in the skincare sector. AI-driven instruments for skin analysis could: Adapt sunscreen formulas to each person's unique skin type, degree of UV exposure, and hydration requirements, Provide on-demand hydrogel mixin, stations so that customers can mix customized antioxidant and SPF mixtures, Utilize smartphone apps to identify when sunscreen is wearing off and send out reminders to reapply (Patil et. al.).

9.8.4. Future Regulatory and Safety Considerations

9.8.4.1. Addressing Skin Penetration and Toxicity Concerns

Safety is important because hydrogel-based sunscreens may improve the absorption of active ingredients. Future advancements will: To regulate the rates of skin penetration, optimize the polymer structure, Create UV filters that are non-toxic and stay on the skin's surface, Long-term dermatological research should be done to confirm safety characteristics.

9.8.4.2. Standardization of Hydrogel Sunscreen Testing

Hydrogel-based formulations might not be completely compatible with the SPF testing techniques now in use. Future studies will: Create innovative procedures for testing the stability and effectiveness of hydrogels, create in vivo testing models that replicate ambient contaminants, water, and perspiration, put into practice global standards for formulations that are reef-safe and biodegradable (Chasanah et. al. 2022).

9.8.5. Integration of Nanotechnology

Nanotechnology breakthroughs have the potential to greatly improve hydrogel-based sunscreens by enhancing the active components' encapsulation, protection, and targeted distribution. Natural antioxidants like anthocyanins and betalains can have their photo stability and skin retention increased by Nano carriers such Nano liposomes and Nano emulsions. Additionally, these technologies allow for controlled, sustained release, guaranteeing continued efficacy while exposed to sunlight. Furthermore, several active chemicals may be co-delivered by nanoparticles in a single hydrogel composition. The medicinal and aesthetic advantages of sunscreens are increased by their multipurpose nature. Therefore, the future generation of sophisticated, performance-driven sun care products will heavily rely on nanotechnology (Shalaby et. al. 2025).

9.8.6. Expansion into Multi-Functional Applications

Future hydrogel sunscreens might be created as multipurpose skincare products that include anti-aging, anti-inflammatory, and healing properties in addition to UV protection. Sunscreens can be transformed into cutting-edge therapeutic skincare products appropriate for everyday use, clinical treatment, and cosmetic enhancement by adding chemicals like vitamin C, niacinamide, ceramides, and hyaluronic acid (Manvi et. al. 2024).

9.8.7. Regulatory and Clinical Advancements

In order to incorporate nanomaterials, encapsulation systems, and biodegradable matrices, regulatory authorities will need to change their requirements as these technologies develop. Standardized procedures will be necessary to assess skin penetration, long-term photostability, and reef safety. Together with this, clinical validation will be required to enable widespread adoption and consumer confidence across a range of skin types, age groups, and medical problems (Aguilera et. al. 2023).

9. CONCLUSION

A promising approach to improving skin protection while guaranteeing better biocompatibility, stability, and application convenience is the creation of hydrogel-based sunscreen formulations. In order to maximize the formulation for broad-spectrum UV protection, improved stability, and extended skin adherence, this study investigated the addition of natural colors, iron nanoparticles, and chitosan-based encapsulation. Because of their high water content, hydrogels have a non-greasy, refreshing feel that makes them ideal for use in sunscreen compositions. Their ability to provide an uninterrupted stream of active chemicals increases the effectiveness of UV filters and reduces the need for frequent reapplications. The use of iron oxide and zinc oxide nanoparticles in this study greatly enhanced multi-spectrum UV protection by efficiently absorbing and dispersing dangerous radiation.

Moreover, pigments were encapsulated in chitosan-based matrices to increase stability, which improved the photo-protective effectiveness of natural pigments and stopped their degradation. One of the main benefits of this product is the decrease in synthetic UV filters, which are frequently linked to skin irritation and environmental toxicity. The addition of chitosan, which offered extra advantages like improved adhesion, improved biocompatibility, and antibacterial qualities, also increased the formulation's effectiveness. The FTIR scan's dominant peaks verified the presence of functional groups necessary for pigment stabilization and the pigments' successful encapsulation in chitosan-based matrices. Additionally, the formulation with exceptional performance characteristics was attained by optimizing the response surface methodology, which maximized pigment loading and encapsulation efficiency. Although this study showed that hydrogel-based sunscreens containing natural pigments and iron nanoparticles are feasible, more investigation is required to evaluate their long-term stability, viability for large-scale manufacturing, and consumer acceptability. Clinical studies will be necessary to confirm the formulation's safety and effectiveness in actual circumstances. In general, hydrogel-based sunscreen formulas that include iron nanoparticles and natural pigments offer a multifaceted approach to UV protection by fusing the advantages of moisturizing, photo-protection, and antioxidants.This work paves the way for further advances in the field of dermatological protection by providing support to the ongoing efforts in developing skin- and environment-friendly substitutes for sunscreens.

REFERENCES

Aguilera, J., Gracia-Cazana, T., & Gilaberte, Y. (2023). New developments in sunscreens. *Photochemical & Photobiological Sciences, 22*(10), 2473–2482. DOI: 10.1007/s43630-023-00453-x PMID: 37543534

Almawash, S., Osman, S. K., Mustafa, G., & El Hamd, M. A. (2022). Current and future prospective of injectable hydrogels—Design challenges and limitations. *Pharmaceuticals (Basel, Switzerland), 15*(3), 371. DOI: 10.3390/ph15030371 PMID: 35337169

Alves, P. L. M., Nieri, V., Moreli, F. D. C., Constantino, E., de Souza, J., Oshima-Franco, Y., & Grotto, D. (2024). Unveiling New Horizons: Advancing Technologies in Cosmeceuticals for Anti-Aging Solutions. *Molecules (Basel, Switzerland), 29*(20), 4890. DOI: 10.3390/molecules29204890 PMID: 39459258

Arnanz, J., Monari, A., Corral, I., & Nogueira, J. J. (2025). Setting the Bases of the Photogenotoxicity of p-Aminobenzoic Acid.

Aungst, B. J. (2012). Absorption enhancers: Applications and advances. *The AAPS Journal, 14*(1), 10–18. DOI: 10.1208/s12248-011-9307-4 PMID: 22105442

Bashir, S., Hina, M., Iqbal, J., Rajpar, A. H., Mujtaba, M. A., Alghamdi, N. A., & Ramesh, S. (2020). Fundamental concepts of hydrogels: Synthesis, properties, and their applications. *Polymers, 12*(11), 2702. DOI: 10.3390/polym12112702 PMID: 33207715

Batista, C. M., de Queiroz, L. A., Alves, Â. V., Reis, E. C., Santos, F. A., Castro, T. N., Lima, B. S., Araújo, A. A. S., Godoy, C. A. P., Severino, P., Cano, A., Santini, A., Capasso, R., de Albuquerque Júnior, R. L. C., Cardoso, J. C., & Souto, E. B. (2022). Photoprotection and skin irritation effect of hydrogels containing hydroalcoholic extract of red propolis: A natural pathway against skin cancer. *Heliyon, 8*(2), e08893. DOI: 10.1016/j.heliyon.2022.e08893 PMID: 35198766

Battistin, M., Bonetto, A., Nicoli, F., Torreggiani, E., Brunetta, A., Cesa, E., & Vertuani, S. (2024). Synthesis and Evaluation of a ZnO-Chitosan Adduct for Safe and Sustainable Enhanced Ultra-Violet (UV) Sunscreens Protection. *Molecules (Basel, Switzerland), 29*(21), 5204. DOI: 10.3390/molecules29215204 PMID: 39519845

Bhardwaj, L. K., Rath, P., Jain, H., & Choudhury, M. (2025). Exploring the effects of e-waste on soil, water quality and human health. *Discover Civil Engineering, 2*(1), 1–16. DOI: 10.1007/s44290-025-00167-2

Cassayre, M., Oline, A., Orneto, C., Wafo, E., Abou, L., Altié, A., & Piccerelle, P. (2024). Optimization of solid lipid nanoparticle formulation for cosmetic application using design of experiments, PART II: Physical characterization and in vitro skin permeation for sesamol skin delivery. *Cosmetics, 11*(4), 120. DOI: 10.3390/cosmetics11040120

Cefali, L. C., Ataide, J. A., Eberlin, S., da Silva Goncalves, F. C., Fernandes, A. R., Marto, J., Ribeiro, H. M., Foglio, M. A., Mazzola, P. G., & Souto, E. B. (2019). In vitro SPF and photostability assays of emulsion containing nanoparticles with vegetable extracts rich in flavonoids. *AAPS PharmSciTech, 20*(1), 1–10. DOI: 10.1208/s12249-018-1217-7 PMID: 30560393

Chasanah, U., Sugiyanto, A. C., Anggraeni, N., Yuniar, V., & Ermawati, D. (2022). Characteristics, Antioxidant Activity, and Sun Protection Factor of Black Pomegranate Peel Ethanolic Extract Hydrogel. *KnE Medicine*, 231-243.

de Araújo, M. M., Schneid, A. C., Oliveira, M. S., Mussi, S. V., de Freitas, M. N., Carvalho, F. C., Bernes Junior, E. A., Faro, R., & Azevedo, H. (2024). NLC-based sunscreen formulations with optimized proportion of encapsulated and free filters exhibit enhanced UVA and UVB photoprotection. *Pharmaceutics*, *16*(3), 427. DOI: 10.3390/pharmaceutics16030427 PMID: 38543321

de Araújo Andrade, T., Heimfarth, L., Dos Santos, D. M., Dos Santos, M. R. V., de Albuquerque-Júnior, R. L. C., dos Santos-Neto, A. G., de Araujo, G. R. S., Lira, A. A. M., Matos, S. S., Frank, L. A., Rabelo, T. K., Quintans-Júnior, L. J., de Souza Siqueira Quintans, J., de Souza Araujo, A. A., & Serafini, M. R. (2022). Hesperetin-based hydrogels protect the skin against UV radiation-induced damage. *AAPS PharmSciTech*, *23*(6), 170. DOI: 10.1208/s12249-022-02323-8 PMID: 35729366

de Souza, I., Grimmelprez, G. C., Yamaguchi, K. K., Schleusener, J., Lohan, S. B., Meinke, M. C., & Gaspar, L. R. (2025). Evaluation of Residues of Amazonian Fruit Piquia (Caryocar villosum) as Sustainable Ingredient for Sunscreen and Cosmetic Formulations. *Antioxidants*, *14*(2), 122. DOI: 10.3390/antiox14020122 PMID: 40002309

Del Rosso, J. Q., & Kircik, L. (2025). Skin 101: Understanding the Fundamentals of Skin Barrier Physiology—Why is This Important for Clinicians? *The Journal of Clinical and Aesthetic Dermatology*, *18*(2), 7. PMID: 40078856

Doan, N. H., Kadokami, K., Le Minh, V., & Duong, H. T. (2024). A preliminary investigation of occurrence of pharmaceuticals and personal care products in total suspended particulate matter in Ha Noi, Viet Nam. *Vietnam Journal of Science and Technology*, *62*(3), 530–538. DOI: 10.15625/2525-2518/16481

Du, L., & Meng, Z. (2024). Carrageenan-induced conjugated oat protein isolate microgel particles as structure modulators in fat analogues and their digestion behaviors. *Journal of Colloid and Interface Science*, *674*, 663–676. DOI: 10.1016/j.jcis.2024.06.107 PMID: 38950465

Fantini, R., Argenziano, M., Cavalli, R., Arletti, R., & Mino, L. (2024). Zeofilters for potentially innovative sunscreen products: Formulation, stability and spectroscopic studies. *Journal of Photochemistry and Photobiology A Chemistry*, *452*, 115585. DOI: 10.1016/j.jphotochem.2024.115585

Gholap, A. D., Pardeshi, S. R., Hatvate, N. T., Dhorkule, N., Sayyad, S. F., Faiyazuddin, M., & Khalid, M. (2024). Environmental implications and nanotechnological advances in octocrylene-enriched sunscreen formulations: A comprehensive review. *Chemosphere*, *358*, 142235. DOI: 10.1016/j.chemosphere.2024.142235 PMID: 38705416

Grant, G. J., Lim, H. W., & Mohammad, T. F. (2025). A review of ultraviolet filters and their impact on aquatic environments. *Photochemical & Photobiological Sciences*, *24*(2), 343–356. DOI: 10.1007/s43630-024-00674-8 PMID: 39704908

Gulati, S., Ansari, N., Moriya, Y., Joshi, K., Prasad, D., Sajwan, G., & Varma, R. S. (2024). Nanobiopolymers in cancer therapeutics: Advancing targeted drug delivery through sustainable and controlled release mechanisms. *Journal of Materials Chemistry. B, 12*(46), 11887–11915. DOI: 10.1039/D4TB00599F PMID: 39502076

Haney, B. (2024). *Aesthetic Procedures: Nurse Practitioner's Guide to Cosmetic Dermatology*. Springer. DOI: 10.1007/978-3-031-51440-1

Heydari, N., Karimi, A. R., Momeni, H. R., Azadikhah, F., & Etemadi, T. (2025). Chitosan Schiff-Base Hydrogel Sunscreen: A Multifunctional Hybrid Network with Antioxidant, Ultraviolet-Shielding, and Self-Healing Properties. *ACS Omega, 10*(8), 8250–8261. DOI: 10.1021/acsomega.4c09976 PMID: 40060864

Hussein, R. S., Bin Dayel, S., Abahussein, O., & El-Sherbiny, A. A. (2025). Influences on skin and intrinsic aging: Biological, environmental, and therapeutic insights. *Journal of Cosmetic Dermatology, 24*(2), e16688. DOI: 10.1111/jocd.16688 PMID: 39604792

Infante, V. H. P., Campos, P. M., Calixto, L. S., Darvin, M. E., Kröger, M., Schanzer, S., & Meinke, M. C. (2021). Influence of physical–mechanical properties on SPF in sunscreen formulations on ex vivo and in vivo skin. *International Journal of Pharmaceutics, 598*, 120262. DOI: 10.1016/j.ijpharm.2021.120262 PMID: 33549814

Jiang, M., Yang, S. Z., Zhang, X. Y., Zhang, L. Z., Gong, J. S., Han, T. T., Chen, Y., Wang, X.-N., & Shi, J. S. (2024). Protective effect of ferulic acid-hyaluronic acid copolymer against UVB irradiation in a human HaCaT cell line. *International Journal of Biological Macromolecules, 279*, 135570. DOI: 10.1016/j.ijbiomac.2024.135570 PMID: 39270908

Kakad, A. V., Sayyad, L. S., Niphade, P. S., Giridhar, B. R., Nalkar, A. S., Laddha, U. D., & Kshirsagar, S. J. (2024). Insight into Suitable Polymers and Materials for the Preparation of Novel Drug Delivery Systems. In *Novel Drug Delivery Systems in the Management of Chronic Diseases* (pp. 55–109). Apple Academic Press. DOI: 10.1201/9781032654881-4

Karisma, V. W., Wu, W., Lei, M., Liu, H., Nisar, M. F., Lloyd, M. D., & Zhong, J. L. (2021). UVA-triggered drug release and photo-protection of skin. *Frontiers in Cell and Developmental Biology, 9*, 598717. DOI: 10.3389/fcell.2021.598717 PMID: 33644041

Karnakar, R. R., & Gite, V. V. (2025). *Hydrogels and Their Derivatives Utilised in Drug Delivery Systems*. CRC Press. DOI: 10.1201/9780367694265-EPPMPT88-1

Korkmaz, N., & El, S. N. (2025). The release of encapsulated phenolics following different in vitro digestion methods. [aop]. *Acta Alimentaria, 54*(2), 1. DOI: 10.1556/066.2017.00061

Li, D. Q., Li, J., & Wang, S. Y. (2025). 1College of Chemistry and Chemical Engineering, Xinjiang Agricultural University, Urumchi, Xinjiang, PR China, 2School of Bioengineering, Dalian University of Technology, Dalian, Liaoning, PR China. *Natural Biopolymers for Drug Delivery*, 163.

Li, H., Cao, J., Wan, R., Feig, V. R., Tringides, C. M., Xu, J., & Lu, B. (2025). PEDOTs-based conductive hydrogels: Design, fabrications, and applications. *Advanced Materials, 37*(7), 2415151. DOI: 10.1002/adma.202415151 PMID: 39711276

Li, X., Tang, J., Guo, W., Dong, X., Cao, K., & Tang, F. (2025). Polydopamine Nanocomposite Hydrogel for Drug Slow-Release in Bone Defect Repair: A Review of Research Advances. *Gels (Basel, Switzerland), 11*(3), 190. DOI: 10.3390/gels11030190 PMID: 40136895

Liu, B., & Chen, K. (2024). Advances in hydrogel-based drug delivery systems. *Gels (Basel, Switzerland), 10*(4), 262. DOI: 10.3390/gels10040262 PMID: 38667681

Lu, Y., Liu, M., Cao, Y., Yin, J., Zhou, H., Yu, W., & Zhao, G. (2021). Hydrogel sunscreen based on yeast/gelatin demonstrates excellent UV-shielding and skin protection performance. *Colloids and Surfaces. B, Biointerfaces, 205*, 111885. DOI: 10.1016/j.colsurfb.2021.111885 PMID: 34091373

Manvi, H., Janhvi, M., Reva, T., Sayali, M., & Rutika, G. (2024). Skin Aura: The Impact of Moisturizers on Maintaining Skin Hydration. *International Journal of Trend in Scientific Research and Development*, *8*(5), 345–351.

Martinović, M., Nešić, I., Bojović, D., Žugić, A., Blagojević, S., Blagojević, S., & Tadić, V. M. (2024). Plant-Based Sunscreen Emulgel: UV Boosting Effect of Bilberry and Green Tea NaDES Extracts. *Gels (Basel, Switzerland), 10*(12), 825. DOI: 10.3390/gels10120825 PMID: 39727583

Michálek, J. I. Ř. Í., Podešva, J., & Dušková-Smrčková, M. (2022). True story of poly (2-hydroxyethyl methacrylate)-based contact lenses: How did it really happen. *Substantia, 6*(2), 79–91.

Miranda, J. A., Cruz, Y. F. D., Girão, Í. C., Souza, F. J. J. D., Oliveira, W. N. D., Alencar, É. D. N., Amaral-Machado, L., & Egito, E. S. T. D. (2024). Beyond traditional sunscreens: A review of liposomal-based systems for photoprotection. *Pharmaceutics, 16*(5), 661. DOI: 10.3390/pharmaceutics16050661 PMID: 38794323

Patil, M. S. V., Patil, M. R., & Gaikwad, M. P. *COSMETIC SCIENCE*. JEC PUBLICATION.

Rennekampff, H. O., Busche, M. N., Knobloch, K., & Tenenhaus, M. (2010). Is UV radiation beneficial in postburn wound healing. *Medical Hypotheses, 75*(5), 436–438. DOI: 10.1016/j.mehy.2010.04.017 PMID: 20447771

Rotaru-Zăvăleanu, A. D., Dinescu, V. C., Aldea, M., & Gresita, A. (2024). Hydrogel-based therapies for ischemic and hemorrhagic stroke: A comprehensive review. *Gels (Basel, Switzerland), 10*(7), 476. DOI: 10.3390/gels10070476 PMID: 39057499

Sahoo, P. (2023). Hydrogen-producing photocatalyst at sunscreen for athletes in preventing and healing muscle-nerve-skin injuries. *Current Topics in Medicinal Chemistry, 23*(4), 249–256. DOI: 10.2174/1568026623666221216142158 PMID: 36529921

Schaldemose, E. L., Andersen, N. T., Finnerup, N. B., & Fardo, F. (2023). When cooling of the skin is perceived as warmth: Enhanced paradoxical heat sensation by pre-cooling of the skin in healthy individuals. *Temperature, 10*(2), 248–263. DOI: 10.1080/23328940.2022.2088028 PMID: 37332303

Schalka, S., Steiner, D., Ravelli, F. N., Steiner, T., Terena, A. C., Marçon, C. R., & Rego, V. R. P. D. A. (2014). Brazilian consensus on photoprotection. *Anais Brasileiros de Dermatologia, 89*(6, suppl 1), 1–74. DOI: 10.1590/abd1806-4841.20143971 PMID: 25761256

Sghier, K., Mur, M., Veiga, F., Paiva-Santos, A. C., & Pires, P. C. (2024). Novel therapeutic hybrid systems using hydrogels and nanotechnology: A focus on nanoemulgels for the treatment of skin diseases. *Gels (Basel, Switzerland), 10*(1), 45. DOI: 10.3390/gels10010045 PMID: 38247768

Shalaby, E. S., Shalaby, S. I., & Ismail, S. A. (2025). Advantages and therapeutic applications of different semisolids as vehicles for nano-based systems. *Therapeutic Delivery, 16*(6), 581–591. DOI: 10.1080/20415990.2025.2483151 PMID: 40118818

Sharma, B., Chauhan, I., & Tiwari, R. K. (2024). Development of NLC-based Sunscreen Gel of Green Tea Extract and its In vitro Characterization. *Current Bioactive Compounds, 20*(5), 31–42. DOI: 10.2174/0115734072260785230920113339

Singh, G., Mohandoss, K., Shah, T., Dixit, R., Arul, V., Verma, D., & Parvez, N. (2024). Potential role of 3D Printing in Cosmeceuticals. *Systematic Reviews, 6*, 824–855.

Singhal, R., & Gupta, K. (2016). A review: Tailor-made hydrogel structures (classifications and synthesis parameters). *Polymer-Plastics Technology and Engineering, 55*(1), 54–70. DOI: 10.1080/03602559.2015.1050520

Staton, J. A. (2021). Sunscreen secondary claims: Market differentiation or market confusion. *Current Problems in Dermatology, 55*, 385–393. DOI: 10.1159/000517646 PMID: 34698041

Thongrueng, M., Sudsakorn, K., Charoenchaitrakool, M., Seubsai, A., Panchan, N., Devahastin, S., & Niamnuy, C. (2024). Synthesis and characterization of environmentally friendly β-cyclodextrin cross-linked cellulose/poly (vinyl alcohol) hydrogels for adsorption of malathion. *ACS Omega, 9*(21), 22635–22649. DOI: 10.1021/acsomega.4c00037 PMID: 38826516

Upadhyay, L. S. B., & Bhagat, P. (2025). Recent Advances and Development in 3D Printable Biosensors. *Soft Materials-Based Biosensing Medical Applications*, 311-339.

Urimubenshi, F., Konditi, D. B., de Dieu Iyakaremye, J., Mpele, P. M., & Munyaneza, A. (2022). A novel approach for low mutual coupling and ultra-compact Two Port MIMO antenna development for UWB wireless application. *Heliyon, 8*(3), e09057. DOI: 10.1016/j.heliyon.2022.e09057 PMID: 35284684

Urimubenshi, F., Konditi, D. B., de Dieu Iyakaremye, J., Mpele, P. M., & Munyaneza, A. (2022). A novel approach for low mutual coupling and ultra-compact Two Port MIMO antenna development for UWB wireless application. *Heliyon, 8*(3), e09057. DOI: 10.1016/j.heliyon.2022.e09057 PMID: 35284684

Urimubenshi, F., Konditi, D. B., de Dieu Iyakaremye, J., Mpele, P. M., & Munyaneza, A. (2022). A novel approach for low mutual coupling and ultra-compact Two Port MIMO antenna development for UWB wireless application. *Heliyon, 8*(3), e09057. DOI: 10.1016/j.heliyon.2022.e09057 PMID: 35284684

Van Der Pols, J. C., Williams, G. M., Pandeya, N., Logan, V., & Green, A. C. (2006). Prolonged prevention of squamous cell carcinoma of the skin by regular sunscreen use. *Cancer Epidemiology, Biomarkers & Prevention, 15*(12), 2546–2548. DOI: 10.1158/1055-9965.EPI-06-0352 PMID: 17132769

Witono, J. R., Setyadi, F. F., Deandra, P. P., Wanta, K. C., Miryanti, A., Santoso, H., Astuti, D. A., & Bulin, C. D. Q. (2024). A Comprehensive Analysis of Chitin Extraction from the Black Soldier Fly for Chitosan Production. *Periodica Polytechnica. Chemical Engineering*, *68*(3), 507–522. DOI: 10.3311/PPch.23714

Yahya, E. B., Ali, S. R., Lalung, J., Zain, M. S. C., Danish, M., & John, A. (2025). Exploring the potential of biopolymers in cosmetic applications: Sustainable, biocompatible, and high-performance materials for future innovations. *Polymer Engineering and Science*, *65*(6), 2789–2802. DOI: 10.1002/pen.27186

Yu, H., Luo, X., Li, Y., Shao, L., Yang, F., Pang, Q., & Hou, R. (2024). Advanced Hybrid Strategies of GelMA Composite Hydrogels in Bone Defect Repair. *Polymers*, *16*(21), 3039. DOI: 10.3390/polym16213039 PMID: 39518248

Zhan, J., Lei, Z., & Zhang, Y. (2022). Non-covalent interactions of graphene surface: Mechanisms and applications. *Chem*, *8*(4), 947–979. DOI: 10.1016/j.chempr.2021.12.015

Zhang, S., & Wang, T. (2023). Preparation of enzymolysis porous corn starch composite microcapsules embedding organic sunscreen agents and its UV protection performance and stability. *Carbohydrate Polymers*, *314*, 120903. DOI: 10.1016/j.carbpol.2023.120903 PMID: 37173013

Zhang, Y., Li, L., Huang, J., He, X., Xuan, X., Shi, W., & Chen, H. (2025). Highly transparent TiO2/cellulose hybrid film with enhanced UV shielding, dye absorption, photocatalytic, and antibacterial properties. *Cellulose (London, England)*, •••, 1–16. DOI: 10.1007/s10570-024-06256-2

Zou, W., Ramanathan, R., Urban, S., Sinclair, C., King, K., Tinker, R., & Bansal, V. (2022). Sunscreen testing: A critical perspective and future roadmap. *Trends in Analytical Chemistry*, *157*, 116724. DOI: 10.1016/j.trac.2022.116724

Chapter 10
Synthesis and Application of Hydrogels of Insect Origin

Victor Ngozi Enwemiwe
https://orcid.org/0000-0002-2116-4448
Delta State University, Abraka, Nigeria

Eric Esiwo
Delta State University, Abraka, Nigeria

Chioma Cynthia Ojianwuna
Delta State University, Abraka, Nigeria

ABSTRACT

Insect hydrogels like other hydrogels are versatile, three-dimensional polymeric networks with significant water-retention capacity, making them valuable in biomedical, agricultural, and environmental applications. This book review explores the techniques for synthesis, and multifaceted applications of insect-derived hydrogels, emphasizing their sustainability, biocompatibility, and functionality. Insects such as darkling beetles/super worm (Zophobas morio), black soldier fly (Hermetia illucens), yellow mealworm (Tenebrio molitor), large brown cicada (Graptopsaltria nigrofuscata), wingless flightless cockroach (Eupolyphaga sinensis), silk worm (Antheraea assama), silk worm (Bombyx mori), and many others serve as point source for hydrogel production. Chitosan, silk fibroin, and hemolymph proteins extracted are transformed into hydrogels via physical, ionic, enzymatic, and chemical crosslinking into valid hydrogels for diverse applications in novel food and feed, drivers for drug delivery, pesticide delivery, drinkers for insects, tissue, skin, bone and organ regeneration and more.

10.1. INTRODUCTION

Hydrogels are water-based, crosslinked gelatinous substances with moisturizing or water-loving properties and soft consistency. They can undergo tensile stress, being stretchable, compatible,

DOI: 10.4018/979-8-3373-2337-4.ch010

and deformable (Dodda et al., 2023). Hydrogels can be processed from a broad spectrum of biomaterials and nonliving polymers. Naturally, they can be extracted from animals, microbes, and plants with classified contents into proteins, tissues that are decellularized (tissues with expunged cells), and polysaccharides (Zhu et al., 2024). Biomaterials from polysaccharides used for hydrogel formulation come from chitosan, cellulose, dextran, agarose, and alginate (Zhao et al., 2023). Protein biomaterials involve gelatin, fiber from silk materials, plant fibers, elastin, and collagen. Animal tissues of decentralized cells are equally biomaterials for hydrogel formation. Naturally occurring biomaterials are sustainable compared to synthetic ones. Of all the potent biomaterials, insects stand out as potential biomaterials considering their ability to be mass-produced in large numbers in a shorter time interval. Insects are quite advantageous because their mass rearing and utilization will serve as an alternative to cushion climate change issues while considering a suitable ingredient for hydrogel production. Insect-derived hydrogels are three-dimensional, hydrophilic polymer networks with water retention capabilities. Living materials occurring in nature, like insects, are endowed with biopolymers like polysaccharides, lipids, carbohydrates, chitin, and insect proteins. Some potent insects that can be extracted include crickets, mealworms, black soldier flies, silk proteins from silk worms, and arthropod-borne hemoproteins. A checklist of insects accompanied by the bio-potent hemoprotein ingredients and exoskeletal materials for hydrogel fabrication is important and shown in Table 10.1.

Insect and insect material synthesis can be done via physical gelation, enzymatic and chemical crosslinking, accompanied by their transformation to hydrogels. Hydrogels by physical means can be classified by swelling, rate of degradation, elasticity, and size. Insect-based hydrogels are derived from chitin transformed to chitosan, silk fibroin, and blood pigment proteins (Zhanget al., 2022). The biologically engineered materials can serve the purpose of the regeneration of organs of the body, the repair of worn-out parts of the body, and several environmental and medical applications. Potential medical applications include culture media for cells, bioengineering of tissues, scaffolding, and delivery media of drugs. Likewise, biomedical applications such as wound healing, drug delivery, the area of agriculture, and environmental biomonitoring are not left out (Ho *et al.,* 2022). Further to their application is the improvement of soil water/moisture retention, biofertilizer delivery system, and water purification and pollutant absorption. The chitin of other arthropod groups other than insects can be bio-converted to chitosan, which is the most recognized carbohydrate biopolymer for hydrogel fabrication. This chapter focuses on the synthesis of insect hydrogels from insect chitin and proteins and their application.

Table 1. Checklist of insects used for hydrogel and substantive biomaterial.

S/N	Common (zoological) name	Family	Insect part	Life stage	Biomaterial	Extraction process	Reference
1.	Darkling beetles/super worm (*Zophobasmorio*)	Tenebrionidae	Cuticle	Adult	Chitosan	deproteination	Zhatkanbayev et al. (2023)
2	Black soldier fly (*Hermetiaillucens*)	Stratiomyidae	Cuticle	Larvae	Chitosan	deproteination	Giani et al. (2025)
	Black soldier fly (*Hermetiaillucens*)	Stratiomyidae	Hemolymph protein	Larvae	Protein	defatting	Lanes et al. (2021)
3	Yellow mealworm (*Tenebrio molitor*)	Tenebroinidae	Cuticle	Adult/larvae	Chitosan	Deproteination	Gkinali et al. (2022)

continued on following page

Table 1. Continued

S/N	Common (zoological) name	Family	Insect part	Life stage	Biomaterial	Extraction process	Reference
4	Large brown cicada (*Graptopsaltrianigrofuscata*)	Cicadidae	Cuticle	Nymph	Chitosan	Deproteination	Zarzosa and Kobayashi, (2025)
5	Wingless flightless cockroach (*Eupolyphagasinensis*)	Corydiidae	Cuticle	nymph/adult	Chitosan	Deproteination	Zhang et al. (2022)
6	Silk worm (*Antheraeaassama*)	Bombycidae	Cuticle and cocoon	Adult/ Larvae	Silk fibroin	Deproteination	Arora et al. (2021)
7	Silk worm (Bombyx mori)	Bombycidae	Cuticle and cocoon	Adult/ Larvae	Silk fibroin	Deproteination	Arora et al. (2021)
8	Japanese rhinoceros beetle (*Trypoxylusdichotomus*)	Scarabaeidae	Cuticle	Adult/larvae	Chitosan	Deproteination	Hauck et al. (2021).
9	Locust (Schistocerca gregaria)	Acrididae	Wings, legs and cuticle	Adult	Resilin	Chromatography, salting-out, coacervation using temperature	Woodrow et al. (2023);
10	Dragonflies	Libellulidae	Wings, legs and cuticle	Adult	Resilin	Chromatography, salting-out, coacervation using temperature	Hauck et al. (2021); Zhou et al. (2024)
11	Cicadas	Cicadidae	Wings, legs and cuticle	Adult	Resilin	Chromatography, salting-out, coacervation using temperature	Hauck et al. (2021) Zhatkanbayev et al. (2023)
12	Sea cockroaches (*Chiton articulatus*)	Blattidae	Wings, legs and cuticle	Adult/ nymph	Resilin	Chromatography, salting-out, coacervation using temperature	Çakmak et al. (2022)
13	Mosquitoes (*An. gambiae*)	Culicidae	Wings, legs, and cuticle	Adult	Resilin	Chromatography, salting-out, coacervation using temperature	Ohkubo et al. (2023)
14	Fruit flies (*Drosophilia*)	Drosophilidae	Cuticle and wings	Adult	Resilin	Chromatography, salting-out, coacervation using temperature	Lerch et al. (2020)

10.2 METHODS OF SYNTHESIZING HYDROGEL

Hydrogels are made from diverse bio-raw materials, including fibrous collagen, chitosan, silk, alginate, hyaluronic acid, and decellularized tissues. The processes of synthesis are known and include physical crosslinking, ionic gelation, chemical, and enzymatic crosslinking. Considering that insects are available in a short time and space, applying these techniques to insect biomaterials would serve as an alternative, thus reversing the reliance on unsustainable sources.

A good number of methods are commonly known for their benefits in aiding the extraction and fabrication of insect hydrogels.

10.2.1. Physical Crosslinking

Insect hydrogels can be physically crosslinked by generating polymer chain links from insect monomers and or complementation via non-covalent interaction. Other physical interaction gateways include electrostatics, bonding between hydrogen ions, Van der Waals forces, and hydrophobic associations. Primary folding into soft protein hydrogels can be formed by random combined effects, leading to networks of aggregation, coacervation, and precipitation. Three-dimensional protein structures can be formed secondarily or through tertiary means. Primary, secondary, or tertiary gelatinous folding caused by physical cross-links can absorb moisture, forming hydrogels. According to Lee et al. (2024), mechanical action crosslinkers are only needed in the synthesis of insect raw materials for hydrogel production. Temperature is a key player in physical cross-links.

Extrusion technique is a physicochemical way of extruding biological pigments or inks from insects and using them to fabricate insect-based hydrogels. The bio-fabricated gels could be referred to as cell-packed hydrogels or biological inks and have broad applications. The extruded gels are presented in layers. There are two extrusion methods: piston-based and screw-based. The piston exerts a vertical force on the biological substrate, causing the embedding ink or pigment to pass through an opening as collected, while the screw system causes a compression as a higher volume of pigments is collected.

Extrusion has been used to harness the functional biomaterials of insects (Alam et al., 2019). Hot-melting extrusion techniques involve the collaborative rotation of screws in the extruding machine (Alam et al., 2019). This aids the extrusion of embedding biomaterials in insect species. The screw extrusion adopting high moisture cooking is another extrusion technique (Kiiru et al.,2020). The screw extrusion techniques adopting high moisture differ from high melting extrusion because higher temperatures are adopted (120-160 °C) in the cooking. The high moisture cooking extruder works at 150 rpm, and it is loaded per hour at 0.4 kg (Kiiru et al., 2020). The twin-screw extruding method is an improved version of the screw extrusion technique.

The inkjet technique is a physical crosslinking printing technique that is used to fabricate cell-packed complex hydrogels (Cheng et al., 2024). The insect morphology is the centre for making the gel from a highly defined and multi-staged process. Under this technique, polymerized structures are formed into hydrogels, adopting a small volume of pigments. The biological ink used in the printing is made up of biological cells that can be manipulated to change their structural conformation to mimic tissues to be repaired or regenerated. Insects, especially in their larval stages, inspire the development of biomaterials for hydrogel formation. The underlying hemolymph, hemocyanin, and other blood pigments of insects can be harnessed as candidates for fabricating hydrogels (Gantumur et al., 2019). It is pertinent to consider the volume of bioink produced, the speed of production, viability, viscosity, and structural design of the extracting equipment (Ghosh and Yi, 2022). Hydrogels extracted from the inkjet technique have been broadly applied in engineering tissues, delivering drugs and biomedicals.

10.2.2. Ionic Gelation

This is a crude technique that encapsulates chitosan from insects and other bioactive ingredients from various sources to biopolymeric functionalities. The differences in charges were previously used as a crude technique for insect hydrogel fabrication. For instance, chitosan has positively charged amino groups that interact with the negatively charged group. The mixture is set up at room temperature with constant stirring. Insect chitosan in sodium tripolyphosphate has been studied to form a hydrogel using the ionotropic gelation technique. Bioactive extracts with medical potentials are mostly incorporated into the hydrogel and are delivered as hydrogel-based drugs. Alginate has been synergized with chitosan, a bioproduct of insects, to form a hydrogel that can encapsulate bioactive ingredients that can help in drug delivery. They have been used to encapsulate insulin.

10.2.3. Chemical Crosslinking

Insect chitosan and many other polymers can be cross-linked chemically by covalent bonding. Synthesis of hydrogels by chemical crosslinking adopts the addition of metallic precursors, including Zinc, Calcium, and Magnesium, to trigger covalent bonding of the biomaterials, leading to gel formation. Chemical synthesis of hydrogel includes the action of enzymes, radicals, temperature, grafting, copper-catalyzed alkyne-azide cycloaddition, and radiation. The solvent, N, N-dimethylacetamide (DMAc), and lithium chloride (LiCl) at various concentrations can be used to harness chitin. The crystalline nature and molecular weight of insect chitin extracted may be reduced due to bleaching, as the solubility of chitin increases in the solvents. Chitin extracted from superworms, crabs, and shrimps was tested for their suitability as a hydrogel by Zarzosa and Kobayashi (2025). Chitin from the nymphal stage of the brown cicada provided a better viscoelastic ingredient for the fabrication of hydrogels than chitin from shrimps and crabs captured from a saltwater environment. Cicada-extracted chitin is considered the best for 3D hydrogels (Guessasma et al., 2024).

10.3. PROTOCOL FOR SYNTHESIS OF FROM INSECT

10.3.1. Chitin and Chitosan-based Hydrogel

Chitin is a basic raw material and ingredient in the exoskeleton of many insects that can be transformed into chitosan under three distinctive steps. They include demineralization, deproteinization, and decolorization. Larvae or adults of insects are dried and blended into fine powders. This fine powder is added to hydrochloric acid in 2 moles. The mixture is stirred, and then filtration and rinsing of residue in deionized water. This occurs in demineralization. Younes and Rinaudo (2015), Maddaloni et al. (2020), and Zhatkanbayev et al. (2023) opined that insects can be deproteinized when the resulting residue is treated with sodium hydroxide in 1 mole at a temperature 70/80 °C without exceeding 6 hours or falling below 4 hours. Then further rinsed in deionized water until neutrality of pH was reached. The formed residue is decolorized by soaking it in ethanol to erase stains or pigments and dried in the purified form. Pure chitin can be converted to chitosan by simply deacetylation. This is done by exposing pure chitin to heat at

40/50% sodium hydroxide at the boiling point or 90 °C. This heating can be done between 3 to 6 hours to transform chitin to chitosan (Figure 10.1).

Hydrogel can be synthesized from chitosan by dissolving chitosan in 1 or 2% acetic acid and stirring for half or a full day at 20-25 °C to form a viscous mixture. Processes of crosslinking can adopt a crosslinker agent depending on the desired form of gel. Tripolyglyceride, glutaraldehyde, and many other crosslinker agents can be used to form a gel depending on the suitable choices (Fujita et al., 2023). Gelatinous liquids can form desirable shapes and forms after controlled cooling or drying. Further rinsing or cleaning can be done in post-hydrogel formation treatment to remove unreactive chemicals.

Hydrogels from chitosan can also be obtained by simple dissolution in phosphoric or lactic acid solution according to Ye et al. (2021). This product can be fortified with compounds of sodium and ammonium, including NH_4NO_3, Na_2HPO_4, or the sodium salt of animal or plant origin. An addition of glutaric aldehyde solution caused hydrogel formation after heating at 40-90 °C for a day. Rinsing and drying of the hydrogel finalizes the production for agricultural applications. Zhatkanbayev et al. (2023) described in detail the production of chitosan hydrogel from insects infesting a pine plantation. The darkling beetle (*Zophobas* sp.) is a potential species for hydrogel production. The insect species is pulverized into chunks and deproteinized in NaOH following standard protocols of stirring constantly for at least 3 hrs. at 80 °C. Details of chitosan extraction can be further read as elucidated in the method by Zhatkanbayev et al. (2023). This method involves neutralization of the hydrogen ion concentration, bleaching in hydrogen peroxide (H_2O_2), washing, deacetylation, incubation, filtration, and drying of the bio-ingredient for hydrogel production. Hydrogels fabricated from chitin using chitin dissolving solution by Zarzosa and Kobayashi (2023) have broad application in biomedicine, wastewater treatment, agriculture, and food (Lv et al., 2023).

The application of the hydrogel is guaranteed after testing. This is severally by whether they are biodegradable, lose weight when in trypsin or protease, retain water, and by weighing of swelling. They are equally checked by dipping in buffer phosphate-saline solution, and calculation of the ratio and tensile strength when mechanically induced to test for resistance to pressure and elasticity.

Figure 1. Process of extraction and synthesis of insect-based hydrogels from chitin

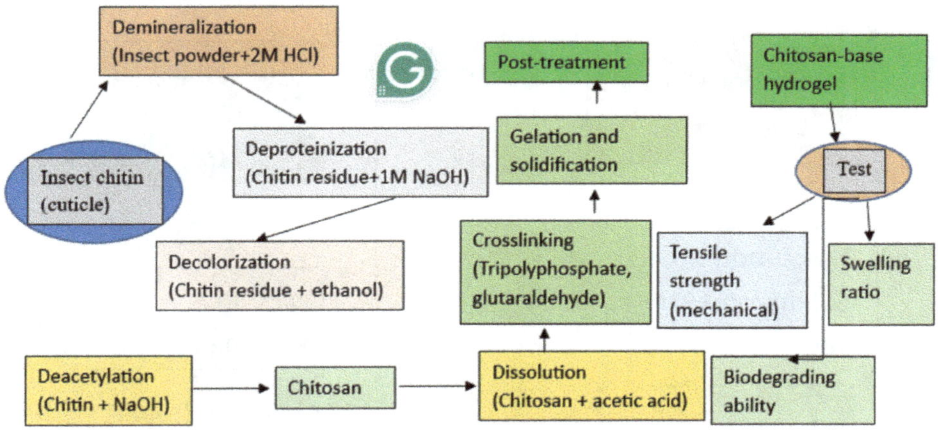

10.3.2. Silk-based Hydrogels

Silk worm (*Bombyx* sp.) and many silk-producing insects, as shown in Table 1, can be harnessed for their silk fibroin and blood lymph or pigment for the synthesis of protein-based hydrogels. The point source of silk is extracted from the cocoon spun by these worms and the blood protein inks of the insects. The silk fibroin and protein extraction process is quite simple compared to the cumbersome process involved in chitin conversion to chitosan, then their fabrication into hydrogels, amongst the cleaning process to get purified hydrogels. The process of silk extraction, as detailed by Liu et al. (2023), starts with degumming, whereby harvested cocoons are heated in a solvent (0.02 mole of Na_2CO_3) between 30 minutes and 1 hour to break loose and eliminate sericin. Rising is done to ward off any gumming impurities. A further step is taken to dissolve the resulting fibroin in 9 moles of Lithium bromide at a temperature of 60 °C for 4 hours. This is to further disassemble the protein structure. The resulting solution is passed through a dialyzing tube and dialyzed for 2 days against deionized water to eliminate lithium bromide and or urea. The solution is filtered, eliminates aggregates, and stored at 4 °C.

To further fabricate hydrogel, the pure fibroin is crosslinked using ethylene glycol, polyethylene glycol diglycidyl ether, genipin, or the enzyme transglutaminase can be adopted as a compatible linking agent to form a gel. Different cross-linkers have different times and temperatures for gelation. According to Lyu et al. (2023) and Wang et al. (2024), constant stirring of pure fibroin added to crosslinkers, including ethylene glycol, polyethylene glycol diglycidyl ether, and genipin, for half or a full day will cause gelation at 0.5-2% addition for genipin and 1-5% for polyethylene glycol diglycidyl ether (Figure 2). Silk-based hydrogels have broad biomedical applications, including their adoption in scaffolding for tissue engineering, drug delivery systems, skin wound healing patches, and biodegradable wound dressings.

Figure 2. Extraction, synthesis, and processing of insect-based hydrogels from silk.

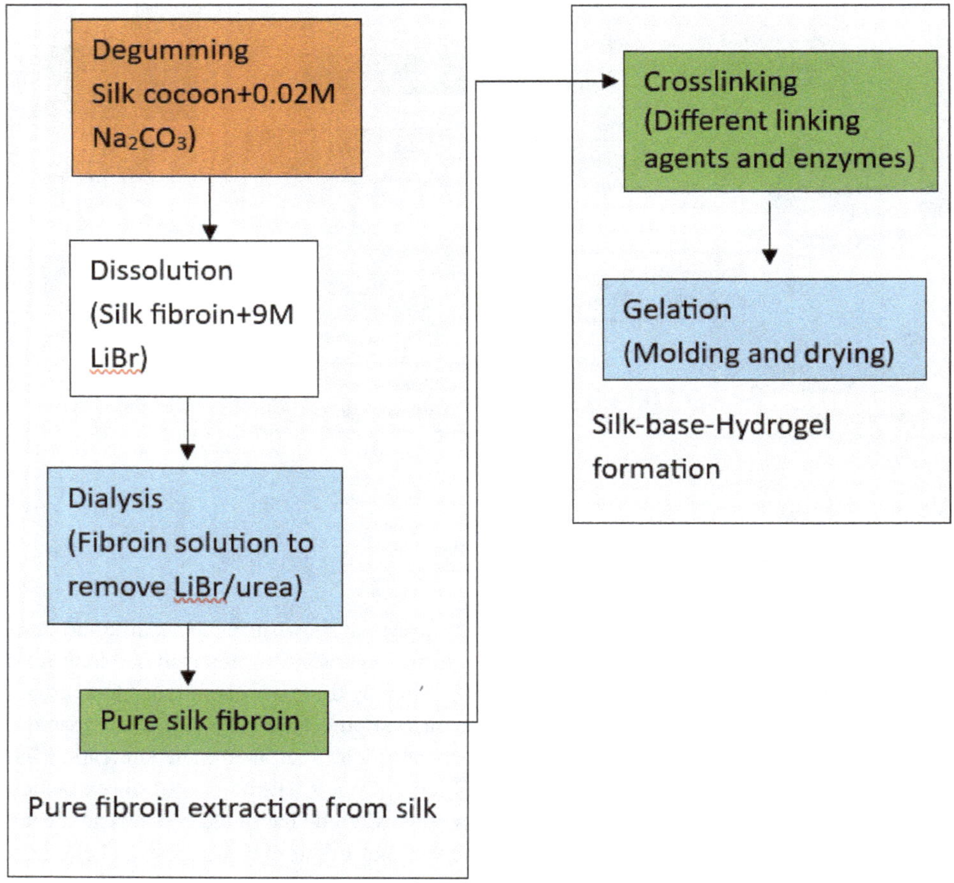

10.3.3. Hydrogels from Insect Blood Fluids

Proteins in insect blood pigment can be harvested by simply piercing the insect exoskeletal covering to allow the pigments to be collected. Three notable steps are involved in extraction and fabrication of a hydrogel based on insect hemolymph. Insect blood fluid collection, purification, and hydrogel fabrication. After the collection of insect blood fluid, sodium citrate, an anticoagulant buffer, is added to allow for free flow of blood pigment into the collecting tube. The purification of the insect blood fluid is done by loading the tube in a centrifuge machine and spinning at 10,000rpm for 10 minutes to expunge residues and cells. The protein can be precipitated using a 50% concentration of ammonium sulphate and centrifuged further. The supernatant is dialyzed in phosphate buffer saline (PBS) to remove excess salt. The pure protein can undergo gelation either by oxidation or by the action of an enzyme. Hydrogen peroxide (0.1-1% concentration) is normally added as a conventional method to oxidize the protein and crosslink them. Stirring is most times applied gently at 20-25 °C until gel is formed. Similarly, the enzymatic pathway to gelation could serve as an alternative. The enzyme peroxidase is added to breakdown protein crosslinks. The mixture is then oven-heated at 37°C through the night (Figure 3).

Figure 3. Extraction, synthesis, and processing of insect-based hydrogels from insect blood fluid.

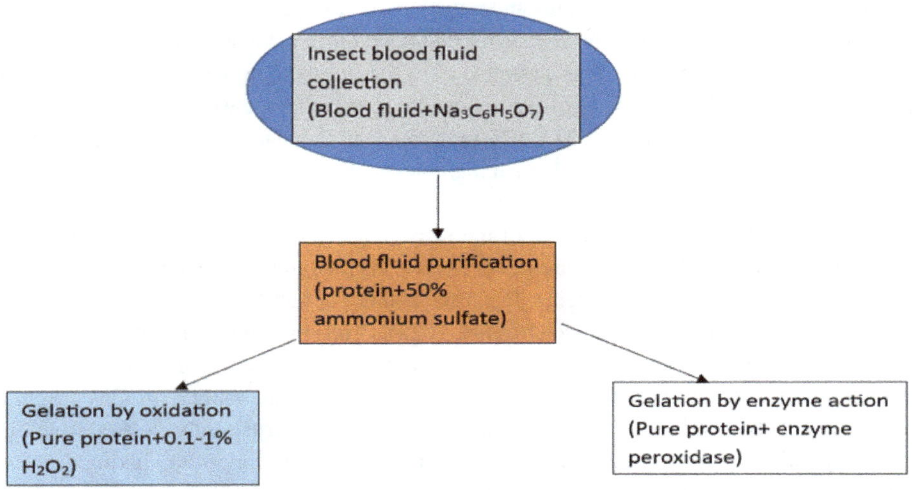

10.4. APPLICATION OF INSECT-DERIVED HYDROGELS

10.4.1 Novel foods and animal feed

Insect hydrogels are applied in foods with the aim of enhancing texture and functionality. Flour from edible cricket has been prepared using the new generation extrusion method to make pellets for snacks (Combrzyńskiet al., 2023). Insect flours have been made from diverse insects and are available in grocery stores as novel foods to combat Westernized phobia of insect incorporation into foods (Ojianwuna et al., 2023). Insects are seen as alternative proteins, and their application of crickets in foods has been studied (Combrzyńskiet al., 2023). The demerits of including extruded materials from cereal, potatoes, and non-insects into foods are quite numerous imbalanced fats, salts, $(CH_2O)_n$ content, and invariable nutrients. However, cricket in 10 to 15% fortified extruded snacks made up of rice showed high iron, protein, fibre, and fat content (Tao et al., 2017). Insect inclusion is done at a low quantity to achieve the best quality of the products, as in cricket inclusion of 7.5% (Igualet al.,2020). Another method of extrusion already tried for cricket cum corn flour is the single screw extrusion method at higher temperatures of 160-180 °C.

A previous study by Sirichocktanasap and Chanput (2022) has implicated the presence of phenol in the wings, cuticles, mid and hind guts of insects. Phenols in insects, especially the edible species, may vary with species due to the feedstocks fed upon and the synthesis of biomaterials (Nino et al., 2021). One resultant benefit of extruded biomaterials into functional food is the gluten-free test bread incorporating cricket (Kowalczewski*et al.,* 2021). Due to the phenol embedded in insect-incorporated foods, most functional foods with insect inclusions portray antioxidant properties, as seen in the study of Suh et al. (2010), whereby *A. dichotoma*, the Japanese rhinoceros beetle, served as a potent candidate. Mealworm-incorporated food hindered the oxidation of certain compounds (Zielińska et al., 2021). Antioxidation may be linked to the corroborative effects of insect proteins and polyphenols (Ruggeri et al., 2023).

Screw extrusion techniques have been used to defat black soldier fly larvae for the application in fish feed meals to formulate desirable diets that measure up with the commercial diets Edah and Owolabi, (2023). The twin-screw hot extrusion technique worked by causing the loss of up to 60% fat in the fly larvae. The 40% protein left in the black soldier flies after defatting the larvae was added as an additive to meal ingredients, including flour from cassava, soybeans, varied minerals, and vitamins, to fabricate fish meals. Feed ingredients, when put together, had 30% moisture. The speed of the motor of the extruding machine was set at 700, 300, and 210 rpm. It is known that adult insects are richer in proteins than in fats, while the immature stages are highly richer in fats than proteins. Defatted soldier fly had 52% of crude proteins, while whole soldier fly had 47% (Edah and Owolabi, 2023). Crude proteins of the defatted soldier fly, when incorporated into other feed diets, were lower than the conventional Danish fish meal. The proteins of the defatted soldier fly fish meal (52%) compared favorably with the conventional meal (72%). In comparing the physical characteristics of the extrusion-formulated float diet of soldier fly with the conventional feed, density (unit and bulk), pellet longevity, sinking/descending velocity, and growth ratio were significant, while buoyancy of pellets, water retention index, and moisture content were insignificant.

10.4.2 Carriers for Drug Delivery

Hydrogels have been used in drug formulations, with insect hydrogels serving as a substrate for drug carrier structures. Chitosan, glutaraldehyde, gelatin, and iota carrageenan were used as agents of cross-linking. Drug delivery systems adopting gelatin, iota carrageenan, or chitosan incorporated ciprofloxacin hydrochloride inhibited bacterial growth. Considering the persistence in the release of the drugs, Abou-Okeil and Taha (2024) exemplified the drug delivery potentials of selected hydrogels and discovered that the fracture rate is between 0.036 and 0.167 Kgf / mm^2, swell rate is between 362 and 412% after 1 day of exposure. Drug-incorporated hydrogel released its active components nearly half a day (10 hours) and gave 98% efficacy in its action. *Escherichia coli* and *Staphylococcus aureus* were reduced following the incorporation of the drug into the hydrogel.

Antibiotics, anticancer, anti-inflammatory, topical anesthetic encapsulations, and lots more drug derivatives can be excellently carried and delivered to patients by insect-derived hydrogels. This is guaranteed for their biocompatible potential, tunable degradation rates, and wide range ability to encapsulate for both hydrophilic and hydrophobic drugs (Liu et al., 2018). Chitosan, being pH-sensitive, allows for targeted release in specific environments such as the acidic conditions of the stomach or slightly alkaline environments of the intestines and blood (Desai et al., 2023). Silk fibroin, another insect derivative, presents the structural potentials of insect-based hydrogels for diverse use and prolonged release profiles (Wani et al., 2022). These hydrogels can be engineered to degrade over specific timescales, ensuring a controlled release of therapeutics. Their mucosal adhesion properties enable the creation of buccal, ocular, nasal, and vaginal delivery systems, making them highly versatile in treating localized and systemic conditions. (Lu et al., 2024).

10.4.3 Novel baits for trapping insects

A bio-derivative lure to kill insect pest innovation utilizing hydrogel was reported by Friuli et al. (2022). The bio-innovative trap adopted the compatible, biodegradable, and naturally occurring hydrogels. The hydrogel acted as an attractant or sex pheromone that attracted insects, possibly the males, for selective mating to mass produce the filial generation. The hydro-gelatinous lures with incorporated pesticides release the active components that kill the insect. Killing the males lured into the sex pheromone trap will reduce the chances of the generational spread of the species and the damage they do to the environment. The bio-derivative lure to kill offers a new cost-effective innovation that aids the trapping of mosquitoes. The functionality can be outdoor or indoor, or even repositioned in or around eaves of houses to trap gravid mosquitoes making their way to lay their eggs, or blood-fed mosquitoes exiting houses at the early hours of the morning after blood meals. The Asian tiger mosquitoes, *Aedes albopictus,* have been tested in the modeled innovative hydrogel lure to kill.

Numerous lure traps face issues of expiring with time or specific durations, repairs or maintenance, cost implications of repairs and maintenance, and abandonment of traps. However, the hydrogel incorporated lure trap to kill possess benefits that outnumber the issues surrounding the conventional traps. One notable feature of the bio-derivative lure to kill trap is that it is biodegradable and can be used after storage when freeze-dried and reused after rehydration, maintaining the properties (Friuli et al., 2021). The lure to kill trap can adopt the flight-interception or sticky-sheet trap to glue insects that contact them since it is a waterless-based trap.

10.4.4 Delivery Channels for Pesticides

Hydrogels have been adopted in delivering insecticides gradually to pests of valuable crops. Chemical insecticides affect nontarget populations and the environment, leading to the invention of micro-robotics that can deliver insecticide encapsulations to specific pest species, causing invasive damage to crops. Maria-Hormigos et al. (2022) designed a micro-robotic system that is capable of delivering hydrogel-incorporated insecticides, which acted as magnetic carriers of the insecticide, ethyl parathion. This specifically wards off the larvae of the mealworms (*Tenebrio molitor*). The activity channel is linked to the degradation of the midgut of the mealworms following responses to the pH of the delivered insecticides to the midgut in 120 minutes. The underlying effectiveness of a rotatory traverse movement of the microrobots in the magnetic field cannot be undermined in causing high mortality in reduced time, especially in less than 1 hour. This hydrogel delivery system using robotics avails a novel potential in the management of insecticide resistance in agricultural fields.

10.4.5 Water-drinkers for Insect Rearing

Hydrogels, according to Adjuik et al. (2023), are hydrophilic gelatinous polymers with superabsorbent properties. They can swell after absorbing water without losing their shape and the biocidal properties. It has never been reported that hydrogels collapsed or dissolved due to water influence (Neethu et al. 2018). Insect rearing is facing a serious hurdle in making the insects drink. Naturally exploited hydrogels are considered a substrate for water storage for insects to get water droplets for drinking. A previous notable study had tried hydrogels as drinkers for bugs by using hydrogels to enhance turgidity in plant tissues and maintain the freshness of the plant (Barbosa et al., 2019). Likewise, the hydrogel associated with the feed served as water storage for crickets (Fricke et al., 2022). It was known over the past decade that the water insects drink is embedded in the food they eat. This has made insect rearing come with uncertainty. This is because supplying insects with flowing water can lead to drowning. Thus, hydrogels offer new hope in their application as insect drinkers (Silva et al., 2023). This is made possible by the controlled release of water for slow drinking while maintaining normal metabolic activities.

Hydrogels are applied in rearing aphids in large numbers. Silva et al. (2023) showed the ability of 0.6 grams to replace 1g of agar in rearing aphids. This did not affect the reproduction and life stages of the insect. Temperature influences the functionality of the hydrogels. Likewise, the substrate, especially the plant host, influences the functionality of hydrogels in rearing insects. It is imperative to adopt a higher dose of hydrogel in insect rearing.

10.4.6 Wound Healing and Tissue Engineering Therapies

Antimicrobial resistance and resistance to antibiotics used for wound healing a public health concerns. A typical example is the Staphylococcus resistant to methicillin. Several other antibiotic-resistant bacteria are common. To mitigate the financial loss resulting from failed treatment, chitosan and pullulan derived from insects are promising alternatives. An innovative fabrication was done by Jiang *et al.* (2024), adopting three ingredients: chitosan from the quaternized-catechol type, pullulan of the methacrylate-dialdehyde type, and gallium ions. These ingredients were used to form various photo-enhanced crosslinked hydrogels capable of killing microbes, halting bleeding from damaged blood vessels, and promoting the regeneration or recuperation of cuts and burns, as well as serving as biomaterial for wound healing.

Silk worm (*Bombyx mori*) silk thread has been reported by Li e*t al.* (2020) for its therapeutic effect on clearing scars in the rabbit's ear. Predominant scars were repaired following the fibre of silk, thus reducing the changes caused by natural lightening of the skin after healing without application of hydrogels. The silk-incorporated hydrogel demonstrated high potential in clearing scars compared to scar-removing cream. Silk worm hydrogel has been reported by Guan et al. (2020) in the repair and healing of wounds using the mouse model. Silkworm hydrogel provided a biofluidic environment that aids wound healing. Likewise, the repair mechanism is possible due to the gluing and resetting of keratin cells and fibroblasts.

Silk worm possesses a natural protein biomaterial called fibrous silk (silk-like fibroin), which benefits diverse biomedical applications (Table 2) (Zhang et al., 2022; Zahra et al., 2023; Li et al., 2023).

Table 2. Summary of Insect-Derived Hydrogels

APPLICATION	INSECT-DERIVED MATERIAL	KEY BENEFITS	REFERENCES
Biomedical Applications			
Wound Healing	Chitosan, Silk Fibroin	Antimicrobial, promotes healing, moist dressing	(Alberts et al., 2025)
Drug Delivery	Chitosan, Silk Fibroin	pH-responsive, controlled release, biocompatible	(Liu et al., 2018)
Tissue Engineering	Silk Fibroin, Resilin	ECM mimicry, 3D scaffolding, biocompatible	(Choi et al., 2024
Agriculture and environment			
Soil Moisture Retention	Chitosan	Improves water retention, biodegradable, soil-safe	Nordin et al., 2024
Agrochemical Release	Chitosan	Sustained delivery, reduces runoff and toxicity	Michalik and Wandzik, (2020)
Bioremediation	Chitosan	Adsorbs heavy metals, safe degradation	Pal et al., 2021
Cosmetic and Personal Care Applications			
Skincare Formulations	Silk Fibroin, Chitosan	Hydrating, anti-aging, antioxidant	Olteanu et al., 2024
Wound Care Patches	Chitosan, Silk Fibroin	Antimicrobial, promotes healing	Kapoor and Kundu, 2016
Hair and Scalp Treatments	Silk Fibroin	Enhances follicle health, reduces hair loss	Chelu et al., 2023

10.4.7 Agricultural and Environmental Applications

10.4.7.1 Hydrogels for water retention, biofertilizer, and improvements

Hydrogel super-absorbents have grown in interest for their application in arid lands and areas with difficulty in water retention. There are no studies on the fabrication of a super-absorbent hydrogel of insect origin. There is a need for insect innovations involving water management systems. Biofertilizers formulated into hydrogels function similarly to water retention hydrogels. Djafaripetroudy et al. (2025) classified biofertilizers incorporated into hydrogel into controlled and slow-release biofertilizers.

10.4.7.2. Soil Moisture Retention

Insect-derived hydrogels are highly effective in enhancing soil water retention, especially in arid and semi-arid regions (Zhu et al., 2024). Chitosan-based hydrogels derived from insect chitin can absorb several hundred times their weight in water, slowly releasing it to plant roots over time. This mechanism helps maintain soil moisture levels, reduce the frequency of irrigation, and support plant growth under water-limited conditions (Agbna and Zaidi, 2025).

Such hydrogels are biodegradable and non-toxic, breaking down into harmless byproducts that enrich soil quality (Nanda et al., 2025). The integration of these materials into soil not only supports sustainable agriculture but also reduces reliance on water-intensive practices, making them indispensable in regions facing water scarcity (Xing & Wang, 2024).

10.4.7.3. Controlled Release of Agrochemicals

The agricultural sector benefits greatly from the ability of insect-derived hydrogels to serve as carriers for fertilizers and pesticides (Mikhailidi et al., 2024). These hydrogels can encapsulate active agents and release them gradually, ensuring prolonged efficacy and reducing the need for repeated applications (Ahmed et al., 2025).

This slow-release mechanism minimizes leaching and runoff, thereby protecting nearby ecosystems from chemical contamination (Li and Li, 2024). Moreover, by releasing nutrients and pesticides in a controlled fashion, these hydrogels improve plant nutrient uptake and pest control efficiency (Vedovello et al., 2024). This dual function supports both crop productivity and environmental safety.

Researchers have also explored incorporating microbial inoculants into these hydrogels, thereby creating a multifunctional matrix that supports plant health through nutrient provisioning and biological pest resistance (Shahwar et al., 2023). These innovations are paving the way for eco-friendly and high-efficiency agricultural inputs.

10.4.8. Bioremediation

Functionalized insect-derived hydrogels show great promise in environmental cleanup. Their porous structure and surface chemistry can be tailored to adsorb heavy metals, dyes, and other industrial pollutants from wastewater (Pal et al., 2021; Visan and Negut, 2025). Chitosan, in particular, exhibits high affinity for metal ions due to its amino groups, making it an effective bio-sorbent (Gamage et al., 2023).

These hydrogels can be deployed in water treatment systems to clean up effluents from textile, mining, and manufacturing industries (Table 10.2). Their biodegradable nature ensures that they do not introduce secondary pollutants during or after the treatment process. In addition, they can be recovered and regenerated, offering a sustainable approach to managing environmental contaminants (Alaghemandi, 2024).

10.4.9 Cosmetic and Personal Care Applications

10.4.9.1 Skincare Formulations

Silk fibroin and chitosan hydrogels derived from insects are increasingly being integrated into cosmetic formulations for their moisturizing, anti-aging, and skin-soothing properties (Table 4). These hydrogels form breathable films on the skin, enhancing hydration and improving skin barrier function (Huang et al., 2022; Olteanu et al., 2024).

They are particularly effective in facial masks and serums, where they deliver active ingredients such as vitamins, peptides, and antioxidants (Januszewski et al., 2023). Their ability to sustain release over several hours ensures long-lasting benefits. Moreover, their biocompatibility reduces the risk of irritation and allergic reactions, making them suitable for sensitive skin types (Ho et al., 2022).

10.4.9.2. Hair and Scalp Treatments

Hydrogels formulated with insect-derived silk proteins are being developed for scalp treatments and hair conditioning (Kapoor and Kundu, 2016). They can encapsulate growth factors or natural extracts that stimulate hair follicles and improve scalp health (Table 2). These hydrogels maintain prolonged contact with the scalp, ensuring effective delivery of active compounds (Chelu et al., 2023).

10.5. ECONOMIC VIABILITY

10.5.1. Production Costs

Farming insects is notably cost-effective compared to other livestock or arthropods (Table 3). Some cost benefits include:

- Low Feed-to-Biomass Ratio (FBR): Insects convert feed into body mass more efficiently.
- Food and/or agricultural waste valorization to biomass is possible with many insects, which helps to drastically cut input costs.
- Space Efficiency: Vertical farming methods for insects reduce land costs.

Despite the processing costs, especially for chitin extraction and conversion into chitosan-forming hydrogel remain relatively high due to the need for chemicals (e.g., NaOH for deacetylation), water, energy, and labor (Younes & Rinaudo, 2015). However, technological improvements such as enzymatic extraction methods, microwave-assisted processing, and biorefinery models are cost-effective extraction methods (Macedo et al., 2023). Insect biomass contains chitin or sericin, which is often treated as a disposal liability. A shift in the conventional raw-material economics into insect can lock in long-term, near-zero-cost feedstock contracts for mills.

Table 3. Feedstock availability and cost

Feedstock	Current scale	Forecast availability	Typical gate price
BSF larvae (spent pupal skins, dead adults)	~80–220 kt (2024) insect biomass worldwide	500–700 kt p.a. by 2030 as insect-protein plants ramp up	Negligible: considered a disposal cost for many farms
Silk degumming liquor (sericin)	50,000 t wet sericin/year (≈ 10,000 t dry) generated as waste from ~109,000 t raw cocoon production in 2019	Waste volume scales linearly with silk output; no feedstock bottleneck foreseen	USD 0–50 t^{-1} (cost of wastewater handling that mills are willing to pay to offload)
Mealworm/frass exoskeleton fines	15–20 kt globally (2024 est.)	Uncertain; depends on the EU feed regulations expansion	Pay-to-take waste stream in most locations

Note: t =tonnes, kt=kilotonnes, p.a=per-annum. Sources: Zhatkanbayev et al. (2023); Giani et al. (2025); Gkinali et al. (2022); Macedo et al. (2023).

10.5.1.1. Conversion Technologies and Processing Costs

Chitin → Chitosan hydrogels (BSF & mealworm routes)

a) Traditional chemistry: Two-step acid demineralization and caustic deproteinization, followed by high-temperature deacetylation, is still the default route and remains energy-intensive. Industry analysts put average variable processing costs at USD 22–36 kg^{-1} for technical-grade chitosan (Riofrio et al., 2021).
b) Green extraction: Enzymatic deproteinization coupled with lactic-acid fermentation cuts chemical inputs by >70% and achieves energy savings of ~30%; pilot lines (10 t y^{-1}) report internal costs of USD 15–18 kg^{-1} but are not yet CAPEX-optimised (Yuan et al., 2024).
c) Hydrogel formation: Once solubilized, insect chitosan behaves identically to crustacean chitosan; thermally or ionically cross-linked hydrogels can be cast or 3-D-printed without further cost penalties — the price driver is the degree of deacetylation (DD).

10.5.1.2 Sericin Hydrogels (Silkworm Route)

a) Recovery stage: Alkali or autoclave degumming recovers 15–36% of cocoon mass as sericin, depending on temperature and liquor ratio (Yuan et al.,2024). A recent techno-economic study using ultrafiltration to simultaneously concentrate and neutralize the liquor reports an all-in operating cost of USD 4.2 kg^{-1} dry sericin for a 2,000 t y^{-1} plant, two-thirds of which is steam and membrane power (Du et al. 2025).
b) Hydrogel cross-linking: Sericin can self-assemble into β-sheet networks above 30% solids, but most industrial prototypes add genipin, silver ions, or enzyme triggers to improve mechanical strength. Resulting gels reach elongation >1 100% and rival commercial PEG dressings (Du et al. 2025).

First-of-a-kind (FOAK) insect chitosan plants of 5,000 t y^{-1} demand an estimated USD 28–32 million installed, ~35% lower than a crustacean shell line of equal capacity because insect feedstock is cleaner (1–2% $CaCO_3$ versus 20% in shrimp) and requires smaller acid reactors. Sericin recovery skids can be containerized and co located at village-scale silk-reeling units for USD 0.5–1 million per 200 t y^{-1} module.

10.5.2. Market Value

The global hydrogel market is projected to exceed $20 billion by 2030, driven by demand in biomedical applications (e.g., wound dressings, tissue engineering), agriculture (e.g., moisture-retaining soil additives), and cosmetics (Alberts et al., 2025). As consumers and industries grow more environmentally conscious, biodegradable and bio-based hydrogels are gaining premium market value (Table 10.4). Insect-based hydrogels can potentially command higher prices than synthetic counterparts, especially when branded as sustainable, cruelty-free, or organic, thereby increasing economic viability.

10.5.2.1. Market Demand Pull

i. **Hydrogel wound dressings:** Global hydrogel dressing sales will top USD 8.38 billion by 2030 (Research and Markets, 2024). Collagen and alginate dominate today, but formulators are seeking animal-free or up-cycled alternatives to meet ESG targets.
ii. **Biomaterials venture funding:** Average biomaterials funding rounds reached USD 11.9 million in 2024 — a proxy that investors perceive a clear exit pathway for novel hydrogel technologies (Bidwai, 2025).
iii. **Sericin niche markets:** The stand-alone sericin market was already valued at USD 2 billion in 2024 and is projected to grow at a 10% CAGR as cosmetic brands switch to *clean-label peptides* (Moharir, 2025). Hydrogels capture only a slice of this, but they raise the price ceiling to USD 120 kg^{-1} for sterile injectable grades (Cocoon Biotech term sheets, 2024, investors' deck).
iv. **Regenerative medicine & bioinks:** Pure sericin and insect chitosan pass cytotoxicity and pyrogenicity tests, enabling their use as *"bio-inks"* for 4D bioprinting. Clinics pay >USD 300 kg-1 for certified printable hydrogels, several orders above bulk polymer pricing, giving innovators a *margin cushion* even at small scale.

Table 4. Comparative economics versus incumbent feedstocks

Attribute	Insect chitosan hydrogel	Crustacean chitosan hydrogel	PEG/PVA hydrogel
Feedstock cost	Near-zero (waste)	USD 200–450 t^{-1} shrimp shells (seasonal)	Oil-linked
Variable processing cost	USD 22–36 kg^{-1} (technical grade)	USD 18–30 kg^{-1}	USD 1.5–2 kg^{-1}
Typical sale price (medical)	USD 35–70 kg^{-1}	USD 30–65 kg^{-1}	USD 60–90 kg^{-1} (PEGDA sterile)
GHG footprint (kg CO_2 eq kg-1)	2.1 (includes larvae rearing credit)	4.6	6–8
Price volatility	Low (year-round insect supply)	High (shrimp seasonality)	Tied to crude oil volatility

Sources: Research and Markets, 2024; Bidwai, (2025); Moharir, 2025

Once plants exceed ~3000 t y^{-1}, insect chitosan hydrogels achieve parity with crustacean-based gels on cost while outperforming them on carbon footprint. They remain dearer than commodity synthetic hydrogels, but in regulated medical niches, price is driven by functionality

10.5.3. Scalability Potential

10.5.3.1 Biomass Supply Chain

Insects like mealworms and black soldier flies can be produced at an industrial scale using modular container systems, enabling decentralized production close to raw material sources. This adaptability makes insect farming highly scalable in both urban and rural settings.

i. **Bioconversion efficiency**: Black soldier flies can convert over 50% of organic waste into harvestable biomass within 15–20 days.
ii. **Lifecycle timing**: Short lifecycles enable multiple harvests per year, ensuring a continuous raw material stream.

10.5.3.2 Processing Infrastructure

Challenges in scalability arise primarily in the **post-harvest processing stage**:

i. **Chitin extraction**: While crustacean chitin extraction is well established, insect chitin extraction is still evolving and often lacks standardized protocols (Mersmann et al.,2025*)*
ii. **Chemical handling**: The use of harsh chemicals (e.g., acid for demineralization) poses safety and environmental concerns on a large scale.
iii. **Consistency**: Ensuring hydrogel consistency and purity across batches can be difficult due to variability in insect species, feed composition, and growth conditions.

Still, integrated biorefinery models, which co-produce insect protein, oil, and chitin in a zero-waste loop, provide a scalable and economically synergistic solution. These models ensure that the revenue from hydrogel production is supplemented by other high-value insect products.

10.5.3.3 Distribution and Logistics

Hydrogel transport is sensitive to environmental conditions (temperature, humidity). However, insect-sourced hydrogels derived from chitosan can often be dried and rehydrated on site, improving shelf life and reducing cold chain dependence. With proper formulation and packaging, distribution logistics can be streamlined.

10.5.4. Ethical and Regulatory Considerations

Considering the disgust factor of insects, large-scale production and processing may raise questions about welfare, especially in regions with less regulatory oversight. International certification systems (e.g., Ecocert, USDA Organic) can help ensure ethical practices. Additionally, the novelty of the source material may present hurdles in regulatory approval, particularly for biomedical applications. However, the use of GRAS (Generally Recognized as Safe) status and biocompatibility testing can mitigate these barriers (Table 5) (Van Loon & Bovenkerk, 2021).

Table 5. Comparative advantages and limitations of insect hydrogels over others

Aspect	Insect-Based Hydrogels	Crustacean-Based Hydrogels	Synthetic Hydrogels
Source sustainability	High	Moderate	Low
Environmental footprint	Low	Medium	High
Production cost (long term)	Medium → Low (with scaling)	Medium	Low
Scalability potential	High	Moderate	High

continued on following page

Table 5. Continued

Aspect	Insect-Based Hydrogels	Crustacean-Based Hydrogels	Synthetic Hydrogels
Market perception	Emerging eco-friendly niche	Established but less sustainable	Widely used, lower-end
Regulatory barriers	High (early-stage)	Low	Low

Economic viability is not speculative, and it is a function of the entrepreneurial interest in scalability. Thus, scalability is hinged on deploying modular extraction capacity alongside insect farms and silk mills, harmonizing quality standards, and securing green premium take-out agreements. This achievement is poised to shift laboratory novelty to mainstream biopolymer platform within the next decade, while delivering both profit and measurable sustainability dividends in insect hydrogel.

10.6 CONCLUSION

Insect-derived hydrogels represent a transformative frontier in sustainable biomaterials research. By harnessing chitosan, silk fibroin, hemolymph proteins, and other polymers from diverse insect species, researchers have developed multifunctional hydrogels with applications across biomedical, agricultural, environmental, and industrial domains. Their inherent biodegradability, biocompatibility, and water retention capacity distinguish them as superior alternatives to many synthetic hydrogels, particularly in addressing global challenges such as wound healing, drug delivery, sustainable agriculture, and pollution remediation. The synthesis methods, ranging from physical and ionic gelation to chemical and enzymatic crosslinking, demonstrate both the versatility and scalability of insect-derived biopolymers. While challenges remain in standardizing extraction processes, lowering production costs, and overcoming consumer perception barriers, the economic viability of insect hydrogels is reinforced by the availability of abundant low-cost feedstocks and their alignment with circular economy models. Ultimately, insect-based hydrogels offer not only functional innovations but also an ethical and ecological pathway toward reducing reliance on unsustainable synthetic polymers and animal-derived materials, intersecting science, sustainability, and societal benefit for the next generation of bio-inspired technologies.

REFERENCES

Abou-Okeil, A., & Taha, G. M. (2024). Investigation and kinetics of hydrogel scaffold with sustained release ciprofloxacin hydrochloride. *Polymer Bulletin*, *81*(18), 17393–17411. DOI: 10.1007/s00289-024-05495-4

Adjuik, T. A., Nokes, S. E., & Montross, M. D. (2023). Biodegradability of bio-based and synthetic hydrogels as sustainable soil amendments: A review. *Journal of Applied Polymer Science*, *140*(12), e53655. DOI: 10.1002/app.53655

Agbna, G. H. D., & Zaidi, S. J. (2025). Hydrogel Performance in Boosting Plant Resilience to Water Stress—A Review. *Gels (Basel, Switzerland)*, *11*(4), 276. DOI: 10.3390/gels11040276 PMID: 40277712

Ahmed, M. S., Yun, S., Kim, H. Y., Ko, S., Islam, M., & Nam, K. W. (2025). Hydrogels and Microgels: Driving Revolutionary Innovations in Targeted Drug Delivery, Strengthening Infection Management, and Advancing Tissue Repair and Regeneration. *Gels (Basel, Switzerland)*, *11*(3), 179. DOI: 10.3390/gels11030179 PMID: 40136884

Alaghemandi, M. (2024). Sustainable Solutions Through Innovative Plastic Waste Recycling Technologies. *Sustainability (Basel)*, *16*(23), 10401. DOI: 10.3390/su162310401

Alam, M. R., Scampicchio, M., Angeli, S., & Ferrentino, G. (2019). Effect of hot melt extrusion on physical and functional properties of insect based extruded products. *Journal of Food Engineering*, *259*, 44–51. DOI: 10.1016/j.jfoodeng.2019.04.021

Alberts, A., Moldoveanu, E.-T., Niculescu, A.-G., & Grumezescu, A. M. (2025). Hydrogels for Wound Dressings: Applications in Burn Treatment and Chronic Wound Care. *Journal of Composites Science*, *9*(3), 133. DOI: 10.3390/jcs9030133

Arora, D., Bhunia, B. K., Janani, G., & Mandal, B. B. (2021). Bioactive three-dimensional silk composite in vitro tumoroid model for high throughput screening of anticancer drugs. *Journal of Colloid and Interface Science*, *589*, 438–452. DOI: 10.1016/j.jcis.2021.01.010 PMID: 33485251

Barbosa, L. R., Santos, F., Soliman, E. P., Rodrigues, A. P., Wilcken, C. F., Campos, J. M., Zanuncio, A. J. V., & Zanuncio, J. C. (2019). Biological parameters, life table and thermal requirements of *Thaumastocorisperegrinus* (Heteroptera: Thaumastocoridae) at different temperatures. *Scientific Reports*, *9*(1), 10174. DOI: 10.1038/s41598-019-45663-5 PMID: 31308394

Bidwai, S. (2025). Biomaterials Market Transforming Healthcare Through Sustainable and Smart Materials. https://www.precedenceresearch.com/biomaterials-market

Çakmak, E., Koc-Bilican, B., Avila-Poveda, O. H., Karaduman, T., Cansaran-Duman, D., Williams, S. T., & Kaya, M. (2022). Discovery of protein-based natural hydrogel from the girdle of the 'sea cockroach' *Chiton articulatus* (Chitonida: Chitonidae). *PeerJ*, *10*, e13386. DOI: 10.7717/peerj.13386 PMID: 35573172

Chelu, M., Musuc, A. M., Popa, M., & Moreno, C. J. (2023). *Aloe vera*-Based Hydrogels for Wound Healing: Properties and Therapeutic Effects. *Gels (Basel, Switzerland)*, *9*(7), 539. DOI: 10.3390/gels9070539 PMID: 37504418

Cheng, C., Williamson, E. J., Chiu, G. T. C., & Han, B. (2024). Engineering biomaterials by inkjet printing of hydrogels with functional particulates. *Med-X, 2*(1), 2024. DOI: 10.1007/s44258-024-00024-4 PMID: 38975024

Choi, H., Choi, W. S., & Jeong, J. O. (2024). A Review of Advanced Hydrogel Applications for Tissue Engineering and Drug Delivery Systems as Biomaterials. *Gels (Basel, Switzerland), 10*(11), 693. DOI: 10.3390/gels10110693 PMID: 39590049

Desai, N., Rana, D., Salave, S., Gupta, R., Patel, P., Karunakaran, B., Sharma, A., Giri, J., Benival, D., & Kommineni, N. (2023). Chitosan: A potential biopolymer in drug delivery and biomedical applications. *Pharmaceutics, 15*(4), 1313. DOI: 10.3390/pharmaceutics15041313 PMID: 37111795

Djafaripetroudy, S., Fatehi, P., El Idrissi, A., Kang, K., Abidi, N., & McLaren, B. (2025). Advancing agricultural efficiency and sustainability: Bio-inspired superabsorbent hydrogels for slow and controlled release fertilizers. *The Science of the Total Environment, 977*, 179366. DOI: 10.1016/j.scitotenv.2025.179366 PMID: 40215637

Dodda, J. M., Deshmukh, K., Bezuidenhout, D., & Yeh, Y. C. (2023). Hydrogels: Definition, History, Classifications, Formation, Constitutive Characteristics, and Applications. In *Multicomponent Hydrogels*. The Royal Society of Chemistry. DOI: 10.1039/BK9781837670055-00001

Du, Z., Yan, Z., Guo, Y., Ye, R., Tehoungue, A., Li, Y., Zhang, G., & Zhang, Y. (2025). A high stretchable, transparent sericin hydrogel with antioxidant activity formed by $CaCl_2$-formic acid dissolution system. *Polymer, 317*, 127929. DOI: 10.1016/j.polymer.2024.127929

Edah, B., & Owolabi, O. D. (2023). Physical properties of defatted and extruded black soldier fly (*Hermetiaillucens*) larvae-based aqua-feed using a twin-screw extruder. *Discover Food, 3*(1), 14. DOI: 10.1007/s44187-023-00056-6

Fricke, A., Psarianos, M., Sabban, J., Fitzner, M., Reipsch, R., Schlüter, O. K., Dreyer, C., Vogt, J. H. M., Schreiner, M., & Baldermann, S. (2022). Composite materials for innovative urban farming of alternative food sources (macroalgae and crickets). *Frontiers in Sustainable Food Systems, 2022*, 1001769. Advance online publication. DOI: 10.3389/fsufs.2022.1001769

Friuli, M., Cafarchia, C., Lia, R. P., Otranto, D., Pombi, M., & Demitri, C. (2022). From tissue engineering to mosquitoes: Biopolymers as tools for developing a novel biomimetic approach to pest management/vector control. *Parasites & Vectors, 15*(1), 79. DOI: 10.1186/s13071-022-05193-y PMID: 35248154

Friuli, M., Nitti, P., Aneke, C. I., Demitri, C., Cafarchia, C., & Otranto, D. (2021). Freeze-drying of *Beauveria bassiana* suspended in hydroxyethyl cellulose-based hydrogel as possible method for storage: Evaluation of survival, growth and stability of conidial concentration before and after processing. *Results in Engineering, 1*, 100283. DOI: 10.1016/j.rineng.2021.100283

Gamage, A., Jayasinghe, N., Thiviya, P., Wasana, M. L. D., Merah, O., Madhujith, T., & Koduru, J. R. (2023). Recent Application Prospects of Chitosan Based Composites for the Metal Contaminated Wastewater Treatment. [Basel]. *Polymers, 15*(6), 1453. DOI: 10.3390/polym15061453 PMID: 36987232

Gantumur, E., Kimura, M., Taya, M., Horie, M., Nakamura, M., & Sakai, S. (2019). Inkjet micropatterning through horseradish peroxidase-mediated hydrogelation for controlled cell immobilization and microtissue fabrication. *Biofabrication*, *12*(1), 011001. DOI: 10.1088/1758-5090/ab3b3c PMID: 31412324

Ghosh, S., & Yi, H. G. (2022). A review on Bioinks and their Application in Plant Bioprinting. *International Journal of Bioprinting*, *8*(4), 612. DOI: 10.18063/ijb.v8i4.612 PMID: 36404783

Giani, M., Valentino, C., Vigani, B., Ruggeri, M., Guarnieri, A., Salvia, R., Scieuzo, C., Falabella, P., & Rossi, S. (2025). *Hermetiaillucens*-derived chitosan as a promising sustainable biomaterial for wound healing applications: Development of sponge-like scaffolds. *International Journal of Biological Macromolecules*, *304*, 140903. DOI: 10.1016/j.ijbiomac.2025.140903 PMID: 39938837

Gkinali, A. A., Matsakidou, A., & Paraskevopoulou, A. (2022). Characterization of *Tenebrio molitor* Larvae Protein Preparations Obtained by Different Extraction Approaches. *Foods*, *11*(23), 3852. DOI: 10.3390/foods11233852 PMID: 36496659

Guan, Y., Sun, F., Zhang, X., Peng, Z., Jiang, B., Liang, M., & Wang, Y. (2020). Silk fibroin hydrogel promote burn wound healing through regulating TLN1 expression and affecting cell dhesion and migration. *Journal of Materials Science. Materials in Medicine*, *31*(5), 48. DOI: 10.1007/s10856-020-06384-8 PMID: 32405818

Guessasma, S., Stephant, N., Durand, S., & Belhabib, S. (2024). Digital light processing route for 3D printing of acrylate-modified PLA/lignin blends: Microstructure and mechanical performance. *Polymers*, *16*(10), 1342. DOI: 10.3390/polym16101342 PMID: 38794534

Hauck, N., Neuendorf, T. A., Männel, M. J., Vogel, L., Liu, P., Stündel, E., Zhang, Y., & Thiele, J. (2021). Processing of fast-gelling hydrogel precursors in microfluidics by electrocoalescence of reactive species. *Soft Matter*, *17*(45), 10312–10321. DOI: 10.1039/D1SM01176F PMID: 34664052

Ho, T. C., Chang, C. C., Chan, H. P., Chung, T. W., Shu, C. W., Chuang, K. P., Duh, T. H., Yang, M. H., & Tyan, Y. C. (2022). Hydrogels: Properties and applications in biomedicine. *Molecules (Basel, Switzerland)*, *27*(9), 2902. DOI: 10.3390/molecules27092902 PMID: 35566251

Huang, C., Dong, L., Zhao, B., Lu, Y., Huang, S., Yuan, Z., Luo, G., Xu, Y., & Qian, W. (2022). Anti-inflammatory hydrogel dressings and skin wound healing. *Clinical and Translational Medicine*, *12*(11), e1094. DOI: 10.1002/ctm2.1094 PMID: 36354147

Igual, M., García-Segovia, P., & Martínez-Monzó, J. (2020). Effect of *Acheta domesticus* (house cricket) addition on protein content, colour, texture, and extrusion parameters of extruded products. *Journal of Food Engineering*, *282*, 110032. DOI: 10.1016/j.jfoodeng.2020.110032

Januszewski, J., Forma, A., Zembala, J., Flieger, M., Tyczyńska, M., Dring, J. C., Dudek, I., Świątek, K., & Baj, J. (2023). Nutritional supplements for skin health—A review of what should be chosen and why. *Medicina*, *60*(1), 68. DOI: 10.3390/medicina60010068 PMID: 38256329

Jiang, F., Duan, Y., Li, Q., Li, X., Li, Y., Wang, Y., Liu, S., Liu, M., Zhang, C., & Pan, X. (2024). Insect chitosan/pullulan/gallium photo-crosslinking hydrogels with multiple bioactivities promote MRSA-infected wound healing. *Carbohydrate Polymers*, *334*, 122045. DOI: 10.1016/j.carbpol.2024.122045 PMID: 38553241

Kapoor, S., & Kundu, S. C. (2016). Silk protein-based hydrogels: Promising advanced materials for biomedical applications. *Acta Biomaterialia*, *31*, 17–32. DOI: 10.1016/j.actbio.2015.11.034 PMID: 26602821

Kiiru, S. M., Kinvuru, J. N., Kiage, B. N., Martin, A., Marel, A. M., & Osen, R. (2020). Extrusion texturization of cricket flour and soy protein isolate: Influence of insect content, extrusion temperature, and moisture-level variation on textural properties. *Food Science & Nutrition*, *8*(8), 4112–4120. DOI: 10.1002/fsn3.1700 PMID: 32884692

Kowalczewski, P. Ł., Gumienna, M., Rybicka, I., Górna, B., Sarbak, P., Dziedzic, K., & Kmiecik, D. (2021). Nutritional value and biological activity of gluten-free bread enriched with cricket powder. *Molecules (Basel, Switzerland)*, *26*(4), 1184. DOI: 10.3390/molecules26041184 PMID: 33672127

Lanes, C. F. C., Pedron, F. A., Bergamin, G. T., Bitencourt, A. L., Dorneles, B. E. R., Villanova, J. C. V., Dias, K. C., Riolo, K., Oliva, S., Savastano, D., & Giannetto, A. (2021). Black soldier fly (*Hermetiaillucens*) larvae and prepupae defatted meals in diets for zebrafish (*Danio rerio*). *Animals (Basel)*, *11*(3), 720. DOI: 10.3390/ani11030720 PMID: 33800826

Lee, H., Kim, J., Myung, S., Jung, T., Han, D., Kim, B., & Lee, J. (2024). Extraction of γ-chitosan from insects and fabrication of PVA/γ-chitosan/kaolin nanofiber wound dressings with hemostatic properties. *Discover Nano*, *19*(1), 77. Advance online publication. DOI: 10.1186/s11671-024-04016-6 PMID: 38693438

Lerch, S., Zuber, R., Gehring, N., Wang, Y., Eckel, B., Klass, K., Lehmann, F., & Moussian, B. (2020). Resilin matrix distribution, variability and function in *Drosophila*. *BMC Biology*, *18*(1), 195. DOI: 10.1186/s12915-020-00902-4 PMID: 33317537

Li, X., & Li, Z. (2024). Global trends and current advances in slow/controlled-release fertilizers: A bibliometric analysis from 1990 to 2023. *Agriculture*, *14*(9), 1502. DOI: 10.3390/agriculture14091502

Li, Z., Song, J., Zhang, J., Hao, K., Liu, L., Wu, B., Zheng, X., Xiao, B., Tong, X., & Dai, F. (2020). Topical application of silk fibroin-based hydrogel in preventing hypertrophic scars. *Colloids and Surfaces. B, Biointerfaces*, *186*, 110735. DOI: 10.1016/j.colsurfb.2019.110735 PMID: 31865120

Li, Z., Zheng, A., Mao, Z., Li, F., Su, T., Cao, L., Wang, W., Liu, Y., & Wang, C. (2023). Silk fibroin–gelatin photo-crosslinked 3D-bioprinted hydrogel with MOF-methylene blue nanoparticles for infected wound healing. *International Journal of Bioprinting*, *9*(5), 459–473. DOI: 10.18063/ijb.773 PMID: 37457928

Liu, H., Wang, C., Li, C., Qin, Y., Wang, Z., Yang, F., Li, Z., & Wang, J. (2018). A functional chitosan-based hydrogel as a wound dressing and drug delivery system in the treatment of wound healing. *RSC Advances*, *8*(14), 7533–7549. DOI: 10.1039/C7RA13510F PMID: 35539132

Liu, X., Qin, S., Xu, L., Fu, G., Huang, Y., Yu, C., Cheng, G., Li, Y., He, Y., Qi, Y., & Sun, D. (2023). A tough and mechanically stable adhesive hydrogel for non-invasive wound repair. *Frontiers in Bioengineering and Biotechnology*, *11*, 1173247. DOI: 10.3389/fbioe.2023.1173247 PMID: 37122868

Lu, P., Ruan, D., Huang, M., Tian, M., Zhu, K., Gan, Z., & Xiao, Z. (2024). Harnessing the potential of hydrogels for advanced therapeutic applications: Current achievements and future directions. *Signal Transduction and Targeted Therapy*, *9*(1), 166. DOI: 10.1038/s41392-024-01852-x PMID: 38945949

Lv, J., Lv, X., Ma, M., Oh, D.-H., Jiang, Z., & Fu, X. (2023). Chitin and chitin-based biomaterials: A review of advances in processing and food applications. *Carbohydrate Polymers*, *299*, 120142. DOI: 10.1016/j.carbpol.2022.120142 PMID: 36876773

Lyu, Y., Liu, Y., He, H., & Wang, H. (2023). Application of silk-fibroin-based hydrogels in tissue engineering. *Gels (Basel, Switzerland)*, *9*(5), 431. DOI: 10.3390/gels9050431 PMID: 37233022

Macedo, G. A., Barbosa, P. P. M., Dias, F. F. G., Crawford, L. M., Wang, S. C., & Bell, J. M. L. N. M. (2023). Optimizing the Integration of Microwave Processing and Enzymatic Extraction to Produce Polyphenol-Rich Extracts from Olive Pomace. *Foods*, *12*(20), 3754. DOI: 10.3390/foods12203754 PMID: 37893645

Maria-Hormigos, R., Mayorga-Martinez, C. C., & Pumera, M. (2022). Magnetic hydrogel microrobots as insecticide carriers for in vivo insect pest control in plants. *Small*, *18*(40), 2204887. DOI: 10.1002/smll.202204887 PMID: 36585370

Mersmann, L., Souza, V. G. L., & Fernando, A. L. (2025). Green Processes for Chitin and Chitosan Production from Insects: Current State, Challenges, and Opportunities. *Polymers*, *17*(9), 1185. DOI: 10.3390/polym17091185 PMID: 40362968

Michalik, R., & Wandzik, I. (2020). A mini-review on chitosan-based hydrogels with potential for sustainable agricultural applications. [Basel]. *Polymers*, *12*(10), 2425. DOI: 10.3390/polym12102425 PMID: 33096639

Mikhailidi, A., Ungureanu, E., Tofanica, B. M., Ungureanu, O. C., Fortună, M. E., Belosinschi, D., & Volf, I. (2024). Agriculture 4.0: Polymer hydrogels as delivery agents of active ingredients. *Gels (Basel, Switzerland)*, *10*(6), 368. DOI: 10.3390/gels10060368 PMID: 38920915

Moharir, A. (2025). Sericin Market Report 2025 (Global Edition). https://www.cognitivemarketresearch.com/sericin-market-report

Nanda, D., Behera, D., Pattnaik, S. S., & Behera, A. K. (2025). Advances in natural polymer-based hydrogels: Synthesis, applications, and future directions in biomedical and environmental fields. *Discover Polymers*, *2*(1), 6. DOI: 10.1007/s44347-025-00017-5

Neethu, T. M., Dubey, P. K., & Kaswala, A. R. (2018). Prospects and applications of hydrogel technology in agriculture. *International Journal of Current Microbiology and Applied Sciences*, *7*(5), 3155–3162. DOI: 10.20546/ijcmas.2018.705.369

Nino, M. C., Reddivari, L., Osorio, C., Kaplan, I., & Liceaga, A. M. (2021). Insects as a source of phenolic compounds and potential health benefits. *Journal of Insects as Food and Feed*, *7*(7), 1077–1087. DOI: 10.3920/JIFF2020.0113

Nordin, N., Farhan, W. A., Afifi, W. M., Majid, S. R., & Bakar, N. A. (2024). Crop resilience enhancement through chitosan-based hydrogels as a sustainable solution for water-limited environments. *International Journal of Biological Macromolecules*, *282*(4), 137202. DOI: 10.1016/j.ijbiomac.2024.137202 PMID: 39489246

Ohkubo, S., Shintaku, T., Mine, S., Yamamoto, D. S., & Togawa, T. (2023). Mosquitoes possess specialized cuticular proteins that are evolutionarily related to the elastic protein resilin. *Insects*, *14*(12), 941. DOI: 10.3390/insects14120941 PMID: 38132614

Ojianwuna, C. C., Enwemiwe, V. N., Esiwo, E., Orji, G. O., & Nkeze, A. J. (2023). Food and feed additive of insects: Economic and environmental impacts. *International Journal of Child Health and Nutrition*, *12*(3), 107–119. DOI: 10.6000/1929-4247.2023.12.03.5

Olteanu, G., Neac u, S. M., Joi a, F. A., Musuc, A. M., Lupu, E. C., Ioni ă-Mîndrican, C. B., Lupuliasa, D., & Mititelu, M. (2024). Advancements in regenerative hydrogels in skin wound treatment: A comprehensive review. *International Journal of Molecular Sciences*, *25*(7), 3849. DOI: 10.3390/ijms25073849 PMID: 38612660

Pal, P., Pal, A., Nakashima, K., & Yadav, B. K. (2021). Application of chitosan in environmental remediation: A review. *Chemosphere*, *266*, 128934. DOI: 10.1016/j.chemosphere.2020.128934 PMID: 33246700

Research and Markets. (2024). **Global Advanced Wound Dressings Market Report, 2024-2030: An $8.38 Billion Opportunity, with Alginate and Collagen Dressings Dominating.** https://www.globenewswire.com/news-release/2024/06/12/2897732/28124/en/Global-Advanced-Wound-Dressings-Market-Report-2024-2030-An-8-38-Billion-Opportunity-with-Alginate-and-Collagen-Dressings-Dominating.html

Riofrio, A., Alcivar, T., & Baykara, H. (2021). Environmental and economic viability of chitosan production in Guayas–Ecuador: A robust investment and life-cycle analysis. *ACS Omega*, *6*(36), 23038–23051. pmc.ncbi.nlm.nih.gov. DOI: 10.1021/acsomega.1c01672 PMID: 34549104

Ruggeri, M., Bianchi, E., Vigani, B., Sánchez-Espejo, R., Spano, M., Totaro Fila, C., Mannina, L., Viseras, C., Rossi, S., & Sandri, G. (2023). Nutritional and functional properties of novel Italian spray-dried cricket powder. *Antioxidants*, *12*(1), 112. DOI: 10.3390/antiox12010112 PMID: 36670974

Shahwar, D., Mushtaq, Z., Mushtaq, H., Alqarawi, A. A., Park, Y., Alshahrani, T. S., & Faizan, S. (2023). Role of microbial inoculants as bio fertilizers for improving crop productivity: A review. *Heliyon*, *9*(6), e16134. DOI: 10.1016/j.heliyon.2023.e16134 PMID: 37255980

Silva, A. P. N., Chagas, C. F., de Andrade Alves, E. L., de Castro Carvalho, V., & Haddi, K. (2023). Hydrogel as an alternative to agar for laboratory rearing of the green peach aphid *Myzuspersicae*. *CABI Agriculture and Bioscience*, *4*(1), 13. DOI: 10.1186/s43170-023-00155-5

Sirichocktanasap, P., & Chanput, W. P. (2022). Phenolic compounds from cricket powder and its protein extract after passing through the simulated in vitro digestion. *Journal of Food Science and Agricultural Technology*, *6*, 62–66.

Suh, H.-J., Kim, S.-R., Lee, K.-S., Park, S., & Kang, S. C. (2010). Antioxidant activity of various solvent extracts from *Allomyrinadichotoma* (Arthropoda: Insecta) larvae. *Journal of Photochemistry and Photobiology. B, Biology*, *99*(2), 67–73. DOI: 10.1016/j.jphotobiol.2010.02.005 PMID: 20236833

Tao, J., Davidov-Pardo, G., Burns-Whitmore, B., Cullen, E. M., & Li, Y. O. (2017). Effects of edible insect ingredients on the physicochemical and sensory properties of extruded rice products. *Journal of Insects as Food and Feed*, *3*(4), 263–278. DOI: 10.3920/JIFF2017.0030

Vedovello, P., Sanches, L. V., da Silva Teodoro, G., Majaron, V. F., Bortoletto-Santos, R., Ribeiro, C., & Putti, F. F. (2024). An overview of polymeric hydrogel applications for sustainable agriculture. *Agriculture*, *14*(6), 840. DOI: 10.3390/agriculture14060840

Visan, A. I., & Negut, I. (2025). Environmental and wastewater treatment applications of stimulus-responsive hydrogels. *Gels (Basel, Switzerland)*, *11*(1), 72. DOI: 10.3390/gels11010072 PMID: 39852043

Wani, S. U. D., Zargar, M. I., Masoodi, M. H., Alshehri, S., Alam, P., Ghoneim, M. M., Alshlowi, A., Shivakumar, H. G., Ali, M., & Shakeel, F. (2022). Silk fibroin as an efficient biomaterial for drug delivery, gene therapy, and wound healing. *International Journal of Molecular Sciences*, *23*(22), 14421. DOI: 10.3390/ijms232214421 PMID: 36430901

Woodrow, C., Cullen, D. A., Montealegre-Z, F., & Gonzalez-Rodriguez, J. (2023). Non-invasive characterization of the elastic protein resilin in insects using Raman spectroscopy. *International Journal of Biological Macromolecules*, *254*(2), 127967. DOI: 10.1016/j.ijbiomac.2023.127967 PMID: 37944738

Xing, Y., & Wang, X. (2024). Precision agriculture and water conservation strategies for sustainable crop production in arid regions. *Plants*, *13*(22), 3184. DOI: 10.3390/plants13223184 PMID: 39599396

Ye, Z., Zhatkanbaeva, Z. H.andRaimbekova, A. (2021). Method for one-step synthesis of biodegradable chitosan hydrogel introduced mineral and/or organic fertilizers. Patent KZ. Sichuan, China: Sichuan Tiannong Agricultural Co.

Younes, I., & Rinaudo, M. (2015). Chitin and chitosan preparation from marine sources. Structure, properties and applications. *Marine drugs, 13*(3), 1133–1174. 10.3390/md13031133

Van Loon, M. S., &Bovenkerk, B. (2021). 32. The ethics and mindedness of insects. In H. Schübel, & I. Walliman-Helmer (Eds.), *Justice and food security in a changing climate* (pp. 218-223). Article 32 Wageningen Academic Publishers. DOI: 10.3920/978-90-8686-915-2_32

Yuan, Y., Nasri, M., Manayi, A., Zhang, J., Wu, C., Jeon, T.-J., & Kang, L. (2024). Sericin coats of silk fibres: A degumming waste or future material? *Materials Today. Bio*, *29*, 101306. DOI: 10.1016/j.mtbio.2024.101306 PMID: 39534681

Zahra, D., Shokat, Z., Ahmad, A., Javaid, A., Khurshid, M., Ashfaq, U. A., & Nashwan, A. J. (2023). Exploring the recent developments of alginate silk fibroin material for hydrogel wound dressing: A review. *International Journal of Biological Macromolecules*, *248*, 125989. DOI: 10.1016/j.ijbiomac.2023.125989 PMID: 37499726

Zarzosa, G. I. G., & Kobayashi, T. (2023). Insect chitins and hydrogels sourced from *Zophobasmorio* in different life stages and their properties. *Chemistry Letters*, *52*(8), 674–677. DOI: 10.1246/cl.230248

Zarzosa, G. I. G., & Kobayashi, T. (2025). Regenerated chitin from insect sources and fabrication of their hydrogel films as an alternative from marine sources. *Biomass Conversion and Biorefinery*, *15*(7), 9835–9848. DOI: 10.1007/s13399-024-05686-z

Zhang, J., Wang, L., Xu, C., Cao, Y., Liu, S., Reis, R. L., Kundu, S. C., Yang, X., Xiao, B., & Duan, L. (2023). Transparent silk fibroin film-facilitated infected-wound healing through antibacterial, improved fibroblast adhesion and immune modulation. *Journal of Materials Chemistry. B*, *12*(2), 475–488. DOI: 10.1039/D3TB02146G PMID: 38099432

Zhang, X., Su, H., Yu, H., Ding, J., Deng, W., Qin, B., Zhou, C., Dou, J., & Guo, M. (2022). A polysaccharide from *Eupolyphagasinensis* Walker with anti-HBV activities in vitro and in vivo. *Frontiers in Pharmacology*, *13*, 827128. DOI: 10.3389/fphar.2022.827128 PMID: 35308231

Zhao, L., Zhou, Y., Zhang, J., Liang, H., Chen, X., & Tan, H. (2023). Natural polymer-based hydrogels: From polymer to biomedical applications. *Pharmaceutics*, *15*(10), 2514. DOI: 10.3390/pharmaceutics15102514 PMID: 37896274

Zhatkanbayev, Y., Zhatkanbayeva, Z., Iskakova, Z., Kolpek, A., Serikov, A., Moldagulova, N., Danlybayeva, G., Sarsenova, A., & Anuarbekova, S. (2023). andAnuarbekova, S. (2023). Application of chitosan-based hydrogel obtained from insects in pine planting. *International Journal of Biomaterials*, *8175405*, 1–6. Advance online publication. DOI: 10.1155/2023/8175405 PMID: 37025976

Zhou, Y., Luo, J., Jing, Q., Ge, S., Chen, S., Guo, Z., Li, J., Liu, Z., He, P., He, X., & Xu, B. B. (2024). Tough, waterproofing, and sustainable bio-adhesive inspired by the dragonfly wing. *Advanced Functional Materials*, *34*(42), 2406557. DOI: 10.1002/adfm.202406557

Zhu, J., Zhang, Z., Wen, Y., Song, X., Tan, W. K., Ong, C. N., & Li, J. (2024). Recent advances in superabsorbent hydrogels derived from agro waste materials for sustainable agriculture: A review. *Journal of Agricultural and Food Chemistry*, *72*(41), 22399–22419. DOI: 10.1021/acs.jafc.4c04970 PMID: 39215710

Zielińska, E., Pankiewicz, U., & Sujka, M. (2021). Nutritional, physiochemical, and biological value of muffins enriched with edible insects flour. *Antioxidants*, *10*(7), 1122. DOI: 10.3390/antiox10071122 PMID: 34356355

Chapter 11
Biomedical Prospects of Hydrogels

Muzammal Hussain
https://orcid.org/0009-0000-9914-8870
Southeast University, China

Aqib Mehmood
University of Wah, Pakistan

Zaheer Ahmad
University of Wah, Pakistan

Munir Hussain
Zhejiang Sci-Tech University, China

Muhammad Ajmal
University of Education, Lahore, Pakistan

ABSTRACT

Hydrogels have revolutionized biomedical applications, evolving from simple water-absorbing materials to multifunctional drug delivery systems. This section highlights their role in drug delivery, wound care, and therapeutic interventions. Their 3D network enables localized and sustained drug release, enhancing efficacy while reducing side effects. pH-sensitive hydrogels facilitate targeted gastrointestinal delivery, while thermosensitive types are effective for nasal and transdermal administration. In ophthalmology, hydrogels improve drug retention, and their ability to bypass the blood-brain barrier makes them a key research focus. Incorporating nanoparticles like silver, zinc oxide, or copper oxide enhances antimicrobial properties. Despite their potential, challenges such as scalability, stability, and precise drug release hinder clinical adaptation. However, advancements in nanoscale hydrogels, smart materials, and innovative fabrication techniques like 3D printing and microfluidics can improve drug loading and controlled release, driving future biomedical breakthroughs.

DOI: 10.4018/979-8-3373-2337-4.ch011

11.1 INTRODUCTION

Hydrogel-driven drug delivery systems (DDS) facilitate the release of therapeutic agents to targeted sites, sensitized by various stimulating factors, including pH, light, redox, electromagnetic potential, temperature, catalysts, and other chemical characteristics (Qiu & Hu, 2013). Their excellent host-response, porous architecture, and volumetric oscillation have enhanced their effectiveness in DDS. The modifiable inherent porosity of hydrogels facilitates the fabrication of drugs within the gel network, resulting in modulated and slow drug release profiles over a prolonged time frame, which proves advantageous for the systematic administration of various therapeutic substances. However, drug inclusion is influenced by the hydrodynamic size and mesh spacing (Hoare & Kohane, 2008; Lin et al., 2019).

The release of the drug is entailed with a multitude of mechanisms: diffusion-mediated, swelling-driven, chemo responsive, and externally activated release. Diffusion-mediated drug delivery from a reservoir DDS allows for drug elution through a concentration differential via hydrogel mesh or water cavities. Hydrogel encapsulates a core enriched with a drug for uniform drug elution. Whereas, hydrogel dispersed with drug, elutes the drug through its mesh or porous sites. The initial burst release is highly dependent on time. Swelling-driven systems involve scattered drugs within a glassy matrix. Upon contact with a biological fluid, the hydrogel swells, initiating the diffusion process (Kasai et al., 2023). Typically, biomedical hydrogels are engineered to degrade or erode through physicochemical hydrolytic or enzymatic cleavage, transforming them into primary units (Correa et al., 2021).

11.1.1 Oral Drug Delivery

For cellular uptake, in intestinal transport, nano-sized particles (50-200nm) are used (Moghimi et al., 2001). pH-sensitized hydrogels have been sorted as cationic and anionic hydrogels. Cationic hydrogels are ionically activated at pH less than the pKa of the polymeric matrix, whereas anionic hydrogels respond to the pH level surpassing the pKa of the polymeric matrix (Kim & Peppas, 2003). An oral ingestion approach was employed to investigate the elution profile of apigenin from a Gellan gum hydrogel matrix over a pH range of 1 to 8. Below pH 4, apigenin was sustainably delivered, achieving 60% over a day. However, at pH 8, it was rapidly dispatched, over 6-8 h (Zhao & Wang, 2019). The release of DIC from MA/HEMA/MAPTAC/alginate-based semi-IPN hydrogel was investigated in a stomach-simulating buffer. At pH 1.2, 4.5% of DIC was eluted over the initial 3 hours. Following a 24h duration, elution reached 95% at pH 7 (Ilgin et al., 2020). Ionically active carboxylic group induced swelling, facilitating DIC liberation.

The diagram depicts oral hydrogel-mediated drug delivery, illustrating the sequential release of therapeutic agents from nanoparticles or hydrogel matrices along the gastrointestinal tract. Within the stomach (low pH), hydrogels remain in a compact or minimally swollen state, resulting in a slow and sustained release profile. Upon transition to the intestine (near-neutral pH), ionizable functional groups undergo dissociation, inducing hydrogel swelling and thereby accelerating drug liberation. Super-porous and gastro-retentive hydrogel systems extend gastric residence time, facilitating controlled drug release over several hours to days. In the colon, enzyme or azo-responsive hydrogels enable site-specific drug delivery, providing a promising strategy for targeted therapies such as colitis treatment.

Recently, a highly porous XG-based hydrogel formulation with rutin encapsulated in an inclusion adduct of HP-β-CD has shown a slightly higher release rate over 48 hours. At pH 1.2, 70% of the elution rate was observed, whereas at pH 7.4, 65% of the elution rate was observed (Naeem et al., 2023). Super-porous hydrogels exhibited prolonged liberation rates as gastro-retentive systems. N, N'-methylenebisacrylamide forms a polymer network of MAA or AM, along with CMC or PVP, using a gas infusion protocol. The fabricated device extended the drug liberation over 14-16 days. PVA/CS hydrogels with a high porosity rate were engineered as gastro-retentive pills, providing prolonged gastric residence of up to 18 hours, which released esomeprazole within 12 hours. In vivo outcomes revealed a peak concentration in plasma (587 ± 4.33 ng/mL) within a time frame of 3.0 ± 0.09 hours. Similarly, a Box-Behnken experimental design was used for the synthesis of a composite hydrogel incorporating Fluvastatin. In simulated analysis, a 10-hour time frame was achieved with 80% of Fluvastatin, while total liberation was completed within 12 hours and 18 minutes. Hydrogels incorporating azo linkages showcased responsiveness to enzymatic stimuli, rendering them an optimal choice for colon-selective drug delivery. 19 Acryloyl chloride-modified osalazine hydrogel delivered 5-FU by 48.13% in 2h (pH=2) and 100% delivery (pH=7.4) in the colon content of the rat. Moreover, it was HCT116 cell-compatible (Ma et al., 2019). Researchers have found the colon to be an optimal site for targeted and broad-spectrum delivery of proteolytic catalysts, enhancing their functionality. Recently, Na-alginate-gallic acid mucoadhesive hybrid matrix was orally administered for the treatment of colitis ulcerosa (CU). The hydrogel demonstrated resistance to stomach acid with sustained durability in the intestinal tract and a reduced severity of CU in mice. Most beneficially, it improved the condition of the colon by reducing inflammation and restoring physiological functions.

Figure 1. **Oral Route of Drug Delivery**

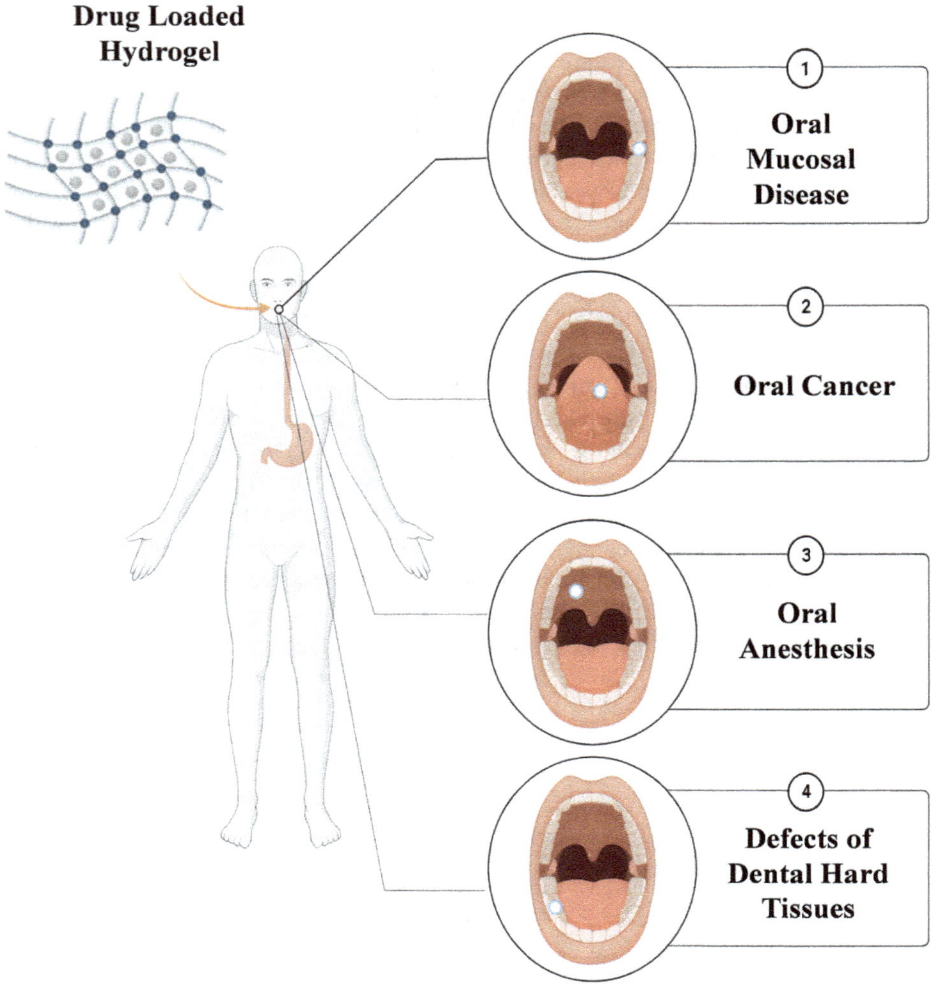

11.1.2 Dermal and Transdermal Drug Delivery

Gel-based therapy has potent topical applications. Indeed, an advantageous route for enhancing the longevity of bioactive ingredients. Hence, granting prolonged drug release. This entails employing polymer material or biopolymers that are responsive to temperature and pH. The diffusion rate is also governed to ensure sustainable delivery and increase exposure time to the bioactive ingredient (Labie & Blanzat, 2023; Weiss, 2011). The curative efficiency hinges on the even distribution of the drug and the consistency of dosage administration. Moreover, it minimizes unintended outcomes from overmedication. Since permeability and dispersibility are linearly related (Cerchiara et al., 2002). Hydrogel's traits are tunable, as drug inclusion in the hydrogel network may induce altered rheological changes. For example, Hydroxyethyl Cellulose (HEC) and Carbopol have demonstrated varying gel properties. Carbopol exhibits flexible solid characteristics. HEC/ Carbopol and HEC itself exemplify viscoelasticity. Calcian and other hydrophobic agents rapidly diffuse out from carbomer-based gels. Moreover, the occlusive property of hydrogels moisturizes the skin through its water-locking ability, helping to alleviate symptoms of eczema.

Vitamin C is an antioxidant compound that prevents skin damage and acts as a lightening agent. To improve its permeability, researchers synthesized vitamin-C-loaded SDEDDS gelled in a xylan gum matrix. Their improved rheological properties minimize the sensitivity of vitamin C to pH, temperature, and oxygen. Thus, increased its permeability into the skin due to the bio-adhesive behavior of xylan gum (Wang et al., 2018). A hydrogel patch composed of oligopeptides, arginine/PEA, PNVCL, and PEG has been developed for transdermal drug delivery. Recently, scientists embedded mesoporous silica nanoparticles (XL-MSNs) into a PAM/PDA hydrogel. Mechanical traits synergistically enhanced its adhesiveness to the skin, which was compromised upon loading with Rhodamine 6G (R6G). Ex vivo analysis revealed that the biocompatibility and healing efficacy of R6G were accelerated over 24 hours, as it had thoroughly permeated the pig skin. Thus, it has the potential for transdermal drug delivery (Jung et al., 2020).

In another study, atopic dermatitis (AD) was medicated with XL-MSNs-PAM/PDA hydrogel incorporated with dexamethasone (DEX). For instance, 32-C alkyl chains were attached to phosphocholine (PC) or dimethylphosphoethanolamine (Me2PE) as the polar group, with methylene blue as the therapeutic ingredient. Release of PC-C_{32}-PC was slightly higher. Additionally, it took a few hours to form a hydrogel. They are intriguing substances with promising potential for use in antimicrobial photodynamic therapy to treat fungal infections, also known as mycoses. An ultrasonically sensitized hydrogel patch was formulated with PEG microcapsules to enhance the penetrability of diclofenac sodium into the skin, thereby reducing epidermal trauma (Huang et al., 2019). A hyaluronic acid/cellulose-based hydrogel was prepared as an acne-controlling agent for iso-liquiritigenin. At a pH of 7, it effectively inhibits the growth of P. Acnes. Liquiritigenin penetrated the skin via follicular hairs. A spongy network of nano-sponges (NS) enables the efficient inclusion and release of NPs or bioactive ingredients through its pores. With this intention, terbinafine hydrochloride (TH) was loaded in NS. In vivo and in vitro investigations demonstrated its enhanced TH penetrability via the topical route. Additionally, Clobetasol propionate (CP) has curative potential against psoriasis, but it also has adverse effects on the skin, including uneven tone, degeneration, and other allergic responses. Researchers have synthesized CP-NS/cyclodextrin hydrogels incorporating Carbopol-934 to counter undesirable outcomes. The CP liberation rate was 86% in a day. In vivo results indicated the inhibitory property of the nanogel against psoriasis. Carbopol-based hydrogel was incorporated with 5, 10, 15, 20-tetrakis(1-methylpyridinium-4-yl)-porphyrin tetra-iodide (TMPTI) polymeric NPs. Without any harm, TMPTI deeply penetrated porcine skin. Further investigation indicated its potency against skin trauma, including skin cancer. Delicate skin has been treated with numerous hydrogel peel-off masks for a cooling and smoothing sensation. Silk sericin-entrapped cellulose-based nanogels have been utilized for this purpose.

Figure 2. **Antibacterial effect of hydrogels for wound recovery**

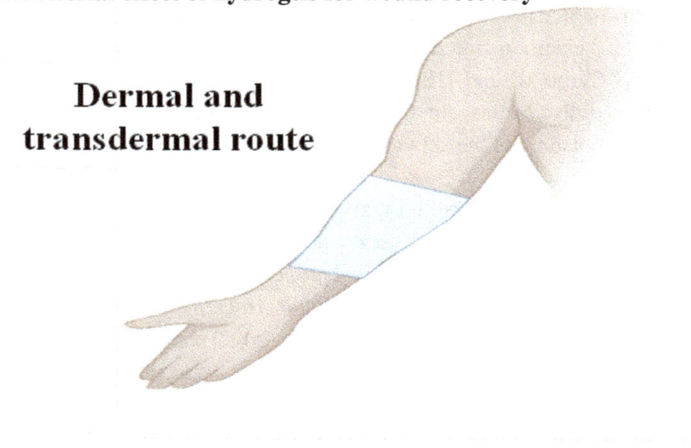

11.1.3 Rectal and Vaginal Drug Delivery

Numerous medicinal substances are delivered via rectal and vaginal pathways, as an alternative route for patients with an impaired oral bioavailability or sensitive metabolism. They are intervened against hemorrhoids, anal abscess and fissures, vaginal candidiasis and vaginosis etc (Dos Santos et al., 2020; Rathi et al., 2022). It offers a range of advantages, including rapid drug uptake, non-gastrointestinal administration, reduced adverse events, and controlled release under relatively stable environmental conditions. Unfortunately, traditional DDS, such as suppositories and enemas, have featured some shortcomings, including brief residency, inconvenience, variable drug absorption rates, expulsion, and patient preferences, which have been overcome by hydrogels with desirable retention rates, suboptimal enzymatic activity, dispersion, and liberation of drug traits (de Araújo Pereira & Bruschi, 2012; Gupta, 2007). More specifically, mucoadhesive hydrogels are well recognized for their versatile formulation options, intimate contact with dosage form and vaginal and rectal mucosa and localized delivery (Purohit et al., 2018).

In a study, 100 patients were implanted with hydrogel spacers (spacer OAR) after undergoing mild-dose 125I brachytherapy, resulting in significant displacement and positive changes in prostate diameter. Additionally, reduced exposure to radiation of the surrounding tissues. These findings support the effectiveness of the spacer to be used for prostate cancer treatment (Morita et al., 2020). A mucoadhesive catechol-functionalized chitosan-based hydrogel enhanced the potency of sulfasalazine through the rectal pathway. Comparatively, equivalent scores were observed for sulfasalazine administered orally or rectally. However, rectally administered hydrogel was more effective against colitis ulcerosa (Xu et al., 2017). Researchers found that a thermally responsive blend of methylcellulose and hyaluronic acid hydrogel entrapped with bovine serum albumin (BSA) exhibited pseudo-plasticity with controlled and extended retention of BSA in the rectum. Moreover, BSA permeates the intestinal barrier, regardless of whether an absorption facilitator accompanies it. Thus, it possesses favorable traits for delivering protein-based therapeutics for the treatment of inflammatory bowel diseases.

Figure 3. **Vaginal route DDS**

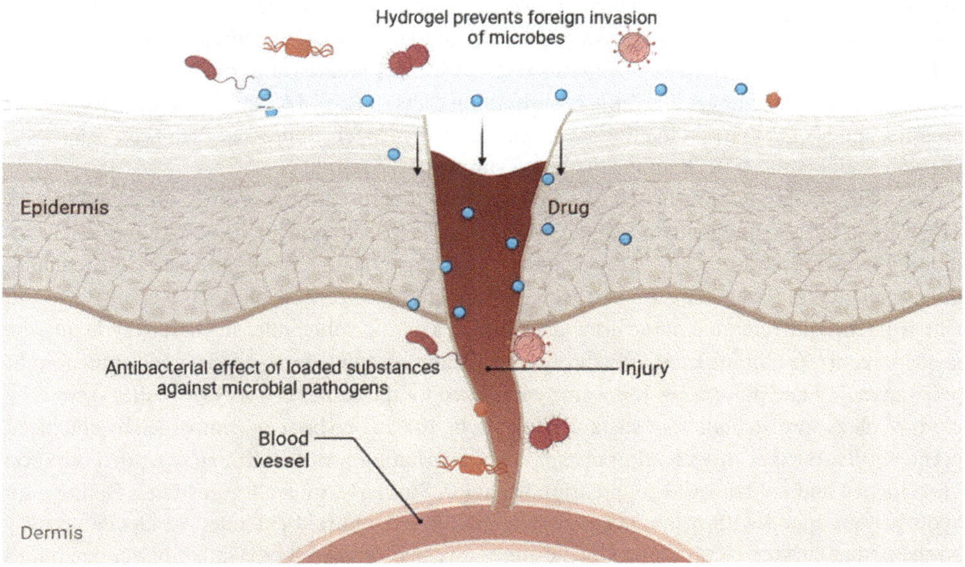

11.1.4 Ocular Drug Delivery/Contact Lenses

Conventionally available ophthalmic medications, such as eye drops, are limited by several factors, including low permeability, tear instability, and blinking of the eye. Hydrogels are physically (due to intermolecular attractions) or chemically (due to covalency) interconnected polymer chains with a network-like structure, arranged three-dimensionally, and are biocompatible when administered intraocularly (Kesharwani et al., 2021; Torres-Luna et al., 2020). Ophthalmic hydrogel formulations are divided based on their behavioral properties. Surface tension allows the adherence of contact lenses to the cornea. Thus, an extended time residence of the drug with sustained liberation enhanced its permeability (Patel et al., 2013). However, visibility, oxygen permeance, and water affinity are key factors to consider when designing contact lens drug delivery systems (DDS). As per traditional practice, contact lens was loaded with the drug by submerging them in a solution. However, its brief release and inefficient loading restricted its scope. Its counter solution was to design a particle-enriched eyeglass lens.

The nanocrystals of Itraconazole (ITZ) were digitally printed on a hydrogel-based soft contact lens by (Tetyczka et al., 2022). Ex vivo analysis demonstrated co-drug release, with 80% liberation achieved within 8 hours. Thus, it serves as a potent DDS for curing glaucoma. As discussed earlier, (Zhou et al., 2024) synthesized a nanozyme-infused probiotic hydrogel, Felab (rGO@FeS2/Lactobacillus@HA), with a generous moisture content. In a mouse, it restricts the scope of the eye lens by allowing an abrupt and unregulated release. Researchers successfully entrapped hydrophobic phenylboronic acid (PBA) into various polymeric lens matrices, including PVP, PHEMA, and PDMA, thereby regulating their moisture content levels. The releases of exemplar ophthalmic agents with contrasting hydrophilic characteristics were investigated. PBA-modified hydrogels strengthened the association of wetting agents and also induced sustained release properties. Similarly, (Zhang et al., 2022) synthesized a hydrogel composed of SBMA, AAc, HEMA, and MMA, exhibiting a zwitterionic nature with a transmittance rate of 94%. The tailored hydrophobic and hydrophilic units resulted in a maintained and controlled water content cycling behavior. Moreover, the built-in antifouling traits of PSBMA facilitated non-adhesion to proteins and bacteria. Upon loading norfloxacin, the hydrogel suppressed the bacteria by 59%. Diabetic retinopathy (DR) restricts the blood supply to the retina.

To address this condition, a PHEMA hydrogel lens incorporating Avastin has been implanted. Throughout 0-1500 minutes, 60% of the drug was released, along with 40-60% of the VEGF inhibiting effect (Goswami et al., 2023). Sometimes, therapeutic ingredients are lost before reaching the cornea due to rapid tear turnover. However, in situ hydrogels develop a bio adhesive mesh, which extends the retention of the drug and facilitates its controlled release, thereby reducing the need for recurrent administration and promoting patient compliance. Recently, nano emulsions of poloxamer 407 and poloxamer 188 were synthesized for the ocular delivery of besifloxacin. The in vitro release profile indicated sustained release of 40% in one hour, accompanied by enhanced permeability. Further investigation reveals its non-irritating profile, with an extended duration of residence and antibacterial suppression behavior. Similarly, ionically-sensitized Gellan gum hydrogel entrapped with brimonidine tartrate (BRT) exemplified extended residence of BRT on the ocular surface. A pharmacokinetic study revealed improved bioavailability compared to BRT drops, (Xu et al., 2023). Resveratrol nanoparticles (NPs) were entrapped in poly (lactic-co-glycolic acid) (PLGA), modified with polyethyleneimine (PEI), and loaded into poloxamer-407. A sustained and extended release of resveratrol was observed for up to 3 days. Furthermore, it exhibits potent antioxidative and anti-inflammatory effects on corneal cells, while also regulating mitochondrial activity (Kolawole & Cook, 2023).

Figure 4. **Ocular routes for DDS**

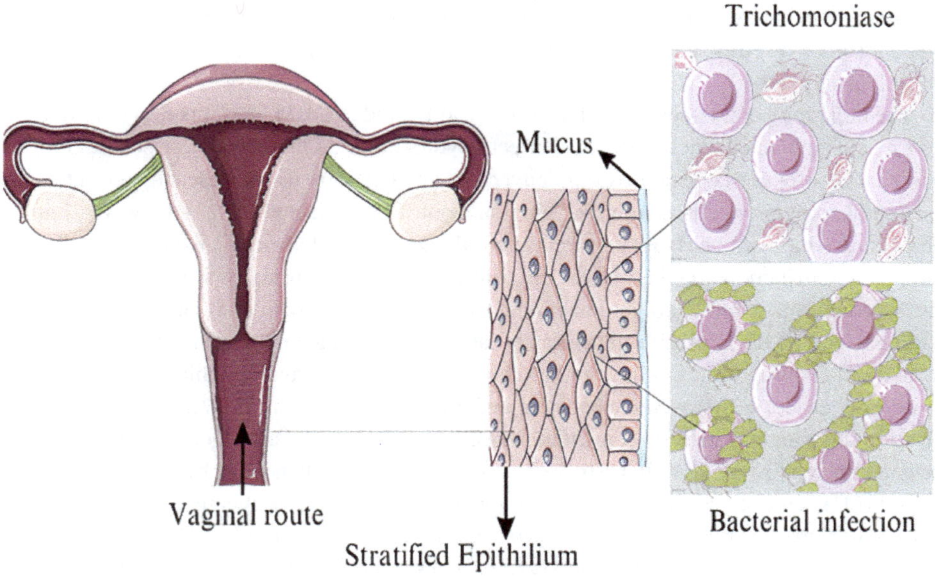

11.1.5 Brain Drug/Epi-cortical Drug Delivery

The intricate barriers within the brain have posed a consistent challenge for delivering effective drugs to their target; foremost among these issues are the arachnoid epithelium, blood-cerebrospinal barrier, choroid plexus epithelium, blood-brain barrier (BBB), and other physiological mechanisms. Due to the physiological complexities of the BBB, the majority of newly synthesized drugs struggle to breach this barrier. Thus, lowers the commercial availability of such medications (Ghasemiyeh & Mohammadi-Samani, 2019; Segal, 2000). However, nanogels have been extensively used for DDS. Over the years, patients diagnosed with Glioblastoma Multiforme (GBM) have not experienced any significant increase in their average survival rate. BBB limits the delivery of drugs to the microenvironment of residual post-surgical tumor cells. (Rajendran & Vadakkepushpakath, 2024) employed site-specific delivery to the surgical cavity using a spraying system with etoposide and Olaparib nanocrystals, along with a pectin-based bio adhesive hydrogel coated with polylactic acid-polyethylene glycol.

The system demonstrated promising results in reducing the viability and adhesion of tumorous cells to the brain, with sustained drug release over 120 hours. The fluorescent-labelled polymeric NPs were sprayed in a mammalian brain (ex vivo). The confocal images revealed the penetration of NPs into brain tissues at significant depths (McCrorie et al., 2020). Similarly, (Kang et al., 2023)have created a soft drug reservoir within the brain's cortex via an injectable thermo-sensitive nanocomposite hydrogel, containing doxorubicin (DOX)-loaded micelles and nanocubes of ferrimagnetic iron oxide with water dispersibility (wFIONs). Under the influence of an alternating magnetic field, the drug was efficiently dispersed through heat generation from wFIONs. Furthermore, the cleavage between the pH-sensitive hydrazone bond of DOX and the hydrogel occurs effectively in the acidic environment surrounding brain tissues or within the endosomes of GBM cells. Thus, release of DOX is accelerated at pH 5.4 as compared to physiological pH 7.4.

The targeting peptide-conjugated micelles readily bind to GBM cells, resulting in increased drug uptake and an enhanced cytotoxic effect. Hydrogel-based therapies offer potential for the treatment of strokes via versatile drug delivery and cell therapy using mesenchymal stem cells and neural progenitor cells to improve functional outcomes (Yu et al., 2023). However, intracerebral hemorrhage stroke (ICH) constrains neural regeneration by various factors, including the brain cavity, inflammation, cell apoptosis, and formation of neural scars. Recently, scientists developed an injectable hydrogel containing micellar chitosan with dual capabilities for drug loading. Upon its injection into the brains of rats, it displayed a combination of swift release of the hydrophilic drug (minocycline) and prolonged release of the hydrophobic drug (edaravone). As a consequence, it improved the behavior and minimized brain atrophy. After 21 days of post-injection, histological analysis confirmed neural regeneration and angiogenesis, thus depicting the potential repair process of ICH stroke in rats treated with a self-healing hydrogel. Parkinson's disease (PD), a prevalent central nervous system (CNS) disorder, result in cell loss in substantia nigra and severe dopaminergic neuron decline. Current treatments merely address the symptoms without slowing down the disease progression. However, hydrogel DDS offers hope for more effective management of PD (Nguyen et al., 2022).

Oxidized tannic acid-stabilized AuNPs with Quinone groups as an effective nano-crosslinker for chitosan (COA). In vitro studies revealed that neural stem cells (NSCs) cultured in the hydrogel exhibited long-term proliferation and differentiated into neurons. Additionally, it effectively recovered inflamed NSCs with a recovery rate of approximately. 90%. Thus, self-healing chitosan hydrogel demonstrated strong anti-inflammatory and anti-oxidative properties. Electrophysiological analysis revealed an improvement in irregular nerve cell discharges in the subthalamic nucleus of PD rats treated with hydrogel. Histological analysis demonstrated that the COA hydrogel can reduce microglial proliferation, leading to an increase in the density of tyrosine hydroxylase-positive neurons and nerve fibers, while decreasing the concentration of Glial fibrillary acidic protein in astrocytes. Thus, COA hydrogel serves as an effective implant for PD therapy (Xu et al., 2023). Epi-cortical delivery using hydrogel offers a brain-localized approach with efficient release of drugs and minimal tissue damage (Ho et al., 2019). In a study, the tunable Cyclosporin A (CsA) released from a hyaluronan-methyl cellulose (HAMC) hydrogel system holds potential as an effective approach to treat strokes by stimulating endogenous neural stem/progenitor cells (NSPCs). The microsphere of poly(lactic-co-glycolic) acid (PLGA) encapsulating CsA has shown prolonged release for 3-4 weeks. In vitro, CsA retained its bioactivity and was effectively delivered to the NSPCs niche in vivo, with a sustained concentration for 24 hours post-implant of 78%.

Figure 5. **Hydrogels administered for the treatment of GBM and another brain tumor.**

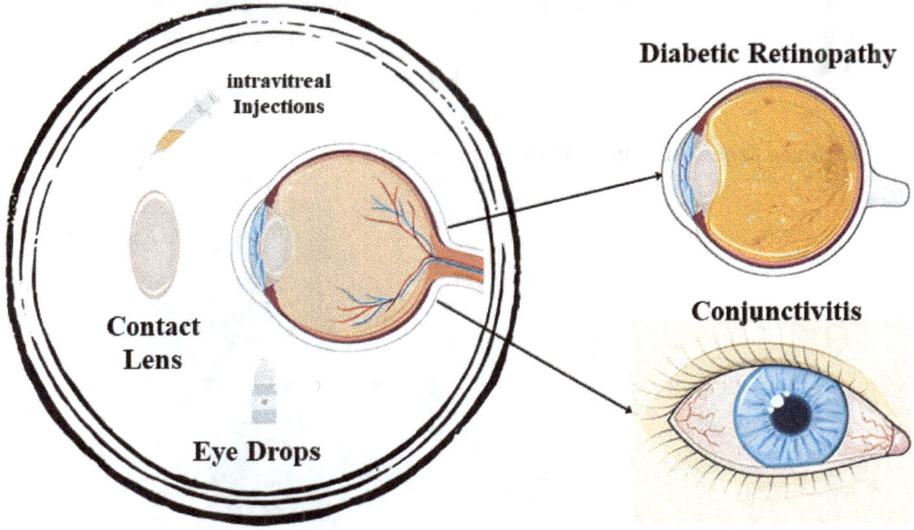

11.1.6 Nasal/Intranasal Drug Delivery

The intranasal pathway, along with the olfactory and trigeminal nerves, plays a crucial role in drug delivery for the treatment of neurodegenerative diseases. By bypassing the BBB, it allows it to target the brain 80 directly. A thermosensitive chitosan-based hydrogel was engineered for the intranasal delivery of ibuprofen, offering efficient spraying characteristics at room temperature and transitioning into a mucoadhesive semi-solid at physiological temperatures. In vitro testing on human epithelial cells unveiled a statistically significant enhancement in the drug delivery rate of ibuprofen with almost no toxic effect. Thus, it serves as a promising candidate for the treatment of brain-affected diseases. In another investigation, a thermo-sensitive hydrogel was synthesized using PF-127/PF-68 grafted on chitosan HCl-co-guar gum. The hydrogel was incorporated with eletriptan hydrobromide-loaded pullulan NPs. In vivo and ex vivo studies substantiated its sustained drug release profile. Moreover, the inclusion of adhesive polymers resulted in the prolonged retention of the formulation in the nasal cavity, thereby evading nasal mucociliary clearance. Thus, the NPs traverse the olfactory mucosa to the brain over more than 24 hours. Icariin is known for its neuroprotective actions (Chen et al., 2016). However, its oral delivery is limited due to hydrophobicity and low brain targeting efficiency. Xu, Dong, et al accepted this challenge. By employing the technique of reverse microemulsion, they synthesized water-soluble icariin nanogels that were loaded into self-assembled thermo-responsive hydrogel systems. In vitro analysis revealed that the release of icariin from the hydrogel system follows zero-order kinetics in phosphate-buffered saline. Fluorescent visualization in vivo confirmed the accelerated brain distribution of NPs within 30 minutes. Compared to oral administration, the nose-to-brain route had a positive emotional influence on depression-induced rats (Xu et al., 2020). Donepezil HCl-loaded liposomes were dispersed in thiolated chitosan hydrogel. The drug release was regulated with a half-life of 3.5 hours. In vivo findings indicated a substantial increase in Donepezil HCl concentration in blood and brain, surpassing that of the oral tablets. Thus, it stands out as a promising candidate for the treatment of Alzheimer's disease (Al Harthi et al., 2019).

Figure 6. **Intranasal route of administration via the BBB**

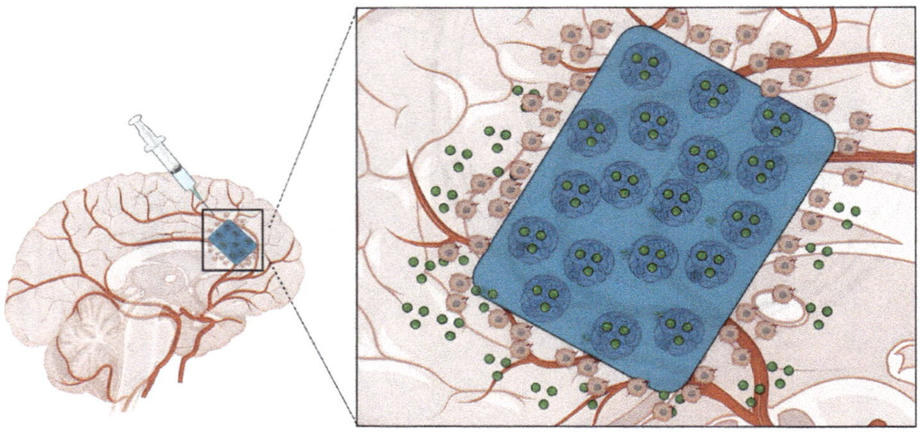

The nasal drug delivery route holds the potential to enhance bioavailability, promote substantial patient compliance, and prevent first-pass metabolism. However, this route is also accompanied by a few setbacks, including barriers to macromolecule absorption across mucosal membranes and a brief period of residence in the nasal region due to the mucosal turnover.

In a study, researchers aimed to enhance insulin absorption via the nasal route by formulating a Hydroxyethyl Cellulose-Based hydrogel (HEC) that incorporated insulin-loaded liposomes modified with cell-penetrating peptides, including HIV-1 peptide and penetrating. Liposomal hydrogel maintained the pseudoplasticity, making it an ideal candidate for nasal drug delivery. Moreover, it exhibited improved drug retention due to an additional barrier for the release. The peptides facilitated a gradual release profile. Formulations, HEC containing insulin and insulin-loaded liposomes, revealed potential as a nasal vehicle for insulin administration, demonstrating anti-hyperglycemic effect. And in vivo release kinetics of insulin conformed to Weibull's mathematical model, with 90% of the release rate of insulin (Von Zuben et al., 2022).

11.1.7 Parenteral Drug Delivery

The Parenteral route of drug administration is favored for the delivery of protein and peptide drugs. Moreover, hydrogels offer a controlled route of drug delivery (Vigata et al., 2020). In a study, a subcutaneous hyaluronic acid hydrogel, chemically and physically cross-linked through polydopamine bonding and forming a polypseudorotaxane structure, was incorporated with donepezil-loaded microspheres. The provided dual crosslinking enhances its bioavailability in the subcutaneous tissue. In vivo studies showed that the drug release profile was controlled, exhibiting longer retention with almost no toxicity (88%). In another investigation, a pH-responsive OE peptide hydrogel was loaded with gemcitabine (GEM) and paclitaxel (PTX). In vitro, the release of PTX from the peptide hydrogel was 96.90% over a week at pH 5.8. For GEM, the release was nearly total at both pH 5.8 and 7.4. Thus, it was concluded that GEM is released in bulk quantities, while PTX is released gradually to prolong the drug's release. In vivo analysis confirms that the peptide hydrogel is less toxic. Thus, they reach the tumor sites by injection and are induced by the tumor microenvironment to release the drugs simultaneously. For intramuscular injection, researchers synthesized goserelin Acetate (GOS)-loaded poloxamer hydrogel polylactic-co-glycolic acid (PLGA) microsphere. An evident core-shell arrangement was observed, responsible for a controlled drug release in vitro. In vivo, GOS was administered for an extended period. The system exhibited a dual-peak plasma concentration pattern, and pH adjustment effectively resolved the delay in drug release, yielding a consistent release profile. In a study, researchers engineered an alendronate-based tetra polyethylene glycol hydrogel for osteoporosis therapy, aiming for intramuscular drug delivery. The tailored tetra-PEG hydrogel exhibited controlled degradation and drug release. The effective drug release promoted anti-osteoporosis effects both in vivo and in vitro. Thus, it made it a promising DDS for treating osteoporosis. In another exploration, researchers utilized host-guest interaction. A polyethyleneimine (PEI)-based hydrogel was intricately crosslinked using PEI branches that carried adamantane and cyclodextrin. Such an engineered hydrogel enhanced the co-encapsulation of protein and nucleic acid, achieved through the use of cationic liposomes containing 1,2-dioleoyl-3-trimethylammonium-propane chloride (DOPTAP). The modified hydrogels released proteins and nucleic acids for over 3 days and exhibited cytotoxic effects against U87MG cells. Upon intravenous administration, it has been shown to suppress tumor growth and improve survival in mice with orthotopic tumors.

Figure 7. **Parenteral Drug Delivery**

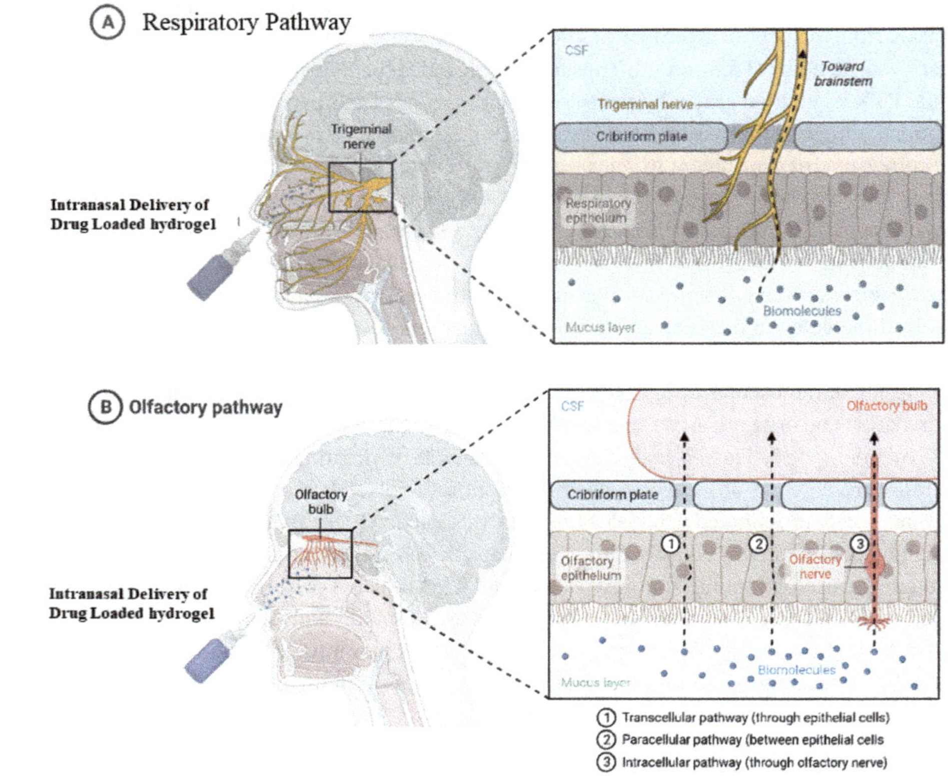

TABLE 1. Hydrogels for Brain/Epi-cortical drug delivery

System Hydrogel	Therapeutic ingredient	Polymer	Applications	References
Angiogenic peptide hydrogels	Self-assembling peptide-based hydrogel	Peptide-based hydrogel	Neuroprotective effect	(Ma et al., 2020)
Liposomal hydrogels	Barbituric acid	Alginate, hyaluronic acid, and methylcellulose	Treatment of GBM	(Han et al., 2020)
PEG-based hydrogels	-	PEG	Activation of microglia and astrocytes	(Bjugstad et al., 2010)
Poloxamer 188-based hydrogel	Curcumin	Poloxamer	Enhance brain-targeted delivery	(Chen et al., 2013)

TABLE 2. Hydrogels for Nasal/Intranasal Drug Delivery

System Hydrogel	Therapeutic ingredient	Polymer	Applications	References
Guanidine -Chitosan hydrogels	Curcumin	Peptide-based hydrogel	Thermo-resposive hydrogel with anti-depressant effect	(Qi et al., 2020)
Pluronic® 127 thermo-sensitive hydrogel	Levetiracetam	Pluronic® 127	Treatment of epilepsy	(Gonçalves et al., 2019)
Chitosan-based hydrogel	Tranexamic acid	Chitosan and glycerol-2-phosphate disodium salt hydrate	Treatment of epistaxis and wound healing of the nasal mucosa	(Gholizadeh et al., 2019)
Paclitaxel-loaded PLGA microspheres	Paclitaxel	PLGA	Treatment of GBM	(Ranganath et al., 2009)

11.2 MECHANISM OF ACTION OF ANTIMICROBIAL AGENTS

The precise mechanism of action of antimicrobial hydrogels is still a topic of debate. Most antimicrobial agents employed against bacterial infections can be grouped based on their functional mechanism. These agents exhibit their action through different principal mechanisms:

i. Perturbation of biosynthetic pathways involved in cell wall formation.
ii. Impediment of the translational process of protein synthesis through the disruption of the ribosomal subunit.
iii. Disruption of nucleotide synthesis sequences.
iv. Disruption of the cellular metabolism (Allison & Lambert, 2024; Tenover, 2006).

Figure 8. Mechanism of action of antibacterial agents

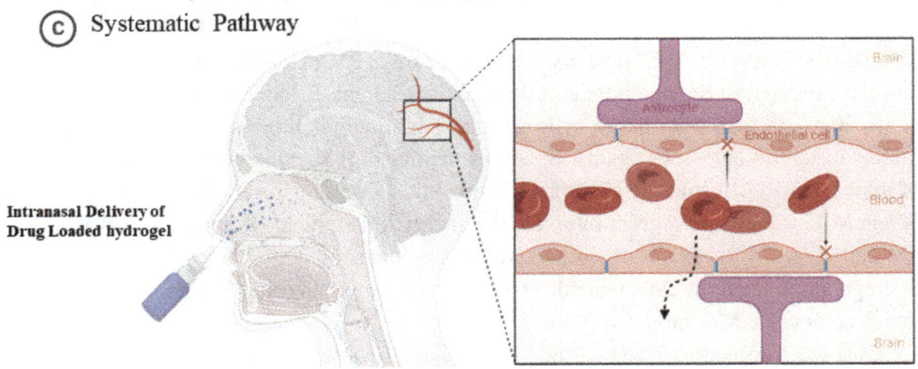

11.2.1 Applications of Hydrogels

11.2.2 AgNPs-Loaded Antibacterial Hydrogel

Ag and Polypyrrole-infused CMC-g-P(AAm-co-AMPS) hydrogel matrices show noteworthy inhibition against microbes E. coli, P. aeruginosa, B. subtilis, C. Albian, A. Niger. The IZD values for the tested sample (100 μg/mL) against microbes are as follows: 22 $\pm$ 0.41, 11 $\pm$ 0.44, 19 $\pm$ 0.45, 14 $\pm$ 0.56, 26 $\pm$ 0.71, respectively. The sample ratio of. However, each dopant, when incorporated separately into the hydrogel, has a mediocre range of inhibition against these microbes (Turky et al., 2021). Ag NPs loaded Am-grafted Cellulose-based hydrogel derived from olive residues (seed, husk). Still, an accelerated rate of release of Ag-NPs was observed for the seed gel, which is attributed to physical crosslinking between them. The stability of the hydrogel impeded the growth of S. aureus, P. aeruginosa, and C. albicans. PVA/CMC hydrogel was infused with AgNPs ranging in size from 8 to 14 nm. The biocidal efficacy of nanocomposite hydrogel (at a concentration of 5mg/ml) was compared to Kanamycin in combating against E. coli, K. pneumonia, P. aeruginosa, P. vulgaris, S. aureus, and P. mirabilis, with IZD values of 16.6, 15.8, 15.6, 15.2, 14.8, 15.0, respectively (Pan et al., 2019).

AgNPs-incorporated GG-g-poly (AMPS-co-AAm) composite hydrogel displayed 18.0mm, 15.0mm, 15.0mm, and 15.0mm clear zones against E. coli, A. hydrophila, S. aureus, and B. subtilis, respectively. The Ag-free gel did not show a significant bacteriostatic effect. Thus, Ag-mediated hydrogel exhibits antibacterial properties (Rayar et al., 2020). AgNPs-incorporated Carrageenan/Lignin hydrogel was networked with bivalent cations (Cu, Ca, and Mg). Within three hours, the Cu-crosslinked hydrogel substantially reduced the viability of both S. aureus and E. coli cells. In contrast, Ca- and Mg-crosslinked hydrogels took 6 hours to achieve similar activity.

Moreover, in vitro animal testing revealed that the Mg-crosslinked hydrogel's wound-healing capability was confirmed by histological analysis (Jaiswal et al., 2020). PVA/starch gelled with AgNPs extracted from Diospyros lotus fruit extract has effective antibacterial activity against E. coli and S. aureus compared to the Ag-unloaded hydrogel (Batool et al., 2019). PVA/corn-starch/Linseed oil polyol hydrogel loaded with LEO/AgNPs and LEO-derived AgNPs has a pronounced effect on the inhibition of C. albicans. At the same time, PVA/CS/LP/LEO/Ag hydrogel film was less effective against K. pneumoniae, A. baumanii, S. aureus, E. faecalis, and P. mirabilis (Sharmin et al., 2020). A guar gum-based hydrogel incorporating curcumin-stabilized AgNPs exhibited more substantial outcomes against S. aureus than against E. coli and P. aeruginosa, with zone of inhibition (ZOI) values of 15 mm, 13.5 mm, and 13 mm, respectively. Cur-AgNPs displayed potent MIC values. Cur-AgNPs inhibited the growth of E. coli and P. aeruginosa at MICs of 1.57 nM and 0.78 nM for S. aureus, respectively (Talodthaisong et al., 2020). In another study, the hydrogel contracted the rat's wounds at a 73% rate, surpassing the commercially available gel, which achieved a 51% rate of wound closure.

Guar-Ag Nano composite hydrogel was efficiently utilized against the inhibition of B. subtilis and S. epidermidis with IZD values of 12.5 mm and 14.1 mm, respectively. The hydrogel resulted in a higher degree of inhibition against S. epidermidis due to the increased electrostatic interactions with the peptidoglycan (Palem et al., 2020). AgNPs/methacrylate-alginate hydrogel and colloidal solution of AgNPs had comparable inhibitory activity against E. coli. However, a hydrogel containing 2.5 nM of AgNPs and colloidal AgNPs at a concentration of 0.50 nM exhibited a strong bactericidal effect. AgNPs have a more substantial bactericidal effect against S. mutans than S-nitroso-MSA. The synergy of S-nitroso-MSA and AgNPs, when incorporated into a hydrogel, resulted in a substantial decrease in the population of S. aureus and E. coli within 2 hours. Overall, the S-nitroso-MSA/AgNPs-incorporated alginate hydrogel exhibits potent antibacterial efficacy against E. coli (Urzedo et al., 2020). PVA/Na-alginate incorporating AgNP synthesized from tea extract exhibited a notable inhibitory effect against S. aureus. However, tea extract has unfavorable outcomes for antibacterial efficacy (Wang et al., 2020). Recently, AgNPs with an average size of 22.31 nm were biologically synthesized from green tea leaf extract. Upon loading into Chitosan-g-PVA hydrogel, a sustained release rate of Ag ions was observed within 24 hours. Hydrogel loaded with 0.8% w/v AgNPs exhibited excellent suppression of E. coli compared to S. aureus. On day 7, wounds demonstrated in vivo skin regeneration with a size ranging between 23% and 30%. Tissue analysis revealed its anti-inflammatory and skin-healing potential, along with antibacterial effectiveness (Aldakheel et al., 2023). The film of Ag-incorporated hydroxyethylacryl chitosan/Na-Alginate hydrogel was insensitive to the concentration of divalent Ca (II) ions; only the Ag-loaded film exhibited inhibitory efficacy against S. aureus and E. coli. However, a decreased concentration of Ca promoted cell proliferation, and the Ag-loaded hydrogel film had a minimal effect on cell viability; however, all samples maintained 90% cell viability (Chalitangkoon et al., 2020).

11.2.2 Zn-/Cu-oxide NPs loaded Antibacterial Hydrogel

PET-corn silk-ZnO based NC hydrogel synthesized with CTAB and non-CTAB hydrogel exhibits substantial impediment to the growth of E. coli, S. aureus, and C. albicans. Non-CTAB hydrogel has shown near-total elimination of pathogens. Whereas CTAB-assisted hydrogel has suppressed E. coli by 90%, S. aureus by 93%, and C. albicans by 98%. A high percentage of cells remain viable at 87% (Amani et al., 2019). Cellulose/ZnO hydrogels have been loaded with curcumin, achieving a productivity of 68.12%. Microbial growth suppression diameter was 12.60±0.40mm against S. aureus and 12.40±0.30mm against T. rubrum (Anagha et al., 2019). XG-PVA/ZnO dressings successfully prevented the foreign invasion of microbes. Hydrogel with 5% of ZnO content exhibited an increase in the clear zone of bacterial growth, efficiently against S. aureus (4cm), C. albicans (3.5cm), and E. coli (2cm) (Raafat et al., 2018). The efficacy of Carrageenan/KCl/ZnO and CuO nanocomposite hydrogels against pathogens was dependent on the type of nanoparticle (NP) employed and its content (in percentage).

Carrageenan/KCl was individually gelled with 1% of Zn and 1% of Cu NPs, and the hydrogel incorporating 0.5% of both metallic-NPs showcased excellent inhibition against E. coli than L. monocytogenes (Oun & Rhim, 2017). PVA-CMC hydrogel was utilized as a carrier matrix for incorporating Cu-NPs with a diameter of 7 nm. It was empirically determined that, with progressive dosage, its inhibition potential exhibited concordance with that of kanamycin against uropathogens. CuNP-Hydrogel (5.0) competently inhibited E. coli, K. pneumoniae, P. aeruginosa, P. vulgaris, S. aureus, and P. mirabilis with IZD values (mm) of 16.0, 15.2, 16.4, 15.8, 15.6, 15.8, respectively (Al-Enizi et al., 2018). Si-CuNPs/Starch hydrogel was tested against E. coli and S. aureus at varying NP concentrations, ranging from 0.36 to 1.27 mmol/g gel, and near-total inhibition was observed for both bacteria. However, the hydrogel was more effective against E. coli (Villanueva et al., 2016). Recently, a ZnO-loaded chitosan-elastin-sodium alginate nanocomposite hydrogel was synthesized by Ramzan, Amna, et al. (2023). Hydrogel was loaded with various compositions. However, notable antibacterial efficacy was observed against E. coli and S. aureus with increasing content of NPs. Cellular compatibility was investigated in human adipose stem cells. The outcomes suggested its biocompatibility for chronic wounds (Ramzan et al., 2023).

11.2.6 Antibacterial activity of Hydrogel Loaded With Antibiotic

Antimicrobial assessment of chloramphenicol (CAP) entrapped in bacterial cellulose and 2,3-dialdehyde cellulose membranes (DABC) against E. coli, S. aureus, and S. pneumoniae reveals that the CAP-loaded DABC exhibits a substantial improvement in adherence and cellular proliferation. The recorded IZD values are 9 mm, 11 mm, and 13 mm, respectively. However, membranes themselves do not exhibit a significant IZD value. Thus, it's a comparable substitute in wound dressing due to its superiority (Laçin, 2014). Cellulose-chalcone nanocomposites with carbon nanotubes were synthesized incorporating various bioactive agents. Among them, an antibiotic named Linezolid was gradually released, showing evidence of potent antimicrobial activity against E. faecium. However, its effectiveness waned following 72 hours. Progression in the zone was demonstrated from 23mm to 30mm over a 48-hour timeframe, resulting in a 7 mm increase. The finest stability and swelling ratio were displayed by the formulation consisting of carboxymethyl chitosan and oxidized gellan gum in a 1:2 proportion, supplemented with 8 mg/mL of GG-based microspheres (GM). By incorporating tetracycline hydrochloride (TH) and silver sulfadiazine (AgSD), it induced inhibiting properties against E. coli and S. aureus. The IZD value (in mm) for TH/gel was measured as 3.37 ± 0.35 and 7.50 ± 2.50. The TH/GMs/Gel formulation showcased values of 2.65 ± 0.35 and 4.05 ± 0.50. For AgSD/Gel, 3.35 ± 0.25 and 3.25 ± 0.50 were observed, while for AgSD/GMs/Gel, 2.50 ± 0.50 and 1.25 ± 0.25 were recorded, respectively, (Zhang et al., 2020). GG-based hydrogel, encapsulated with 5% ibuprofen, successfully delivered the drug within 2 hours, inhibited S. aureus, while E. coli remained unaffected (Sebri & Amin, 2016). In the evaluation against S. aureus and E. coli, the Xylan gum/hydroxypropyl β-cyclodextrin (XG/HPCD) hydrogel encapsulating levofloxacin (LVFX) drug lengthened the release rate. Thus, precise zone percentages of 66.67% and 53.33%, respectively, were demonstrated, indicating significant suppression of bacteria (Wang et al., 2016). Succinic anhydride-modified xanthan (XG–SA) hydrogel encapsulating gentamicin (GS) has been reported to exhibit antibacterial efficacy against S. aureus.

11.2.7 Antibacterial Hydrogel With a Synergistic Effect

Quercetin (QE) extracted from onion scale leftovers was encapsulated into a carrier system consisting of Chitosan/Dialdehyde cellulose/Zn NPs. Quercetin/onion peel drug (QE/OPD) entrapped in nanohybrid hydrogel (CHGZ) exhibits potent antimicrobial capabilities. The drug liberation was proportional to the efficiency of drug encapsulation. Where QE has antifungal potency against T. rubrum, and OPD loaded in respective gel matrices have shown remarkable S. aureus inhibition. However, the incorporation of QE/OPD into the hydrogel system yields a substantial increase in the diameter of the zone of inhibition, indicating a synergistic effect of the loaded drug and ZnO. PVA/Alginate and PVA/Alginate-reduced GO hydrogels themselves were ineffective against planktonic S. aureus and P. aeruginosa. However, the inclusion of Tb^{3+} decreased the adhesion of S. aureus and P. aeruginosa by 79% and 99%, respectively. The notable impediment to biofilm populations of S. aureus and P. aeruginosa was observed at magnitudes of

2 and 1 orders, respectively. At the same time, the synergistic effect of Tb-reduced GO-loaded PVA/Alginate hydrogel elevated the biofilm suppression by a factor of 1.5 to 2 orders, respectively (Wang et al., 2019). The synergistically enhanced antibacterial action of Sultamicillin tosylate (ST) and AgNPs/ZnONPs loaded onto niosomes incorporated with carboxymethyl chitosan hydrogel was observed against various bacterial strains. ST-Nio-AgNPs@CMC and ST-Nio -ZnONPs@CMC were more effective against S. aureus, P. aeruginosa, K. pneumoniae, and E. coli. The synergistic effect of ST-Nio-AgNPs and ST-Nio-ZnONPs was confirmed by the MBEC assay, indicating their potential to impede the biofilm of E. coli by ~4 and 5 log, respectively (Ashkezari et al., 2023).

11.3 CONCLUSIONS

Hydrogels have offered solutions to the longstanding challenges to revolutionized drug delivery systems. Their unique physicochemical characteristic, biocompatibility and ability to be tailored for the specific routes make them an attractive candidate for biomedical research. This chapter highlights the hydrogel-mediated delivery of therapeutic agents across the oral, dermal and transdermal, ocular, brain and epi-cortical, nasal and intranasal, paternal, rectal and vaginal routes. The ability of hydrogels to combat different bacterial infections, their wound healing, and tissue regeneration properties has also been discussed. The versatility of hydrogels is evident across various routes. Their biocompatibility and stimuli-responsiveness coupled with controlled and targeted release behavior makes them a promising platform for therapeutic applications.

REFERENCES

Al-Enizi, A. M., Ahamad, T., Al-Hajji, A. B., Ahmed, J., Chaudhary, A. A., & Alshehri, S. M. (2018). Cellulose gum and copper nanoparticles based hydrogel as antimicrobial agents against urinary tract infection (UTI) pathogens. *International Journal of Biological Macromolecules, 109*, 803–809. DOI: 10.1016/j.ijbiomac.2017.11.057 PMID: 29133089

Al Harthi, S., Alavi, S. E., Radwan, M. A., El Khatib, M. M., & AlSarra, I. A. (2019). Nasal delivery of donepezil HCl-loaded hydrogels for the treatment of Alzheimer's disease. *Scientific Reports, 9*(1), 9563. DOI: 10.1038/s41598-019-46032-y PMID: 31266990

Aldakheel, F. M., Mohsen, D., El Sayed, M. M., Alawam, K. A., Binshaya, A. S., & Alduraywish, S. A. (2023). Silver nanoparticles loaded on chitosan-g-PVA hydrogel for the wound-healing applications. *Molecules (Basel, Switzerland), 28*(7), 3241. DOI: 10.3390/molecules28073241 PMID: 37050003

Allison, D. G., & Lambert, P. A. (2024). Modes of action of antibacterial agents. In *Molecular medical microbiology* (pp. 597–614). Elsevier. DOI: 10.1016/B978-0-12-818619-0.00133-7

Amani, A., Montazer, M., & Mahmoudirad, M. (2019). Synthesis of applicable hydrogel corn silk/ZnO nanocomposites on polyester fabric with antimicrobial properties and low cytotoxicity. *International Journal of Biological Macromolecules, 123*, 1079–1090. DOI: 10.1016/j.ijbiomac.2018.11.093 PMID: 30447368

Anagha, B., George, D., Maheswari, P. U., & Begum, K. M. S. (2019). Biomass derived antimicrobial hybrid cellulose hydrogel with green ZnO nanoparticles for curcumin delivery and its kinetic modelling. *Journal of Polymers and the Environment, 27*(9), 2054–2067. DOI: 10.1007/s10924-019-01495-y

Ashkezari, S., Abtahi, M. S., Sattari, Z., Yaraki, M. T., Hosseini, F., Salehi, R. I., Afzali, E., Hajihosseini, S., & Mousavi-Niri, N. (2023). Antibiotic and inorganic nanoparticles co-loaded into carboxymethyl chitosan-functionalized niosome: Synergistic enhanced antibacterial and anti-biofilm activities. *Journal of Drug Delivery Science and Technology, 83*, 104386. DOI: 10.1016/j.jddst.2023.104386

Batool, S., Hussain, Z., Niazi, M. B. K., Liaqat, U., & Afzal, M. (2019). Biogenic synthesis of silver nanoparticles and evaluation of physical and antimicrobial properties of Ag/PVA/starch nanocomposites hydrogel membranes for wound dressing application. *Journal of Drug Delivery Science and Technology, 52*, 403–414. DOI: 10.1016/j.jddst.2019.05.016

Bjugstad, K., Lampe, K., Kern, D., & Mahoney, M. (2010). Biocompatibility of poly (ethylene glycol)-based hydrogels in the brain: An analysis of the glial response across space and time. *Journal of Biomedical Materials Research. Part A, 95*(1), 79–91. DOI: 10.1002/jbm.a.32809 PMID: 20740603

Cerchiara, T., Luppi, B., Bigucci, F., Orienti, I., & Zecchi, V. (2002). Physically cross-linked chitosan hydrogels as topical vehicles for hydrophilic drugs. *The Journal of Pharmacy and Pharmacology, 54*(11), 1453–1459. DOI: 10.1211/00223570281 PMID: 12495547

Chalitangkoon, J., Wongkittisin, M., & Monvisade, P. (2020). Silver loaded hydroxyethylacryl chitosan/sodium alginate hydrogel films for controlled drug release wound dressings. *International Journal of Biological Macromolecules*, *159*, 194–203. DOI: 10.1016/j.ijbiomac.2020.05.061 PMID: 32416295

Chen, X., Zhi, F., Jia, X., Zhang, X., Ambardekar, R., Meng, Z., Paradkar, A. R., Hu, Y., & Yang, Y. (2013). Enhanced brain targeting of curcumin by intranasal administration of a thermosensitive poloxamer hydrogel. *The Journal of Pharmacy and Pharmacology*, *65*(6), 807–816. DOI: 10.1111/jphp.12043 PMID: 23647674

Chen, Y. J., Zheng, H. Y., Huang, X. X., Han, S. X., Zhang, D. S., Ni, J. Z., & He, X. Y. (2016). Neuroprotective effects of icariin on brain metabolism, mitochondrial functions, and cognition in triple-transgenic Alzheimer's disease mice. *CNS Neuroscience & Therapeutics*, *22*(1), 63–73. DOI: 10.1111/cns.12473 PMID: 26584824

Correa, S., Grosskopf, A. K., Lopez Hernandez, H., Chan, D., Yu, A. C., Stapleton, L. M., & Appel, E. A. (2021). Translational applications of hydrogels. *Chemical Reviews*, *121*(18), 11385–11457. DOI: 10.1021/acs.chemrev.0c01177 PMID: 33938724

de Araújo Pereira, R. R., & Bruschi, M. L. (2012). Vaginal mucoadhesive drug delivery systems. *Drug Development and Industrial Pharmacy*, *38*(6), 643–652. DOI: 10.3109/03639045.2011.623355 PMID: 21999572

Dos Santos, A. M., Carvalho, S. G., Araujo, V. H. S., Carvalho, G. C., Gremião, M. P. D., & Chorilli, M. (2020). Recent advances in hydrogels as strategy for drug delivery intended to vaginal infections. *International Journal of Pharmaceutics*, *590*, 119867. DOI: 10.1016/j.ijpharm.2020.119867 PMID: 32919001

Ghasemiyeh, P., & Mohammadi-Samani, S. (2019). Hydrogels as drug delivery systems; pros and cons. *Trends in Pharmaceutical Sciences and Technologies*, *5*(1), 7–24.

Gholizadeh, H., Messerotti, E., Pozzoli, M., Cheng, S., Traini, D., Young, P., Kourmatzis, A., Caramella, C., & Ong, H. X. (2019). Application of a thermosensitive in situ gel of chitosan-based nasal spray loaded with tranexamic acid for localised treatment of nasal wounds. *AAPS PharmSciTech*, *20*(7), 299. DOI: 10.1208/s12249-019-1517-6 PMID: 31482286

Gonçalves, J., Bicker, J., Gouveia, F., Liberal, J., Oliveira, R. C., Alves, G., Falcão, A., & Fortuna, A. (2019). Nose-to-brain delivery of levetiracetam after intranasal administration to mice. *International Journal of Pharmaceutics*, *564*, 329–339. DOI: 10.1016/j.ijpharm.2019.04.047 PMID: 31015006

Goswami, M., Sadasivam, R., & Packirisamy, G. (2023). Viability studies of hydrogel contact lens on a 3D printed platform as ocular drug delivery carrier for diabetic retinopathy. *Materials Letters*, *333*, 133636. DOI: 10.1016/j.matlet.2022.133636

Gupta, P. (2007). Suppositories in anal disorders: A review. *European Review for Medical and Pharmacological Sciences*, *11*(3). PMID: 17970232

Han, X., Huang, J., To, A. K., Lai, J. H., Xiao, P., Wu, E. X., Xu, J., & Chan, K. W. (2020). CEST MRI detectable liposomal hydrogels for multiparametric monitoring in the brain at 3T. *Theranostics*, *10*(5), 2215–2228. DOI: 10.7150/thno.40146 PMID: 32089739

Ho, M. T., Teal, C. J., & Shoichet, M. S. (2019). A hyaluronan/methylcellulose-based hydrogel for local cell and biomolecule delivery to the central nervous system. *Brain Research Bulletin*, *148*, 46–54. DOI: 10.1016/j.brainresbull.2019.03.005 PMID: 30898580

Hoare, T. R., & Kohane, D. S. (2008). Hydrogels in drug delivery: Progress and challenges. *polymer, 49*(8), 1993-2007.

Huang, D., Sun, M., Bu, Y., Luo, F., Lin, C., Lin, Z., Weng, Z., Yang, F., & Wu, D. (2019). Microcapsule-embedded hydrogel patches for ultrasound responsive and enhanced transdermal delivery of diclofenac sodium. *Journal of Materials Chemistry. B*, *7*(14), 2330–2337. DOI: 10.1039/C8TB02928H PMID: 32254681

Ilgin, P., Ozay, H., & Ozay, O. (2020). Synthesis and characterization of pH responsive alginate based-hydrogels as oral drug delivery carrier. *Journal of Polymer Research*, *27*(9), 251. DOI: 10.1007/s10965-020-02231-0

Jaiswal, L., Shankar, S., Rhim, J.-W., & Hahm, D.-H. (2020). Lignin-mediated green synthesis of AgNPs in carrageenan matrix for wound dressing applications. *International Journal of Biological Macromolecules*, *159*, 859–869. DOI: 10.1016/j.ijbiomac.2020.05.145 PMID: 32445819

Jung, H., Kim, M. K., Lee, J. Y., Choi, S. W., & Kim, J. (2020). Adhesive hydrogel patch with enhanced strength and adhesiveness to skin for transdermal drug delivery. *Advanced Functional Materials*, *30*(42), 2004407. DOI: 10.1002/adfm.202004407

Kang, T., Cha, G. D., Park, O. K., Cho, H. R., Kim, M., Lee, J., Kim, D., Lee, B., Chu, J., Koo, S., Hyeon, T., Kim, D.-H., & Choi, S. H. (2023). Penetrative and sustained drug delivery using injectable hydrogel nanocomposites for postsurgical brain tumor treatment. *ACS Nano*, *17*(6), 5435–5447. DOI: 10.1021/acsnano.2c10094 PMID: 36926815

Kasai, R. D., Radhika, D., Archana, S., Shanavaz, H., Koutavarapu, R., Lee, D.-Y., & Shim, J. (2023). A review on hydrogels classification and recent developments in biomedical applications. *International Journal of Polymeric Materials*, *72*(13), 1059–1069. DOI: 10.1080/00914037.2022.2075872

Kesharwani, P., Bisht, A., Alexander, A., Dave, V., & Sharma, S. (2021). Biomedical applications of hydrogels in drug delivery system: An update. *Journal of Drug Delivery Science and Technology*, *66*, 102914. DOI: 10.1016/j.jddst.2021.102914

Kim, B., & Peppas, N. A. (2003). Poly (ethylene glycol)-containing hydrogels for oral protein delivery applications. *Biomedical Microdevices*, *5*(4), 333–341. DOI: 10.1023/A:1027313931273

Kolawole, O. M., & Cook, M. T. (2023). In situ gelling drug delivery systems for topical drug delivery. *European Journal of Pharmaceutics and Biopharmaceutics*, *184*, 36–49. DOI: 10.1016/j.ejpb.2023.01.007 PMID: 36642283

Labie, H., & Blanzat, M. (2023). Hydrogels for dermal and transdermal drug delivery. *Biomaterials Science*, *11*(12), 4073–4093. DOI: 10.1039/D2BM02070J PMID: 37060109

Laçin, N. T. (2014). Development of biodegradable antibacterial cellulose based hydrogel membranes for wound healing. *International Journal of Biological Macromolecules, 67*, 22–27. DOI: 10.1016/j.ijbiomac.2014.03.003 PMID: 24631550

Lin, H. A., Varma, D. M., Hom, W. W., Cruz, M. A., Nasser, P. R., Phelps, R. G., Iatridis, J. C., & Nicoll, S. B. (2019). Injectable cellulose-based hydrogels as nucleus pulposus replacements: Assessment of in vitro structural stability, ex vivo herniation risk, and in vivo biocompatibility. *Journal of the Mechanical Behavior of Biomedical Materials, 96*, 204–213. DOI: 10.1016/j.jmbbm.2019.04.021 PMID: 31054515

Ma, X., Agas, A., Siddiqui, Z., Kim, K., Iglesias-Montoro, P., Kalluru, J., Kumar, V., & Haorah, J. (2020). Angiogenic peptide hydrogels for treatment of traumatic brain injury. *Bioactive Materials, 5*(1), 124–132. DOI: 10.1016/j.bioactmat.2020.01.005 PMID: 32128463

Ma, Z., Ma, R., Wang, X., Gao, J., Zheng, Y., & Sun, Z. (2019). Enzyme and PH responsive 5-flurouracil (5-FU) loaded hydrogels based on olsalazine derivatives for colon-specific drug delivery. *European Polymer Journal, 118*, 64–70. DOI: 10.1016/j.eurpolymj.2019.05.017

McCrorie, P., Mistry, J., Taresco, V., Lovato, T., Fay, M., Ward, I., Ritchie, A. A., Clarke, P. A., Smith, S. J., Marlow, M., & Rahman, R. (2020). Etoposide and olaparib polymer-coated nanoparticles within a bioadhesive sprayable hydrogel for post-surgical localised delivery to brain tumours. *European Journal of Pharmaceutics and Biopharmaceutics, 157*, 108–120. DOI: 10.1016/j.ejpb.2020.10.005 PMID: 33068736

Moghimi, S. M., Hunter, A. C., & Murray, J. C. (2001). Long-circulating and target-specific nanoparticles: Theory to practice. *Pharmacological Reviews, 53*(2), 283–318. DOI: 10.1016/S0031-6997(24)01494-7 PMID: 11356986

Morita, M., Fukagai, T., Hirayama, K., Yamatoya, J., Noguchi, T., Ogawa, Y., Igarashi, A., Niiya, A., Kato, M., Morota, M., Oshinomi, K., Ogawa, Y., & Lederer, J. L. (2020). Placement of Space OAR hydrogel spacer for prostate cancer patients treated with iodine-125 low-dose-rate brachytherapy. *International Journal of Urology, 27*(1), 60–66. DOI: 10.1111/iju.14123 PMID: 31587417

Naeem, A., Yu, C., Zang, Z., Zhu, W., Deng, X., & Guan, Y. (2023). Synthesis and evaluation of rutin–hydroxypropyl β-cyclodextrin inclusion complexes embedded in xanthan gum-based (HPMC-g-AMPS) hydrogels for oral controlled drug delivery. *Antioxidants, 12*(3), 552. DOI: 10.3390/antiox12030552 PMID: 36978800

Nguyen, T. T., Bao, N. S., & Van Vo, G. (2022). Advances in hydrogel-based drug delivery systems for Parkinson's disease. *Neurochemical Research, 47*(8), 2129–2141. DOI: 10.1007/s11064-022-03617-w PMID: 35596041

Oun, A. A., & Rhim, J.-W. (2017). Carrageenan-based hydrogels and films: Effect of ZnO and CuO nanoparticles on the physical, mechanical, and antimicrobial properties. *Food Hydrocolloids, 67*, 45–53. DOI: 10.1016/j.foodhyd.2016.12.040

Palem, R. R., Shimoga, G., Rao, K. K., Lee, S.-H., & Kang, T. J. (2020). Guar gum graft polymer-based silver nanocomposite hydrogels: Synthesis, characterization and its biomedical applications. *Journal of Polymer Research*, *27*(3), 68. DOI: 10.1007/s10965-020-2026-8

Pan, Y., Zhao, X., Li, X., & Cai, P. (2019). Green-based antimicrobial hydrogels prepared from bagasse cellulose as 3D-scaffolds for wound dressing. *Polymers*, *11*(11), 1846. DOI: 10.3390/polym11111846 PMID: 31717464

Patel, A., Cholkar, K., Agrahari, V., & Mitra, A. K. (2013). Ocular drug delivery systems: An overview. *World Journal of Pharmacology*, *2*(2), 47. DOI: 10.5497/wjp.v2.i2.47 PMID: 25590022

Purohit, T. J., Hanning, S. M., & Wu, Z. (2018). Advances in rectal drug delivery systems. *Pharmaceutical Development and Technology*, *23*(10), 942–952. DOI: 10.1080/10837450.2018.1484766 PMID: 29888992

Qi, X.-J., Liu, X.-Y., Tang, L.-M.-Y., Li, P.-F., Qiu, F., & Yang, A.-H. (2020). Anti-depressant effect of curcumin-loaded guanidine-chitosan thermo-sensitive hydrogel by nasal delivery. *Pharmaceutical Development and Technology*, *25*(3), 316–325. DOI: 10.1080/10837450.2019.1686524 PMID: 31661648

Qiu, X., & Hu, S. (2013). "Smart" materials based on cellulose: A review of the preparations, properties, and applications. *Materials (Basel)*, *6*(3), 738–781. DOI: 10.3390/ma6030738 PMID: 28809338

Raafat, A. I., El-Sawy, N. M., Badawy, N. A., Mousa, E. A., & Mohamed, A. M. (2018). Radiation fabrication of Xanthan-based wound dressing hydrogels embedded ZnO nanoparticles: In vitro evaluation. *International Journal of Biological Macromolecules*, *118*, 1892–1902. DOI: 10.1016/j.ijbiomac.2018.07.031 PMID: 30017987

Rajendran, A. T., & Vadakkepushpakath, A. N. (2024). Natural Food Components as Biocompatible Carriers: A Novel Approach to Glioblastoma Drug Delivery. *Foods*, *13*(17), 2812. DOI: 10.3390/foods13172812 PMID: 39272576

Ramzan, A., Mehmood, A., Ashfaq, R., Andleeb, A., Butt, H., Zulfiqar, S., Nasir, M., Hasan, A., Yar, M., & Akram, S. J. (2023). Freeze Gelation (Chitosan/elastin/sodium alginate)/Zinc Oxide Nanocomposite gel with enhanced Adipose Stem Cells proliferation and Antibacterial properties. WOUND REPAIR AND REGENERATION, Ranganath, S. H., Kee, I., Krantz, W. B., Chow, P. K.-H., & Wang, C.-H. (2009). Hydrogel matrix entrapping PLGA-paclitaxel microspheres: Drug delivery with near zero-order release and implantability advantages for malignant brain tumour chemotherapy. *Pharmaceutical Research*, *26*(9), 2101–2114.

Rathi, R., Sanshita, , Kumar, A., Vishvakarma, V., Huanbutta, K., Singh, I., & Sangnim, T. (2022). Advancements in rectal drug delivery systems: Clinical trials, and patents perspective. *Pharmaceutics*, *14*(10), 2210. DOI: 10.3390/pharmaceutics14102210 PMID: 36297645

Rayar, A., Babaladimath, G., Ambalgi, A., & Chapi, S. (2020). An eco-friendly synthesis, characterisation and antibacterial applications of gellan gum based silver nanocomposite hydrogel. *Materials Today: Proceedings*, *23*, 211–220. DOI: 10.1016/j.matpr.2020.02.019

Sebri, N. M., & Amin, K. A. M. (2016). Gellan gum/ibuprofen hydrogel for dressing application: Mechanical properties, release activity and biocompatibility studies.

Segal, M. B. (2000). The choroid plexuses and the barriers between the blood and the cerebrospinal fluid. *Cellular and Molecular Neurobiology*, *20*(2), 183–196. DOI: 10.1023/A:1007045605751 PMID: 10696509

Sharmin, E., Kafyah, M. T., Alzaydi, A. A., Fatani, A. A., Hazazzi, F. A., Babgi, S. K., Alqarhi, N. M., Sindi, A. A. H., Akram, D., Alam, M., Alam, J., Al-Madboly, L. A., Shoeib, N. A., Alqahtani, A. M., & Mojally, M. (2020). Synthesis and characterization of polyvinyl alcohol/corn starch/linseed polyol-based hydrogel loaded with biosynthesized silver nanoparticles. *International Journal of Biological Macromolecules*, *163*, 2236–2247. DOI: 10.1016/j.ijbiomac.2020.09.044 PMID: 32931835

Talodthaisong, C., Boonta, W., Thammawithan, S., Patramanon, R., Kamonsutthipaijit, N., Hutchison, J. A., & Kulchat, S. (2020). Composite guar gum-silver nanoparticle hydrogels as self-healing, injectable, and antibacterial biomaterials. *Materials Today. Communications*, *24*, 100992. DOI: 10.1016/j.mtcomm.2020.100992

Tenover, F. C. (2006). Mechanisms of antimicrobial resistance in bacteria. *The American Journal of Medicine*, *119*(6), S3–S10. DOI: 10.1016/j.amjmed.2006.03.011 PMID: 16735149

Tetyczka, C., Brisberger, K., Reiser, M., Zettl, M., Jeitler, R., Winter, C., Kolb, D., Leitinger, G., Spoerk, M., & Roblegg, E. (2022). Itraconazole nanocrystals on hydrogel contact lenses via inkjet printing: Implications for ophthalmic drug delivery. *ACS Applied Nano Materials*, *5*(7), 9435–9446. DOI: 10.1021/acsanm.2c01715

Torres-Luna, C., Fan, X., Domszy, R., Hu, N., Wang, N. S., & Yang, A. (2020). Hydrogel-based ocular drug delivery systems for hydrophobic drugs. *European Journal of Pharmaceutical Sciences*, *154*, 105503. DOI: 10.1016/j.ejps.2020.105503 PMID: 32745587

Turky, G., Moussa, M. A., Hasanin, M., El-Sayed, N. S., & Kamel, S. (2021). Carboxymethyl cellulose-based hydrogel: Dielectric study, antimicrobial activity and biocompatibility. *Arabian Journal for Science and Engineering*, *46*(1), 17–30. DOI: 10.1007/s13369-020-04655-8

Urzedo, A. L., Goncalves, M. C., Nascimento, M. H., Lombello, C. B., Nakazato, G., & Seabra, A. B. (2020). Cytotoxicity and antibacterial activity of alginate hydrogel containing nitric oxide donor and silver nanoparticles for topical applications. *ACS Biomaterials Science & Engineering*, *6*(4), 2117–2134. DOI: 10.1021/acsbiomaterials.9b01685 PMID: 33455338

Vigata, M., Meinert, C., Hutmacher, D. W., & Bock, N. (2020). Hydrogels as drug delivery systems: A review of current characterization and evaluation techniques. *Pharmaceutics*, *12*(12), 1188. DOI: 10.3390/pharmaceutics12121188 PMID: 33297493

Villanueva, M. E., Diez, A. M. R., González, J. A., Pérez, C. J., Orrego, M., Piehl, L., Teves, S., & Copello, G. J. (2016). Antimicrobial activity of starch hydrogel incorporated with copper nanoparticles. *ACS Applied Materials & Interfaces*, *8*(25), 16280–16288. DOI: 10.1021/acsami.6b02955 PMID: 27295333

Von Zuben, E. S., Eloy, J. O., Inácio, M. D., Araujo, V. H. S., Baviera, A. M., Gremião, M. P. D., & Chorilli, M. (2022). Hydroxyethylcellulose-based hydrogels containing liposomes functionalized with cell-penetrating peptides for nasal delivery of insulin in the treatment of diabetes. *Pharmaceutics, 14*(11), 2492. DOI: 10.3390/pharmaceutics14112492 PMID: 36432681

Wang, B., Han, Y., Lin, Q., Liu, H., Shen, C., Nan, K., & Chen, H. (2016). In vitro and in vivo evaluation of xanthan gum–succinic anhydride hydrogels for the ionic strength-sensitive release of antibacterial agents. *Journal of Materials Chemistry. B, 4*(10), 1853–1861. DOI: 10.1039/C5TB02046H PMID: 32263062

Wang, Q., Zhang, H., Huang, J., Xia, N., Li, T., & Xia, Q. (2018). Self-double-emulsifying drug delivery system incorporated in natural hydrogels: A new way for topical application of vitamin C. *Journal of Microencapsulation, 35*(1), 90–101. DOI: 10.1080/02652048.2018.1425752 PMID: 29308686

Wang, T., Zhang, F., Zhao, R., Wang, C., Hu, K., Sun, Y., Politis, C., Shavandi, A., & Nie, L. (2020). Polyvinyl alcohol/sodium alginate hydrogels incorporated with silver nanoclusters via green tea extract for antibacterial applications. *Designed Monomers and Polymers, 23*(1), 118–133. DOI: 10.1080/15685551.2020.1804183 PMID: 33029080

Wang, Y., Lu, Y., Zhang, J., Hu, X., Yang, Z., Guo, Y., & Wang, Y. (2019). A synergistic antibacterial effect between terbium ions and reduced graphene oxide in a poly (vinyl alcohol)–alginate hydrogel for treating infected chronic wounds. *Journal of Materials Chemistry. B, 7*(4), 538–547. DOI: 10.1039/C8TB02679C PMID: 32254787

Weiss, S. C. (2011). Conventional topical delivery systems. *Dermatologic Therapy, 24*(5), 471–476. DOI: 10.1111/j.1529-8019.2012.01458.x PMID: 22353153

Xu, D., Lu, Y.-R., Kou, N., Hu, M.-J., Wang, Q.-S., & Cui, Y.-L. (2020). Intranasal delivery of icariin via a nanogel-thermoresponsive hydrogel compound system to improve its antidepressant-like activity. *International Journal of Pharmaceutics, 586*, 119550. DOI: 10.1016/j.ijpharm.2020.119550 PMID: 32554031

Xu, H., Liu, Y., Jin, L., Chen, X., Chen, X., Wang, Q., & Tang, Z. (2023). Preparation and characterization of ion-sensitive brimonidine tartrate in situ gel for ocular delivery. *Pharmaceuticals (Basel, Switzerland), 16*(1), 90. DOI: 10.3390/ph16010090 PMID: 36678587

Xu, J., Tam, M., Samaei, S., Lerouge, S., Barralet, J., Stevenson, M. M., & Cerruti, M. (2017). Mucoadhesive chitosan hydrogels as rectal drug delivery vessels to treat ulcerative colitis. *Acta Biomaterialia, 48*, 247–257. DOI: 10.1016/j.actbio.2016.10.026 PMID: 27769943

Yu, W., Gong, E., Liu, B., Zhou, L., Che, C., Hu, S., Zhang, Z., Liu, J., & Shi, J. (2023). Hydrogel-mediated drug delivery for treating stroke. *Chinese Chemical Letters, 34*(9), 108205. DOI: 10.1016/j.cclet.2023.108205

Zhang, J., Qian, S., Chen, L., Wu, M., Cai, Y., Mou, X., & Feng, J. (2022). Antifouling and antibacterial zwitterionic hydrogels as soft contact lens against ocular bacterial infections. *European Polymer Journal, 167*, 111037. DOI: 10.1016/j.eurpolymj.2022.111037

Zhang, X., Pan, Y., Li, S., Xing, L., Du, S., Yuan, G., Li, J., Zhou, T., Xiong, D., Tan, H., Ling, Z., Chen, Y., Hu, X., & Niu, X. (2020). Doubly crosslinked biodegradable hydrogels based on gellan gum and chitosan for drug delivery and wound dressing. *International Journal of Biological Macromolecules*, *164*, 2204–2214. DOI: 10.1016/j.ijbiomac.2020.08.093 PMID: 32798543

Zhao, X., & Wang, Z. (2019). A pH-sensitive microemulsion-filled gellan gum hydrogel encapsulated apigenin: Characterization and in vitro release kinetics. *Colloids and Surfaces. B, Biointerfaces*, *178*, 245–252. DOI: 10.1016/j.colsurfb.2019.03.015 PMID: 30875583

Zhou, C., Wang, Q., Cao, H., Jiang, J., & Gao, L. (2024). Nanozybiotics: Advancing antimicrobial strategies through biomimetic mechanisms. *Advanced Materials*, *36*(33), 2403362. DOI: 10.1002/adma.202403362 PMID: 38874860

Chapter 12
Super-Absorbent Hydrogels for Environmental Remediation

Peter Olusakin Oladoye
https://orcid.org/0000-0003-2872-9197
Department of Chemistry and Biochemistry, Florida International University, USA

Elizabeth O. Omotola
Department of Chemical Sciences, Tai Solarin University of Education, Nigeria

Ivan Oyege
Department of Earth and Environment, Florida International University, USA

Mayowa E. Oladipo
https://orcid.org/0000-0002-9799-7104
Department of Chemistry and Biochemistry, Florida International University, USA

Wycliffe C. Wanyonyi
https://orcid.org/0000-0002-4055-0176
Department of Mathematics, University of Kabianga, Kenya

ABSTRACT

The removal of water and wastewater contaminants by adsorptive means has been shown to be very promising. Amongst different adsorbent materials, hydrogels have been found to be a good tool for eliminating both inorganic and organic toxic chemical substances from water and wastewater, because of swelling characteristics, bio-compatibility, non-toxicity, re-usability and high functionalities. From previous studies, it has been shown that hydrogels can be re-used and regenerated for up to 5 times, with only about 10-15% adsorption capacity loss. This chapter focuses on the viability of hydrogels and its modified forms for the treatment of water and wastewater that contains micro-pollutants like pharmaceuticals, pesticides, dyes and heavy metals. Specifically, the mechanisms of interaction/uptake, adsorption isotherm modelling, adsorption capacities and regeneration potentials of (modified) hydrogels for these selected water contaminants shall be discussed.

DOI: 10.4018/979-8-3373-2337-4.ch012

12.1 INTRODUCTION

Daily human-induced activities have caused significant harm to the air, soil, water, and biota. For survival's sake, humans must engage in processes such as leather production, textile manufacturing, foundry operations, mining, smelting, and other industrial or chemically related activities that ultimately result in the release of micro-pollutants into the environment. Directly or indirectly, these harmful chemicals that are released contaminate our drinking water and natural water. Worst still, the micro-pollutants bio-accumulate in crops, aquatic organisms, and biota. For instance, recent studies have indicated a surprisingly high level of heavy metals in food crops grown on heavy metal-contaminated soils (N.-W. Hu et al., 2023; Tagumira et al., 2022). Dolu et al. study reported the presence of paracetamol in barley and corn, with a concentration of 16.8 ng/kg and 43.3 ng/kg, respectively, as a result of irrigation with wastewater from a treatment plant (Dolu & Nas, 2023). This gives a concern and an urgent need to remediate.

Adsorption of inorganic and (colored) organic micro-pollutants has been identified to have more pros than cons, compared to all other remediation techniques (Ahmad et al., 2020; Oladoye et al., 2022). To achieve effective pollutant uptake from (waste)water, suitable adsorbent materials are required, like hydrogels. Hydrogels are three-dimensional crosslinked polymer networks with tunable properties that can absorb and retain excess amounts of water (Wang et al., 2020). Their tunable properties make them fit for diverse applications in engineering, agriculture, and biomedical sectors (Caló & Khutoryanskiy, 2015; Wang et al., 2020). Hydrogels have also been used in soft robotics (Lee et al., 2020) and wastewater treatment (Chelu et al., 2023). In terms of structures, hydrogels have been described as porous materials with different chemical moieties suitable for capturing dyes, pharmaceuticals, heavy metal ions, pesticides, etc, from wastewater, and to later chemically release the adsorbed pollutants into aqueous solutions by changing the surface chemistry (Bharali et al., 2003).

Hydrogels as adsorbent materials have a strong capacity for the remediation of ionic toxic substances as a result of hydrophilic chemical functionalities like sulfonic, carboxylic, hydroxyl, and amine groups as complexing agents for the uptake of toxic inorganic and organic compounds from wastewater (Shalla et al., 2018). Upon interaction with and adsorption of inorganic and organic toxic substances, hydrogels swell up to several times their initial volume, which can become accessible via the opening of the hydrogel. This chapter focuses on the treatment or adsorption prowess of hydrogels with respect to pesticides, dyes, heavy metal ions, drugs, etc, as pollutants. Moreover, it specifically seeks to provide a detailed update on the adsorption mechanism, capacity, isotherm model, and re-usability studies of the toxic substances afore listed.

12.2 FABRICATION STRATEGIES OF HYDROGELS FOR ENVIRONMENTAL REMEDIATION

The primary fabrication strategies fall under physical, chemical, and enzymatic crosslinking, with several advanced techniques now complementing these methods. Physical crosslinking employs non-covalent interactions such as ionic bonding, hydrogen bonding, hydrophobic interactions, and freeze-thaw cycling. These methods are free from toxic chemicals and enable reversible, biocompatible hydrogels. For instance, PVA forms gels through crystallite formation during repeated freeze-thaw cycles, while gelatin and carrageenan undergo thermal gelation via temperature-induced transitions (Dodda et al., 2023; Song et al., 2022). Chemical crosslinking involves covalent bond formation, resulting in superior durability and structural control of hydrogels. Common approaches include using small-molecule crosslinkers like genipin, free-radical polymerization, and click chemistry reactions. Advanced methods such as photopolymerization offer spatial control, while ATRP (Atom Transfer Radical Polymerization) enables precise chain design for enhanced performance (Visan & Negut, 2025). Enzymatic crosslinking uses biological catalysts such as horseradish peroxidase and transglutaminase to form stable hydrogel networks under mild conditions. These are ideal for applications with critical biocompatibility and selectivity (S. Das et al., 2018; Dodda et al., 2023). Hybrid systems like interpenetrating polymer networks (IPNs) are developed to meet complex environmental requirements by combining physical and chemical crosslinking, enhancing strength, flexibility, and degradation control (Sonwane & Ingle, 2024).

Emerging fabrication technologies include 3D printing, which allows for customized, multi-functional hydrogel architectures; microfluidics for producing uniform, tunable hydrogel particles; electrospinning for creating high surface area nanofibrous gels; and photolithography for microscale patterning. These techniques offer precision and adaptability for pollutant-specific applications (Sonwane & Ingle, 2024). Additional enhancements involve embedding magnetic nanoparticles, photocatalysts, or carbon nanomaterials to create multi-functional, stimuli-responsive hydrogels (Nille et al., 2021; Visan & Negut, 2025).

12.3 PHYSICOCHEMICAL PROPERTIES AND ADSORPTION CAPACITY OF HYDROGELS

12.3.1 Particle Size

The particle size of hydrogels is an important morphology to be considered for their practical use. Generally, the pore sizes of various hydrogels can vary significantly (Tan et al., 2024). This property aids the understanding of the retention capacity of various hydrogels. For instance, smaller particles have a larger surface area and interstitial gaps and absorb/retain more water (Rizwan et al., 2021). Also, hydrogels' smaller particle sizes enhance their mechanical properties, including hardness, complex modulus, work of shear, and adhesiveness (Fasolin et al., 2021). The greater interface area and smaller particles' capacity to function as active fillers, strengthening the structure, are the causes of this (Fasolin et al., 2021). Moreover, the swelling properties of hydrogels, particularly their kinetics, are influenced by particle size; therefore, the size of the hydrogel particles should be appropriate for the particular target application (Ahmed, 2015). Particles larger than 80 mesh cause swelling to occur slowly and the liquid to remain in the system for extended periods; these sizes may be suitable for use in agriculture, where the slow release and swelling are necessary to produce sufficient soil humidity (Ahmed, 2015). While larger particles swell more slowly and hold onto liquid for longer, making them appropriate for uses like agriculture, where gradual release is advantageous, smaller particles initially increase the swelling rate, but the swelling ratio at equilibrium falls with increasing particle size (Batista et al., 2019; Razmjou et al., 2013).

12.3.2 Surface area

Hydrogels' functionality and use in a variety of sectors are greatly influenced by their surface area. Water retention of hydrogels can be increased by approximately 60 by altering their surface characteristics, such as by forming milliscale patterns (Chhetry & Gooh Pattader, 2024). This advancement is important for applications in drug delivery, bio-separations, wound healing, tissue engineering, pharmaceuticals, and sustainable agriculture (Chhetry & Gooh Pattader, 2024). The dynamic modulus of embedded materials, such as graphene in hydrogel composites, is influenced by their surface area. For bio-actuator applications, a higher dynamic modulus results from a larger graphene surface area within the hydrogel composite (Tungkavet et al., 2015). In hydrogels, a wider interface radius expands the area that can be used to store charge, improving sensitivity and capacitance (L. Wu, 2025). This is especially helpful for applications that need constant observation since it gives a steady signal even when the hydrogel is only slightly swelled (L. Wu, 2025). Summarily, hydrogels' surface area has a major impact on their mechanical characteristics, charge storage capacity, protein binding and release, and water retention, making it an essential component in maximizing their performance for a range of applications.

12.3.3 Porosity

Higher porosity of hydrogels generally correlates with increased water absorption capacity and swelling ability, which is essential for applications like drug delivery systems (Londhe et al., 2025). Porosity mediates interactions with cells and surrounding tissues, facilitating cellular permeation and new tissue formation (D. Das et al., 2019); it additionally ensures the removal of cellular debris and the supply of oxygen and nutrients to cells, both of which are critical for cell migration and tissue integration (Rybachuk et al., 2024). Increased porosity may decrease mechanical strength and stiffness while increasing flexibility and deformability. For applications needing particular mechanical qualities, this balance is essential (Kim et al., 2023; Londhe et al., 2025). Porosity can improve mechanical stiffness by promoting fluid drainage and releasing internal pressure, or it can decrease it by creating defects (Kim et al., 2023). Generally, hydrogels' mechanical characteristics, biocompatibility, drug delivery effectiveness, and water treatment capabilities are all greatly impacted by their porosity, and it is crucial to optimize porosity in order to customize hydrogels for particular uses (Le et al., 2023; Londhe et al., 2025).

12.4 MECHANISM OF MICRO-POLLUTANT UPTAKE

12.4.1 Heavy metals

Several mechanisms occur when hydrogels bind with heavy metals, which are electrostatic interaction, surface complexation, ion exchange, and coordination. These mechanisms were brought about by the association of the hydrogel's functional groups with metal ions at the set adsorptive conditions (pH, initial metal concentration, temperature, etc.). The process of assigning the adsorption mechanisms has been quite understood with the utilization of the analytical techniques, e.g., Energy-dispersive X-ray spectroscopy (EDS), X-ray photoelectron spectroscopy (XPS), Fourier transform infrared spectroscopy (FTIR), supplemented with adsorption energies and thermodynamic data (i.e., changes in entropy and enthalpy) (Lin et al., 2021). Table 1 shows some applications of hydrogels for the adsorption of heavy metal uptake with their corresponding mechanisms.

The heavy metals uptake by hydrogels via electrostatic interaction occurs when the opposite charge of the hydrogel's surface functional groups and that of the metal ions create an attraction force. In the medium, electrostatic interaction can be assessed by the Zeta Potential (ZP) – the potential difference between the particle and its ionic atmosphere surrounding the medium, which is strongly dependent on pH. The point of zero charge (pH_{PZC}) has been used to interpret the power of the interaction in relation to pH. Consequently, at $pH > pH_{PZC}$ the heavy metals' adsorption on hydrogel is governed by electrostatic interaction because the functional groups, such as -COOH, -OH, etc., deprotonate to form $-COO^-$, $-O^-$, respectively, on the surface of the hydrogel, which are readily available to bind electrostatically with the cationic metal ions. For instance, Aljar et al. (2023) observed that at $pH > pH_{PZC}$ the functional group present in Poly(vinyl alcohol)-Chitosan/Cellulose hydrogel beads (PVA-CS/CE), i.e., -OH deprotonates to form $-O^-$ on the PVA-CS/CE surface, resulting in the electrostatic interaction of cationic Co(II), Zn(II), Cd(II), and Pb(II) on the surface of the PVA-CS/CE hydrogel. Similarly, in the investigation of a magnetic calcium-alginate hydrogel-sphere encapsulated with Fe-Mn-Zr ternary metal composite (MCHETC) for the copper uptake in aqueous solutions. Yuan et al. (2023) carried out FT-IR analysis to explore the mechanism study and observed that electrostatic attraction occurred between copper and the deprotonated carboxyl group ($-COO^-$) on the MCHETC surface at $pH > pH_{PZC}$ (5.9), resulting in the shifts in the adsorption band of C=O and C-O and C=O in the carboxyl group from 1435 to 1385 cm^{-1} and from 1639 to 1616 cm^{-1}. In another study, poly(Hydroxyethyl methacrylate/Maleamic acid) (p(HEA/MALA)) hydrogel removed the utmost amount of Pb^{2+}, Cu^{2+}, Cd^{2+}, Ni^{2+} from aqueous solution via electrostatic interactions at higher pH (N. Wu & Li, 2013). Chen et al. (2023) also observed enhanced electronegativity of the magnetic sodium lignosulfonate hydrogel (Fe_3O_4@LS) surface at $pH > pH_{PZC}$ (3.2), resulting in the electrostatic interaction between Cd^{2+} and the Fe_3O_4@LS surface.

In the adsorption of metal ions by hydrogels through ion exchange, metal ions are exchanged with H^+ on oxygen in hydroxyl and carboxyl groups containing hydrogels, as detailed below:

$$M^{n+} + \text{-COOH} \rightarrow \text{-COO...M} + nH^+ \quad (1)$$

$$M^{n+} + \text{-OH} \rightarrow \text{-O...M} + nH^+ \quad (2)$$

$$M^{n+} + \text{-2COOH} \rightarrow \text{-COO...M...OOC-} + nH^+ \quad (3)$$

$$M^{n+} + \text{-2OH} \rightarrow \text{-O...M...O-} + nH^+ \quad (4)$$

M^{n+}-OH + -COOH → -COO…M…O- + nH^+ (5)

Also, the metal ions (M_i^{n+}) to remove from the aqueous solution can exchange the existing metal ion (M) with n ionic charges bound with the oxygen in hydroxyl and carboxyl functional groups containing hydrogels, in which this process is cation exchange.

$$-COOM + M_i^{n+} \rightarrow -COO...M_i + M^{n+} \quad (6)$$

In a similar manner to electrostatic interaction, the ion exchange mechanism also shows a strong dependency on the solution pH. On the other hand, the mechanism of ion exchange for the heavy metal uptake by hydrogels occurs at $pH < pH_{PZC}$, resulting in the protonation of the hydrogel surface and subsequent exchange of metal ions with H^+ from -COOH and/or -OH groups (Akter et al. 2021). Hence, XPS spectra reveal the presence and number of oxygen-containing functional groups available in the adsorbent and the oxygen-containing groups taking part in the adsorption process (Ma et al., 2018). Wu and Li (2013) conducted XPS measurement of p(HEA/MALA) hydrogels before and after metal uptake and observed ion exchange between the carboxyl groups and the positively charged ions (Cu^{2+}, Ni^{2+}, Pb^{2+}, and Cd^{2+}) in the pH range of 1-6.1, 1-7.0, 1-5.2, and 1-7.1, respectively. A similar ion exchange mechanism was found for the adsorption of Ni^{2+}, Cu^{2+}, Cd^{2+}, Fe^{3+}, and Cd^{2+} onto the active -COOH regions of polyacrylic acid (PAA) based porous hydrogel (Gokmen et al., 2021). In the study of polyvinyl alcohol/alginate (PVA/ALG) hydrogel for the adsorption of Pb^{2+} from aqueous solution, the XPS spectra of PVA/ALG hydrogel indicated the reduction of the Ca signal after adsorption. This means that Pb^{2+} replaced Ca^{2+} from $-COO^-$ on the PVA/ALG surface through an ion-exchange mechanism (P. Zhang et al., 2022). In the same vein, M. Zhang et al. (2023) observed the disappearance of K2p and K2s peaks from the ETS-PAA signal peaks i.e., C1s, K2p, K2s, and O1s, after adsorption of Cd^{2+}, Pb^{2+}, and Cu^{2+}. The result showed that K bound to $-COO^-$ was exchanged by Pb^{2+}, Cu^{2+}, and Cd^{2+}. Likewise, Cu^{2+} exchanged with Ca^{2+} that were initially bound onto the carboxylic and hydroxyl ions on MCHETC hydrogel (Yuan et al., 2023).

Surface complexation is another important mechanism controlling the uptake of metal ions by the hydrogels. In this mechanism, the interaction occurs between the electron acceptor and electron pair donor, resulting in the formation of various complexes. Rahmatpour and Alijani (2023) prepared chitosan and carboxymethyl gear gum (CS/CMGG) hydrogel for removing Cu^{2+} from aqueous solution. With the aid of FTIR analysis, it was revealed that $-NH_2$, -OH, and $-COO^-$ groups present in CS/CMGG hydrogel are involved in complexation with Cu^{2+}, suggesting that Cu^{2+} adsorption occurred via surface complexation. Also, sodium lignosulfonate-guar gum composite (GG/SLS) hydrogel was utilized for Co^{2+} and Cu^{2+} uptake, and calculated adsorption capacities were 601 mg/g and 709 mg/g, respectively. The adsorption of Co^{2+} and Cu^{2+} by GG/SLS hydrogel was mainly via the surface complexation between the lone pair electron of the oxygen atom present in the $-COO^-$ group and the vacant Cu^{2+} and Co^{2+} electron orbital (Li et al., 2021). In another study, XPS spectra of Cd^{2+} adsorption on Fe_3O_4@LS hydrogel showed that Cd^{2+} undergoes surface complexation on iron oxides, resulting in coprecipitation with iron oxides to form Cd^{2+} hydroxide (Ding et al., 2023). Moreover, cations (Pb^{2+}, Cd^{2+}, Zn^{2+}, and Co^{2+}) combine with the atoms having lone pair electrons on the PVA-CS/CE hydrogel in the outer orbitals through coordination interaction (Aljar et al. 2023).

Table 1. Adsorption mechanisms for heavy metal uptake using hydrogels

Adsorbent	Heavy metal	Governing Mechanism	References
Xylan-rich hemicellulose-based hydrogel	Pb^{2+}, Cd^{2+}, and Zn^{2+}	Ion mechanism	Peng et al. (2012)
Microcrystalline cellulose/bentonite grafted polyacrylic acid (MCC/Ben-g-PAA) hydrogel	Cd^{2+}	Ion complexation, physical adsorption hydrogen bonding, and electrostatic attraction	Ding et al. (2023)
Double network peanut shells-polyacrylic acid PS(H)-PAA hydrogel	Pb^{2+} and Cd^{2+}	Chelation/coordination, and physical adsorption	Yang et al. (2021)
Lignin-g-PAAc/MMT hydrogel	Cu^{2+}	Ion exchange, electrostatic attraction, and chemisorption	Sun et al. (2019)
Licorice residue-epichlorohydrin (LR-EPI) hydrogel	Pb^{2+}, Cr^{3+}, and Cu^{2+}	Ion exchange, hydrophobic, coordination, and electrostatic interactions	Yin et al. (2023)
Fe_3O_4@LS hydrogel	Cd^{2+}	Surface complexation, electronic attraction, and coprecipitation	Chen et al. (2023)
Multi-engineered alginate-based porous nanocomposite (GO@PAN-PPy) hydrogel	Cr(VI) and Cu(II)	Ion exchange, electrostatic interactions, hydrogen bonding, reduction, and complexation	W. Zhang et al. (2021)
CTS/SA/Ca^{2+} PCDNH hydrogel	Pb^{2+} and Cd^{2+}	Electrostatic interactions and coordination interactions	Tang et al. (2020)
Sodium acrylate and acrylamide copolymer/graphene oxide (P(AANa-co-AM)/GO) hydrogels	Pb^{2+} and Cd^{2+}	Chelation and ion exchange	He et al. (2016)
Poly(AN)-g-starch composite hydrogel	Cd^{2+}	Coordination/chelation	Hashem et al. (2021)
Alginate-polyvinyl alcohol double network hydrogels	Pb(II) and Cr(VI)	Electrostatic interaction, complexation, cation exchange, reduction, and chelation	P. Zhang et al. (2022)
p(HEA/MALA) hydrogel	Pb^{2+}, Cd^{2+}, Ni^{2+} and Cu^{2+}	Electrostatic interaction, chelation, and ion exchange	Wu and Li (2013)
PVA-CS/CE hydrogel	Pb(II), Cd(II), Zn(II) and Co(II)	Electrostatic interactions and coordination interactions	Aljar et al. (2023)
Esterified tobacco straw-based (ETS-PAA) hydrogels	Pb^{2+}, Cu^{2+} and Cd^{2+}	Ion exchange and coordination interactions	M. Zhang et al. (2023)
Carboxymethyl chitosan-nanoclay hydrogel	Cr(VI) and Pb(II)	Chelation and electrostatic interactions	El et al. (2023)
Graphene-based porous composite (GO-g -P4VP@PAA) hydrogel	Cd^{2+} and Pb^{2+}	Electrostatic interaction and chelation	L. Zhang et al. (2023)
Porous kappa-carrageenan/cellulose hydrogels	Pb^{2+}	Ion exchange and chelation	Kalaiselvi et al. (2023)
Dual-sensitive chitosan hydrogel	Cu^{2+}	Cation-π interaction and electrostatic interaction	Romal and Ong (2023)

12.4.2 Dyes

Dye uptake mechanism, also known as adsorption, is a surface occurrence where dye (adsorbate) molecules are attracted to the surface of the adsorbent through chemical bonds, electrostatic attraction, or van der Waals forces of attraction. This manifestation finds several important industrial applications such as dyeing textiles, heterogeneous catalysis, softening and deionization of water, wastewater treatment, removal of colouring matter from aqueous solutions, and chromatography analysis. The mechanism of dye removal from the aqueous phase greatly depends on the type of dye and the nature of the solid material. For instance, the maximum adsorption capacity of an adsorbent is impelled by porosity, surface area, distinctive bond interactions, including electrostatic forces, H-bonding, and hydrophobic interactions between the surface of the adsorbent that contains different functionality like hydroxyl groups, carboxylate, imidazole moieties, and N,N-dimethyl, N-quaternized amino with groups on the dye molecule (Abu Elella et al., 2019). Electrostatic interactions significantly influence dye uptake of ionic compounds/dyes because of the surface charge on the adsorbent (Gupta et al., 2013). The prevailing dye solution conditions, like temperature, pH, dye concentration, adsorbent dosage, ionic concentration, and presence of impurities, control the rate of adsorption and quantity of dye adsorbed.

High dye concentration provides a greater concentration slope and powerful driving force between the adsorbent and aqueous phase and thereby increasing dye uptake (Ghaedi et al., 2014). However, as the adsorbent gets saturated, the rate of adsorption drops due to reduced dye concentration and active adsorption sites. Higher surface area and porosity generally result in higher adsorption capacities because they provide more active sites for the dye molecule to stick to the adsorbent. For example, adsorption of Congo red dye on *E. crassipes* roots greatly depended on dye concentration, *E. crassipes* dosage, and surface area (Wanyonyi et al., 2014). Temperature greatly impacts the adsorption phenomenon by increasing the diffusion rate of the dye molecules through the boundary layer outside and inside the pores of the solid adsorbent, due to decreased solution viscosity and increased dye solubility (Felista et al., 2020). In addition, increasing temperature may upsurge both the swelling of the adsorbent and kinetic energy of the dye molecules causing an increase in dye uptake (Abu Elella et al., 2019). pH significantly controls the characteristic surface charge on the adsorbent, ionization of dye species, and morphology of the adsorbate (Muinde et al., 2020). Furthermore, zeta potential reveals how particles interact with other molecules or particles in the surrounding solution through surface charges.

12.4.3 Pesticides

Hydrogels' adsorption of pesticides is a complex process involving multiple interactions, such as chemical bonding, physical entrapment, and electrostatic interactions. Chemical bonding occurs between the pesticides and the molecular backbone of hydrogels, facilitated by hydrogen bonding. This bonding strengthens the attachment of pesticides to the hydrogel matrix, enhancing their adsorption capacity. Additionally, due to their porous structure, physical entrapment occurs as pesticides get trapped within the gel-water phase of hydrogels. Hydrogels like wet superabsorbent polymers (w-SAP) exhibit high water absorption capacity, creating a buffer zone for handling high pesticide concentrations. The porous nature of hydrogels allows for increased surface area, enabling efficient diffusion of pesticides into the gel-water phase, thus improving overall adsorption efficiency. On the other hand, electrostatic interactions play a role in the adsorption process, with anionic pesticides potentially experiencing lower retention due to electrostatic repulsion between them and the anionic backbone of the hydrogel (Jing et al., 2021).

Moreover, hydrogels are compatible with constructed wetlands (CWs), further enhancing pesticide removal efficiency. The gel-water phase in hydrogels facilitates retention and release processes, while the increased pore volume improves hydraulic retention time (HRT), contributing to overall removal efficiency (Jing et al., 2021).

The unique properties of hydrogels, such as water absorption capacity, porous structure, and compatibility with constructed wetlands, contribute to their effectiveness in pesticide removal. Characterization techniques and investigations of pH, temperature, and kinetic and isothermal behavior support a comprehensive understanding of the adsorption mechanisms. Further research in this area will develop sustainable and efficient materials for water remediation and environmental protection.

12.5 KINETICS OF MICRO-POLLUTANT UPTAKE

The kinetic data on dye uptake can assist in establishing whether the adsorption process is caused by diffusion through internal pores or external surface (Shi et al., 2014). Adsorption kinetics can further be used to forecast the speed at which a contaminant is adsorbed from solutions, thus providing important data for understanding the mechanism of dye uptake (Alpatova et al., 2010). The rate at which the adsorption process occurs is essentially determined by the characteristic nature a particular adsorbent for a specific dye, which influences the design of a cost-efficient adsorption system. Pseudo-first-order and Pseudo-second-order kinetic models are commonly applied in describing the kinetics of dye uptake:

I. Pseudo-first-order kinetics

Pseudo-first-order kinetic model assumes that the rate of adsorption is directly proportional to the difference between the initial concentration of the dye and the concentration at any given time during the adsorption process. The linearized integral form of the pseudo-first-order model (Lagergren, 1898) is denoted by:

$$\log\left(q_e - q_t\right) = \log q_e - \frac{k_1}{2.303}t \tag{7}$$

Where q_e and q_t are the amounts of adsorbed adsorbate (mgg^{-1}) at equilibrium and at any instant of time (min), respectively, while k_1 is the rate constant of pseudo-first-order adsorption (min^{-1}). The values of k_1 and $q_{e,cal,}$ are computed from the slope and intercept of log (q_e - q_t) against t. The pseudo-first-order model is easy to apply, but it may not accurately represent the complex nature of the adsorption process in many cases. The high value of R^2 and concurrence between theoretical and experimental adsorption capacities values depict the conformity of the model in describing the sorption process.

II. Pseudo-second-order kinetics

Pseudo-second-order kinetic model assumes that the adsorption rate is proportional to the square of the difference between the equilibrium adsorption and the quantity of dye removed at a given time. The differential equation is generally described as Ho and Mckay, 1999:

$$\frac{dq_e}{dt} = k_2(q_e - q_t)^2 \tag{8}$$

Integrating equation when $q_t = 0$ at $t = 0$, gives:

$$\frac{t}{q_t} = \frac{1}{k_2 q_e^2} + \frac{t}{q_e} \quad (9)$$

Where k_2 is the adsorption rate constant (g/mg min), q_e and q_t are the dye removed at equilibrium and time t(mgg^{-1}) respectively. Values of k_2 and q_e are calculated from the intercepts and slope of t/q_e vs t. High R^2 values and concurrence between the theoretical $q_{e,cal}$ values and experimental $q_{e.,exp}$ values imply that the pseudo-second order model is appropriate in describing the sorption process. Conformity to the equation further suggests that dye uptake is boosted by chemisorption, comprising exchange and sharing of electrons between the adsorbent particles and adsorbate molecules Ahmad and Mirza, 2018. The pseudo-second-order model often gives a good fit to experimental data, especially when the sorption process is complex and involves multiple steps.

III. **Other models explored for heavy metals**

To understand the adsorption kinetics of the heavy metal uptake by hydrogels, studies have extensively employed kinetic models including pseudo-first-order (PFO), pseudo-second-order (PSO), Elovich, and internal diffusion to model the experimental kinetic data of heavy metal uptake by hydrogels, and the best-fitted kinetic model to each adsorption is listed in Table 2.

Table 2. Adsorption isotherms, kinetic models, and the adsorption capacity of the heavy metal uptake by hydrogels.

Adsorbent	Heavy metal	Adsorption capacity	Adsorption isotherm	Kinetic model	Reference
Poly(vinyl alcohol)-chitosan/ cellulose (PVA-CS/CE) hydrogel	Pb^{2+} Cd^{2+} Zn^{2+} Co^{2+}	50.54 mg/g 41.82 mg/g 26.74 mg/g 20.53 mg/g	Langmuir	PSO	Aljar et al. (2023)
Magnetic sodium lignosulfonate (Fe_3O_4@LS) hydrogel	Cd^{2+}	88.00 mg/g	Temkin	PSO	Chen et al. (2023)
Poly(*N*-isopropylacrylamide-*co*-acrylic acid) hydrogels	Pb(II) Cr(VI) Mn(II)	588.23 mg/g 443.89 mg/g 232.30 mg/g	Langmuir	PSO	Zdravkovic, Nikolic, and Petrovic (2018)
Super-adsorbent poly (AN)-g-starch composite hydrogel	Cd^{2+}	834.53 mg/g	Langmuir	Intraparticle diffusion	Hashem et al. (2021)
Amalgamating chitosan and carboxymethyl guar gum (CS/ CMGG) hydrogels	Cu^{2+}	155.51 mg/g	Langmuir	PSO	Rahmatpour and Alijani (2023)
Acrylamide/acrylic acid cellulose hydrogels	Cu(II) Pb(II) Cd(II)	157.51 mg/g 393.28 mg/g 289.97 mg/g	Langmuir	PSO	Zhao et al. (2019)
MCHETC hydrogel	Cu(II)	103.1 mg/g	Langmuir	Intraparticle surface diffusion	Yuan et al. (2023)
Chitosan/sodium alginate/calcium ion (CTS/SA/Ca^{2+}) hydrogel	Cu^{2+} Cd^{2+}	174.82 mg/g 157.72 mg/g	Langmuir	PSO	(H. Sun et al., 2023)

continued on following page

Table 2. Continued

Adsorbent	Heavy metal	Adsorption capacity	Adsorption isotherm	Kinetic model	Reference
Sodium lignosulfonate-guar gum composite hydrogel	Cu^{2+} Co^{2+}	709 mg/g 601 mg/g	Langmuir	PSO	Li et al. (2021)
Polyacrylate-coated kaolin/zeolite composite hydrogels	Cr^{3+} Co^{2+} Ni^{2+} Pb^{2+}	84.5 mg/g 73.3 mg/g 61 mg/g 97 mg/g	Langmuir	PSO	Sorour et al. (2019)
Macroporous Xylan-rich hemicellulose-based hydrogel	Pb^{2+} Cd^{2+} Zn^{2+}	859 mg/g 495 mg/g 274 mg/g	Langmuir	PSO	Peng et al. (2012)
Carboxymethyl chitosan-nanoclay hydrogel	Cr(VI) Pb(II)	205 mg/g 125 mg/g	Langmuir	PFO	El et al. (2023)
ETS-PAA hydrogels	Pb^{2+} Cu^{2+} Cd^{2+}	1.93 mmol/g 1.93 mmol/g 1.77 mmol/g	Freundlich	PSO	(M. Zhang, Zhou, Yang, Lu, Zhao, Chen, Duan, et al., 2023)
p(HEA/MALA) hydrogel	Pb^{2+} Cd^{2+} Ni^{2+} Cu^{2+}	0.166 mmol/g 0.142 mmol/g 0.168 mmol/g 0.161 mmol/g	Langmuir	PSO	Wu and Li (2013)
Ethyl acetoacetate modified chitosan/organo-functionalized alginate hydrogel beads	Cu(II)	390.21 mg/g	Langmuir	PSO	Abid et al. (2023)
PVA/ALG hydrogel	Pb(II)	139.4 mg/g	Langmuir	PSO	(P. Zhang et al., 2022)
Polyvinyl alcohol/alginate/chitosan	Cr(VI)	86.1 mg/g	Langmuir	PFO	(P. Zhang et al., 2022)
Xanthan Gum/Poly(acrylic acid)/ Cloisite 15A Semi-IPN hydrogel	Co^{2+} Cu^{2+} Ni^{2+}	436.62 mg/g 530.14 mg/g 511.74 mg/g	Co^{2+} - Langmuir Cu^{2+} - Temkin Ni^{2+} - Freundlich	PSO	Esmaeildoost et al. (2022)
Sodium acrylate and acrylamide copolymer/graphene oxide (P(AANa-*co*-AM)/GO) hydrogels	Pb^{2+} Cd^{2+}	452.3 mg/g 196.4 mg/g	Langmuir	PFO	He et al. (2016)
Physically crosslinked chitosan/ sodium alginate/calcium ion double-network hydrogel	Pb^{2+} Cu^{2+} Cd^{2+}	176.50 mg/g 70.83 mg/g 81.25 mg/g	Langmuir-Freundlich	Pb^{2+}, Cu^{2+} and Cd^{2+} (20 mg/L and 100 mg/L) – PFO and PSO Pb^{2+}, Cu^{2+} and Cd^{2+} (200 mg/L) - Elovich	Tang et al. (2020)
Cellulose/bentonite grafted polyacrylic acid hydrogel	Cd(II)	242.53 mg/g	Freundlich	PSO	Ding et al. (2023)
Alginate-based porous nanocomposite hydrogels	Cr(VI) Cu(II)	133.7 mg/g 87.2 mg/g	Freundlich	PSO	(W. Zhang et al., 2021)
Licorice residue-based hydrogel	Pb^{2+} Cr^{3+} Cu^{2+}	591.75 mg/g 458.3 mg/g 121.4 mg/g	Pb^{2+}, Cu^{2+} - Langmuir Cr^{3+} - Freundlich	PSO	(Yin et al., 2023)
Superabsorbent hydrogel based on lignin and montmorillonite	Cu(II)	1.17 mmol/g	Freundlich	PSO	(X.-F. Sun et al., 2019)

12.6 ADSORPTION ISOTHERM MODELLING

Adsorption capacity refers to the maximum amount of a substance (e.g., dye) that an adsorbent material can adsorb or hold per unit mass or unit volume. It is a measure of the adsorbent's efficiency in capturing the target substance from a solution or gas mixture. Adsorption capacity is typically expressed in units such as milligrams of adsorbate per gram of adsorbent (mgg^{-1}) or milligrams of adsorbate per unit volume of adsorbent (mg/L). Adsorption capacity at diverse equilibrium concentrations can be exemplified by an adsorption isotherm. Adsorption isotherms are graphical representations of the equilibrium relationships between the concentration of the dye in the aqueous phase and the quantity of the dye adsorbed on the surface of the adsorbent material at a constant temperature. Adsorption isotherms provide critical evidence about the interaction between molecules of the dye and the adsorbent surface at equilibrium time. The most commonly used adsorption isotherms are the Langmuir isotherm and the Freundlich isotherm.

Langmuir adsorption isotherm (Langmuir, 1918) presumes that dye uptake happens on a uniform, homogeneous surface with sites of indistinguishable nature on the surface of the adsorbent. It is applicable for monolayer adsorption, where one adsorbate molecule is adsorbed on one active site. The nonlinear and linearized forms of the Langmuir isotherm equation are denoted by:

$$q_e = \frac{(Q_m bC_e)}{1 + bC_e} \tag{10}$$

$$\frac{1}{q_e} = \frac{1}{Q_m} + \left(\frac{1}{Q_m b}\right)\frac{1}{C_e} \tag{11}$$

Where q_e is the adsorption density (mgg^{-1}) of dye at equilibrium, C_e is the equilibrium concentration of the dye, Q_m is the one-layer adsorption capacity mgg^{-1} and b is the Langmuir constant (L mg). Q_m and b are calculated from the intercepts and gradient of the plots $1/q_e$ vs $1/C_{e.}$ A dimensionless constant termed the separation factor (R_L) can be applied to evaluate the suitability of an adsorption process. R_L denoted by:

$$R_L = \frac{1}{1 + b(C_o)} \tag{12}$$

The R_L values demonstrate that sorption isotherms are either unfavorable ($R_L > 1$), linear ($R_L = 1$), favorable ($0< R_L <1$), and irreversible ($R_L = 0$).

Freundlich isotherm (Freundlich, 1907) is valid for non-ideal sorption on heterogeneous surfaces and is suitable for multilayer adsorption, where multiple layers of adsorbate molecules are adsorbed on the adsorbent surface. It assumes a heterogeneous surface with different energies of adsorption. The nonlinear and linearized forms of the equation are represented as:

$$q_{e=K_f} \cdot C_e^{1/n} \tag{13}$$

In

$$q_e = In K_f + \left(\frac{1}{n}\right) In C_e \tag{14}$$

Where k_f and n are the Freundlich constants related to adsorption capacity and adsorption intensity of the adsorbent. k_f and n values are attained from the intercepts (lnk_f) and gradient (1/n) of the plot of lnq_e versus lnC_e. 1/n value indicates how favorable the sorption process and values of n>1 point to a favorable condition of adsorption. Experimental determination of adsorption isotherms allows researchers to understand the adsorption behavior of the adsorbent and the adsorbate and to calculate the adsorption capacity under different conditions. This information is crucial in designing and optimizing adsorption processes for various environmental and industrial applications.

Experimental determination of adsorption modelling entails carrying out adsorption experiments carefully to generate critical data. The accuracy of experimental adsorption data is vital for establishing adsorption modelling. Sample selection for analysis should be homogeneous and representative of the adsorbent. Prior to performing adsorption experiments, it is necessary to prepare the adsorbent material to avoid analytical errors. Some of the preparatory measures include thorough washing to remove dirt, foreign particles, and soluble impurities, drying to remove moisture content, and finally, proper storage to avert moisture reabsorption. During the adsorption process, it is important to ensure that the adsorbent weight is accurately measured and the difference between the initial time and each subsequent measurement time is not excessively small to avert analytical errors. Experiments should be performed in triplicate and the average value presented to minimize errors, improve accuracy, reliability, and reproducibility (Mutunga et al., 2020). A set of experimental bottles containing a predetermined amount of adsorbent is vital for determining isotherm points in parallel. A minimum of 8-10 experimental bottles of different solute concentrations is recommended to obtain adequate data points for the isotherm equation fitting. The performance of adsorbents is significantly controlled by critical factors such as aqueous phase temperature, agitation, contact time, pH, dosage, surface area/particle size, and initial concentration, which ought to be optimized and controlled during the sorption test. Equilibrium data evaluation of adsorption is crucial for determining the capacity of specific solute-adsorbent pairs. The amount of solute adsorbed per unit mass of the adsorbent q_e (mg g^{-1}) can be calculated using eq 1.

$$q_e = \frac{V(C_o - C_e)}{W} \tag{15}$$

Where C_o and C_e are the initial and the equilibrium adsorbate concentrations (mg L^{-1}), V is the volume of solution (L), and W is the amount of adsorbent used (g). Experimental adsorption data are fitted to the linear and the nonlinear form of the models. The adsorption isotherm model that gives a better fit to the experimental data with a higher correlation coefficient (R^2) is adopted in describing the adsorption process

12.7 REUSABILITY/REGENERATION STUDIES

Reusability/regeneration studies are imperative considerations in determining the potential use of adsorbents as a prototype-adsorption design in water and wastewater applications. A good adsorbent is known for its ability to maintain the adsorption capacity through the number of adsorption-desorption cycles and has the property of removing contaminants, minimizing the process cost (Sherlala et al., 2018).

The potential of an adsorbent to be regenerated is important because it allows recycling of both adsorbent and adsorbate for multiple uses, making the adsorption process economically efficient and helping to reduce environmental pollution from saturated adsorbent. Adsorbents with higher sorption capacity and exceptionally high desorption properties are economically feasible and environmentally sustainable (Baskar et al., 2022). The phenomenon of desorption is predominantly found in physical adsorption. In chemical adsorption, binding may involve the formation of covalent bonds, coordination complexes, or other chemical interactions, depending on the nature of the dye and the material that makes dye separation irreversible. Desorption mechanism is highly controlled by a combination of factors, which include dye concentration, temperature, contact time, pH, surface area, and nature of the adsorbent (Momina et al., 2019). Several methods have been tested and used for adsorbent regeneration, including thermal treatment, chemical extraction, bio-regeneration, supercritical regeneration, microwave irradiation, ultrasonic regeneration, oxidative regeneration, and photocatalytic (Baskar et al., 2022).

Chemical and thermal techniques have been widely applied in adsorbent regeneration (Baskar et al., 2022). Chemical regeneration involves dissolving dye molecules in the eluent or removing them from the adsorbent surface by an ion-exchange process. In some cases, varying solution pH can facilitate dye removal from the adsorbent. pH strongly influences the electrostatic forces of attraction between the dye molecules and the adsorbent surface and can weaken or promote dye release. Up to 100% regeneration efficiency can be obtained by washing the adsorbent with acid or alkali(Dhodapkar et al., 2007). Introduction of a highly concentrated solution of another species that has a stronger affinity for the adsorbent sites can competitively catalyze the desorption process. The selection of specific solvents that have a higher affinity for the dye than the adsorbent facilitates the desorption process. For certain adsorbents and dye systems that are held together by electrostatic forces of attraction, applying an electrical potential or using an electrochemical cell can facilitate the desorption of dyes and regenerate the adsorbent. Thermal regeneration involves heating the adsorbent to a higher temperature to desorb the dyes. This method is effective for certain adsorbent-dye systems where weak interactions are involved, and the dye can be easily desorbed upon heating. It is important to monitor and characterize the adsorbent after each cycle's regeneration because repeated cycles of adsorption and desorption change the surface properties of the adsorbent, and stability affects its functioning over time.

12.8 COMPETITIVE ADSORPTION OF OTHER POLLUTANTS ON HYDROGEL

Wastewater contains a wide range of inorganic ions and organic substances that can negatively affect the performance of hydrogels by causing competitive adsorption dynamics between various constituents and target pollutants on the hydrogel's surface (Zheng et al., 2024; Zhou et al., 2019). The diverse ionic species in wastewater significantly influence the hydrogel's capacity to adsorb ions. Certain pollutants may hinder the adsorption of others through competition, while others may facilitate co-adsorption (Antić et al., 2016; Van Tran et al., 2018; Zheng et al., 2010).

Research conducted by Ma and co-workers demonstrated that cations such as K^+, Na^+, Mg^{2+}, and Ca^{2+} in solution compete for chelation sites on the hydrogel. This competition results in cations sharing binding sites with target pollutants, including Cd^{2+}, Cu^{2+}, and Pb^{2+}, thereby reducing the efficiency of their removal (Ma et al., 2017). Zheng et al. (2024) studied the effect of varying concentrations of Na^+, K^+, and Ca^{2+} on the adsorption of Cu(II) by sodium alginate hydrogel (SAH). They found that when the concentrations of Na^+ and K^+ were between 0 and 0.01 M, there was minimal effect on Cu(II) adsorption. However, Ca^{2+} exhibited a notable influence due to its lower charge and the lack of available orbitals outside the nucleus for Na^+ and K^+. At concentrations above 0.01 M, competitive binding occurred at the active sites of the hydrogel, leading to a reduced capacity for Cu(II) adsorption. Furthermore, increasing the concentrations of anions like SO_4^{2-} and NO_3^- to 0.1 M led to sediment formation from their interaction with Ca^{2+}, which occupied the hydrogel's active sites, further limiting its ability to adsorb Cu^{2+}.

Humic acids (HAs), which are a significant component of natural organic matter in water, also affect hydrogel efficiency. HAs have polar structural groups such as carboxyls and hydroxyls that bind hydrophilic solutes, while their hydrophobic components help remove non-polar compounds (Guan et al., 2006; Liu & Gonzalez, 2000; Ran et al., 2007; Smilek et al., 2015). However, this interaction can compete with the hydrogel for binding to pollutants, thereby reducing the overall effectiveness of the hydrogel (Zheng et al., 2024).

Clay minerals, such as kaolin, montmorillonite, and bentonite, can also impact the performance of hydrogels. These minerals have small particle sizes and high specific surface areas, and their Si-OH groups can ionize H^+ in aqueous solutions, resulting in a negative charge on the particles. This negative charge facilitates electrostatic interactions with pollutants, especially heavy metals, leading to competitive adsorption (Hu et al., 2017; Huang et al., 2018; Zheng et al., 2024). Additionally, chelating agents like ethylenediaminetetraacetic acid (EDTA), citrate, gluconic acid, nitrile-tri-acetic acid (NTA), and organophosphonates, which are commonly found in industrial wastewater, can pose a challenge for large-scale hydrogel application. These chelating agents compete with hydrogels by binding to and masking metal ions (Knepper, 2003; Zheng et al., 2024).

To minimize competitive adsorption, researchers suggest studying ion properties, including electronegativity, radius, and ionization potential. Observations indicate that ions with larger radii generally have a higher affinity for hydrogels than for other ions, while ions with smaller hydration radii are less affected by competing cations (Antić et al., 2016; Dwivedi et al., 2015). Introducing functional groups such as R-COOH, R-OH, R-SH, R-NH_2, and R-SO_3H onto the hydrogel surface can enhance its preferential binding to specific pollutants (Badsha et al., 2021). Additionally, increasing the number of active sites on the hydrogel surface can help reduce competitive adsorption and improve pollutant removal efficiency (Badsha & Lo, 2020; Hua & Li, 2014).

12.9 CONCLUSION

Findings gathered from literature have proved hydrogels and its modified form to be a suitable and competitive alternative for dye, heavy metals, pharmaceuticals and pesticide removal from wastewater. Both the hydrogels and its derivative indicate potential for re-usability. Mechanistically, for heavy metal uptakes, ion exchange, electrostatic attraction, and chemisorption are the dominant mechanisms. Electrostatic and hydrophobic interactions are common for pharmaceutical, pesticides and dye uptake. This is further corroborated from the kinetic isotherm models whereby pseudo-second order models are the most common kinetic model applicable for heavy metals, pesticides, pharmaceuticals and dyes.

In adsorption isotherm modelling of experimental data, Langmuir isotherm is mainly reported, which again reveals chemisorption and monolayer adsorption. This thus suggests one can rule out physisorption as a mechanism for heavy metal, pesticide, pharmaceutical and dyes. Furthermore, several factors influence the adsorption behavior and uptake rates, including pollutant concentration, pH, temperature, hydrogel properties, and the presence of nanocomposites. The findings underscore the potential of hydrogels as effective adsorbents for pesticide removal, contributing to the development of sustainable and efficient materials for water remediation. Lastly, hydrogel composition and structure optimization can enhance adsorption performance and sustainability, facilitating their practical application in water treatment systems.

REFERENCES

Abid, O., Ahmed, E., Shehzad, H., Sharif, A., Farooqi, Z. H., Liu, Z., Zhou, L., Ouyang, J., Begum, R., Irfan, A., Rasool, A., & Imran, M. (2023). Competitive recovery of copper ions using ethyl acetoacetate modified chitosan / organo-functionalized alginate hydrogel beads : Kinetics and isothermal sorption studies. *Colloids and Surfaces. A, Physicochemical and Engineering Aspects*, *675*(July), 132019. DOI: 10.1016/j.colsurfa.2023.132019

Abu Elella, M. H., Sabaa, M. W., ElHafeez, E. A., & Mohamed, R. R. (2019). Crystal violet dye removal using crosslinked grafted xanthan gum. *International Journal of Biological Macromolecules*, *137*, 1086–1101. DOI: 10.1016/j.ijbiomac.2019.06.243

Ahmad, M. A., Eusoff, M. A., Oladoye, P. O., Adegoke, K. A., & Bello, O. S. (2020). Statistical optimization of Remazol Brilliant Blue R dye adsorption onto activated carbon prepared from pomegranate fruit peel. *Chemical Data Collections*, *28*, 100426. Advance online publication. DOI: 10.1016/j.cdc.2020.100426

Ahmed, E. M. (2015). Hydrogel: Preparation, characterization, and applications: A review. *Journal of Advanced Research*, *6*(2), 105–121. DOI: 10.1016/j.jare.2013.07.006

Akter, M., Bhattacharjee, M., Dhar, A. K., Rahman, F. B. A., Haque, S., Rashid, T. U., & Kabir, S. M. F. (2021). Cellulose-Based Hydrogels for Wastewater Treatment: A Concise Review. *Gels (Basel, Switzerland)*, *7*(30), 1–28. DOI: 10.3390/gels7010030

Aljar, M. A. A., Rashdan, S., Almutawah, A., & El-Fattah, A. A. (2023). Synthesis and Characterization of Biodegradable Poly(vinyl alcohol)-Chitosan/Cellulose Hydrogel Beads for Efficient Removal of Pb (II), Cd (II), Zn (II), and Co (II) from Water. *Gels (Basel, Switzerland)*, *9*(328), 1–21. DOI: 10.3390/gels9040328

Alpatova, A. L., Shan, W., Babica, P., Upham, B. L., Rogensues, A. R., Masten, S. J., Drown, E., Mohanty, A. K., Alocilja, E. C., & Tarabara, V. V. (2010). Single-walled carbon nanotubes dispersed in aqueous media via non-covalent functionalization: Effect of dispersant on the stability, cytotoxicity, and epigenetic toxicity of nanotube suspensions. *Water Research*, *44*(2), 505–520. DOI: 10.1016/j.watres.2009.09.042

Antić, K. M., Babić, M. M., Vuković, J. S., Onija, A. E., Filipović, J. M., & Tomić, S. L. (2016). Removal of Pb2+ from aqueous solution by P(HEA/IA)hydrogels. *Hemijska Industrija*, *70*(6), 695–705. DOI: 10.2298/HEMIND151225006A

Badsha, M. A., & Lo, I. M. C. (2020). An innovative pH-independent magnetically separable hydrogel for the removal of Cu(II) and Ni(II) ions from electroplating wastewater. *Journal of Hazardous Materials*, *381*, 121000. DOI: 10.1016/j.jhazmat.2019.121000

Badsha, M. A. H., Khan, M., Wu, B., Kumar, A., & Lo, I. M. C. (2021). Role of surface functional groups of hydrogels in metal adsorption: From performance to mechanism. *Journal of Hazardous Materials*, *408*, 124463. DOI: 10.1016/j.jhazmat.2020.124463

Baskar, A. V., Bolan, N., Hoang, S. A., Sooriyakumar, P., Kumar, M., Singh, L., Jasemizad, T., Padhye, L. P., Singh, G., Vinu, A., Sarkar, B., Kirkham, M. B., Rinklebe, J., Wang, S., Wang, H., Balasubramanian, R., & Siddique, K. H. M. (2022). Recovery, regeneration and sustainable management of spent adsorbents from wastewater treatment streams: A review. *The Science of the Total Environment*, *822*, 153555. DOI: 10.1016/j.scitotenv.2022.153555

Batista, R. A., Espitia, P. J. P., Quintans, J., Freitas, M. M., Cerqueira, M. Â., Teixeira, J. A., & Cardoso, J. C. (2019). Hydrogel as an alternative structure for food packaging systems. *Carbohydrate Polymers*, *205*, 106–116. DOI: 10.1016/j.carbpol.2018.10.006

Bharali, D. J., Sahoo, S. K., Mozumdar, S., & Maitra, A. (2003). Cross-linked polyvinylpyrrolidone nanoparticles: A potential carrier for hydrophilic drugs. *Journal of Colloid and Interface Science*, *258*(2), 415–423. DOI: 10.1016/S0021-9797(02)00099-1

Caló, E., & Khutoryanskiy, V. V. (2015). Biomedical applications of hydrogels: A review of patents and commercial products. *European Polymer Journal*, *65*, 252–267. DOI: 10.1016/j.eurpolymj.2014.11.024

Chelu, M., Popa, M., Calderon Moreno, J., Leonties, A. R., Ozon, E. A., Pandele Cusu, J., Surdu, V. A., Aricov, L., & Musuc, A. M. (2023). Green Synthesis of Hydrogel-Based Adsorbent Material for the Effective Removal of Diclofenac Sodium from Wastewater. *Gels (Basel, Switzerland)*, *9*(6), 454. Advance online publication. DOI: 10.3390/gels9060454

Chen, W., Xie, H., Jiang, N., Guo, X., & Liu, Z. (2023). Synthesis of magnetic sodium lignosulfonate hydrogel (Fe 3 O 4 @ LS) and its adsorption behavior for Cd 2 + in wastewater. *International Journal of Biological Macromolecules*, *245*(June), 125498. DOI: 10.1016/j.ijbiomac.2023.125498

Chhetry, G., & Gooh Pattader, P. S. (2024). Effect of patterns on Polyacrylamide hydrogel surface towards enhancement of water retention. *Polymer*, *308*, 127362. DOI: 10.1016/j.polymer.2024.127362

Das, D., Pham, H. T. T., Lee, S., & Noh, I. (2019). Fabrication of alginate-based stimuli-responsive, non-cytotoxic, terpolymric semi-IPN hydrogel as a carrier for controlled release of bovine albumin serum and 5-amino salicylic acid. *Materials Science and Engineering C*, *98*, 42–53. DOI: 10.1016/j.msec.2018.12.127

Das, S., Kumar, V., Tiwari, R., Singh, L., & Singh, S. (2018). Recent advances in hydrogels for biomedical applications. *Asian Journal of Pharmaceutical and Clinical Research*, *11*(11), 62. DOI: 10.22159/ajpcr.2018.v11i11.27921

Dhodapkar, R., Rao, N. N., Pande, S. P., Nandy, T., & Devotta, S. (2007). Adsorption of cationic dyes on Jalshakti®, super absorbent polymer and photocatalytic regeneration of the adsorbent. *Reactive & Functional Polymers*, *67*(6), 540–548. DOI: 10.1016/j.reactfunctpolym.2007.03.007

Ding, W., Liang, H., Zhang, H., Sun, H., Geng, Z., & Xu, C. (2023). A cellulose / bentonite grafted polyacrylic acid hydrogel for highly-efficient removal of Cd (II). *Journal of Water Process Engineering*, *51*(3), 103414. DOI: 10.1016/j.jwpe.2022.103414

Dodda, J. M., Deshmukh, K., Bezuidenhout, D., & Yeh, Y.-C. (2023). Hydrogels: Definition, History, Classifications, Formation, Constitutive Characteristics, and Applications. In *Multicomponent Hydrogels* (pp. 1–25). The Royal Society of Chemistry. DOI: 10.1039/BK9781837670055-00001

Dolu, T., & Nas, B. (2023). Dissemination of nonsteroidal anti-inflammatory drugs (NSAIDs) and metabolites from wastewater treatment plant to soils and agricultural crops via real-scale different agronomic practices. *Environmental Research, 227*, 115731. DOI: 10.1016/j.envres.2023.115731

Dwivedi, C., Pathak, S. K., Kumar, M., Tripathi, S. C., & Bajaj, P. N. (2015). Preparation and characterization of potassium nickel hexacyanoferrate-loaded hydrogel beads for the removal of cesium ions. *Environmental Science. Water Research & Technology, 1*(2), 153–160. DOI: 10.1039/C4EW00021H

El, M., Abdel, S., Rasha, R., Dalia, S. K., & Asmaa, E. H. (2023). Gamma Irradiation Synthesis of Carboxymethyl Chitosan - Nanoclay Hydrogel for the Removal of Cr (VI) and Pb (II) from Aqueous Media. *Journal of Inorganic and Organometallic Polymers and Materials, 33*(4), 895–913. DOI: 10.1007/s10904-023-02543-w

Esmaeildoost, F., Shahrousvand, M., Goudarzi, A., & Bagherieh, M. B. (2022). Optimization of Xanthan Gum / Poly (acrylic acid)/ Cloisite 15A Semi - IPN Hydrogels for Heavy Metals Removal. *Journal of Polymers and the Environment, 30*(10), 4271–4286. DOI: 10.1007/s10924-022-02501-6

Fasolin, L. H., Martins, A. J., Cerqueira, M. A., & Vicente, A. A. (2021). Modulating process parameters to change physical properties of bigels for food applications. *Food Structure, 28*, 100173. DOI: 10.1016/j.foostr.2020.100173

Felista, M. M., Wanyonyi, W. C., & Ongera, G. (2020). Adsorption of anionic dye (Reactive black 5) using macadamia seed Husks: Kinetics and equilibrium studies. *Scientific African, 7*, e00283. Advance online publication. DOI: 10.1016/j.sciaf.2020.e00283

Freundlich, H. (1907). Über die Adsorption in Lösungen. *Zeitschrift für Physikalische Chemie, 57U*(1), 385–470. DOI: 10.1515/zpch-1907-5723

Ghaedi, M., Nasab, A. G., Khodadoust, S., Rajabi, M., & Azizian, S. (2014). Application of activated carbon as adsorbents for efficient removal of methylene blue: Kinetics and equilibrium study. *Journal of Industrial and Engineering Chemistry, 20*(4), 2317–2324. DOI: 10.1016/j.jiec.2013.10.007

Gokmen, F. O., Yaman, E., & Temel, S. (2021). Eco-friendly polyacrylic acid based porous hydrogel for heavy metal ions adsorption : Characterization, adsorption behavior, thermodynamic and reusability studies. *Microchemical Journal, 168*(May), 106357. DOI: 10.1016/j.microc.2021.106357

Guan, X.-H., Chen, G.-H., & Shang, C. (2006). Combining kinetic investigation with surface spectroscopic examination to study the role of aromatic carboxyl groups in NOM adsorption by aluminum hydroxide. *Journal of Colloid and Interface Science, 301*(2), 419–427. DOI: 10.1016/j.jcis.2006.05.031

Gupta, V. K., Kumar, R., Nayak, A., Saleh, T. A., & Barakat, M. A. (2013). Adsorptive removal of dyes from aqueous solution onto carbon nanotubes: A review. *Advances in Colloid and Interface Science, 193–194*, 24–34. DOI: 10.1016/j.cis.2013.03.003

Hashem, A., Aniagor, C. O., Afifi, M. A., Abou-okeil, A., & Samaha, S. H. (2021). *Synthesis of super-absorbent poly (AN) -g-starch composite hydrogel and its modelling for aqueous sorption of cadmium ions. 38*(10), 2157–2170. DOI: 10.1007/s11814-021-0856-7

He, S., Zhang, F., Cheng, S., & Wang, W. (2016). *Synthesis of Sodium Acrylate and Acrylamide Copolymer/GO Hydrogels and Their Eff ective Adsorption for Pb 2+ and Cd 2+*. DOI: 10.1021/acssuschemeng.6b00796

Hu, C., Zhu, P., Cai, M., Hu, H., & Fu, Q. (2017). Comparative adsorption of Pb(II), Cu(II) and Cd(II) on chitosan saturated montmorillonite: Kinetic, thermodynamic and equilibrium studies. *Applied Clay Science, 143*, 320–326. DOI: 10.1016/j.clay.2017.04.005

Hu, N.-W., Yu, H.-W., Deng, B.-L., Hu, B., Zhu, G.-P., Yang, X.-T., Wang, T.-Y., Zeng, Y., & Wang, Q.-Y. (2023). Levels of heavy metal in soil and vegetable and associated health risk in peri-urban areas across China. *Ecotoxicology and Environmental Safety, 259*, 115037. DOI: 10.1016/j.ecoenv.2023.115037

Hua, R., & Li, Z. (2014). Sulfhydryl functionalized hydrogel with magnetism: Synthesis, characterization, and adsorption behavior study for heavy metal removal. *Chemical Engineering Journal, 249*, 189–200. DOI: 10.1016/j.cej.2014.03.097

Huang, L., Jin, Q., Tandon, P., Li, A., Shan, A., & Du, J. (2018). High-resolution insight into the competitive adsorption of heavy metals on natural sediment by site energy distribution. *Chemosphere, 197*, 411–419. DOI: 10.1016/j.chemosphere.2018.01.056

Jing, Y., Krauss, M., Zschieschang, S., Miltner, A., Butkovskyi, A., Eggen, T., Kästner, M., & Nowak, K. M. (2021). Superabsorbent polymer as a supplement substrate of constructed wetland to retain pesticides from agricultural runoff. *Water Research, 207*, 117776. DOI: 10.1016/j.watres.2021.117776

Kalaiselvi, K., Mohandoss, S., Ahmad, N., & Khan, M. R. (2023). *Adsorption of Pb 2 + Ions from Aqueous Solution onto Porous Kappa-Carrageenan / Cellulose Hydrogels : Isotherm and Kinetics Study.*

Kim, S. D., Kim, K., & Shin, M. (2023). Recent advances in 3D printable conductive hydrogel inks for neural engineering. *Nano Convergence, 10*(1), 41. DOI: 10.1186/s40580-023-00389-z

Knepper, T. P. (2003). Synthetic chelating agents and compounds exhibiting complexing properties in the aquatic environment. *Trends in Analytical Chemistry, 22*(10), 708–724. DOI: 10.1016/S0165-9936(03)01008-2

Lagergren, S. (1898). About the theory of so-called adsorption of soluble substances, Kungliga Svenska Vetenskapsakademiens. *Handlingar, 24*, 1.

Langmuir, I. (1918). THE ADSORPTION OF GASES ON PLANE SURFACES OF GLASS, MICA AND PLATINUM. *Journal of the American Chemical Society*, *40*(9), 1361–1403. DOI: 10.1021/ja02242a004

Le, V. T., Joo, S.-W., Berkani, M., Mashifana, T., Kamyab, H., Wang, C., & Vasseghian, Y. (2023). Sustainable cellulose-based hydrogels for water treatment and purification. *Industrial Crops and Products*, *205*, 117525. DOI: 10.1016/j.indcrop.2023.117525

Li, X., Wang, X., Han, T., Hao, C., Han, S., & Fan, X. (2021). Synthesis of sodium lignosulfonate-guar gum composite hydrogel for the removal of Cu 2 + and Co 2 +. *International Journal of Biological Macromolecules*, *175*, 459–472. DOI: 10.1016/j.ijbiomac.2021.02.018

Lin, Z., Yang, Y., Liang, Z., Zeng, L., & Zhang, A. (2021). Preparation of Chitosan / Calcium Alginate / Bentonite Composite Hydrogel and Its Heavy Metal Ions Adsorption Properties. *Polymers, 13*(1891), 1–19.

Liu, A., & Gonzalez, R. D. (2000). Modeling Adsorption of Copper(II), Cadmium(II) and Lead(II) on Purified Humic Acid. *Langmuir*, *16*(8), 3902–3909. DOI: 10.1021/la990607x

Londhe, P. V., Londhe, M. V., Salunkhe, A. B., Laha, S. S., Mefford, O. T., Thorat, N. D., & Khot, V. M. (2025). Magnetic hydrogel (MagGel): An evolutionary pedestal for anticancer therapy. *Coordination Chemistry Reviews*, *522*, 216228. DOI: 10.1016/j.ccr.2024.216228

Ma, J., Liu, Y., Ali, O., Wei, Y., Zhang, S., Zhang, Y., Cai, T., Liu, C., & Luo, S. (2018). Fast adsorption of heavy metal ions by waste cotton fabrics based double network hydrogel and influencing factors insight. *Journal of Hazardous Materials*, *344*, 1034–1042. DOI: 10.1016/j.jhazmat.2017.11.041

Ma, J., Zhou, G., Chu, L., Liu, Y., Liu, C., Luo, S., & Wei, Y. (2017). Efficient Removal of Heavy Metal Ions with An EDTA Functionalized Chitosan/Polyacrylamide Double Network Hydrogel. *ACS Sustainable Chemistry & Engineering*, *5*(1), 843–851. DOI: 10.1021/acssuschemeng.6b02181

Momina, R., Rafatullah, M., Ismail, S., & Ahmad, A. (2019). Optimization Study for the Desorption of Methylene Blue Dye from Clay Based Adsorbent Coating. *Water (Basel)*, *11*(6), 1304. DOI: 10.3390/w11061304

Muinde, V. M., Onyari, J. M., Wamalwa, B., & Wabomba, J. N. (2020). Adsorption of malachite green dye from aqueous solutions using mesoporous chitosan–zinc oxide composite material. *Environmental Chemistry and Ecotoxicology*, *2*, 115–125. DOI: 10.1016/j.enceco.2020.07.005

Mutunga, M. F., Wanyonyi, W. C., & Ongera, G. (2020). Utilization of Macadamia seed husks as a low-cost sorbent for removing cationic dye (basic blue 3 dye) from aqueous solution. *Environmental Chemistry and Ecotoxicology*, *2*, 194–200. DOI: 10.1016/j.enceco.2020.09.005

Nille, O. S., Patil, A. S., Kolekar, G. B., & Gore, A. H. (2021). *Carbon-Based Composite Hydrogels for Environmental Remediation.*, DOI: 10.1007/978-981-15-6699-8_20

Oladoye, P. O., Ajiboye, T. O., Omotola, E. O., & Oyewola, O. J. (2022). Methylene blue dye: Toxicity and potential elimination technology from wastewater. *Results in Engineering*, *16*, 100678. DOI: 10.1016/j.rineng.2022.100678

Peng, X., Zhong, L., Ren, J., & Sun, R. (2012). *Highly Effective Adsorption of Heavy Metal Ions from Aqueous Solutions by Macroporous Xylan-Rich Hemicelluloses-Based Hydrogel.*

Rahmatpour, A., & Alijani, N. (2023). An all-biopolymer self-assembling hydrogel film consisting of chitosan and carboxymethyl guar gum : A novel bio-based composite adsorbent for Cu 2 + adsorption from aqueous solution. *International Journal of Biological Macromolecules, 242*(P2), 124878. DOI: 10.1016/j.ijbiomac.2023.124878

Ran, Y., Sun, K., Yang, Y., Xing, B., & Zeng, E. (2007). Strong Sorption of Phenanthrene by Condensed Organic Matter in Soils and Sediments. *Environmental Science & Technology, 41*(11), 3952–3958. DOI: 10.1021/es062928i

Razmjou, A., Simon, G. P., & Wang, H. (2013). Effect of particle size on the performance of forward osmosis desalination by stimuli-responsive polymer hydrogels as a draw agent. *Chemical Engineering Journal, 215–216*, 913–920. DOI: 10.1016/j.cej.2012.11.088

Rizwan, M., Rubina Gilani, S., Iqbal Durani, A., & Naseem, S. (2021). Materials diversity of hydrogel: Synthesis, polymerization process and soil conditioning properties in agricultural field. *Journal of Advanced Research, 33*, 15–40. DOI: 10.1016/j.jare.2021.03.007

Romal, J. R. A., & Ong, S. K. (2023). Single-Step Fabrication of a Dual-Sensitive Chitosan Hydrogel by C-Mannich Reaction: Synthesis, Physicochemical Properties, and Screening of its Cu2+ Uptake. *Processes (Basel, Switzerland), 11*(354), 1–18. DOI: 10.3390/pr11020354

Rybachuk, O., Nesterenko, Y., & Zhovannyk, V. (2024). Modern advances in spinal cord regeneration: Hydrogel combined with neural stem cells. *Frontiers in Pharmacology, 15*, 1419797. Advance online publication. DOI: 10.3389/fphar.2024.1419797

Shalla, A. H., Bhat, M. A., & Yaseen, Z. (2018). Hydrogels for removal of recalcitrant organic dyes: A conceptual overview. *Journal of Environmental Chemical Engineering, 6*(5), 5938–5949. DOI: 10.1016/j.jece.2018.08.063

Sherlala, A. I. A., Raman, A. A. A., Bello, M. M., & Asghar, A. (2018). A review of the applications of organo-functionalized magnetic graphene oxide nanocomposites for heavy metal adsorption. *Chemosphere, 193*, 1004–1017. DOI: 10.1016/j.chemosphere.2017.11.093

Shi, L., Zhang, G., Wei, D., Yan, T., Xue, X., Shi, S., & Wei, Q. (2014). Preparation and utilization of anaerobic granular sludge-based biochar for the adsorption of methylene blue from aqueous solutions. *Journal of Molecular Liquids, 198*, 334–340. DOI: 10.1016/j.molliq.2014.07.023

Smilek, J., Sedláček, P., Kalina, M., & Klučáková, M. (2015). On the role of humic acids' carboxyl groups in the binding of charged organic compounds. *Chemosphere, 138*, 503–510. DOI: 10.1016/j.chemosphere.2015.06.093

Song, M., Wang, J., He, J., Kan, D., Chen, K., & Lu, J. (2022). Synthesis of Hydrogels and Their Progress in Environmental Remediation and Antimicrobial Application. *Gels (Basel, Switzerland), 9*(1), 16. DOI: 10.3390/gels9010016

Sonwane, S., & Ingle, R. (2024). Recent Advances in Hydrogel Drug Delivery Systems: Innovations and Applications. *INTERNATIONAL JOURNAL OF DRUG DELIVERY TECHNOLOGY, 14*(04), 945–947. DOI: 10.25258/ijddt.14.4.67

Sorour, M. H., Hani, H. A., Shaalan, H. F., Marwa, M., Sayed, E., & El-toukhy, M. A. (2019). Adsorption isotherms and kinetics pertinent to modified composite hydrogel adsorbents adopted for heavy metals removal. *International Journal of Applied Science and Engineering, 18*(5), 2021013.

Sun, H., Zhan, J., Chen, L., & Zhao, Y. (2023). Preparation of CTS/PAMAM/SA/Ca2+ hydrogel and its adsorption performance for heavy metal ions. *Applied Surface Science, 607*, 155135. DOI: 10.1016/j.apsusc.2022.155135

Sun, X.-F., Hao, Y., Cao, Y., & Zeng, Q. (2019). Superadsorbent hydrogel based on lignin and montmorillonite for Cu(II) ions removal from aqueous solution. *International Journal of Biological Macromolecules, 127*, 511–519. DOI: 10.1016/j.ijbiomac.2019.01.058

Tagumira, A., Biira, S., & Amabayo, E. B. (2022). Concentrations and human health risk assessment of selected heavy metals in soils and food crops around Osukuru phosphate mine, Tororo District, Uganda. *Toxicology Reports, 9*, 2042–2049. DOI: 10.1016/j.toxrep.2022.11.007

Tan, J., Li, J., Zhang, Y., Li, X., Han, S., Li, Z., & Zhou, X. (2024). Application of photocrosslinked gelatin, alginate and dextran hydrogels in the in vitro culture of testicular tissue. *International Journal of Biological Macromolecules, 260*, 129498. DOI: 10.1016/j.ijbiomac.2024.129498

Tang, S., Yang, J., Lin, L., Peng, K., Chen, Y., Jin, S., & Yao, W. (2020). Construction of physically crosslinked chitosan / sodium alginate / calcium ion double-network hydrogel and its application to heavy metal ions removal. *Chemical Engineering Journal, 393*(January), 124728. DOI: 10.1016/j.cej.2020.124728

Tungkavet, T., Seetapan, N., Pattavarakorn, D., & Sirivat, A. (2015). Graphene/gelatin hydrogel composites with high storage modulus sensitivity for using as electroactive actuator: Effects of surface area and electric field strength. *Polymer, 70*, 242–251. DOI: 10.1016/j.polymer.2015.06.027

Van Tran, V., Park, D., & Lee, Y.-C. (2018). Hydrogel applications for adsorption of contaminants in water and wastewater treatment. *Environmental Science and Pollution Research International, 25*(25), 24569–24599. DOI: 10.1007/s11356-018-2605-y

Visan, A. I., & Negut, I. (2025). Environmental and Wastewater Treatment Applications of Stimulus-Responsive Hydrogels. *Gels (Basel, Switzerland), 11*(1), 72. DOI: 10.3390/gels11010072

Wang, W., Narain, R., & Zeng, H. (2020). Hydrogels. *Polymer Science and Nanotechnology: Fundamentals and Applications*, 203–244. DOI: 10.1016/B978-0-12-816806-6.00010-8

Wanyonyi, W. C., Onyari, J. M., & Shiundu, P. M. (2014). Adsorption of Congo Red Dye from Aqueous Solutions Using Roots of Eichhornia Crassipes: Kinetic and Equilibrium Studies. *Energy Procedia, 50*, 862–869. DOI: 10.1016/j.egypro.2014.06.105

Wu, L. (2025). Coupled swelling and capacitive behavior in hydrogel-elastomer composite cylinders for sensing applications. *Materials Today. Communications, 42*, 111426. DOI: 10.1016/j.mtcomm.2024.111426

Wu, N., & Li, Z. (2013). Synthesis and characterization of poly (HEA / MALA) hydrogel and its application in removal of heavy metal ions from water. *Chemical Engineering Journal, 215–216*, 894–902. DOI: 10.1016/j.cej.2012.11.084

Yang, Z., Yang, T., Yang, Y., Yi, X., Hao, X., Xie, T., & Liao, C. J. (2021). The behavior and mechanism of the adsorption of Pb (II) and Cd (II) by a porous double network porous hydrogel derived from peanut shells. *Materials Today. Communications*, *27*(May), 102449. DOI: 10.1016/j.mtcomm.2021.102449

Yin, X., Ke, T., Zhu, H., Xu, P., & Wang, H. (2023). Efficient Removal of Heavy Metals from Aqueous Solution Using Licorice Residue-Based Hydrogel Adsorbent. *Gels (Basel, Switzerland)*, *9*(559), 1–21. DOI: 10.3390/gels9070559

Yin, X., Ke, T., Zhu, H., Xu, P., & Wang, H. (2023). Efficient Removal of Heavy Metals from Aqueous Solution Using Licorice Residue-Based Hydrogel Adsorbent. *Gels (Basel, Switzerland)*, *9*(7), 559. DOI: 10.3390/gels9070559

Yuan, B., Huang, X., Yang, S., Yang, Y., Lin, Z., Semiat, R., & Chen, J. P. (2023). Development of a magnetic calcium-alginate hydrogel-sphere encapsulated with Fe – Mn – Zr ternary metal composite for heavy metal adsorption. *Separation and Purification Technology, 306*(PA), 122531. DOI: 10.1016/j.seppur.2022.122531

Zdravkovic, A., Nikolic, V., & Petrovic, S. (2018). *The removal of heavy metal ions from aqueous solutions by hydrogels based on N -isopropylacrylamide and acrylic acid.* 4797–4821. DOI: 10.1007/s00289-018-2295-0

Zhang, L., Su, T., Luo, Z., Xu, B., Yao, W., Zhou, M., & Yang, W. (2023). A graphene-based porous composite hydrogel for efficient heavy metal ions removal from wastewater. *Separation and Purification Technology, 305*(September 2022), 122484. DOI: 10.1016/j.seppur.2022.122484

Zhang, M., Zhou, Y., Yang, X., Lu, X., Zhao, X., Chen, Z., Duan, W., Li, J., Zhao, M., & Yin, Q. (2023). Preparation of esterified biomass waste hydrogels and their removal. *Environmental Science and Pollution Research International, 30*(19), 56580–56593. DOI: 10.1007/s11356-023-26350-5

Zhang, M., Zhou, Y., Yang, X., Lu, X., Zhao, X., Chen, Z., Duan, W., Li, J., Zhao, M., & Yin, Q. (2023). Preparation of esterified biomass waste hydrogels and their removal of Pb2+, Cu2+ and Cd2+ from aqueous solution. *Environmental Science and Pollution Research International, 30*(19), 56580–56593. DOI: 10.1007/s11356-023-26350-5

Zhang, P., Zou, K., Yuan, L., Liu, J., Liu, B., Qing, T., & Feng, B. (2022). A biomass resource strategy for alginate-polyvinyl alcohol double network hydrogels and their adsorption to heavy metals. *Separation and Purification Technology*, *301*(August), 122050. DOI: 10.1016/j.seppur.2022.122050

Zhang, W., Ou, J., Wang, B., Wang, H., He, Q., Song, J., Zhang, H., Tang, M., Zhou, L., Gao, Y., & Sun, S. (2021). Efficient heavy metal removal from water by alginate-based porous nanocomposite hydrogels: The enhanced removal mechanism and influencing factor insight. *Journal of Hazardous Materials*, *418*, 126358. DOI: 10.1016/j.jhazmat.2021.126358

Zhao, B., Jiang, H., Lin, Z., Xu, S., Xie, J., & Zhang, A. (2019). Preparation of acrylamide / acrylic acid cellulose hydrogels for the adsorption of heavy metal ions. *Carbohydrate Polymers*, *224*(June), 115022. DOI: 10.1016/j.carbpol.2019.115022

Zheng, C., Wu, Q., Sun, K., Xu, B., Sun, Y., & Zheng, H. (2024). Insight into the impact of environmental factors on heavy metal adsorption by sodium alginate hydrogel: Inspiration on applicable scenarios. *Environmental Research*, *262*, 119878. DOI: 10.1016/j.envres.2024.119878

Zheng, Y., Hua, S., & Wang, A. (2010). Adsorption behavior of Cu2+ from aqueous solutions onto starch-g-poly(acrylic acid)/sodium humate hydrogels. *Desalination*, *263*(1-3), 170–175. DOI: 10.1016/j.desal.2010.06.054

Zhou, A., Chen, W., Liao, L., Xie, P., Zhang, T. C., Wu, X., & Feng, X. (2019). Comparative adsorption of emerging contaminants in water by functional designed magnetic poly (N-isopropylacrylamide)/chitosan hydrogels. *The Science of the Total Environment*, *671*, 377–387. DOI: 10.1016/j.scitotenv.2019.03.183

Chapter 13
Catalytic Hydrogels and Hydrogel-Based Composites:
Applications and Innovations

Irum Fatima
https://orcid.org/0009-0008-4202-4170
University of Wah, Pakistan

Abbas Faheem
https://orcid.org/0009-0002-8465-4492
University of Wah, Pakistan

Sana Javaid
University of Wah, Pakistan

Gul E. Nayyab
Universiti Tun Hussein Onn Malaysia, Malaysia

ABSTRACT

Catalytic hydrogels represent a groundbreaking class of materials that integrate the unique properties of hydrogels with catalytic functionality, offering transformative solutions across environmental remediation, biotechnology, and industrial processes. This chapter highlights the synthesis and utilization of hydrogel-nanoparticle composites, demonstrating their efficacy in pollutant degradation, energy conversion, and chemical synthesis. Photocatalytic hydrogels, particularly those incorporating TiO_2 and ZnO, exhibit remarkable efficiency in degrading organic contaminants and reducing heavy metals like CrVI. Electrocatalytic hydrogels are explored for their role in oxygen reduction reactions (ORR), hydrogen evolution reactions (HER), and nitrogen reduction reactions (NRR). Additionally, heterogeneous catalyst hydrogels, including Ag-based and Fe_3O_4-hydrogel systems, showcase high reusability and stability in treating organic and inorganic pollutants.

DOI: 10.4018/979-8-3373-2337-4.ch013

13.1 EXPLORING THE VERSATILITY OF CATALYTIC HYDROGELS

Catalytic hydrogels represent an innovative class of materials that combine the structural and functional properties of hydrogels with catalytic capabilities (Dadashi et al., 2022). These materials feature hydrophilic polymer networks that retain substantial amounts of water while maintaining their three-dimensional integrity, simultaneously housing active catalytic sites that facilitate chemical transformations. This dual-functionality enables catalytic hydrogels to operate efficiently in aqueous environments, making them particularly valuable for applications ranging from environmental remediation to biomedical interventions and industrial processing.

The fundamental architecture of catalytic hydrogels consists of a hydrophilic polymer framework with integrated catalytic centers. These catalytic sites accelerate chemical reactions by lowering activation energy barriers, increasing reaction rates while remaining unchanged throughout the process. The hydrogel matrix offers several advantages as a catalyst support: it provides high surface area for reactant interaction, enables controlled diffusion of substrates and products, stabilizes catalytic species against aggregation or leaching, and offers a tunable microenvironment that can influence reaction selectivity and efficiency (Dadashi et al., 2022).

Catalytic hydrogels facilitate numerous chemical transformations critical to various applications. In environmental contexts, these materials decompose hazardous pollutants into benign compounds, removing contaminants like dyes, heavy metals, and organic toxins from water sources (Dadashi et al., 2022). In biotechnology, they enable controlled enzymatic reactions for biosensing or therapeutic purposes (Hernandez Monroy et al., 2025). Industrial applications include flow reactors for continuous chemical synthesis, where the hydrogel structure ensures catalyst retention while allowing reactant and product diffusion (Abou Zeid & Leprince-Wang, 2024).

Among the diverse applications of catalytic hydrogels, photocatalytic systems have attracted significant attention due to their ability to harness light energy for chemical transformations. When embedded within hydrogel matrices, photocatalysts like titanium dioxide (TiO_2) or zinc oxide (ZnO) efficiently degrade organic pollutants upon light activation. For instance, photocatalytic hydrogels demonstrate impressive efficiency in decomposing common water pollutants such as methylene blue, crystal violet, malachite green, and rhodamine B. The mechanism involves light absorption by the photocatalyst, generating electron-hole pairs that produce reactive oxygen species (ROS), particularly hydroxyl radicals (·OH). These ROS subsequently oxidize organic contaminants, converting them into harmless molecules such as carbon dioxide and water (Katzenberg et al., 2020).

Figure 1. Adsorption and photocatalytic degradation efficiency of MB, including stability. This figure includes two parts: (a) a graph showing the adsorption and photocatalytic degradation efficiency of methylene blue (MB) over different hydrogels with and without TiO_2, with a grey area indicating the adsorption phase (120 minutes, light off) and a white area for the degradation phase under UV light (180 minutes, light on); (b) a graph demonstrating the long-term stability of photocatalytic degradation for the hydrogel with TiO_2 (C15E1Ti15). It highlights the effectiveness of TiO_2-loaded hydrogels in removing organic pollutants, with data showing high degradation efficiency and stability over multiple cycles.

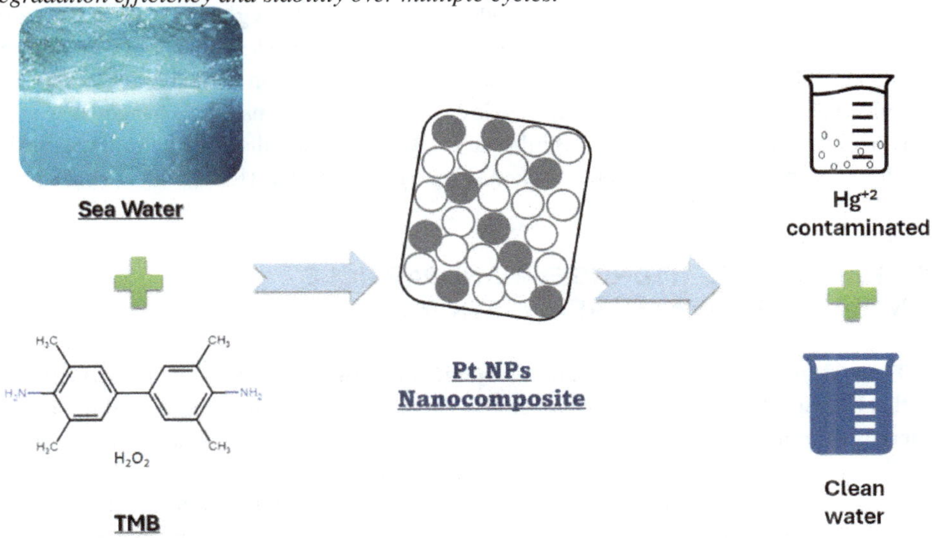

Due to its unique properties, zinc oxide has emerged as a particularly effective photocatalyst for environmental applications. ZnO exhibits high electron mobility (200–300 $cm^2/V{\cdot}s$), high photocatalytic efficiency, excellent chemical stability, and economic viability regarding production costs. However, its susceptibility to photo-induced corrosion in prolonged UV exposure can limit performance, though recent modifications like doping enhance its stability (Aaron J. Clasky et al., 2021). Moreover, ZnO's band gap energy (~3.37 eV) allows it to absorb UV light efficiently, generating the reactive species necessary for pollutant degradation. This combination of characteristics makes ZnO-loaded hydrogels particularly promising for water treatment applications where durability and cost-effectiveness are essential considerations.

The integration of photocatalysts within hydrogels addresses several challenges associated with conventional photocatalytic systems. By immobilizing nanoscale catalysts within the hydrogel network, particle aggregation is minimized, active surface areas are maximized, and catalyst recovery is simplified. Furthermore, the hydrogel matrix can be engineered to enhance light penetration, improving photocatalytic efficiency throughout the material volume. This represents a significant advancement over traditional suspended catalyst systems, where light scattering and catalyst recovery present persistent operational challenges (Liao et al., 2025).

Beyond photocatalysis, catalytic hydrogels incorporate various other active species, including metal nanoparticles (e.g., Pt, Pd, Au), enzymes, organocatalysts, and metal-organic frameworks. Each catalytic system offers unique advantages for specific applications, expanding the versatility of these materials across different reaction types and environmental conditions. The polymeric network of the hydrogel can be tailored to optimize performance for particular catalytic systems, adjusting properties such as mesh size, hydrophilicity/hydrophobicity balance, and functional group density to enhance substrate binding or product release (Dadashi et al., 2022).

Recent advances in catalytic hydrogel design have focused on developing stimuli-responsive systems, where catalytic activity can be modulated by external triggers such as temperature, pH, light, or electrical stimuli. These smart materials offer unprecedented control over reaction kinetics and selectivity, enabling on-demand activation or deactivation of catalytic processes. For example, thermoresponsive catalytic hydrogels can undergo conformational changes at specific temperatures, altering substrate accessibility to catalytic sites and thereby controlling reaction rates without changing catalyst concentration or reaction conditions (Karchoubi et al., 2024).

13.2 HARNESSING HYDROGEL-NANOPARTICLE SYNERGY FOR DIVERSE APPLICATIONS

The strategic integration of nanoparticles within hydrogel matrices creates hybrid nanocomposites that leverage the complementary properties of both components, resulting in multifunctional materials with enhanced performance characteristics. This synergistic combination has generated significant interest across scientific and industrial communities, as it enables the development of advanced materials with tailored properties for specific applications (Bertolacci et al., 2020).

Hydrogels contribute several crucial attributes to these nanocomposites, including high water content, structural flexibility, biocompatibility, and tunable mechanical properties. Additionally, the three-dimensional polymer network of hydrogels provides spatial confinement for nanoparticles, preventing their aggregation and maintaining their dispersed state—a critical factor for preserving their unique size-dependent properties. The hydrogel matrix also facilitates the controlled diffusion of reactants and products, creating an optimal microenvironment for catalytic reactions while ensuring long-term stability of the embedded nanoparticles (Bertolacci et al., 2020).

Nanoparticles, depending on their composition and structure, endow the nanocomposites with distinctive functional properties. Metallic nanoparticles (e.g., gold, silver, platinum) introduce plasmonic effects, catalytic activity, and antimicrobial properties. Magnetic nanoparticles (e.g., iron oxide, cobalt, nickel) enable responsiveness to external magnetic fields, facilitating remote manipulation and heating capabilities. Semiconductor nanoparticles (e.g., TiO_2, ZnO, quantum dots) contribute photocatalytic and optoelectronic properties, including light absorption and luminescence. Additionally, polymeric nanoparticles can enhance mechanical strength or introduce specific functional groups for molecular recognition or stimuli-responsiveness (Shim et al., 2024b).

The selection of appropriate nanoparticles depends on the intended application and desired functionality. Gold and silver nanoparticles are frequently employed for their excellent antimicrobial properties and surface plasmon resonance effects, making them valuable for biosensing and therapeutic applications. Platinum nanoparticles exhibit superior catalytic activity, particularly for oxidation and reduction reactions, positioning them as excellent candidates for chemical synthesis and environmental remediation. Magnetic nanoparticles enable the development of remotely controllable systems for targeted drug delivery or magnetic separation processes. Quantum dots and semiconductor nanoparticles provide tunable optical properties for imaging, sensing, and photocatalysis (Bertolacci et al., 2020).

The methods for incorporating nanoparticles into hydrogels fall into two main categories: in-situ synthesis and ex-situ incorporation. In-situ approaches involve the formation of nanoparticles directly within the hydrogel matrix, typically through reduction of metal precursors or precipitation reactions. This method often results in uniform nanoparticle distribution and strong interactions between the particles and polymer network. Ex-situ methods involve the separate synthesis of nanoparticles followed by their incorporation into the hydrogel during or after gelation. This approach offers greater control over nanoparticle size, shape, and surface properties, but may result in less homogeneous distribution or weaker interactions with the hydrogel matrix (Bertolacci et al., 2020).

The resulting hydrogel-nanoparticle nanocomposites find applications across diverse fields. In healthcare, these materials serve as advanced platforms for controlled drug delivery, tissue engineering scaffolds, wound dressings with antimicrobial properties, and diagnostic sensors (Shim et al., 2024b). Environmental applications include water purification systems, where catalytic nanoparticles embedded in hydrogels degrade pollutants while the hydrogel matrix ensures easy catalyst recovery and reuse (Dadashi et al., 2022). In electronics and energy storage, conductive hydrogel-nanoparticle composites function as flexible electrodes, supercapacitors, and components of wearable devices (Bertolacci et al., 2020). Chemical synthesis applications leverage the catalytic properties of embedded nanoparticles for selective transformations, with the hydrogel matrix facilitating catalyst recovery and product separation (Dadashi et al., 2022).

Figure 2. Schematic depiction of the detection mechanism: embedding platinum nanoparticles (PtNPs) within nanocomposites (center) ensures their stability in complex environmental settings (such as seawater) while maintaining their catalytic capabilities. As a result, PtNPs can function as nanozymes to catalyze the chromogenic reaction of TMB and act as a sensor for Hg^{2+} contamination in water. Trace levels of Hg^{2+} suppress the chromogenic reaction by forming an amalgam with PtNPs, passivating their surface.

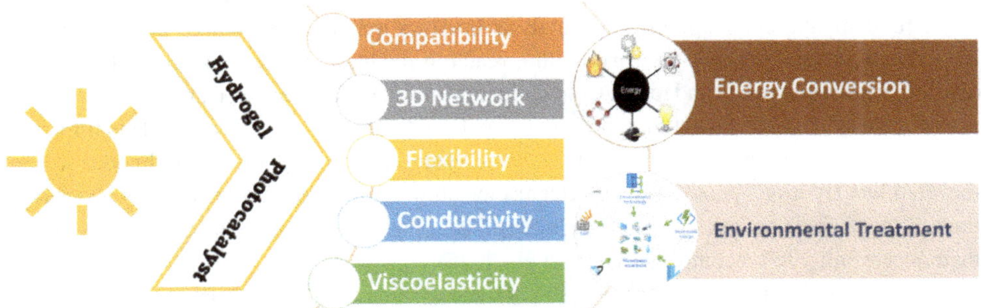

A particularly notable example of hydrogel-nanoparticle synergy is illustrated in Figure 13.2, which depicts platinum nanoparticles (PtNPs) embedded within a hydrogel nanocomposite for mercury ion (Hg^{2+}) detection in complex aqueous environments (Bertolacci et al., 2020). In this system, the hydrogel matrix provides stability and protection for the platinum nanoparticles, allowing them to maintain their nanozyme activity even in challenging media such as seawater. The PtNPs catalyze the chromogenic reaction of 3,3',5,5'-tetramethylbenzidine (TMB), producing a visible color change. When mercury ions are present in the water sample, they form an amalgam with the platinum surface, effectively passivating the catalytic sites and inhibiting the chromogenic reaction. This inhibition results in a concentration-dependent reduction in color intensity, enabling quantitative detection of mercury contamination. The hydrogel matrix is crucial for this application, as it prevents nanoparticle aggregation while allowing diffusion of both the analyte (Hg^{2+}) and substrate (TMB), maintaining the system's sensitivity and selectivity even in complex environmental matrices (Bertolacci et al., 2020).

The development of hydrogel-nanoparticle composites continues to evolve, with current research focusing on enhancing the interface between organic and inorganic components, developing more sophisticated stimuli-responsive behaviors, and improving the long-term stability of these materials. Advanced characterization techniques, including in-situ spectroscopy and microscopy methods, provide deeper insights into the nanoscale interactions governing the performance of these composite materials. These fundamental studies are essential for the rational design of next-generation hydrogel-nanoparticle systems with optimized properties and functionalities for specific applications (Bertolacci et al., 2020).

13.3 PHOTOCATALYTIC APPLICATIONS OF HYDROGELS AND HYDROGEL-BASED COMPOSITES

Integrating photocatalytic materials within hydrogel matrices represents a significant advancement in addressing the dual challenges of sustainable energy generation and environmental remediation. Photocatalysts have attracted substantial research interest due to their ability to harness light energy to drive chemical transformations, potentially addressing both energy requirements and environmental pollution simultaneously (T. Sandu et al., 2025). However, the practical implementation of photocatalytic systems, particularly those employing nanoscale materials, faces several challenges that hydrogel-based approaches can effectively address.

A fundamental challenge in photocatalysis involves maintaining optimal contact between photocatalytic particles and the surrounding reaction medium while simultaneously enabling efficient catalyst recovery and reuse. Conventional nanoscale photocatalysts, while offering high surface area and enhanced activity, present separation difficulties post-reaction. This limitation can lead to secondary pollution through catalyst loss and reduced economic viability due to poor recyclability. Hydrogel photocatalysts offer an elegant solution to these challenges by providing a three-dimensional network structure that immobilizes photocatalytic materials while maintaining their accessibility to reactants (T. Sandu et al., 2025).

As illustrated in Figure 13.3, hydrogel photocatalysts facilitate both energy conversion and environmental treatment processes through their unique structural and functional properties. The porous nature of hydrogels enables efficient mass transport of reactants and products, while their high-water content provides an ideal medium for aqueous-phase reactions. The polymer network stabilizes photocatalytic particles, preventing aggregation and enhancing their long-term performance. The transparent or translucent nature of many hydrogels allows effective light penetration, ensuring activation of embedded photocatalysts throughout the material volume. Characterization via SEM and porosimetry often reveals pore diameters of 10–100 μm and swelling ratios of 500–1,300% (wet/dry) (Y. Yue et al., 2020). Visible-range transmittance typically exceeds 95% (400–800 nm), and elastic moduli in the swollen state span 10^4–10^7 Pa depending on crosslink density (Y. Yue et al., 2020). These parameters critically influence both light delivery and pollutant ingress.

Figure 3. Hydrogel photocatalysts as an effective strategy for energy conversion and environmental treatment

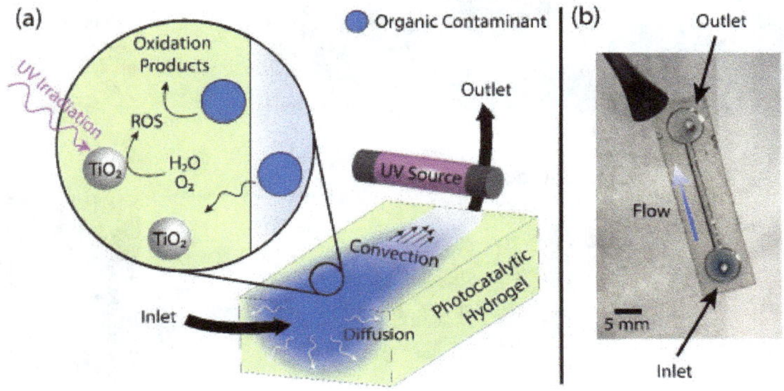

The performance of hydrogel photocatalysts depends on several critical factors, including hydrogel composition, crosslinking density, photocatalyst loading, light penetration depth, and mass transport characteristics. These parameters can be systematically tuned to optimize catalytic efficiency for specific applications. For instance, hydrogels with lower crosslinking density typically exhibit higher swelling ratios and consequently improved mass transport, facilitating the diffusion of reactants to catalytic sites. However, excessive swelling may dilute photocatalyst concentration and reduce mechanical stability. Similarly, higher photocatalyst loading generally enhances catalytic activity but may compromise light penetration and mechanical properties. Therefore, careful balancing of these parameters is essential for designing efficient hydrogel photocatalytic systems (Hernandez Monroy et al., 2025).

13.3.1 Harnessing Photocatalytic Hydrogels to Eliminate Contaminants From Water Solutions

Water pollution by recalcitrant organic contaminants presents a significant environmental and public health challenge worldwide. Municipal and industrial wastewater treatment facilities play a crucial role in managing these pollutants, yet conventional treatment methods often prove inadequate for complete removal or degradation of certain persistent compounds. Organic contaminants of particular concern include synthetic dyes, pharmaceuticals, personal care products, pesticides, and endocrine-disrupting compounds, many of which resist biodegradation and conventional physicochemical treatment approaches (Ren et al., 2021).

Figure 4. Multiple methods for the elimination or reduction of inorganic/organic pollutants.

(a)

(b)

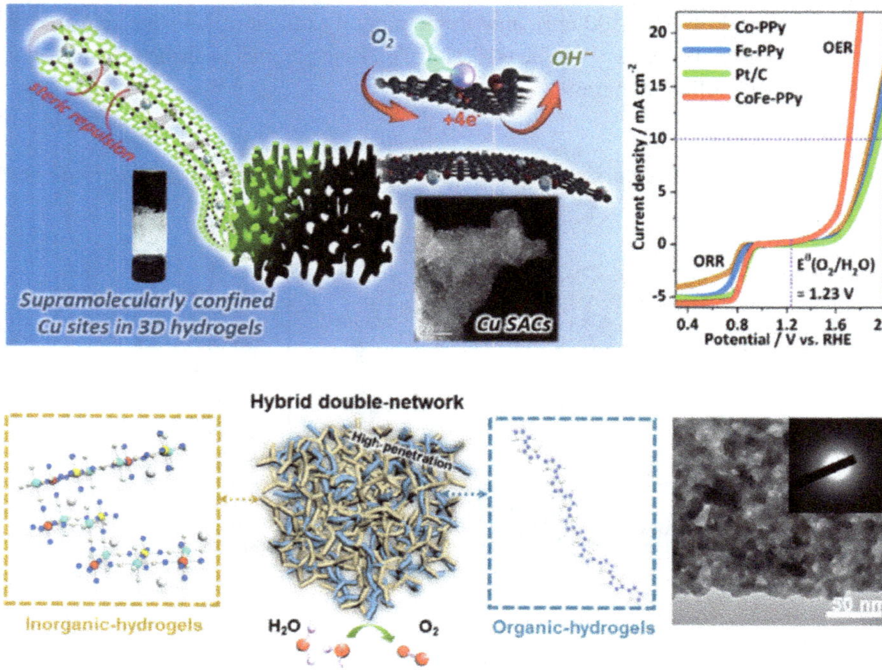

Due to their low environmental concentrations, these contaminants, often classified as trace organic compounds (TrOCs), can persist through traditional water treatment processes and consequently appear in treated wastewater, natural water bodies, and even drinking water. Their persistence raises significant ecological and human health concerns, as many TrOCs exhibit bioactive properties even at extremely low concentrations. This situation has intensified research efforts to develop advanced water treatment technologies capable of efficiently removing or degrading these recalcitrant compounds (Ren et al., 2021).

Photocatalytic systems have emerged as promising solutions for addressing these persistent pollutants through advanced oxidation processes. When semiconductor photocatalysts absorb light with energy equal to or greater than their band gap, electron-hole pairs are generated on or near their surfaces. These charge carriers can then participate in redox reactions with water and dissolved oxygen, producing reactive oxygen species (ROS) such as hydrogen peroxide (H_2O_2), singlet oxygen ($^1O_2$), hydroxyl radicals (OH·), and superoxide anions (O_2^-). These highly reactive species can attack organic molecules through various oxidation pathways, ultimately leading to their mineralization into harmless products like carbon dioxide and water (Kumar et al., 2025). Kinetic analysis generally follows pseudo-first-order decay with activation energies of 10–40 kJ/mol (e.g. 13.8 kJ/mol for phenol on TiO_2) and time-resolved spectroscopy shows electron-hole recombination on ns–μs scales versus pollutant oxidation on ms–s scales (Li et al., 2020).

Recent research has focused on designing photocatalysts with enhanced light absorption properties and tunable band gaps using earth-abundant materials. Common inorganic semiconductors employed as photocatalytic nanoparticles include titanium dioxide (TiO_2), cadmium sulfide (CdS), tungsten trioxide (WO_3), tin sulfide (SnS), and zinc oxide (ZnO). Among these, TiO_2 has received the most intensive investigation due to its photochemical stability, high activity, controllable particle size distribution, and economic viability. Comparative tests under identical UV illumination have shown TiO_2 achieving apparent first-order rate constants roughly 2× greater than ZnO—for example, 0.045 min^{-1} vs. 0.023 min^{-1} degradation of methylene blue at 20 mg/L—while ZnO still removes >85% of dye in 60 min, offering lower cost and faster electron mobility (200–300 $cm^2/V{\cdot}s$) (Abou Zeid & Leprince-Wang, 2024; A. J. Clasky et al., 2021). However, TiO_2 and many other efficient photocatalysts present challenges related to catalyst recovery and recycling when used as free nanoparticles in water treatment applications (Lee et al., 2023).

Incorporating photocatalytic nanoparticles within hydrogel matrices represents a strategic approach to overcome these limitations. Recent studies have investigated TiO_2-hydrogel composites for the removal of organic dye molecules from aqueous solutions through combined adsorption and photocatalytic degradation mechanisms. These processes are interdependent and significantly influenced by the structural and chemical properties of the photocatalytic hydrogel (Hernandez Monroy et al., 2025).

The efficiency of contaminant removal depends on various material characteristics, including the chemical composition of the hydrogel (e.g., acrylic acid content and crosslinking degree), which determines the solubility and diffusivity of solute species within the gel matrix. Additional influential factors include the hydrogel's swelling behavior, ionic strength, TiO_2 loading, light transmission characteristics, and potential nanoparticle release dynamics. These parameters interact in complex ways to determine the overall performance of the photocatalytic hydrogel system (Katzenberg et al., 2020).

Figure 13.5 illustrates the operational principle of a photocatalytic hydrogel reactor designed for water purification. In this system, contaminated water containing organic pollutants (exemplified by methylene blue) flows through a channel embedded within a TiO_2-loaded hydrogel. The pollutant molecules diffuse into the hydrogel matrix, where they encounter reactive oxygen species generated by UV-activated TiO_2 nanoparticles, resulting in their degradation. This configuration offers several advantages: the flow-through design ensures continuous operation, the hydrogel matrix prevents catalyst loss, and the intimate contact between pollutants and photocatalyst enhances degradation efficiency.

Figure 5. Operational principle of a photocatalytic hydrogel reactor for water purification. (a) Organic molecules (e.g., methylene blue) were introduced as solutes in water flowing through a small channel embedded in a photocatalyst-loaded hydrogel. In theory, methylene blue would permeate the gel and be broken down by reactive oxygen species (ROS) produced by UV-excited TiO_2 within the hydrogel. (b) A photocatalytic hydrogel reactor after two hours of operation.

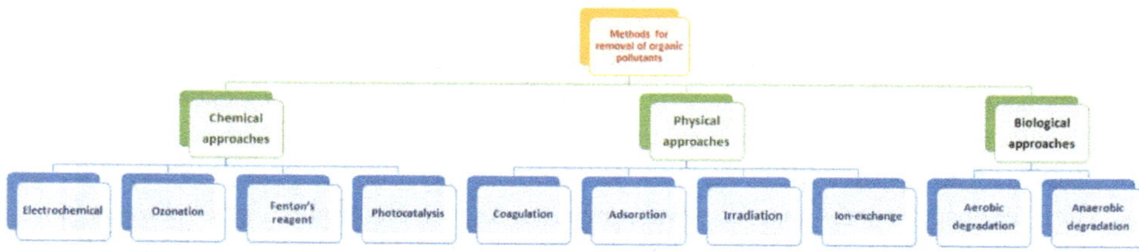

Beyond traditional metal oxide photocatalysts, recent innovations have explored organic photocatalytic units integrated directly into the hydrogel polymer network. Kuckhoff and colleagues developed highly transparent photocatalytic hydrogels incorporating benzothiadiazole acrylamide monomers with photocatalytic properties. These hydrogels were synthesized through radical copolymerization with water-compatible N,N-dimethylacrylamide, creating a system where only the photocatalytic component absorbs visible light. This approach offers several advantages: the transparent nature of the polymer network maximizes light utilization, the covalent attachment of photocatalytic units prevents leaching, and the hydrophilic hydrogel structure facilitates pollutant diffusion to active sites (Kuckhoff et al., 2021). However, one potential drawback is that organic chromophores like benzothiadiazole may undergo photobleaching or photooxidative degradation under prolonged illumination, which can reduce long-term activity (Cigáň et al., 2010).

In real wastewater, light attenuation by suspended solids and colored dissolved organics can reduce incident UV–vis intensity to <1% within a few centimeters, and diffusion lengths for bulky organics may be limited by pore radii; this 'self-screening' effect has been quantified in model turbid media (sand:water 1:10 w/v) demonstrating a 90% drop in 365 nm intensity in 5 cm (Teodor Sandu et al., 2025; Shim et al., 2024a). These photocatalytic hydrogels demonstrated impressive versatility in environmental remediation, degrading both organic pollutants and heavy metals. For example, glyphosate (N-(phosphonomethyl)glycine), a widely used herbicide, was effectively photodegraded. In a half-liter batch under visible light, the hydrogel achieved substantial glyphosate conversion, demonstrating practical potential (Kuckhoff et al., 2021). Importantly, time-resolved ^31P NMR showed that glyphosate concentration decayed nearly linearly (pseudo–first-order kinetics). The NMR spectra revealed phosphate (2.5 ppm) as the main product and an intermediate at ~19.3 ppm, identified as (aminomethyl)phosphonic acid (AMPA) (Kuckhoff et al., 2021). This confirms the C–P bond cleavage pathway in glyphosate degradation and highlights that AMPA (a persistent, potentially toxic metabolite) is formed during the process.

Heavy metal remediation represents another critical application area for photocatalytic hydrogels. Heavy metals like chromium (VI) are prevalent pollutants in industrial wastewater, presenting serious environmental and health risks due to their non-biodegradability, high toxicity, and carcinogenic properties. The high-transparency photocatalytic hydrogels developed by Kuckhoff et al. effectively photoreduced toxic Cr (VI) to the less harmful Cr (III) under aqueous conditions.

Figure 6. Schematic representation of Cr (VI) photoreduction by hydrogel-supported photocatalysts under visible or blue light. Reduction leads to the precipitation of Cr(III), facilitating removal from aqueous systems

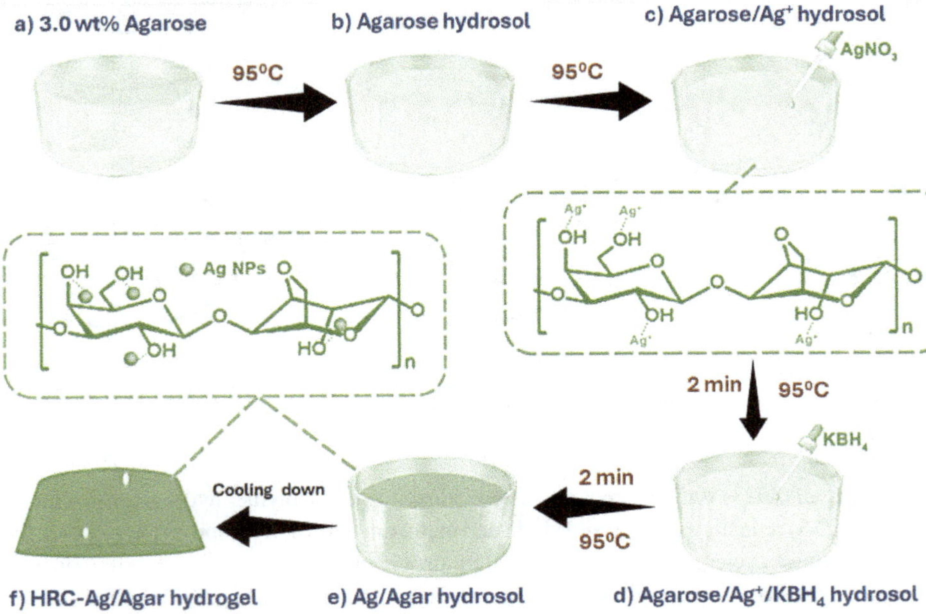

The water-soluble Cr (VI) was converted to water-insoluble Cr (III), facilitating easy precipitation and separation from the treated water. This photoreduction process, monitored through complexation with diphenylcarbazide and UV/Vis spectroscopy, achieved over 92% conversion within approximately 60 minutes under blue light illumination. Control experiments without light exposure or without the photocatalyst showed no significant conversion, confirming the photocatalytic nature of the process (Kuckhoff et al., 2021).

Specifically, Cr(VI) remained 0% reduced in dark or catalyst-free controls, whereas photocatalytic runs achieved >92% reduction in 60 min; dye adsorption alone removed <10% of methylene blue vs. >95% under UV with catalyst (Kuckhoff et al., 2021; Yiying Yue et al., 2020). In line with these findings, related photocatalytic hydrogels exhibit excellent recyclability under operational conditions. Yue et al. (2020) showed that a TEMPO-chitin/polyacrylamide hydrogel loaded with TiO_2 nanoparticles removed ~97.3% of methyl orange under UV in 90 min—while adsorption alone removed only ~8.5% in 150 min—and retained ≈ 96% of that removal efficiency after 10 reuse cycles, demonstrating robust activity with negligible loss over repeated runs (Yiying Yue et al., 2020).

Figure 7. Photocatalytic degradation mechanism of methyl orange using a TiO_2-loaded chitin/polyacrylamide hydrogel under UV irradiation. The network structure promotes efficient dye breakdown and recyclability

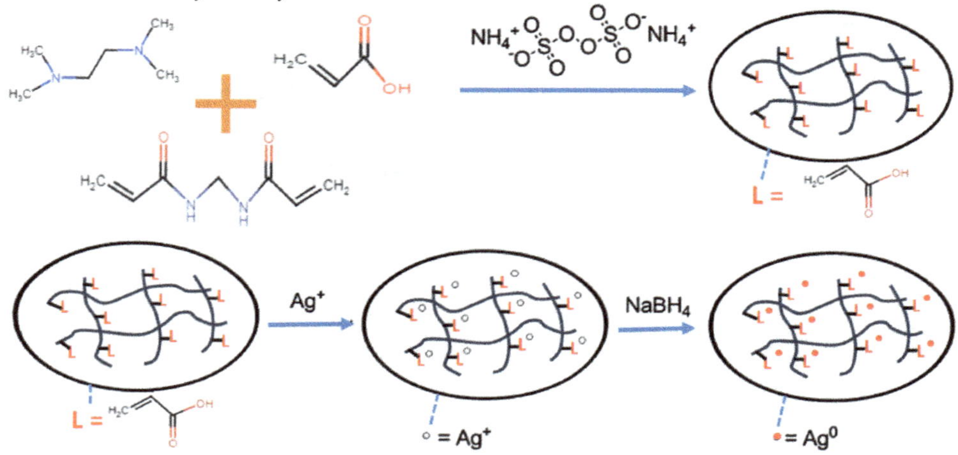

(You et al., 2021) Copyright 2021 Springer Nature Creative Commons Licenses.

Zhang et al. (2023) similarly reported a benzothiadiazole-copolymer hydrogel that achieved complete Cr(VI) reduction under visible light and maintained full conversion efficiency over five consecutive reuse cycles; control tests in the dark or without the hydrogel gave zero reduction, highlighting the strict photocatalytic mechanism (Kuckhoff et al., 2021). A ZnO/natural-rubber–PAA hydrogel tested by Sabuad et al. (2024) removed ~97% of methylene blue via combined adsorption and photodegradation, and still degraded >80% of the dye after 10 reuse cycles under solar illumination (Sabuad et al., 2024). Together, these benchmarks underscore the durability and practical reusability of hydrogel-supported photocatalysts.

Comparative performance analysis revealed that these photocatalytic hydrogels demonstrated superior efficiency compared to fully conjugated microporous polyazulene photocatalysts, requiring significantly less active material to achieve comparable results. This efficiency advantage can be attributed to the highly accessible photocatalytic sites within the hydrogel's hydrophilic matrix, which facilitates intimate contact between the catalyst and aqueous-phase pollutants. Additionally, when compared with conventional metal oxide photocatalysts like TiO_2 and ZnO, these organic photocatalytic hydrogels offer advantages in terms of visible light utilization and integration into existing water treatment infrastructure (Kuckhoff et al., 2021).

Despite these promising results, several challenges remain. Photocatalytic hydrogels must still overcome light penetration limits (especially in turbid water), diffusion constraints for large molecules, and mechanical stability under prolonged use. The formation of toxic byproducts during incomplete degradation is also a concern. For example, incomplete oxidation of glyphosate produces AMPA (as noted above) (Kuckhoff et al., 2021). Similarly, incomplete Cr (VI) reduction would leave residual Cr (VI) (highly toxic) instead of benign Cr (III). Ensuring complete mineralization to innocuous end-products (CO_2, water, PO_4^{3-}, etc.) is critical; kinetic analyses and intermediate identification (via NMR or GC–MS) are needed to verify complete degradation. Finally, the long-term photostability of the catalyst (to prevent photobleaching) must be evaluated under extended illumination. Addressing these challenges requires interdisciplinary approaches combining materials science, photochemistry, reaction engineering, and toxicology to develop next-generation photocatalytic hydrogels with enhanced performance and practical applicability (Beil et al., 2024).

Integrating computational modeling with experimental studies represents another promising avenue for advancing photocatalytic hydrogel technologies. Multiphysics simulations incorporating light transport, reaction kinetics, and mass transfer phenomena can provide valuable insights for optimizing material composition and reactor configuration. Additionally, artificial intelligence and machine learning approaches offer potential for accelerating materials discovery and process optimization in this field (Melchionna & Fornasiero, 2020).

13.3.2 Catalytic Silver–Poly(acrylic acid) Hydrogel Composites for Redox transformations

Silver nanoparticles (Ag NPs) embedded in poly(acrylic acid) (PAA) hydrogels form efficient heterogeneous catalysts for redox transformations. The PAA network (rich in –COOH groups) binds Ag^+ ions and allows in situ reduction to Ag^0, yielding well-dispersed NPs (Berraho & Martín Andreu, 2025; Zhang et al., 2023). For example, Zhang et al. (2023) observed that the carboxyl and amide groups in a PAA-based hydrogel "provide abundant active sites for complexing silver ions, facilitating the in situ reduction and confinement of silver nanoparticles"(Zhang et al., 2023). In practice this stabilization prevents NP agglomeration: Berraho & Martín Andreu (2025) showed that Ag/PAA composites are thermally stable above 100 °C and "prevent aggregation and oxidation" of Ag NPs(Berraho & Martín Andreu, 2025). Dadashi et al. (2022) noted that silver-hydrogel catalysts combine "high catalytic ability, reusability, easy work-up, and simple synthesis" (Dadashi et al., 2022).

Figure 8. A one-pot UV polymerization can prepare Ag-NP/PAA hydrogel composites. Part (a) shows liquid-phase exfoliation of graphite into multilayer graphene platelets (MLGPs), and part (b) depicts mixing acrylic acid (AA), crosslinker (MBAA), and MLGPs (with $AgNO_3$), followed by UV irradiation to form the PAA hydrogel with embedded Ag^0 nanoparticles

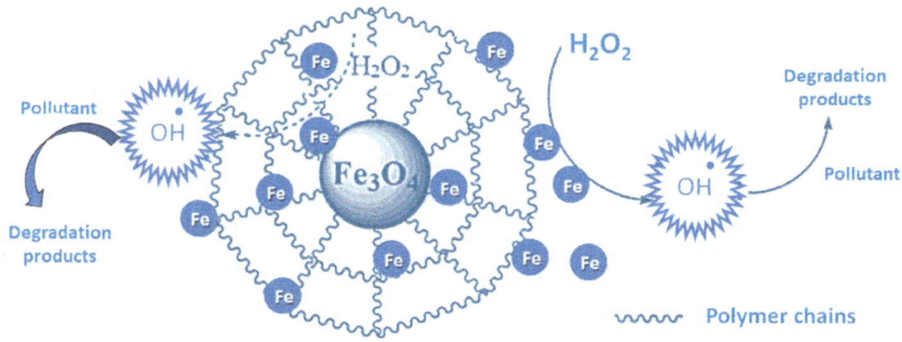

A) Reduction of nitroarenes

A canonical test of Ag/PAA catalysts is the $NaBH_4$-mediated reduction of 4-nitrophenol (4-NP) to 4-aminophenol. In a recent study, Zhang et al. achieved complete conversion of 5 mM 4-NP in just 113 s (rate constant $k \approx 4.06 \times 10^{-2}\ s^{-1}$) under ambient conditions (Zhang et al., 2023). Notably, >99% conversion was attained even in tap or seawater, and >98% of the activity was retained after eight successive runs. Earlier work by Heidari et al. (2018) similarly showed that Ag/PAA gels, with $NaBH_4$ present, rapidly degrade model pollutants (e.g. methyl orange, rhodamine B, methylene blue) in minutes (Heidari et al., 2018). These results confirm that Ag/PAA hydrogels serve as highly active reduction catalysts, effectively treating nitroaromatics and related pollutants with excellent turnover and recyclability.

Figure 9. Schematic mechanism for $NaBH_4$-assisted reduction of 4-nitrophenol on Ag nanoparticles (green arrows indicate electron transfer) and concurrent reduction of methylene blue (MB) to leucomethylene blue. Ag NPs adsorb BH_4^- and the nitrophenolate, mediating sequential $2e^-/2H^+$ steps.

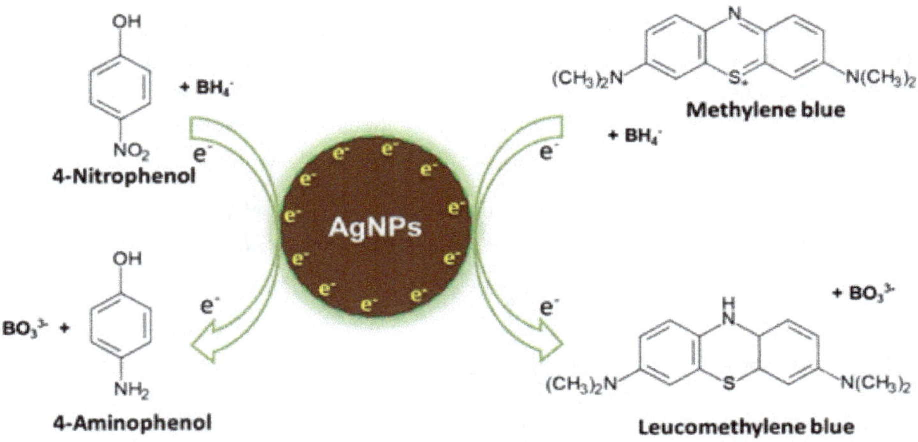

B) Oxidation of alcohols

Ag/PAA catalysts are also active for aerobic oxidation of alcohols, e.g., benzyl alcohol → benzaldehyde. Rajabi et al. (2022) demonstrated that nitroxide-modified Ag NPs oxidize benzyl alcohol under mild conditions using O_2 as oxidant, yielding aldehydes in near-quantitative yield and ~100% selectivity. (Krogul-Sobczak et al., 2022). In their system, the nitroxyl-radical functionality promotes O_2 activation, but the key point is that Ag NPs can drive the conversion of benzylic alcohols to benzaldehyde with high efficiency. These findings align with general trends in Ag catalysis – simple base-free protocols or base-assisted O_2 oxidation can achieve high benzaldehyde selectivity. Thus, Ag/PAA composites can serve as eco-friendly oxidation catalysts, complementing the reductive applications above.

In the presence of molecular oxygen, Ag/PAA catalysts can dehydrogenate benzyl alcohol ($PhCH_2OH$) to benzaldehyde (PhCHO) with water as the only by-product. Mechanistically, silver surfaces activate O_2 to form surface oxygen species (e.g., Ag–O or Ag–O_2^-) that abstract hydrogen from the benzylic C–H. The reaction is essentially an oxidative dehydrogenation: one H atom (as H· or H^-) is transferred to O_2 on Ag, yielding benzaldehyde and H_2O. Notably, the carboxylate groups of the PAA matrix help disperse and stabilize the Ag NPs, and may provide acid–base sites to assist proton transfers. An "oxygen spillover" effect has been invoked in supported systems: activated O atoms migrate from Ag to the support, facilitating hydrogen removal from the alcohol. In practice, the oxidation is highly selective to benzaldehyde (negligible benzoic acid byproduct) under mild conditions. Golub et al. review that Ag catalysts exhibit near-100% selectivity in benzyl alcohol oxidation due to the low activation barrier for dehydrogenation on Ag(111) compared to further oxidation (Torbina et al., 2018).

Ag/PAA catalysts quantitatively give high benzyl alcohol conversions with excellent selectivity. Gorey et al. demonstrated that their nitroxide-grafted Ag NPs in water gave 96% benzaldehyde yield (TON ≈114) under an oxygen balloon in just one cycle (Under air, conversion was slightly lower but still ~90+%.) Even after multiple uses, the Ag/PAA system retained substantial activity: for example, yields remained above 50% after seven cycles. These figures align with other reports of supported Ag oxidations; typical turnover frequencies are on the order of 10–100 h^{-1} at 60–80 °C, with benzoic acid negligible under controlled conditions(Torbina et al., 2018). In sum, Ag/PAA catalysts drive benzyl alcohol → benzaldehyde with turnovers in the 10^2 range and selectivity ~100%, demonstrating clear utility for green alcohol oxidation.

Figure 10. Benzaldehyde yields over consecutive oxidation cycles with Ag-based catalysts. The initial cycle reaches ~95–100% benzaldehyde yield, confirming the high single-pass conversion.

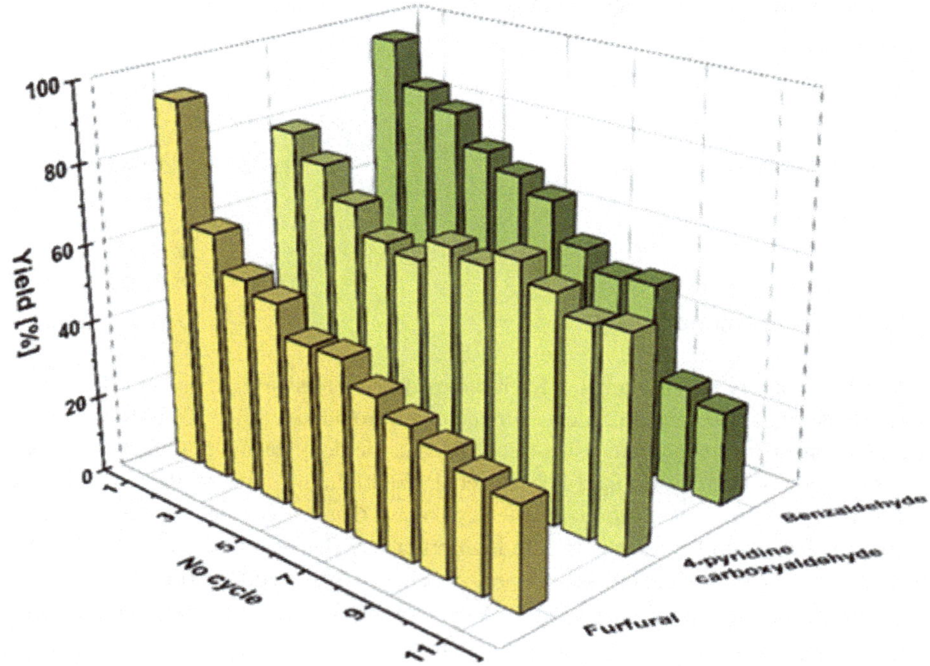

Adapted from (Krogul-Sobczak et al., 2022) Copyright 2022 Springer Nature Creative Commons Licenses.

13.3.3 Hydrogel-Coated Fe_3O_4 Nanoparticles as Heterogeneous Fenton Catalysts for Phenol Degradation

Phenol is a high-production aromatic chemical (global output on the order of 10^7 t/yr) primarily synthesized by the cumene process from benzene and propylene. It enters wastewater from numerous industries (e.g., petrochemical, resin, pharmaceuticals) and is highly persistent and toxic (Tutić et al., 2025). Due to its potent biological effects, the U.S. EPA classifies phenol as a priority pollutant. In water, phenol exists primarily in its non-ionized form (pK_A ≈9.99) and can cause severe irritation to skin, eyes, and mucous membranes. Prolonged or high-level exposure leads to systemic toxicity (lethal oral dose for humans ≈50–500 mg/kg) (Tutić et al., 2025). Elevated phenol levels have been documented worldwide in industrial effluents (plastics, dyes, coking, etc.) and surface waters, posing significant environmental and health risks. Thus, effective removal of phenol from wastewater remains an urgent challenge.

The homogeneous Fenton process (Fe^{2+}/H_2O_2) generates •OH radicals that can mineralize phenol but suffer critical drawbacks. In practice, optimal degradation requires strongly acidic conditions (pH~2–4) to keep iron soluble (B. Bouzayani & M. Á. Sanromán, 2024). Even slight increases in pH cause Fe $(OH)_3$ precipitation and copious iron sludge, creating secondary waste and operational complexity. Moreover, the reaction consumes large quantities of H_2O_2 and iron (i.e., high reagent demand), further limiting its sustainability. Traditional Fenton is efficient but costly and impractical for large-scale treatment without extensive pH control and sludge handling. These limitations have spurred interest in heterogeneous Fenton catalysts that achieve similar •OH production under milder, more practical conditions.

Magnetite (Fe_3O_4) nanoparticles are widely studied as heterogeneous Fenton catalysts because they provide surface Fe^{2+}/Fe^{3+} sites for H_2O_2 activation and can be easily recovered by magnetic separation (Shen et al., 2019). An ideal Fe_3O_4-based catalyst would match the high activity of homogeneous Fenton while operating at a broader pH and exhibiting minimal iron leaching. Encapsulating Fe_3O_4 within hydrophilic polymer networks (hydrogels) markedly improves catalytic performance and stability. Natural and synthetic hydrogels (e.g. cellulose derivatives, chitosan, alginate, carboxymethyl cellulose) offer a 3D water-swollen matrix that disperses magnetite nanoparticles and enhances mass transfer. The polymer swelling is often pH-responsive: as the hydrogel network swells, it increases porosity and allows pollutant molecules to more readily contact the catalyst (Shen et al., 2019).

Figure 11. Fenton-like catalytic mechanism of polymer-coated Fe_3O_4 nanoparticles. The Fe_3O_4 core is shown at center, surrounded by polymeric hydrogel chains that contain Fe^{2+}/Fe^{3+} sites. Hydrogen peroxide diffuses through the polymer network and reacts at Fe sites to generate •OH radicals, which then oxidize nearby pollutant molecules. The polymer coating provides hydrophilic channels and functional groups (–COOH, etc.) that promote H_2O_2 diffusion, pollutant adsorption, and retention of iron on the particle surface

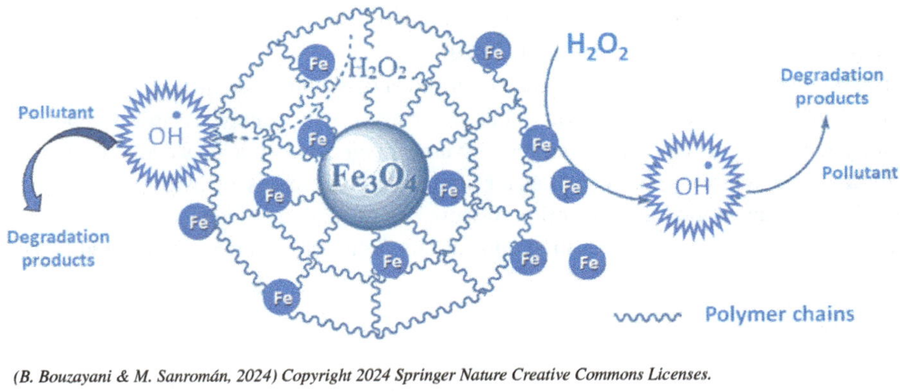

(B. Bouzayani & M. Sanromán, 2024) Copyright 2024 Springer Nature Creative Commons Licenses.

In practice, Shen et al. (2019) coated Fe_3O_4 with a carboxymethyl cellulose/poly(acrylic acid–acrylamide) hydrogel. The resulting Fe_3O_4@hydrogel particles (~10 nm) remained well-dispersed and magnetically recoverable. Crucially, the hydrogel network provided p^H-sensitive swelling and high hydrophilicity, which enhanced phenol uptake and reaction kinetics. Under optimal conditions, Shen et al. reported ~98.2% phenol removal and ~80.4% COD reduction in 180 min using Fe_3O_4@hydrogel. Even at pH 5 (higher than classical Fenton), the catalyst achieved ≈50% COD removal reflecting improved activity at milder pH (Shen et al., 2019). Other studies similarly highlight this synergistic effect of hydrogel supports: Chen et al. (2019) grew Fe_3O_4 on a cellulose/graphene-oxide hydrogel, achieving ~98% removal of an organic dye over 20 cycles (Chen et al., 2019). The graphene oxide helped maintain the Fe^{2+}/Fe^{3+} balance on the surface, while the cellulose matrix prevented aggregation. Notably, Fang et al. (2022) showed that adding Fe_3O_4 to an agar/κ-carrageenan hydrogel containing phenol-degrading bacteria increased the phenol oxidation rate by over 10%, owing to Fe_3O_4-induced •OH generation and improved mass/electron transfer (Fang et al., 2022). These examples illustrate that hydrogels not only stabilize the magnetite but also actively contribute to faster pollutant oxidation and catalyst reuse.

13.4 ELECTROCATALYTIC PROPERTIES OF HYDROGELS

Electrocatalytic hydrogels represent an innovative class of functional materials that combine the structural attributes of hydrogels with electrochemical catalytic capabilities. These advanced materials integrate the three-dimensional polymer network structure of hydrogels—capable of retaining substantial amounts of water or electrolyte solutions—with catalytically active components that facilitate electron transfer processes at electrochemical interfaces. This synergistic combination creates multifunctional materials with significant potential for applications in energy conversion and storage, environmental remediation, sensing, and beyond (Fang et al., 2020).

The fundamental architecture of electrocatalytic hydrogels typically involves a hydrophilic polymer network that provides structural support and a hydrated environment, combined with catalytically active sites that promote specific electrochemical reactions. These catalytic centers can comprise various entities, including metal nanoparticles, metal-organic complexes, conductive polymers, carbon nanomaterials, or combinations thereof. The hydrogel component offers several advantages for electrochemical applications: it facilitates ion transport through its water-filled channels, provides a high surface area for catalytic reactions, stabilizes catalytic species against aggregation or detachment, and can be engineered with specific functional groups to enhance catalyst-substrate interactions (Fang et al., 2020).

The porous structure of hydrogels enables efficient mass transport of reactants and products to and from catalytic sites. At the same time, their hydrophilic nature ensures excellent compatibility with aqueous electrolytes commonly used in electrochemical systems. Furthermore, the mechanical flexibility and potential for stimuli-responsiveness of hydrogel matrices add dimensions of adaptability and tunability to these electrocatalytic systems, enabling operation under diverse conditions and integration into various device architectures (Fang et al., 2020).

13.4.1 Applications of Electrocatalyst Hydrogels

Electrocatalysis, which involves using electricity to lower energy barriers and reduce activation energy for chemical reactions, has become increasingly important in addressing global challenges related to energy security, climate change, and sustainable development. As conventional fossil fuel resources diminish and the impacts of global warming intensify, efficient electrocatalytic processes offer promising pathways for developing alternative energy systems and mitigating environmental impacts.

Hydrogel materials, with their diverse physicochemical properties and structural advantages, have attracted substantial research interest for their potential roles in various electrocatalytic reactions, including the oxygen reduction reaction (ORR), oxygen evolution reaction (OER), hydrogen evolution reaction (HER), nitrogen reduction reaction (NRR), and carbon dioxide reduction reaction (CO_2RR). The high compositional flexibility and ability to uniformly distribute metal species make hydrogels and their derivatives particularly valuable as precursors or active materials in electrocatalysis (Fang et al., 2020).

In recent years, significant research efforts have focused on developing noble-metal-free electrocatalysts as alternatives to traditional noble-metal-based systems, which face limitations related to high cost and scarcity. Within this context, catalysts composed of transition metals (particularly Fe, Co, and Ni) co-doped with nitrogen on carbon supports (M-N/C) have emerged as promising next-generation candidates, especially for the oxygen reduction reaction. The structural properties of hydrogels make them excellent precursors for synthesizing these catalysts, as the appropriate selection of crosslinkers and monomers enables uniform distribution and fixation of M-N-C active sites onto carbon frameworks. This approach provides a scalable pathway for producing single-atom catalysts with maximized atom utilization and catalytic efficiency (You et al., 2015).

Supramolecular hydrogels have demonstrated particular utility as precursors for synthesizing highly efficient electrocatalysts. For instance, researchers have successfully developed copper and cobalt-containing polypyrrole hydrogels, achieving a remarkable atom utilization rate of 96% for Cu(I)-N_2 sites in oxygen reduction reactions. The supramolecular confinement provided by the polymer backbones effectively prevents metal aggregation during the thermal treatment (calcination) process, preserving the integrity of the three-dimensional interconnected structures essential for efficient electrocatalysis (Meng et al., 2020).

Similarly, dual crosslinked polypyrrole (PPy) hydrogels have been employed as precursors to generate cobalt-iron sub nanoclusters evenly dispersed on nanotubular carbon structures. These Co/Fe-PPy derived catalysts demonstrate enhanced electrocatalytic activity in both oxygen evolution and oxygen reduction reactions compared to their monometallic counterparts. This superior performance can be attributed to the higher density of active sites and faster oxygen-binding kinetics facilitated by the bimetallic composition and the hierarchical structure inherited from the hydrogel precursor (Yang et al., 2017).

Beyond serving as precursors for carbon-based electrocatalysts, hydrogels also function effectively as templates for synthesizing alloy electrocatalysts with precisely controlled composition and crystallinity. For instance, Fang and colleagues developed bismuth-nickel (Bi-Ni) alloy electrocatalysts derived from Bi-Ni inorganic hydrogels specifically for the nitrogen reduction reaction. The amorphous Bi-Ni alloys inherit the structural characteristics of their hydrogel precursors, exhibiting disordered atomic arrangements and defect-rich surfaces that provide numerous active sites for nitrogen adsorption. Furthermore, the three-dimensional bimetallic framework functions as an effective conductive network, enhancing electronic and mass transport properties. These structural advantages translate to improved electrochemical surface area and enhanced stability during operation, resulting in superior electrocatalytic performance for nitrogen fixation processes with a yield rate of 17.5 μg h^{-1} mg_cat^{-1} and Faradaic efficiency of 13.8% (Fang et al., 2021).

Figure 12. Hydrogels utilized in the context of electrocatalysis. (b) Atomic Fe–N-C sites and Pt/C derived from hydrogels, employed for the electrocatalytic processes of oxygen evolution reaction (OER) and oxygen reduction reaction (ORR). (c) Hybrid double-network hydrogels exhibiting stability in acidic water oxidation for electrocatalytic applications. (e) A graphical representation illustrating the aforementioned concepts. (f) Fe-N-C synthesized on a gram scale. (g) The Fe-N-C catalytic material. (h) Co9S8NSC as an alternative catalyst. (i) Transmission electron microscopy (TEM) image depicting structural characteristics. (j, k) LSC measurements conducted at varying temperatures. (l) K-I plots demonstrating the relationship between current and potential

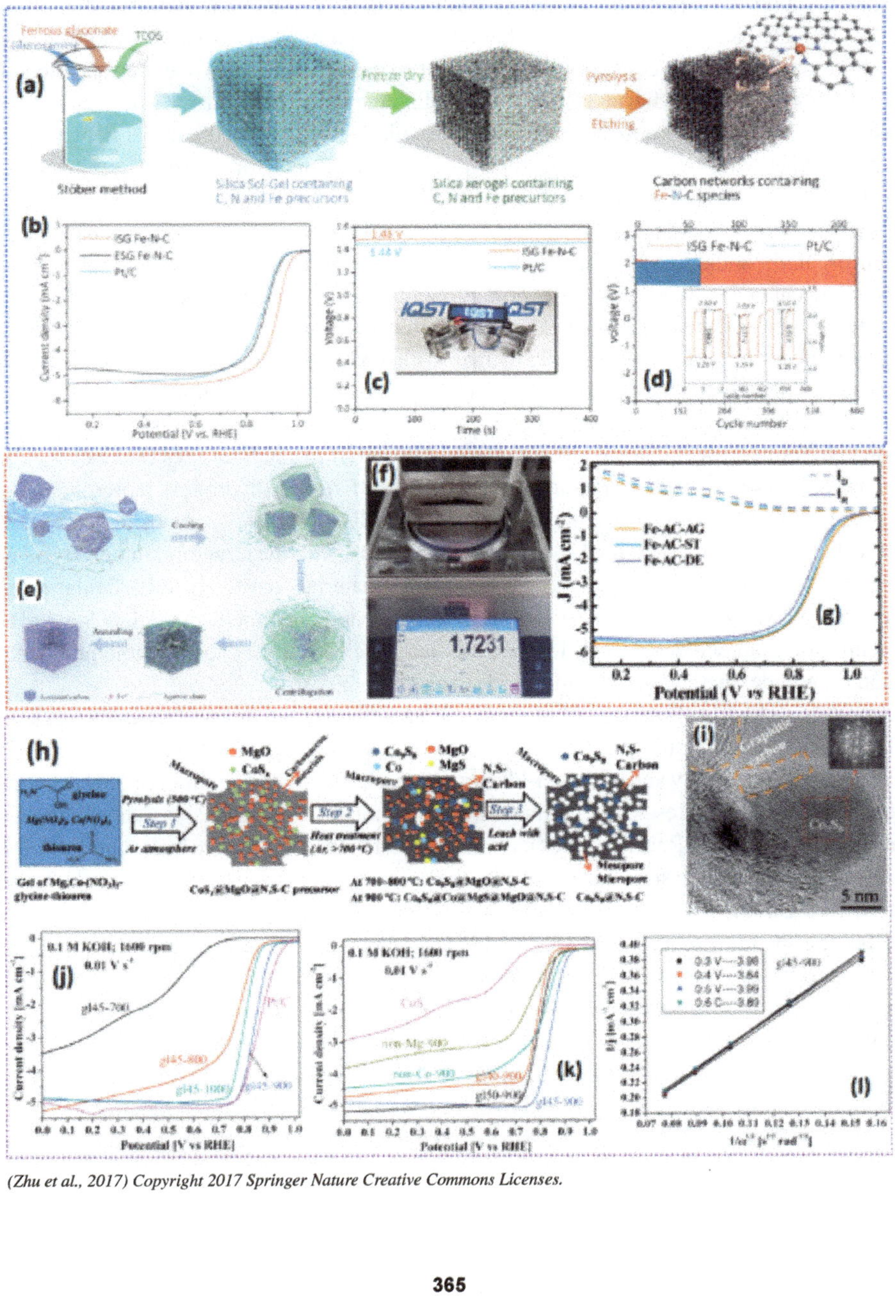

(Zhu et al., 2017) Copyright 2017 Springer Nature Creative Commons Licenses.

Recent research has highlighted the efficacy of nanostructured hydrogel electrocatalysts, including both metal-containing and metal-free variants, for driving various electrochemical reactions. Metal-containing cyanogels—a specialized category of inorganic hydrogels—have demonstrated particular utility for acidic water oxidation applications due to their robust metal-cyano structural units. For example, Fang and colleagues developed an indium-iron-cobalt cyanogel/polypyrrole (InFeCo-cyano/PPy) hybrid hydrogel catalyst specifically for water oxidation in strongly acidic electrolytes. This material exhibited remarkable stability, showing no significant degradation even after 3,000 operational cycles (Fang et al., 2019). Beyond photocatalysis, electrocatalytic hydrogel electrodes also demonstrate outstanding durability. For example, Shi et al. (2023) fabricated a crosslinked poly (sodium acrylate) hydrogel on carbon cloth for the oxygen evolution reaction. Under continuous operation, it sustained 10 mA cm^{-2} at an overpotential of ≈316 mV for 48 hours and over 15,000 cyclic voltammetry cycles with <5% performance loss; post-test XPS and SEM confirmed negligible structural change in the hydrogel network (Pei et al., 2023). Such long-term stability further validates the promise of hydrogel architectures for robust electrochemical applications.

A particularly innovative approach involves the in-situ polymerization of inorganic hydrogels within organic conductive hydrogel matrices, resulting in highly porous nanopolymers with amorphous characteristics. This methodology facilitates rapid charge and mass transfer while providing abundant unsaturated metal atoms within the inorganic hydrogel component. The resulting materials exhibit significantly increased electrochemically active surface areas, which translates to enhanced catalytic performance. The extensive surface area supports efficient adsorption and desorption processes, while the porous structures enable rapid mass transport, and the conductive frameworks facilitate electron transfer to and from catalytic sites (Fang et al., 2020).

The integration of metal organic frameworks (MOFs) with hydrogels represents another promising direction in electrocatalyst development. MOF-hydrogel composites combine the high surface area and well-defined catalytic sites of MOFs with the structural integrity and mass transport capabilities of hydrogels. These composite materials have shown particular promise for electrocatalytic CO_2 reduction, where the hydrogel component helps modulate local pH and concentration gradients near catalytic sites, enhancing selectivity toward valuable products like carbon monoxide, methanol, or formic acid (Fang et al., 2020).

Conductive hydrogels incorporating polyaniline, polypyrrole, or poly(3,4-ethylenedioxythiophene) (PEDOT) offer unique advantages as electrocatalyst supports due to their combination of electronic conductivity and hydrogel-like properties. These materials provide direct electronic pathways to catalytic sites while maintaining the beneficial mass transport characteristics of conventional hydrogels. Recent studies have demonstrated that conductive hydrogels loaded with bifunctional catalysts can significantly reduce the overpotentials required for water splitting reactions, offering improved energy efficiency for hydrogen production systems (Fang et al., 2020).

Looking forward, several emerging trends are shaping the development of hydrogel-based electrocatalysts: (1) hierarchical structures with multi-scale architectural design from nano to macro dimensions to optimize both catalytic site accessibility and bulk material properties; (2) stimuli-responsive systems that can adapt their properties in response to external stimuli for dynamic control of catalytic performance; (3) multi-catalytic systems integrating multiple catalytic functionalities within a single hydrogel matrix to enable cascade or tandem reactions; (4) in-operando characterization techniques for studying electrocatalytic processes under realistic conditions; and (5) computational design approaches using molecular modeling and machine learning to predict and optimize hydrogel-catalyst interactions (Fang et al., 2020).

13.5 CONCLUSION

Catalytic hydrogels and their composites represent a diverse and rapidly evolving class of materials that bridge multiple scientific disciplines, including polymer chemistry, catalysis, materials science, and environmental engineering. This comprehensive exploration has examined how these advanced materials harness the synergistic combination of hydrogel networks and catalytic components to enable efficient chemical transformations across various applications.

Integrating nanoparticles within hydrogel matrices creates multifunctional composites with enhanced performance characteristics, where the hydrogel component provides structural support, stabilization, and a favorable microenvironment for catalytic processes. The resulting nanocomposites exhibit tunable properties suitable for diverse applications ranging from environmental remediation to healthcare and energy conversion systems.

Photocatalytic hydrogels offer promising solutions for water treatment challenges, particularly recalcitrant organic pollutants and heavy metal contaminants that resist conventional treatment methods. These materials address practical limitations of nanoscale photocatalysts by providing a structured support that prevents catalyst loss while maintaining accessibility to pollutants and light. Despite significant progress, challenges remain in optimizing light penetration, mass transport, long-term stability, and scaling these systems for industrial implementation. Future efforts should focus on identifying more abundant/cheaper chromophores, optimizing synthetic routes, and conducting techno-economic analyses to ensure these hydrogels can be produced affordably and at scale. Bulk prices for polymer precursors like acrylamide or alginate are <\$10/kg and TiO_2/ZnO photocatalysts \$10–50/kg; a techno-economic study estimated treatment costs of \$0.7–62 per kg pollutant removed depending on reusability and process integration (Abou Zeid & Leprince-Wang, 2024; Beil et al., 2024).

Electrocatalytic hydrogels demonstrate remarkable potential for energy conversion applications, particularly for reactions central to renewable energy technologies. The hydrogel architecture facilitates efficient ion transport and catalyst stability while providing a platform for precise control over catalytic site distribution and accessibility. From noble-metal-free oxygen reduction catalysts to bimetallic alloys for nitrogen reduction, hydrogel-derived catalysts are expanding the boundaries of electrocatalytic performance and durability.

Looking forward, several research directions merit focused attention: (1) developing mechanistic understanding of catalyst-hydrogel interactions at the molecular level to guide rational material design; (2) enhancing the long-term stability and mechanical robustness of catalytic hydrogels under operational conditions; (3) exploring bio-inspired approaches that mimic natural catalytic systems within hydrogel environments; (4) implementing advanced in-situ and operando characterization techniques to study catalytic processes within hydrogel matrices; (5) establishing standardized testing protocols for fair comparison of different catalytic hydrogel systems; and (6) addressing scalability and manufacturing challenges for transitioning from laboratory demonstrations to practical applications.

The continued advancement of catalytic hydrogels will likely contribute significantly to addressing pressing global challenges related to environmental sustainability, renewable energy implementation, and resource efficiency. By leveraging the structural versatility of hydrogels with the functional capabilities of various catalytic components, these materials offer promising pathways toward more efficient and environmentally friendly chemical processes.

REFERENCES

Abou Zeid, S., & Leprince-Wang, Y. (2024). Advancements in ZnO-Based Photocatalysts for Water Treatment: A Comprehensive Review. *Crystals, 14*(7), 611. https://www.mdpi.com/2073-4352/14/7/611. DOI: 10.3390/cryst14070611

Beil, S. B., Bonnet, S., Casadevall, C., Detz, R. J., Eisenreich, F., Glover, S. D., Kerzig, C., Næsborg, L., Pullen, S., Storch, G., Wei, N., & Zeymer, C. (2024). Challenges and Future Perspectives in Photocatalysis: Conclusions from an Interdisciplinary Workshop. *JACS Au, 4*(8), 2746–2766. DOI: 10.1021/jacsau.4c00527 PMID: 39211583

Berraho, M., & Martín Andreu, C. (2025). Ecofriendly Synthesis of A Hybrid Hydrogel Incorporating Silver Nanoparticles: Biological and Photocatalytic Activities. *Chemistry (Weinheim an der Bergstrasse, Germany), 31*(22), e202404623. DOI: 10.1002/chem.202404623 PMID: 40042110

Bertolacci, L., Valentini, P., & Pompa, P. P. (2020). A nanocomposite hydrogel with catalytic properties for trace-element detection in real-world samples. *Scientific Reports, 10*(1), 18340. DOI: 10.1038/s41598-020-75103-8 PMID: 33110162

Bouzayani, B., & Sanromán, M. (2024). Polymer-Supported Heterogeneous Fenton Catalysts for the Environmental Remediation of Wastewater. *Molecules (Basel, Switzerland), 29*(10), 2188. Advance online publication. DOI: 10.3390/molecules29102188 PMID: 38792049

Bouzayani, B., & Sanromán, M. Á. (2024). Polymer-Supported Heterogeneous Fenton Catalysts for the Environmental Remediation of Wastewater. *Molecules (Basel, Switzerland), 29*(10), 2188. https://www.mdpi.com/1420-3049/29/10/2188. DOI: 10.3390/molecules29102188 PMID: 38792049

Chen, Y., Pötschke, P., Pionteck, J., Voit, B., & Qi, H. (2019). RETRACTED: Fe3O4 Nanoparticles Grown on Cellulose/GO Hydrogels as Advanced Catalytic Materials for the Heterogeneous Fenton-like Reaction. *ACS Omega, 4*(3), 5117–5125. DOI: 10.1021/acsomega.9b00170 PMID: 31459688

Cigáň, M., Gáplovský, A., Gajdoš, P., Magdolen, P., Zahradník, P., & Vetríková, Z. (2010). Photophysical properties and photostability of novel benzothiazole-based D-π-A-π-D systems. *Spectrochimica Acta. Part A: Molecular and Biomolecular Spectroscopy, 77*(5), 984–993. DOI: 10.1016/j.saa.2010.08.035 PMID: 20863742

Clasky, A. J., Watchorn, J. D., Chen, P. Z., & Gu, F. X. (2021). From prevention to diagnosis and treatment: Biomedical applications of metal nanoparticle-hydrogel composites. *Acta Biomaterialia, 122*, 1–25. https://doi.org/https://doi.org/10.1016/j.actbio.2020.12.030. DOI: 10.1016/j.actbio.2020.12.030 PMID: 33352300

Clasky, A. J., Watchorn, J. D., Chen, P. Z., & Gu, F. X. (2021). From prevention to diagnosis and treatment: Biomedical applications of metal nanoparticle-hydrogel composites. *Acta Biomaterialia, 122*, 1–25. DOI: 10.1016/j.actbio.2020.12.030 PMID: 33352300

Dadashi, J., Ghasemzadeh, M. A., & Salavati-Niasari, M. (2022). Recent developments in hydrogels containing copper and palladium for the catalytic reduction/degradation of organic pollutants []. *RSC Advances, 12*(36), 23481-23502. https://doi.org/DOI: 10.1039/D2RA03418B

Deng, A., Wu, S., Hao, J., Pan, H., Li, M., & Gao, X. (2022). Photocatalytic Removal of Cr(VI) by Thiourea Modified Sodium Alginate/Biochar Composite Gel. *Gels (Basel, Switzerland), 8*(5), 293. https://www.mdpi.com/2310-2861/8/5/293. DOI: 10.3390/gels8050293 PMID: 35621591

Fang, Y., Hu, J., Wang, H., Chen, D., Zhang, A., Wang, X., & Ni, Y. (2022). Development of stable agar/carrageenan-Fe(3)O(4)-Klebsiella pneumoniae composite beads for efficient phenol degradation. *Environmental Research, 205*, 112454. DOI: 10.1016/j.envres.2021.112454 PMID: 34856163

Fang, Z., Li, P., & Yu, G. (2020). Gel Electrocatalysts: An Emerging Material Platform for Electrochemical Energy Conversion. *Advanced Materials, 32*(39), 2003191. https://doi.org/https://doi.org/10.1002/adma.202003191. DOI: 10.1002/adma.202003191 PMID: 32830391

Fang, Z., Wu, P., Qian, Y., & Yu, G. (2021). Gel-Derived Amorphous Bismuth–Nickel Alloy Promotes Electrocatalytic Nitrogen Fixation via Optimizing Nitrogen Adsorption and Activation. *Angewandte Chemie International Edition, 60*(8), 4275–4281. https://doi.org/https://doi.org/10.1002/anie.202014302. DOI: 10.1002/anie.202014302 PMID: 33197124

Fang, Z., Wu, P., Yu, K., Li, Y., Zhu, Y., Ferreira, P. J., Liu, Y., & Yu, G. (2019). Hybrid Organic–Inorganic Gel Electrocatalyst for Stable Acidic Water Oxidation. *ACS Nano, 13*(12), 14368–14376. DOI: 10.1021/acsnano.9b07826 PMID: 31800208

Heidari, A., Moghimi, H., Rashidiani, J., & Taheri, R. A. (2018). Molecular development of silver nanoparticles-loaded poly acrylic acid hydrogel as a catalyst for dye degradation. *Physical Chemistry Research, 6*, 857–869. DOI: 10.22036/pcr.2018.145269.1528

Hernandez Monroy, L., Tavares, J. R., & Dumont, M.-J. (2025). Application of photocatalytic hydrogels used for the degradation of dyes, pharmaceuticals and other contaminants in water. *Adsorption, 31*(3), 52. DOI: 10.1007/s10450-025-00602-x

Karchoubi, F., Afshar Ghotli, R., Pahlevani, H., & Baghban Salehi, M. (2024). New insights into nanocomposite hydrogels; a review on recent advances in characteristics and applications. *Advanced Industrial and Engineering Polymer Research, 7*(1), 54-78. https://doi.org/https://doi.org/10.1016/j.aiepr.2023.06.002

Katzenberg, A., Raman, A., Schnabel, N. L., Quispe, A. L., Silverman, A. I., & Modestino, M. A. (2020). Photocatalytic hydrogels for removal of organic contaminants from aqueous solution in continuous flow reactors []. *Reaction Chemistry & Engineering, 5*(2), 377-386. https://doi.org/DOI: 10.1039/C9RE00456D

Krogul-Sobczak, A., Pisarek, N., Cieciórski, P., & Megiel, E. (2022). Silver Nanoparticles Densely Grafted with Nitroxides as a Recyclable Green Catalyst in the Selective Oxidation of Alcohols. *Nanomaterials (Basel, Switzerland), 12*(15), 2542. Advance online publication. DOI: 10.3390/nano12152542 PMID: 35893509

Kuckhoff, T., Landfester, K., Zhang, K. A. I., & Ferguson, C. T. J. (2021). Photocatalytic Hydrogels with a High Transmission Polymer Network for Pollutant Remediation. *Chemistry of Materials*, *33*(23), 9131–9138. DOI: 10.1021/acs.chemmater.1c02180

Kumar, A. S., Pabba, D. P., Reddy, N. R., & Joo, S. W. (2025). Emerging photocatalytic systems for environmental and energy applications: A comprehensive review of mechanisms, materials, and future directions. *Journal of Alloys and Compounds*, *1031*, 180934. https://doi.org/https://doi.org/10.1016/j.jallcom.2025.180934. DOI: 10.1016/j.jallcom.2025.180934

Lee, D.-E., Kim, M.-K., Danish, M., & Jo, W.-K. (2023). State-of-the-art review on photocatalysis for efficient wastewater treatment: Attractive approach in photocatalyst design and parameters affecting the photocatalytic degradation. *Catalysis Communications*, *183*, 106764. https://doi.org/https://doi.org/10.1016/j.catcom.2023.106764. DOI: 10.1016/j.catcom.2023.106764

Li, P., Jin, Z., Qian, Y., Fang, Z., Xiao, D., & Yu, G. (2020). Supramolecular confinement of single Cu atoms in hydrogel frameworks for oxygen reduction electrocatalysis with high atom utilization. *Materials Today*, *35*, 78–86. https://doi.org/https://doi.org/10.1016/j.mattod.2019.10.006. DOI: 10.1016/j.mattod.2019.10.006

Liao, J., Sun, J., Jia, W., He, W., Wang, H., Huang, W., Wang, Y., Yu, M., Xie, Y., & Chen, Y. (2025). External stimuli-driven catalytic hydrogels for biomedical applications []. *Chemical Communications*, *61*(20), 3946-3966. https://doi.org/DOI: 10.1039/D4CC05256K

Melchionna, M., & Fornasiero, P. (2020). Updates on the Roadmap for Photocatalysis. *ACS Catalysis*, *10*(10), 5493–5501. DOI: 10.1021/acscatal.0c01204

Meng, Y., Yin, J., Jiao, T., Bai, J., Zhang, L., Su, J., Liu, S., Bai, Z., Cao, M., & Peng, Q. (2020). Self-assembled copper/cobalt-containing polypyrrole hydrogels for highly efficient ORR electrocatalysts. *Journal of Molecular Liquids*, *298*, 112010. https://doi.org/https://doi.org/10.1016/j.molliq.2019.112010. DOI: 10.1016/j.molliq.2019.112010

Nguyen, B. C., Truong, T. M., Nguyen, N. T., Dinh, D. N., Hollmann, D., & Nguyen, M. N. (2024). Advanced cellulose-based hydrogel TiO2 catalyst composites for efficient photocatalytic degradation of organic dye methylene blue. *Scientific Reports*, *14*(1), 10935. DOI: 10.1038/s41598-024-61724-w PMID: 38740877

Pei, Z., Tan, H., Gu, J., Lu, L., Zeng, X., Zhang, T., Wang, C., Ding, L., Cullen, P. J., Chen, Z., & Zhao, S. (2023). A polymeric hydrogel electrocatalyst for direct water oxidation. *Nature Communications*, *14*(1), 818. DOI: 10.1038/s41467-023-36532-x PMID: 36781856

Ren, G., Han, H., Wang, Y., Liu, S., Zhao, J., Meng, X., & Li, Z. (2021). Recent Advances of Photocatalytic Application in Water Treatment: A Review. *Nanomaterials (Basel, Switzerland)*, *11*(7), 1804. Advance online publication. DOI: 10.3390/nano11071804 PMID: 34361190

Sabuad, A., Khaokong, C., Kongseng, P., & Chantarak, S. (2024). Superabsorbent ZnO/rubber-based hydrogel composite for removal and photocatalytic degradation of methylene blue. *International Journal of Biological Macromolecules*, *275*(Pt 1), 133421. DOI: 10.1016/j.ijbiomac.2024.133421 PMID: 38945320

Sandu, T., Chiriac, A.-L., Zaharia, A., Iordache, T.-V., & Sarbu, A. (2025). New Trends in Preparation and Use of Hydrogels for Water Treatment. *Gels (Basel, Switzerland)*, *11*(4), 238. https://www.mdpi.com/2310-2861/11/4/238. DOI: 10.3390/gels11040238 PMID: 40277674

Sandu, T., Chiriac, A. L., Zaharia, A., Iordache, T. V., & Sarbu, A. (2025). New Trends in Preparation and Use of Hydrogels for Water Treatment. *Gels (Basel, Switzerland)*, *11*(4), 238. Advance online publication. DOI: 10.3390/gels11040238 PMID: 40277674

Shen, J., Zhou, Y., Li, S., Gu, P., & Xue, G. (2019). Hydrogel-coated Fe3O4 nanoparticles as an efficient heterogeneous Fenton catalyst for degradation of phenol. *Journal of Materials Science*, *54*(15), 10684–10694. DOI: 10.1007/s10853-019-03661-y

Shim, H., Kim, J., Im, S., Kim, Y.-S., & Kim, M. (2024a). Continuous-flow catalytic reactors based on 1d hydrogels for catalytic conversion and photocatalytic removal of organic materials in water. *Journal of Water Process Engineering*, *59*, 105055. DOI: 10.1016/j.jwpe.2024.105055

Shim, H., Kim, J., Im, S. H., Kim, Y.-S., & Kim, M. H. (2024b). Continuous-flow catalytic reactors based on 1d hydrogels for catalytic conversion and photocatalytic removal of organic materials in water. *Journal of Water Process Engineering*, *59*, 105055. https://doi.org/https://doi.org/10.1016/j.jwpe.2024.105055. DOI: 10.1016/j.jwpe.2024.105055

Torbina, V. V., Vodyankin, A. A., Ten, S., Mamontov, G. V., Salaev, M. A., Sobolev, V. I., & Vodyankina, O. V. (2018). Ag-Based Catalysts in Heterogeneous Selective Oxidation of Alcohols: A Review. *Catalysts*, *8*(10), 447. https://www.mdpi.com/2073-4344/8/10/447. DOI: 10.3390/catal8100447

Tutić, A., Miloloža, M., Cvetnić, M., Ukić, Š., & Kučić Grgić, D. (2025). Phenol, Cyanide, and Thiocyanate in Aquatic Media: The Ecotoxicity of Individual Substances and Their Mixtures. *Environments (Basel, Switzerland)*, *12*(4), 128. https://www.mdpi.com/2076-3298/12/4/128. DOI: 10.3390/environments12040128

Yang, J., Wang, X., Li, B., Ma, L., Shi, L., Xiong, Y., & Xu, H. (2017). Novel Iron/Cobalt-Containing Polypyrrole Hydrogel-Derived Trifunctional Electrocatalyst for Self-Powered Overall Water Splitting. *Advanced Functional Materials*, *27*(17), 1606497. https://doi.org/https://doi.org/10.1002/adfm.201606497. DOI: 10.1002/adfm.201606497

You, B., Yin, P., Zhang, J., He, D., Chen, G., Kang, F., Wang, H., Deng, Z., & Li, Y. (2015). Hydrogel-derived non-precious electrocatalysts for efficient oxygen reduction. *Scientific Reports*, *5*(1), 11739. DOI: 10.1038/srep11739 PMID: 26130371

You, X., Huang, H., Zhang, R., Yang, Z., Xu, M., Wang, X., & Yao, Y. (2021). Immobilization of TiO_2 Nanoparticles in Hydrogels Based on Poly(Methyl Acrylate) and Succinamide Acid for the Photodegradation of Organic Dyes. *Catalysts*, *11*(5), 613. DOI: 10.3390/catal11050613

Yue, Y., Wang, X., Wu, Q., Han, J., & Jiang, J. (2020). Highly recyclable and super-tough hydrogel mediated by dual-functional TiO_2 nanoparticles toward efficient photodegradation of organic water pollutants. *Journal of Colloid and Interface Science*, *564*, 99–112. https://doi.org/https://doi.org/10.1016/j.jcis.2019.12.069. DOI: 10.1016/j.jcis.2019.12.069 PMID: 31911232

Yue, Y., Wang, X., Wu, Q., Han, J., & Jiang, J. (2020). Highly recyclable and super-tough hydrogel mediated by dual-functional TiO_2 nanoparticles toward efficient photodegradation of organic water pollutants. *Journal of Colloid and Interface Science, 564*, 99–112. DOI: 10.1016/j.jcis.2019.12.069 PMID: 31911232

Zhang, F., Gao, C., Zhai, S.-R., & An, Q.-D. (2023). Nanosilver anchored alginate/poly(acrylic acid/acrylamide) double-network hydrogel composites for efficient catalytic degradation of organic dyes. *Frontiers of Chemical Science and Engineering, 17*(7), 893–905. DOI: 10.1007/s11705-022-2290-8

Zhu, C., Aoki, Y., & Habazaki, H. (2017). Co9 S8 Nanoparticles Incorporated in Hierarchically Porous 3D Few-Layer Graphene-Like Carbon with S,N-Doping as Superior Electrocatalyst for Oxygen Reduction Reaction. *Particle & Particle Systems Characterization, 34*(11), 1700296. DOI: 10.1002/ppsc.201700296

Chapter 14
Hydrogel-Based Sensor Devices

Moses Kumi
https://orcid.org/0000-0002-7971-8802
Northwestern Polytechnical University, China

Shephrah Olubusola Ogungbesan
https://orcid.org/0009-0003-4032-283X
Southeast University, China

Muritala Adeniyi Olusola
https://orcid.org/0000-0002-4826-9661
Federal University of Agriculture, Abeokuta, Nigeria

Frederick Nii Ofei Bruce
https://orcid.org/0000-0003-3928-3928
Northwestern Polytechnical University, China

ABSTRACT

Sensor devices integrate physical and digital systems for real-time monitoring in healthcare, agriculture, and environmental sectors. Meeting rising demands requires enhanced sensitivity, durability, and compatibility. Hydrogels—hydrophilic, biocompatible, stimuli-responsive, and selective—serve as adaptable sensing platforms. This chapter analyzes structural properties governing hydrogel sensing efficacy. It explores optical, electrochemical, and mechanical mechanisms for transducing stimuli like pH, temperature, and analytes into measurable signals. Fabrication methods, including photopolymerization and 3D printing, enable tailored designs for wearable health monitors, precision agriculture, and pollution detection. Correlating hydrogel properties to performance establishes design principles for optimizing sensitivity and selectivity. Challenges such as spatial inhomogeneity, stability under variable conditions, and biocompatibility validation are critically addressed. This chapter highlights hydrogels' transformative role in next-generation sensors.

DOI: 10.4018/979-8-3373-2337-4.ch014

14.1. INTRODUCTION

In today's data-driven world, perceiving and interpreting environmental information in real-time is more critical than ever. Sensor devices, the sentinels of modern industries, seamlessly bridge the gap between the tangible and the digital realms, enabling us to monitor, analyse, and respond to a myriad of dynamic processes (Wang et al. 2023). From healthcare and environmental monitoring to robotics and consumer electronics, sensor technologies have catalysed breakthroughs that redefine the boundaries of human interaction with the world around us (Xu et al., 2023). The widespread adoption of sensor devices across various industries indicates their growing importance in enabling real-time monitoring and intelligent decision-making. In healthcare, sensors empower medical practitioners with real-time insights into physiological parameters, transforming diagnosis and treatment strategies (Ahmed et al. 2024; Damre et al. 2024; Malakhov 2023; Chen et al. 2023). In agriculture, they optimize resource utilization, ushering in an era of precision farming (Hesham, 2023). Environmental monitoring leverages sensors to track climate patterns, pollution levels, and natural disasters, facilitating data-driven decisions for a sustainable future (Mohammed et al. 2023). From smart cities to manufacturing, transportation to defence, sensor devices lay the foundation for innovation, efficiency, and informed decision-making.

Central to the evolution of sensor technologies is the innovative fusion of responsive materials with engineering prowess. As the demand for smarter, more adaptive, and user-friendly sensor systems increases, the role of advanced functional materials becomes ever more critical. Among these materials, hydrogels have emerged as a promising platform, uniquely positioned to meet the growing need for flexible, biocompatible, and highly sensitive sensors. These three-dimensional networks of hydrophilic polymers are capable of absorbing and releasing water without compromising their structural integrity, a property that underpins their functional versatility (Li et al., 2023). This dynamic behavior, when coupled with their inherent biocompatibility, renders hydrogels exceptionally well-suited for a wide spectrum of sensing applications (Vijayakanth et al., 2023). From wearable health monitors to environmental sensors, the responsiveness of hydrogels to subtle physicochemical changes in their environment forms the foundation of their efficacy as transducers (Wang et al., 2023).

As sensor devices evolve, the quest to enhance their performance characteristics intensifies. Sensitivity, the ability to detect minute changes, stands as a sentinel in the realm of sensor development. Hydrogels, owing to their remarkable responsiveness, offer a canvas to amplify sensitivity, pushing the boundaries of detection to new horizons. Selectivity, the capability to discriminate between target analytes and interfering substances, further challenges sensor designers. By engineering hydrogels with tailored recognition elements, the selectivity of sensor devices can be finely tuned, resulting in higher accuracy and reduced false positives (Heydari et al., 2023). Biocompatibility, a paramount consideration in medical and biological applications, harmonizes seamlessly with the natural properties of hydrogels (Shi et al., 2023). Their compatibility with biological tissues and fluids endows hydrogel-based sensor devices with the potential to revolutionize healthcare diagnostics and monitoring, paving the way for patient-centric approaches.

In the sections that follow, we delve into the intricate world of hydrogel-based sensor devices. Exploring their sensing principles, fabrication techniques, and real-world applications, we navigate the interplay of material science, engineering ingenuity, and sensory perception. As we embark on this journey, we uncover the tapestry of possibilities woven by hydrogels, unravelling their potential to redefine how we sense and interact with the world.

14.2. GENERAL PROPERTIES OF HYDROGELS SENSORS

In the practical realm of polymer hydrogels, a thorough understanding of their molecular makeup and elasticity is indispensable. The remarkable elastic characteristics of hydrogels are influenced by various intriguing molecular attributes. In this section, we delve into the fundamental attributes of hydrogels, with a particular focus on examining the intricate connections between the circumstances under which hydrogels form and their attributes, including their swelling behavior, elastic resilience, and spatial variability.

14.2.1. Elasticity and Swelling of Hydrogels Sensors

The elasticity of hydrogels is determined by their molecular structure, specifically the arrangement of polymer chains within the hydrogel network (Zhang et al., 2023). Hydrogels' elastic modulus and equilibrium swelling degree are largely dependent on molecular interactions among polymer chains, crosslinking density, and ion concentration (Hoque et al., 2023). These characteristics rely on the concentration of cross-linked polymer during gel production, as well as the charge and cross-link densities of the polymer network. Although theoretical forecasts can provide information on how hydrogels would swell and behave elastically in various scenarios, there is only a limited degree of agreement between theory and experimental findings.

We can investigate the synthesis of polyacrylamide (PAAm) hydrogels utilizing AAm and N,N′-methylenebis(acrylamide) (MBAAm) in aqueous solutions to obtain a better knowledge of the influence of elasticity on hydrogel properties (Islamkulov et al., 2023). The impact of the initial monomer concentration on the hydrogels' post-synthesis elastic modulus which normally range from 0.01 to 10 kPa (Bose et al., 2020), are much smaller than the values determined by their cross-linking. This discovery demonstrates that the threshold concentration of MBAAm for infinite network formation increases with increasing dilution in the reaction system. Because of cyclization and numerous cross-linking processes and shows that the cross-linker concentration drops as the starting monomer concentration lowers.

It's commonly known that hydrogels with more ionic groups have a greater potential to swell. This impact is mainly caused by the simultaneous increase in the number of counterions in the gel, which results in more osmotic pressure and encourages gel expansion (Flory, 1953). It is possible to reduce the excessive swelling in comparison to non-ionic hydrogels by increasing the amount of salt present in the external solution. For ionic PAAm hydrogels with different charge densities in water and aqueous NaCl solutions, this action lessens the concentration differential of counterions between the inner and outer phases of the gel

Because AMPS Na units disintegrate completely over the whole pH range, ionic comonomers, such the sodium salt of 2-acrylamido-2-methylpropane sulfonic acid (AMPS Na), produce hydrogels with pH-independent swelling (Kalaithong et al., 2021). An increase in AMPS Na content from 0 to 80 mol% causes the hydrogel's volume in water to grow 27 times. However, because of the hydrogel's screening of charge interactions, the swelling ratio in a 1.0 M NaCl solution becomes almost independent of the ionic group level.

Due to their temperature sensitivity, which is linked to the behavior of hydrogen bonding and hydrophobic interactions that is temperature-dependent, hydrophobically modified hydrogels have attracted a lot of attention (Liu et al., 2024). Immersing these hydrogels in aqueous solutions of organic solvents or linear polymers can cause them to show a reentrant swelling transition. The gel undergoes a phase change during which it first compresses and then expands again under the influence of several external factors like the concentration of linear polymer or an organic solvent (Zhang et al., 2023).

Hydrogels' ability to hold and release nutrients and biological products is largely dependent on their water content. As a dry hydrogel absorbs water, primary bound water initially hydrates the polar hydrophilic regions, leading to network swelling. Further interactions between the hydrophilic regions and water molecules result in secondary bound water. Together, these constitute total bound water (Almajidi et al., 2023). Free water, which fills the spaces between network chains and large pores in the hydrogel network, is absorbed more easily into the network due to osmotic forces (Damiri et al., 2023). The hydrogel's final state—disintegration or dissolution—is determined by how quickly the degradable network is consumed.

It is crucial to take into account the swelling capacity when examining the swelling properties of hydrogels. This capacity is impacted by various factors, including the concentration, cross-link density, number of ionic groups, ionizable group pKa or pKb, hydrophobicity or hydrophilicity, and degree of ionization (Boldrini, 2023). Greater swelling capacity is a result of having more ionic groups in the hydrogel. This is mainly because more counterions are present, which raises the osmotic pressure and increases the gel's capacity to swell (Chaudhuri et al., 2023).

14.2.2. Permeation and Porosity of Hydrogels Sensors

The permeability and porosity of hydrogels play a central role in defining their performance as sensor platforms. These properties influence not only the transport of analytes but also the material's responsiveness, sensitivity, and mechanical adaptability. Understanding the origins of pore structures and the strategies to tailor them for specific functions is essential for advancing hydrogel-based sensor design.

14.2.2.1. Pore Formation and Influencing Factors

In hydrogels, pores may emerge during phase separation or they may already be tiny pores that are present throughout the network. The hydrogel's tortuosity (El Idrissi et al., 2023) refers to the critical elements of the hydrogel matrix, such as pore interconnectivity, pore size distribution, and average pore size. Three main factors determine the distribution of pore diameters in the hydrogel:

- The concentration of the polymer matrix that is chemically cross-linked, as measured by its ratio to the concentration of the monomer (Le et al., 2023).
- The initial concentrations of polymerizable monomers in the aqueous solution, which determine the concentrations of physical entanglements in the polymer matrix (Pięta et al., 2023).
- The polyelectrolyte hydrogel's net charge accumulation, which is determined by taking the initial concentration of cationic or anionic monomer units (Bujok et al., 2023).

Hydrogel porosity is further modulated by the nature of the surrounding fluid. Both ionic and nonionic solutes can interact with the polymer network, altering the effective porosity and permeability of the hydrogel. Environmental stimuli such as temperature variations can also induce swelling or shrinking, dynamically affecting pore characteristics and leading to time-dependent porosity profiles.

14.2.2.2. Modulation of Porosity and Functional Innovations

The tunability of porosity and permeability is a crucial asset of hydrogels, enabling the customization of their physical properties for diverse sensing applications. Recent innovations have aimed to overcome the traditional trade-off between swelling capacity and mechanical robustness. Notably, Y. Alsaid *et al* (Alsaid et al., 2021). introduced a cononsolvency-based photopolymerization strategy to engineer hydrogel networks with enhanced structural and functional attributes. Their development—a poly(N-isopropylacrylamide) hydrogel synthesized via mixed-solvent photopolymerization—features a highly porous architecture with interconnected microchannels (Figure 1).

This hierarchical pore structure facilitates ultra-fast volumetric swelling and deswelling, nearly ten times faster than previously reported hydrogels, while simultaneously maintaining superior stiffness and toughness. Unlike conventional approaches that rely on a single solvent system, this method capitalizes on the cononsolvency effect, where competitive adsorption among mixed solvents yields uniform and hierarchical porosity. Impressively, this approach has been successfully extended to other hydrogel systems, including poly(N-tert-butylacrylamide-co-acrylamide) and polyacrylamide, even those that are non-thermoresponsive. Such versatility opens avenues for advanced 3D printing technologies and reinforces the broader potential of cononsolvency as a universal mechanism for hydrogel design.

Figure 1. Schematic of the mixed solvent photopolymerization technique, generating porous hydrogels through the cononsolvency effect, resulting in a hierarchical pore structure. Reproduced from ref (Alsaid et al., 2021). copyright 2021, John Wiley and Sons

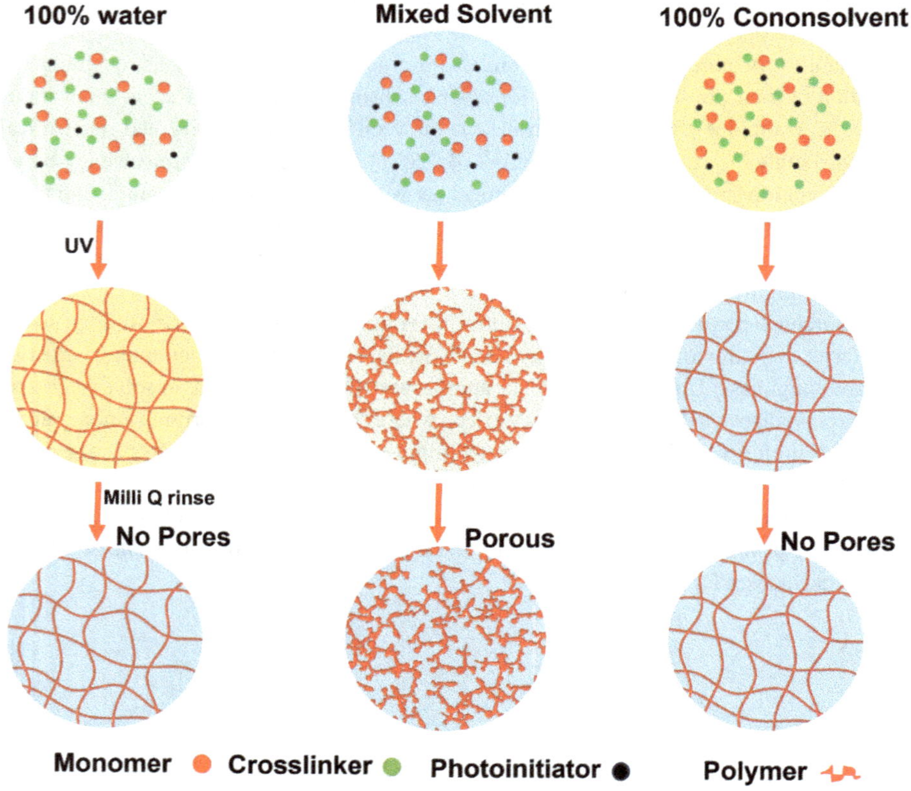

14.2.3. Inhomogeneity of Hydrogels Sensors

Ideal and nonideal hydrogels are the two groups into which hydrogels can be divided (Maryam et al., 2024). In contrast to nonideal hydrogels, which have an uneven cross-link distribution leading to what is referred to as spatial gel inhomogeneity, ideal hydrogels have a uniform distribution of cross-links inside their matrix. This inhomogeneity, which is associated with spatial concentration changes, has a negative impact on the optical clarity and strength of hydrogels. Scattering techniques such as light scattering, small angle X-ray scattering, and small angle neutron scattering have been used to analyze these inhomogeneities. Because more network defects result in places with different cross-link densities, spatial inhomogeneity grows with higher gel cross-link density. On the other hand, when gel ionization increases, inhomogeneity diminishes due to the influence of electrostatic repulsion, mobile counter ions, and the Donnan potential. The Scattering intensities (Wilcox et al., 2023) are strongly influenced by the initial monomer concentration employed in gel production; an inflection point is shown where inhomogeneity reaches its highest value. The concentration of monomers determines the effective cross-link density of hydrogels; an increase in monomer concentration both increases spatial inhomogeneity and the concentration differential between densely and loosely cross-linked regions (Hazra et al., 2023). Reducing the concentration of monomers causes the gels to swell less after preparation, which gradually lessens the concentration difference and apparent inhomogeneity. A maximum inhomogeneity at a critical monomer concentration arises from the interaction of these opposing effects, which determines the inhomogeneity in hydrogels.

Khodambashi *et al* (Khodambashi et al., 2021). have developed heterogeneous hydrogel structures that can be reconfigured over time and space by employing adjustable and addressable hydrogel units, termed soft voxel actuators (SVAs). These SVAs, created through a mixed-solvent photopolymerization technique, leverage the rapid reaction kinetics and the cononsolvency effect of poly(N-isopropylacrylamide) (PNIPAAm). This process yields hydrogels with highly connected pore networks, allowing for the control of their swelling ratio, rate, and mechanical stiffness. The production method is straightforward, involving a single-step casting that is also scalable for mass production. The strategic placement and swelling characteristics of each voxel, along with the integration of Joule heaters within them, enable the hydrogel structures to perform complex movements. These movements include grasping and moving objects, navigating around barriers, creating wave motions, and altering shape. Such advancements are crucial for the creation of adaptable, untethered hydrogel-based robots with a high degree of freedom, capable of operating in unpredictable environments. This work also demonstrates the potential for creating structural diversity within hydrogels by patterning them with varying swelling capacities.

14.2.4. Biocompatibility of Hydrogels Sensors

Nanohydrogels must pass cytotoxicity testing and demonstrate biocompatibility in order to be used in biomedicine. The term "biocompatibility" describes a biomaterial's capacity to trigger a suitable host response (Anderson, 2019). It includes the material's ability to interact positively with host cells and surrounding tissues as well as biosafety. Because continuous interactions with the body throughout processes like healing and cellular regeneration inform the safety of the biomaterial and monitor scaffold degradation, biocompatibility research is essential. A hydrogel can be considered harmful to neighboring or vasculature-connected tissues if it is not biocompatible. Since it is not always possible to achieve 100% conversion, the thorough investigation of potentially harmful compounds frequently utilized in the polymerization of synthetic hydrogels poses a barrier for in vivo biocompatibility (Li et al., 2023). In addition, certain substances used in the polymerization and hydrogel formation process, such as initiators, organic solvents, emulsifiers, stabilizers, unreacted monomers, and cross-linkers, may be harmful to host cells, particularly if they contaminate encapsulated cells or seep into tissues (Sinad et al., 2023).

Biocompatibility of hydrogels is critical in the context of clinical applications. The work of Qi *et al* (Qi et al., 2020)., who developed a novel hybrid hydrogel made of curdlan and salecan, serves as an example of this. In this composition, curdlan acts as a thickening and salecan as the base network matrix. This hydrogel's capacity to adjust its chemical and physical characteristics, including shape, swelling behavior, mechanical robustness, and heat resistance, by adjusting the polysaccharide concentration, is one of its main features. Crucially, assessments of cytotoxicity have validated the hydrogel's safety for application in biological systems. Moreover, the hydrogel system exhibits exceptional proficiency in facilitating cellular processes such as migration, development, and functionality. The cooperative study led by Qi *et al* (Qi et al., 2020). opens the door for a novel strategy in the creation of hydrogel scaffolds based on polysaccharides. The potential of these scaffolds for tissue engineering is enormous, especially in terms of improving cell implantation, survival, proliferation, and functioning.

Hydrogels' biocompatibility is an important consideration in the fields of tissue engineering and cell adhesion. Using the live/dead experiment, Qi *et al* (Qi et al., 2020). carried out a comprehensive examination of the possible toxicity of their created salecan/curdlan hydrogel; the findings are shows L929 cells proliferated inside and on the hydrogel's surface, as was to be predicted. In addition, the CCK-8 assay was used to evaluate the cytotoxicity of the hydrogel these panels amply demonstrated the lack of discernible cytotoxicity, hence confirming the hydrogel's high degree of biocompatibility.

Furthermore, utilizing L929 cells in a CCK-8 experiment, the proliferation of cells in the salecan/curdlan hydrogels was investigated. The optical density (OD) values in the salecan/curdlan hydrogels were substantially greater than those in the pure curdlan hydrogels. This implies that the hydrogel scaffold's development and regulation are significantly influenced by the salecan polysaccharide. In addition to providing space for cells, the scaffold's three-dimensional shape makes it easier for nutrients and metabolic waste to be transported, which promotes increased cell proliferation.

In conclusion, these findings clearly show that the salecan/curdlan hydrogel, because of its exceptional biocompatibility and growth-promoting environment, is a suitable choice for tissue engineering applications.

14.2.5. Mechanical Properties of Hydrogels Sensors

The mechanical characteristics of hydrogels are essential from a pharmaceutical and biomedical standpoint for applications like tissue engineering, drug delivery matrices, ligament and tendon regeneration, wound dressings, and cartilage replacement (Chen et al., 2023). They support the preservation of the structural integrity while therapeutic substances are released under strict control. The degree of cross-linking can be changed to modify these characteristics, providing flexibility and control in hydrogel design (Thang et al., 2023). Stronger hydrogels are produced by increasing cross-linking because it increases stiffness; however, excessive cross-linking can decrease elongation and increase brittleness (Far et al., 2023). Therefore, to create robust yet flexible hydrogels, the ideal cross-linking degree is crucial.

Miao *et al* (Miao et al., 2022)., conducted research that led to the development of a novel natural polymer hydrogel called Chitosan/polymethylacrylic acid (CMA). This hydrogel is remarkable for its remarkable flexibility, self-healing properties, and recyclability. With its tensile strain ranging from 1357% to 3012% at pressures between 26 kPa and 125 kPa, and its ability to tolerate 50% compressive strain at pressures between 32 kPa and 176 kPa, this hydrogel exhibits remarkable flexibility and modifiable mechanical properties. The remarkable capacity for self-healing exhibited by the CMA-3 version of this hydrogel is worth mentioning. In just five minutes, it can restore up to 92.9% of its initial mechanical strength and quickly fix mechanical damage in just one minute. Furthermore, because of its dynamic and reversible interactions, the CMA-3 hydrogel performs exceptionally well in powder self-healing. This enables it to repair damages brought on by desiccation or part loss with ease. After a drying-powdering procedure and rehydration, an entire hydrogel may be rebuilt in the desired shape in under a minute, demonstrating its exceptional recyclability and amazing mechanical strength recovery. The CMA hydrogel exhibits promising qualities that render it a material of choice for diverse applications such as tissue engineering, stretchy electronics, and implantable or wearable devices (Ding et al., 2023). As seen in Figure 2, all of the CMA hydrogels had excellent stretching properties. These hydrogels are made to have regulated mechanical characteristics that meet a range of application needs. With different MA concentrations, the CMA-3 hydrogel exhibits the best mechanical properties, reaching 50% compressive strain at 176 kPa and 3012% stretching strain at 125 kPa. This outperforms the previously documented self-healing hydrogels, especially when it comes to hydrogel electrolytes and hydrogel biomaterials. Moreover, the CMA-3 hydrogel exhibits remarkable elasticity. The hydrogel has remarkable self-recovery capabilities as seen by its quick return to its original shape upon stress release. Its strong reversible hydrogen interactions and continuous porous structure, which successfully oppose outside forces and improve its mechanical properties, are responsible for this resilience. Sticks and disks made of prepared CMA-3 hydrogel were utilized to illustrate the gel's elasticity. Figures 2 demonstrates how the CMA-3 hydrogel keeps its good flexibility without breaking even during repeated stretching and bending operations. It also has outstanding compressive strength. The hydrogel does not break even when subjected to the tension of human weight, and it quickly reverts to its initial state when the load is released (Menezes et al., 2023). These exceptional mechanical qualities led to the selection of the CMA-3 hydrogel for further testing.

Figure 2. The CMA-3 hydrogel compresses. flexibility and recovers. Reproduced from ref (Miao et al., 2022). Open Access.

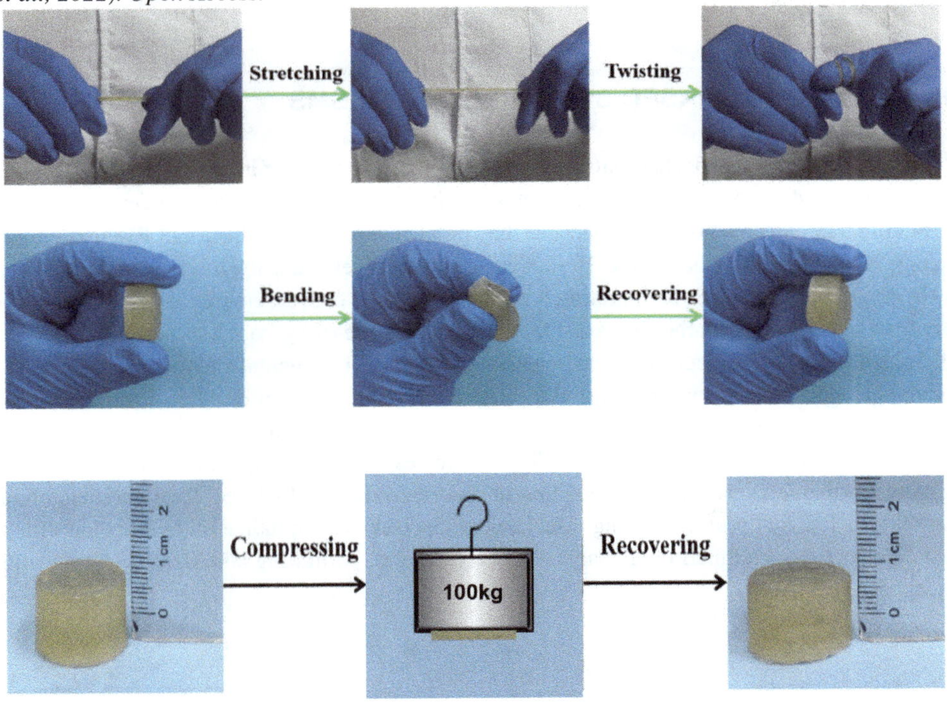

Dryness poses a significant challenge for hydrogels, often leading to their rapid desiccation and shrinkage, which in turn results in performance degradation and limits their applications since it affects the mechanical properties (Feng & Wang, 2023). However, the CMA-3 hydrogel developed by Miao *et al.*, (Miao et al., 2022) benefiting from its strong reversible interactions, showcases an exceptional powder self-healing property that effectively combats this issue. When the CMA-3 hydrogel fails and dries, it can be reduced to a powder form. This powder, when placed in a round-shaped mold and mixed with a small amount of water at room temperature, swiftly reforms into a hydrogel, taking the shape of the mold within just one minute (SI-V1). The hydrogel's dynamic hydrogen bonds are responsible for this amazing change, and the water that was introduced was an essential medium for these dynamic interactions, which restored the hydrogel's structural framework. In order to showcase the remarkable ability of the recycled CMA-3 hydrogel to mend itself, it was once more dried, powdered, and then inserted into molds of different forms. The regenerated hydrogel doesn't lose its flexibility even when it is continuously stretched and released. This remarkable powder self-healing property not only solves the dryness problem but also allows the hydrogel to self-repair in the event that any sections are lost. In approximately five minutes, a fully formed hydrogel can be restored by quickly adding a small amount of prepared hydrogel powder and water. Even with an obvious crack at the point of contact, the self-healing hydrogel exhibits strong mechanical characteristics, able to bear tensile pressures without cracking. Characterized by exceptional flexibility, powder self-healing capabilities, and self-healing ability, the CMA hydrogel successfully addresses common hydrogel problems such mechanical damage, dryness, and missing pieces. These problems frequently cause hydrogel materials to fail and then be abandoned. Most importantly, the hydrogel's exceptional capacity for recycling assures sustainability, prolonging its useful life and improving its suitability as a material for a range of uses.

14.3. HYDROGEL-BASED SENSING MECHANISMS

Hydrogel-based sensors have attracted a lot of interest because of its exceptional sensitivity, adaptability, and promise for a wide range of uses. Hydrogels are fascinating wonders in the field of biomimicry and novel materials because they mimic the sensitive behavior of live tissues and change how we think about sensing. Here, we explore the diverse sensing mechanisms employed by hydrogel-based sensors, highlighting how these materials convert environmental stimuli into measurable signals with precision and functionality. The use of hydrogels as sensing materials exploits a spectrum of mechanisms to detect analytes or environmental changes. These mechanisms span optical, electrochemical, and mechanical domains, each offering distinct advantages and sensitivities. Optical mechanisms leverage the hydrogel's ability to modulate light properties, such as absorbance, fluorescence, or reflection, in response to analyte interactions. Electrochemical principles rely on the modulation of redox reactions at hydrogel interfaces, enabling highly selective and precise sensing. Mechanical mechanisms capitalize on hydrogel swelling and deformation as a function of analyte-induced changes, permitting label-free detection with exceptional sensitivity.

14.3.1. Optical Sensing Mechanism of Hydrogels Sensors

The optical sensing mechanism of hydrogels is a remarkable phenomenon that capitalizes on their capacity to undergo changes in various optical properties, including absorbance, fluorescence, and refractive index, in response to interactions with specific analytes (Song et al., 2023). Embedded with dyes or nanoparticles, hydrogels transform their appearance as analyte molecules permeate their matrix, inducing changes in the local refractive index and causing the material to alter its coloration, akin to a chameleon's skin. This visual metamorphosis serves as a captivating scientific display, revealing the concentration of the analyte. The foundation of the optical sensing mechanism rests upon the refractive index changes incurred by the hydrogel when interacting with analytes. Variations in the molecular environment upon analyte binding cause changes in the refractive index, which controls the passage of light through the material. Furthermore, hydrogels can exhibit alterations in absorbance properties when analytes bind to them, causing a decrease in the transmitted light intensity that correlates with analyte concentration. Some hydrogels incorporate fluorophores, molecules that emit light when excited by a specific wavelength, and analyte interactions can alter their fluorescence properties, including emission intensity, wavelength, or decay kinetics, which are harnessed in fluorescence-based sensors (Ganguly & Margel, 2023). Sensor design and fabrication require careful selection of the hydrogel material with tunable optical properties and compatibility with the analyte of interest. Functionalizing the hydrogel with recognition elements, such as antibodies or molecular imprints, ensures specific interactions with the target analyte. The sensing mechanism encompasses three steps: analyte interaction with the hydrogel, leading to structural or conformational changes within the hydrogel matrix; subsequent optical responses, resulting in measurable signals; and quantification of these signals using various optical techniques, such as spectrophotometry, fluorometry, or interferometry. This comprehensive understanding of the optical sensing mechanism of hydrogels holds great promise for diverse applications in environmental monitoring, medical diagnostics, and chemical analysis.

One prevalent method for thermal sensing in materials involves observing color shifts. As shown in Figure 3a, this phenomenon is caused by the redistribution and/or reorientation of materials and is frequently brought on by the temperature-induced phase separation of hydrophilic components or water from the polymer matrix. It has been found that PNIPAm hydrogels work especially well in this situation when combined with other doping agents. For example, it has been demonstrated that the incorporation of cellulosic materials can fine-tune PNIPAm hydrogels by reducing their native Lower Critical Solution Temperature (LCST) by 8°C, from 32°C to roughly 15°C. Additionally, the incorporation of plasmonic nanoparticles, like gold nanoparticles, into these hydrogels has enabled the creation of temperature-responsive microgels. These gels are capable of detecting temperature variations with a high degree of precision (0.2°C resolution), changing color in the range of 25°C to 45°C. Another aspect of thermal sensing in hydrogels relates to the electrical conduction of functional components within them, as depicted in Figure 3b. For example, reversible PVA and humectant hydrogels, when doped with silver nanotubes, exhibit enhanced moisture retention over a broad temperature spectrum. These gels show an increase in electrical conductivity when the temperature ranges from 0 to 60°C. Excitonic hopping between isolated silver nanotubes in conjunction with the nanotubes' expansion or contraction is thought to be the fundamental mechanism causing this phenomenon. Moreover, hydrogels made of poly(acrylic acid) (PAA) and PANI nanofiber hydrogels have shown an electrical reaction to heat stimulation. The addition of glycerol as an antifreeze ingredient in these gels allows them to endure extensive temperature fluctuations and serve as fever detection sensors with a resolution of 2.7°C. Similarly, thermally controlled electrical signals can be produced by PNIPAm hydrogels embedded with adsorbed or punctured PANI. These gels are effective in a temperature detection range from 2 to 50°C, showcasing their versatility and potential in various thermal sensing applications.

Incorporating ion-sensitive moieties into hydrogels presents an effective strategy for detecting ions, their concentrations, and pH levels (Asadi-Zaki et al., 2024). The key to this detection lies in the hydrogel's ability to undergo volumetric changes through swelling and deswelling, which are crucial for monitoring pH and ions (Yaacob et al., 2023). Functionalized hydrogels that chelate ingested metal ions experience a volumetric reduction, a phenomenon that can be harnessed to measure ion concentrations, as shown in Figure 3c. PAM hydrogels functionalized with 5,6-dicarboxylic fluorescein cross-linked amine are useful for detecting and quantifying cupric species because they prefer Cu2+ ions over other metals. Polymerized colloidal photonic crystals (CPCs) can combine volumetric alterations with diffraction-based visual color shift. Figure 3d shows how PAM hydrogel expansion alters the color of immobilized Polystyrene (PS) CPCs. After benzo-9-crown-3 functionalization, these hydrogels choose beryllium ions (Be^{2+}). Depending on ion electronegativity, CPCs in poly(acrylamide-hydroxyethyl methacrylate) hydrogels chose Pb^{2+} over Cu^2+, Ni^2+, Zn^2+, and Cd^2+. Using optical waveguides, PEGDA hydrogels may detect harmful heavy-metal ions like Hg^{2+}. These hydrogels form flat surfaces with low light scattering. Carbon dots (CDs) in these hydrogels glow to identify absorbed metal ions. Electron transport from heavy-metal ions quenches CD fluorescence. By washing in chelators, hydrogels can be regenerated to sense Hg ions linearly at 5nM. Hg2 + -induced fluorescence red shifts are exploited in thymine-rich DNA-polyacrylamide hydrogels. The sensitive fluorescence red shift of SYBR Green I can visibly identify 10nM Hg^{2+} ions in the gel. Figure 3e shows how fiber Bragg gratings (FBG) can alter wavelength attenuation using chelation swelling hydrogel strain. The wavelength shift from swelling relies on ion concentration. A FBG coating of diverse chemicals affects Cr (VI) ions in solution. Hydrogel-coated fiber-optic cables can monitor pH and water for bridge maintenance. Fiber losses can be detected at various distances using optical time domain reflectometry (OTDR). Hydrogel swelling at specific areas can cause optical losses in fibers, which OTDR can quantify (Figure 3f). Due to pH-dependent swelling or deswelling, hydrogels loaded with conductive polymers like PEDOT doped with PSS have different electrical currents. Also discussed was a pH-sensitive hydrogel device for a diffraction grating that demonstrated pH and ion detection utilizing hydrogels. Hydrogel swelling alters diffracted light patterns, as shown in Figure 3g.

Figure 3. Hydrogel sensing methods. (a) In PNIPAm, phase separation takes place above the LCST, forming crystalline areas that cause light scattering. (b)Ionic movements are enhanced and the electrical current is altered as a result of swelling and temperature changes. (d) Chelation can result in deswelling, which lowers volume and creates cross-linking. The consequence of CPCs is diffraction grating. Light diffraction is affected by variations in the grating frequency in relation to the swelling. (e) Swelling causes strain on Fiber Bragg Grating (FBG) sensors coated in hydrogel. As a result, the grating frequency and reflected light are altered. (f) OTDR detecting fiber strain as a result of the hydrogel swelling. (g) Hydrogen grating alters diffraction following swelling. Reproduced from ref (Kaliaraj et al., 2023). Open Access.

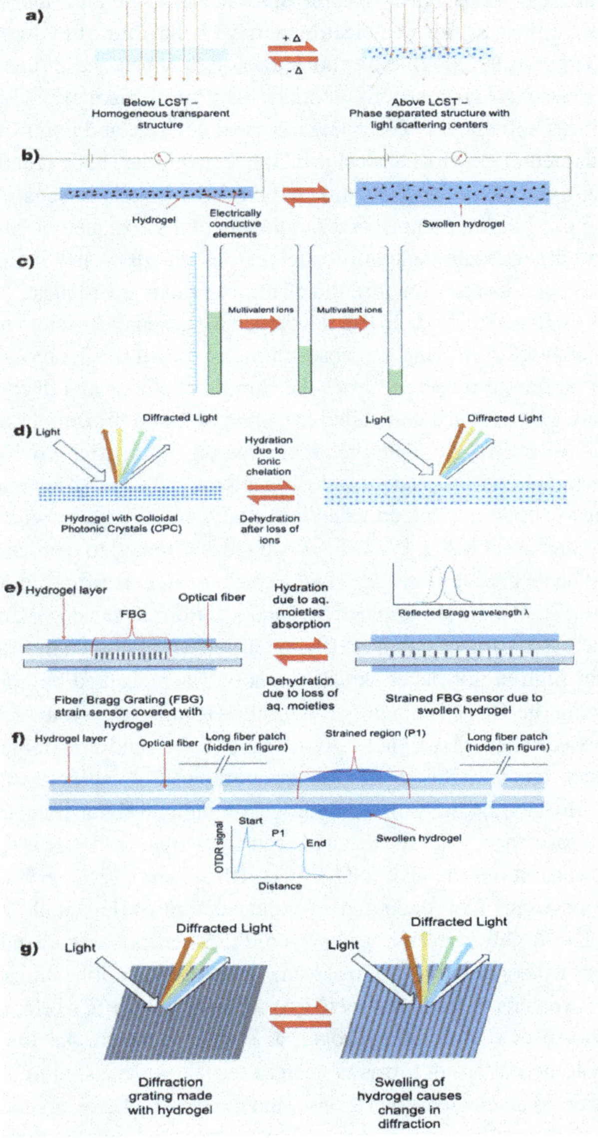

14.3.2. Electrochemical Sensing Mechanism of Hydrogels Sensors

The electrochemical sensing mechanism of hydrogels unveils a captivating molecular dialogue within their matrices, where they act as intricate electrochemical canvases, translating molecular language into electrical symphonies. This mechanism harnesses the capacity of hydrogels to facilitate and regulate redox reactions at their interfaces, creating sensors that detect and quantify analytes through changes in electrical currents or potentials (Shafique et al., 2023). Hydrogels, endowed with inherent conductivity, transform into electrochemical transducers, allowing subtle shifts in analyte concentration to modulate electron flow within their structure, generating measurable electrical signals that mirror the intricate chemical interplay. Electrochemical hydrogel-based sensors exploit the electrochemical behaviour of hydrogels when exposed to analytes. By incorporating redox-active molecules or enzymes, these sensors detect analyte-induced changes in electrical currents or potentials. This mechanism offers high specificity and sensitivity, rendering it suitable for biosensing, medical devices, and point-of-care diagnostics. It builds upon fundamental electrochemical principles, including redox reactions and electrode interfaces, utilizing redox-active species and enzymatic hydrogels in sensor design and fabrication. The sensing mechanism comprises two fundamental steps: analyte interaction inducing redox reactions, and the resultant generation of electrical signals, which manifest as changes in current flow or potential, quantifiable through electrochemical techniques.

Puttasakul *et al* (Puttasakul et al. 2019). developed a gas sensing system to evaluate electrochemical property changes following gas absorption, as shown in Figure 4a. This system was activated by a diaphragm pump placed after the sample chamber and drying unit. The pump delivered air through silica gel at a controlled flow rate of 1,000 mL/min in order to maintain a relative humidity of 40–50%. After that, the air entered the pump after passing past a sample in a vessel with temperature control. The air was directed toward a measurement box that held the gas sensor in order to conduct an electrochemical study, either with or without the sample. A screen-printed gold electrode with a PAAM gel applied and linked to a potentiostat was used to create this sensor. The technology was designed to pick up signals from samples weighing 10 g or 10 mL. A transducer (the conversion system) and a solidified sensing element (the recognition system) are the basic parts of an electrochemical sensor, as shown in Figure 4b. A general design paradigm for high-performance dc-IRS sensors was presented by Fan *et al* (Fan et al., 2019). By substituting the HNaCl α hydrogel with the $HCu(NO_3)_2 - \alpha$ hydrogel, these sensors demonstrated improved sensitivity, signal-to-noise ratio (SNR), and signal stability. Based on the electrochemical redox reaction, the greater reduction rate of Cu^{2+} at the cathode and the stable chemical composition of the $HCu(NO_3)_2 - \alpha$ hydrogel throughout signal transmission are credited for the increased performance. With its ultrawide sensing range (from 0.005 to over 1500%) and ultrahigh detection resolution, the $HCu(NO_3)_2 - \alpha$ IRS-sensor offers very sensitive, accurate, and dependable monitoring of a broad range of human activities (Fan et al., 2019). It opens the door for the creation of highly effective, self-powered IRS-sensors with adaptable features, thus expanding the use of hydrogel-based strain sensors in real-world applications.

Connecting the two ends of the hydrogel to two separate copper (Cu) electrodes is necessary to integrate an IRS-sensor with ionic hydrogel, as shown in Figure 4c. Ionic hydrogels have the ability to promote two different forms of current signal transmission in direct current (DC) circuits: the formation of an electric double layer and the electrochemical redox reaction. Under the impact of DC voltage, anions and cations within the hydrogel migrate towards the anode and cathode, respectively, when the IRS-sensor is seen as a supercapacitor in the context of electric double layer formation. An electric double layer forms at the electrode-ionic hydrogel interface as a result of this migration. The duration of current transfer, denoted as t_m, can be calculated using the formula $t_m = \frac{CU}{I}$, where C represents capacitance, U the applied voltage, and I the electric current. Given the relatively small capacitance of the IRS-sensor (approximately 80nF for the $H_{NaCl\text{-}0.5}$ sensor), t_m is significantly less than 1 second (Figure 4d). Thus, it is deduced that an electrochemical redox process plays a major role in the current transit in the IRS-sensor. Cu electrodes are oxidized at the anode to produce Cu^{2+}/Cu^{+} species, as shown in Figure 1c. This happens because Cu has a lower redox potential than typical anions like Cl^{-}, OH^{-}, and NO^{3-}. H^{+} is electrochemically reduced to H_2 at the cathode due to common cations such as Na^{+} and Al^{3+} having a lower redox potential than the H^{+} ions ionized by water. Copper was selected due to its favorable redox potential and electrochemical stability, making it a suitable electrode material for ionic hydrogel-based sensor applications. The following formula (as illustrated in Figure 4e) can be used to characterize the electrochemical reaction that takes place during the current transfer of the IRS-sensor, especially when the anions in the ionic hydrogel are unable to produce Cu+ species and react with Cu^{2+}:

$$Anode\ Reaction: Cu \rightarrow Cu^{2+} + 2e^{-} \tag{1}$$

$$Cathode\ Raection: 2H^{+} + 2e^{-} \rightarrow H_2 \tag{2}$$

This representation highlights the fundamental electrochemical processes underpinning the operation of the IRS-sensor, emphasizing its potential in various applications where precise and efficient current signal transmission is required.

Figure 4. Electrochemical Gas Sensing System and Reactions in Hydrogel-Based IRS Sensor (a) The specially designed gas flow system for detecting explosive gasses. The explosive materials chamber, diaphragm pump, measurement chamber, sensing electrode, and potentiostat instrument are the five basic parts of this system. (b) The structure and basic principle of electrochemical sensor (c) The IRS sensor's optical image and schematic diagram. (d) $H_{Na}Cl^{-}_{0.5}$ IRS-sensor current as a function of time when a 2 V applied voltage is used. Optical pictures of the $H_{Na}Cl^{-}_{0.5}$ IRS-sensor are displayed in the insets following varying periods of operation at 2 V. (e) Diagram showing the electrochemical redox reaction that occurs during the IRS-sensor's current transit, in which the cations have a lower redox potential than H^{+}. Reproduced from ref (Fan et al., 2019). copyright 2019, American Chemical Society

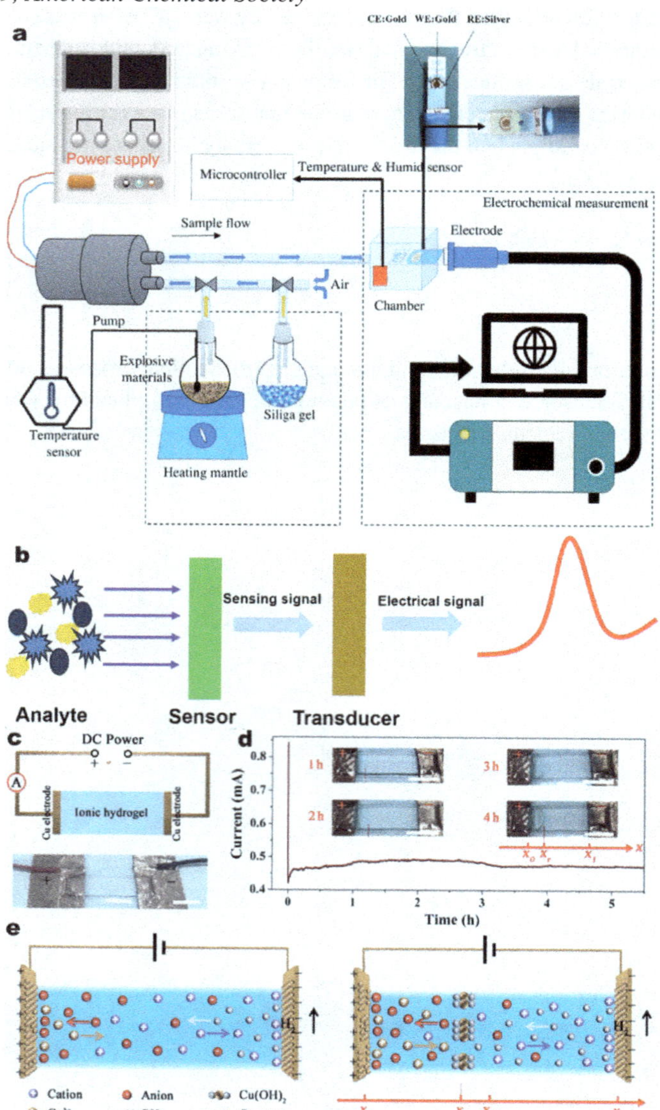

14.3.3. Mechanical Sensing Mechanism of Hydrogels Sensors

The mechanical sensing mechanism of hydrogels represents a captivating realm where science and imagination entwine, revealing the eloquent verses of molecular interactions. Hydrogels express the presence and impact of analytes through their delicate mechanical symphonies. As analytes bind to hydrogel networks, subtle alterations in their swelling properties orchestrate a silent conversation with observers, conveying the subtlest fluctuations in the surrounding milieu. This mechanical sensing mechanism, akin to a delicate dance of molecular interactions, unveils the hidden language of hydrogels, allowing us to discern the presence and influence of analytes in a tangible and intuitive manner. This change in volume can lead to mechanical deformations, such as bending or stretching, in the hydrogel structure (Qian et al., 2023). The hydrogel's exceptional capacity to alter its volume and deform in response to analyte interactions or environmental changes forms the basis of the mechanical sensing mechanism and is used in the creation of sensors that use mechanical measures to quantify analytes. It relies on fundamental principles of swelling and volume changes, wherein hydrogels, as highly water-absorbing materials, undergo swelling when exposed to aqueous analyte solutions, resulting in increased volume and dimensions (Cui et al., 2023). This swelling-induced mechanical deformation can manifest as bending, stretching, or compression, depending on the hydrogel's geometry and design (Inobeme et al., 2023). The choice of hydrogel material is crucial for achieving the desired mechanical responsiveness, favouring hydrogels with high swelling capacity and defined mechanical behaviour. Hydrogels can be integrated into various sensing configurations, such as cantilever arrays, membranes, or microfluidic channels, to enhance and detect mechanical deformations more effectively (Chapi, 2023). The sensing mechanism unfolds in two critical steps: the interaction of the target analyte with the hydrogel, inducing changes in the hydrogel's osmotic balance or specific binding interactions, leading to the absorption of solvent molecules and subsequent swelling; this swelling results in volume changes and mechanical deformations, which are measurable as alterations in the hydrogel's shape or position (Jansen-van Vuuren et al., 2023; Mahardika et al., 2023; Ye et al., 2023).

Wearable sensor technology has become increasingly vital in the realms of health monitoring and gesture control for machinery (Saeidi et al., 2023). This technique makes use of the idea that strain caused by physical motion can be converted into electrical impulses and then analyzed intelligently (Alshangiti et al., 2023). For example, strain transducers for motion sensing have been effectively employed in hydrogels made from okara cellulose. These hydrogels can be used in strain and motion detection applications because they become electrically conductive when doped with conductive ions. The usage of acrylic acid-crosslinked allyl cellulose hydrogels is a prominent illustration of this application. These hydrogels become more electrically conductive when ammonium persulfate, urea, and sodium hydroxide are added. This combination has been proven effective in facilitating repeatable motion sensing. The hydrogels are capable of detecting changes in electrical current over a significant number of cycles, maintaining reliable performance for up to 1000 cycles. This advancement in hydrogel technology underscores the potential of these materials in developing sophisticated, wearable sensors for various applications, including real-time health monitoring and interactive control of machines through gesture recognition. Liu *et al* (Liu et al., 2020). created a graphene-enhanced hydrogel conductor by polymerizing methoxyethyl acrylate and acrylic acid in a DMSO-aqueous solution (Figure 5). This hydrogel, with its unique hydrophobic-hydrophilic network, is resistant to solvents, non-swelling, and strongly adhesive in various liquids. It demonstrates excellent mechanical strength, freeze resistance, and high sensitivity and durability in strain and pressure sensing. These hydrogel sensors can reliably and incredibly sensitively identify various human motions and health states in any kind of weather. Furthermore, even in a variety of liquids such as water, chloroform, hexane, and dodecane, they are able to precisely detect mechanical changes such as pressure, stretching, and bending.

Figure 5. Schematic of the mechanical sensing hydrogel. Thee schematic of the solvent-resistant graphene-assisted hydrogel toward mechanical sensors in diverse solvents. Also, illustration of the real-time resistance variations of wearable hydrogel sensors for monitoring human motions of beathing, speaking and knee bending. Reproduced from ref (Liu et al., 2020). copyright 2020, American Chemical Society.

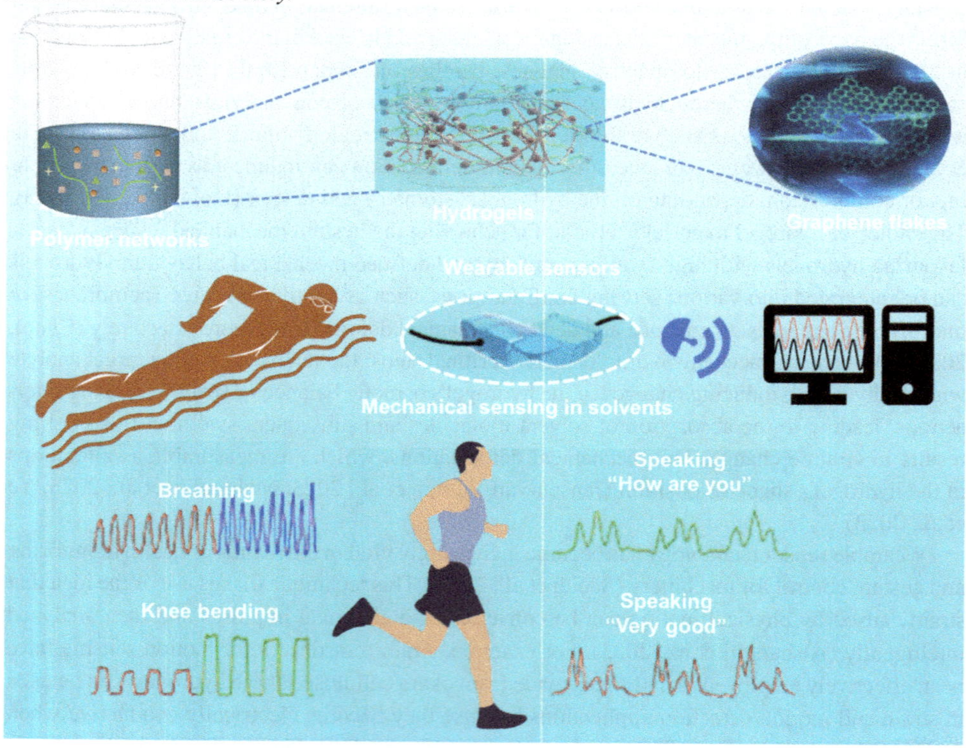

14.3.4. Correlating Hydrogel Sensor Properties to Analyte Concentration

Here, we unmask the underlying choreography of hydrogel behaviour, where the seemingly abstract changes in properties gain tangible significance. The behaviour of hydrogels, characterized by properties like swelling, volume change, and refractive index, forms the basis for their sensing capabilities. Swelling occurs as hydrogels absorb analytes, resulting in volumetric changes that can be correlated to analyte concentration. The volume change often leads to mechanical deformation, which influences the hydrogel's optical properties, such as refractive index and light scattering behaviour. Consequently, these changes establish a direct link between the analyte's presence and measurable alterations in hydrogel properties. This intrinsic correlation facilitates the design of accurate and responsive sensing systems. The success of hydrogel-based sensors hinges on their ability to correlate changes in hydrogel properties to analyte concentration or environmental variations. Here's an exploration of how this correlation works:

- **Swelling and Volume Changes:** Hydrogels react to their surroundings by expanding and contracting. By measuring these intricate changes in volume, scientists decipher the rhythm of analyte concentration (Feng & Wang, 2023). The intimate connection between swelling kinetics and analyte interactions becomes a language that hydrogel-based sensors fluently speak, translating molecular dialogues into observable phenomena (Ming et al., 2023). The amount of the analyte present in the surrounding environment is directly correlated with the degree of hydrogel swelling and volume change (Dehkordi et al., 2023). This swelling-induced response serves as a quantitative indicator of analyte concentration, allowing for accurate detection and quantification.
- **Refractive Index Changes:** In the realm of optical sensing, refractive index takes centre stage. The interaction between light and hydrogel density becomes a prism through which analyte concentration can be read (Pokusaev et al., 2023). As analytes bind and alter the hydrogel's refractive index, the material refracts light in distinct ways, painting a vivid picture of the unseen world. This prismatic language unveils the secrets that analytes whisper to the hydrogel matrix (Cao et al., 2023). Changes in refractive index due to analyte binding alter the propagation of light through the hydrogel (Ma et al., 2023). This modulation can be precisely measured and related to analyte concentration, enabling label-free and real-time detection.
- **Electrical Index Changes:** In electrochemical hydrogel-based sensors, the analyte's presence leads to changes in electrical properties. This can involve alterations in conductivity, impedance, or redox potential (Yue, 2023). The interaction between the analyte and redox-active molecules within the hydrogel can induce electron transfer, which forms the basis for quantification.
- **Mechanical Deformation:** The analyte's influence can cause the hydrogel to swell or deform, altering its mechanical properties (Wu et al., 2023). As the hydrogel responds to the analyte's presence, it may bend, stretch, or compress. These mechanical changes can be measured and correlated to analyte concentration through careful calibration (Wang et al., 2023).
- **Calibration and Validation:** To establish the correlation between hydrogel properties and analyte concentration, calibration is essential. This involves subjecting the hydrogel sensor to known concentrations of the analyte and recording the resulting changes in properties. A calibration curve or equation is then constructed, mapping the relationship between property changes and analyte concentration (Hervas-Raluy et al., 2023).
- **Selectivity and Specificity:** Ensuring the accuracy of the correlation requires addressing selectivity and specificity. Hydrogel sensors must respond predominantly to the target analyte rather than interferents (Falandt et al., 2023). Designing hydrogels with high selectivity, through molecular recognition elements like antibodies or enzymes, minimizes false readings due to cross-reactivity (Wang et al., 2023).

- **Dynamic Range and Sensitivity:** The dynamic range of the sensor refers to the concentration range over which it can accurately measure analytes. The sensitivity of the sensor determines how well it can detect small changes in analyte concentration within this range (Zhou et al., 2023). Optimizing the hydrogel's properties and sensor configuration can enhance the dynamic range and sensitivity.

Correlating hydrogel properties to analyte concentration involves understanding how the analyte-induced changes in hydrogel's physical, optical, electrical, or mechanical properties translate into measurable signals. This correlation forms the basis for quantitative measurements and relies on careful calibration, selectivity enhancement, and sensor optimization. Through this correlation, hydrogel-based sensors become valuable tools for detecting and quantifying analytes across a wide range of applications.

14.4. CONCLUSION

Hydrogel-based sensor devices exemplify the convergence of material science, engineering, and biotechnology, offering exceptional versatility across diverse sensing applications. Their unique properties—such as elasticity, swelling dynamics, porosity, and biocompatibility—are not passive traits but active enablers of dynamic responsiveness to environmental stimuli. Through optical, electrochemical, and mechanical mechanisms, hydrogels effectively translate molecular interactions into measurable signals, enabling applications from wearable health monitors to environmental sensing systems. Significant progress has been made in enhancing hydrogel functionality through innovations like self-healing structures, heterogeneous architectures, and improved mechanical resilience. However, challenges remain, including spatial inhomogeneity, long-term stability, and limitations in scalability. Biocompatibility and the safe integration of responsive elements also require continued attention for clinical and wearable applications. Looking ahead, research should prioritize scalable fabrication techniques, the development of multifunctional and self-adaptive hydrogels, and the integration of AI-driven analytics for real-time feedback. Sustainability and eco-friendly synthesis will be essential to align hydrogel sensor technology with broader environmental goals. Hydrogel-based sensors are poised to redefine how we monitor, interpret, and interact with the world—bridging the physical and digital with unprecedented precision, adaptability, and intelligence.

REFERENCES

Ahmed, S. F., Alam, M. S., Afrin, S., Rafa, S. J., Rafa, N., & Gandomi, A. H. (2024). Insights into Internet of Medical Things (IoMT): Data fusion, security issues and potential solutions. *Information Fusion*, *102*, 102060. DOI: 10.1016/j.inffus.2023.102060

Almajidi, Y. Q., Gupta, J., Sheri, F. S., Zabibah, R. S., Faisal, A., Ruzibayev, A., Adil, M., Saadh, M. J., Jawad, M. J., Alsaikhan, F., Narmani, A., & Farhood, B. (2023). Advances in chitosan-based hydrogels for pharmaceutical and biomedical applications: A comprehensive review. *International Journal of Biological Macromolecules*, *253*, 127278. DOI: 10.1016/j.ijbiomac.2023.127278

Alsaid, Y., Wu, S., Wu, D., Du, Y., Shi, L., Khodambashi, R., Rico, R., Hua, M., Yan, Y., Zhao, Y., Aukes, D., & He, X. (2021). Tunable sponge-like hierarchically porous hydrogels with simultaneously enhanced diffusivity and mechanical properties. *Advanced Materials*, *33*(20), 2008235. DOI: 10.1002/adma.202008235

Alshangiti, D. M., El-Damhougy, T. K., Zaher, A., Madani, M., & Mohamady ghobashy, M. (2023). Revolutionizing biomedicine: advancements, applications, and prospects of nanocomposite macromolecular carbohydrate-based hydrogel biomaterials: a review. *RSC Advances*, *13*(50), 35251–35291. DOI: 10.1039/D3RA07391B

Anderson, J. M. (2019). Biocompatibility and bioresponse to biomaterials. In *Principles of regenerative medicine* (pp. 675–694). Elsevier. DOI: 10.1016/B978-0-12-809880-6.00039-4

Asadi-Zaki, N., Mardani, H., Roghani-Mamaqani, H., & Wang, F. (2024). Stimuli-induced adjustment of spatial distribution of fluorescence resonance energy transfer dyads in smart polymers. *Coordination Chemistry Reviews*, *500*, 215518. DOI: 10.1016/j.ccr.2023.215518

Boldrini, D. E. (2023). Starch-based materials for drug delivery in the gastrointestinal tract-A review. *Carbohydrate Polymers*, *320*, 121258. DOI: 10.1016/j.carbpol.2023.121258

Bose, S., Koski, C., & Vu, A. A. (2020). Additive manufacturing of natural biopolymers and composites for bone tissue engineering. *Materials Horizons*, *7*(8), 2011–2027. DOI: 10.1039/D0MH00277A

Bujok, S., Konefał, R., Nevoralová, M., Bednarz, S., Mielczarek, K., & Beneš, H. (2023). Cation identity in clay-polyelectrolyte self-assembled hydrogels: Rheological and NMR study of the polyitaconate-counterion interactions. *Colloids and Surfaces. A, Physicochemical and Engineering Aspects*, *656*, 130346. DOI: 10.1016/j.colsurfa.2022.130346

Cao, X., Sun, L., Xu, D., Miao, S., Li, N., & Zhao, Y. (2023). Melanin-Integrated Structural Color Hybrid Hydrogels for Wound Healing. *Advanced Science (Weinheim, Baden-Wurttemberg, Germany)*, *10*(22), 2300902. DOI: 10.1002/advs.202300902

Chapi, S. (2023). Preparation, characterization and swelling studies of bio-waste-derived absorbent hydrogels. *Polymers from Renewable Resources*, *14*(4), 237–249. DOI: 10.1177/20412479231200323

Chaudhuri, A., Sandha, K. K., Agrawal, A. K., & Gupta, P. N. (2023). Introduction to smart polymers and their application. In *Smart Polymeric Nano-Constructs in Drug Delivery* (pp. 1–46). Elsevier. DOI: 10.1016/B978-0-323-91248-8.00002-7

Chen, R., Chen, F., Chen, K., & Xu, J. (2023). Advances in the application of hydrogel-based scaffolds for tendon repair. *Genes & Diseases.*

Cui, J., Varma, J., Emrick, T., Bien, C., Preda, D., & Gamliel, D. (2023). Carboxymethyl cellulose foams: Fabrication, aqueous stability, and water capture. *Journal of Materials Science*, *58*(19), 8230–8240. DOI: 10.1007/s10853-023-08514-3

Damiri, F., Salave, S., Vitore, J., Bachra, Y., Jadhav, R., Kommineni, N., Karouach, F., Paiva-Santos, A. C., Varma, R. S., & Berrada, M. (2023). Properties and valuable applications of superabsorbent polymers: A comprehensive review. *Polymer Bulletin*, •••, 1–31.

Damre, S. S., Shendkar, B. D., Kulkarni, N., Chandre, P. R., & Deshmukh, S. (2024). Smart Healthcare Wearable Device for Early Disease Detection Using Machine Learning. *International Journal of Intelligent Systems and Applications in Engineering*, *12*(4s), 158–166.

Dehkordi, N. K., Shojaei, S., Asefnejad, A., Hassani, K., & Benisi, S. Z. (2023). The effect of three types of cross-linked hydrogels and volume fraction of polyacrylamide on the swelling and thermal behavior using molecular dynamics simulation. *Journal of Materials Research and Technology*, *24*, 4627–4638. DOI: 10.1016/j.jmrt.2023.04.102

Ding, Q., Zhang, S., Liu, X., Zhao, Y., Yang, J., Chai, G., Wang, N., Ma, S., Liu, W., & Ding, C. (2023). Hydrogel Tissue Bioengineered Scaffolds in Bone Repair: A Review. *Molecules (Basel, Switzerland)*, *28*(20), 7039. DOI: 10.3390/molecules28207039

El Idrissi, A., Channab, B., Essamlali, Y., & Zahouily, M. (2023). Superabsorbent hydrogels based on natural polysaccharides: Classification, synthesis, physicochemical properties, and agronomic efficacy under abiotic stress conditions: A review. *International Journal of Biological Macromolecules*, •••, 128909.

Falandt, M., Bernal, P. N., Dudaryeva, O., Florczak, S., Größbacher, G., Schweiger, M., Longoni, A., Greant, C., Assunção, M., Nijssen, O., van Vlierberghe, S., Malda, J., Vermonden, T., & Levato, R. (2023). Spatial-Selective Volumetric 4D Printing and Single-Photon Grafting of Biomolecules within Centimeter-Scale Hydrogels via Tomographic Manufacturing. *Advanced Materials Technologies*, *8*(15), 2300026. DOI: 10.1002/admt.202300026

Fan, W., Zhang, X., Cui, H., Liu, C., Li, Y., Xia, Y., & Sui, K. (2019). Direct Current-Powered High-Performance Ionic Hydrogel Strain Sensor Based on Electrochemical Redox Reaction. *ACS Applied Materials & Interfaces*, *11*(27), 24289–24297. DOI: 10.1021/acsami.9b06523

Far, B. F., Naimi-Jamal, M. R., Jahanbakhshi, M., Rostamani, H., Karimi, M., & Keihankhadiv, S. (2023). Synthesis and characterization of chitosan/collagen/polycaprolactone hydrogel films with enhanced biocompatibility and hydrophilicity for artificial tendon applications. *International Journal of Biological Macromolecules*, *253*, 127448. DOI: 10.1016/j.ijbiomac.2023.127448

Feng, W., & Wang, Z. (2023). Tailoring the Swelling-Shrinkable Behavior of Hydrogels for Biomedical Applications. *Advanced Science (Weinheim, Baden-Wurttemberg, Germany), 10*(28), 2303326. DOI: 10.1002/advs.202303326

Flory, P. J. (1953). *Principles of polymer chemistry.* Cornell university press.

Ganguly, S., & Margel, S. (2023). Fluorescent quantum dots-based hydrogels: Synthesis, Fabrication and multimodal biosensing. *Talanta Open, 8*, 100243. DOI: 10.1016/j.talo.2023.100243

Hazra, N., Ninarello, A., Scotti, A., Houston, J. E., Mota-Santiago, P., Zaccarelli, E., & Crassous, J. J. (2023). Structure of Responsive Microgels down to Ultralow Cross-Linkings. *Macromolecules.*

Hervas-Raluy, S., Wirthl, B., Guerrero, P. E., Rei, G. R., Nitzler, J., Coronado, E., de Mora Sainz, J. F., Schrefler, B. A., Gomez-Benito, M. J., & Garcia-Aznar, J. M. (2023). Tumour growth: An approach to calibrate parameters of a multiphase porous media model based on in vitro observations of Neuroblastoma spheroid growth in a hydrogel microenvironment. *Computers in Biology and Medicine, 159*, 106895. DOI: 10.1016/j.compbiomed.2023.106895

Hesham, M. (2023). *Unveiling the Potential of Artificial Intelligence in Agriculture.*

Heydari, A., Khatibi, M., & Ashrafizadeh, S. N. (2023). Fine-tuning ionic transport through hybrid soft nanochannels: The role of polyelectrolyte charge density distribution. *Physics of Fluids, 35*(8), 082006. DOI: 10.1063/5.0160678

Hoque, M., Alam, M., Wang, S., Zaman, J. U., Rahman, M. S., Johir, M. A. H., Tian, L., Choi, J.-G., Ahmed, M. B., & Yoon, M.-H. (2023). Interaction chemistry of functional groups for natural biopolymer-based hydrogel design. *Materials Science and Engineering R Reports, 156*, 100758. DOI: 10.1016/j.mser.2023.100758

Inobeme, A., Ikechukwu Ajai, A., Inobeme, J., Adetunji, C. O., Obar, A., Mathew, J. T., Jacob, J. O., & Nwakife, N. (2023). Superabsorbent Polymers for the Development of Nanofiltration. In *Properties and Applications of Superabsorbent Polymers: Smart Applications with Smart Polymers* (pp. 157–170). Springer. DOI: 10.1007/978-981-99-1102-8_7

Islamkulov, M., Karakuş, S., & Özeroğlu, C. (2023). Design artificial intelligence-based optimization and swelling behavior of novel crosslinked polymeric network hydrogels based on acrylamide-2-hydroxyethyl methacrylate and acrylamide-N-isopropylacrylamide. *Colloid & Polymer Science, 301*(3), 259–272. DOI: 10.1007/s00396-023-05064-7

Jansen-van Vuuren, R. D., Naficy, S., Ramezani, M., Cunningham, M., & Jessop, P. (2023). CO 2-responsive gels. *Chemical Society Reviews.*

Kalaithong, W., Molloy, R., Nalampang, K., & Somsunan, R. (2021). Design and optimization of polymerization parameters of carboxymethyl chitosan and sodium 2-acrylamido-2-methylpropane sulfonate hydrogels as wound dressing materials. *European Polymer Journal, 143*, 110186. DOI: 10.1016/j.eurpolymj.2020.110186

Kaliaraj, G., Shanmugam, D., Dasan, A., & Mosas, K. (2023). Hydrogels—A Promising Materials for 3D Printing Technology. *Gels (Basel, Switzerland), 9*(3), 260. DOI: 10.3390/gels9030260

Khodambashi, R., Alsaid, Y., Rico, R., Marvi, H., Peet, M. M., Fisher, R. E., Berman, S., He, X., & Aukes, D. M. (2021). Heterogeneous Hydrogel Structures with Spatiotemporal Reconfigurability using Addressable and Tunable Voxels. *Advanced Materials*, *33*(10), 2005906. Advance online publication. DOI: 10.1002/adma.202005906

Le, T. N., Kim, H., Elumalai, R., Kim, J., Lee, J., & Suh, M. C. (2023). Highly efficient organic light emitting diodes fabricated by solution process with new hole transport materials cross-linked at 120° C. *Chemical Engineering Journal*, *471*, 144540. DOI: 10.1016/j.cej.2023.144540

Li, W., Liu, J., Wei, J., Yang, Z., Ren, C., & Li, B. (2023). Recent progress of conductive hydrogel fibers for flexible electronics: Fabrications, applications, and perspectives. *Advanced Functional Materials*, *33*(17), 2213485. DOI: 10.1002/adfm.202213485

Li, Y., Saiding, Q., Wang, Z., & Cui, W. (2023). Engineered Biomimetic Hydrogels for Organoids. *Progress in Materials Science*, •••, 101216.

Liu, J., Du, C., Huang, W., & Lei, Y. (2024). Injectable smart stimuli-responsive hydrogels: Pioneering advancements in biomedical applications. *Biomaterials Science*.

Liu, X., Zhang, Q., & Gao, G. (2020). Solvent-Resistant and Nonswellable Hydrogel Conductor toward Mechanical Perception in Diverse Liquid Media. *ACS Nano*, *14*(10), 13709–13717. DOI: 10.1021/acsnano.0c05932

Ma, Z., Sun, J., Zhou, S., Shan, W., Yan, Y., & Liu, Y. (2023). Compact fiber sensor for pH measurement based on the composite effect of hydrogel deformation and LC refractive index variation. *Optics Letters*, *48*(1), 139–142. DOI: 10.1364/OL.479025

Mahardika, I. H., Kin, S., Kwon, O.-S., & Shin, K. (2023). Extended functionalities of paper-based analytical devices using hydrogels for biomedical applications. *Flexible and Printed Electronics*, *8*(3), 33001. DOI: 10.1088/2058-8585/ace4da

Malakhov, K. S. (2023). Insight into the digital health system of Ukraine (ehealth): Trends, definitions, standards, and legislative revisions. *International Journal of Telerehabilitation*, *15*(2). Advance online publication. DOI: 10.5195/ijt.2023.6599

Maryam, L., Gulzar, A., & Iqbal, M. (2024). An overview of the mechanical behavior of hydrogels. *Hydrogels: Antimicrobial Characteristics, Tissue Engineering, Drug Delivery Vehicle*, 17.

Menezes, R., Vincent, R., Osorno, L., Hu, P., & Arinzeh, T. L. (2023). Biomaterials and tissue engineering approaches using glycosaminoglycans for tissue repair: Lessons learned from the native extracellular matrix. *Acta Biomaterialia*, *163*, 210–227. DOI: 10.1016/j.actbio.2022.09.064

Miao, H., Hao, W., Liu, H., Liu, Y., Fu, X., Huang, H., Ge, M., & Qian, Y. (2022). Highly Flexibility, Powder Self-Healing, and Recyclable Natural Polymer Hydrogels. *Gels (Basel, Switzerland)*, *8*(2), 89. DOI: 10.3390/gels8020089

Ming, X., Sheng, Y., Yao, L., Li, X., Huang, Y., Zhu, H., Zhang, Q., & Zhu, S. (2023). Anti-swelling conductive polyampholyte hydrogels via ionic complexations for underwater motion sensors and dynamic information storage. *Chemical Engineering Journal*, *463*, 142439. DOI: 10.1016/j.cej.2023.142439

Mohammed, M. A., Ahmed, M. A., & Hacimahmud, A. V. (2023). Data-Driven Sustainability: Leveraging Big Data and Machine Learning to Build a Greener Future. *Babylonian Journal of Artificial Intelligence, 2023*, 17–23. DOI: 10.58496/BJAI/2023/005

Pięta, M., Purohit, V. B., Pietrasik, J., & Plummer, C. M. (2023). Disulfide-containing monomers in chain-growth polymerization. *Polymer Chemistry*. Advance online publication. DOI: 10.1039/D2PY01291J

Pokusaev, B., Zakharov, N., Vyazmin, A., Nekrasov, D., Volkova, E., & Moshin, A. (2023). Study of hydrogel materials thermophysical properties. *Thermal Science, 00*, 71.

Puttasakul, T., Pintavirooj, C., Sangma, C., & Sukjee, W. (2019). Hydrogel Based-Electrochemical Gas Sensor for Explosive Material Detection. *IEEE Sensors Journal, 19*(19), 8556–8562. DOI: 10.1109/JSEN.2019.2922170

Qi, X., Su, T., Zhang, M., Tong, X., Pan, W., Zeng, Q., & Shen, J. (2020). Sustainable, flexible and biocompatible hydrogels derived from microbial polysaccharides with tailorable structures for tissue engineering. *Carbohydrate Polymers, 237*, 116160. DOI: 10.1016/j.carbpol.2020.116160

Qian, C., Li, Y., Liu, L., Chen, C., & Han, L. (2023). NIR responsive and conductive PNIPAM/PANI nanocomposite hydrogels with high stretchability for self-sensing actuators. *Journal of Materials Chemistry. C, Materials for Optical and Electronic Devices, 11*(20), 6741–6749. DOI: 10.1039/D3TC00990D

Saeidi, M., Chenani, H., Orouji, M., Adel Rastkhiz, M., Bolghanabadi, N., Vakili, S., Mohamadnia, Z., Hatamie, A., & Simchi, A. (2023). Electrochemical Wearable Biosensors and Bioelectronic Devices Based on Hydrogels: Mechanical Properties and Electrochemical Behavior. *Biosensors (Basel), 13*(8), 823. DOI: 10.3390/bios13080823

Shafique, H., de Vries, J., Strauss, J., Khorrami Jahromi, A., Siavash Moakhar, R., & Mahshid, S. (2023). Advances in the Translation of Electrochemical Hydrogel-Based Sensors. *Advanced Healthcare Materials, 12*(1), 2201501. DOI: 10.1002/adhm.202201501

Shi, S., Ou, X., & Cheng, D. (2023). How Advancing is Peripheral Nerve Regeneration Using Nanofiber Scaffolds? A Comprehensive Review of the Literature. *International Journal of Nanomedicine, 18*, 6763–6779. DOI: 10.2147/IJN.S436871

Sinad, K. V. G., Ebubechukwu, R. C., & Chu, C. K. (2023). Recent advances in double network hydrogels based on naturally-derived polymers: Synthesis, properties, and biological applications. *Journal of Materials Chemistry. B, 11*(48), 11460–11482. DOI: 10.1039/D3TB00773A

Song, Q., Wang, Z., Xu, D., Liu, S., Liu, H., & Zhang, K. (2023). Self-assembly of polysaccharide nanocrystals: From aggregation in suspensions to optical materials. *Progress in Polymer Science*, •••, 101768.

Thang, N. H., Chien, T. B., & Cuong, D. X. (2023). Polymer-based hydrogels applied in drug delivery: An overview. *Gels (Basel, Switzerland), 9*(7), 523. DOI: 10.3390/gels9070523

Vijayakanth, T., Shankar, S., Finkelstein-Zuta, G., Rencus-Lazar, S., Gilead, S., & Gazit, E. (2023). Perspectives on recent advancements in energy harvesting, sensing and bio-medical applications of piezoelectric gels. *Chemical Society Reviews*, *52*(17), 6191–6220. DOI: 10.1039/D3CS00202K

Wang, C., He, T., Zhou, H., Zhang, Z., & Lee, C. (2023). Artificial intelligence enhanced sensors-enabling technologies to next-generation healthcare and biomedical platform. *Bioelectronic Medicine*, *9*(1), 17. DOI: 10.1186/s42234-023-00118-1

Wang, S., Gerlach, G., & Körner, J. (2023). A study of smart hydrogels as sensing elements in gaseous environment for VOC detection. *Polymer*, *278*, 126009. DOI: 10.1016/j.polymer.2023.126009

Wang, Y., Jiang, X., Li, X., Ding, K., Liu, X., Huang, B., Ding, J., Qu, K., Sun, W., Xue, Z., & Xu, W. (2023). Bionic ordered structured hydrogels: Structure types, design strategies, optimization mechanism of mechanical properties and applications. *Materials Horizons*, *10*(10), 4033–4058. DOI: 10.1039/D3MH00326D

Wang, Y., Ren, Y., & Chen, T. (2023). From kirigami to hydrogels: A tutorial on designing conformally transformable surfaces. *Journal of Applied Mechanics*, *90*(4), 44801. DOI: 10.1115/1.4056350

Wang, Z., Wei, H., Huang, Y., Wei, Y., & Chen, J. (2023). Naturally sourced hydrogels: Emerging fundamental materials for next-generation healthcare sensing. *Chemical Society Reviews*, *52*(9), 2992–3034. DOI: 10.1039/D2CS00813K

Wilcox, K. G., Grinevich, A., Linscott, A., Singh, A., & Morozova, S. (2023). Properties of Cross-Linked Poly-l-lysine Hydrogels across the Random Coil–Helix Transition. *Macromolecular Chemistry and Physics*, *224*(24), 2300227. DOI: 10.1002/macp.202300227

Wu, J., Li, Y., Duan, S., Wang, Z., Jing, X., Lin, Y., Zhu, D., Lei, W., Shi, Q., & Tao, L. (2023). Bioinspired Stretchable MXene Deformation-Insensitive Hydrogel Temperature Sensors for Plant and Skin Electronics. *Research*, *6*, 106. DOI: 10.34133/research.0106

Xu, J., Zhu, X., Zhao, J., Ling, G., & Zhang, P. (2023). Biomedical applications of supramolecular hydrogels with enhanced mechanical properties. *Advances in Colloid and Interface Science*, *321*, 103000. DOI: 10.1016/j.cis.2023.103000

Yaacob, S. F. F. S., Olasupo, A., & Suah, F. B. M. (2023). Polymer inclusion membranes based Optode: Recent advances and perspectives. *Trends in Analytical Chemistry*, •••, 117498.

Ye, R., Liu, S., Zhu, W., Li, Y., Huang, L., Zhang, G., & Zhang, Y. (2023). Synthesis, Characterization, Properties, and Biomedical Application of Chitosan-Based Hydrogels. *Polymers*, *15*(11), 2482. DOI: 10.3390/polym15112482

Yue, Y. (2023). Nanocomposite Hydrogels for Strain Sensing Based on Optical and Electrical Signals: A Review. *Chemical Communications (Cambridge)*, *59*(58), 8894–8910. DOI: 10.1039/D3CC01580G

Zhang, D., Liu, Y., Gross, W., Tang, Y., & Zheng, J. (2023). Designing Antifreezing Hydrogels with Enhanced Mechanical Properties Using a Simple Crosslinker. *Giant (Oxford, England)*, *16*, 100203. DOI: 10.1016/j.giant.2023.100203

Zhang, H., Shi, L. W. E., & Zhou, J. (2023). Recent developments of polysaccharide-based double-network hydrogels. *Journal of Polymer Science*, *61*(1), 7–43. DOI: 10.1002/pol.20220510

Zhou, B., Fan, K., Li, T., Luan, G., & Kong, L. (2023). A biocompatible hydrogel-coated fiber-optic probe for monitoring pH dynamics in mammalian brains in vivo. *Sensors and Actuators. B, Chemical*, *380*, 133334. DOI: 10.1016/j.snb.2023.133334

Chapter 15
Hydrogel-Based Bioelectronics

Akintomiwa Olumide Esan
https://orcid.org/0000-0003-0578-7458
Ladoke Akintola University of Technology, Ogbomoso, Nigeria

Ibidotun Theophilus Olawoore
https://orcid.org/0009-0003-1093-4579
Ladoke Akintola University of Technology, Ogbomoso, Nigeria

Blessing Ayomide Olafimihan
Ladoke Akintola University of Technology, Ogbomoso, Nigeria

Promise Chima Ubah
https://orcid.org/0009-0004-4924-9330
Federal University of Technology, Akure, Nigeria

Siwaporn Meejoo Smith
Mahidol University, Thailand

Adewusi John Adepoju
Ladoke Akintola University of Technology, Ogbomoso, Nigeria

ABSTRACT

Hydrogel-based bioelectronics represent a groundbreaking advancement in biomedical engineering, merging the biocompatibility of hydrogels with the functional capabilities of electronic devices. These materials offer superior flexibility, conductivity, and biocompatibility, making them ideal for applications in biosensors, neural interfaces, cardiac monitoring, and drug delivery systems. This chapter explores the fundamental properties of hydrogels, highlighting their advantages over traditional bioelectronic materials. Innovations in soft robotics and prosthetics also leverage hydrogel properties to improve artificial muscle function and sensory feedback. Despite these advancements, challenges remain, particularly regarding material stability and long-term biocompatibility. Addressing these concerns through scalable manufacturing and interdisciplinary research will be crucial for use in clinical applications. Future perspectives suggest the integration of artificial intelligence and machine learning to enhance bioelectronic performance for neural and cardiac applications.

DOI: 10.4018/979-8-3373-2337-4.ch015

15.1 INTRODUCTION

15.1.1 Definition of Bioelectronics

Bioelectronics is the integration of electronic systems and biological systems to develop devices that detect, interact with, or even control physiology. A cross-disciplinary science seems to blend biology, materials science, and electrical engineering to offer new technologies such as biosensors, neural interfaces, or the development of wearable technology devices. Such devices enable real-time sensing, focal stimulation and therapeutic intervention with emphasis placed on precision and minimal invasiveness. Being a rapidly evolving discipline, bioelectronics is promising to drive diagnostics, personalized medicine, and restoration of perturbed biological processes (Xu, 2024).

15.1.2 Overview and Importance of Hydrogel-Based Bioelectronics

Hydrogels are 3D hydrophilic polymer networks that retain water while maintaining structure, making them ideal for drug delivery, tissue engineering, and wound healing (Aswathy *et al.*, 2020). They allow stimuli-responsive drug release (Jahanbekam *et al.*, 2025) and mimic the extracellular matrix (El-Sherbiny & Yacoub, 2013). Conductive hydrogels, following graphene or polymer incorporation, allow for flexible electronics (Shin *et al.*, 2024), interfacing hard devices with soft tissue (Wang *et al.*, 2024). In medicine, they enable real-time, noninvasive monitoring (Hua *et al.*, 2023; Mukherjee *et al.*, 2025) and improve brain-computer interfaces (Qian & Liao, 2023). Their stretchability is suitable for prosthetics and wearables (Shin *et al.*, 2024), and their combination with nanotech and 3D printing drives personalized treatments (Dave *et al.*, 2024).

Figure 1. Comparison between Traditional Bioelectronic Materials and Hydrogel-Based Bioelectronics

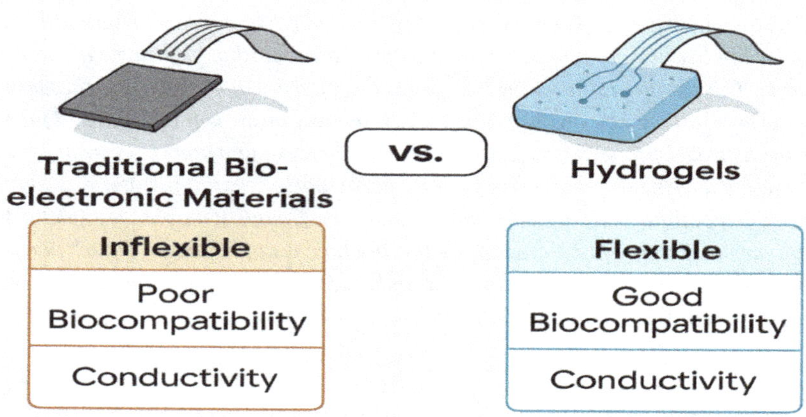

15.1.3 Advantages of Hydrogels over Traditional Bioelectronic Materials

Hydrogels possess various benefits over traditional materials like metals and silicon due to their physicochemical properties. They exhibit good biocompatibility, causing low immune responses and reducing irritation to tissues due to their soft, tissue-like nature (Figure 1) (Xu *et al.*, 2022a). Their flexibility and stretchability enable higher integration with dynamic biological surfaces as opposed to inflexible substitutes like silicon (Shin *et al.*, 2024). The high water content replicates the environment of the native tissue and allows for good ion conduction upon mixing with conductive fillers (Mo *et al.*, 2025). The tunable properties of hydrogels such as porosity and mechanical properties allow for application-based modification (Jahanbekam *et al.*, 2025). The ease of functionalization allows for compatibility with bioactive agents and sensors (Khan *et al.*, 2024). Moreover, their sustainability and biodegradability reduce environmental footprints and facilitate safe in-body degradation (Zhang *et al.*, 2023), and thus they are apt for next-generation bioelectronics.

15.2 FUNDAMENTALS OF HYDROGELS IN BIOELECTRONICS

15.2.1 Composition and Structure of Hydrogels

Hydrogels are three-dimensional networks of hydrophilic polymers that can capture and retain large volumes of water, often more than 90% of their weight, while maintaining structural integrity. Their composition and architecture are central to their role in bioelectronics, as they dictate mechanical, electrical, and biological behaviors. In this context, different classifications of hydrogels based on network formation and source materials provide distinct design advantages for specific applications.

15.2.1.1 Polymer-Based vs. Supramolecular Hydrogels

Polymer-based hydrogels, such as polyacrylamide and poly(ethylene glycol), are covalently crosslinked to form stable, mechanically strong networks, making them suitable for long-term applications like implantable biosensors (Irmukhametova *et al.*, 2025; Shin *et al.*, 2024). In contrast, supramolecular hydrogels rely on non-covalent interactions, including hydrogen bonding, hydrophobic forces, or host–guest complexation, to create reversible, stimuli-responsive structures (Sánchez-Fernández, 2023). Although less robust, their flexibility makes them well-suited for wearable sensors. For example, cyclodextrin-based hydrogels can release conductive nanoparticles upon environmental triggers, enabling dynamic bioelectronic functionality (El-Husseiny *et al.*, 2021). This classification highlights the spectrum of hydrogel stabilityfrom permanent networks to reversible assemblies laying the foundation for understanding how different sources and compositions further influence their performance.

15.2.1.2 Classification: Synthetic vs. Natural Hydrogels

Hydrogels are also broadly classified as synthetic or natural. Synthetic hydrogels such as polyacrylamide, PVA, and PEG offer tunable physicochemical properties and precise functionalization for bioelectronics, though they lack inherent biocompatibility without modification (Ho *et al.*, 2022). Natural hydrogels, including alginate, collagen, and hyaluronic acid, provide excellent biocompatibility and support for tissue integration, but exhibit weaker mechanics and batch variability (Kim & Cha, 2025). For instance, collagen mimics the extracellular matrix, promoting cell adhesion (Kim & Cha, 2025). Increasingly, hybrid hydrogels that combine synthetic and natural components are explored to balance mechanical robustness, functional tunability, and biological performance (Ullah *et al.*, 2025).

Taken together, both structural (polymer vs. supramolecular) and source-based (synthetic vs. natural) classifications reveal how design choices influence hydrogel function. These compositional foundations directly impact their physicochemical properties, which in turn govern their ability to interface effectively with electronic systems.

15.2.1.3 Physicochemical Properties Related to Bioelectronics

The physicochemical properties of hydrogels including hydrophilicity, porosity, and swellability are crucial for their use in bioelectronics. High water content (70–99%) enables ionic conductivity (Shin *et al.*, 2024), while porosity facilitates functionalization with biomolecules or nanomaterials. Swellability affects both mechanical response and sensitivity, especially in sensing applications (Barhoum *et al.*, 2023). Furthermore, their tunable surface chemistry allows for seamless integration with electronic components and biological tissues (Vo *et al.*, 2020).

Understanding these physicochemical parameters provides natural transition into the discussion of conductivity mechanisms, since hydrogel structure directly influences whether ionic, electronic, or mixed pathways dominate their electrical performance.

15.2.2 Mechanisms of Conductivity in Hydrogels

Conductivity underpins the function of hydrogel-based bioelectronics, enabling signal transmission in devices ranging from biosensors to neural interfaces. Depending on composition and functionalization, hydrogels exhibit ionic, electronic, or mixed conduction mechanisms, each suited to different device requirements.

15.2.2.1 Ionic Conductivity

Ionic conductivity arises from the migration of ions within the hydrogel's water-rich matrix. High water content and porosity facilitate the diffusion of ions such as sodium and potassium, either inherent to the hydrogel or sourced from surrounding biological fluids (Mo *et al.*, 2025). Polyelectrolyte-derived hydrogels like poly(acrylic acid) exemplify this property, offering biocompatible, flexible platforms for skin-interfaced devices that detect bioelectric signals, such as electrocardiograms (Shin *et al.*, 2024). However, ionic conduction can be sensitive to environmental conditions, particularly dehydration (Mo *et al.*, 2025).

15.2.2.2 Electronic Conductivity

Electronic conductivity is introduced by incorporating conductive nanomaterials such as carbon nanotubes (CNTs), graphene, or conductive polymers like polyaniline (PANI) and PEDOT (Mo *et al.*, 2025). These components establish percolating networks for electron transport. PEDOT:PSS-based hydrogels, for instance, combine softness and high conductivity, making them promising for neural electrodes (Bhat *et al.*, 2021). While such modifications enhance conductivity, they may compromise biocompatibility or mechanical properties, necessitating strategies such as in-situ polymerization or layer-by-layer assembly to optimize performance (Musa *et al.*, 2025).

15.2.2.3 Mixed Conduction Systems

Mixed conduction hydrogels integrate both ionic and electronic pathways to mimic the signal transmission characteristics of biological tissues (Ghosh *et al.*, 2024). For example, PEDOT-based hydrogels doped with ionic salts simultaneously support electron and ion conduction, enhancing neural interface performance (Zhang *et al.*, 2025). These dual-mode systems are particularly valuable for brain–machine interfaces that require concurrent sensing and stimulation, though balancing conductivity with biocompatibility and mechanical strength remains a key challenge (Shin *et al.*, 2024).

15.2.3 Mechanical Properties and Biocompatibility as Design Constraints

Beyond conductivity, mechanical compatibility and biocompatibility are critical design considerations that directly complement the electrical functionality of hydrogel-based devices. Hydrogels' softness and elasticity (Young's modulus: ~1 kPa to 1 MPa) allow them to mimic tissues such as skin and neural structures, reducing mechanical mismatch and minimizing tissue damage during long-term integration (Wang *et al.*, 2024; Sun *et al.*, 2023). Biocompatibility is equally important: natural hydrogels like hyaluronic acid exhibit low cytotoxicity (Ullah *et al.*, 2025), while synthetic systems can be modified for example, by PEGylation to reduce immune responses (Xu *et al.*, 2022a).

However, durability remains a challenge. Swelling, dehydration, and degradation must be controlled through crosslinking strategies or the inclusion of biodegradable components to achieve reliable long-term performance (Jiao *et al.*, 2025; Ahmad *et al.*, 2022). When considered alongside electrical performance, these mechanical and biological constraints form an integrated design framework that ensures both functionality and safety in hydrogel-based bioelectronics.

15.3 HYDROGEL-BASED BIOELECTRONICS FABRICATION TECHNIQUES

15.3.1 Traditional Hydrogel Fabrication Techniques

Traditional methods offer scalable means of engineering hydrogels with specificity in structure, mechanical properties, and conductivity for bioelectronic uses.

15.3.1.1 Sol-Gel Processing

Sol-gel processing converts a sol to a gel through pH, temperature, or ionic stimuli, incorporating conductive or bioactive species (Bokov *et al.*, 2021). Conductive and suitable mechanical properties are maintained by graphene oxide-doped silica hydrogels (Ma *et al.*, 2017). Disadvantages include nonuniformity of the filler and the gelation rate (Sakka, 2007).

15.3.1.2 Free-Radical Polymerization

This method forms electroactive hydrogels with initiators like ammonium persulfate or UV (Reddy *et al.*, 2020). Conducting polymers like PEDOT or PANI are utilized for enhanced performance (Deng *et al.*, 2022), though cytotoxicity and uncontrolled microstructure may result (Bas *et al.*, 2015).

15.3.1.3 Crosslinking Strategies

16.3.1.3.1 Physical Crosslinking

Physical crosslinking is done through the application of non-covalent interactions like hydrogen bonding or van der Waals interaction to form reversible networks (Yang *et al.*, 2022). Freeze-thaw-crosslinked PVA hydrogels are biocompatible and flexible, suitable for wearables (Khan & Rumon, 2025). Though self-healing and responsive, they are usually deficient in mechanical strength.

16.3.1.3.2 Chemical Crosslinking

Chemical methods create covalent bonds via Michael addition or click chemistry and produce strong hydrogels like PEG diacrylate networks, ideal for neural electrodes (Pal *et al.*, 2017). Crosslinkers that are toxic need strict purification (De Siqueira *et al.*, 2023).

16.3.1.3.3 Ionic Crosslinking

This technique relies on electrostatic interaction, e.g., in calcium-ion crosslinked alginate gels (Ručigaj *et al.*, 2024). These gels are easy to prepare and biocompatible but not stable in the physiological environment due to ion exchange (Shin *et al.*, 2024).

15.3.2 Advanced Microfabrication Techniques

15.3.2.1 3D Printing of Hydrogels

3D printing allows for high-resolution production of hydrogel structures. Extrusion-based, SLA, and DLP methods allow for incorporation of conductive fillers like CNTs into GelMA inks (Dong *et al.*, 2021). SLA and DLP allow for delicate microfeatures for neural interfaces (Wang *et al.*, 2024). Though multifunctionality is accommodated, printability must be optimized and mechanical and biological compatibility ensured (Tuli *et al.*, 2024). Bioprinting with live cells also opens up prospects (Persaud *et al.*, 2022).

15.3.2.2 Lithography-Based Patterning

Soft lithography uses PDMS molds to create microchannels or arrays at low expense (Qin *et al.*, 2010). It suits microfluidics or wearable sensors (Zhang *et al.*, 2024). Photolithography, light-based patterning of PEG-based hydrogels, allows ultra-fine structures like neural microelectrodes (Hahn *et al.*, 2005). Highly resolved, it is costly and biocompatibility can be compromised (Tanwar *et al.*, 2022).

15.3.2.3 Electrospinning and Nanostructuring

Electrospinning produces nanofibrous hydrogel scaffolds that are apt for tissue engineering or biosensors (Zhong *et al.*, 2021). PEDOT or graphene-based electrospun scaffolds are flexible and conductive for neural devices (Adabavazeh *et al.*, 2025). Although nanostructured morphology is better for cell interaction, scalability remains an issue (Mirshafiei *et al.*, 2024). Co-electrospinning increases capability by using multimaterial scaffolds (Xing *et al.*, 2023).

15.3.3 Integration with Electronic Components

Hydrogel incorporation into electronic components is crucial in the fabrication of functional bioelectronic devices that bridge biological and electronic systems. Incorporation enables conductivity, mechanical compatibility, and biocompatibility for applications in sensing, stimulation, and drug delivery.

15.3.3.1 Hybrid Hydrogel-Electronic Interfaces

Hybrid hydrogel-electronic interfaces combine the soft, biocompatible character of hydrogels with the conductivity of traditional electronic materials, such as metals or silicon. These interfaces are achieved by coating rigid electrodes with conducting hydrogels or encapsulating devices within hydrogel matrices (Shin *et al.*, 2024). For example, hydrogel-coated gold electrodes reduce impedance and improve signal quality in neural recording by offering a soft, tissue-like interface (Liu et al., 2020). Techniques like layer-by-layer assembly or in-situ polymerization enable robust adhesion between hydrogels and electronic substrates to avoid delamination when in use (Zou et al., 2025). Hybrid interfaces are critical in uses like implantable neural probes, where hydrogels reduce tissue damage without sacrificing electrical performance. Challenges include long-term stability and preventing degradation at the hydrogel-electrode interface (Ullah *et al.*, 2025).

15.3.3.2 Conductive Fillers: Nanoparticles, Polymers, and Carbon-Based Materials

Conductive fillers are employed to strengthen hydrogels with enhanced electrical performance, enabling electron or ion transport for bioelectronic uses. Common fillers are:

16.3.3.2.1 Nanoparticles

Metal nanocubes (e.g., gold, silver) or carbon nanomaterials (e.g., graphene, carbon nanotubes) enhance electronic conductivity (Khan *et al.*, 2017). For instance, silver nanoparticles-doped hydrogels impart excellent biosensor conductivity but can be a cytotoxicity concern (Fahim et al., 2024).

16.3.3.2.2 Conductive Polymers

Polymers like PEDOT:PSS or PANI are widely used due to their high conductivity and compatibility with hydrogel matrix (Bhat *et al.*, 2021). PEDOT:PSS hydrogels, for example, provide low impedance and flexibility, favorable for applications in neural electrodes (Cysewska & Pawłowska, 2024).

16.3.3.2.3 Carbon-Based Materials

Graphene and carbon nanotubes offer good conductivity and mechanical reinforcement. Graphene oxide hydrogels, for instance, show ionic and electronic conductivity, suitable for wearable sensors (Alqaderi & Ramakrishnan, 2025).

The preferred filler depends on the application, where there are trade-offs between electrical conductivity, biocompatibility, and cost. Dispersion techniques, including ultrasonication or chemical functionalization, allow for even filler distribution in the hydrogel matrix (Shin *et al.*, 2024).

15.3.3.3 Enhancing Conductivity and Mechanical Performance

Enhancement of conductivity and mechanical performance is required for long-lasting hydrogel-based bioelectronics. Approaches include:

16.3.3.3.1 Double-Network Hydrogels

Combining two interpenetrating polymer networks, e.g., one rigid network to introduce strength and one flexible network to introduce elasticity, increases toughness and extensibility (Chen *et al.,* 2015). Double-network hydrogels with PEDOT:PSS, for example, is conductive even under high strain, rendering it applicable in wearable devices.

16.3.3.3.2 Nanocomposite Reinforcement

Incorporation of nanomaterials like graphene or silica nanoparticles imparts mechanical strength and conductivity. The composites are resistant to cyclic deformation with electrical performance maintained (Rashid et al., 2024).

16.3.3.3.3 Stimuli-Responsive Design

Hydrogels may be engineered to change conductivity or mechanical behavior with respect to stimuli like pH or temperature, which enables adaptive bioelectronic devices (El-Husseiny *et al.*, 2021). pH-responsive hydrogels that are doped with conducting polymers, for instance, can switch the conductivity of drug delivery systems. Such methods guarantee hydrogel-based bioelectronics meet the demands of flexibility, conductivity, and stamina in dynamic biology.

15.4 CURRENT AND EMERGING APPLICATIONS OF HYDROGEL-BASED BIOELECTRONICS

Hydrogel-based bioelectronics have transformed implantable and wearable devices due to their softness, biocompatibility, and flexibility. They enable the real-time monitoring of physiological signals non-invasively.

15.4.1 Implantable and Wearable Biosensors

Sweat, Saliva, and Tear Sensors: Hydrogels' porosity and water content enable the detection of analytes in body fluids. PEDOT:PSS-doped polyacrylamide hydrogels are used for sweat sensors to detect sodium and potassium (Moreno *et al.*, 2025). Glucose oxidase embedded in hyaluronic acid hydrogels enables saliva-based glucose sensing (Ameen *et al.*, 2025). Tear sensor-based contact lenses enable ocular biomarkers like lactoferrin (Shetty *et al.*, 2024) to be monitored, though long-term stability remains to be accomplished (Xiao *et al.*, 2023).

Continuous Glucose Monitoring (CGM): Hydrogels swell or alter optical signals when exposed to glucose. PEG-phenylboronic acid hydrogels are ideal for optical/electrochemical CGMs (Efstathiou *et al.*, 2025), while alginate hydrogels reduce immune responses in implants (Xu & Lee, 2020). Microneedle patches collect interstitial fluid painlessly (Dai *et al.*, 2023).

Electrochemical and Optical Biosensing: Analytes are sensed electrochemically by conductive hydrogels, such as lactate during exercise (Saeidi *et al.*, 2023; Deng, 2024). Transparence of the hydrogel is utilized by optical biosensors for pH/fluorescence sensing (Youssef *et al.*, 2023), but is associated with fouling and complexity issues (Janik *et al.*, 2024).

15.4.2 Neural Interfaces and Brain-Machine Interfaces

EEG, ECoG, and Neural Stimulation: Hydrogel EEG electrodes improve skin adhesion and signal clarity (Hsieh *et al.*, 2022). GelMA ECoG electrodes offer low inflammation and chronical recordings (Oribe *et al.*, 2019). PEDOT:PSS-doped hydrogels offer neural stimulation in Parkinson's treatment (Zhu *et al.*, 2025; Shah *et al.*, 2023). Non-Invasive vs. Invasive Applications: EEG hydrogel patches find usage in sleep research (Yadav & Maini, 2023). Invasive hydrogel-coated electrodes reduce scarring (Tang *et al.*, 2024) and help restore mobility (Jorfi *et al.*, 2014; Zhang *et al.*, 2024). Signal Fidelity and Biocompatibility: Hydrogels must compromise on flexibility and impedance (Chen *et al.*, 2024). PEG coatings mitigate biofouling (Wang *et al.*, 2025), and hyaluronic acid hydrogels mitigate immune responses (Revete *et al.*, 2022). Self-healing hydrogels improve durability (Xue *et al.*, 2025).

15.4.3 Cardiac and Muscle Bioelectronics

Monitoring and Stimulation: KCl-doped hydrogels reduce ECG noise (Shin *et al.*, 2024). PVA-based ECG patches are good at detecting arrhythmia (Cui *et al.*, 2022), while graphene hydrogels enhance EMG monitoring (Munasir *et al.*, 2025), though it is disrupted by sweat (Wang *et al.*, 2024). Pacemakers and FES: Hydrogel coatings of defibrillators reduce tissue damage (Hua *et al.*, 2023). PEDOT:PSS-alginate hydrogels improve electronic function in pacemakers

(Wang *et al.*, 2025). FES systems enable rehabilitation through ionic hydrogels (Cooper *et al.*, 2011; Linh *et al.*, 2024).

15.4.4 Smart Therapeutics

Electroresponsive hydrogels deliver drugs like insulin and chemotherapy agents (Ghosh *et al.*, 2024; Dias *et al.*, 2024). Degradation of hydrogels and drug release remain concerns (Huanbutta *et al.*, 2024). Smart patches for cancer and wound healing are imminent (Mariello, 2025). Microfluidic-integrated hydrogels ensure precise dosing (Zhang & Staples, 2024; Omidian & Chowdhury, 2025).

15.4.5 Prosthetics and Soft Robotics

Hydrogel actuators mimic muscle movements of soft robots (Banerjee *et al.*, 2018; Wang *et al.*, 2025), and hydrogels provide sensory feedback through pressure and temperature sensing (Rahmani & Shojaei, 2021; Munasir *et al.*, 2025). AI-hydrogel systems improve robotic control as well as prosthetic responsiveness (Wang *et al.*, 2025; Li, 2024).

15.5 FUTURE HORIZONS AND CHALLENGES OF HYDROGEL-BASED BIOELECTRONICS

Hydrogel-based bioelectronics promise much in wearable and implantable medical devices but several technical, biological, and commercial challenges must be overcome to achieve clinical and industrial relevance.

15.5.1 Technical Challenges

15.5.1.1 Stability and Durability

Soft and biocompatible hydrogels experience extreme stability issues. Environmental conditions like pH and temperature fluctuations, and variations in water content lead to swelling, dehydration, or degradation, degrading performance (Shin *et al.*, 2024; Li *et al.*, 2024). Biodegradable hydrogels like gelatin and chitosan decompose over time, limiting reusability (Garcia-Garcia *et al.*, 2024). Double-network structure development and enhanced crosslinking (e.g., polyacrylamide-alginate systems) are stable but may compromise flexibility or biocompatibility (Khan *et al.*, 2024; Nezhad-Mokhtari *et al.*, 2024).

15.5.1.2 Electrical Performance

Hydrogels are less conductive than metals and therefore are not ideal for high-performance electronics (Shin *et al.*, 2024). Doping with conductivity (e.g., CNTs, PEDOT:PSS, graphene) enhances conductivity but induces heterogeneity and particle clustering (Li *et al.*, 2009; Munasir *et al.*, 2025). Layer-by-layer assembly and in-situ polymerization reduce but scaling up is challenging (Mumtaz *et al.*, 2024). Research on mixed ionic-electronic conductive systems for the future is promising but at an early stage (Hasan *et al.*, 2025).

15.5.1.3 Mechanical-Electronic Integration

Mechanical incompatibility between soft hydrogels (Young's modulus ~1 kPa–1 MPa) and hard electronics (>100 MPa) often leads to delamination or failure (Sunwoo *et al.*, 2020). Wearable devices suffer from disconnection due to motion stress (Ying & Liu, 2021). Partial solutions based on covalent bonding and bio-inspired adhesives but with a fabrication complexity (Wan *et al.*, 2024). Power sources and wireless modules integration is currently hindered by biocompatibility and miniaturization limitations (Shin *et al.*, 2024; Obaideen *et al.*, 2024).

15.5.2 Biocompatibility and Safety

15.5.2.1 Immune Response

Though the native hydrogels, e.g., collagen and hyaluronic acid, have low immunogenicity, their synthetic analogs like polyacrylamide have been reported to cause inflammatory responses (Ho *et al.*, 2022). Immunogenicity is reduced by PEGylation (Isaac *et al.*, 2024), yet persistent inflammation remains problematic. Hydrogels reduce mechanical irritation but may cause protein adsorption and fibrotic encapsulation (Norahan *et al.*, 2022). Anti-inflammatory and bioactive incorporation are being explored for improving biocompatibility (Raina *et al.*, 2022; Bu *et al.*, 2022).

15.5.2.2 Degradation and Toxicity

Biodegradable hydrogels like chitosan hydrolyze to innocuous products, with certain fillers (e.g., silver nanoparticles) even being cytotoxic (Naghib *et al.*, 2024). Long-term safety evidence is limited to short-term animal tests, and chronic toxicity has been inadequately investigated (Van Norman, 2019). PEG hydrogels are more stable but risk accumulation (Agbna & Zaidi, 2025). In-depth genotoxicity and carcinogenicity assessments are required, particularly among sensitive populations (Hartwig *et al.*, 2020).

15.5.2.3 Regulatory Issues

As hybrid devices (electronic + biomaterial), hydrogel devices pose formidable regulatory hurdles. Preclinical requirements by bodies like FDA and EMA, including ISO 10993 testing, are stringent (Al-Zyoud *et al.*, 2023; Lin *et al.*, 2025). Addition of conductive substances or drugs in hydrogels increases the challenge of approval (Sultana *et al.*, 2022). Clinical trial data are scarce, and manufacturing variability discourages compliance (Catoira *et al.*, 2020). Reproducible testing and predictive in vitro models are badly needed (Masarone *et al.*, 2024).

15.5.3 Scalability and Commercialization

15.5.3.1 Manufacturing Cost

Hydrogels incorporate expensive materials (e.g., PEDOT:PSS, graphene) and advanced manufacturing processes like in-situ polymerization and 3D printing (Shin *et al.*, 2024). Double-network hydrogel production is energy-intensive owing to precise crosslinking demands (Li *et al.*, 2022). Sol-gel and electrospinning scale-up is constrained by equipment needs (Owens *et al.*, 2016). Other natural polymers are investigated but without optimal performance (Hiwrale *et al.*, 2023). Roll-to-roll and high-throughput printing methods may optimize costs (Tuli *et al.*, 2024).

15.5.3.2 Industrial production

Batch-to-batch variability in natural polymers like collagen affects device performance (She *et al.*, 2024; Ullah *et al.*, 2025). Conductive filler dispersion defects affect quality (Király & Ronkay, 2013). Sterilization damages hydrogel structures (Bento *et al.*, 2023). Huge capital is needed for mass production of hydrogel bioelectronics (Asin-Garcia *et al.*, 2024).

15.5.3.3 Commercialization Trends

The commercial market for hydrogel-based bioelectronics increases, driven by neural interfaces and continuous glucose monitoring (Javaid *et al.*, 2024). Collaboration between universities and industry is a main channel for commercialization (Perkmann *et al.*, 2012). Profusa and GraphWear are strong investor-supported companies (Shang *et al.*, 2021). However, competition from silicon-based sensors and the high R&D cost are persistent entry barriers (Mahalakshmi *et al.*, 2025). Public-private finance and strategic partnerships are essential (Prain *et al.*, 2020).

15.5.4 Future Horizons

15.5.4.1 Next-Generation Hydrogels

Future hydrogels will be multifunctional, enabling sensing, actuation, and therapy. Self-healing, stimuli-responsive, and bioinspired hydrogels in the future will enhance toughness and integration (El-Husseiny *et al.*, 2021; Revete *et al.*, 2022). Conductive degradable polymers show promise; clinical validation is awaited (Liu & Echeverry-Rendón, 2025; Zhang *et al.*, 2024).

15.5.4.2 AI and Machine Learning

AI/ML can transform hydrogel biosensor applications by enabling personalized diagnostics and results interpretation in automation. For example, machine learning enhances glucose prediction in diabetes management (Van Doorn *et al.*, 2021). In BMIs, ML translates hydrogel electrode signals into precise prosthetic control (Vidal *et al.*, 2016). Integration challenges are low-power processing and ensuring confidentiality of data (Lewin *et al.*, 2024). Edge computing and cloud-AI hardware are being explored for this (Ficili *et al.*, 2025).

15.5.4.3 Brain-Machine Interfaces (BMIs)

Hydrogels enable more effective BMIs by improved neural signal quality and tissue compatibility. PEDOT:PSS hydrogel electrodes have shown promise in electrocorticography (ECoG) for epilepsy treatment (Wang *et al.*, 2022; Oribe *et al.*, 2019). Upcoming BMI systems will integrate hydrogels with wireless, closed-loop systems for AI-guided adaptive stimulation (Zhang *et al.*, 2025). Miniaturization and longevity remain challenges (Ognard *et al.*, 2025), but innovations like 3D-printed hydrogel scaffolds and nanotech are advancing the field (Nguyen *et al.*, 2025).

15.6 CONCLUSION

15.6.1 Summary of Key Findings

Hydrogel-based bioelectronics represent a convergence of materials science, biomedical engineering, and electronics with transformative potential in healthcare. Their hydrophilic polymer networks have high water content, flexibility, and biocompatibility, enabling implantation with biological tissues (Hua *et al.*, 2023). Tunable hydrogels are critical in wearable biosensors, neural electrodes, cardiac monitors, drug delivery systems, and soft robotics (Khan *et al.*, 2024). They mimic tissue-like mechanics (Young's modulus ~1 kPa–1 MPa), minimizing injury and supporting physiological integration (Chen *et al.*, 2023). Ionic, electronic, or mixed conductivity further supports signal transduction in neuro-electronic devices (Shin *et al.*, 2024).

Advanced synthesis methods, including sol–gel processing, polymerization, 3D printing, and electrospinning, yield multifunctional adaptive systems (Anusiya & Jaiganesh, 2022). Saliva, sweat, and tear-based biosensors are emerging (Mukherjee *et al.*, 2025), while low impedance improves neural electrodes (Chen *et al.*, 2024) and cardiac monitors (Li *et al.*, 2023). Electro-responsive hydrogels enable smart drug delivery (Ghosh *et al.*, 2024). AI enhances hydrogel robotics and prosthetics (Chen *et al.*, 2023). Remaining challenges include swelling, low conductivity, immune responses, production cost, and regulation (Zhang *et al.*, 2023; Virumbrales *et al.*, 2024). Future targets include self-healing, stimuli-responsive hydrogels and AI-driven neural interfaces (Ullah *et al.*, 2025).

15.6.2 Medicine, Healthcare, and Bioengineering Implications

Future hydrogel bioelectronics will be characterized by self-healing, mechanical recovery, and multifunctionality, improving device durability and enabling closed-loop therapies (Hua *et al.*, 2023). Integration with AI will advance predictive analytics, precision neural stimulation, and robotic control (Oh *et al.*, 2025). Hydrogel electrodes offer intuitive brain–machine interfaces for neurological disorders (Kılıç, 2024), while energy efficiency, data security, and privacy remain critical (Mohammed *et al.*, 2023). Promising applications include epilepsy, Parkinson's disease, and spinal cord injury through low-impedance neural interfaces (Lu *et al.*, 2024; Ullah *et al.*, 2025). Nanostructuring and 3D printing enhance precision and accessibility (Javaid *et al.*, 2022). Key challenges include durability, immunosafety, and scalability, with roll-to-roll fabrication and regulatory flexibility offering solutions (Al-Zyoud *et al.*, 2023; Boutiette *et al.*, 2020; Duarte *et al.*, 2024).

15.6.3 Future of Hydrogel-Based Bioelectronics

Future hydrogel bioelectronics will be defined by self-healing, mechanical recovery, and multifunctionality, extending device life and enabling closed-loop therapies (Hua *et al.,* 2023). AI integration will enhance predictive analytics, precision neural stimulation, and robotic control (Oh *et al.*, 2025). Hydrogel electrodes promise brain–machine interfaces for neurological disorders (Kılıç, 2024), while energy efficiency, data security, and privacy remain essential (Mohammed *et al.*, 2023). Applications include epilepsy, Parkinson's disease, and spinal cord injury, where hydrogels provide low-impedance neural interfaces (Lu *et al.*, 2024; Ullah *et al.*, 2025). Nanostructuring and 3D printing support precision and accessibility (Javaid *et al.*, 2022). Remaining challenges include durability, immunosafety, and scalability, with roll-to-roll fabrication and regulatory flexibility as solutions (Al-Zyoud *et al.*, 2023; Boutiette *et al.*, 2020; Duarte *et al.*, 2024).

REFERENCES

Adabavazeh, Z., Johari, N., & Baino, F. (2025). Electrospun Conductive polymer scaffolds: Tailoring fiber diameter and electrical properties for tissue engineering applications. *Materials Today. Communications, 112596*, 112596. Advance online publication. DOI: 10.1016/j.mtcomm.2025.112596

Agbna, G. H. D., & Zaidi, S. J. (2025). Hydrogel Performance in Boosting plant resilience to Water Stress—A Review. *Gels (Basel, Switzerland), 11*(4), 276. DOI: 10.3390/gels11040276 PMID: 40277712

Ahmad, Z., Salman, S., Khan, S. A., Amin, A., Rahman, Z. U., Al-Ghamdi, Y. O., Akhtar, K., Bakhsh, E. M., & Khan, S. B. (2022). Versatility of hydrogels: From synthetic strategies, classification, and properties to biomedical applications. *Gels (Basel, Switzerland), 8*(3), 167. DOI: 10.3390/gels8030167 PMID: 35323280

Al-Zyoud, W., Haddadin, D., Hasan, S. A., Jaradat, H., & Kanoun, O. (2023). Biocompatibility Testing for Implants: A Novel tool for selection and Characterization. *Materials (Basel), 16*(21), 6881. DOI: 10.3390/ma16216881 PMID: 37959478

Alqaderi, A. I., & Ramakrishnan, N. (2025). Carbon-based flexible strain sensors: Recent advances and performance insights in human motion detection. *Chemical Engineering Journal, 162609*, 162609. Advance online publication. DOI: 10.1016/j.cej.2025.162609

Ameen, S. S. M., Omer, K. M., Mansour, F. R., Bedair, A., & Hamed, M. (2025). Non-invasive wearable electrochemical sensors for continuous glucose monitoring. *Electrochemistry Communications, 107894*, 107894. Advance online publication. DOI: 10.1016/j.elecom.2025.107894

Anusiya, G., & Jaiganesh, R. (2022). A review on fabrication methods of nanofibers and a special focus on application of cellulose nanofibers. *Carbohydrate Polymer Technologies and Applications, 4*, 100262. DOI: 10.1016/j.carpta.2022.100262

Asin-Garcia, E., Fawcett, J. D., Batianis, C., & Santos, V. P. M. D. (2024). A snapshot of biomanufacturing and the need for enabling research infrastructure. *Trends in Biotechnology*. Advance online publication. DOI: 10.1016/j.tibtech.2024.10.014 PMID: 39592270

Aswathy, S., Narendrakumar, U., & Manjubala, I. (2020). Commercial hydrogels for biomedical applications. *Heliyon, 6*(4), e03719. DOI: 10.1016/j.heliyon.2020.e03719 PMID: 32280802

Banerjee, H., Suhail, M., & Ren, H. (2018). Hydrogel Actuators and Sensors for Biomedical Soft Robots: Brief Overview with Impending Challenges. *Biomimetics, 3*(3), 15. DOI: 10.3390/biomimetics3030015 PMID: 31105237

Barhoum, A., Sadak, O., Ramirez, I. A., & Iverson, N. (2023). Stimuli-bioresponsive hydrogels as new generation materials for implantable, wearable, and disposable biosensors for medical diagnostics: Principles, opportunities, and challenges. *Advances in Colloid and Interface Science, 317*, 102920. DOI: 10.1016/j.cis.2023.102920 PMID: 37207377

Bas, O., De-Juan-Pardo, E. M., Chhaya, M. P., Wunner, F. M., Jeon, J. E., Klein, T. J., & Hutmacher, D. W. (2015). Enhancing structural integrity of hydrogels by using highly organised melt electrospun fibre constructs. *European Polymer Journal, 72*, 451–463. DOI: 10.1016/j.eurpolymj.2015.07.034

Bento, C. S. A., Gaspar, M. C., Coimbra, P., De Sousa, H. C., & Braga, M. E. M. (2023). A review of conventional and emerging technologies for hydrogels sterilization. *International Journal of Pharmaceutics, 634*, 122671. DOI: 10.1016/j.ijpharm.2023.122671 PMID: 36736965

Bhat, M. A., Rather, R. A., & Shalla, A. H. (2021a). PEDOT and PEDOT:PSS conducting polymeric hydrogels: A report on their emerging applications. *Synthetic Metals, 273*, 116709. DOI: 10.1016/j.synthmet.2021.116709

Bhat, M. A., Rather, R. A., & Shalla, A. H. (2021b). PEDOT and PEDOT:PSS conducting polymeric hydrogels: A report on their emerging applications. *Synthetic Metals, 273*, 116709. DOI: 10.1016/j.synthmet.2021.116709

Bokov, D., Jalil, A. T., Chupradit, S., Suksatan, W., Ansari, M. J., Shewael, I. H., Valiev, G. H., & Kianfar, E. (2021). Nanomaterial by Sol-Gel Method: synthesis and application. *Advances in Materials Science and Engineering, 2021*(1). https://doi.org/DOI: 10.1155/2021/5102014

Boutiette, A. L., Toothaker, C., Corless, B., Boukaftane, C., & Howell, C. (2020). 3D printing direct to industrial roll-to-roll casting for fast prototyping of scalable microfluidic systems. *PLoS One, 15*(12), e0244324. DOI: 10.1371/journal.pone.0244324 PMID: 33370381

Bu, W., Wu, Y., Ghaemmaghami, A. M., Sun, H., & Mata, A. (2022). Rational design of hydrogels for immunomodulation. *Regenerative Biomaterials, 9*, rbac009. Advance online publication. DOI: 10.1093/rb/rbac009 PMID: 35668923

Catoira, M. C., González-Payo, J., Fusaro, L., Ramella, M., & Boccafoschi, F. (2020). Natural hydrogels R&D process: Technical and regulatory aspects for industrial implementation. *Journal of Materials Science. Materials in Medicine, 31*(8), 64. Advance online publication. DOI: 10.1007/s10856-020-06401-w PMID: 32696261

Chen, Q., Chen, H., Zhu, L., & Zheng, J. (2015). Fundamentals of double network hydrogels. *Journal of Materials Chemistry. B, 3*(18), 3654–3676. DOI: 10.1039/C5TB00123D PMID: 32262840

Chen, R., Gao, M., Chu, D., Cheng, W., & Lu, Y. (2024). Self-powered hydrogel wearable bioelectronics. *Nano Energy, 128*, 109960. DOI: 10.1016/j.nanoen.2024.109960

Chen, Y., Zhang, Y., Li, H., Shen, J., Zhang, F., He, J., Lin, J., Wang, B., Niu, S., Han, Z., & Guo, Z. (2023a). Bioinspired hydrogel actuator for soft robotics: Opportunity and challenges. *Nano Today, 49*, 101764. DOI: 10.1016/j.nantod.2023.101764

Chen, Y., Zhang, Y., Li, H., Shen, J., Zhang, F., He, J., Lin, J., Wang, B., Niu, S., Han, Z., & Guo, Z. (2023b). Bioinspired hydrogel actuator for soft robotics: Opportunity and challenges. *Nano Today, 49*, 101764. DOI: 10.1016/j.nantod.2023.101764

Chen, Y., Zhou, Y., Hu, Z., Lu, W., Li, Z., Gao, N., Liu, N., Li, Y., He, J., Gao, Q., Xie, Z., Li, J., & He, Y. (2023). Gelatin-Based Metamaterial Hydrogel Films with High Conformality for Ultra-Soft Tissue Monitoring. *Nano-Micro Letters, 16*(1), 34. Advance online publication. DOI: 10.1007/s40820-023-01225-z PMID: 38019305

Cooper, G., Barker, A. T., Heller, B. W., Good, T., Kenney, L. P., & Howard, D. (2011). The use of hydrogel as an electrode–skin interface for electrode array FES applications. *Medical Engineering & Physics, 33*(8), 967–972. DOI: 10.1016/j.medengphy.2011.03.008 PMID: 21482167

Cui, T., Li, D., Huang, X., Yan, A., Dong, Y., Xu, J., Guo, Y., Wang, Y., Chen, Z., Shao, W., Tang, Z., Tian, H., Yang, Y., & Ren, T. (2022). Graphene-Based flexible electrode for electrocardiogram signal monitoring. *Applied Sciences (Basel, Switzerland), 12*(9), 4526. DOI: 10.3390/app12094526

Cysewska, K., & Pawłowska, S. (2024). A simple route of providing a soft interface for PEDOT:PSS film metallic electrodes without loss of their electrical interface parameters. *Electrochimica Acta, 145115*, 145115. Advance online publication. DOI: 10.1016/j.electacta.2024.145115

Dai, Y., Nolan, J., Madsen, E., Fratus, M., Lee, J., Zhang, J., Lim, J., Hong, S., Alam, M. A., Linnes, J. C., Lee, H., & Lee, C. H. (2023). Wearable Sensor Patch with Hydrogel Microneedles for In Situ Analysis of Interstitial Fluid. *ACS Applied Materials & Interfaces*, acsami.3c12740. Advance online publication. DOI: 10.1021/acsami.3c12740 PMID: 38041570

Dave, S., Churi, H. S., Vishwakarma, P. A., Krishnamoorthy, A., & Jagtap, U. P. (2024). Advancements in Healthcare through 3D- Printed Micro and nanosensors: Innovation, Application, and Prospects. *Hybrid Advances, 100311*, 100311. Advance online publication. DOI: 10.1016/j.hybadv.2024.100311

De Siqueira, E. C., De França, J. A., De Souza, R. F. M., Da Silva Leoterio, D. M., Cordeiro, J. N., & Doboszewski, B. (2023). Mecanisms of the chemical crosslinking to obtain the hydrogels: Synthesis, conditions of crosslinking and biopharmaceutical applications. *Research Society and Development, 12*(8), e18312943072. DOI: 10.33448/rsd-v12i8.43072

Deng, S. (2024). Application of graphene oxide nanosheet lactate biosensors in continuous assessment of athlete fitness. *Alexandria Engineering Journal, 88*, 31–35. DOI: 10.1016/j.aej.2024.01.017

Deng, Z., Guo, Y., Zhao, X., Du, T., Zhu, J., Xie, Y., Wu, F., Wang, Y., & Guan, M. (2022). Poly(N-Isopropylacrylamide) based electrically conductive hydrogels and their applications. *Gels (Basel, Switzerland), 8*(5), 280. DOI: 10.3390/gels8050280 PMID: 35621578

Dias, D., Resina, L., Ferreira, F. C., Sanjuan-Alberte, P., & Esteves, T. (2024). Synthesis strategies and cancer therapy applications of PEDOT nanoparticles. *Materials Advances, 5*(19), 7561–7583. DOI: 10.1039/D4MA00260A

Dong, L., Bu, Z., Xiong, Y., Zhang, H., Fang, J., Hu, H., Liu, Z., & Li, X. (2021). Facile extrusion 3D printing of gelatine methacrylate/Laponite nanocomposite hydrogel with high concentration nanoclay for bone tissue regeneration. *International Journal of Biological Macromolecules, 188*, 72–81. DOI: 10.1016/j.ijbiomac.2021.07.199 PMID: 34364938

Duarte, J., Mascarenhas-Melo, F., Pires, P. C., Veiga, F., & Paiva-Santos, A. C. (2024). Multifunctional hydrogels-based therapies for chronic diabetic wound healing. *European Polymer Journal, 211*, 113026. DOI: 10.1016/j.eurpolymj.2024.113026

Efstathiou, S., Wemyss, A. M., Coursari, D., Hand, R. A., Tinley, E. C., Ford, J., Edwards, S. E., Bates, S., Evans, R. L., Khoshdel, E., & Haddleton, D. M. (2025). Development of colorimetric PEG-based hydrogel sensors for urea detection. *Soft Matter, 21*(12), 2268–2281. Advance online publication. DOI: 10.1039/D4SM01500B PMID: 39992679

El-Husseiny, H. M., Mady, E. A., Hamabe, L., Abugomaa, A., Shimada, K., Yoshida, T., Tanaka, T., Yokoi, A., Elbadawy, M., & Tanaka, R. (2021). Smart/stimuli-responsive hydrogels: Cutting-edge platforms for tissue engineering and other biomedical applications. *Materials Today. Bio, 13*, 100186. DOI: 10.1016/j.mtbio.2021.100186 PMID: 34917924

El-Sherbiny, I. M., & Yacoub, M. H. (2013). Hydrogel scaffolds for tissue engineering: Progress and challenges. *Global Cardiology Science & Practice, 2013*(3), 38. DOI: 10.5339/gcsp.2013.38 PMID: 24689032

Fahim, M., Shahzaib, A., Nishat, N., Jahan, A., Bhat, T. A., & Inam, A. (2024). Green Synthesis of Silver Nanoparticles: A Comprehensive review of methods, influencing factors, and applications. *JCIS Open, 16*, 100125. DOI: 10.1016/j.jciso.2024.100125

Ficili, I., Giacobbe, M., Tricomi, G., & Puliafito, A. (2025). From Sensors to Data Intelligence: Leveraging IoT, Cloud, and Edge Computing with AI. *Sensors (Basel), 25*(6), 1763. DOI: 10.3390/s25061763 PMID: 40292910

Garcia-Garcia, A., Muñana-González, S., Lanceros-Mendez, S., Ruiz-Rubio, L., Alvarez, L. P., & Vilas-Vilela, J. L. (2024). Biodegradable natural hydrogels for tissue engineering, controlled release, and soil remediation. *Polymers, 16*(18), 2599. DOI: 10.3390/polym16182599 PMID: 39339063

Ghosh, S., Kumar, N., & Chattopadhyay, S. (2024). ELECTRICALLY CONDUCTIVE "SMART" HYDROGELS FOR ON-DEMAND DRUG DELIVERY. *Asian Journal of Pharmaceutical Sciences, 20*(1), 101007. DOI: 10.1016/j.ajps.2024.101007 PMID: 39935975

Hahn, M., Taite, L., Moon, J., Rowland, M., Ruffino, K., & West, J. (2005). Photolithographic patterning of polyethylene glycol hydrogels. *Biomaterials, 27*(12), 2519–2524. DOI: 10.1016/j.biomaterials.2005.11.045 PMID: 16375965

Hartwig, A., Arand, M., Epe, B., Guth, S., Jahnke, G., Lampen, A., Martus, H., Monien, B., Rietjens, I. M. C. M., Schmitz-Spanke, S., Schriever-Schwemmer, G., Steinberg, P., & Eisenbrand, G. (2020). Mode of action-based risk assessment of genotoxic carcinogens. *Archives of Toxicology, 94*(6), 1787–1877. DOI: 10.1007/s00204-020-02733-2 PMID: 32542409

Hasan, M. B., Parvez, M. M., Abir, A. Y., & Ahmad, M. F. (2025). A review on conducting organic polymers: Concepts, applications, and potential environmental benefits. *Heliyon, 42375*(3), e42375. Advance online publication. DOI: 10.1016/j.heliyon.2025.e42375 PMID: 39975833

Hiwrale, A., Bharati, S., Pingale, P., & Rajput, A. (2023). Nanofibers: A current era in drug delivery system. *Heliyon, 9*(9), e18917. DOI: 10.1016/j.heliyon.2023.e18917 PMID: 37674834

Ho, T., Chang, C., Chan, H., Chung, T., Shu, C., Chuang, K., Duh, T., Yang, M., & Tyan, Y. (2022). Hydrogels: Properties and applications in biomedicine. *Molecules (Basel, Switzerland), 27*(9), 2902. DOI: 10.3390/molecules27092902 PMID: 35566251

Hsieh, J., Alawieh, H., Li, Y., Iwane, F., Zhao, L., Anderson, R., Abdullah, S. I., Tang, K. W. K., Wang, W., Pyatnitskiy, I., Jia, Y., Del, R., Millán, J., & Wang, H. (2022). A highly stable electrode with low electrode-skin impedance for wearable brain-computer interface. *Biosensors & Bioelectronics, 218*, 114756. DOI: 10.1016/j.bios.2022.114756 PMID: 36209529

Hua, J., Su, M., Sun, X., Li, J., Sun, Y., Qiu, H., Shi, Y., & Pan, L. (2023). Hydrogel-Based Bioelectronics and their applications in health monitoring. *Biosensors (Basel), 13*(7), 696. DOI: 10.3390/bios13070696 PMID: 37504095

Huanbutta, K., Puri, V., Sharma, A., Singh, I., Sriamornsak, P., & Sangnim, T. (2024). The rise of implantable drugs: A chronicle of breakthroughs in drug delivery systems. *Saudi Pharmaceutical Journal : SPJ : the Official Publication of the Saudi Pharmaceutical Society, 32*(12), 102193. DOI: 10.1016/j.jsps.2024.102193 PMID: 39564378

Irmukhametova, G., Kazybayeva, D., Mun, G., & Khutoryanskiy, V. (2025). Synthesis and chemistry of hydrogels. In *Elsevier eBooks* (pp. 1–37). https://doi.org/DOI: 10.1016/B978-0-443-22017-3.00004-4

Isaac, A. H., Phillips, S. Y. R., Ruben, E., Estes, M., Rajavel, V., Baig, T., Paleti, C., Landsgaard, K., Lee, R. H., Guda, T., Criscitiello, M. F., Gregory, C., & Alge, D. L. (2024). Impact of PEG sensitization on the efficacy of PEG hydrogel-mediated tissue engineering. *Nature Communications, 15*(1), 3283. Advance online publication. DOI: 10.1038/s41467-024-46327-3 PMID: 38637507

Jahanbekam, S., Asare-Addo, K., Alipour, S., & Nokhodchi, A. (2025). Smart hydrogels and the promise of Multi-Responsive In-Situ systems. *Journal of Drug Delivery Science and Technology, 107*, 106758. DOI: 10.1016/j.jddst.2025.106758

Janik, M., Koba, M., & Śmietana, M. (2024). Optical fiber chemo and biosensors operating in the electrochemical domain – A review. *Trends in Analytical Chemistry, 178*, 117829. DOI: 10.1016/j.trac.2024.117829

Javaid, M., Haleem, A., & Singh, R. P. (2024). Health informatics to enhance the healthcare industry's culture: An extensive analysis of its features, contributions, applications and limitations. *Informatics and Health, 1*(2), 123–148. DOI: 10.1016/j.infoh.2024.05.001

Javaid, M., Haleem, A., Singh, R. P., & Suman, R. (2022). 3D printing applications for healthcare research and development. *Global Health Journal (Amsterdam, Netherlands), 6*(4), 217–226. DOI: 10.1016/j.glohj.2022.11.001

Jiao, C., Liu, J., Yan, S., Xu, Z., Hou, Z., & Xu, W. (2025). Hydrogel soft bioelectronic interfaces and their applications. *Journal of Materials Chemistry. C, Materials for Optical and Electronic Devices, 13*(6), 2620–2645. Advance online publication. DOI: 10.1039/D4TC04150J

Jorfi, M., Skousen, J. L., Weder, C., & Capadona, J. R. (2014). Progress towards biocompatible intracortical microelectrodes for neural interfacing applications. *Journal of Neural Engineering, 12*(1), 011001. DOI: 10.1088/1741-2560/12/1/011001 PMID: 25460808

Khan, I., Saeed, K., & Khan, I. (2017). Nanoparticles: Properties, applications and toxicities. *Arabian Journal of Chemistry*, *12*(7), 908–931. DOI: 10.1016/j.arabjc.2017.05.011

Khan, M. M. R., & Rumon, M. M. H. (2025). Synthesis of PVA-Based Hydrogels for Biomedical applications: Recent trends and advances. *Gels (Basel, Switzerland)*, *11*(2), 88. DOI: 10.3390/gels11020088 PMID: 39996631

Khan, M. U. A., Aslam, M. A., Abdullah, M. F. B., Al-Arjan, W. S., Stojanovic, G. M., & Hasan, A. (2024). Hydrogels: Classifications, fundamental properties, applications, and scopes in recent advances in tissue engineering and regenerative medicine – A comprehensive review. *Arabian Journal of Chemistry*, *17*(10), 105968. DOI: 10.1016/j.arabjc.2024.105968

Kılıç, S. (2024). Brain-Computer Interfaces enhanced by AI: Applications in rehabilitation and assistive technology. *Next Frontier.*, *8*(1), 207. DOI: 10.62802/m89avz38

Kim, D. G., & Cha, G. D. (2025). Extracellular-Matrix-Mimetic hydrogels by using nanomaterials. *International Journal of Molecular Sciences*, *26*(11), 4987. DOI: 10.3390/ijms26114987 PMID: 40507800

Király, A., & Ronkay, F. (2013). Effect of filler dispersion on the electrical conductivity and mechanical properties of carbon/polypropylene composites. *Polymer Composites*, *34*(7), 1195–1203. DOI: 10.1002/pc.22530

Lewin, S., Chetty, R., Ihdayhid, A. R., & Dwivedi, G. (2024). Ethical challenges and opportunities in applying artificial intelligence to cardiovascular medicine. *The Canadian Journal of Cardiology*, *40*(10), 1897–1906. DOI: 10.1016/j.cjca.2024.06.029 PMID: 39038648

Li, F. (2024). Application and challenges of artificial intelligence in cybersecurity. *Applied and Computational Engineering*, *47*(1), 262–268. DOI: 10.54254/2755-2721/47/20241480

Li, J., Liu, J., & Gao, C. (2009). On the mechanism of conductivity enhancement in PEDOT/PSS film doped with multi-walled carbon nanotubes. *Journal of Polymer Research*, *17*(5), 713–718. DOI: 10.1007/s10965-009-9360-1

Li, L., Wu, P., Yu, F., & Ma, J. (2022). Double network hydrogels for energy/environmental applications: Challenges and opportunities. *Journal of Materials Chemistry. A, Materials for Energy and Sustainability*, *10*(17), 9215–9247. DOI: 10.1039/D2TA00540A

Li, P., Hu, J., Wang, J., Zhang, J., Wang, L., & Zhang, C. (2023). The role of hydrogel in cardiac repair and regeneration for myocardial infarction: Recent advances and future perspectives. *Bioengineering (Basel, Switzerland)*, *10*(2), 165. DOI: 10.3390/bioengineering10020165 PMID: 36829659

Li, Y., Cheng, Q., Deng, Z., Zhang, T., Luo, M., Huang, X., Wang, Y., Wang, W., & Zhao, X. (2024). Recent progress of Anti-Freezing, Anti-Drying, and Anti-Swelling conductive hydrogels and their applications. *Polymers*, *16*(7), 971. DOI: 10.3390/polym16070971 PMID: 38611229

Li, Z., Song, P., Li, G., Han, Y., Ren, X., Bai, L., & Su, J. (2024). AI energized hydrogel design, optimization and application in biomedicine. *Materials Today. Bio*, *25*, 101014. DOI: 10.1016/j.mtbio.2024.101014 PMID: 38464497

Liang, X., Chen, Z., Ruan, H., & Li, Y. (2025). Fishnet-Inspired Dual-Crosslinked Nanocomposite Hydrogels with Core-Shell SiO2@ZnO for Self-Powered Sensing and Intelligent Mobility Assistance Systems. *Colloids and Surfaces. A, Physicochemical and Engineering Aspects, 137365*, 137365. Advance online publication. DOI: 10.1016/j.colsurfa.2025.137365

Lin, C., Huang, X., Xue, Y., Jiang, S., Chen, C., Liu, Y., & Chen, K. (2025). Advances in medical devices using nanomaterials and nanotechnology: Innovation and regulatory science. *Bioactive Materials*, *48*, 353–369. DOI: 10.1016/j.bioactmat.2025.02.017 PMID: 40060145

Linh, V. T. N., Han, S., Koh, E., Kim, S., Jung, H. S., & Koo, J. (2024). Advances in wearable electronics for monitoring human organs: Bridging external and internal health assessments. *Biomaterials*, *314*, 122865. DOI: 10.1016/j.biomaterials.2024.122865 PMID: 39357153

Liu, S., Zhao, Y., Hao, W., Zhang, X., & Ming, D. (2020). Micro- and nanotechnology for neural electrode-tissue interfaces. *Biosensors & Bioelectronics*, *170*, 112645. DOI: 10.1016/j.bios.2020.112645 PMID: 33010703

Liu, Y., & Echeverry-Rendón, M. (2025). 3D-printed biodegradable polymer scaffolds for tissue engineering: An overview, current stage and future perspectives. *Next Materials*, *8*, 100647. DOI: 10.1016/j.nxmate.2025.100647

Lu, X., Luo, Y., Hu, X., Chu, J., Li, S., Hao, M., Zhuang, J., Liu, Y., Gao, J., & Yin, Y. (2024). Emerging hydrogel therapies for translating brain disease: Materials, mechanisms, and recent research. *Materials Today. Advances*, *22*, 100490. DOI: 10.1016/j.mtadv.2024.100490

Ma, Z., Jiang, Y., Xiao, H., Jiang, B., Zhang, H., Peng, M., Dong, G., Yu, X., & Yang, J. (2017). Sol-gel preparation of Ag-silica nanocomposite with high electrical conductivity. *Applied Surface Science*, *436*, 732–738. DOI: 10.1016/j.apsusc.2017.12.101

Mahalakshmi, K., Palanivelu, V. R., & Kirubakaran, D. (2025). *Global market trends in biomedical sensors: materials, device engineering, and healthcare applications*. Deleted Journal., DOI: 10.1007/s44174-025-00362-w

Mariello, M. (2025). Advances and perspectives on bioelectronic and atomic nanogenerators for anticancer therapy. *Nanoenergy Advances*, *5*(2), 4. DOI: 10.3390/nanoenergyadv5020004

Masarone, S., Beckwith, K. V. V., Wilkinson, M., Tuli, S., Lane, A., Windsor, S., Lane, J., & Hosseini-Gerami, L. (2024). Advancing predictive toxicology: Overcoming hurdles and shaping the future. *Digital Discovery*. Advance online publication. DOI: 10.1039/D4DD00257A

Mirshafiei, M., Rashedi, H., Yazdian, F., Rahdar, A., & Baino, F. (2024). Advancements in tissue and organ 3D bioprinting: Current techniques, applications, and future perspectives. *Materials & Design*, *240*, 112853. DOI: 10.1016/j.matdes.2024.112853

Mo, F., Lin, Y., Liu, Y., Zhou, P., Yang, J., Ji, Z., & Wang, Y. (2025). Advances in ionic conductive hydrogels for skin sensor applications. *Materials Science and Engineering R Reports*, *165*, 100989. DOI: 10.1016/j.mser.2025.100989

Mohammed, S., Nanthini, S., Krishna, N. B., Srinivas, I. V., Rajagopal, M., & Kumar, M. A. (2023). A new lightweight data security system for data security in the cloud computing. *Measurement. Sensors*, *29*, 100856. DOI: 10.1016/j.measen.2023.100856

Moreno, J. M. C., Chelu, M., & Popa, M. (2025). Eco-Friendly Conductive hydrogels: Towards green wearable electronics. *Gels (Basel, Switzerland)*, *11*(4), 220. DOI: 10.3390/gels11040220 PMID: 40277656

Mukherjee, M. D., Gupta, P., Kumari, V., Rana, I., Jindal, D., Sagar, N., Singh, J., & Dhand, C. (2025). Wearable Biosensors in Modern Healthcare: Emerging Trends and Practical applications. *Talanta Open*, *100486*, 100486. Advance online publication. DOI: 10.1016/j.talo.2025.100486

Mumtaz, N., Li, Y., Artiaga, R., Farooq, Z., Mumtaz, A., Guo, Q., & Nisa, F. (2024). Fillers and methods to improve the effective (out-plane) thermal conductivity of polymeric thermal interface materials – A review. *Heliyon*, *10*(3), e25381. DOI: 10.1016/j.heliyon.2024.e25381 PMID: 38352797

Munasir, N., Prapanca, A., Aliansah, M. F., Paramudhita, F. A., Faaizatunnisa, N., Ariesta, M. N., & Taufiq, A. (2025). Self-Healing Graphene-Based Composite Hydrogels for Motion sensing: Source, fabrication, and Applications in Assistive Technologies – A review. *Sensors International*, *100338*, 100338. Advance online publication. DOI: 10.1016/j.sintl.2025.100338

Musa, A. A., Bello, A., Adams, S. M., Onwualu, A. P., Anye, V. C., Bello, K. A., & Obianyo, I. I. (2025). Nano-Enhanced Polymer Composite Materials: A review of current advancements and challenges. *Polymers*, *17*(7), 893. DOI: 10.3390/polym17070893 PMID: 40219283

Naghib, S. M., Amiri, S., & Mozafari, M. R. (2024). Stimuli-responsive chitosan-based nanocarriers for drug delivery in wound dressing applications: A review. *Carbohydrate Polymer Technologies and Applications*, *7*, 100497. DOI: 10.1016/j.carpta.2024.100497

Nezhad-Mokhtari, P., Hasany, M., Kohestanian, M., Dolatshahi-Pirouz, A., Milani, M., & Mehrali, M. (2024). Recent advancements in bioadhesive self-healing hydrogels for effective chronic wound care. *Advances in Colloid and Interface Science*, *334*, 103306. DOI: 10.1016/j.cis.2024.103306 PMID: 39423587

Nguyen, T. D., Nguyen, T., Vo, V. T., & Nguyen, T. (2025). Advances in three-dimensional printing of hydrogel formulations for vascularized tissue and organ regeneration. *Journal of Biomaterials Science. Polymer Edition*, •••, 1–43. DOI: 10.1080/09205063.2025.2582702 PMID: 39899080

Norahan, M. H., Pedroza-González, S. C., Sánchez-Salazar, M. G., Álvarez, M. M., & De Santiago, G. T. (2022). Structural and biological engineering of 3D hydrogels for wound healing. *Bioactive Materials*, *24*, 197–235. DOI: 10.1016/j.bioactmat.2022.11.019 PMID: 36606250

Obaideen, K., Albasha, L., Iqbal, U., & Mir, H. (2024). Wireless power transfer: Applications, challenges, barriers, and the role of AI in achieving sustainable development goals - A bibliometric analysis. *Energy Strategy Reviews*, *53*, 101376. DOI: 10.1016/j.esr.2024.101376

Ognard, J., Hajj, G. E., Verma, O., Ghozy, S., Kadirvel, R., Kallmes, D. F., & Brinjikji, W. (2025). Advances in Endovascular Brain Computer Interface: Systematic review and future implications. *Journal of Neuroscience Methods, 420,* 110471. DOI: 10.1016/j.jneumeth.2025.110471 PMID: 40355001

Oh, M., Kim, E., Lee, J., Jeong, I., Kim, E., Paek, J., Lee, T., Kim, D., An, S. H., Kim, S., Lim, J. A., & Park, J. (2025). Machine Learning Enhanced multimodal bioelectronics: Advancement toward intelligent healthcare systems. *Advanced Sensor Research, 4*(7), e00028. Advance online publication. DOI: 10.1002/adsr.202500028

Omidian, H., & Chowdhury, S. D. (2025). Multifunctional Hydrogel microneedles (HMNs) in drug delivery and diagnostics. *Gels (Basel, Switzerland), 11*(3), 206. DOI: 10.3390/gels11030206 PMID: 40136911

Oribe, S., Yoshida, S., Kusama, S., Osawa, S., Nakagawa, A., Iwasaki, M., Tominaga, T., & Nishizawa, M. (2019). Hydrogel-Based Organic Subdural Electrode with High Conformability to Brain Surface. *Scientific Reports, 9*(1), 13379. Advance online publication. DOI: 10.1038/s41598-019-49772-z PMID: 31527626

Owens, G. J., Singh, R. K., Foroutan, F., Alqaysi, M., Han, C., Mahapatra, C., Kim, H., & Knowles, J. C. (2016). Sol–gel based materials for biomedical applications. *Progress in Materials Science, 77,* 1–79. DOI: 10.1016/j.pmatsci.2015.12.001

Pal, R. K., Turner, E. E., Chalfant, B. H., & Yadavalli, V. K. (2017). Mechanically robust, photopatternable conductive hydrogel composites. *Reactive & Functional Polymers, 120,* 66–73. DOI: 10.1016/j.reactfunctpolym.2017.09.006

Perkmann, M., Tartari, V., McKelvey, M., Autio, E., Broström, A., D'Este, P., Fini, R., Geuna, A., Grimaldi, R., Hughes, A., Krabel, S., Kitson, M., Llerena, P., Lissoni, F., Salter, A., & Sobrero, M. (2012). Academic engagement and commercialisation: A review of the literature on university–industry relations. *Research Policy, 42*(2), 423–442. DOI: 10.1016/j.respol.2012.09.007

Persaud, A., Maus, A., Strait, L., & Zhu, D. (2022). 3D Bioprinting with Live Cells. *Engineered Regeneration, 3*(3), 292–309. DOI: 10.1016/j.engreg.2022.07.002

Prain, G., Wheatley, C., Odsey, C., Verzola, L., Bertuso, A., Roa, J., & Naziri, D. (2020). Research-development partnerships for scaling complex innovation: Lessons from the Farmer Business School in IFAD-supported loan-grant collaborations in Asia. *Agricultural Systems, 182,* 102834. DOI: 10.1016/j.agsy.2020.102834

Qian, X., & Liao, C. (2023). Engineering liquid metal-based implantable electrodes toward brain-machine interfaces. *Health Sciences Review (Oxford, England), 9,* 100118. DOI: 10.1016/j.hsr.2023.100118

Qin, D., Xia, Y., & Whitesides, G. M. (2010). Soft lithography for micro- and nanoscale patterning. *Nature Protocols, 5*(3), 491–502. DOI: 10.1038/nprot.2009.234 PMID: 20203666

Rahmani, P., & Shojaei, A. (2021). A review on the features, performance and potential applications of hydrogel-based wearable strain/pressure sensors. *Advances in Colloid and Interface Science, 298,* 102553. DOI: 10.1016/j.cis.2021.102553 PMID: 34768136

Raina, N., Pahwa, R., Bhattacharya, J., Paul, A. K., Nissapatorn, V., De Lourdes Pereira, M., Oliveira, S. M. R., Dolma, K. G., Rahmatullah, M., Wilairatana, P., & Gupta, M. (2022). Drug delivery Strategies and Biomedical Significance of hydrogels: Translational considerations. *Pharmaceutics*, *14*(3), 574. DOI: 10.3390/pharmaceutics14030574 PMID: 35335950

Rashid, A. B., Haque, M., Islam, S. M. M., & Labib, K. R. U. (2024). Nanotechnology-enhanced fiber-reinforced polymer composites: Recent advancements on processing techniques and applications. *Heliyon*, *10*(2), e24692. DOI: 10.1016/j.heliyon.2024.e24692 PMID: 38298690

Reddy, K. R., El-Zein, A., Airey, D. W., Alonso-Marroquin, F., Schubel, P., & Manalo, A. (2020). Self-healing polymers: Synthesis methods and applications. *Nano-Structures & Nano-Objects*, *23*, 100500. DOI: 10.1016/j.nanoso.2020.100500

Revete, A., Aparicio, A., Cisterna, B. A., Revete, J., Luis, L., Ibarra, E., González, E. S., Molino, J., & Reginensi, D. (2022). Advancements in the use of hydrogels for Regenerative Medicine: Properties and biomedical applications. *International Journal of Biomaterials*, *2022*, 1–16. DOI: 10.1155/2022/3606765 PMID: 36387956

Ručigaj, A., Golobič, J., & Kopač, T. (2024). The role of multivalent cations in determining the cross-linking affinity of alginate hydrogels: A combined experimental and modeling study. *Chemical Engineering Journal Advances*, *20*, 100678. DOI: 10.1016/j.ceja.2024.100678

Saeidi, M., Chenani, H., Orouji, M., Rastkhiz, M. A., Bolghanabadi, N., Vakili, S., Mohamadnia, Z., Hatamie, A., & Simchi, A. (2023). Electrochemical wearable biosensors and bioelectronic devices based on hydrogels: Mechanical properties and electrochemical behavior. *Biosensors (Basel)*, *13*(8), 823. DOI: 10.3390/bios13080823 PMID: 37622909

Sakka, S. (2007). Sol–gel technology as representative processing for nanomaterials: Case studies on the starting solution. *Journal of Sol-Gel Science and Technology*, *46*(3), 241–249. DOI: 10.1007/s10971-007-1651-6

Sánchez-Fernández, J. A. (2023). Structural strategies for supramolecular hydrogels and their applications. *Polymers*, *15*(6), 1365. DOI: 10.3390/polym15061365 PMID: 36987146

Shah, S. S., Niaz, F., Ehsan, M. A., Das, H. T., Younas, M., Khan, A. S., Rahman, H. U., Nayem, S. A., Oyama, M., & Aziz, M. A. (2023). Advanced strategies in electrode engineering and nanomaterial modifications for supercapacitor performance enhancement: A comprehensive review. *Journal of Energy Storage*, *79*, 110152. DOI: 10.1016/j.est.2023.110152

Shang, T., Zhang, J. Y., Thomas, A., Arnold, M. A., Vetter, B. N., Heinemann, L., & Klonoff, D. C. (2021). Products for monitoring glucose levels in the human body with noninvasive optical, noninvasive fluid sampling, or minimally invasive technologies. *Journal of Diabetes Science and Technology*, *16*(1), 168–214. DOI: 10.1177/19322968211007212 PMID: 34120487

She, J., Liu, J., Mu, Y., Lv, S., Tong, J., Liu, L., He, T., Wang, J., & Wei, D. (2024). Recent advances in collagen-based hydrogels: Materials, preparation and applications. *Reactive & Functional Polymers*, *207*, 106136. DOI: 10.1016/j.reactfunctpolym.2024.106136

Shetty, K. H., Desai, D. T., Patel, H. P., Shah, D. O., Willcox, M. D., & Maulvi, F. A. (2024). Contact lens as an emerging platform for non-invasive bio-sensing: A review. *Sensors and Actuators. A, Physical, 376*, 115617. DOI: 10.1016/j.sna.2024.115617

Shin, Y., Lee, H. S., Jeong, H., & Kim, D. (2024). *Recent advances in conductive hydrogels for soft biointegrated electronics: materials, properties, and device applications*. Wearable Electronics., DOI: 10.1016/j.wees.2024.10.004

Sultana, A., Zare, M., Thomas, V., Kumar, T. S., & Ramakrishna, S. (2022). Nano-based drug delivery systems: Conventional drug delivery routes, recent developments and future prospects. *Medicine in Drug Discovery, 15*, 100134. DOI: 10.1016/j.medidd.2022.100134

Sun, C., Cheng, Z., Abu-Halimah, J., & Tian, B. (2023). Perspectives on tissue-like bioelectronics for neural modulation. *iScience, 26*(5), 106715. DOI: 10.1016/j.isci.2023.106715 PMID: 37216128

Sunwoo, S., Han, S. I., Joo, H., Cha, G. D., Kim, D., Choi, S. H., Hyeon, T., & Kim, D. (2020). Advances in soft bioelectronics for brain research and clinical neuroengineering. *Matter, 3*(6), 1923–1947. DOI: 10.1016/j.matt.2020.10.020

Tang, H., Li, Y., Liao, S., Liu, H., Qiao, Y., & Zhou, J. (2024). Multifunctional conductive hydrogel interface for bioelectronic recording and stimulation. *Advanced Healthcare Materials, 13*(22), 2400562. Advance online publication. DOI: 10.1002/adhm.202400562 PMID: 38773929

Tanwar, A., Gandhi, H., Kushwaha, D., & Bhattacharya, J. (2022). A review on microelectrode array fabrication techniques and their applications. *Materials Today. Chemistry, 26*, 101153. DOI: 10.1016/j.mtchem.2022.101153

Tuli, N. T., Khatun, S., & Rashid, A. B. (2024). Unlocking the future of precision manufacturing: A comprehensive exploration of 3D printing with fiber-reinforced composites in aerospace, automotive, medical, and consumer industries. *Heliyon, 10*(5), e27328. DOI: 10.1016/j.heliyon.2024.e27328 PMID: 38495162

Ullah, A., Kim, D. Y., Lim, S. I., & Lim, H. (2025). Hydrogel-Based Biointerfaces: Recent advances, challenges, and future directions in Human–Machine Integration. *Gels (Basel, Switzerland), 11*(4), 232. DOI: 10.3390/gels11040232 PMID: 40277668

Van Doorn, W. P. T. M., Foreman, Y. D., Schaper, N. C., Savelberg, H. H. C. M., Koster, A., Van Der Kallen, C. J. H., Wesselius, A., Schram, M. T., Henry, R. M. A., Dagnelie, P. C., De Galan, B. E., Bekers, O., Stehouwer, C. D. A., Meex, S. J. R., & Brouwers, M. C. G. J. (2021). Machine learning-based glucose prediction with use of continuous glucose and physical activity monitoring data: The Maastricht Study. *PLoS One, 16*(6), e0253125. DOI: 10.1371/journal.pone.0253125 PMID: 34166426

Van Norman, G. A. (2019). Limitations of animal studies for predicting toxicity in clinical trials. *JACC. Basic to Translational Science, 4*(7), 845–854. DOI: 10.1016/j.jacbts.2019.10.008 PMID: 31998852

Vidal, G. W. V., Rynes, M. L., Kelliher, Z., & Goodwin, S. J. (2016). Review of Brain-Machine Interfaces Used in Neural Prosthetics with New Perspective on Somatosensory Feedback through Method of Signal Breakdown. *Scientifica, 2016*, 1–10. DOI: 10.1155/2016/8956432 PMID: 27313959

Virumbrales, C., Hernández-Ruiz, R., Trigo-López, M., Vallejos, S., & García, J. M. (2024). Sensory Polymers: Trends, challenges, and Prospects ahead. *Sensors (Basel)*, *24*(12), 3852. DOI: 10.3390/s24123852 PMID: 38931634

Vo, R., Hsu, H., & Jiang, X. (2020). Hydrogel facilitated bioelectronic integration. *Biomaterials Science*, *9*(1), 23–37. DOI: 10.1039/D0BM01373K PMID: 33094761

Wan, X., Wang, Z., Liu, M., Zhang, F., & Wang, S. (2024). Bioinspired structural adhesives: A decades-old science but emerging materials. *Matter*, *7*(5), 1710–1723. DOI: 10.1016/j.matt.2024.02.015

Wang, H., Du, J., & Mao, Y. (2025). Hydrogel-Based continuum soft robots. *Gels (Basel, Switzerland)*, *11*(4), 254. DOI: 10.3390/gels11040254 PMID: 40277689

Wang, J., Li, L., Wu, Y., & Liu, Y. (2025). Design and application of Antifouling Bio-Coatings. *Polymers*, *17*(6), 793. DOI: 10.3390/polym17060793 PMID: 40292673

Wang, J., Yang, B., Jiang, Z., Liu, Y., Zhou, L., Liu, Z., & Tang, L. (2024). Recent advances of conductive hydrogels for flexible electronics. *Electronic Materials*, *5*(3), 101–131. DOI: 10.3390/electronicmat5030008

Wang, S., Wei, L., Wang, F., Wang, L., & Mao, J. (2025). Advanced implantable energy storage for powering medical devices. *eScience*, 100409. https://doi.org/DOI: 10.1016/j.esci.2025.100409

Wang, X., Liu, J., Dong, R., Gilchrist, M. D., & Zhang, N. (2024). High-precision digital light processing (DLP) printing of microstructures for microfluidics applications based on a machine learning approach. *Virtual and Physical Prototyping*, *19*(1), e2318774. Advance online publication. DOI: 10.1080/17452759.2024.2318774

Wang, X., Sun, X., Gan, D., Soubrier, M., Chiang, H., Yan, L., Li, Y., Li, J., Yu, S., Xia, Y., Wang, K., Qin, Q., Jiang, X., Han, L., Pan, T., Xie, C., & Lu, X. (2022). Bioadhesive and conductive hydrogel-integrated brain-machine interfaces for conformal and immune-evasive contact with brain tissue. *Matter*, *5*(4), 1204–1223. DOI: 10.1016/j.matt.2022.01.012

Wang, Y., Zhang, Y., Zhang, Q., Li, X., Yan, Q., & Zhu, Y. (2024). Mechanical properties modulation and biological applications of DNA hydrogels. *Advanced Sensor and Energy Materials*, *3*(3), 100113. DOI: 10.1016/j.asems.2024.100113

Xiao, L., Jiang, H., Zhang, D., Ou, C., Lai, J., Wang, M., Ma, Y., & Huang, Y. (2023). Designing anti-dehydration and ion-conductive tough hydrogels as environment-adaptable strain sensors for e-skin. *Chemical Engineering Journal*, *474*, 145944. DOI: 10.1016/j.cej.2023.145944

Xing, J., Zhang, M., Liu, X., Wang, C., Xu, N., & Xing, D. (2023). Multi-material electrospinning: From methods to biomedical applications. *Materials Today. Bio*, *21*, 100710. DOI: 10.1016/j.mtbio.2023.100710 PMID: 37545561

Xu, F., Dawson, C., Lamb, M., Mueller, E., Stefanek, E., Akbari, M., & Hoare, T. (2022a). Hydrogels for Tissue Engineering: Addressing key design needs toward Clinical translation. *Frontiers in Bioengineering and Biotechnology*, *10*, 849831. Advance online publication. DOI: 10.3389/fbioe.2022.849831 PMID: 35600900

Xu, J., & Lee, H. (2020). Anti-Biofouling strategies for Long-Term continuous use of implantable biosensors. *Chemosensors (Basel, Switzerland)*, *8*(3), 66. DOI: 10.3390/chemosensors8030066

Xu, Y. (2024). Integrating Biological Systems with Electronic Technologies in Bioelectronics Advancements. *Highlights in Science Engineering and Technology*, *119*, 110–116. DOI: 10.54097/vq0xt606

Xue, L., An, R., Zhao, J., Qiu, M., Wang, Z., Ren, H., Yu, D., & Zhu, X. (2025). Self-Healing hydrogels: Mechanisms and biomedical applications. *MedComm*, *6*(5), e70181. Advance online publication. DOI: 10.1002/mco2.70181 PMID: 40276645

Yadav, H., & Maini, S. (2023). Electroencephalogram based brain-computer interface: Applications, challenges, and opportunities. *Multimedia Tools and Applications*, *82*(30), 47003–47047. DOI: 10.1007/s11042-023-15653-x PMID: 37362726

Yang, J., Chen, Y., Zhao, L., Zhang, J., & Luo, H. (2022). Constructions and properties of physically Cross-Linked hydrogels based on natural polymers. *Polymer Reviews (Philadelphia, Pa.)*, *63*(3), 574–612. DOI: 10.1080/15583724.2022.2137525

Ying, B., & Liu, X. (2021). Skin-like hydrogel devices for wearable sensing, soft robotics and beyond. *iScience*, *24*(11), 103174. DOI: 10.1016/j.isci.2021.103174 PMID: 34755087

Youssef, K., Ullah, A., Rezai, P., Hasan, A., & Amirfazli, A. (2023). Recent advances in biosensors for real time monitoring of pH, temperature, and oxygen in chronic wounds. *Materials Today. Bio*, *22*, 100764. DOI: 10.1016/j.mtbio.2023.100764 PMID: 37674780

Zhang, C. W., Chen, C., Duan, S., Yan, Y., He, P., & He, X. (2024). Hydrogel-based soft bioelectronics for personalized healthcare. *Med-X*, *2*(1), 20. Advance online publication. DOI: 10.1007/s44258-024-00036-0

Zhang, H., Jiao, L., Yang, S., Li, H., Jiang, X., Feng, J., Zou, S., Xu, Q., Gu, J., Wang, X., & Wei, B. (2024). Brain-computer interfaces: The innovative key to unlocking neurological conditions. *International Journal of Surgery*, *110*(9), 5745–5762. DOI: 10.1097/JS9.0000000000002022 PMID: 39166947

Zhang, H., Lin, X., Cao, X., Wang, Y., Wang, J., & Zhao, Y. (2023). Developing natural polymers for skin wound healing. *Bioactive Materials*, *33*, 355–376. DOI: 10.1016/j.bioactmat.2023.11.012 PMID: 38282639

Zhang, P., Yang, Y., Li, Z., Xue, Y., Wang, F., Shan, L., Wang, Y., Shi, X., Wu, K., & Liu, J. (2025). Conducting Hydrogel-Based neural biointerfacing technologies. *Advanced Functional Materials*, *35*(25), 2422869. Advance online publication. DOI: 10.1002/adfm.202422869

Zhang, Q., Shang, J., Qiao, H., Liu, J., Liu, X., & Yang, Q. (2025). Synergistic Ion-Electron dual conduction in mechanically robust hydrogels composites for High-Fidelity bioelectronics. *Composites Communications*, *102470*, 102470. Advance online publication. DOI: 10.1016/j.coco.2025.102470

Zhang, S., & Staples, A. E. (2024). Microfluidic-based systems for the management of diabetes. *Drug Delivery and Translational Research*, *14*(11), 2989–3008. DOI: 10.1007/s13346-024-01569-y PMID: 38509342

Zhang, Y., Li, J., Jiao, S., Li, Y., Zhou, Y., Zhang, X., Maryam, B., & Liu, X. (2024). Microfluidic sensors for the detection of emerging contaminants in water: A review. *The Science of the Total Environment*, *929*, 172734. DOI: 10.1016/j.scitotenv.2024.172734 PMID: 38663621

Zhang, Y., Tan, Y., Lao, J., Gao, H., & Yu, J. (2023). Hydrogels for flexible electronics. *ACS Nano*, *17*(11), 9681–9693. DOI: 10.1021/acsnano.3c02897 PMID: 37199537

Zhong, H., Huang, J., Wu, J., & Du, J. (2021). Electrospinning nanofibers to 1D, 2D, and 3D scaffolds and their biomedical applications. *Nano Research*, *15*(2), 787–804. DOI: 10.1007/s12274-021-3593-7

Zhu, P., Zhao, Z., & Gao, Y. (2025). Advances in hydrogel for diagnosis and treatment for Parkinson's disease. *Frontiers in Pharmacology*, *16*, 1552586. Advance online publication. DOI: 10.3389/fphar.2025.1552586 PMID: 40012627

Zou, X., Wang, X., Xie, L., Yang, Y., Wei, C., Zhou, Y., Huo, P., Yue, O., & Liu, X. (2025). Scaleable asymmetric "Janus" patch with seamless elastomer-hydrogel integration for wet on-skin electronic interfaces. *Chemical Engineering Journal*, *162244*, 162244. Advance online publication. DOI: 10.1016/j.cej.2025.162244

Chapter 16
The Biotechnology of Hydrogel-Based Biocomposites

Omoremime Elizabeth Dania
https://orcid.org/0000-0002-1713-3999
Covenant University, Ota, Nigeria

Emeka E. J. Iweala
Covenant University, Ota, Nigeria

ABSTRACT

Hydrogels are a unique type of polymer that is hydrophilic. This chapter explores the applications of hydrogels in tissue engineering, skin regeneration, soft robots, and artificial muscle production. For tissue engineering, there is a need to maintain proper hardening and desirable strength of hydrogels at physiological temperatures under biological conditions for proper formation of scaffolds with complex shapes. The use of hydrogels for skin regeneration stems from the need to correct skin defects without scarring by enhancing reparative properties and regeneration. Hydrogels promote angiogenesis, cell proliferation, and wound closure effectively. Hydrogels have been used to construct actuators, sensors, communicators, power sources, and computational circuits for soft robots. They are responsive to ambient stimuli, increasing their sensitivity; they are quite flexible and potentially enduring large deformations, making them good substrates for innovative research in artificial muscle generation. While hydrogels offer significant advantages in wound healing and tissue engineering, challenges such as optimising mechanical strength, biocompatibility, and production scalability remain.

16.1 INTRODUCTION TO HYDROGELS

As science and technology evolve and discoveries are made in material science and the life sciences, the need for a better understanding of the three-dimensional arrangement of biological systems to aid a better understanding of the basic biological processes, so that more effective treatments and improved quality of life can be made available as a result of these new developments,

DOI: 10.4018/979-8-3373-2337-4.ch016

is important. This chapter aims to explore the applications of hydrogel-based bio composites in life science for tissue engineering, regeneration of skin and soft tissues, development of artificial muscles and in hydrogel robotics. For such biocomposites to be effective in the mission they are proposed for, they need to be biocompatible, be durable to withstand mechanical stress, have a high propensity for cellular attachment and allow for cellular migration and proliferation. The biocomposite materials used in the field of medicine need to have properties that make their handling easier and be easily sterilised to prevent contamination, which can result in life-threatening infections when they are administered.

Hydrogels have existed for over 50 years. They are a unique set of polymers with the ability to absorb and retain large amounts of water. When the hydrophilic groups present in the polymer become hydrated, the hydrogel structure is formed (Akhtar et al., 2015). Their viscoelastic network permits the diffusion and attachment of molecules and substances. Hydrogels were first synthesised in 1960 and used in contact lens manufacturing based on their ability to absorb and hold moisture due to the presence of hydrophilic molecules like carboxylic acid, hydroxy, and amine groups, to mention just a few, in the polymer network. Hydrogels also contain crosslinking groups that are employed in the interactions between the hydrogel and biological tissues.

16.2 SYNTHESIS AND CLASSIFICATION OF HYDROGELS

Hydrogels are synthesised by crosslinking natural polymers, synthetic polymers or a combination of both. They are classified based on the mode of synthesis or the overall charge on the network. Hydrogels composed of natural polymers such as polysaccharides, polyamides, and polyphenols, and biological polymers such as polynucleotides and polypeptides are similar to the extracellular matrix and, therefore, are biocompatible but very delicate; therefore, they usually require crosslinking with monomers or mixing with synthetic polymers to improve their mechanical qualities. They can also be synthetic, such as polyvinyl alcohol. Natural hydrogels, mostly polysaccharides, have biological activities. As a result of their physiological functions, they are employed in various biomedical applications, while synthetic polymer hydrogels have more mechanical strength but are not environmentally friendly.

Hydrogels are formulated based on their proposed use. This is achieved by selecting the monomer or polymer and determining the formation reactions, either by chemical or physical crosslinking (Wang & Han, 2017). Chemical crosslinking involves the use of optical polymerisation, enzymatic reactions, or click reaction, which involves the use of catalysts, while physical crosslinking can be achieved through ionic crosslinking, temperature-dependent methods, or pH-dependent methods. Hydrogels are ideal for creating scaffolds that support cell growth and tissue regeneration. In wound healing, hydrogels provide a moist environment that promotes cell growth and nutrient exchange, facilitating tissue formation and regeneration. They can encapsulate and deliver therapeutic agents, such as antibiotics and growth factors, directly to the wound site, promoting healing and reducing infection risk, thereby enhancing the healing process. Also, in wound healing, electrospun hydrogel fibres, which mimic the extracellular matrix, can be used to provide an optimal environment for cellular proliferation, which is required for the healing process. These types of hydrogels can be customised to control drug release and incorporate smart features for real-time monitoring of the entire process. They can monitor wound status, providing real-time feedback on healing progress. Another group of smart hydrogels is the stimulus-responsive hydrogel. These hydrogels can respond to environmental changes, allowing for controlled drug release and adaptation to dynamic biological environments. The development of these intelligent and stimuli-responsive hydrogels, as well as nanocomposite hydrogels, which exhibit improved functional properties, has further expanded their potential in clinical applications.

Figure 1. Hydrogel classification

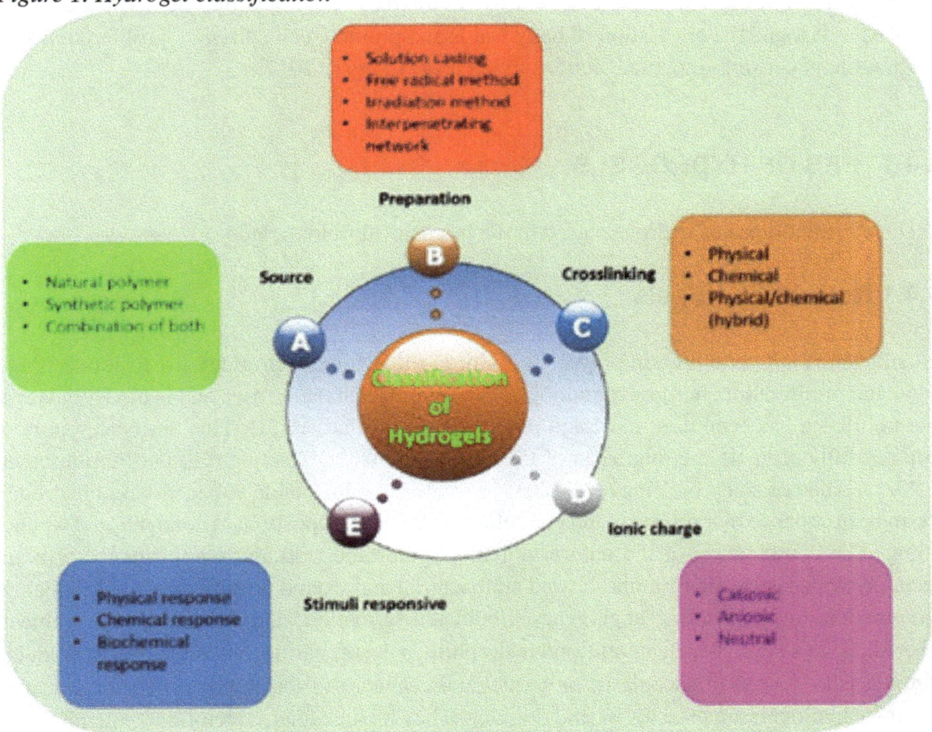

(Bashir et al., 2020) Copyright 2020 MDPI. https://www.mdpi.com/openaccess

16.3 PROPERTIES OF HYDROGELS

The properties of hydrogel depend on the type. The polymers for natural hydrogels are obtained from natural sources such as polysaccharides, e.g., fibrin, collagen and gelatine. Natural polymers are usually biocompatible, non-toxic, highly durable and susceptible to enzymatic degradation (Cicha et al., 2017). The polymer sources for synthetic hydrogels include polyacrylamides and polyethylene glycol (PEG). These are more hydrophilic, with more gel strength and therefore have better mechanical strength (Gyles et al., 2017). Synthetic hydrogels have better mechanical properties than natural hydrogels. Therefore, they are more suitable for tissue engineering. It is difficult to compare the mechanical parameters of hydrogels because these parameters are measured using time-dependent and time-independent viscoelastic theories. The factors that determine the mechanical properties of such hydrogels include monomer concentration, polymerisation temperature, quantity of crosslinking agent, degree of swelling at the time of measurement and method of measurement. For example, when harder substrates are used in culturing cells, such cells tend to have a faster rate of proliferation and migration compared to cells cultured with softer substrates (Vedadghavami et al., 2017).

In the case of biological properties, hydrogels should be biocompatible and non-toxic. They should be able to aid the circulation of body fluids and promote cell adhesion, proliferation and growth without causing immune reactions in the host (Li et al., 2017).

16.4 TYPES OF HYDROGELS

The different types of hydrogels used in the biomedical field include

16.4.1 Natural Hydrogels

Collagen is a natural protein polymer present in the extracellular matrix. It fosters cell adhesion and proliferation, has low immunogenicity, promotes haemostasis, and is biodegradable. Type I collagen has been used to design porous dressings that are 3D. This has been reported to successfully promote the migration of fibroblasts and wound healing (Gaspar-Pintiliescu et al., 2021). Another study by Ying et al. (2019) revealed that injectable collagen-based hydrogel dressing can enhance blood vessels and epithelium formation, deposition of collagen, and wound healing. The combination of the natural polymer hyaluronic acid and ε-polysine to form an injectable wound dressing formula showed antibacterial and wound healing effects by reducing inflammation, and promoting angiogenesis, collagen formation, and keratinocyte movement (Hu et al., 2023). Alginate, an anionic and hydrophilic polysaccharide, has been used to produce hydrogels with very good viscoelastic properties and a capacity to absorb proteins. Various forms of alginate hydrogels are used for wound dressing. They include foams, nanofibers, membranes, and reinforced composites (Ahmad et al., 2023). Wang et al. (2023) developed a hydrogel from sodium alginate and platelet-rich plasma containing high amounts of vascular endothelial growth factor and epidermal growth factor to help in wound healing. This hydrogel was found to promote angiogenesis, cell proliferation, and wound closure effectively.

16.4.2 Synthetic Hydrogels

These have some advantages over their natural counterparts due to their modifiable molecular weight and better mechanical strength. The functionalities of synthetic polymers can also be modified for better outcomes. Polyethylene glycol provides an excellent scaffold for nanoparticles and drug loading because it is biocompatible, soluble in water, and has low protein absorption. Polyethylene glycol has been modified extensively to form polymers used in wound healing. For example, polyethylene glycol-thioketal bonds with a sensitivity to reactive oxygen species, which helps to transport epidermal growth factor (Lei & Fan, 2021); some polyethylene glycol-modified polymers improve wound healing through the creation of micro-negative pressure. This is achieved through its potential to prevent protein and bacterial absorption. The work of Chen et al. (2023), which combined polyurethane, polyethylene glycol, and triethoxysilane, showed enhanced wound healing. Polyethylene glycol and triethoxysilane were combined using self-foaming reactions, and the dressing formed was seen to prevent adhesion, enhance the inflammatory phase of the wound healing process, and promote the deposition of collagen. Polyacrylamide-based hydrogels, when used as scaffolds, responded readily to pH changes, and their mechanical properties were adjustable; therefore, they could be used to produce structures with entrapped hollow particles. All these features make them good for wound healing. When polyvinyl alcohol was used by Chen et al. (2020) to culture hematopoietic stem cells, it generated a thousand-fold colony of the cells that maintained their stemness. Polyvinyl alcohol has also been used to produce sponges that are employed in surgeries as clinical haemostats and to reduce the risks of bleeding after surgery (Wang et al., 2020). One observed disadvantage of polyvinyl alcohol for wound dressing is the potential for bacterial infection as a result of the formation of a thick water layer at secretion sites due to the accumulation of hydrophilic hydroxyl groups in the molecular chains of polyvinyl alcohol (Nakayama et al., 2010).

16.4.3 Hybrid Hydrogels

These are formed from the combination of natural and synthetic polymers. Qian et al. (2021) produced a hybrid gel from silk fibroin cross-linked with haemoglobin and gallium, which successfully enhanced wound healing and resolved infection in diabetic wounds of the model animals used in the research within 15 days. A hybrid hydrogel formed from the combination of conjugated collagen, oxidised sodium alginate, and antimicrobial peptides enhanced the healing of very thick wounds through its antibacterial effect against *Escherichia coli* and *Staphylococcus aureus*. It also promoted angiogenesis and proliferation. In the animal model used, this hybrid hydrogel successfully enhanced the healing process through collagen buildup and skin regrowth enhancement (Feng et al., 2020).

16.4.4 Biomimetic Hydrogels

These are produced from polymeric materials with similar features to the extracellular matrix, which enables them to mimic the biochemical functions of the extracellular matrix (Dutta, Patel & Lim, 2019). To obtain this, decellularised extracellular matrix is usually combined with a natural or synthetic polymer. Decellularised extracellular matrix is obtained when cellular tissues are removed from the matrix, leaving behind the structural integrity of the extracellular matrix, which enhances its biocompatibility (Zhang et al., 2022). Xia et al. (2023) produced a biomimetic hydrogel with decellularised extracellular matrix from human umbilical cord tissues that showed excellent biocompatibility and an abundance of growth factors and structural proteins that aid wound healing. This biomimetic hydrogel was designed to be injectable within a 30-minute timeframe for hydrogel formation. When nanofibers were produced from it through electrospinning, its product was structurally similar to the extracellular matrix present in the dermal layer.

16.4.5 Composite Hydrogels (Nanocomposites)

Biological tissues are composed of composite materials with anisotropic properties that aid their function. Hydrogels used for regenerating or repairing natural tissues need to have similar properties, with the right elastic modulus and mechanical strength (Afewerki et al., 2019). To achieve this, composite hydrogels are synthesised. Composite hydrogels can be polymer composite hydrogels from natural-natural, natural-synthetic and synthetic-synthetic polymer hydrogels or nanocomposite hydrogels. Nanocomposite hydrogels are synthesised from a blend of organic-inorganic nanocomposites similar to what is found in nature in bones and teeth. The inorganic nanomaterials are embedded in the matrices of the organic polymer network. Nanocomposite hydrogels can be used for controlled drug delivery, remote-controlled, stimuli-responsive, self-healable and biological applications. Commonly used nanomaterials include metal nanoparticles, mineral, silica, magnetic, carbon and polymer nanoparticles, such as silver, natural silicate clay, mesoporous silica, iron oxide, graphene oxide and nanocellulose, respectively. Other forms of composite hydrogels include supramolecular and stimuli-responsive hydrogels. Supramolecular hydrogels have a dynamic nature and therefore, exhibit interesting and unique properties such as self-healing, molecular recognition and bioadhesiveness, while stimuli-responsive hydrogels are sensitive to various environmental stimuli such as pH, temperature, light, enzyme and glucose, to mention a few (Li et al., 2020)

16.5 BIOMEDICAL APPLICATIONS OF HYDROGELS

Hydrogels also have the ability to absorb biological fluids due to the presence of hydrophilic groups such as amine, carboxyl, and hydroxyl groups (Chamkouri & Chamkouri, 2021). Hydrogels can be produced within the body at the desired location based on biological conditions. This is achieved by the injection of the polymer solution into the body, which is then converted into solid hydrogel by physical and chemical crosslinking agents (Riederer et al., 2016). The factors that stimulate the formation of hydrogels include the presence of ions, ultraviolet radiation, pH, and temperature changes (Alexander et al., 2014).

16.5.1 Tissue Engineering

In tissue engineering, hydrogels serve as scaffolds that mimic the natural extracellular matrix, supporting the growth and differentiation of cells necessary for tissue regeneration. Hydrogels are used in cartilage regeneration and complex tissue engineering due to their mechanical properties and bioactivity, which can be employed for vascular tissue engineering and organ-on-a-chip systems, thereby enhancing the development of functional tissues and aiding in repairing damaged areas. Their tunable properties and ability to respond to physiological stimuli further enable the controlled release of therapeutics, fostering tissue regeneration and repair in various biomedical applications. Both natural and synthetic polymers are used to form hydrogels, which are employed in tissue engineering. For example, to reconstruct defective joints and tissues. Hydrogels from natural sources provide good scaffolds for tissue engineering. This is because they can provide biochemical signals that aid the proliferation and differentiation of cells through the stimulation of the extracellular matrix of body tissues. Maintaining proper hardening and desirable strength at physiological temperatures under biological conditions is important so that the hydrogel can form a scaffold with complex shapes. Also important is the need for the mechanical properties of such a hydrogel to be similar to what is obtainable in natural tissues (Zhou et al., 2018).

Collagen is the main component of the extracellular matrix. It forms 25 to 35% of the protein in the body (Pollot et al., 2018). Type I collagen can be obtained in cartilage, bone, ligament, and skin through enzymatic and acid-base processes. The synthesis of collagen-based hydrogels occurs at 37 °C, which is the physiological temperature, resulting in the assembly of type I collagen fibrils, which are solubilised (Liu et al., 2019). Through the hydrolysis of collagen, gelatin can be synthesised. Gelatin is less immunogenic than collagen due to a reduction in the aromatic groups and also cross-links at 30- 35 °C. Alginate obtained from seaweed easily forms a hydrogel through physical crosslinking in the presence of divalent cations like calcium at room temperature. It is biocompatible and supports the growth and differentiation of encapsulated chondrocytes while maintaining their phenotype. Type II collagen is formed when such encapsulated chondrocytes are inoculated, forming cartilage genes. Chitosan can also be used in tissue engineering; it is obtained from oyster shells, fungi, and marine waste.

Natural hydrogels make it easier to reproduce the chemical, structural, and biological complexities of the natural microenvironment, which can be used for muscle development (Choi et al., 2016). Although natural hydrogels have a high degree of compatibility with the microenvironment, it is difficult to control their biofunctionality, properties, and degradation (Mancha Sanchez et al., 2020). Synthetic hydrogels can be used to balance these deficiencies. These have better mechanical and physical properties, such as elastic modulus, permeability, and viscoelasticity, which can be controlled by adjusting molecular chain length and cross-linking density. Synthetic hydrogels also have reduced risk of batch-to-batch variability and pathogen transfection (Gyles et al., 2017). Synthetic hydrogels are hydrophilic with high chain mobility; they can be modified with proteins or short peptides to aid cell adhesion (Madduma-Banderage & Madihally, 2021). For example, gelatin methacryloyl (GelMA) is formed from the photo crosslinking of methacrylic anhydride with gelatin. Gelatin still preserves its original cell adhesive properties when GelMA is formed because only approximately 5% of its amino acid residues are used (Xiao et al., 2019).

Due to the biocompatibility of hydrogels and their amenability, they can quickly be processed into various conFigureurational designs using advanced microfabrication techniques (Xie et al.,2020; Vedadghavami et al., 2017). This can be used to fabricate hydrogel microfibers that can either serve as individual self-standing structures or as part of the building block for the assembling of a full-scale 3-dimensional skeletal muscle tissue construct (Onoe et al., 2013). This will ultimately provide a very conducive, cell-compatible microenvironment and structure for proper muscle tissue development (Shang et al., 2019). For example, Narayanan et al. (2021) reported that hyaluronic acid hydrogels supported myoblast differentiation when conjugated with chondroitin sulfate. it also allowed proper integration of the hydrogel into the host tissue when implanted at 4 weeks.

For skeletal muscle tissue engineering, myoblasts need to be aided to align properly, instead of the entangled, disordered structure formed when myoblasts proliferate. To achieve this, various methods are employed, such as;

a. Microgrooves or Micropillars

Cellular proliferation is physiologically regulated by contact. This idea is used to streamline the growth of seeded myoblasts using micromolding. Hydrogels are patterned within microgrooves.

The microgroove structures regulate the boundary of growth for myoblasts, resulting in the formation of muscle-like structures (Denes et al., 2019). Smaller grooves are encouraged because they induce higher cell alignment, while deeper channels aid cell proliferation and multilayering, thereby improving cell alignment (Hume et al., 2012). Micropillars, on the other hand, help in designing hydrogels with musculoskeletal micro-arrangement that mimic the transmission of force from muscle to bone as seen *in vivo* (Furuhashi et al., 2021). This is because myoblasts will end up exerting isotropic contractile force on the hydrogel in a micropillar, resulting in the generation of passive longitudinal tension.

b. Fibre Biofabrication

This is important because it covers the deficiency of the microgroove or micropillar method. In this process, it provides some level of control in the production of scaffold geometry necessary for myoblast alignment and tissue development. The biomimetic scaffold provides a platform for muscle cell elongation and alignment. This can be achieved through electrospinning, microfluidic spinning, 3D bioprinting, moulding and extrusion. Fibre-based biofabrication for skeletal muscle tissue engineering using moulding was reported by Szymanski and Feinberg (2013). The hydrogel was composed of alginate and fibrinogen, and the muscle cells used were C2C12; the fibre diameter was approximately 20 μm. C2C12 are myoblast cell lines that differentiate rapidly to form contractile myotubes and muscle proteins. At the end of the experiment, the C2C12 myoblasts remained attached when the substrate was dissolved. Using indirect 3D bioprinting, Kim et al. (2018) combined fibrinogen, gelatin, and hyaluronic acid in human mesenchymal progenitor cells (hMPC) using 300 μm fibres to obtain 82% functional skeletal muscle recovery at the end of an *in vivo* implantation using a rat model for a duration of 8 weeks. Carnes and Pins (2020) used an extrusion mechanism to seed fibrinogen in C2C12 cell lines using MES-based chemical treatments. The result was cells aligned along the superficial microgrooves.

Figure 2. Hydrogel-based fiber biofabrication techniques used for the development of advanced skeletal muscle tissue constructs

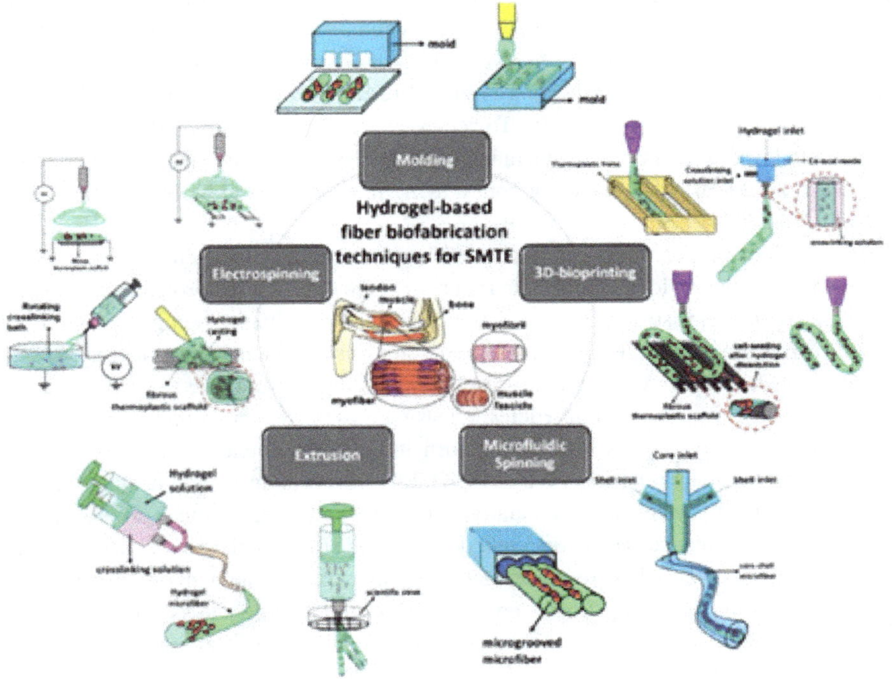

c. Organ-on-Chip

The organ-on-chip systems are systems that have undergone extensive design, modelling and applications in tissue engineering because they are made to mimic *in vivo* physiological environments of cells, tissues and organs. These systems are designed as microchambers where cells or organs are sealed and provided with sufficient nutrients and the timely removal of wastes. Most organ-on-chips are a bit different from natural organs because, in most cases, they contain other materials aside from the natural biocomponents found in nature, which make them perform better (Sutterby et al., 2020) Organ-on-chip systems help to solve, to some extent, the challenges of insufficient supply of organs and the time-consuming nature of animal studies because they can be conveniently manipulated, cost less and accurately replicate single-organ functions and multi-organ interactions (Chen et al., 2023). Organ-on-chips are mostly integrated with a sensor to help mimic the organ's function using the micro-physiological unit and monitor static or dynamic parameters using sensory units. More recently, 3D bioprinting technology has been used in the microfabrication of organ-on-chips (Li et al., 2020). Sensors integrated in organ-on-chips are used to monitor the movement of cells, record the concentration of metabolites and the landscape forms of the tissues.

16.5.2 Skin Regeneration

The human skin is composed of three layers: the epidermis, the dermis, and the hypodermis, all interconnected and serving as the largest organ in the human body (Monteiro-Riviere, 2020). Research is ongoing to find ways to correct skin defects without scarring. This requires approaches that will support wound healing by successfully regenerating the extracellular matrix's microenvironment (Sultana et al., 2020). To achieve this, various natural and synthetic biomaterials are employed to assist the regenerative capacities of skin tissue and its growth through the construction of scaffolds containing both polymeric materials and biological molecules or cells (Fani et al., 2024).

The hydrophilic property of hydrogel makes it very permeable to oxygen and readily absorbs wound exudate while providing a moist environment at the interface. Hydrogels can also mimic the mechanical characteristics of natural tissues, providing mechanical support and sufficient space for the movement of cells and the regeneration of tissue (Vedadghavami et al., 2017). A unique set of hydrogels known as stimuli-responsive hydrogels can serve as carriers for drugs and other substances, such as nanoparticles, releasing these active substances gradually or on demand. They can also serve as sources of information for treatment planning by monitoring the wound site's microenvironment (Wu et al., 2022). Advances in wound healing treatments aim to prevent scarring by enhancing reparative properties and regeneration. This is achieved by engineering scaffolds that can correctly repair damaged skin through the provision of supportive structure and signals that help cells to navigate the extracellular matrix microenvironment for regrowth (Chen et al., 2019). Hydrogels are used for wound healing and skin regeneration because of their characteristic properties. Their elasticity is similar to natural tissues, with moisture retention, which can help in the removal of wound debris through autolysis, and porosity, which helps to facilitate gaseous exchange and enhancement of migration and proliferation of cells (Abasalizadeh et al., 2020).

Skin regeneration hydrogels are designed based on the stage of wound healing for which they will be applied. The first stage of wound healing requires the control of bleeding, and hydrogels designed for this stage need to have the potential of attracting platelets so as to facilitate coagulation and also adhere to tissues efficiently to seal the opening and reduce bleeding (Cheng et al., 2021). But it is not always easy for hydrogels to stick to the moist surfaces of a wound. Therefore, techniques such as electrostatic interaction, topological adhesion, bionic adhesion and dynamic covalent bonding can be used to enhance adhesion (Cui & Liu, 2021). Liu et al. (2020) produced a hydrogel aldohydroxyethyl starch/amino carboxymethyl chitosan by combining ethylene diamine and carboxymethyl chitosan. This injectable hydrogel enhanced blood clotting by the interaction of cationic amino groups and the anionic charged platelets, thereby reducing blood loss.

For wounds to heal effectively, infections need to be avoided to prevent secondary wounds and scarring to achieve this, antibacterial agents have been incorporated into hydrogel network for example, hyaluronic acid was combined with iron (III)-ethylenediamine tetra acetic acid and the hydrogel formed possessed antibacterial properties due to the presence of iron (III) complexes (Tian et al., 2018). Qi et al. (2022) synthesised anti-inflammatory hydrogels that help to reduce pro-inflammatory cytokines and enhance macrophage polarisation so as to control the inflammation and aid proper healing. This can also be achieved through the introduction of anti-inflammatory agents or antioxidants into hydrogels. For example, Zhang et al. (2020) added antioxidants to the hydrogel to enhance the degradation of reactive oxygen species and reduce the inflammatory response. Natural antioxidants such as polyphenols, vitamins C, and vitamin A can be incorporated into hydrogels. Ahmadian et al. (2022) combined tannic acid and gelatin to form a hydrogel with good radical scavenging and antibacterial potential. It also helped to reduce bleeding.

16.5.3 Hydrogel Robotics

Robots made from hydrogels are known as soft robots. Traditionally, robots are made to assist humans, thereby increasing productivity or efficiency. Such robots are usually rigid-bodied. Also, robots have been attached to living organisms to assist with deformities. Recently, concerns about the safety of such robotic components have increased, giving rise to the invention of soft-bodied robots that are more compliant (Rus & Tolley, 2015). Since hydrogels are polymers with cross-linked chains, they can form 3D elastic structures with adjustable conFigureuration when varied mechanical strain is applied (Yang & Suo, 2018). Other features of hydrogels that make them good for robot construction include their high level of transparency even in the bulk state (Keplinger et al., 2013) enabling the efficient transmission of optical information and their conductivity (Lee, Kim & Sun, 2018) due to the ability for ions to dissolve in the water present in hydrogels, thereby acting as an ionic conductor. Hydrogels have been used in the construction of actuators, sensors, communicators, power sources and computational circuits needed for the construction of robots.

a. Actuator

A robot needs to have the capacity to generate motion. The component responsible for this feat is called an actuator. An actuator can be controlled by different stimuli. It can be a thermal actuator, chemical actuator (Guo et al., 2019), or optically regulated actuator (Kuwahara et al., 2013).

b. Sensors

For hydrogel robots to respond to stimuli, sensors are needed to measure the changes in the environment for appropriate responses. Such sensors can be chemical, temperature, volumetric responses, stress or strain sensors (Shin et al., 2019).

c. Communicators

The communicator of a hydrogel robot helps to transmit information in a way that humans can understand. This is mostly controlled electrically and is classified according to the target sensory organ. Optical communicator expresses information based on light intensity, colour change or transmittance (Alesanco et al., 2016), acoustic communicators transmit information via mechanical vibration, which is a good way of sending information to various targets (Keplinger et al., 2013), and tactile communicators transfer mechanical stimuli to the skin of humans (Tiwari et al., 2018).

d. Power Source

Hydrogel robots require a power source for the supply of energy to component parts. This could either be an energy storage device or an energy generator. An energy storage device stores either chemical or electrical energy (Daunert et al., 2000) or, in some cases, both. Energy generators harvest wasted mechanical energy and turn it into electrical energy (Zhou et al., 2017). These generators have hydrogels as electrodes because hydrogels are stretchable and have a high ionic conductivity. They can either be dielectric elastomer generators or triboelectric nanogenerators (Wirthl et al., 2017).

e. Computational Circuit

Soft robots need computational circuits to help process information and control the components of the robot. The computational circuit could be a conductive hydrogel with ions or electrons as charge carriers (Ren et al., 2019), a semiconductor that processes ionic signals directly from biological systems, acting either as a diode for ionic signals rectification or a transistor to regulate and amplify ionic signals (Lee et al., 2019). Others are circuit boards and bio-computational circuits, which involve the processing of chemical signals by genetically programmed cells cultured in a hydrogel matrix (Liu et al., 2017).

Though all these components of soft robots have been synthesised from hydrogels, there is still a need to systematically assemble them together for proper interaction and applicability. There is also a need to improve their durability because functionality is greatly diminished as a result of dehydration of hydrogels (Lee et al., 2020).

16.5.4 Artificial Muscle

The muscle is a natural example of an actuator. Once it detects a stimulus, it responds adequately, giving rise to varied possible movement options such as twitching, twisting, gripping, or running. Synthetically replicating the dynamics of a muscle has been the focus of research to produce artificial muscles for medical devices or soft robots (Spinks, 2020). Since hydrogels can swell, they are good substrates for innovative research in artificial muscle generation through the alteration of the swelling ratio by applied external stimuli to create motion (Oveissi et al., 2021). Hydrogels are also responsive to ambient stimuli, increasing their sensitivity; they are quite flexible, with the potential of enduring large deformations. Dicker et al. (2017) experimentally replicated the molecular-level behaviour of muscles using a pH-responsive hydrogel and a photoacid generator. Exposure to light triggers an acid autocatalytic reaction that results in changes to the hydrogel length. Certain physico-mechanical parameters need to be taken into consideration in the synthesis of artificial muscles from hydrogels. They include swelling, which can be directed to different types of motions, either linear, bending or twisting (Giraud et al., 2008); actuation response time and diffusion, which measures the time it takes the hydrogel to swell and deswell; the characteristic of the hydrogel when uniaxial and biaxial stress is applied and the toughness of the hydrogel. Toughness of a hydrogel is the measure of the energy required to cause mechanical failure.

One major disadvantage of using regular hydrogels for artificial muscle production is their isotropic structure, which restricts their contraction and expansion options (Sano, Ishida & Aida,

2018). This is because they are composed of monomers that disperse homogenously in aqueous media. Artificial muscles need to be produced from anisotropic hydrogels so as to produce complex motions such as those seen in natural muscles. Anisotropic hydrogels have a multi-layered structure with varied characteristics and unique roles (Itai et al., 2018). Such hydrogels are produced with polymers and materials that aid responsiveness, speed, and extent of motion, and the operational conditions of the artificial muscle. The bending motion of artificial muscles will require multiple layers with different levels of responsiveness. To achieve this, hydrogel blocks can be assembled and immersed in solvent (Qin et al., 2019), which will ultimately result in the generation of an integrated hydrogel that can carry out composite motion. Another means is when stimuli-responsive materials and anisotropic structures are arranged together by design to give rise to delicately programmed motions (Wu et al., 2013). Artificial muscles are mainly used to lift or push objects (Greene et al., 2017). For example, a gripper was produced with a finger-like bending strip (Li et al., 2017). The gripper was successfully used to catch and hold on to a live goldfish without causing any damage to the fish as a result of the softness of the gripper (Yuk et al., 2017). Hydrogel locomotors can also be produced (Francis et al., 2017).

Figure 3. Artificial muscle

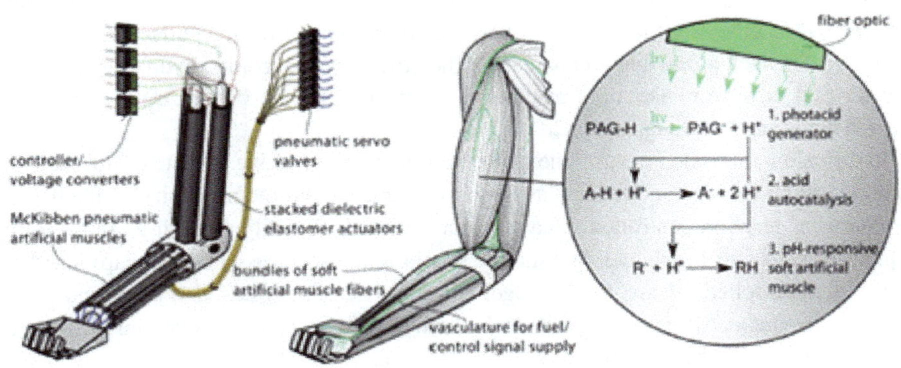

(Dicker et al., 2017) Copyright 2017 Springer Nature. Creative Commons Licence.

16.5.5 Drug Delivery System

Smart hydrogels are used in advanced drug delivery due to their sensitive nature. smart hydrogels respond to changes in pH, light, temperature, and concentration of biomolecules, releasing the drug component they carry, improving pharmacotherapy (Bordbar-Khiabani & Gasik, 2022). Smart hydrogels have the ability to return to their original shape and state once the trigger no longer acts on them (Wang et al., 2020). Bao et al., (2019) reported the fabrication of customised nanogel carriers from a UV-triggered dynamic self-assembly strategy. When the irradiation time was varied, and the photochemical method used, different attractive structures of disulphide-linked nanoparticles were formed in the dilute aqueous solutions, which had excellent capacities for encapsulation of active drug components, good separation and excellent release of the drug component they carry in a controlled manner. The removal of the UV lamp resulted in the suspension of the exchange reaction and immediate deactivation of the active sites of the nanogel carriers. Subsequent reactivation of the stable inert sites by turning on the UV light restores the nanoparticle formation process. This method of fabrication has been used for the screening of components for cancer therapy. One unique property of this fabrication method is that different morphologies and size evolutions can be achieved. Also, the formation of desired shapes can be suspended and reactivated as the need arises, just by using the external UV switcher.

Hydrogels can be used in drug delivery via various routes such as oral, ocular, transdermal, injectables and drug implantations (Vigata et al., 2020). Their hydrophilic nature enables them to serve as good drug carriers. They also efficiently release the drugs encapsulated when present in a hydrated environment. Innovations in drug delivery are essential to prevent off-target effects of drugs that are distributed throughout the body and also enhance stability, solubility and bioavailability of the active drug component. Because hydrogels have high water content, it is easy to load the drugs onto their matrix by dispersing or dissolving them within the hydrogel matrix, thereby ensuring uniform drug distribution, which can be controlled over extended periods. Since it is possible to adjust the cross-linking density of hydrogels, thereby customising their porosity, the drug diffusion rate can also be controlled. In situations where sustained drug release is required, biodegradable hydrogels are used. As the hydrogel breaks down, it releases the encapsulated drug. For targeted drug deliveries, specific ligands can be attached to hydrogel surfaces to direct the drugs to the target sites, thereby reducing off-target effects and improving efficiency. This is a promising area in cancer therapy. For pH-sensitive hydrogels, they can be used for oral drug administration, resulting in controlled release based on the acidity of the stomach environment (Suhag et al., 2024).

In diffusion-controlled release, the concentration gradients drive the rate of release via diffusion based on the size and molecular weight of the drug and the mesh size of the hydrogel. For chemical reaction-controlled release, specific chemical reactions such as those initiated by pH cause changes in the hydrogel matrix, resulting in the release of the encapsulated drug. In swelling-controlled release, the hydrogel absorbs water and swells, and pores and channels are created in the matrix, providing channels through which the drug is released (Mortensen et al., 2022), while for matrix erosion-controlled release, as the hydrogel undergoes degradation over time, the encapsulated drug is released. This mechanism is mostly seen in biodegradable hydrogels (Wang et al., 2013).

16.5.6 Hydrogels as Biosensors for Highly Sensitive Detections

Since hydrogels have been used in the design of wearable sensors, nanocomposite hydrogels can be used for monitoring physiological signals such as electrocardiogram readings and the monitoring of movement. A nanocomposite hydrogel synthesised from graphene oxide, polyvinyl alcohol and polydopamine acted as a biosensor with self-healing properties and good adhesive features (Zhang et al., 2020). Wearable biosensors for the detection of glucose and tryptophan in sweat have been developed. The glucose biosensor contains polyaniline and N-doped graphene quantum dots in its matrix, while the tryptophan biosensor is composed of a hydrogel of PEDOT: SS, dopamine and poly (2-(*N*-3-sulfopropyl-*N, N*-dimethyl ammonium) ethyl methacrylate) (Lin et al., 2023; Xu et al., 2023). Biosensors have been used for personalised health monitoring, such as heart rate, blood pressure, and blood oxygen levels. Xu et al., (2020) developed a chip from plasmonic gold nano-island (pGold) assay, which had high sensitivity and enhanced quantitative measurement of biomarkers in serum. the pGold assay chip had almost 96% specificity and 100% sensitivity for myocardial infarction.

Multistimuli-responsive hydrogels are used in the production of eye drops for the treatment of glaucoma. Timolol maleate, the medication for glaucoma, requires longer ocular residency and bioavailability to achieve effectiveness, *in situ* forming gel formulations that are responsive to pH, temperature and ionic strength are required (Patel et al., 2023). Patel et al. (2019) developed a multistimuli hydrogel drug delivery system based on pH and ion-sensitive gelling. It contained deacetylated gellan gum- a polysaccharide with the ability to form a translucent gel when exposed to cations, both monovalent and divalent.

16.6 SUSTAINABILITY OF HYDROGELS AND ENVIRONMENTAL IMPLICATIONS OF THEIR USAGE

Natural plant-based biopolymers for hydrogel synthesis are sustainable. For instance, cellulose is a promising option due to its abundance and biodegradability. Cellulose-based hydrogels could be obtained from cellulose, chitosan or chitin. The use of hydrogel in tissue engineering for tissue regeneration, repair and transplantation is dependent on the availability of biocompatible material with properties similar to the extracellular matrix and which will favour cell proliferation. Cellulose is a sustainable alternative to synthetic polymers (Jasni et al., 2023). The synthesis of hydrogels from cellulose is eco-friendly and less energy-intensive. The life cycle assessment (LCA) of cellulose-derived hydrogels from manufacturing to use to disposal shows that the use of sustainable forestry practices and efficient resource management will reduce the environmental impact of cellulose consumption (Jasni et al., 2025). Xie et al. (2025) developed a sustainable assessment model using the LCA and concluded that optimisation of materials and energy inputs can collectively reduce environmental impact and cost by approximately 73% and 89% respectively.

16.7 CONCLUSION AND FUTURE PERSPECTIVE

The biomedical field has gained much from innovations in hydrogels. Further improvements of hydrogels with inorganic materials will produce a synergistic effect that can improve the degradation profile, mechanical behaviour, and interaction of the hydrogel with biological species for tissue engineering. The combination of multifunctional materials during hydrogel processing can be used to improve the efficiency of hydrogel robots to obtain faster and more precise deformation, especially when new assembly technologies like nano and microtechnologies are used. Finally, for skin regeneration, more research is needed for the effective utilisation of hydrogel in the treatment of chronic and deep wounds by enhancing its mechanical features so that it can navigate the changing nature of the wound microenvironment, better deliver therapeutic agents to prevent infection and enhance healing and effectively produce hydrogel formulations tailored to the needs of individual patient to reduce side effects. The possible applications of hydrogels in the biotechnology and biomedical field are enormous and require more research for large-scale production and optimisation.

16.8 REFERENCES

Abasalizadeh, F., Moghaddam, S. V., Alizadeh, E., Akbari, E., Kashani, E., Fazljou, S. M. B., Torbati, M., & Akbarzadeh, A. (2020). Alginate-based hydrogels as drug delivery vehicles in cancer treatment and their applications in wound dressing and 3D bioprinting. *Journal of Biological Engineering*, *14*, 1–22.

Afewerki, S., Amir, S., Kannan, S., Ahadian, S., & Khademhosseini, A. (2019). Gelatin-polysaccharide composite scaffolds for 3D cell culture and tissue engineering: Towards natural therapeutics. *Bioengineering & Translational Medicine*, *4*(1), 96–115. DOI: 10.1002/btm2.10124 PMID: 30680322

Ahmad, F., Mushtaq, B., Ahmad, S., Rasheed, A., & Nawab, Y. (2023). A novel composite of hemp fiber and alginate hydrogel for wound dressings. *Journal of Polymers and the Environment*, *31*(6), 2294–2305. DOI: 10.1007/s10924-023-02756-7

Ahmadian, Z., Gheybi, H., & Adeli, M. (2022). Efficient wound healing by antibacterial property: Advances and trends of hydrogels, hydrogel-metal NP composites and photothermal therapy platforms. *Journal of Drug Delivery Science and Technology*, *73*, 103458. DOI: 10.1016/j.jddst.2022.103458

Alesanco, Y., Viñuales, A., Cabañero, G., Rodriguez, J., & Tena-Zaera, R. (2016). Colorless to neutral color electrochromic devices based on asymmetric viologens. *ACS Applied Materials & Interfaces*, *8*(43), 29619–29627. DOI: 10.1021/acsami.6b11321 PMID: 27723968

Bao, W., Lyu, J., Li, C., Zhang, J., Sun, T., Wang, X., Zhou, J., & Li, D. (2019). Fabrication of customized nanogel carriers from a UV-triggered dynamic self-assembly strategy. *Frontiers in Chemistry*, *7*, 489556. DOI: 10.3389/fchem.2019.00769 PMID: 31781547

Bashir, S., Hina, M., Iqbal, J., Rajpar, A. H., Mujtaba, M. A., Alghamdi, N. A., Wageh, S., Ramesh, K., & Ramesh, S. (2020). Fundamental concepts of hydrogels: Synthesis, properties, and their applications. *Polymers*, *12*(11), 2702. DOI: 10.3390/polym12112702 PMID: 33207715

Bordbar-Khiabani, A., & Gasik, M. (2022). Smart hydrogels for advanced drug delivery systems. *International Journal of Molecular Sciences*, *23*(7), 3665. DOI: 10.3390/ijms23073665 PMID: 35409025

Carnes, M. E., & Pins, G. D. (2020). Etching anisotropic surface topography onto fibrin microthread scaffolds for guiding myoblast alignment. *Journal of Biomedical Materials Research. Part B, Applied Biomaterials*, *108*(5), 2308–2319. DOI: 10.1002/jbm.b.34566 PMID: 31967415

Chen, C. F., Chen, S. H., Chen, R. F., Liu, K. F., Kuo, Y. R., Wang, C. K., Lee, T.-M., & Wang, Y. H. (2023). A multifunctional polyethylene glycol/triethoxysilane-modified polyurethane foam dressing with high absorbency and antiadhesion properties promotes diabetic wound healing. *International Journal of Molecular Sciences*, *24*(15), 12506. DOI: 10.3390/ijms241512506 PMID: 37569881

Chen, D., Hou, Q., Zhong, L., Zhao, Y., Li, M., & Fu, X. (2019). Bioactive molecules for skin repair and regeneration: Progress and perspectives. *Stem Cells International*, *2019*(1), 6789823. DOI: 10.1155/2019/6789823 PMID: 32082386

Chen, H., Luo, Z., Lin, X., Zhu, Y., & Zhao, Y. (2023). Sensors-integrated organ-on-a-chip for biomedical applications. *Nano Research*, *16*(7), 10072–10099. DOI: 10.1007/s12274-023-5651-9 PMID: 37359077

Chen, X., Lu, B., Zhou, D., Shao, M., Xu, W., & Zhou, Y. (2020). Photocrosslinking maleilated hyaluronate/methacrylated poly (vinyl alcohol) nanofibrous mats for hydrogel wound dressings. *International Journal of Biological Macromolecules*, *155*, 903–910. DOI: 10.1016/j.ijbiomac.2019.11.048 PMID: 31730992

Cheng, J., Liu, J., Li, M., Liu, Z., Wang, X., Zhang, L., & Wang, Z. (2021). Hydrogel-based biomaterials engineered from natural-derived polysaccharides and proteins for hemostasis and wound healing. *Frontiers in Bioengineering and Biotechnology*, *9*, 780187. DOI: 10.3389/fbioe.2021.780187 PMID: 34881238

Choi, Y. J., Kim, T. G., Jeong, J., Yi, H. G., Park, J. W., Hwang, W., & Cho, D. W. (2016). 3D cell printing of functional skeletal muscle constructs using skeletal muscle-derived bioink. *Advanced Healthcare Materials*, *5*(20), 2636–2645. DOI: 10.1002/adhm.201600483 PMID: 27529631

Cicha, I., Detsch, R., Singh, R., Reakasame, S., Alexiou, C., & Boccaccini, A. R. (2017). Biofabrication of vessel grafts based on natural hydrogels. *Current Opinion in Biomedical Engineering*, *2*, 83–89. DOI: 10.1016/j.cobme.2017.05.003

Cui, C., & Liu, W. (2021). Recent advances in wet adhesives: Adhesion mechanism, design principle and applications. *Progress in Polymer Science*, *116*, 101388. DOI: 10.1016/j.progpolymsci.2021.101388

Daunert, S., Barrett, G., Feliciano, J. S., Shetty, R. S., Shrestha, S., & Smith-Spencer, W. (2000). Genetically engineered whole-cell sensing systems: Coupling biological recognition with reporter genes. *Chemical Reviews*, *100*(7), 2705–2738. DOI: 10.1021/cr990115p PMID: 11749302

Denes, L. T., Riley, L. A., Mijares, J. R., Arboleda, J. D., McKee, K., Esser, K. A., & Wang, E. T. (2019). Culturing C2C12 myotubes on micromolded gelatin hydrogels accelerates myotube maturation. *Skeletal Muscle*, *9*(1), 1–17. DOI: 10.1186/s13395-019-0203-4 PMID: 31174599

Dicker, M. P., Baker, A. B., Iredale, R. J., Naficy, S., Bond, I. P., Faul, C. F., Rossiter, J. M., Spinks, G. M., & Weaver, P. M. (2017). Light-triggered soft artificial muscles: Molecular-level amplification of actuation control signals. *Scientific Reports*, *7*(1), 9197. DOI: 10.1038/s41598-017-08777-2 PMID: 28835614

Dutta, S. D., Patel, D. K., & Lim, K. T. (2019). Functional cellulose-based hydrogels as extracellular matrices for tissue engineering. *Journal of Biological Engineering*, *13*(1), 55. DOI: 10.1186/s13036-019-0177-0 PMID: 31249615

Fani, N., Moradi, M., Zavari, R., Parvizpour, F., Soltani, A., Arabpour, Z., & Jafarian, A. (2024). Current advances in wound healing and regenerative medicine. *Current Stem Cell Research & Therapy*, *19*(3), 277–291. DOI: 10.2174/1574888X18666230301140659 PMID: 36856176

Feng, X., Zhang, X., Li, S., Zheng, Y., Shi, X., Li, F., Guo, S., & Yang, J. (2020). Preparation of aminated fish scale collagen and oxidized sodium alginate hybrid hydrogel for enhanced full-thickness wound healing. *International Journal of Biological Macromolecules, 164*, 626–637. DOI: 10.1016/j.ijbiomac.2020.07.058 PMID: 32668308

Francis, W., Dunne, A., Delaney, C., Florea, L., & Diamond, D. (2017). Spiropyran-based hydrogel actuators—Walking in the light. *Sensors and Actuators. B, Chemical, 250*, 608–616. DOI: 10.1016/j.snb.2017.05.005

Furuhashi, M., Morimoto, Y., Shima, A., Nakamura, F., Ishikawa, H., & Takeuchi, S. (2021). Formation of contractile 3D bovine muscle tissue for construction of millimetre-thick cultured steak. *npj Science of Food, 5*(1), 6.

Gaspar-Pintiliescu, A., Anton, E. D., Iosageanu, A., Berger, D., Matei, C., Mitran, R. A., Negreanu-Pirjol, T., Craciunescu, O., & Moldovan, L. (2021). Enhanced wound healing activity of undenatured type I collagen isolated from discarded skin of Black Sea gilthead bream (*Sparus aurata*) conditioned as 3D porous dressing. *Chemistry & Biodiversity, 18*(8), e2100293. DOI: 10.1002/cbdv.202100293 PMID: 34156756

Giraud, M. N., Ayuni, E., Cook, S., Siepe, M., Carrel, T. P., & Tevaearai, H. T. (2008). Hydrogel-based engineered skeletal muscle grafts normalize heart function early after myocardial infarction. *Artificial Organs, 32*(9), 692–700. DOI: 10.1111/j.1525-1594.2008.00595.x PMID: 18684206

Greene, A. F., Danielson, M. K., Delawder, A. O., Liles, K. P., Li, X., Natraj, A., Wellen, A., & Barnes, J. C. (2017). Redox-responsive artificial molecular muscles: Reversible radical-based self-assembly for actuating hydrogels. *Chemistry of Materials, 29*(21), 9498–9508. DOI: 10.1021/acs.chemmater.7b03635

Guo, M., Wu, Y., Xue, S., Xia, Y., Yang, X., Dzenis, Y., Li, Z., Lei, W., Smith, A. T., & Sun, L. (2019). A highly stretchable, ultra-tough, remarkably tolerant, and robust self-healing glycerol-hydrogel for a dual-responsive soft actuator. *Journal of Materials Chemistry. A, Materials for Energy and Sustainability, 7*(45), 25969–25977. DOI: 10.1039/C9TA10183G

Gyles, D. A., Castro, L. D., Silva, J. O. C.Jr, & Ribeiro-Costa, R. M. (2017). A review of the designs and prominent biomedical advances of natural and synthetic hydrogel formulations. *European Polymer Journal, 88*, 373–392. DOI: 10.1016/j.eurpolymj.2017.01.027

Hu, D., Wen, J., Zhao, X., Liu, K., Zhang, Y., Bu, Y., & Wang, K. (2023). A wound-friendly antibacterial hyaluronic acid dressing with on-demand removability for infected wound healing. *Biomaterials Research, 27*(1), 38. DOI: 10.1186/s40824-023-00340-7 PMID: 37127622

Hume, S. L., Hoyt, S. M., Walker, J. S., Sridhar, B. V., Ashley, J. F., Bowman, C. N., & Bryant, S. J. (2012). Alignment of multi-layered muscle cells within three-dimensional hydrogel macrochannels. *Acta Biomaterialia, 8*(6), 2193–2202. DOI: 10.1016/j.actbio.2012.02.001 PMID: 22326973

Itai, S., Tajima, H., & Onoe, H. (2018). Double-layer perfusable collagen microtube device for heterogeneous cell culture. *Biofabrication, 11*(1), 015010. DOI: 10.1088/1758-5090/aaf09b PMID: 30499456

Jasni, A. H., Azmi, A. S., Puad, N. I. M., Ali, F., & Nor, Y. A. (2023). Naturally derived biomaterials: advances and opportunities. In *Engineered Biomaterials: Synthesis and Applications* (pp. 3–41). Springer Nature Singapore. DOI: 10.1007/978-981-99-6698-1_1

Jasni, A. H., Azmi, A. S., Puad, N. I. P. M., Ali, F., & Nor, Y. A. (2025). *Sustainable synthesis of cellulose-derived hydrogels for tissue engineering. in sustainable green biomaterials as drug delivery systems.* Springer Nature Switzerland.

Keplinger, C., Sun, J. Y., Foo, C. C., Rothemund, P., Whitesides, G. M., & Suo, Z. (2013). Stretchable, transparent, ionic conductors. *Science*, *341*(6149), 984–987. DOI: 10.1126/science.1240228 PMID: 23990555

Kim, J. H., Seol, Y. J., Ko, I. K., Kang, H. W., Lee, Y. K., Yoo, J. J., Atala, A., & Lee, S. J. (2018). 3D bioprinted human skeletal muscle constructs for muscle function restoration. *Scientific Reports*, *8*(1), 12307. DOI: 10.1038/s41598-018-29968-5 PMID: 30120282

Kuwahara, Y., Kaji, M., Okada, J., Kim, S., Ogata, T., & Kurihara, S. (2013). Self-alignment and photomechanical properties of alternative multi-layered films containing azobenzene polymer liquid crystal and polyvinyl alcohol layers. *Materials Letters*, *113*, 202–205. DOI: 10.1016/j.matlet.2013.09.085

Lee, H. R., Kim, C. C., & Sun, J. Y. (2018). Stretchable ionics–a promising candidate for upcoming wearable devices. *Advanced Materials*, *30*(42), 1704403. DOI: 10.1002/adma.201704403

Lee, H. R., Woo, J., Han, S. H., Lim, S. M., Lim, S., Kang, Y. W., Song, W. J., Park, J.-M., Chung, T. D., Joo, Y.-C., & Sun, J. Y. (2019). A stretchable ionic diode from copolyelectrolyte hydrogels with methacrylated polysaccharides. *Advanced Functional Materials*, *29*(4), 1806909. DOI: 10.1002/adfm.201806909

Lee, S. S., Choi, K. H., Kim, S. H., & Lee, S. Y. (2018). Wearable supercapacitors printed on garments. *Advanced Functional Materials*, *28*(11), 1705571. DOI: 10.1002/adfm.201705571

Lei, H., & Fan, D. (2021). Conductive, adaptive, multifunctional hydrogel combined with electrical stimulation for deep wound repair. *Chemical Engineering Journal*, *421*, 129578. DOI: 10.1016/j.cej.2021.129578

Li, H., Song, X., Li, B., Kang, J., Liang, C., Wang, H., Yu, Z., & Qiao, Z. (2017). Carbon nanotube-reinforced mesoporous hydroxyapatite composites with excellent mechanical and biological properties for bone replacement material application. *Materials Science and Engineering C*, *77*, 1078–1087. DOI: 10.1016/j.msec.2017.04.048 PMID: 28531981

Li, J., Wu, C., Chu, P. K., & Gelinsky, M. (2020). 3D printing of hydrogels: Rational design strategies and emerging biomedical applications. *Materials Science and Engineering R Reports*, *140*, 100543. DOI: 10.1016/j.mser.2020.100543

Li, X., Cai, X., Gao, Y., & Serpe, M. J. (2017). Reversible bidirectional bending of hydrogel-based bilayer actuators. *Journal of Materials Chemistry. B*, *5*(15), 2804–2812. DOI: 10.1039/C7TB00426E PMID: 32264167

Lin, Y. C., Rinawati, M., Chang, L. Y., Wang, Y. X., Wu, Y. T., Yen, Y. H., Chen, K. J., Ho, K. C., & Yeh, M. H. (2023). A non-invasive wearable sweat biosensor with a flexible N-GQDs/PANI nanocomposite layer for glucose monitoring. *Sensors and Actuators. B, Chemical, 383*, 133617. DOI: 10.1016/j.snb.2023.133617

Liu, J., Li, J., Yu, F., Zhao, Y. X., Mo, X. M., & Pan, J. F. (2020). In situ forming hydrogel of natural polysaccharides through Schiff base reaction for soft tissue adhesive and hemostasis. *International Journal of Biological Macromolecules, 147*, 653–666. DOI: 10.1016/j.ijbiomac.2020.01.005 PMID: 31923505

Liu, X., Tang, T. C., Tham, E., Yuk, H., Lin, S., Lu, T. K., & Zhao, X. (2017). Stretchable living materials and devices with hydrogel–elastomer hybrids hosting programmed cells. *Proceedings of the National Academy of Sciences of the United States of America, 114*(9), 2200–2205. DOI: 10.1073/pnas.1618307114 PMID: 28202725

Madduma-Bandarage, U. S., & Madihally, S. V. (2021). Synthetic hydrogels: Synthesis, novel trends, and applications. *Journal of Applied Polymer Science, 138*(19), 50376. DOI: 10.1002/app.50376

Mancha Sánchez, E., Gómez-Blanco, J. C., López Nieto, E., Casado, J. G., Macías-García, A., Díaz Díez, M. A., Carrasco-Amador, J. P., Torrejón Martín, D., Sánchez-Margallo, F. M., & Pagador, J. B. (2020). Hydrogels for bioprinting: A systematic review of hydrogels synthesis, bioprinting parameters, and bioprinted structures behavior. *Frontiers in Bioengineering and Biotechnology, 8*, 776. DOI: 10.3389/fbioe.2020.00776 PMID: 32850697

Monteiro-Riviere, N. A. (2020). Comparative anatomy, physiology, and biochemistry of mammalian skin. *Dermal and Ocular Toxicology*, 3-71.

Mortensen, N., Toews, P., & Bates, J. (2022). Crosslinking-dependent drug kinetics in hydrogels for ophthalmic delivery. *Polymers, 14*(2), 248. DOI: 10.3390/polym14020248 PMID: 35054655

Nakayama, K. H., Batchelder, C. A., Lee, C. I., & Tarantal, A. F. (2010). Decellularized rhesus monkey kidney as a three-dimensional scaffold for renal tissue engineering. *Tissue Engineering. Part A, 16*(7), 2207–2216. DOI: 10.1089/ten.tea.2009.0602 PMID: 20156112

Narayanan, N., Jia, Z., Kim, K. H., Kuang, L., Lengemann, P., Shafer, G., Bernal-Crespo, V., Kuang, S., & Deng, M. (2021). Biomimetic glycosaminoglycan-based scaffolds improve skeletal muscle regeneration in a Murine volumetric muscle loss model. *Bioactive Materials, 6*(4), 1201–1213. DOI: 10.1016/j.bioactmat.2020.10.012 PMID: 33163701

Onoe, H., Okitsu, T., Itou, A., Kato-Negishi, M., Gojo, R., Kiriya, D., Sato, K., Miura, S., Iwanaga, S., Kuribayashi-Shigetomi, K., Matsunaga, Y. T., Shimoyama, Y., & Takeuchi, S. (2013). Metre-long cell-laden microfibres exhibit tissue morphologies and functions. *Nature Materials, 12*(6), 584–590. DOI: 10.1038/nmat3606 PMID: 23542870

Oveissi, F., Fletcher, D. F., Dehghani, F., & Naficy, S. (2021). Tough hydrogels for soft artificial muscles. *Materials & Design, 203*, 109609. DOI: 10.1016/j.matdes.2021.109609

Patel, G. C., Parmar, V. K., & Patel, P. S. (2019). Stimuli-responsive polymers for ocular therapy. In *Stimuli Responsive Polymeric Nanocarriers for Drug Delivery Applications* (pp. 463–489). Woodhead Publishing. DOI: 10.1016/B978-0-08-101995-5.00023-4

Patel, P., Patel, P., & Patel, G. (2023). Container closure selection and stability studies of developed multistimuli-responsive ocular sustained in situ hydrogel formulation of timolol maleate. *Journal of Sol-Gel Science and Technology*, *105*(2), 443–450. DOI: 10.1007/s10971-022-06006-5

Pollot, B. E., Rathbone, C. R., Wenke, J. C., & Guda, T. (2018). Natural polymeric hydrogel evaluation for skeletal muscle tissue engineering. *Journal of Biomedical Materials Research. Part B, Applied Biomaterials*, *106*(2), 672–679. DOI: 10.1002/jbm.b.33859 PMID: 28306190

Qi, L., Zhang, C., Wang, B., Yin, J., & Yan, S. (2022). Progress in hydrogels for skin wound repair. *Macromolecular Bioscience*, *22*(7), 2100475. DOI: 10.1002/mabi.202100475 PMID: 35388605

Qian, Y., Xu, C., Xiong, W., Jiang, N., Zheng, Y., He, X., Ding, F., Lu, X., & Shen, J. (2021). Dual cross-linked organic-inorganic hybrid hydrogels accelerate diabetic skin wound healing. *Chemical Engineering Journal*, *417*, 129335. DOI: 10.1016/j.cej.2021.129335

Qin, H., Zhang, T., Li, N., Cong, H. P., & Yu, S. H. (2019). Anisotropic and self-healing hydrogels with multi-responsive actuating capability. *Nature Communications*, *10*(1), 2202. DOI: 10.1038/s41467-019-10243-8 PMID: 31101823

Ren, Y., Guo, J., Liu, Z., Sun, Z., Wu, Y., Liu, L., & Yan, F. (2019). Ionic liquid–based click-ionogels. *Science Advances*, *5*(8), eaax0648. DOI: 10.1126/sciadv.aax0648 PMID: 31467977

Rus, D., & Tolley, M. T. (2015). Design, fabrication and control of soft robots. *Nature*, *521*(7553), 467–475. DOI: 10.1038/nature14543 PMID: 26017446

Sano, K., Ishida, Y., & Aida, T. (2018). Synthesis of anisotropic hydrogels and their applications. *Angewandte Chemie International Edition*, *57*(10), 2532–2543. DOI: 10.1002/anie.201708196 PMID: 29034553

Shang, L., Yu, Y., Liu, Y., Chen, Z., Kong, T., & Zhao, Y. (2019). Spinning and applications of bioinspired fiber systems. *ACS Nano*, *13*(3), 2749–2772. DOI: 10.1021/acsnano.8b09651 PMID: 30768903

Shin, S. H., Lee, W., Kim, S. M., Lee, M., Koo, J. M., Hwang, S. Y., Oh, D. X., & Park, J. (2019). Ion-conductive self-healing hydrogels based on an interpenetrating polymer network for a multimodal sensor. *Chemical Engineering Journal*, *371*, 452–460. DOI: 10.1016/j.cej.2019.04.077

Spinks, G. M. (2020). Advanced actuator materials powered by biomimetic helical fiber topologies. *Advanced Materials*, *32*(18), 1904093. DOI: 10.1002/adma.201904093 PMID: 31793710

Suhag, D., Kaushik, S., & Taxak, V. B. (2024). Hydrogels in drug delivery. In Handbook of Biomaterials for Medical Applications, Volume 1: Fundamentals (213-246). Singapore: Springer Nature Singapore. DOI: 10.1007/978-981-97-4818-1_8

Sultana, N., Bandyopadhyay-Ghosh, S., & Soon, C. F. (Eds.). (2020). Tissue engineering strategies for organ regeneration. CRC Press, Taylor & Francis Group. DOI: 10.1201/9780429422652

Sutterby, E., Thurgood, P., Baratchi, S., Khoshmanesh, K., & Pirogova, E. (2020). Microfluidic skin-on-a-chip models: Toward biomimetic artificial skin. *Small, 16*(39), 2002515. DOI: 10.1002/smll.202002515 PMID: 33460277

Szymanski, J. M., & Feinberg, A. W. (2013). Fabrication of freestanding alginate microfibers and microstructures for tissue engineering applications. *Biofabrication, 6*(2), 024104. DOI: 10.1088/1758-5082/6/2/024104 PMID: 24695323

Tian, R., Qiu, X., Yuan, P., Lei, K., Wang, L., Bai, Y., Liu, S., & Chen, X. (2018). Fabrication of self-healing hydrogels with on-demand antimicrobial activity and sustained biomolecule release for infected skin regeneration. *ACS Applied Materials & Interfaces, 10*(20), 17018–17027. DOI: 10.1021/acsami.8b01740 PMID: 29693373

Tiwari, N., Rajput, M., Chien, N. A., & Mathews, N. (2018). Highly transparent and integrable surface texture change device for localized tactile feedback. *Small, 14*(1), 1702312. DOI: 10.1002/smll.201702312 PMID: 29140605

Vedadghavami, A., Minooei, F., Mohammadi, M. H., Khetani, S., Kolahchi, A. R., Mashayekhan, S., & Sanati-Nezhad, A. (2017). Manufacturing of hydrogel biomaterials with controlled mechanical properties for tissue engineering applications. *Acta Biomaterialia, 62*, 42–63. DOI: 10.1016/j.actbio.2017.07.028 PMID: 28736220

Vigata, M., Meinert, C., Hutmacher, D. W., & Bock, N. (2020). Hydrogels as drug delivery systems: A review of current characterization and evaluation techniques. *Pharmaceutics, 12*(12), 1188. DOI: 10.3390/pharmaceutics12121188 PMID: 33297493

Volpi, M., Paradiso, A., Costantini, M., & Świ szkowski, W. (2022). Hydrogel-based fiber biofabrication techniques for skeletal muscle tissue engineering. *ACS Biomaterials Science & Engineering, 8*(2), 379–405. DOI: 10.1021/acsbiomaterials.1c01145 PMID: 35084836

Wang, T., Yi, W., Zhang, Y., Wu, H., Fan, H., Zhao, J., & Wang, S. (2023). Sodium alginate hydrogel containing platelet-rich plasma for wound healing. *Colloids and Surfaces. B, Biointerfaces, 222*, 113096. DOI: 10.1016/j.colsurfb.2022.113096 PMID: 36542954

Wang, X., Yang, Y., Shi, Y., & Jia, F. (2020). Smart hydrogels in tissue engineering and regenerative medicine. *Frontiers in Chemistry, 8*, 245. DOI: 10.3389/fchem.2020.00245 PMID: 32351934

Wang, Y., Tu, S., Pinchuk, A. N., & Xiong, M. P. (2013). Active drug encapsulation and release kinetics from hydrogel-in-liposome nanoparticles. *Journal of Colloid and Interface Science, 406*, 247–255. DOI: 10.1016/j.jcis.2013.05.081 PMID: 23809875

Wang, Z., Hu, W., Du, Y., Xiao, Y., Wang, X., Zhang, S., Wang, J., & Mao, C. (2020). Green gas-mediated cross-linking generates biomolecular hydrogels with enhanced strength and excellent hemostasis for wound healing. *ACS Applied Materials & Interfaces, 12*(12), 13622–13633. DOI: 10.1021/acsami.9b21325 PMID: 32163261

Wirthl, D., Pichler, R., Drack, M., Kettlguber, G., Moser, R., Gerstmayr, R., Hartmann, F., Bradt, E., Kaltseis, R., Siket, C. M., Schausberger, S. E., Hild, S., Bauer, S., & Kaltenbrunner, M. (2017). Instant tough bonding of hydrogels for soft machines and electronics. *Science Advances, 3*(6), e1700053. DOI: 10.1126/sciadv.1700053 PMID: 28691092

Wu, Y., Wang, Y., Long, L., Hu, C., Kong, Q., & Wang, Y. (2022). A spatiotemporal release platform based on pH/ROS stimuli-responsive hydrogel in wound repairing. *Journal of Controlled Release*, *341*, 147–165. DOI: 10.1016/j.jconrel.2021.11.027 PMID: 34813880

Wu, Z. L., Moshe, M., Greener, J., Therien-Aubin, H., Nie, Z., Sharon, E., & Kumacheva, E. (2013). Three-dimensional shape transformations of hydrogel sheets induced by small-scale modulation of internal stresses. *Nature Communications*, *4*(1), 1586. DOI: 10.1038/ncomms2549 PMID: 23481394

Xia, W., Jin, M., Feng, Z., Zhang, J., Rong, Y., Zhang, Y., Zhang, S., Yu, Y., Yang, H., & Wang, T. (2023). Injectable decellularzied extracellular matrix hydrogel derived from human umbilical cord: A novel perspective to deal with refractory wound via medical wastes. *Materials & Design*, *229*, 111877. DOI: 10.1016/j.matdes.2023.111877

Xiao, S., Zhao, T., Wang, J., Wang, C., Du, J., Ying, L., Lin, J., Zhang, C., Hu, W., Wang, L., & Xu, K. (2019). Gelatin methacrylate (GelMA)-based hydrogels for cell transplantation: An effective strategy for tissue engineering. *Stem Cell Reviews and Reports*, *15*(5), 664–679. DOI: 10.1007/s12015-019-09893-4 PMID: 31154619

Xie, R., Zheng, W., Guan, L., Ai, Y., & Liang, Q. (2020). Engineering of hydrogel materials with perfusable microchannels for building vascularized tissues. *Small*, *16*(15), 1902838. DOI: 10.1002/smll.201902838 PMID: 31559675

Xie, Y., Guo, X., Zhang, D., Shen, X., Sun, Q., & Chu, F. (2025). Life cycle assessment and environmental cost evaluation of biomimetic cellulose-enhanced hydrogels. *ACS Sustainable Chemistry & Engineering*, *13*(24), 9186–9194. DOI: 10.1021/acssuschemeng.5c02547

Xu, W., Wang, L., Zhang, R., Sun, X., Huang, L., Su, H., Wei, X., Chen, C.-C., Lou, J., Dai, H., & Qian, K. (2020). Diagnosis and prognosis of myocardial infarction on a plasmonic chip. *Nature Communications*, *11*(1), 1654. DOI: 10.1038/s41467-020-15487-3 PMID: 32245966

Xu, Z., Liu, Y., Lv, M., Qiao, X., Fan, G. C., & Luo, X. (2023). An anti-fouling wearable molecular imprinting sensor based on semi-interpenetrating network hydrogel for the detection of tryptophan in sweat. *Analytica Chimica Acta*, *1283*, 341948. DOI: 10.1016/j.aca.2023.341948 PMID: 37977778

Yang, C., & Suo, Z. (2018). Hydrogel ionotronics. *Nature Reviews. Materials*, *3*(6), 125–142. DOI: 10.1038/s41578-018-0018-7

Ying, H., Zhou, J., Wang, M., Su, D., Ma, Q., Lv, G., & Chen, J. (2019). In situ formed collagen-hyaluronic acid hydrogel as biomimetic dressing for promoting spontaneous wound healing. *Materials Science and Engineering C*, *101*, 487–498. DOI: 10.1016/j.msec.2019.03.093 PMID: 31029343

Yuk, H., Lin, S., Ma, C., Takaffoli, M., Fang, N. X., & Zhao, X. (2017). Hydraulic hydrogel actuators and robots optically and sonically camouflaged in water. *Nature Communications*, *8*(1), 14230. DOI: 10.1038/ncomms14230 PMID: 28145412

Zhang, S., Hou, J., Yuan, Q., Xin, P., Cheng, H., Gu, Z., & Wu, J. (2020). Arginine derivatives assist dopamine-hyaluronic acid hybrid hydrogels to have enhanced antioxidant activity for wound healing. *Chemical Engineering Journal*, *392*, 123775. DOI: 10.1016/j.cej.2019.123775

Zhang, X., Chen, X., Hong, H., Hu, R., Liu, J., & Liu, C. (2022). Decellularized extracellular matrix scaffolds: Recent trends and emerging strategies in tissue engineering. *Bioactive Materials*, *10*, 15–31. DOI: 10.1016/j.bioactmat.2021.09.014 PMID: 34901526

Zhang, Y., Liang, B., Jiang, Q., Li, Y., Feng, Y., Zhang, L., Zhao, Y., & Xiong, X. (2020). Flexible and wearable sensor based on graphene nanocomposite hydrogels. *Smart Materials and Structures*, *29*(7), 075027. DOI: 10.1088/1361-665X/ab89ff

Zhou, Y., Hou, Y., Li, Q., Yang, L., Cao, Y., Choi, K. H., Wang, Q., & Zhang, Q. M. (2017). Biocompatible and flexible hydrogel diode-based mechanical energy harvesting. *Advanced Materials Technologies*, *2*(9), 1700118. DOI: 10.1002/admt.201700118

Zhou, Y., Zhao, S., Zhang, C., Liang, K., Li, J., Yang, H., Gu, S., Bai, Z., Ye, D., & Xu, W. (2018). Photopolymerized maleilated chitosan/thiol-terminated poly (vinyl alcohol) hydrogels as potential tissue engineering scaffolds. *Carbohydrate Polymers*, *184*, 383–389. DOI: 10.1016/j.carbpol.2018.01.009 PMID: 29352933

Chapter 17
Hydrogels for 3D/4D Printing and Additive Manufacturing

Akintomiwa Olumide Esan
https://orcid.org/0000-0003-0578-7458
Ladoke Akintola University of Technology, Ogbomoso, Nigeria

Olawale Samuel Dabo
https://orcid.org/0000-0002-4252-3315
Ladoke Akintola University of Technology, Ogbomoso, Nigeria

Ibironke Ayobami
https://orcid.org/0009-0003-4925-6525
Ladoke Akintola University of Technology, Ogbomoso, Nigeria

Ibidotun Theophilus Olawoore
https://orcid.org/0009-0003-1093-4579
Ladoke Akintola University of Technology, Ogbomoso, Nigeria

ABSTRACT

Recent advancements in printing and additive manufacturing have positioned three-dimensional (3D) and four-dimensional (4D) printing at the forefront of cutting-edge research. These technologies have revolutionized material fabrication, offering significant advantages across various industries, including healthcare, aerospace, and bioengineering. This chapter reviews recent developments in the use of hydrogels for 3D and 4D printing, their role in additive manufacturing, their advantages and limitations in 3D printing, and the latest innovations in 4D printing technologies that leverage hydrogels as stimuli-responsive materials are reviewed.

17.1 INTRODUCTION

The field of additive manufacturing has been completely transformed in the last few decades due to the tremendous growth and development in Three-dimensional (3D) printing technologies.

DOI: 10.4018/979-8-3373-2337-4.ch017

There are several advantages that accompany 3D printing technology, these include customized printing prototyping, different design methods among others. Furthermore, batch customization is also an added advantage provided by this technology. This gives the end-users (consumers) more options regarding the design and the shape (specifics) of the end-product. This ensures that the technology can be more employed and adopted across different industries and at home for personal usage (Iftekar *et al.* 2023; Li *et al.* 2024).

According to the International Standard ISO/ASTM 52900, additive manufacturing also known as 3D printing is a process that involves combining materials to produce parts from the 3D model data. This technology involves placing the substances 'layer over layer.' The final object is designed with the use of Computer-aided Design (CAD) software, while the final design is sent to the 3D printer (Srinivasan *et al.* 2021; Jayswal & Adanur, 2022). As mentioned by Parupelli and Desai (2019), there are eight steps involved in the 3D printing process, which include design and modelling, file conversion to stereolithography (.STL) format, equipment preparation and transfer of STL file to the additive manufacturing (AM) equipment, machine configuration, part fabrication, part removal and cleaning, post processing and end-use application.

Various materials, including ceramics and metals, have been extensively utilized in different 3D printing techniques. However, 4D printing remains in its early developmental stages, with its ability to fabricate highly functional 3D structures hinged on the choice of stimuli-responsive materials and the printing methods employed (Zhao *et al.* 2022). Various smart materials, including shape memory polymers, liquid crystal elastomers (LCE), and hydrogels, have been investigated for use in 4D printing. Among these, hydrogels are particularly notable for their ability to reversibly hydrate or dehydrate without dissolving in water. This distinctive property makes them especially suitable for printing and additive manufacturing applications (Zhao *et al.* 2022). In contrast to traditional three-dimensional (3D) printing, which fabricates static structures, four-dimensional (4D) printing integrates stimuli-responsive materials to produce dynamic objects capable of undergoing controlled shape transformations due to the influence of environmental triggers such as temperature, light, chemicals, or moisture (Wang *et al.* 2024).

The selection of a suitable stimuli-responsive material is crucial, as it directly determines the "fourth dimension;" the time-dependent behavior that enables programmed deformations such as folding, curling, twisting, or color changes upon exposure to stimuli (Tamay *et al.* 2019). Based on ISO/ASTM guidelines, additive manufacturing is categorized into seven main technologies: material extrusion, material jetting, powder bed fusion, binder jetting, vat photopolymerization, sheet lamination, and directed energy deposition. Common 3D printing techniques include stereolithography, fused deposition modeling, polymer jetting, digital light processing, inkjet printing, and direct ink writing. Hydrogels have demonstrated significant potential in these techniques, particularly in applications requiring high biocompatibility, flexibility, and responsiveness to environmental conditions (Zhao *et al.* 2022; Li *et al.* 2024).

Additive manufacturing has found diverse applications in several fields including in personalized medicine. For example, drug delivery could be personalized for individual patients leading to higher efficiency (Alzoubi *et al.* 2023). In a study, Liu *et al.* (2024), developed a 4D printed hydrogel scaffold which is triggered based on body temperature. This is particularly useful in creating minimally invasive implants that could be triggered in the human body depending on the temperature. In another study, Nain *et al.* (2024), examined the applicability of 4D printing to produce complex vascular structures, this holds great potential for tissue engineering.

17.2 3D PRINTING: HISTORICAL PERSPECTIVE, ADVANCEMENTS AND LIMITATIONS

17.2.1 Historical Perspective and Advancements

The works of Charles W. Hull laid the foundation for the development of additive manufacturing. In March 1983, Hull was able to successfully print a teacup on stereolithography (STL) apparatus. This was the 3D printer developed. In the years that followed, interest in this field grew rapidly leading to accessibility, prompting more research. Between the 1990s and 2000s, attention was focused on new functional materials leading to applications of this technology in fields such as aviation, healthcare, etc. (Attaran 2017; Jimenez *et al.* 2019; Iftekar *et al.* 2023). After the work of Hull in 1983, Carl Deckard worked on 'methods and apparatus for producing parts by selective sintering' in 1986. This was the approach in developing additive manufacturing through selective sintering. Deckard's work was followed by Michael Feygin and his colleagues' efforts in 1988 for developing integral object formation from lamination. In 1989, Scott Crump worked on apparatus and methods for the creation of 3D objects. This represents a first in creating additive manufacturing through the use of fused deposition modelling (FDM). Emmanuel Sachs and co-workers in 1989 also developed a 3D printing technique of 'integrating binding agent and coloured ink.' (Jimenez *et al.* 2019).

Traditional manufacturing techniques including casting, forging, subtractive machining are usually time consuming and generate wastes. Consequently, leading to increased cost of fabrication. Additive manufacturing on the hand consumes time with minimal wastes produced. Diverse material availability also makes additive manufacturing gain widespread adoption in various industries (Srinivasan *et al.* 2021; Jayswal and Adanur, 2022). In addition, the increased demand for 3D printing technology across these fields is as a result of its ability to generate intricate structures without compromising on precision and accuracy. Different smart materials such as hydrogels, metals, extracellular matrix materials, ceramics, polymers etc. are widely employed in additive manufacturing (Lu *et al.* 2015; Iftekar *et al.* 2023).

Several works such as Solid creation system (SCS), Solid Object Ultraviolet Laser Printer (SOULP), Inkjet Rapid Prototyping were all advancements of Charles Hull's works while the advancements on Deckard's work include Direct Metal Laser Sintering (DMLS) and Selective Laser Melting (SLM). Metal Deposition (MD) and Fused Filament Fabrication (FFF) were the progresses made after Scott Crump's works. Multijet Modelling System (MJS), Model Maker and Pattern Master and Prometal were the advancements made through Sach and co-workers' efforts (Jimenez *et al.* 2019).

The working principle behind 3D printing is quite similar to the standard inkjet printing, although, in the place of printing several sheets of ink on paper, 3D printing in contrast produces 3D objects using materials. Although, 3D printing and additive manufacturing are used alternately, additive manufacturing is often regarded as the industrial form of 3D printing while the use of 3D printing technique is referred to as rapid prototyping (Attaran 2017; Framinan *et al.* 2022). To better visualize the chronological evolution of additive manufacturing techniques, key historical milestones and their contributors are summarized in Table 1.

Table 1. Summary of landmark dates in additive manufacturing

Author and Date	Key Advancement Recorded	References
Charles Hull, 1983	He developed the first 3D printer and applied successfully in printing a teacup on a stereolithography apparatus	Iftekar *et al.* 2019
Carl Deckard, 1986	He was the first to develop selective sintering. A method of creating 3D objects	Jimenez *et al.* 2019
Feygin and Co-workers, 1988	Integral object formation from lamination was developed	Jimenez *et al.* 2019
Scott Crump, 1989	He created the first technique for printing 3D objects through fused deposition modeling (FDM)	Jandyal *et al.* 2022
Sachs and Co-workers, 1989	3D printing method of injecting binding and coloured ink was created	Gross *et al.* 2014

17.3 MAJOR 3D PRINTING TECHNIQUES

Several 3D printing techniques have been created recently, each with its different functionalities and intended applications. This technology (3D printing) has been massively scaled from its early beginning of prototyping to its present state of widespread adoption for printing in different areas of human endeavours (Shahrubudin *et al.* 2019).

17.3.1 Vat Photopolymerization

The printing techniques in this category are Stereolithography (SLA) and Digital Light Processing. In stereolithography, a chain reaction is initiated with the aid of Ultraviolet (UV) light on a resin layer or solution of monomers. Acrylic or epoxy-based monomers which are active to UV-radiation can easily be converted to chains of polymers when activated (Ngo *et al.* 2018; Shahrubudin *et al.* 2019). The selection of a UV light source is determined by the type of resin being used. Commonly utilized sources include HeCd lasers and xenon lamps. Furthermore, two photon polymerization has been utilized in Stereolithography (SLA) fabrication to attain enhanced resolution in the resulting printed material. However, resins present significant challenges in SLA technology. These include high costs and the limitation of using only one resin at a time, which restricts the complexity of device designs (Gross *et al.* 2014). To obtain an object of the required mechanical strength, the object printed could be processed further through heating. The main advantages of SLA is exceptional print quality with resolutions as fine as 10μm while slow process, high printing cost and restricted range of printable materials are its drawbacks (Gross *et al.* 2014; Ngo *et al.* 2018).

Figure 1. Vat Photopolymerization printing technique adapted from Li et al. (2022), BioMed Research International, reused under the Creative Common CC-BY license

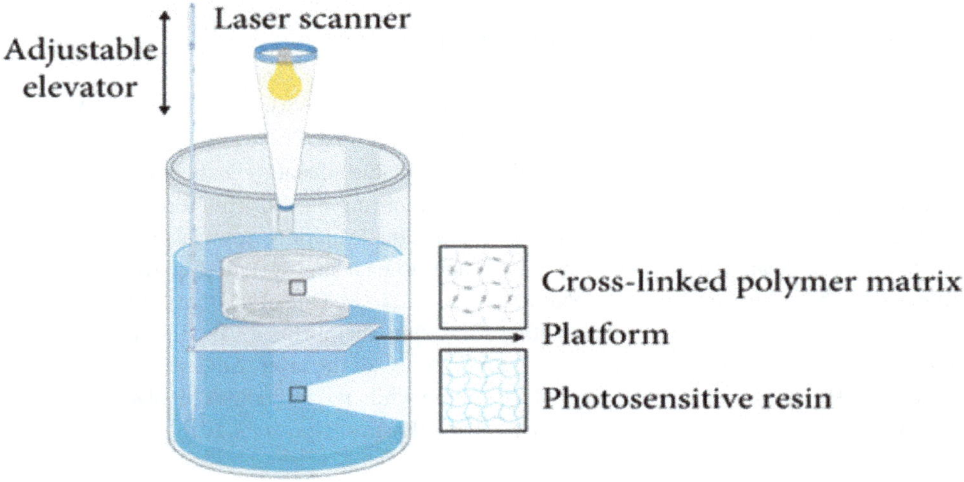

17.3.2 Direct Energy Deposition (DED)

This is a more intricate 3D printing method that is normally utilized for repairing or adding more materials to 'already existing components.' Direct Energy Deposition is also beneficial because of its ability to produce prints of fine quality (Shahrubudin *et al.* 2019). DED also called Laser Engineered Net Shaping (LENS), Laser Solid Forming (LSF), Direct Metal Deposition e.t.c. In this process, a concentrated energy source, typically a laser, is precisely directed at a small, specific area of the substrate. Simultaneously, this energy source melts the initial material, usually in the form of powder or wire. Following the movement of the energy source used, the melted starting material is directly placed onto the substrate (Ngo *et al.* 2018; Jandyal *et al.* 2022). The main benefits of DED are high speed and its ability to combine with traditional subtractive processes. Its drawbacks are reduced accuracy and surface quality. It also has limitations in printing intricate parts when compared to powder bed fusion methods (Ngo *et al.* 2018).

17.3.3 Powder Bed Fusion

This 3D printing technique involves methods such as Electron Beam Melting (EBM), Selective Laser Sintering and Selective Heat Sintering (Jayakrishna *et al.* 2023). In this method fine powder coatings are distributed and tightly fixed on a printing platform. The thin layers of powder in each step are joined together with the aid of a 'laser beam or a binder.' The final 3D object is produced by stacking different layers of powder sequentially. Final clean-up of the created 3D object can be done using a vacuum and could be further processed through coating or sintering (Ngo *et al.*, 2018; Shahrubudin *et al.* 2019; Jayakrishna *et al.* 2023). The most critical parameters that affect this 3D printing process are size distribution of powder and fixing. The major advantages are high resolution, superior print quality and the limitations are increased costs, and significant porosity resulting from fusing the powder with a binder (Ngo *et al.* 2018).

17.3.4 Extrusion-Based 3D Printing Technology

Extrusion-based 3D printing technologies, such as Fused Deposition Modeling (FDM) and Direct Ink Writing (DIW), have significantly advanced over time. These technologies are known for their simplicity, affordability, and versatility. One notable advantage is their ability to incorporate active ingredients directly into the printing ink during the fabrication process (Song *et al.* 2025). FDM involves the use of elongated filament made up of thermoplastic polymer. The process includes heating the filament 'nozzle' to approach an almost fluid (liquid) state. This is followed by thrusting it on a printing platform (Li *et al.* 2022). One of the most critical features of the filament is its flexibility. Vital printing parameters that influence final properties of the printed objects are filament direction, the width/thickness of the layer. Advantages of FDM include cost efficiency, fast speed and simple operation while the limitations involve diminished mechanical properties and low-quality surface characteristics (Ngo *et al.* 2018).

Figure 2. Fused deposition modelling printing technique adapted from Li et al. (2022), BioMed Research International, reused under the Creative Common CC-BY license.

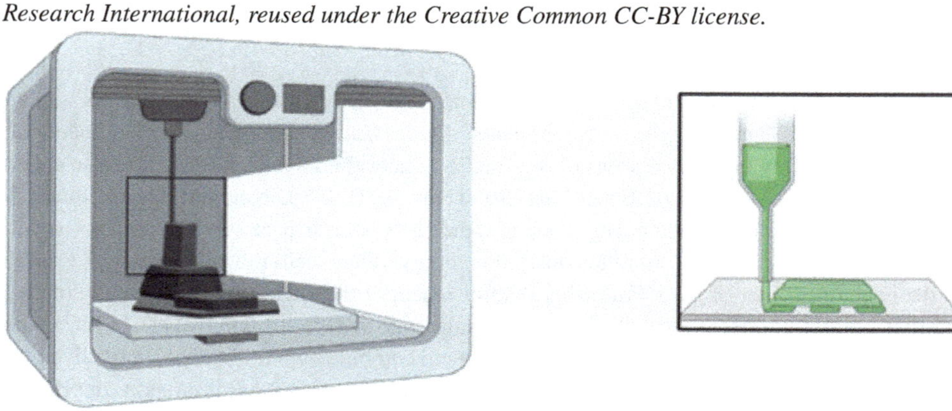

17.3.5 Sheet Lamination

In this method, sheets of materials are joined together in order to print parts of a material. Laminated Object Manufacturing (LOM) and Ultrasound Additive Manufacturing (UAM) are the 3D printing techniques that make use of this technology (Shahrubudin *et al.* 2019). In LOM, materials or sheets are cut and laminated step-wisely. Sequential layers of materials are cut with precision and are joined together (either through form-then-bond or bond-then-form) (Jandyal *et al.* 2022). Post-processing of printed objects includes high temperature treatment. Ceramics, paper, and metal filled tapes can be printed using LOM. Advantages of LOM are lowered instrumental costs and production time. Larger structures can be effectively printed through this technique. Its drawbacks are lower surface quality, especially without undergoing post-processing, lowered accuracy and difficulty in removing excess parts (Wang *et al.* 2012).

The summary of different 3D printing methods, materials used, applications, and advantages is shown in Table 2.

Table 2. Analysis of the benefits and challenges associated with several 3D printing techniques.

3D Printing Method	Materials Used	Applications	Advantages	Limitations	Reference(s)
Vat Photopolymerization	Acrylic or epoxy-based monomers	It is widely applied in printing biomedical objects	Exceptional print quality	Slow process, High cost, Restricted range of printable materials	Ngo *et al.* 2018; Shahrubudin *et al.* 2019.
Direct Energy Deposition	Usually powder or wire	It can be applied for printing ceramics and polymers but used more for metals	High speed Ability to be combined with traditional subtractive processes	Reduced accuracy Reduced surface quality and difficulty in creating intricate parts	Ngo *et al.* 2018; Shahrubudin *et al.* 2019
Powder Bed Fusion	Powder	For printing metals, plastics, and, ceramics e.t.c	High resolution Superior print quality	Increased costs Significant porosity	Ngo *et al.* 2018; Shahrubudin *et al.* 2019
Fused Deposition Modelling	Filament made up thermoplastic polymer	Printing complex composite materials and toys	Cost efficiency Fast speed Simple operation	Diminished mechanical properties Low-quality surface characteristics	Ngo *et al.* 2018; Li *et al.* 2022
Inkjet Printing	Ceramics (Wax-based inks and liquid suspension)	For printing ceramics	Speed Efficiency	Reduced workability Poor interlayer adhesion Coarse resolution	Gross *et al.* 2014; Ngo *et al.* 2018; Park *et al.* 2022
Sheet Lamination	Paper, plastic or metal	Paper and complex structures	Reduced instrumental cost and production time Ability to print large structures	Lower surface quality Reduced accuracy Difficulty in removing excess parts	Wong and Hernandez, 2012; Jandyal *et al.* 2022

17.4 LIMITATIONS AND CHALLENGES OF 3D PRINTING

Presently, there are two main drawbacks to using 3D printers; these are increased cost for printers with enhanced capabilities or lower capabilities accompanied with cheaper prices. This gap has been bridged recently and more affordable printers with higher capabilities are expected with increased research efforts in this area (Attaran 2017). Some of the other challenges that require further attention include limitations in part size, issues like the 'elephant foot effect,' 'over-extrusion', misaligned layers, 'gaps in top layers,' 'under-extrusion,' 'pillowing,' 'warping,' and inaccuracies. Additionally, there are concerns about low production efficiency, high costs of production, difficulties in building overhang structures, anisotropic mechanical properties, and material limitations. (Praveena *et al.* 2022).

17.5 EMERGENCE OF 4D PRINTING: INNOVATIONS AND CHALLENGES

The field of four-dimensional (4D) printing was opened up by Skylar Tibbits in 2011 through his works at Massachusetts Institute of Technology (MIT) and subsequently through his TEDTalk (Imam *et al.* 2021). Traditional 3D techniques lead to the production of 'static' materials, but due to research driven mainly by researchers both in the industry and academia, this technology has been vastly improved. There has however been interest towards 4D printing as a result of its ability to create objects that react to environmental triggers after printing (Champeau *et al.* 2020). 4D printing represents an advancement to 3D printing through the addition of the fourth dimension; 'time.' This technique allows the printed object to react to external influence, such as light, temperature, and pH, over time. This concept offers significant possibilities for use in fields like biomedical engineering, robotics, aerospace, and 'smart materials.' There are several crucial factors that influence the 4D printing process. These factors include; printing method utilized, the kind of 'stimuli-responsive' polymers used, the kind of stimuli, mechanism of interaction and the mathematical modelling employed (Champeau *et al.* 2020; Ghosh *et al.* 2023; Li *et al.* 2025).

17.5.1 Hydrogels: Properties and Suitability for 3D/4D Printing

Hydrogels are 3D networks composed of natural or synthetic polymers, distinguished by their exceptionally high water content, which gives them remarkable flexibility (Ho *et al.* 2022). These polymer networks are formed through chemical or physical bonds between macromolecules (Abel *et al.* 2023). A better understanding of hydrogel chemistry can be visualized through their 3D network structure, as shown in Figure 18.3. When exposed to physiological conditions, these structures can absorb a significant quantity of water or biological fluids, displaying a soft, pliable texture akin to living tissues. This unique property renders hydrogels suitable for various applications. Natural polymers such as polypeptides, polysaccharides, and DNA, as well as synthetic options like polyacrylamide and poly(vinyl alcohol), have been widely researched for hydrogel production due to their beneficial properties (Khan *et al.* 2013; Correa *et al.* 2021). Hydrogels possess properties such as specific utility, adaptability, capacity for sterilization, and compatibility with living organisms, which enable them to meet both material and biological needs for treating or replacing tissues and organs, as well as interacting with biological systems. In addition, hydrogel's 'shape-memory' capabilities can be fine-tuned to meet the specific needs of applications across diverse fields like tissue engineering and biomedicine. (Peppas & Hoffman, 2020; Zhu *et al.* 2023). Gelatin and agar have been utilized in various fields for centuries, but the use of hydrogels in biomedical applications began to take shape in 1936. During this time, researchers at DuPont introduced methacrylic polymers, including poly(2-hydroxyethyl methacrylate) (polyHEMA), as described in the work of Wichterle and Lim in 1960. Initially, polyHEMA was regarded as a hard, brittle, and glassy material with limited significance. However, in 1960, Wichterle and Lim discovered that polymerizing and crosslinking polyHEMA in the presence of water and other solvents produced a completely different outcome. Crosslinking refers to the process of chemically bonding polymer chains together to form a three-dimensional network, which transforms a liquid monomer into a soft, water-retaining gel (Li & Gong, 2024). The development of hydrogel formulations has grown steadily over time. However, issues such as limited solubility, excessive crystallinity, lack of biodegradability, suboptimal mechanical and thermal features, unreacted monomer residuals, and the presence of toxic crosslinkers continue to pose challenges (Joshi *et al.* 2022). To overcome these obstacles, integrating natural and synthetic polymers with tailored features, including biodegradability, improved solubility, controlled crystallinity, and specific biological activities, presents promising opportunities for innovation. (Kuzina *et al.* 2023). Crosslinking can take place in two distinct environments: in vitro, during the preparation of the hydrogel, or in vivo (in situ), after it has been applied to a specific location within the human body (Chen *et al.* 2022). Chemical crosslinking requires the addition of a low molecular weight crosslinking agent to the polymer within the reaction mixture (Nasution *et al.* 2022). Without crosslinking, hydrophilic linear polymer chains solubilize in water as a result of their natural thermodynamic compatibility with it (Gao *et al.* 2021). When crosslinking points are present, the balance between solubility and the retractive force of the network's elasticity ensures that swelling stabilizes when these forces are in equilibrium (Jiang *et al.* 2021). The hydrophilicity of the network arises from hydrophilic groups like $-NH_2$, and $-COOH$, in addition to osmotic pressure and the capillary effect (Nasution *et al.* 2022).

Chemical and physical crosslinking points are vital in preserving the 3D structure of hydrogels in their swollen state. In chemical crosslinking, polymer chains form covalent bonds through crosslinking agents, hydrogen bonding, hydrophobic interactions, or ionic complexation. This process allows for applications like solvent casting, bulk modifications, reshaping, and biodegradability, though chemically crosslinked hydrogels often lack non-toxic properties (Ahmad *et al.* 2022). A representation of natural and synthetic hydrogel is shown in Figure 18.4

Figure 3. Schematic diagram of a hydrogel network structure, showing the crosslinked polymer matrix. Reproduced from Wang et al. (2018), RSC Advances, 8, 16773–16780, licensed under CC BY.

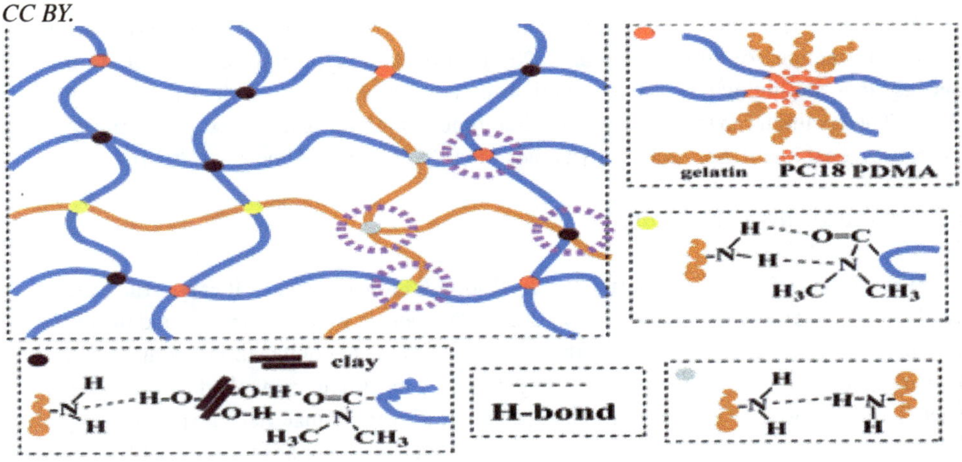

Figure 4. Sources of hydrogel materials. Natural types include hyaluronic acid, alginate, chitosan, and gelatin, while synthetic types include PNIPAAm, PVA, PEG, and PAA. Adapted from Gan et al. (2024), under CC BY licence

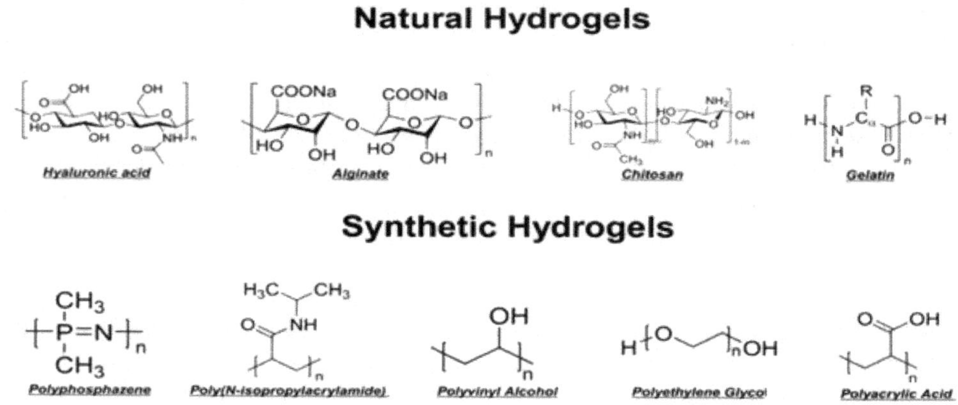

17.6 DIFFERENT SHAPE MEMORY EFFECT IN 4D PRINTING OF HYDROGELS

A vital characteristic of 4D printing is the use of external stimuli to cause variations in the produced structure over time (Chu *et al.* 2020). Hydrogels are especially suitable for 4D printing because they are highly responsive to various external factors (Rastogi & Kandasubramanian, 2019). These stimuli can induce reversible alterations in the hydrogel's shape, size, or mechanical properties, enabling the printing of dynamic, 'smart' structures (Khoo *et al.* 2015).

17.6.1 Water and Humidity

Water is one of the most important external stimuli for hydrogels, as it causes the material to swell and alter its shape (Al Nahali *et al.* 2023). The rate of water absorption and the extent of swelling can be modified by altering the hydrogel's chemical composition (Dong *et al.* 2020). For instance, hydrogels that react to varying humidity levels can be applied in areas like smart packaging or sensors, where environmental changes trigger the material's response (Jiang *et al.* 2021). The capacity to control both the timing and extent of water absorption enables designers to create objects that evolve over time, leading to potential usage in areas including responsive architecture, biomedical implants, and robotics (Deng *et al.* 2021).

17.6.2 Temperature

Temperature is another key external stimulus that can induce changes in hydrogel behavior (Abdullah & Okay, 2023). Some hydrogels are thermos-responsive, meaning they can swell, shrink, or alter their mechanical features as a result of temperature fluctuations (Li *et al.* 2022).

For example, certain hydrogels undergo a phase transition at specific temperatures, shifting between a hydrated, swollen state and a dehydrated, collapsed state (Solis & Czekanski, 2022). In 4D printing, temperature changes can be leveraged to control the deformation of hydrogel structures (Kuang *et al.* 2019). This is very beneficial in areas where materials must adapt to varying environmental temperatures, such as in self-healing materials, shape-memory devices, or smart textiles (Aydin *et al.* 2025).

Liu *et al.* (2024), developed a 4D printed hydrogel scaffold which is triggered based on body temperature. Through multidimensional morphing from 1D to 3D using Dynamic Thermoset Polyurethane (DTPU). The dry samples of DTPU displayed remarkable strength at 42.5 MPa, Young's Modulus of 1.1 GPA and low breaking strain of 4.6 percent, while the toughness was reported to be 1.2 MJm^{-3}. The energy properties of DTPU was attributed to the cross-linked structure. In addition, the material (DTPU), also displayed a remarkable heat-induced shape morphing effect. This was shown in the swift changes in the Young's Modulus with respect to changes in body temperature. Changes in temperature between 10 °C to 50 °C led to decreased tensile strength (14.6 to 0.73). The temperature-influenced mechanical properties of DTPU was further tested by holding 100 g of material at 4 °C without morphing. However, when the temperature was increased to 37 °C, the support flattened. This innovation has implications for minimally invasive implants and devices that deploy automatically inside the human body at physiological temperatures.

Browser *et al.* (2023), on the other hand worked on developing thermo-sensitive hydrogels using stereolithography (SLA) for 4D printing applications. The authors synthesised two hydrogels systems (PiPrOx-netPAM and PMOx-net-PNIPAM) which demonstrated significant shape-morphing behaviour in response to temperature changes. These hydrogels underwent swelling below their Volume Phase Transition Temperature (VPTT) and Shrinkage above it, thereby showcasing contraction and expansion effects. Other key findings from the study included the ability to control thermal responsiveness by varying the ratio of thermo-sensitive units to hydrophilic components resulting in morphological changes. Such systems are promising for temperature-controlled microfluidic devices or actuators in soft robotics where structural transformations are essential.

Abdullah and Okay (2023), achieved SLA printing of PAAc-based hydrogels with self-healing and shape-memory properties. The hydrogels were fabricated using solvent-free co-polymerization of hydrophilic AAc and hydrophobic C16A monomers, eliminating the need for surfactants or chemical cross-linkers. The hydrogels demonstrated significant thermal responsiveness, with modulus changes that varied based on C16A content. Moreover, the mechanical features of the hydrogels revealed that higher C16A content increased Young's modulus and the number of crystalline domains acting as physical cross-links, though at the expense of brittleness. The study also highlighted the hydrogels' self-healing capabilities, demonstrating high recovery efficiencies for mechanical performance metrics such as fracture strength and Young's modulus, exceeding 90% after thermal healing at 65 °C. These properties make them suitable candidates for reusable wearable devices or artificial skin with integrated damage-recovery features.

17.6.3 pH and Ionic Strength

Hydrogels can also respond to changes in pH or ionic strength, which can alter their chemical structure and behavior (Andrade *et al.* 2021). For instance, hydrogels made from polyacrylic acid or similar materials may swell significantly in an alkaline environment, while shrinking in an acidic one (Vigata *et al.* 2021). Through 4D printing, the pH sensitivity of hydrogels can be used to print structures that can change shape or function when exposed to different chemical environments (Stano *et al.* 2021). By precisely adjusting the hydrogel's composition, designers can develop highly responsive materials for diverse applications (Pourjavadi *et al.* 2021).

Lai *et al.* (2021), explored the shape-morphing behaviour of a hydrogel composite comprising alginate and methylcellulose (Alg_3/MC_9) under external stimuli. They focused on how printed patterns and structural configurations influence their deformation. As reported by the authors, the hydrogels exhibited shape-morphing behaviour both in DI water and 0.1 M $CaCl_2$ solution. However, uncross-linked hydrogels in DI water were mechanically weak and prone to collapse, while cross-linked hydrogels in 0.1 M $CaCl_2$ solution showed improved structural stability. This study demonstrates the feasibility of designing hydrogel-based actuators with programmable motion, which could be applied in biomedical stents, self-assembling scaffolds, or soft robotic limbs with tailored motion paths.

Nain *et al.* (2024), also investigated the application of 4D printing to produce complex vascular structures using a nanoengineered dual-component hydrogel system. The hydrogel system consists of carbonized alginate (CAlg) and alginate/methylcellulose (Alg/MC) which exhibited excellent and anti-oxidant, anti-inflammatory and anti-thrombotic properties. The 3D printed structures showed remarkable cell viability and morphology, with both HUVECs and NIH3T3 cells maintaining their morphology. This case underscores the potential of 4D printed hydrogels in vascular tissue engineering, enabling the creation of biologically functional vessels with advanced bioactive properties for regenerative medicine.

Li and Co-workers on the other hand proposed a strategy for achieving 4D printing of programmable shape-morphing hydrogel structures using highly viscous hydrogels with smartly pre-designed patterns to create desirable 4D printed flat structures. The hydrogels are printed using Direct Ink Writing (DIW) and are able to alter their shape in response to soaking in $CaCl_2$ solution. The rheological properties (shear thinning and thixotropic) of the hydrogels were excellent making it suitable for DIW printing.

17.6.4 Electric and Magnetic Fields

Some hydrogels are responsive to electric or magnetic fields, which can be used to induce deformation (Wei *et al.* 2023). For example, conductive hydrogels can alter their shape or mechanical properties when exposed to an 'electric field,' making them well-suited for usage in soft robotics and actuators (He *et al.* 2023). Magnetic hydrogels, which contain magnetic particles embedded within the polymer matrix, can respond to external magnetic fields, enabling their use in targeted drug delivery, biomedical imaging, and robotics (Deng *et al.* 2021). The ability to design hydrogels that react to electric and magnetic fields offers exciting possibilities for printing dynamic, reconfigurable objects in 4D printing (Hu *et al.* 2023).

Cho *et al.* (2024), demonstrated a novel approach to 4D printing by utilizing Functionally Graded Additive Manufacturing (FGAM) to create Functionally Graded Multi-Materials (FGMMs) for advanced soft robotics. By leveraging Binder Jetting 4D Printing (BJ4DP), the researchers fabricated porous hygroscopic networks capable of responding dynamically to changes in humidity, as well as structures with magnetic-responsive properties. These functionally graded materials exhibited a wide range of customizable responses, including swelling, bending, and crawling motions under different stimuli. For instance, the magneto-hygrobot demonstrated dual responsiveness to humidity and magnetism, enabling complex tasks like directional locomotion and autonomous navigation through track-like environments. Table 3 provides an overview of 4D-printed hydrogels, detailing the printing techniques employed, the stimuli triggering shape-morphing effects, and their respective applications. This case clearly illustrates the real-world applicability of 4D-printed hydrogels in the development of autonomous soft robots, wearable sensors, and eco-friendly bioactuators capable of intelligent, stimulus-driven navigation.

Table 3. Summary of 4D-printed hydrogels, the printing methods utilized, the stimulus causing shape-morphing effect, and the intended applications

Stimulus	Composition of Hydrogel Used	Printing Methods	Intended Application	References
Temperature	Diels-alder diol + Polyurethane	Fused deposition modelling	Minimally invasive treatment of tissue defects	Liu *et al.* 2024
Ionic Strength and Water	Alginate + Methylcellulose	Direct Ink Writing	Biomedical engineering	Lai *et al.* 2021
Ion Strength	Carbonized Alginate and Alginate + Methylcellulose	Extrusion-based	Vascular tissue engineering	Nain *et al.* 2024
Ion Strength	Alginate + Methylcellulose, Alginate + PVA and GELMA	Direct Ink Writing	Bioprinting and Smart devices	Li *et al.* 2025
Temperature	1,6-hexaneditosylate + acetonitrile + 2-Methyl-2-oxazoline and 1,6-hexaneditosylate + 2-isopropyl-2-oxazoline + acetonitrile	Stereolithography	Biomedical devices	Browser *et al.* 2023
Humidity and Magnetism	Polyvinyl alcohol (PVA) + Sodium tetraborate and Polyvinyl alcohol (PVA) + Carbonyl iron powder	Binder jetting	Robotics	Cho *et al.* 2024
Temperature	Diphenyl phosphine oxide + n-hexadecyl acrylate	Stereolithography	Soft robotics and drug delivery systems	Abdullah & Okay 2023

As reflected in Table 18.3, alginate-based hydrogels often blended with methylcellulose or polyvinyl alcohol are prominently featured due to their excellent biocompatibility and ion-sensitive behavior, which make them well-suited for applications in tissue scaffolding and smart devices. Temperature-responsive systems dominate applications in soft robotics and drug delivery, given their reversible responsiveness and ease of external control (Karimi *et al.* 2016; Abasalizadeh *et al.* 2020; Kavand *et al.* 2024).

In terms of fabrication methods, Direct Ink Writing (DIW) and extrusion-based printing are frequently utilized due to their compatibility with high-viscosity hydrogels and their ability to maintain structural fidelity post-printing. Conversely, stereolithography, though less flexible in terms of material compatibility, offers superior resolution, making it ideal for producing fine biomedical components (Baniasadi *et al.* 2024).

Importantly, while stimulus-responsive printing strategies offer dynamic functionality, they also introduce challenges in long-term durability, especially under physiological conditions (Khonakdar *et al.* 2025). Materials must retain mechanical integrity and responsiveness over time without causing adverse biological responses. Moreover, the absence of standardized metrics for comparing performance between hydrogel systems limits cross-study validation and scalability toward clinical applications.

17.7 OTHER MATERIALS USED IN 3D/4D PRINTING AND ADDITIVE MANUFACTURING

Various materials are employed for different usage in multiple industries in 3D and 4D printing (Bhatia & Seghal, 2023). While hydrogels are gaining significant popularity due to their unique properties, other materials including 'shape-memory polymers (SMPs),' 'liquid crystal elastomers (LCEs),' and metal-based materials also play crucial roles (Enyan *et al.* 2025). Table 4 offers a comparative overview of the key properties, applications, limitations, and relative advantages of hydrogels, SMPs, LCEs, and metal-based materials in 3D/4D printing. By comparing these materials, it becomes evident how each one provides distinct benefits for specific applications, illustrating how they either complement or differ from hydrogels in terms of performance and suitability across industries like biomedical engineering, soft robotics, and structural applications.

Table 4. Comparative summary of these materials, outlining their properties, applications, limitations, and how they compare to hydrogels

Material	Shape-Memory Polymers (SMPs)	Liquid Crystal Elastomers (LCEs)	Metal-Based Materials
Properties	- Stimuli-responsive (heat, light, pH)	- Shape-changing behavior due to heat/light stimuli	- High strength, durability, and corrosion resistance
	- Reversible shape transition	- Elasticity and flexibility	- Excellent thermal and electrical conductivity
	- High flexibility and reusability	- Light-responsive properties	- High-quality surface finish in 3D-printed parts
	- Capable of large, reversible deformations	- Combination of liquid crystal and elastomer properties	- Suitable for high-performance structural applications
Applications	- Biomedical Devices: Self-expanding stents, scaffolds	- Soft Robotics: Shape-changing actuators for robots	- Aerospace & Automotive: Lightweight, high-performance components
	- Soft Robotics: Actuators, self-deploying structures	- Wearable Devices: Smart clothing, adaptive lenses	- Biomedical Implants: Custom implants, prosthetics, orthopedic parts
	- Aerospace: Deployable structures like antennas or solar arrays	- Biomedical Applications: Sensors, adaptive prosthetics	- Tooling & Manufacturing: Molds, production aids
Limitations	- Limited Stimuli Response: Sensitivity mainly to temperature/light	- Thermal Sensitivity: Limited range of temperature responses	- High Cost: Expensive metal powders and printing processes
	- Mechanical Properties: Weaker compared to metals or ceramics	- Processing Complexity: Requires precise control in fabrication	- Post-Processing: Requires significant finishing steps
	- Processing Complexity: Requires specific processing conditions	- Material Constraints: Limited range of LCEs available	- Material Constraints: Limited number of printable metals

continued on following page

Table 4. Continued

Material	Shape-Memory Polymers (SMPs)	Liquid Crystal Elastomers (LCEs)	Metal-Based Materials
Comparison with Hydrogels	- SMPs are more robust and suitable for structural applications	- LCEs are ideal for dynamic soft robotics and wearables	- Metals offer high strength but are unsuitable for biological applications
	- Hydrogels offer biocompatibility and flexibility for biomedical uses	- Hydrogels excel in moisture-rich environments, better for tissue engineering	- Hydrogels are better for applications requiring flexibility and biocompatibility

17.8 CHALLENGES OF 4D PRINTING

In conventional 3D printing, the primary goal is to ensure the stability of the printed material so that it retains its shape without undergoing morphological changes in response to environmental stimuli. While this stability is beneficial in many applications, it also imposes limitations on the functionality of 3D-printed objects. In contrast, 4D printing introduces a transformative advantage by allowing objects to change shape dynamically in response to environmental triggers, expanding its potential applications (Shen & Shen, 2025). Despite its advantages, 4D printing faces several challenges that hinder its widespread adoption. The challenges are highlighted below:

17.1. Limited Smart Material Compatibility and Material Selection

One major obstacle is the compatibility between smart materials and existing 3D printing technologies. Most conventional 3D printing materials are unsuitable for 4D applications because they do not exhibit stimuli-responsive behavior. This significantly narrows the range of materials that can be used, limiting innovation and design flexibility. Additionally, the few smart materials currently available for 4D printing typically respond to only specific stimuli, further restricting their applications. To advance the field, researchers must develop new smart materials that are not only highly responsive but also durable, adaptable, and seamlessly compatible with 3D printing processes (Ahmed *et al.* 2021; Sahafnejad Mohammadi *et al.* 2023).

Material selection remains a central challenge in 4D printing. For instance, printing 4D structures using shape memory alloys (SMAs) such as Nickel-Titanium (NiTi) requires strict control over composition to prevent defects. Some materials degrade under repeated shape transformations, while others, such as hydrogels, may deteriorate when subjected to multiple hydration and dehydration cycles. In certain cases, materials must be blended with additional components to enhance their printability and stability (Ahmed *et al.* 2021).

17.2. Inadequate Printing Resolution for Biomedical Applications

Achieving high-resolution and precise printing is particularly critical in biomedical applications, where complex designs and accurate dimensions are essential for functional performance. However, current printing technologies struggle to meet these demands, making it difficult to fabricate intricate 4D structures with the necessary level of detail and reliability (Ramezani & Mohd Ripin, 2023).

17.3. Technical Limitations of Printing Methods

Another obstacle in 4D printing is the lack of widely available 3D printing techniques that are compatible with smart materials. At present, only a few research centres possess the necessary 3D printers capable of handling 4D printing processes. Currently, PolyJet and Selective Laser Melting (SLM) are the primary techniques employed, each with stringent material requirements. PolyJet printing necessitates liquid-state materials that are UV-light-responsive, whereas SLM requires powder-based materials. These constraints underscore the need for more advanced printing technologies (Ahmed *et al.* 2021; Navaf *et al.* 2022; Aldawood, 2023). While extrusion-based 4D printing is more cost-effective and allows for multi-material printing, it suffers from lower resolution. In contrast, laser-assisted 4D printing offers higher precision but is restricted to single-material applications and comes at a higher cost. To realize the full potential of 4D printing, it is imperative to optimize existing 3D printing techniques and develop new methods that align with the unique requirements of 4D fabrication (Sahafnejad Mohammadi *et al.* 2023).

17.4. Durability and Environmental Sustainability

A major hindrance to advancement of 4D printing technology is the inadequate comprehension of the long-term durability and performance of the materials presently employed in this field. In particular, the environmental impact of these materials remains an open question, necessitating further studies on their sustainability and biodegradability (Ahmed *et al.* 2021; Kantaros *et al.* 2023). Furthermore, environmental sustainability of 4D printing materials and processes remains an underexplored area. Given the increasing emphasis on eco-friendly manufacturing, addressing the environmental footprint of 4D printing will be crucial for its long-term adoption and integration into various industries. More research is needed to assess the recyclability, biodegradability, and overall life-cycle impact of smart materials used in 4D printing. Addressing these sustainability concerns will be a key step in ensuring the responsible and widespread use of this emerging technology.

17.5. Structural Design Complexity

Designing smart structures for 4D printing also presents unique challenges. These structures must be carefully engineered to achieve predictable and controlled shape transformations. For example, smart actuators perform best when pre-strained before printing, yet no existing 3D printing technology can directly produce pre-strained membranes. This limitation highlights the urgent need for advancements in material design and fabrication techniques (Ahmed *et al.* 2021; Sahafnejad Mohammadi *et al.* 2023). The complexity of designing 4D-printed structures requires specialized expertise, limiting the number of individuals who can effectively work with the technology (Ahmed *et al.* 2021; Kantaros *et al.* 2023).

17.6. External Barriers: Policy, Cost, and Accessibility

Despite 4D printing advantages, policy constraints, high production costs, and difficulties in scaling from laboratory prototypes to large-scale manufacturing remain significant obstacles. Additionally, the choice and design of intelligent materials used in 4D printing are critical, as these materials must possess specific characteristics, including mechanical strength and stimuli-responsiveness, tailored to the intended application. Achieving high-resolution prints is another challenge, as 4D-printed objects undergo intricate shape transformations that require precise material properties and fine-printing accuracy (Shen & Shen, 2025).

17.7. Stimuli Control Limitations

Controlling environmental stimuli such as pH, temperature, and humidity poses technical limitations, though ongoing research aims to develop more reliable and efficient control mechanisms (Shen & Shen, 2025). In Table 5, the various challenges of 4D printing of hydrogels are summarized and ranked.

Table 5. Different challenges and Rank of 4D Printing of Hydrogels

Rank	Challenge	Brief Description	References
1	Material compatibility	Limited availability of smart materials compatible with existing 3D printing technologies. Most conventional materials lack stimuli-responsive behavior.	Ahmed *et al.* 2021; Sahafnejad Mohammadi *et al.* 2023
2	High-resolution printing	Biomedical applications require intricate designs and dimensional accuracy, which current printing technologies struggle to deliver.	Ramezani & Mohd Ripin, 2023
3	Environmental sustainability	Sustainability and long-term durability of smart materials remain underexplored, with limited knowledge on recyclability and biodegradability.	Ahmed *et al.* 2021; Kantaros *et al.* 2023
4	Limited smart material types	Existing smart materials typically respond to only one stimulus, restricting multifunctional applications.	Ahmed *et al.* 2021
5	Complexity in structure design	Engineering smart structures with predictable shape transformations is limited by printing constraints (e.g., pre-straining actuators).	Sahafnejad Mohammadi *et al.* 2023
6	Printing technology limitations	Few printing techniques (e.g., PolyJet, SLM) can handle 4D materials; others like extrusion are cost-effective but low in resolution.	Navaf *et al.* 2022; Aldawood, 2023
7	Expertise and accessibility	Specialized knowledge is needed to work with smart materials and compatible printers, limiting broader adoption.	Shen & Shen, 2025
8	Policy and cost barriers	High production cost, policy constraints, and scale-up difficulties limit the industrialization of 4D printing.	Shen & Shen, 2025
9	Stimuli control	Technical challenges remain in controlling environmental triggers like temperature and pH with precision.	Shen & Shen, 2025

17.9 CONCLUSION AND FUTURE PERSPECTIVES

The rapid technological advancement of 3D and 4D printing has transformed additive manufacturing, offering new opportunities across various industries, particularly in biomedical engineering, soft robotics, and advanced materials design. Among the diverse materials explored for these technologies, hydrogels have stood-out as a strong candidate as result of their exceptional water retention capacity, biocompatibility, and stimuli-responsive nature. Their ability to undergo reversible shape transformations renders them ideal for applications requiring adaptability and precision. However, despite these benefits, challenges such as limited mechanical strength, printability issues, and durability concerns continue to hinder their widespread adoption. Additionally, existing 3D printing techniques must be refined to accommodate the complex requirements of hydrogel-based 4D printing, ensuring precise control over material behavior and external triggering mechanisms.

Overcoming these challenges will require continued innovation in material science, printing techniques, and computational modeling. The strategic incorporation of artificial intelligence and predictive modeling has the capacity to improve the design and fabrication of hydrogel based structures. By leveraging these advanced tools, researchers can optimize the performance and functionality of these structures, better aligning them with the demands of real-world applications. Furthermore, as the demand for environmentally friendly manufacturing grows, research into biodegradable and eco-friendly hydrogel materials will become increasingly important. With ongoing developments in these areas, hydrogel based 3D and 4D printing is positioned to play a transformative role in shaping the future of additive manufacturing, offering smarter, more adaptive, and environmentally sustainable solutions for a wide range of industries.

REFERENCES

Abasalizadeh, F., Moghaddam, S. V., Alizadeh, E., akbari, E., Kashani, E., Fazljou, S. M. B., Torbati, M., & Akbarzadeh, A. (2020). Alginate-based hydrogels as drug delivery vehicles in cancer treatment and their applications in wound dressing and 3D bioprinting. *Journal of Biological Engineering*, *14*(1), 8. DOI: 10.1186/s13036-020-0227-7 PMID: 32190110

Abdullah, T., & Okay, O. (2023). 4D printing of body temperature-responsive hydrogels based on poly(acrylic acid) with shape-memory and self-healing abilities. *ACS Applied Bio Materials*, *6*(2), 703–711. DOI: 10.1021/acsabm.2c00939 PMID: 36700540

Abel, S., Busatto, C. A., Karp, F., Estenoz, D., & Calderón, M. (2023). Weaving the next generation of (bio)materials: Semi-interpenetrated and interpenetrated polymeric networks for biomedical applications. *Advances in Colloid and Interface Science*, *321*, 103026. DOI: 10.1016/j.cis.2023.103026 PMID: 39491440

Ahmad, Z., Salman, S., Khan, S. A., Amin, A., Rahman, Z. U., Al-Ghamdi, Y. O., Akhtar, K., Bakhsh, E. M., & Khan, S. B. (2022). Versatility of hydrogels: From synthetic strategies, classification, and properties to biomedical applications. *Gels (Basel, Switzerland)*, *8*(3), 167. DOI: 10.3390/gels8030167 PMID: 35323280

Ahmed, A., Arya, S., Gupta, V., Furukawa, H., & Khosla, A. (2021). 4D printing: Fundamentals, materials, applications and challenges. *Polymer*, *228*, 123926. DOI: 10.1016/j.polymer.2021.123926

Aldawood, F. K. (2023). A comprehensive review of 4D printing: State of the arts, opportunities, and challenges. *Actuators*, *12*(3), 101. DOI: 10.3390/act12030101

Alzoubi, L., Aljabali, A. A., & Tambuwala, M. M. (2023). Empowering precision medicine: The impact of 3D printing on personalized therapeutic. *AAPS PharmSciTech*, *24*(8), 228. Advance online publication. DOI: 10.1208/s12249-023-02682-w PMID: 37964180

Andrade, F., Roca-Melendres, M. M., Durán-Lara, E. F., Rafael, D., & Schwartz, S.Jr. (2021). Stimuli-responsive hydrogels for cancer treatment: The role of pH, light, ionic strength and magnetic field. *Cancers (Basel)*, *13*(5), 1164. DOI: 10.3390/cancers13051164 PMID: 33803133

Attaran, M. (2017). The rise of 3-D printing: The advantages of additive manufacturing over traditional manufacturing. *Business Horizons*, *60*(5), 677–688. DOI: 10.1016/j.bushor.2017.05.011

Aydin, G., Abdullah, T., & Okay, O. (2025). 4D printing of self-healing and shape-memory hydrogels sensitive to body temperature. *European Polymer Journal*, *223*, 113651. DOI: 10.1016/j.eurpolymj.2024.113651

Baniasadi, H., Abidnejad, R., Fazeli, M., Lipponen, J., Niskanen, J., Kontturi, E., Seppälä, J., & Rojas, O. J. (2024). Innovations in hydrogel-based manufacturing: A comprehensive review of direct ink writing technique for biomedical applications. *Advances in Colloid and Interface Science*, *324*, 103095. DOI: 10.1016/j.cis.2024.103095 PMID: 38301316

Bhatia, A., & Sehgal, A. K. (2023). Additive manufacturing materials, methods and applications: A review. *Materials Today: Proceedings*, *81*, 1060–1067. DOI: 10.1016/j.matpr.2021.04.379

Champeau, M., Heinze, D. A., Viana, T. N., de Souza, E. R., Chinellato, A. C., & Titotto, S. (2020). 4D printing of hydrogels: A review. *Advanced Functional Materials*, *30*(31), 1910606. DOI: 10.1002/adfm.201910606

Chen, C., Xi, Y., & Weng, Y. (2022). Recent advances in cellulose-based hydrogels for tissue engineering applications. *Polymers*, *14*(16), 3335. DOI: 10.3390/polym14163335 PMID: 36015592

Cho, S. Y., Ho, D. H., Jo, S. B., & Cho, J. H. (2024). Direct 4D printing of functionally graded hydrogel networks for biodegradable, untethered, and multimorphic soft robots. *International Journal of Extreme Manufacturing*, *6*(2), 025002. DOI: 10.1088/2631-7990/ad1574

Chu, H., Yang, W., Sun, L., Cai, S., Yang, R., Liang, W., & Liu, L. (2020). 4D printing: A review on recent progresses. *Micromachines*, *11*(9), 796. DOI: 10.3390/mi11090796 PMID: 32842588

Correa, S., Grosskopf, A. K., Lopez Hernandez, H., Chan, D., Yu, A. C., Stapleton, L. M., & Appel, E. A. (2021). Translational applications of hydrogels. *Chemical Reviews*, *121*(18), 11385–11457. DOI: 10.1021/acs.chemrev.0c01177 PMID: 33938724

Deng, Z., Yu, R., & Guo, B. (2021). Stimuli-responsive conductive hydrogels: Design, properties, and applications. *Materials Chemistry Frontiers*, *5*(5), 2092–2123. DOI: 10.1039/D0QM00868K

Deng, Z., Yu, R., & Guo, B. (2021). Stimuli-responsive conductive hydrogels: Design, properties, and applications. *Materials Chemistry Frontiers*, 5 (5), 2092–2123.

Dong, Y., Wang, S., Ke, Y., Ding, L., Zeng, X., Magdassi, S., & Long, Y. (2020). 4D printed hydrogels: Fabrication, materials, and applications. *Advanced Materials Technologies*, *5*(6), 2000034. DOI: 10.1002/admt.202000034

Enyan, M., Bing, Z., Amu-Darko, J. N. O., Issaka, E., Otoo, S. L., & Agyemang, M. F. (2025). Advances in smart materials soft actuators on mechanisms, fabrication, materials, and multifaceted applications: A review. *Journal of Thermoplastic Composite Materials*, *38*(1), 302–370. DOI: 10.1177/08927057241248028

Framinan, J. M., Perez-Gonzalez, P., & Fernandez-Viagas, V. (2022). An overview on the use of operations research in additive manufacturing. *Annals of Operations Research*, *322*(1), 5–40. DOI: 10.1007/s10479-022-05040-4 PMID: 33583990

Gan, X., Wang, X., Huang, Y., Li, G., & Kang, H. (2024). Applications of hydrogels in osteoarthritis treatment. *Biomedicines*, *12*(4), 923. DOI: 10.3390/biomedicines12040923 PMID: 38672277

Gao, Y., Peng, K., & Mitragotri, S. (2021). Covalently crosslinked hydrogels via step-growth reactions: Crosslinking chemistries, polymers, and clinical impact. *Advanced Materials*, *33*(25), 2006362. DOI: 10.1002/adma.202006362 PMID: 33988273

Ghosh, S., Chaudhuri, S., Roy, P., & Lahiri, D. (2023). 4D Printing in biomedical engineering: A state-of-the-art review of technologies, biomaterials, and application. *Regenerative Engineering and Translational Medicine*, *9*(3), 339–365. DOI: 10.1007/s40883-022-00288-5

Gross, B. C., Erkal, J. L., Lockwood, S. Y., Chen, C., & Spence, D. M. (2014). Evaluation of 3D printing and its potential impact on biotechnology and the Chemical Sciences. *Analytical Chemistry, 86*(7), 3240–3253. DOI: 10.1021/ac403397r PMID: 24432804

He, Y., Tang, J., Hu, Y., Yang, S., Xu, F., Zrínyi, M., & Chen, Y. M. (2023). Magnetic hydrogel-based flexible actuators: A comprehensive review on design, properties, and applications. *Chemical Engineering Journal, 462*, 142193. DOI: 10.1016/j.cej.2023.142193

Ho, T. C., Chang, C. C., Chan, H. P., Chung, T. W., Shu, C. W., Chuang, K. P., Duh, T.-H., Yang, M.-H., & Tyan, Y. C. (2022). Hydrogels: Properties and applications in biomedicine. *Molecules (Basel, Switzerland), 27*(9), 2902. DOI: 10.3390/molecules27092902 PMID: 35566251

Hu, L., Chee, P. L., Sugiarto, S., Yu, Y., Shi, C., Yan, R., & Huang, W. (2023). Hydrogel-based flexible electronics. *Advanced Materials, 35*(14), 2205326. DOI: 10.1002/adma.202205326 PMID: 36037508

Iftekar, S. F., Aabid, A., Amir, A., & Baig, M. (2023). Advancements and limitations in 3D printing materials and technologies: A critical review. *Polymers, 15*(11), 2519. DOI: 10.3390/polym15112519 PMID: 37299318

Imam, S. S., Hussain, A., Altamimi, M. A., & Alshehri, S. (2021). Four-dimensional printing for hydrogel: Theoretical concept, 4D materials, shape-morphing way, and future perspectives. *Polymers, 13*(21), 3858. DOI: 10.3390/polym13213858 PMID: 34771414

Jandyal, A., Chaturvedi, I., Wazir, I., Raina, A., & Ul Haq, M. I. (2022). 3D printing – A review of processes, materials and applications in industry 4.0. *Sustainable Operations and Computers, 3*, 33–42. DOI: 10.1016/j.susoc.2021.09.004

Jayakrishna, M., Vijay, M., & Khan, B. (2023). An overview of extensive analysis of 3D printing applications in the manufacturing sector. *Journal of Engineering, 2023*, 1–23. DOI: 10.1155/2023/7465737

Jayswal, A., & Adanur, S. (2022). An overview of additive manufacturing methods, materials, and applications for flexible structures. *Journal of Industrial Textiles, 52*, 15280837221114638. Advance online publication. DOI: 10.1177/15280837221114638

Jiang, K., Yan, H., Rickert, C., Marczynski, M., Sixtensson, K., Vilaplana, F., & Crouzier, T. (2021). Modulating the bioactivity of mucin hydrogels with crosslinking architecture. *Advanced Functional Materials, 31*(10), 2008428. DOI: 10.1002/adfm.202008428

Jiang, Z., Shen, P., Tan, M. L., Yan, Q., Viktorova, J., Cementon, C., & Connal, L. A. (2021). 3D and 4D printable dual cross-linked polymers with high strength and humidity-triggered reversible actuation. *Materials Advances, 2*(15), 5124–5134. DOI: 10.1039/D1MA00223F

Jiménez, M., Romero, L., Domínguez, I. A., Espinosa, M. del, & Domínguez, M. (2019). Additive Manufacturing Technologies: An overview about 3D printing methods and future prospects. *Complexity*, 2019 (1). DOI: 10.1155/2019/9656938

Joshi, J., Homburg, S. V., & Ehrmann, A. (2022). Atomic force microscopy (AFM) on biopolymers and hydrogels for biotechnological applications—Possibilities and limits. *Polymers*, *14*(6), 1267. DOI: 10.3390/polym14061267 PMID: 35335597

Kantaros, A., Ganetsos, T., & Piromalis, D. (2023). 4D printing: Technology Overview and smart materials utilized. *Journal of Mechatronics and Robotics*, *7*(1), 1–14. DOI: 10.3844/jmrsp.2023.1.14

Karimi, M., Sahandi Zangabad, P., Ghasemi, A., Amiri, M., Bahrami, M., Malekzad, H., Ghahramanzadeh Asl, H., Mahdieh, Z., Bozorgomid, M., Ghasemi, A., Rahmani Taji Boyuk, M. R., & Hamblin, M. R. (2016). Temperature-Responsive Smart Nanocarriers for Delivery Of Therapeutic Agents: Applications and Recent Advances. *ACS Applied Materials & Interfaces*, *8*(33), 21107–21133. DOI: 10.1021/acsami.6b00371 PMID: 27349465

Kavand, A., Noverraz, F., & Gerber-Lemaire, S. (2024). Recent Advances in Alginate-Based Hydrogels for Cell Transplantation Applications. *Pharmaceutics*, *16*(4), 469. DOI: 10.3390/pharmaceutics16040469 PMID: 38675129

Khan, A., Othman, M. B. H., Razak, K. A., & Akil, H. M. (2013). Synthesis and physicochemical investigation of chitosan-PMAA-based dual-responsive hydrogels. *Journal of Polymer Research*, *20*(10), 1–8. DOI: 10.1007/s10965-013-0273-7

Khonakdar, H., Ehsani, M., Naderi, G., Shokrolahi, F., & Khonakdar, H. A. (2025). Recent Advances in Injectable Stimuli-Responsive Hydrogels for Biomedical Applications. *Journal of Polymers and the Environment*, *33*(8), 3512–3554. Advance online publication. DOI: 10.1007/s10924-025-03620-6

Kuzina, M. A., Kartsev, D. D., Stratonovich, A. V., & Levkin, P. A. (2023). Organogels versus hydrogels: Advantages, challenges, and applications. *Advanced Functional Materials*, *33*(27), 2301421. DOI: 10.1002/adfm.202301421

Lai, J., Ye, X., Liu, J., Wang, C., Li, J., Wang, X., Ma, M., & Wang, M. (2021). 4D printing of highly printable and shape morphing hydrogels composed of alginate and methylcellulose. *Materials & Design*, *205*, 109699. DOI: 10.1016/j.matdes.2021.109699

Li, B., Zhang, M., Lu, Q., Zhang, B., Miao, Z., Li, L., Zheng, T., & Liu, P. (2022). Application and development of modern 3D printing technology in the field of Orthopedics. *BioMed Research International*, 2022 (1). DOI: 10.1155/2022/8759060

Li, H., Bartolo, P. J., & Zhou, K. (2025). Direct 4D printing of hydrogels driven by structural topology. *Virtual and Physical Prototyping*, *20*(1), e2462962. Advance online publication. DOI: 10.1080/17452759.2025.2462962

Li, T., Huang, Z., Tsui, G., Tang, C., & Deng, Y. (2024). Recent advances in 4D printing of hydrogels. *Reviews on Advanced Materials Science*, *63*(1), 20240028. DOI: 10.1515/rams-2024-0028

Li, X., & Gong, J. P. (2024). Design principles for strong and tough hydrogels. *Nature Reviews. Materials*, *9*(6), 380–398. DOI: 10.1038/s41578-024-00672-3

Li, Y., Zheng, W., Li, B., Dong, J., Gao, G., & Jiang, Z. (2022). Double-layer temperature-sensitive hydrogel fabricated by 4D printing with fast shape deformation. *Colloids and Surfaces. A, Physicochemical and Engineering Aspects*, *648*, 129307. DOI: 10.1016/j.colsurfa.2022.129307

Liu, B., Li, H., Meng, F., Xu, Z., Hao, L., Yao, Y., Zhu, H., Wang, C., Wu, J., Bian, S., Lu, W. W., Liu, W., Pan, H., & Zhao, X. (2024). 4D printed hydrogel scaffold with swelling-stiffening properties and programmable deformation for minimally invasive implantation. *Nature Communications*, *15*(1), 1587. Advance online publication. DOI: 10.1038/s41467-024-45938-0 PMID: 38383668

Lu, B., Li, D., & Tian, X. (2015). Development trends in additive manufacturing and 3D printing. *Engineering*, 1 (1), 085–089. DOI: 10.15302/J-ENG-2015012

Nain, A., Chakraborty, S., Jain, N., Choudhury, S., Chattopadhyay, S., Chatterjee, K., & Debnath, S. (2024). 4D hydrogels: Fabrication strategies, stimulation mechanisms, and biomedical applications. *Biomaterials Science*, *12*(13), 3249–3272. DOI: 10.1039/D3BM02044D PMID: 38742277

Nain, A., Joshi, A., Debnath, S., Choudhury, S., Thomas, J., Satija, J., Huang, C.-C., & Chatterjee, K. (2024). A 4D printed nanoengineered super bioactive hydrogel scaffold with programmable deformation for potential bifurcated vascular channel construction. *Journal of Materials Chemistry. B*, *12*(31), 7604–7617. DOI: 10.1039/D4TB00498A PMID: 38984474

Nasution, H., Harahap, H., Dalimunthe, N. F., Ginting, M. H. S., Jaafar, M., Tan, O. O., & Herfananda, A. L. (2022). Hydrogel and effects of crosslinking agent on cellulose-based hydrogels: A review. *Gels (Basel, Switzerland)*, *8*(9), 568. DOI: 10.3390/gels8090568 PMID: 36135281

Navaf, M., Sunooj, K. V., Aaliya, B., Akhila, P. P., Sudheesh, C., Mir, S. A., & George, J. (2022). 4D printing: A new approach for food printing; effect of various stimuli on 4D printed food properties. A comprehensive review. *Applied Food Research*, *2*(2), 100150. DOI: 10.1016/j.afres.2022.100150

Ngo, T. D., Kashani, A., Imbalzano, G., Nguyen, K. T. Q., & Hui, D. (2018). Additive Manufacturing (3D printing): A review of materials, methods, applications and challenges. *Composites. Part B, Engineering*, *143*, 172–196. DOI: 10.1016/j.compositesb.2018.02.012

Park, Y., Yun, I., Chung, W. G., Park, W., Lee, D. H., & Park, J. (2022). High-resolution 3D printing for Electronics. *Advancement of Science*, *9*(8), 2104623. Advance online publication. DOI: 10.1002/advs.202104623 PMID: 35038249

Parupelli, S. K., & Desai, S. (2019). A comprehensive review of additive manufacturing (3D printing): Processes, applications and future potential. *American Journal of Applied Sciences*, *16*(8), 244–272. DOI: 10.3844/ajassp.2019.244.272

Peppas, N. A., & Hoffman, A. S. (2020). Hydrogels. In *Biomaterials science* (pp. 153–166). Academic Press. DOI: 10.1016/B978-0-12-816137-1.00014-3

Pourjavadi, A., Heydarpour, R., & Tehrani, Z. M. (2021). Multi-stimuli-responsive hydrogels and their medical applications. *New Journal of Chemistry*, *45*(35), 15705–15717. DOI: 10.1039/D1NJ02260A

Praveena, N, L., Buradi, A., N, S., B L, P., & R, V. (. (2022). A comprehensive review of emerging additive manufacturing (3D printing technology): Methods, materials, applications, challenges, trends and future potential. *Materials Today: Proceedings, 52*, 1309–1313. DOI: 10.1016/j.matpr.2021.11.059

Ramezani, M., & Mohd Ripin, Z. (2023). 4D printing in biomedical engineering: Advancements, challenges, and future directions. *Journal of Functional Biomaterials, 14*(7), 347. DOI: 10.3390/jfb14070347 PMID: 37504842

Rastogi, P., & Kandasubramanian, B. (2019). Breakthrough in the printing tactics for stimuli-responsive materials: 4D printing. *Chemical Engineering Journal, 366*, 264–304. DOI: 10.1016/j.cej.2019.02.085

Sahafnejad-Mohammadi, I., Karamimoghadam, M., Zolfagharian, A., Akrami, M., & Bodaghi, M. (2023). 4D printing technology in Medical Engineering: A narrative review. *Journal of the Brazilian Society of Mechanical Sciences and Engineering, 44*(6), 233. Advance online publication. DOI: 10.1007/s40430-022-03514-x

Shahrubudin, N., Lee, T. C., & Ramlan, R. (2019). An overview on 3D printing technology: Technological, materials, and applications. *Procedia Manufacturing, 35*, 1286–1296. DOI: 10.1016/j.promfg.2019.06.089

Shen, C., & Shen, A. (2025). 4D printing: Innovative Solutions and technological advances in orthopedic repair and reconstruction, personalized treatment and drug delivery. *Biomedical Engineering Online, 24*(1), 5. Advance online publication. DOI: 10.1186/s12938-025-01334-3 PMID: 39838448

Solis, D. M., & Czekanski, A. (2022). The effect of the printing temperature on 4D DLP printed pNIPAM hydrogels. *Soft Matter, 18*(17), 3422–3429. DOI: 10.1039/D2SM00201A PMID: 35437561

Song, W., Li, H., & Li, F. (2025). The integration of environmental governance and 3D printing technology: Current situation and prospects. *Chemical Engineering Journal, 505*, 159498. DOI: 10.1016/j.cej.2025.159498

Srinivasan, D., Meignanamoorthy, M., Ravichandran, M., Mohanavel, V., Alagarsamy, S. V., Chanakyan, C., Sakthivelu, S., Karthick, A., Prabhu, T. R., & Rajkumar, S. (2021). 3D printing manufacturing techniques, materials, and applications: An overview. *Advances in Materials Science and Engineering*, 2021 (1). DOI: 10.1155/2021/5756563

Staňo, R., Košovan, P., Tagliabue, A., & Holm, C. (2021). Electrostatically cross-linked reversible gels—Effects of pH and ionic strength. *Macromolecules*, 54 (10), 4769–4781.

Tamay, D. G., Dursun Usal, T., Alagoz, A. S., Yucel, D., Hasirci, N., & Hasirci, V. (2019). 3D and 4D printing of polymers for tissue engineering applications. *Frontiers in Bioengineering and Biotechnology, 7*, 164. Advance online publication. DOI: 10.3389/fbioe.2019.00164 PMID: 31338366

Vigata, M., Meinert, C., Bock, N., Dargaville, B. L., & Hutmacher, D. W. (2021). Deciphering the molecular mechanism of water interaction with gelatin methacryloyl hydrogels: Role of ionic strength, pH, drug loading and hydrogel network characteristics. *Biomedicines*, *9*(5), 574. DOI: 10.3390/biomedicines9050574 PMID: 34069533

Wang, F., Yong, X., Deng, J., & Wu, Y. (2018). Poly(N,N-dimethylacrylamide-octadecyl acrylate)-clay hydrogels with high mechanical properties and shape memory ability. *RSC Advances*, *8*(30), 16773–16780. DOI: 10.1039/C8RA01167B PMID: 35540536

Wang, R., Yuan, C., Cheng, J., He, X., Ye, H., Jian, B., Li, H., Bai, J., & Ge, Q. (2024). Direct 4D printing of ceramics driven by hydrogel dehydration. *Nature Communications*, *15*(1), 758. DOI: 10.1038/s41467-024-45039-y PMID: 38272972

Wichterle, O., & Lim, D. (1960). Hydrophilic gels for biological use. *Nature*, *185*(4706), 117–118. DOI: 10.1038/185117a0

Wong, K. V., & Hernandez, A. (2012). A review of Additive Manufacturing. *ISRN Mechanical Engineering*, *2012*, 1–10. DOI: 10.5402/2012/208760

Zhao, C., Lv, Q., & Wu, W. (2022). Application and prospects of Hydrogel Additive Manufacturing. *Gels (Basel, Switzerland)*, *8*(5), 297. DOI: 10.3390/gels8050297 PMID: 35621595

Zhu, T., Ni, Y., Biesold, G. M., Cheng, Y., Ge, M., Li, H., & Lai, Y. (2023). Recent advances in conductive hydrogels: Classifications, properties, and applications. *Chemical Society Reviews*, *52*(2), 473–509. DOI: 10.1039/D2CS00173J PMID: 36484322

KEY TERMS AND DEFINITIONS

Additive Manufacturing: This is also known as 3D printing, it is a process that involves combining materials to produce parts from the 3D model data.

Hydrogel: These are 3D networks composed of natural or synthetic polymers, distinguished by their exceptionally high water content, which gives them remarkable flexibility.

4D Printing: This technique involve the printing of 3D objects through the addition of the fourth dimension called 'time.' Through this technique, the printed object reacts to external influence, such as light, temperature, and pH, over time.

3D Printing Techniques: These are additive manufacturing methods that creates physical objects from digital designs through layering materials such as plastics, metals and hydrogels, allowing for complex shapes and structures to be produced.

Smart Materials: These are materials capable of sensing and responding to external stimuli such as temperature, light or moisture, with a programmed action, leading to changes in their properties or behavior.

Crosslinking of Hydrogels: This refers to the formation of bonds between polymer chains within the hydrogel network, resulting in improved mechanical strength, stability etc.

Shape-memory Effect: This a property of certain materials that enables them to return to their original shape after being deformed. They are often triggered by external stimuli such as temperature, light, pH, ionic strength etc.

Chapter 18
Theoretical and 3D Simulation of Hydrogel Composites

Ephriam Akor
https://orcid.org/0009-0009-8825-9087
Redeemer's University, Nigeria

Excel Oluwafemi Anjorin
https://orcid.org/0009-0001-7738-2727
Redeemer's University, Nigeria

Victor Inioluwa Olaoye
University of Ibadan, Nigeria

Jumoke Elizabeth Oyaniyi
Redeemer's University, Nigeria

ABSTRACT

Hydrogel composites, known for their mechano-tunability, biocompatibility, and stimuli-sensitivity, are revolutionizing biomedical, environmental, and industrial fields. This study explores theoretical modeling (molecular dynamics (MD), coarse-grained simulations, finite element analysis (FEA)) and its synergy with advanced 3D printing (stereolithography, extrusion) to decipher the nuances of hydrogel composites design principles. MD unravels atomic-scale mechanics, designing fracture-resistant networks, whereas FEA engineers' macroscale mechanics towards tissue scaffolds or body sensors. Machine learning speeds up material design, predicting optimal mixtures for drug release or pollutant capture. 3D-printed hydrogels facilitate precision structures: tailored wound healing, enzymes-based biosensors, and aerogels that capture heavy metals. Scalability challenges, as well as fragility, are met with hybrid modeling-experiment approaches, confirmed with AFM and rheometry. AI-based multi-physics models are new trends toward smart hydrogels as well as greener bioinks.

DOI: 10.4018/979-8-3373-2337-4.ch018

18.1 INTRODUCTION

In the last decade or so, hydrogels have transitioned from specialized laboratory curiosities to base materials in several industries based on their unique blend of tunable water content, mechanical flexibility, and biocompatibility (Park et al., 2025; Shukur et al., 2025). The international market for hydrogel stood at USD 27.62 billion in 2023 and is likely to increase to USD 56.81 billion by 2034, growing at a compound annual growth rate of 6.7 percent, with a growing demand for biomedical, environmental, and consumer-care uses (Luthfianti et al., 2024; Raeisi & Farjadian, 2024). This rapid market development is facilitated by hydrogels' ability to mimic soft living tissue, and that is why they are in great demand as dressings for wounds, tissue scaffolds, and contact lenses, as well as how they fare in agriculture and wastewater treatment with water retension and selective adsorption (Solanki et al., 2023).

This rapid market development is facilitated by hydrogels' ability to mimic soft living tissue, and that is why they are in great demand as dressings for wounds, tissue scaffolds, and contact lenses, and how they fare in agriculture and wastewater treatment with water holding and selective adsorption (Majcher & Hoare, 2018). Conventional trial-and-error experimental methods are inadequate to keep up with the increasing demand for customized materials optimized for particular performance specifications, i.e., swelling ratio, elastic modulus, or mesh size. Computational modeling and three-dimensional simulation technologies have therefore come to the fore as effective enablers, enabling researchers to study hydrogel dynamics in silico, minimize development cycles, and tailor materials with hitherto unmatched accuracy (Danole et al., 2018; Matole & Digge, 2022).

The combination of theoretical models, such as thermodynamic swelling theory and molecular dynamics, together with digital fabrication tools has yielded new possibilities for high-throughput design of hydrogel composites (Murakami, 2015). Furthermore, the application of artificial intelligence and machine learning in hydrogel research may demystify complex structure–property relationships elucidated from big experimental and simulation data sets, hence enabling the rational design of future hydrogels with tailored functionalities at an accelerated rate (Shukur et al., 2025). The foundational theory for understanding hydrogel swelling hinges on the Flory–Rehner framework, which balances the elastic free energy of a cross-linked polymer network against its mixing free energy with solvent (Luthfianti et al., 2024; Matole & Digge, 2022). While classical applications of the model adequately describe equilibrium swelling ratios and mesh sizes for neutral hydrogels, they become unsuitable when systems with highly charged or copolymers are addressed. In order to remedy these issues, scientists have extended the Flory–Rehner equation to incorporate ionic effects through the Gibbs–Donnan effect and the finite extensibility of polymer chains, and in this way they have been able to attain an excellent agreement with experimental swelling data over broad ranges in both polymer volume fractions and charge densities (Cascone & Lamberti, 2020; Park et al., 2025; Solanki et al., 2023).

In addition to continuum-level theory, mesoscale methods, i.e., the generalized Flory–Rehner copolymer model, allow for quantitative predictions of binary copolymer microgel swelling behavior by defining individual interaction parameters for each of the monomer species. At the molecular level, both coarse-grained and atomistic simulations have yielded atomistic detail to network formation processes, crosslinker distribution, and dynamic response to stress (Singh, 2018). These investigations show how the length of polymer chains, crosslink density, and solvent quality influence mesh size distributions, which dictate the diffusion of drugs or nutrients through the hydrogel matrix (Cascone & Lamberti, 2020; Singh, 2018).

At the same time, simulations of fluid-structure interaction have started to clarify the effects of perfusion and mechanical stimulation on cell behavior in three-dimensional hydrogel scaffolds, guiding design of bioreactors and tissue constructs (Murakami, 2015). Concurrently with mechanistic model, machine learning data-driven methodologies have emerged to summarize insights from vast amounts of data gathered from experiment and simulation in common, rapidly predicting parameter regimes maximizing beneficial properties such as elastic modulus, permeability, or stimulus sensitivity (Park et al., 2025; Raeisi & Farjadian, 2024). Collectively, these theoretical and computational advances form an integrated suite of tools for rational hydrogel composite design with predictive ability spanning scales from molecular networks to macroscopic devices.

This chapter was aimed at developing a comprehensive framework to understand the recent technological advancement in understanding unified polymer network theory with three-dimensional simulation of hydrogel composites. It was also intended to understand several applications of models including the modelling of drug release kinetics from printed matrices by extending Merino et al.'s (2015) sequenced delivery approach, and those predictions with continuum simulations of scaffold mechanics to optimize stiffness for tissue engineering (W. Dai et al., 2020; Merino et al., 2015). Further objectives included understanding of simulation need in analyte–hydrogel interactions following Rebelo et al.'s (2019) sensitivity enhancements, thereby guiding wearable sensor fabrication, and other advanced applications (Rebelo et al., 2019; Wang et al., 2021). These goals rested on a theoretical foundation combining the different theoretical models to capture coupled deformation, diffusion, and reaction phenomena within the hydrogel composite constructs.

18.2 THEORETICAL MODELLING OF HYDROGEL COMPOSITES

In a broad classification of the different theoretical modelling techniques applied in literature for the prediction of the physiochemical properties of hydrogel composites, five (5) classes stood out. These classes include molecular-level models (Utech & Boccaccini, 2016; L. Wu et al., 2018; T. Wu et al., 2019), statistical mechanics models (Chen, 2022), continuum mechanism models (Castilho et al., 2018; Y. Liu et al., 2022), Multiphysics models (Chinthala et al., 2021) and machine learning models (Gholami et al., 2023; Shokrollahi et al., 2022). Each of these model types utilizes different approaches in the prediction of physical and/or chemical properties.

18.3 MOLECULAR-LEVEL MODELLING TECHNIQUE

18.3.1 Molecular Dynamics Simulations (MD)

This model entails the utilization of numerical integration of Newton's laws of motion to study interacting molecules and atoms within a given period (Filipecka-Szymczyk et al., 2023). MD as a theoretical modelling technique provides information on the structural arrangement of the material studied and also the dynamics of the hydrogel atoms (including its composite's atoms) differentiated by colours as shown in figure 19.1 and 19.2 (Filipecka-Szymczyk et al., 2023). The molecular computational model has been applied for the modelling of hydrogel composites, which are materials made of a reinforcing filler and polymer network which is usually hydrophilic. The study of these hydrogel composite's structure and interactive properties is done following certain laid-down principles of this theoretical modelling technique (Hu et al., 2022; Ou et al., 2015).

Figure 1. A snapshot of MDS simulated hydrogel-copolymer; with carbon atoms represented in grey, Oxygen atoms represented in white, Nitrogen atoms represented in green and Phosphorus represented in yellow.

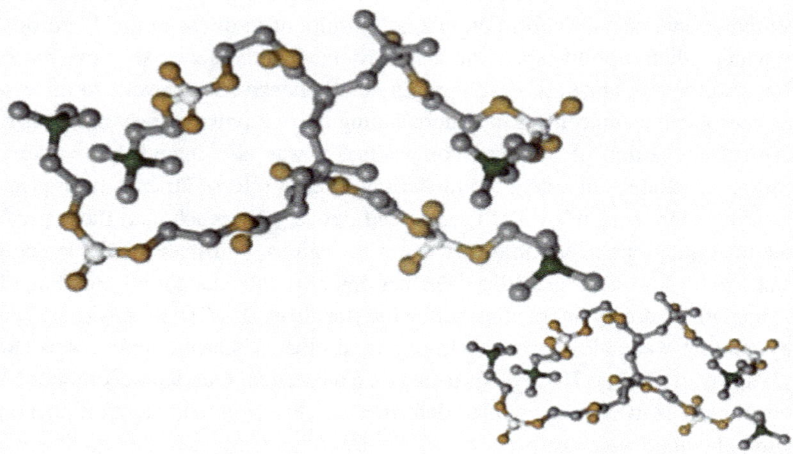

Following the MD principles; the hydrogel composite system is represented holistically using a set of particles interacting with themselves via a potential energy function. The potential energy function is referred to as the 'forcefield'. With the force field, non-bonded and boned interactions within the system (Hydrogel composite) are identified and described, this includes information on bonding such as the van der Waals force, covalent bonds, dihedral and electrostatic bonds (Hu et al., 2022). The system configuration using the MD model relies on experimental data first generated or a random or regular arrangement of data with the assumption of being void of boundary effects and, thus, considered in a closed system; this makes the simulation to infinity possible. The valet and leapfrog methods are employed in the calculation of the particle motion in the system to determine effectively variables such as the position, velocity and acceleration of the particles in the system at specific time intervals (Gao et al., 2019; Zheng et al., 2017).

For a successful MD simulation, parameters considered during this modelling process for hydrogel composites at the molecular level involve the choice of the force field (the potential energy), the size and shape of the simulation box for the modelling process, the system pre-configuration process, the kinetic/thermodynamic conditions and, the duration of the simulation (Dai et al., 2023). Different forces field can be compatible to simulate a particular property, for example, the COMPASS force field has been reported to simulate the water-absorbing capacity of the hydrogel composite (Ou et al., 2015) and can describe the bonding and non-bonding molecules of the hydrogel together with their interactions (Chen et al., 2013; Wang et al., 2021).

Additionally, in pricking the duration of the simulation, it should be long enough to accommodate the points where the system attains equilibrium, so it is relevant to the study. The optimization of the time step of the MD simulation process can be done using conditions such as the auto-correlation function, energy conservation or the convergence of the properties (Ou et al., 2015).

The MD simulation model also addresses both the temporal and spatial measure of time during the modelling process of hydrogel composite, this aspect of the study connects the macroscopic behaviours of the composite hydrogel to its molecular interactions. It also provides direct insight into the multi-component interaction of the hydrogel composite (Filipecka-Szymczyk et al., 2023). The time steps of this modelling technique range from femto-scale to the nano-scale of seconds, this aids in the calculation of the hydrogel atom's potion, velocity and the force acting on each particle making up the composite (Moreno & Mendoza, 2015). Despite the minute scale in time steps, the MD simulation model gives insightful information on the macroscopic behaviour of the composite hydrogel. It is from the accumulation of these smaller time scales to form a larger one that this information is extrapolated. Macroscopic properties such as diffusion and structural changes are then addressed (Filipecka-Szymczyk et al., 2023; Moreno & Mendoza, 2015).

Figure 2. Simulated free volume of a hydrogel composite showing (a) <10% (dried co-polymer) (b) 20%, and (c) 60% water by mass. Water molecule in this schematic is marked in black.

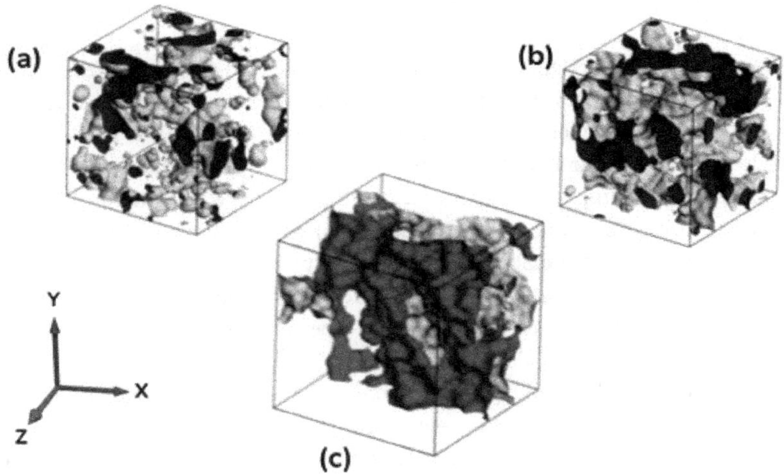

The validation of the MD simulation modelling technique for accuracy and reliability in the prediction of hydrogel composite properties is an important part of the model study (Lee et al., 2009; Wong et al., 2021). To accurately validate the MD simulation model, several methods have been designed to yield success; this includes assessing different molecular theories surrounding the model (Childers & Dagget, 2018). Methods have been developed to assess the accuracy of the intermolecular interactions between atoms measured, the degree of sampling, the convergence of the simulation reached and of course, the quality of software used for the modelling. Additionally, techniques such as the solid-state NMR have been used for the characterization process to validate measured results together with experimental mechanical testing (Wang et al., 2014; Xiao et al., 2013). All of these variables form a standard checklist, comprising theoretical, experimental and computational tests which on proper checking pronounce the process as reliable and accurate or otherwise (Heydari Dokoohaki et al., 2020).

Despite the incredible successes ascribed to the use of the MD simulation modelling techniques and the valuable processed data that its approach carries; it faces several limiting challenges that are still called for scientific advancement (Hou et al., 2019). One of these challenges is the limitation posed by the computational timescales available to the model. Phenomenons such as swelling and polymer degradation are not well observable within the limited timescale made available, therefore acting as a strain on the efficient representation of some critical properties of the composite hydrogel (Hou et al., 2019). Another limitation to this model is the difficulty experienced in the choosing of a valid force field which encompasses all of the diverse and complex interactions of the composite hydrogel simulations (Hou et al., 2019; Li et al., 2016). Additionally, in an attempt to include many parameters to simply and bring to reality the MD simulations layers and layers of complexity are added making the process tedious in accounting for a particular hydrogel behaviour; therefore, there is a need for continuous all-round development and synergy with other computational models to actualize seamless and efficient modelling (Chen et al., 2013; Lee et al., 2009).

18.3.2 Coarse-Grained Models (CG)

The CG model for the modelling of hydrogel composites reduces the represented models of atoms available to the hydrogel structure to only retain their molecular essence and properties. Those properties which are then of interest are then focused on (Singh & Li, 2019). This simulation model reduces the number of atoms making up the large-scale system of the composite hydrogel material simulating a few atom-to-atom interactions as depicted in Figure 19.3 (Jin et al., 2022). The principle of this model has been applied in the literature to model composite hydrogels, investigating and predicting the structural model of polymeric materials such as chitosan modified with varieties of chemical species (acetyl, heptanoyl, butanoyl etc.) which has shown viable applications in medical sciences, especially in drug delivery and engineering (Filipecka-Szymczyk et al., 2023; Singhal et al., 2020). This model is also important in the study of composite hydrogel's self-assembly phenomenon, via simulating the interaction between the different constituents' interactions with each other at the molecular level (Zhao et al., 2020). The CG model because of its efficiency has also been applied to the investigation of nanosized composite materials to great precision and accuracy of data (Khan et al., 2022).

The CG model enjoys popularity in literature as a meso-level theoretical modelling technique for both biological and chemical material, this is because of some significant advantages this technique possesses. Part of these advantages is the ease and speed the modelling process takes to execute and predict properties of their complex systems. This is because of the reduced degrees of freedom of the actual complex systems as fewer atomic particles are considered in the simulation (Singh & Li, 2019; Singhal et al., 2020). Another advantage to this scheme is the flexible time scale which greater length as compared to the atomistic models. It is also well suited for large and complex systems such as the composite hydrogel system, providing insight into the macroscopic behaviour of the system (Mehandzhiyski & Zozoulenko, 2021), but this model is dented with the limitation of lost details to the complex system, which to some property is considered significant. This also has a negative effect on the accuracy of properties results presented of a complex system especially when compared with the atomistic modelling technique (Mehandzhiyski & Zozoulenko, 2021; Pak & Voth, 2018; Singhal et al., 2020).

The CG model utilizes a couple of techniques for its development and validation. One of the most common techniques for the model's validation is the use of CG model simulations of the polymeric structure, where a replica model of the actual composite hydrogel gets reproduced to investigate different structural, thermodynamical and mechanical parameters of the hybrid system (Khan et al., 2022). This technique has well been employed in the characterization of protein-DNA composite systems and also for their optimization (Narayanan et al., 2022). Another common technique utilized for the validation and development of the CG model is a keen investigation of the molecular interaction and the properties portrayed by the macro-sized system. This is done in research by studying these materials near a surface made of silica and using the CG model to then understand the different modes of molecular interaction with the surface, therefore creating a study link from the material's nanosized molecules to its macroscopic behaviour (Dannert et al., 2019).

Other techniques evident in literature for validated CG simulations include the contact mechanics modelling -which focuses on the contact mechanics of the hydrogel composite (Müser et al., 2019), the bottom-up coarse-graining technique (Jin et al., 2022)- which validates the CG model from it reverse operational principle, the multiscale CG model (Singhal et al., 2020)- which focuses on the mechanical and transport properties of the composite system and the CG molecular dynamics simulations (Liu et al., 2023), which combines the features of the MD simulations in investigating fracture mechanism of the interaction molecules of the hybrid structure. All of these techniques and methods provide a framework for comprehensive investigation and validation of the developing CG model for predicting the properties of composite hydrogel materials and, also allow for a connecting link between the molecular level and the visible macro-level behaviours of the hybrid material.

Figure 3. (A) Shows the Coarse-grain model of PAAm hydrogel; where grey beads are the AAm atoms, the blue are the water beads and the red are the MBAA beads (B) The model without water beads.

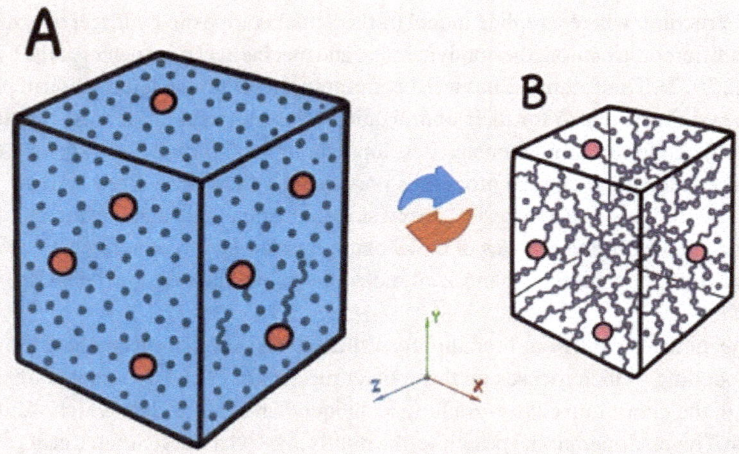

18.4 STATISTICAL MECHANICS MODELS

18.4.1 Flory-Rehner Theory

The Flory-Rehner theory which is a model type under the statistical framework has its principle reliant on the fundamentals such as the polymer-solvent interaction, swelling equilibrium of hydrogel materials and the mesh size of the polymeric material (hydrogel composite) (Casalini & Perale, 2019). This model type stems from the principle of Flory Huggins' mean field theory and the chain concept of ideality within a given solvent system. In the Flory Huggins mean field theory, the elastic energy of the polymer chain is harmonized with the mixing entropy of the polymer-solvent system (Casalini & Perale, 2019; DiBenedetto, 1987). This theory is infused into the Flory-rehner model which utilizes a pivotal equation correlating the swelling degree of the polymer, the polymer-solvent interaction parameter and the polymer concentration in the solvent (Casalini & Perale, 2019). The Flory- rehner model also leverages thermodynamical insights into the polymer material predicting not only the composite's swelling properties but also its mechanical attributes (Borges et al., 2020; Chen, 2020; Friesen et al., 2022; Lai & Hu, 2018; Lopez & Richtering, 2017; Tanimoto et al., 2023). Therefore, with these parameters in control, the tailoring of hydrogel composites with desired functionalities and properties is designed effectively.

In employing the Flory-rehner model, factors such as temperature, pH and other external stimuli are considered greatly (Alziyadi & Denton, 2021). This multifactored theory creates a balance between the elastic forces and electrostatic forces within the simulated system (Alziyadi

& Denton, 2021). This has aided its utilization in the study of ionic hydrogel composites. An example of this is seen in the investigation of poly(N-t-butylacrylamide-co-acrylamide) (P(TBA-co-AAm)) hydrogels swelling properties (Turan et al., 2006). The investigation of this study focused explicitly on the swelling property of the hydrogels under different pH conditions buffered for meticulous analysis. As a result, from the swelling studies, the derivation of the polymer-solvent interaction parameter (χ) and that of the average molecular weight between the different crosslinks was carried out and calculated. Results from this study aligned with the Flory Rehner equation grounded in the affine network model and the Donnan ideal theory (Turan et al., 2006).

The Flory-rehner theory which has proven very useful in literature still holds her limitations together with held assumptions needed for the theory to function. This flaw is significantly notable especially when applied to non-ideal or intricate composite polymeric systems (Rodriques, 2016). One of the assumptions the model goes by is the random walk scenario of polymer chain movements between crosslinks. Though this assumption aids in the simplification of its predictions and understanding but becomes void of the true complexity of the naturally occurring hydrogel systems (Rodriques, 2016). Another assumption lies in the ideal behaviours of the hydrogel molecules without any deviation, this is true universally (Denton & Tang, 2016). The assumption of ideality, following the ideal-gas model shows uncorrelations between experimental data and the theoretically generated data (Denton & Tang, 2016). Despite factors such as pH, temperature, ionic strength and other external stimuli that can be controlled, the incorporation of these constant assumptions creates a hinge on the modelling process precision (Alziyadi & Denton, 2021; Denton & Tang, 2016).

There exists limited evidence of the Flory-rehner theory utilization for modelling hydrogel composites in literature but, recently a proposed Flory-rehner equation was developed for the calculation of the molecular mass of the used crosslinker and that of the branched segments. An example of this is illustrated in Figure 19.4. This model provided a more viable calculation of the mesh dimension for a complex polymeric network. This network contained macromeric cross-linkers with a high degree of crosslinks forming the biocompatible hydrogel. This investigation was carried out as a case study comparing experimental data and theoretical data from the state-of-the-art approach (Borges et al., 2020). A more recent study calculated the porosity and pore size of hydrogel composites and their elasticity using the Flory-Rehner theory and compared its results with the Rubber elasticity theory which is another research model for predicting hydrogel properties. Both theories take into consideration thermodynamic variables such as enthalpy and entropy that could affect the hydrogel pore size or porosity (Mahmood et al., 2022).

Figure 4. The Flory-rehner theory expressing a crosslinker's tetragonal cell making up a hydrogel (i) Crosslinker at negligible size (ii) Crosslinker is 2.5 times than its backbone polymeric structure.

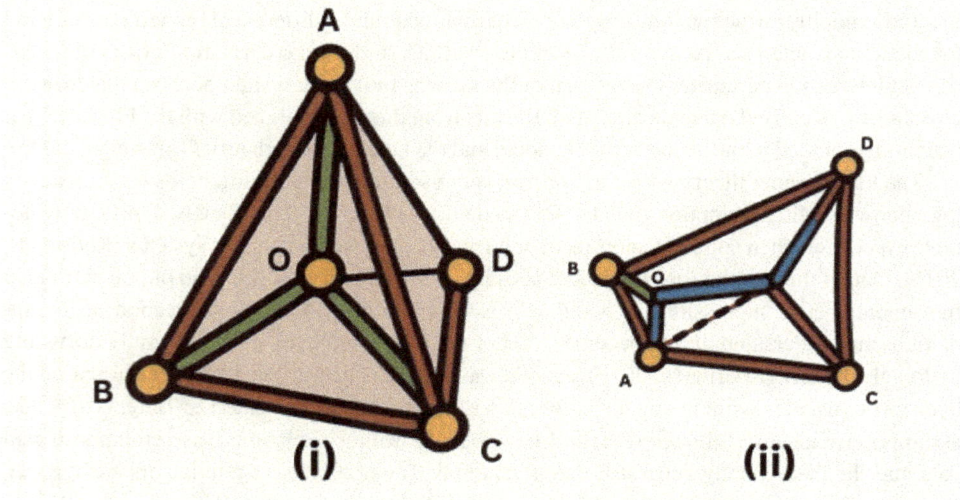

18.5 CONTINUUM MECHANICS MODEL

The continuum modelling technique is another tool in the arsenal of theoretical modelling, that effectively aids in the prediction of intricate properties of hydrogel composites. This form of theoretical modelling capitalizes on the fact that materials can be treated as continuous mediums (Figure 5), with variables that can be expressed in partial differential equations (Prakash & Ghosh, 2023). The continuum mechanics model which embodies properties of the material such as its microstructure, deformation and damage evolution reveals much information on the material's behaviour. An example of this is expressed in the Parametrically Upscaled Continuum Damage Mechanics (PUCDM) model. This exemplar was formed to predict the deformation by high strain stress of materials accounting for its damage progressively in the macro-scale. Its coefficients reveal information on the material microstructure and general morphology, it also reveals information on the material's micro-inertia (Prakash & Ghosh, 2023).

Figure 5. The scale expression of continuum mechanics showing that the length of matter (hydrogel) can be treated as a continuum medium.

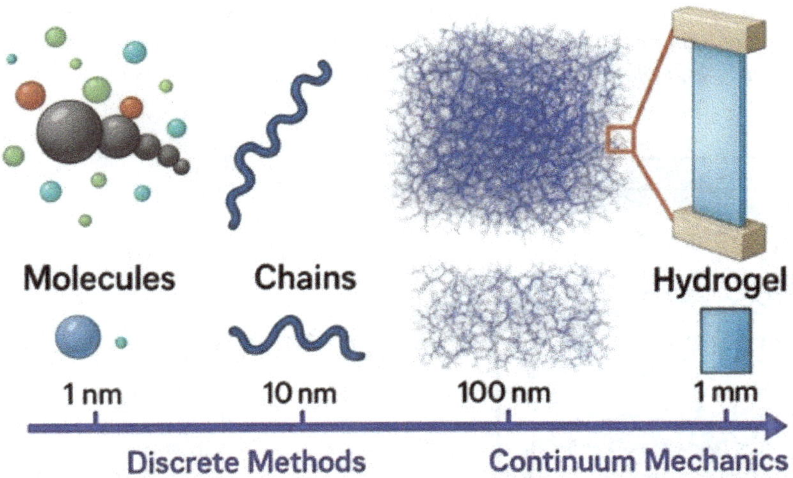

The continuum damage mechanics (CDM) and multiscale constitutive modelling (MCM) methods are also reported models in literature with successful applications, with the CDM method finding application in the prediction of mechanical properties of fibre-reinforced polymeric composites. It has also been reported for the investigation of woven composites' impact damage parameters (Hinton, n.d.; Völkerink et al., 2023). To add to the announced models under the continuum mechanics model type the multi-mechanism-based phenomenological model is another with is held in the principle and framework of the finite deformation kinematics. This modelling technique has emerged as a viable tool in the prediction of thermodynamic properties of polymers especially those with shape memory during their deformation process (Shojaei et al., 2014). A more recent and viable model type under this classification is the Finite Element Analysis (FEA) (Völkerink et al., 2023). This technique emerges as indispensable since its partial differential equations explore approximations of different materials' governing boundaries (Shreedhar, 2012; Völkerink et al., 2023).

Figure 6. shows a schematic of continuum mechanics model of the PAAm hydrogel using the mapping method (a) One CG particle represents two AAm molecules, or eight water molecules, or one MBAA molecule (b) shows the undisplay Bead-spring models of PAAm hydrogel without displaying with colour blue. Grey, magenta and red represent polymer chains networks, polymer branched networks, isolated polymer networks and crosslinkers respectively.

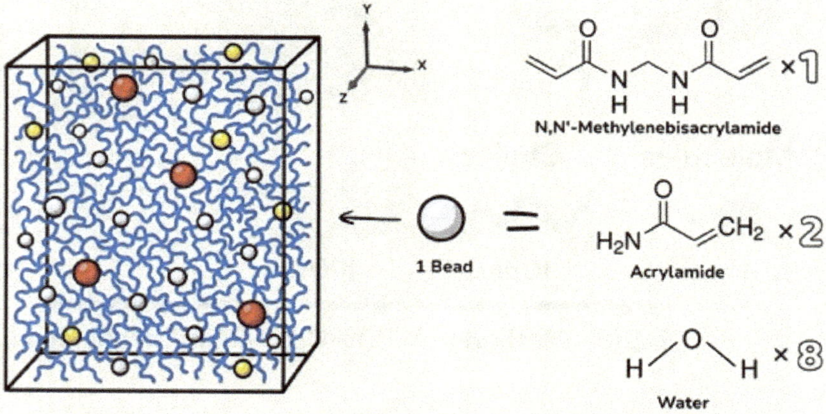

18.5.1 Finite Element Analysis (FEA)

The finite element analysis (FEA) an exemplar under the continuum mechanics model is a pivotal tool in the generation of microstructural images and many other diverse aspects of simulating and predicting the intrinsic behaviours of the hydrogel composites (Gholami et al., 2023b). The FEA model fragments bulk materials into smaller elements to carry out these studies- as exemplified in Figure 19.6, making the investigation of different load conditions and intricate behaviours easy. An example of its application is in a recent study which was directly coupled with the deep-learning technique for the investigation of the composite's mechanical properties (Gholami et al., 2023b). The FEA technique also goes further in the literature to predict the tensile fracture property of polymeric composites. The study of a stochastic tow-based discontinuous composite material was used for this study (Ryatt & Ramulu, 2022). It has also been deployed in the discerning of cohesive parameter fibre-reinforced interfaces (Yuan et al., 2022). Furthermore, the FEA technique has also been proven to be used in the optimization process of residual stresses in multi-arch welding and the defining of stress distributions in nano-reinforced composite gear teeth (Mohsenzadeh, 2022). These advances highlight how flexible FEA is in understanding and forecasting the behaviour of hydrogel composites by simulating its structure (figure 8).

The FEA simulation of hydrogel composites involves the step process of modelling, data generation, and machine learning process (Yuan et al., 2022; Zhou et al., 2022). The 2D and 3D simulation of these materials using the FEA technique has nexuses with machine learning as data generated from the FEA model can serve as a database for the developed machine learning algorithm. In literature, this has been reported in a recent study involving the harnessing of data using the FEA modelling technique for the development of a machine learning database. This built system was reported effective in the determination of cohesive parameters of fibre-reinforced composite materials (Yuan et al., 2022). The FEA model has also been reported effective in simulation for the investigation of stress distribution in polymeric composite materials, an example of this is the nano-sized CaCO3 composite gear teeth modelled for its stress distributions (Mohsenzadeh, 2022). The significance of FEA surrogate models for the development of composites' high-fidelity simulations has proven to be pivotal in the development of the machine learning models; since it allows for a comprehensive component of the hydrogel composite's feasibility assessment and the optimization of the different parameters and controllable factors involved in the process (Zhou et al., 2022).

In the construction of the FEA model for the investigation and prediction of hydrogel composite properties involves many parameter considerations. This multifaceted process must be contained and treated uniquely for different applications of the modelled hydrogel composites (Gholami et al., 2023b). Examples of these properties include the young modulus coefficient, particle attributes, interfacial properties and the swelling parameters of the hydrogel material (Gholami et al., 2023b). This methodology has been adopted for different composite hydrogel materials employed in different applications, e.g., modeling of swelling capacity of saltwater systems using the response surface methodology (RSM) optimized hydrogel (Heidari et al., 2021; Yuan et al., 2022). The heterogeneity of conditions in these processes implied the specificity of this modelling process with the FEA method used on hydrogel composites. Despite this, experimental validation of FEA simulations is still required no matter how powerful and versatile they are. Despite this, experimental validation of FEA simulations is still required no matter how powerful and versatile they are. Though Finite Element Analysis (FEA) is a potent tool for uncovering the behaviours of composite materials, empirical verification of its predictions is a necessary prerequisite. Experiments have confirmed this to be so by merging FEA-based calculations of stress patterns or cohesive parameters with experimental testing on composite materials, testing for equivalence between simulated and real-world behaviors (Mohsenzadeh, 2022; Yuan et al., 2022). Such equivalence emphasizes the importance of conducting experiments to verify the accuracy and reliability of FEA predictions that were made based on continuum mechanics of hydrogel composites.

It should be pointed out that the use of the FEA model for the prediction of hydrogel composites has a line of limitation and challenge. This zone of limitation must draw a keen scientific eye to painstakingly cut and nurture innovatively. Among them is the system processing complexity -of composites; that is, the degree to which it is difficult to effectively mimic hydrogel behavior in FEA models, such as swelling response and compression mode depicted in Figure 7. This also means that the material's microstructure and inherent material properties should be well addressed (Abalymov et al., 2020). Modelling interfacial properties between hydrogels and other materials is yet another challenge, owing to the intricate nature of hydrogel-particle interactions, demanding precision in FEA modelling (Yuan et al., 2022). Further research data-related uncertainties add to the challenges to the precision of FEA calibration, requiring further tests for increased dependability and calibration of simulation parameters (Kuhlmann et al., 2014). The intricate hydrogel composites' microstructures pose computational challenges in controlling computational effectiveness due to the size and nature of the models, where novel strategies and resource management are required (Gholami et al., 2023b). Concurrently, model verification remains necessary but problematic, involving reliable experimental data to compare FEA predictions with, highlighting the necessity of coordination of simulated and actual behaviors (Heidari et al., 2021).

Figure 7. Schematic representing the analysis of stress using the Finite Element Analysis (FEA), on a polymeric structure with different densities measured in von Mises stress (MPa).

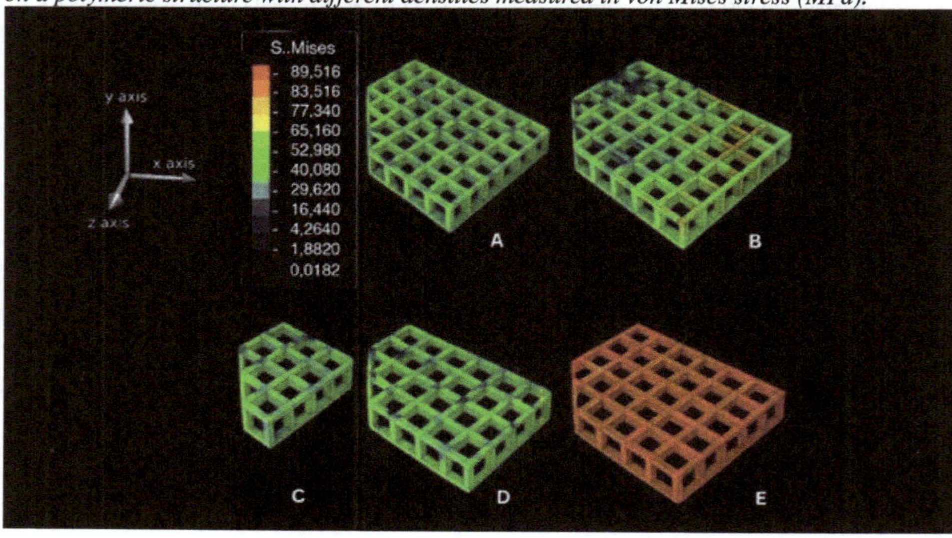

In literature, several developments have been used and proposed to improve these limitations. Most of these developments rely on the cross-combination of different domains for attaining probable development and success (Yuan et al., 2022; Zhou et al., 2022). Use of Image-based Machine Learning Methods (IBMLMs) has also enabled the creation of surrogate models of fidelitous FEA simulations, and feasibility analysis, as well as process optimization, became more efficient (Zhou et al., 2022). Furthermore, the application of FEA and Machine Learning coupling has allowed for the estimation of cohesive parameters of fibre-reinforced interfaces by predicting interfacial shear strength and fracture toughness without extra assumptions, and it has improved simulation accuracy (Yuan et al., 2022). Moreover, the incorporation of Deep Learning (DL) in controlled environment agriculture (CEA) has pushed the horizon further, implementing convolutional neural networks (CNNs) for diverse applications, demonstrating the viability of DL across an extensive array of industries (Ojo & Zahid, 2022).

Figure 8. Schematic of a FEA model used to simulate a porous scaffold.

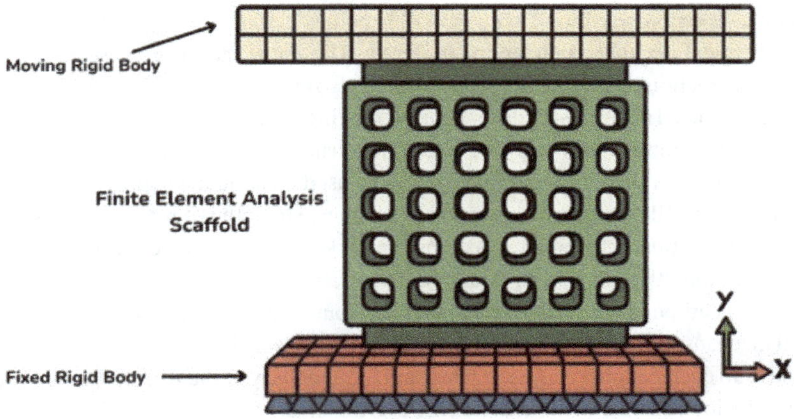

18.6 MULTIPHYSICS MODEL

The multi-physics model's large-scale modelling of the hydrogel composites gives prognoses of swelling parameters, mechanical properties and electrochemical stimulus response of the material. This theoretical model captures the complicated properties of the hydrogel composites as a function of the governing scientific principles or existing theories (Xu et al., 2018). Mechanical modelling of behaviour incorporating nonlinear viscoplastic and viscoelastic models accounts for swelling effects as well as interactions in the composite (van Leeuwen, 2018). The application of a Fickian model of swelling kinetics enables extrinsic modelling of factors such as electric potentials on swelling and corresponding volume changes (van Leeuwen, 2018). Mechanical behaviour of a hydrogel composite system is associated with its charge-discharge cycles in electrochemical response modelling. Thus, the model can provide a general view of the system by bringing together different phenomena that act together to create its performance. Through this, it closes the gaps between these different phenomena to achieve a more efficient knowledge of how the composite system functions (Xu et al., 2018).

The different phenomena bring about different equations in theory that are interrelated to give the complete picture of the system as a unit. The holistic study ties the interaction of different components such as taking an approach of a degradation model from a glass transition surface to outline multi-state behaviour (van Leeuwen, 2018). Geometrical parameters such as fibre volume fraction and matrix porosity, in turn, affect multifunctional composite performance during swelling/shrinking and are revealed through metric analysis (Xu et al., 2018). Numerical simulations, such as the finite element approach, allow for the visualization and understanding of physical events that occur within hydrogel composites (Xu et al., 2018). This method suggested aims at demystifying hydrogel composite complex mechanisms, suggesting a pathway towards the high-performance material development and diverse application. Finally, research culture on hydrogel composites is concerned with the need for a unified Multiphysics modelling strategy. Greater comprehension of such materials comes by integrating mechanical, swelling, and electrochemical parameters under one umbrella. These that have been described go beyond the different interaction between different factors but also provide possibilities for the development of tailored, optimized materials for different uses, ranging from biomedical to structural engineering.

The filler complexity in terms of size, filler structure and polymeric material form plays a key role in the creation of the Multiphysics model used in modeling hydrogel composite systems. The impact such filler properties exert on mechanical strength is supported by using fillers with specific aspect ratios that promote load transfer and slow down crack growth, thereby increasing the mechanical strength (Hong et al., 2021). Moreover, the composition of the filler, i.e., the amount of modified carbon fibres, gets oriented and distributed in the hydrogel matrix, thus influencing overall mechanical strength (Hong et al., 2021). Smaller-sized fillers with a higher surface area to volume ratio enable higher diffusion rates, and the geometry of the filler influences the tortuosity of the diffusion channel, hence influencing overall diffusivity. Also, the filller composition, i.e., Si-C composites, itself interacts with the hydrogel matrix, altering transport properties and, in turn, diffusivity in the composite material (Gao et al., 2020). Such above-discussed properties i.e., size effect, composition and shape collectively constituted a pivotal integral viewpoint to explain the phenomenon of swelling/shrinkage in charging/discharging (Xu et al., 2018).

The experimentally obtained data's verification and calibration of processed data from the multiphysics model is not an aspect worth neglecting in this study. Numerical theory and simulation that are backed by experimentally derived data make the model more accurate and reliable (Fagerstr¨om et al., 2023). The experimentally drawn data's correlation with such numerical theory is the method's need for verification to be up-to-date. An example of this correlational study has been carried out on the structure of material properties, with numerical simulations on the mesoscale complemented by spatial distribution from experimental tests (Fagerstr¨om et al., 2023). Using experimental data, parametric analysis aids in revealing the impact of various parameters on multifunctional performance and refines the model for better agreement with experimental observations (Hong et al., 2021; Xu et al., 2018).

18.7 MACHINE LEARNING MODEL

The modelling of hydrogel composites via the integration of machine learning into different theoretical modelling techniques holds a great prospect for better understanding, optimizing and predicting the properties of these polymeric materials (Shokrollahi et al., 2022). Machine learning also aids in the predictive process after training with bulk data. It uses types of materials and the way they are made to make designs better and perform well (Sai et al., 2020). This forecast ability allows for special use in many areas, from medical engineering to material science. In addition, the models when combined with the machine learning program simplify complex systems with numerous parameters for consideration, therefore revealing mechanisms and designs considered to be superior in design (Zobeiry & Poursartip, 2021). One significant thing that machine learning does in this area is to use generated data to make predictions (Kumar et al., 2022). By studying data from experiments, ML helps scientists find important factors affecting the performance and behaviour of hydrogel composites which further help better design plans. This allows for more informed design solutions, where insights from numerous data sources, including mechanical, geometrical, and fatigue loading stresses, improve forecast accuracy (Sai et al., 2020). Also, machine learning's skill to handle skewed data in sets (Hassanat et al., 2022) is a powerful feature. It helps make predictions better and makes the design and performance of polymeric materials more effective.

There exist literature examples of the symbiotic relationship between the theoretical modelling technique and machine learning technology. Studies from (Kek et al., 2022; Potočnik et al., 2022) investigated and evaluated the effectiveness of utilising machine learning techniques such as neural networks (NN), convolutional autoencoder (CAE), discriminant analysis (DA) in the retrieval of data and the analysis of it via the acoustic emission measurement (AE). This literature shows the robustness of the ML technique in the classification of materials and the modelling accuracy over a wide range of applications. There are other instances where this model has also been reported with success in the application, others are; the artificial neural network (ANNs), convolutional neural networks (CNN) shown in Figure 9, support vector machine (SVM) and the support vector regression (SVR). Quantities such as mechanical properties and responses to different stimuli for hydrogel composites have been reported (Kek et al., 2022; Potočnik et al., 2022).

Encouragement of a systemic approach in the development of consistently accurate hydrogel composites is important. This systematic approach which ensures a direct trajectory to accuracy involves first the collection of data, primarily or secondarily, via material 2D and 3D simulations or laboratory-generated data (Bhattacharyya et al., 2013; Omidian & Chowdhury, 2023). This is then followed by the selection of the desired feature exhibited by the material, which is then identified based on preference and utilized in the formation of ML algorithms for the modelling of material properties. Finally, the validation of the modelling results obtained from the ML algorithm ensures the results' application in the real world (Xu et al., 2023).

As excellent as this model type appears, its continuous development and advancement are required to keep it relevant. This technique is also plagued with its share of challenges and limitations, which have been contemporary issues of research. Data security, scarcity and privacy are common examples of its challenges (Ezugwu et al., 2022; S. Wang & Sun, 2023). The limited availability of data deters the training process of the machine learning program, therefore limiting its application in different areas particularly in sensitive domains such as medical sciences making the handling of generated data and its privacy a major issue (Aouedi et al., 2023). Moreover, the multifaceted nature of hydrogel composites complicates the development of accurate theoretical models, demanding more advanced algorithms capable of handling such complexity (Jiang et al., 2022).

The contemporary efforts scientifically to advance the course of this symbiotic relationship between theoretical models and machine learning algorithms fuel the prospects of this study (Wang & Sun, 2023). The foremost area of this advancement requires the transparency and explainability of designed ML models to foster understanding and reliability of the model operations (Patel, 2023). To effectively mask out biases and unequal outcomes of the ML models is of paramount interest to stakeholder scientists in the different fields of material application, this is followed suit by curbing privacy issues and heightening the security of data generated using theoretical models, experimentally or via simulations for the training of ML algorithms (Aouedi et al., 2023). This impacts and refines the Internet of Medical Things (IoMT) an example of the ML model application development in modern healthcare systems (Aouedi et al., 2023).

Figure 9. Schematic of data representation in the Deep Learning Machine learning model expressed in the (a) 3-D SAW and (b) 3D- CNN model (Self-avoiding walk network model and Convolutional Neural Network).

18.8 CONCLUSION

This chapter provides a comprehensive synthesis of theoretical frameworks and computational methodologies driving advancements in hydrogel composite research, underscoring their transformative role in bridging molecular design with macroscopic functionality. By integrating molecular dynamics (MD), coarse-grained (CG) models, Flory-Rehner theory, continuum mechanics, multiphysics simulations, and machine learning (ML), the text establishes a multiscale paradigm for predicting and optimizing hydrogel behavior. Molecular-level models elucidate atomic interactions and network dynamics, while statistical mechanics offers thermodynamic insights into swelling and elasticity. Continuum approaches, such as finite element analysis (FEA), enable macroscopic mechanical predictions, and multiphysics models capture coupled phenomena like diffusion and electrochemical responses.

The integration of ML further accelerates data-driven discovery, decoding intricate structure-property relationships and enabling high-throughput material design. However, challenges persist: MD faces temporal and force field limitations, CG models sacrifice atomic precision, and ML grapples with data scarcity and interpretability. Experimental validation remains critical to ensure model fidelity, particularly for complex phenomena like interfacial interactions and degradation. Looking ahead, advancements in computational efficiency, hybrid multiscale modeling, and explainable AI promise to overcome these barriers, fostering tailored hydrogels for biomedical, environmental, and wearable technologies. This interdisciplinary synergy not only refines predictive accuracy but also democratizes access to bespoke material innovation, positioning hydrogels as pivotal tools in addressing global challenges in healthcare, sustainability, and smart material engineering. The chapter thus serves as both a foundational guide and a catalyst for future research, advocating for continued collaboration across computational, experimental, and theoretical domains to unlock the full potential of hydrogel composites.

REFERENCES

Abalymov, A., Van der Meeren, L., Skirtach, A. G., & Parakhonskiy, B. V. (2020). Identification and Analysis of Key Parameters for the Ossification on Particle Functionalized Composites Hydrogel Materials. *ACS Applied Materials & Interfaces*, *12*(35), 38862–38872. DOI: 10.1021/acsami.0c06641 PMID: 32539334

Alziyadi, M. O., & Denton, A. R. (2021). Osmotic pressure and swelling behavior of ionic microcapsules. *The Journal of Chemical Physics*, *155*(21), 214904. Advance online publication. DOI: 10.1063/5.0064282 PMID: 34879668

Aouedi, O., Sacco, A., Piamrat, K., & Marchetto, G. (2023). Handling Privacy-Sensitive Medical Data With Federated Learning: Challenges and Future Directions. *IEEE Journal of Biomedical and Health Informatics*, *27*(2), 790–803. DOI: 10.1109/JBHI.2022.3185673 PMID: 35737624

Bhattacharyya, R., Ray, S. K., & Mandal, B. (2013). A systematic method of synthesizing composite superabsorbent hydrogels from crosslink copolymer for removal of textile dyes from water. *Journal of Industrial and Engineering Chemistry*, *19*(4), 1191–1203. DOI: 10.1016/j.jiec.2012.12.017

Borges, F. T. P., Papavasiliou, G., & Teymour, F. (2020). Characterizing the Molecular Architecture of Hydrogels and Crosslinked Polymer Networks beyond Flory–Rehner—I. Theory. *Biomacromolecules*, *21*(12), 5104–5118. DOI: 10.1021/acs.biomac.0c01256 PMID: 33253542

Casalini, T., & Perale, G. (2019). From Microscale to Macroscale: Nine Orders of Magnitude for a Comprehensive Modeling of Hydrogels for Controlled Drug Delivery. *Gels (Basel, Switzerland)*, *5*(2), 28. DOI: 10.3390/gels5020028 PMID: 31096685

Cascone, S., & Lamberti, G. (2020). Hydrogel-based commercial products for biomedical applications: A review. *International Journal of Pharmaceutics*, *118803*, 118803. Advance online publication. DOI: 10.1016/j.ijpharm.2019.118803 PMID: 31682963

Castilho, M., Hochleitner, G., Wilson, W., van Rietbergen, B., Dalton, P. D., Groll, J., Malda, J., & Ito, K. (2018). Mechanical behavior of a soft hydrogel reinforced with three-dimensional printed microfibre scaffolds. *Scientific Reports*, *8*(1), 1245. DOI: 10.1038/s41598-018-19502-y PMID: 29352189

Chen, G. (2022). Thermodynamics of hydrogels for applications in atmospheric water harvesting, evaporation, and desalination. *Physical Chemistry Chemical Physics*, *24*(20), 12329–12345. DOI: 10.1039/D2CP00356B PMID: 35545966

Chen, J., Zhou, S.-M., Ma, B.-G., Zhang, L.-M., & Yi, J.-Z. (2013). Molecular dynamics simulations on dextran hydrogels. *E-Polymers*, *13*(1), 004. Advance online publication. DOI: 10.1515/epoly-2013-0104

Chen, Z. (2020). Modeling of Kerogen Swelling by Solvents with Flory-Rehner and Regular Solution Model. Retrieved from https://api.semanticscholar.org/CorpusID:226404653

Childers, M. C., & Daggett, V. (2018). Validating Molecular Dynamics Simulations against Experimental Observables in Light of Underlying Conformational Ensembles. *The Journal of Physical Chemistry B, 122*(26), 6673–6689. DOI: 10.1021/acs.jpcb.8b02144 PMID: 29864281

Chinthala, V. S. R. K., Mulay, S. S., & Harish, A. B. (2021). Constitutive modeling of pH-sensitive hydrogel: Multi-physics coupling of electromagnetics with mechanics and thermodynamics. *Mechanics of Materials, 161*, 104002. DOI: 10.1016/j.mechmat.2021.104002

Dai, G., Zhan, L., Ma, B., Guan, C., & Huang, M. (2023). Effect of process parameters on interlaminar properties of thermoplastic composite: Molecular dynamics simulation and experimental verification. *Polymer, 280*, 126062. DOI: 10.1016/j.polymer.2023.126062

Dai, W., Sun, M., Leng, X., Hu, X., & Ao, Y. (2020). Recent Progress in 3D Printing of Elastic and High-Strength Hydrogels for the Treatment of Osteochondral and Cartilage Diseases. *Frontiers in Bioengineering and Biotechnology, 8*, 604814. Advance online publication. DOI: 10.3389/fbioe.2020.604814 PMID: 33330436

Dannert, C., Stokke, B. T., & Dias, R. S. (2019). Nanoparticle-Hydrogel Composites: From Molecular Interactions to Macroscopic Behavior. *Polymers, 11*(2), 275. DOI: 10.3390/polym11020275 PMID: 30960260

Danole, A., Kothali, B., Apte, A., Kulkarni, A., Khot, V., Patil, A., & Upadhye, S. (2018). *A Review On- Hydrogel*. American Journal of PharmTech Research., DOI: 10.46624/ajptr.2018.v8.i3.005

Denton, A. R., & Tang, Q. (2016). Counterion-induced swelling of ionic microgels. *The Journal of Chemical Physics, 145*(16), 164901. Advance online publication. DOI: 10.1063/1.4964864 PMID: 27802663

DiBenedetto, A. T. (1987). Prediction of the glass transition temperature of polymers: A model based on the principle of corresponding states. *Journal of Polymer Science. Part B, Polymer Physics, 25*(9), 1949–1969. DOI: 10.1002/polb.1987.090250914

Ezugwu, A. E., Ikotun, A. M., Oyelade, O. O., Abualigah, L., Agushaka, J. O., Eke, C. I., & Akinyelu, A. A. (2022). A comprehensive survey of clustering algorithms: State-of-the-art machine learning applications, taxonomy, challenges, and future research prospects. *Engineering Applications of Artificial Intelligence, 110*, 104743. DOI: 10.1016/j.engappai.2022.104743

Fagerstr¨om, M., Catalanotti, G., Rauter, N., Widera, I. M., & Lammering, R. (2023). NUMERICAL - EXPERIMENTAL APPROACH FOR THE CORRELATION STRUCTURE DETERMINATION OF SHORT FIBER-REINFORCED COMPOSITES. Retrieved from https://api.semanticscholar.org/CorpusID:261749771

Filipecka-Szymczyk, K., Makowska-Janusik, M., & Marczak, W. (2023). Molecular Dynamics Simulation of Hydrogels Based on Phosphorylcholine-Containing Copolymers for Soft Contact Lens Applications. *Molecules (Basel, Switzerland), 28*(18), 6562. DOI: 10.3390/molecules28186562 PMID: 37764338

Friesen, S., Kakorin, S., & Hellweg, T. (2022). Modified Flory–Rehner Theory Describes Thermotropic Swelling Transition of Smart Copolymer Microgels. *Polymers, 14*(10), 1999. DOI: 10.3390/polym14101999 PMID: 35631881

Gao, M., Li, X., Ren, C., Wang, Z., Pan, Y., & Guo, L. (2019). Construction of a Multicomponent Molecular Model of Fugu Coal for ReaxFF-MD Pyrolysis Simulation. *Energy & Fuels*, *33*(4), 2848–2858. DOI: 10.1021/acs.energyfuels.8b04434

Gao, X., Lu, W., & Xu, J. (2020). Modeling framework for multiphysics-multiscale behavior of Si–C composite anode. *Journal of Power Sources*, *449*, 227501. DOI: 10.1016/j.jpowsour.2019.227501

Gholami, K., Ege, F., & Barzegar, R. (2023a). Prediction of Composite Mechanical Properties: Integration of Deep Neural Network Methods and Finite Element Analysis. *Journal of Composites Science*, *7*(2), 54. DOI: 10.3390/jcs7020054

Gholami, K., Ege, F., & Barzegar, R. (2023b). Prediction of Composite Mechanical Properties: Integration of Deep Neural Network Methods and Finite Element Analysis. *Journal of Composites Science*, *7*(2), 54. DOI: 10.3390/jcs7020054

Hassanat, A. B., Tarawneh, A. S., Abed, S. S., Altarawneh, G. A., Alrashidi, M., & Alghamdi, M. (2022). RDPVR: Random Data Partitioning with Voting Rule for Machine Learning from Class-Imbalanced Datasets. *Electronics (Basel)*, *11*(2), 228. DOI: 10.3390/electronics11020228

Heidari, S., Esmaeilzadeh, F., Mowla, D., & Ghasemi, S. (2021). Optimization of Key Parameters Affecting Swelling Capacity of Poly(acrylamide-co-acrylic acid) Hydrogels in Salt Water Using Response Surface Methodology (RSM). *Journal of Macromolecular Science, Part B: Physics*, *60*(12), 971–988. DOI: 10.1080/00222348.2021.1932950

Heydari Dokoohaki, M., Zolghadr, A. R., Ghatee, M. H., & Klein, A. (2020). Aqueous solutions of binary ionic liquids: Insight into structure, dynamics, and interface properties by molecular dynamics simulations and DFT methods. *Physical Chemistry Chemical Physics*, *22*(47), 27882–27895. DOI: 10.1039/D0CP04303F PMID: 33284294

Hinton, M. J. (n.d.). Micromechanics Based Continuum Damage Model for Ply Failure in Unidirectional Composites Synopsis. Retrieved from https://api.semanticscholar.org/CorpusID:614306

Hong, Z., Hu, Z., Yang, R., You, J., Fu, Y., Zhou, L., & Yin, S. (2021). Multiphysics modeling framework for composite structural batteries with modified carbon fibers as electrodes. *Composites Communications*, *27*, 100853. DOI: 10.1016/j.coco.2021.100853

Hou, D., Xu, J., Zhang, Y., & Sun, G. (2019). Insights into the molecular structure and reinforcement mechanism of the hydrogel-cement nanocomposite: An experimental and molecular dynamics study. *Composites. Part B, Engineering*, *177*, 107421. DOI: 10.1016/j.compositesb.2019.107421

Hu, N., Wang, Y., Ma, R., Zhang, W., Li, B., Zhao, X., Zhang, L., & Gao, Y. (2022). Optimizing the fracture toughness of a dual cross-linked hydrogel via molecular dynamics simulation. *Physical Chemistry Chemical Physics*, *24*(29), 17605–17614. DOI: 10.1039/D2CP02478K PMID: 35829708

Jiang, Y., Li, G., Ge, H., Wang, F., Li, L., Chen, X., Lu, M., & Zhang, Y. (2022). Machine Learning and Application in Terahertz Technology: A Review on Achievements and Future Challenges. *IEEE Access : Practical Innovations, Open Solutions*, *10*, 53761–53776. DOI: 10.1109/ACCESS.2022.3174595

Jin, J., Pak, A. J., Durumeric, A. E. P., Loose, T. D., & Voth, G. A. (2022). Bottom-up Coarse-Graining: Principles and Perspectives. *Journal of Chemical Theory and Computation, 18*(10), 5759–5791. DOI: 10.1021/acs.jctc.2c00643 PMID: 36070494

Kek, T., Potočnik, P., Misson, M., Bergant, Z., Sorgente, M., Govekar, E., & Šturm, R. (2022). Characterization of Biocomposites and Glass Fiber Epoxy Composites Based on Acoustic Emission Signals, Deep Feature Extraction, and Machine Learning. *Sensors (Basel), 22*(18), 6886. DOI: 10.3390/s22186886 PMID: 36146236

Khan, P., Kaushik, R., & Jayaraj, A. (2022). Approaches and Perspective of Coarse-Grained Modeling and Simulation for Polymer–Nanoparticle Hybrid Systems. *ACS Omega, 7*(51), 47567–47586. DOI: 10.1021/acsomega.2c06248 PMID: 36591142

Kuhlmann, H., Volgers, P., & Zhang, Z. (2014). Application and CAE Simulation of Over Molded Short and Continuous Fiber Thermoplastic Composite. Retrieved from https://api.semanticscholar.org/CorpusID:198923257

Kumar, S., Singh, K. S. K., & Singh, K. K. (2022). Data-driven modeling for predicting tribo-performance of graphene-incorporated glass-fabric reinforced epoxy composites using machine learning algorithms. *Polymer Composites, 43*(9), 6599–6610. DOI: 10.1002/pc.26974

Lai, Y., & Hu, Y. (2018). Probing the swelling-dependent mechanical and transport properties of polyacrylamide hydrogels through AFM-based dynamic nanoindentation. *Soft Matter, 14*(14), 2619–2627. DOI: 10.1039/C7SM02351K PMID: 29577116

Lee, S. G., Brunello, G. F., Jang, S. S., & Bucknall, D. G. (2009). Molecular dynamics simulation study of P (VP-co-HEMA) hydrogels: Effect of water content on equilibrium structures and mechanical properties. *Biomaterials, 30*(30), 6130–6141. DOI: 10.1016/j.biomaterials.2009.07.035 PMID: 19656562

Li, X., Murthy, N. S., Becker, M. L., & Latour, R. A. (2016). Multiscale approach for the construction of equilibrated all-atom models of a poly(ethylene glycol)-based hydrogel. *Biointerphases, 11*(2), 021002. Advance online publication. DOI: 10.1116/1.4944829 PMID: 27013229

Liu, F., Wu, D., & Hong, W. (2023). Mechanism study on structural and mechanical properties of triblock copolymer hydrogels by coarse-grained molecular dynamics simulations. *Journal of Molecular Liquids, 382*, 121910. DOI: 10.1016/j.molliq.2023.121910

Liu, Y., Zhang, Z., Liang, Z., Yong, Y., Yang, C., & Li, Z. (2022). Multifunctional polyurethane hydrogel based on a phenol–carbamate network and an Fe 3+ –polyphenol coordination bond toward NIR light triggered actuators and strain sensors. *Journal of Materials Chemistry. A, Materials for Energy and Sustainability, 10*(32), 16928–16940. DOI: 10.1039/D2TA04837J

Lopez, C. G., & Richtering, W. (2017). Does Flory–Rehner theory quantitatively describe the swelling of thermoresponsive microgels? *Soft Matter, 13*(44), 8271–8280. DOI: 10.1039/C7SM01274H PMID: 29071323

Luthfianti, H. R., Waresindo, W. X., Edikresnha, D., Hapidin, D., Noor, F., Elfahmi, E., & Khairurrijal, K. (2024). Bioactive compounds-loaded polyvinyl alcohol hydrogels: Advancements in smart delivery media for biomedical applications. *Materials Research Express*, *11*(6), 062002. Advance online publication. DOI: 10.1088/2053-1591/ad4fdd

Mahmood, A., Patel, D., Hickson, B., DesRochers, J., & Hu, X. (2022). Recent Progress in Biopolymer-Based Hydrogel Materials for Biomedical Applications. *International Journal of Molecular Sciences*, *23*(3), 1415. DOI: 10.3390/ijms23031415 PMID: 35163339

Majcher, M., & Hoare, T. (2018). Applications of Hydrogels. 1–39. DOI: 10.1007/978-3-319-92066-5_17-1

Matole, V., & Digge, P. (2022). A Brief Review on Hydrogel. *Research Journal of Topical and Cosmetic Sciences*, 99–100. Advance online publication. DOI: 10.52711/2321-5844.2022.00016

Mehandzhiyski, A. Y., & Zozoulenko, I. (2021). A Review of Cellulose Coarse-Grained Models and Their Applications. *Polysaccharides*, *2*(2), 257–270. DOI: 10.3390/polysaccharides2020018

Merino, S., Martín, C., Kostarelos, K., Prato, M., & Vázquez, E. (2015). Nanocomposite Hydrogels: 3D Polymer–Nanoparticle Synergies for On-Demand Drug Delivery. *ACS Nano*, *9*(5), 4686–4697. DOI: 10.1021/acsnano.5b01433 PMID: 25938172

Mohsenzadeh, R. (2022). Development of Stress Distribution of Composite Gear Tooth Reinforced by Nano-CaCO3, using Finite Element Analysis and its Correlation with Experience. *Journal of Failure Analysis and Prevention*, *22*(4), 1495–1503. DOI: 10.1007/s11668-022-01433-6

Moreno, A., & Mendoza, M. E. (2015). Crystallization in Gels. In Handbook of Crystal Growth (pp. 1277–1315). Elsevier. DOI: 10.1016/B978-0-444-63303-3.00031-6

Murakami, T. (2015). Hydrogels in Biomedical Use. *Proceedings of the Institution of Mechanical Engineers. Part H, Journal of Engineering in Medicine*, *229*(12), 825–827. DOI: 10.1177/0954411915616528 PMID: 26614796

Müser, M. H., Li, H., & Bennewitz, R. (2019). Modeling the Contact Mechanics of Hydrogels. *Lubricants (Basel, Switzerland)*, *7*(4), 35. DOI: 10.3390/lubricants7040035

Narayanan, R. P., Procyk, J., Nandi, P., Prasad, A., Xu, Y., Poppleton, E., Williams, D., Zhang, F., Yan, H., Chiu, P.-L., Stephanopoulos, N., & Šulc, P. (2022). Coarse-Grained Simulations for the Characterization and Optimization of Hybrid Protein–DNA Nanostructures. *ACS Nano*, *16*(9), 14086–14096. DOI: 10.1021/acsnano.2c04013 PMID: 35980981

Ojo, M. O., & Zahid, A. (2022). Deep Learning in Controlled Environment Agriculture: A Review of Recent Advancements, Challenges and Prospects. *Sensors (Basel)*, *22*(20), 7965. DOI: 10.3390/s22207965 PMID: 36298316

Omidian, H., & Chowdhury, S. D. (2023). Advancements and Applications of Injectable Hydrogel Composites in Biomedical Research and Therapy. *Gels (Basel, Switzerland)*, *9*(7), 533. DOI: 10.3390/gels9070533 PMID: 37504412

Ou, X., Han, Q., Dai, H.-H., & Wang, J. (2015). Molecular dynamic simulations of the water absorbency of hydrogels. *Journal of Molecular Modeling*, *21*(9), 231. DOI: 10.1007/s00894-015-2784-0 PMID: 26271733

Pak, A. J., & Voth, G. A. (2018). Advances in coarse-grained modeling of macromolecular complexes. *Current Opinion in Structural Biology*, *52*, 119–126. DOI: 10.1016/j.sbi.2018.11.005 PMID: 30508766

Park, S.-J., Hwang, T., Jo, S., Wooh, S., Lee, H., Jung, Y., & Yoo, J. (2025). Unveiling the Diverse Principles for Developing Sprayable Hydrogels for Biomedical Applications. *Biomacromolecules*, *26*(2), 753–772. Advance online publication. DOI: 10.1021/acs.biomac.4c01312 PMID: 39760352

Patel, J. (2023). A Comprehensive Review of Innovations in Artificial Intelligence and Enhancing Machine Learning Model Precision and Efficiency. 2023 14th International Conference on Computing Communication and Networking Technologies (ICCCNT), 1–4. DOI: 10.1109/ICCCNT56998.2023.10306844

Potočnik, P., Misson, M., Šturm, R., Govekar, E., & Kek, T. (2022). Deep Feature Extraction Based on AE Signals for the Characterization of Loaded Carbon Fiber Epoxy and Glass Fiber Epoxy Composites. *Applied Sciences (Basel, Switzerland)*, *12*(4), 1867. DOI: 10.3390/app12041867

Prakash, C., & Ghosh, S. (2023). Self-consistent homogenization-based parametrically upscaled continuum damage mechanics model for composites subjected to high strain-rate loading. *Journal of Composite Materials*, *57*(4), 545–563. DOI: 10.1177/00219983221140564

Raeisi, A., & Farjadian, F. (2024). Commercial hydrogel product for drug delivery based on route of administration. *Frontiers in Chemistry*, *12*, 1336717. Advance online publication. DOI: 10.3389/fchem.2024.1336717 PMID: 38476651

Rebelo, R., Barbosa, A. I., Caballero, D., Kwon, I. K., Oliveira, J. M., Kundu, S. C., Reis, R. L., & Correlo, V. M. (2019). 3D biosensors in advanced medical diagnostics of high mortality diseases. *Biosensors & Bioelectronics*, *130*, 20–39. DOI: 10.1016/j.bios.2018.12.057 PMID: 30716590

Rodriques, S. G. (2016). Swelling of Hydrogels. Retrieved from https://api.semanticscholar.org/CorpusID:245331538

Ryatt, J., & Ramulu, M. (2022). Prediction of tensile failure of stochastic tow-based discontinuous composites via mesoscale finite element analysis. *Composite Structures*, *279*, 114769. DOI: 10.1016/j.compstruct.2021.114769

Sai, W., Chai, G. B., & Srikanth, N. (2020). Fatigue Life Prediction of GLARE Composites Using Regression Tree Ensemble-Based Machine Learning Model. *Advanced Theory and Simulations*, *3*(6), 2000048. Advance online publication. DOI: 10.1002/adts.202000048

Shojaei, A., Li, G., Fish, J., & Tan, P. J. (2014). Multi-scale constitutive modeling of Ceramic Matrix Composites by Continuum Damage Mechanics. *International Journal of Solids and Structures*, *51*(23–24), 4068–4081. DOI: 10.1016/j.ijsolstr.2014.07.026

Shokrollahi, Y., Dong, P., Gamage, P. T., Patrawalla, N., Kishore, V., Mozafari, H., & Gu, L. (2022). Finite Element-Based Machine Learning Model for Predicting the Mechanical Properties of Composite Hydrogels. *Applied Sciences (Basel, Switzerland)*, *12*(21), 10835. DOI: 10.3390/app122110835

Shreedhar, R. (2012). Analysis of T-beam Bridge Using Finite Element Method. Retrieved from https://api.semanticscholar.org/CorpusID:56114110

Shukur, M. F. A., Makhijani, S., Ingle, R., & Saifee, M. (2025). Natural Hydrogel-based Drug Delivery System: A Global Scenario, Current Development, and Future Prospective. *Current Drug Delivery*, *22*(10), 1376–1392. Advance online publication. DOI: 10.2174/0115672018320746241101052039 PMID: 39781715

Singh, N., & Li, W. (2019). Recent Advances in Coarse-Grained Models for Biomolecules and Their Applications. *International Journal of Molecular Sciences*, *20*(15), 3774. DOI: 10.3390/ijms20153774 PMID: 31375023

Singh, S. (2018). Advanced Hydrogels for Biomedical Applications. *Biomedical Journal of Scientific & Technical Research*, *5*(1). Advance online publication. DOI: 10.26717/BJSTR.2018.05.001144

Singhal, A., Schneible, J. D., Lilova, R. L., Hall, C. K., Menegatti, S., & Grafmüller, A. (2020). A multiscale coarse-grained model to predict the molecular architecture and drug transport properties of modified chitosan hydrogels. *Soft Matter*, *16*(47), 10591–10610. DOI: 10.1039/D0SM01243B PMID: 33156313

Solanki, D., Yadav, P., Soni, A., & Jaiswal, R. (2023). *HYDROGEL- A REVIEW. International Journal of Pharma Professional's Research.* IJPPR., DOI: 10.48165/ijppronline.2023.14112

Tanimoto, R., Ebara, M., & Uto, K. (2023). Tunable enzymatically degradable hydrogels for controlled cargo release with dynamic mechanical properties. *Soft Matter*, *19*(33), 6224–6233. DOI: 10.1039/D3SM00475A PMID: 37493066

Turan, E., Demirel, G., Zengin, M., Şimsek, D., & Çaykara, T. (2006). PH-dependent swelling behavior and network parametes of ionic poly(N - t -butylacrylamide- co -acrylamide) hydrogels. *Journal of Applied Polymer Science*, *102*(2), 1624–1630. DOI: 10.1002/app.23945

Utech, S., & Boccaccini, A. R. (2016). A review of hydrogel-based composites for biomedical applications: Enhancement of hydrogel properties by addition of rigid inorganic fillers. *Journal of Materials Science*, *51*(1), 271–310. DOI: 10.1007/s10853-015-9382-5

van Leeuwen, R. A. (2018). A Multiphysics Numerical Framework for Epoxy Resins: Investigating Hygrothermal Ageing in Laminated Composites. Retrieved from https://api.semanticscholar.org/CorpusID:139687327

Völkerink, O., Koord, J., Petersen, E., & Hühne, C. (2023). Comparison of Continuum Shell and Solid Element-Based Modeling Strategies for Mesoscale Progressive Damage Analysis of Fiber Composites. *Mechanics of Composite Materials*, *59*(2), 219–238. DOI: 10.1007/s11029-023-10093-y

Wang, J., Ma, Z., Wang, Y., Shao, J., & Yan, L. (2021). Ultra-Stretchable, Self-Healing, Conductive, and Transparent PAA/DES Ionic Gel. *Macromolecular Rapid Communications*, *42*(2), 2000445. Advance online publication. DOI: 10.1002/marc.202000445 PMID: 33191561

Wang, M., Zhu, H., & Shen, J. (2021). Synthesis and molecular dynamics simulation of CuS@ GO–CS hydrogel for enhanced photothermal antibacterial effect. *New Journal of Chemistry*, *45*(15), 6895–6903. DOI: 10.1039/D0NJ05891B

Wang, S., & Sun, Z. (2023). Hydrogel and Machine Learning for Soft Robots' Sensing and Signal Processing: A Review. *Journal of Bionic Engineering*, *20*(3), 845–857. DOI: 10.1007/s42235-022-00320-y PMID: 36466727

Wang, Y., Li, X., Wei, Q., Chai, W., Yang, M., & Wei, S. (2014). Molecular Dynamics Simulation of Mechanical Properties for Poly(vinyl pyrrolidone)/Poly(vinyl alcohol) Hydrogel. *Asian Journal of Chemistry*, *26*(17), 5378–5382. DOI: 10.14233/ajchem.2014.18116

Wong, C. Y., Wong, W. Y., Liu, L., Shibutani, Y., & Loh, K. S. (2021). Molecular dynamic simulation approach to understand the physical and proton transport properties of chitosan/sulfonated Poly(Vinyl alcohol) composite membranes. *Polymer*, *217*, 123458. DOI: 10.1016/j.polymer.2021.123458

Wu, L., Mao, G., Nian, G., Xiang, Y., Qian, J., & Qu, S. (2018). Mechanical characterization and modeling of sponge-reinforced hydrogel composites under compression. *Soft Matter*, *14*(21), 4355–4363. DOI: 10.1039/C8SM00678D PMID: 29767186

Wu, T., Li, H., & Deng, Z. (2019). A review on recent development of theoretical modeling of hydrogel phase behavior subject to mechanics and multiphysics coupled effects. *Mechanics of Soft Materials*, *1*(1), 11. DOI: 10.1007/s42558-019-0011-3

Xiao, Y., Friis, E. A., Gehrke, S. H., & Detamore, M. S. (2013). Mechanical Testing of Hydrogels in Cartilage Tissue Engineering: Beyond the Compressive Modulus. *Tissue Engineering. Part B, Reviews*, *19*(5), 403–412. DOI: 10.1089/ten.teb.2012.0461 PMID: 23448091

Xu, J., Lindbergh, G., & Varna, J. (2018). Multiphysics modeling of mechanical and electrochemical phenomena in structural composites for energy storage: Single carbon fiber micro-battery. *Journal of Reinforced Plastics and Composites*, *37*(10), 701–715. DOI: 10.1177/0731684418760207

Xu, P., Ji, X., Li, M., & Lu, W. (2023). Small data machine learning in materials science. *npj Computational Materials*, *9*(1), 42. DOI: 10.1038/s41524-023-01000-z

Yuan, M., Zhao, H., Liu, S., Ren, H., Zhang, B., Sun, X., & Chen, J. (2022). Determination of cohesive parameters for fibre-reinforced composite interfaces based on finite element analysis and machine learning. *Journal of Composite Materials*, *56*(27), 4113–4122. DOI: 10.1177/00219983221127400

Zhao, Z., Vizetto-Duarte, C., Moay, Z. K., Setyawati, M. I., Rakshit, M., Kathawala, M. H., & Ng, K. W. (2020). Composite Hydrogels in Three-Dimensional in vitro Models. *Frontiers in Bioengineering and Biotechnology*, *8*, 611. Advance online publication. DOI: 10.3389/fbioe.2020.00611 PMID: 32656197

Zheng, M., Li, X., Nie, F., & Guo, L. (2017). Investigation of model scale effects on coal pyrolysis using ReaxFF MD simulation. *Molecular Simulation*, *43*(13–16), 1081–1088. DOI: 10.1080/08927022.2017.1356456

Zhou, H., Xu, Q., Nie, Z., & Li, N. (2022). A Study on Using Image-Based Machine Learning Methods to Develop Surrogate Models of Stamp Forming Simulations. *Journal of Manufacturing Science and Engineering*, *144*(2), 021012. Advance online publication. DOI: 10.1115/1.4051604

Zobeiry, N., & Poursartip, A. (2021). THEORY-GUIDED MACHINE LEARNING FOR PROCESS SIMULATION OF ADVANCED COMPOSITES A PREPRINT.

About the Contributors

Onome Ejeromedoghene is a post-doctoral fellow at Soochow University, Suzhou, China. He obtained his Ph.D. degree in Materials Physics and Chemistry in 2023 from Southeast University, Nanjing, China. He has authored and co-authored over 50 peer-reviewed articles in international scholarly journal of high impact factor. He has also served as reviewer for many reputable journals like ACS Applied Materials and Interfaces, Chemical Engineering Journal, Separation and Purification Technology, Journal of Applied Polymer Science, Macromolecular Research etc. His current research focus is on the development of biofunctional and stimuli-responsive inorganic-polymeric soft materials (hydrogel composites) as smart materials for biological applications, digital technologies, and photochromic devices.

Guodong Fu is a distinguished professor and faculty member at the School of Chemistry and Chemical Engineering, Southeast University. Renowned for his expertise in polymer science, he specializes in polymerization, polymeric materials, and polymer synthesis. His research encompasses nanomaterials, polymeric biomaterials, and polymer nanoscience, with a focus on innovative applications like hydrogel actuators. His advancements in polymer processing and engineering have made him a leading figure in polymer research, significantly contributing to both academic and industrial progress in polymer technology.

Olayinka Kehinde Oderinde is an Assistant Professor at Nile University of Nigeria, Abuja, Nigeria. He is a PhD graduate of Chemical Engineering and Technology from the prestigious Southeast University, Nanjing City, PR China. His area of expertise revolves around functional nanomaterials synthesis and characterization, especially in photoresponsive inorganic-organic hybrids, hydrogel, metal-organic frameworks, and air pollution assessment and monitoring. His awards include Chinese Government Scholarship (CSC) (2015-2020) and Top Downloaded Paper 2018-2019 in Polymers for Advanced Technology by John Wiley & Sons. Inc. (2020) and Royal Society of Chemistry (RSC) Travel Grant to attend RSC Chemical Nanoscience and Nanotechnology Network Annual Symposium (2023). He is a member of the Royal Society of Chemistry (RSC, UK) with membership ID 731933 and also a founding member of the Organization of African Academic Doctors (OAAD). Dr. Oderinde is a Reviewer for several top journals and also a Co-Guest Editor for Fermentation Journal.

***

Aqeel Ahmad earned a Ph.D. in chemical engineering from Universiti Teknologi PETRONAS (UTP), Malaysia. Currently, he is serving as a post-doctoral fellow at King Fahd University of Petroleum and Minerals (KFUPM), Saudi Arabia. His research primarily focuses on the synthesis and development of novel adsorbent materials aimed at environmental remediation, addressing challenges such as water purification and pollutant removal. With a strong background in chemical engineering and material science, he is dedicated to advancing sustainable solutions for environmental protection.

Zaheer Ahmad is a distinguished researcher in the field of chemistry, specializing in Organic Synthesis and Medicinal Chemistry. With over 15 years of academic and research experience, he has made significant contributions to Natural Product Chemistry, Medicinal Chemistry and Nanochemistry. Dr.Ahmad learned his Ph.D. in Chemistry from University of Karachi, Karachi-Pakistan in 2009, focusing on Natural Product Chemistry. Following his doctoral studies, he completed a research work at university of Karachi, Karachi where he worked on Phytochemistry. Currently, he serves as a Associate Professor at Department of Chemistry, University of Wah, Wah Cantt where he leads a research group investigating Medicinal Chemistry, Phytochemistry and Hydrogel Materials. His work has been published in numerous high-impact journals, and he has been recognized with several prestigious awards, including different Projects. In addition to research, Professor Dr. Ahmad is dedicated to teaching and mentoring students at both undergraduate and graduate levels. He has supervised 29 MS and 5 PhD students are in progress, many of whom have gone on to successful careers in academia and industry.

Muhammad Ajmal graduated from Quaid-i-Azam University Islamabad Pakistan and currently serving as Assistant Professor of Chemistry at department of Chemistry University of Education Lahore Pakistan. He is working on hydrogels from the last 13 years and published 50 research articles in international journals.

Arooba Binte Amir is a dedicated researcher and educator with a strong background in chemistry, having specialization in inorganic chemistry.With a Bachelor's degree in chemistry, Arooba has developed a keen interest in the field of hydrogels, exploring their potential applications in various scientific and industrial contexts.Her research focuses on synthesis, characterization and application of hydrogels,partricularly emphasis on enhancing functional properties of hydrogels, used in biomedical and environmental fields. Her work aims to contribute to the development of novel hydrogel-based materials and technologies. She has taught various chemistry and science courses, developing a passion for sharing knowledge and inspring young scientists.She aims to contribute to advancement of scientific knowledge and inspire future researchers.

Victor Ngozi Enwemiwe is an Entomologist and Research Assistant, Department of Animal and Environmental Biology, Delta State University, Abraka, Nigeria. Enwemiwe is a PhD holder in the same Institution. He holds a Master of Science in Zoology (Entomology Option) from the University of Ibadan, Ibadan, Nigeria and a Bachelor Degree from the Department of Animal and Environmental Biology, Delta State University, Abraka. He was one of the best graduating students of the 2011/2012 academic session with a Second Class Upper Honours. Till date, Enwemiwe is an Entomology Technician for Delta State Entomological Surveillance Project, National Malaria Elimination & Control, sponsored by Global Fund and have served as Supervisor for Blackfly

Entomology for NTD (Onchocerciasis) in Delta North Senatorial District, Delta State by Carter Center, Nigeria. He has reviewed for a number of journals; International Journal of Tropical Insect Science, BMJ Open Ecology Journal, PLoS Neglected Tropical Disease, and EcoHealth. He is equally a member of Entomological Society of Nigeria, Zoological society of Nigeria, International Society of Plant Protection and Royal Society of Tropical Medicine and Hygiene, London. Enwemiwe have published over 40 research articles in International, National and Local journals. His research interest includes but not limited to Stored Product, Forest, Aquatic, Urban, Medical and Molecular Entomology.

Moses Kumi is currently a Research Assistant in the Research Institute for Intelligent Wearable Systems, School of Fashion and Textiles, The Hong Kong Polytechnic University,Hong Kong, under the supervision of Profs. Tao Xiaoming and Pu Junhong. His current research focuses on the development of Electronically Controllable Fibrous Actuators for Soft Robotics and Magnetic Textiles for Smart Wearables

Shazia Naheed is Working as an Assistant Professor at the Department of Chemistry, University of Wah. She completed PhD in Chemistry from Bahauddin Zakariya University Multan in 2016. Her research interests include Synthesis, Characterization and Applications of Organic-Inorganic Hybrid Materials.

Frederick Nii Ofei Bruce, a distinguished scholar in Chemical Engineering and Aerospace Sciences, earned his Bachelor of Science degree with honors from Kwame Nkrumah University of Science and Technology (KNUST), Ghana. During his undergraduate tenure, Bruce conducted pioneering research in product formulation, chemical plant design, organic synthesis, energy management, waste disposal, and cosmetic development with institutions such as the IDIN network and MIT. Expanding his expertise, Bruce pursued a Master's degree in Aviation and Astronavigation Science and Technology at the Beijing Institute of Technology, focusing on interfacial fluid dynamics and 3D modeling of low Reynolds number fluid flows. He is a PhD Postgraduate Researcher at Northwestern Polytechnical University studying Aeronautical and Science and Technology within the School of Astronautics, delving into combustion chemistry of alternative aviation fuels, flame retardants, and Li-ion battery electrolytes. He addresses energy sustainability challenges by employing quantum calculations, machine learning, molecular dynamics, and chemical kinetic modeling. He is also the research lead for the multinational health company Quanbio, which focuses on developing well-advanced algorithms for the early detection and screening of cardiovascular diseases. Bruce doubles as a creative writer with several published novels, plays, and other genres of creative arts and a serial entrepreneur in many areas of green food and cosmetics production, energy sustainability, real estate, and education. Bruce is a research fellow at the Nesvard Institute of Molecular Sciences, affiliated with the South African Department of Science and Innovation (DSI), South African National Research Network (SANReN), and Data Intensive Research Infrastructure of South Africa (DIRISA). His research contributions, published in esteemed journals, underscore his eminence in Chemical Engineering and Aerospace Sciences.

Adebayo Ponle Oduwole is a recent graduate with a Bachelor's degree in Chemistry] from Lagos State University. With a strong interest in inorganic and materials chemistry, he focused on the synthesis and characterization of metal complexes during his undergraduate studies. Passionate about renewable energy and catalysis, Adebayo Ponle Oduwole aims to pursue graduate research in electrocatalysis, specifically the development of transition metal-based complexes for hydrogen production. He is eager to contribute to innovative research that addresses global energy and environmental challenges.

Shephrah Olubusola Ogungbesan, a Doctor of Material Physics and Chemistry, has her research focus on the fabrication of biocompatible, photoresponsive nanomaterials for applications in the biomedical field. Her areas of expertise encompass both organic and inorganic synthesis, as well as nanocomposites, intelligent hydrogels, and other functionalized polymeric matrices. Among her most recent endeavors is the development of an antibacterial, recyclable photochromic nanohybrid film. She has had the opportunity to publish her findings in peer-reviewed journals and has presented her work at various international conferences. Strong communication skills and a collaborative spirit fuel her passion for furthering research in the use of nanomaterials in healthcare and sustainable applications.

Peter Olusakin Oladoye is currently a doctoral researcher at Florida International University, Miami, FL. He obtained his B.Sc in Industrial Chemistry and M.Sc in Analytical Chemistry in 2010 and 2014, respectively, from University of Ibadan, Ibadan, Nigeria. He was a faculty member at Ladoke Akintola University of Technology, Ogbomoso, Nigeria between 2019 and 2021. His research focuses on valorization of biomass into useful materials for wastewater treatment and energy sources. Additionally, he is interested in understanding the biogeochemical cycling of mercury in the environment. He is currently a member of Editorial board of Journal of Trace Element and Minerals (Elsevier) and Desalination and Wastewater Treatment (Elsevier).

Adefarati Oloruntoba is a Postdoctoral Associate at the University of Calgary, Canada, where he has been working on process intensification and life cycle assessment of existing and emerging energy technologies and systems since 2023. He received his PhD in Power Engineering and Thermophysics in 2023, with a focus on process intensification. Prior to his current role, Adefarati completed his doctoral studies at China University of Petroleum Beijing. He also holds a Master's degree in Energy and Environment from the University of Leeds, UK. His professional experience includes working as a process engineer at the Bioresource Development Centre, Nigeria, for over 5 years, where he focused on biomass valorisation. Dr Oloruntoba is passionate about driving innovative, industry-leading advancements in cleaner energy and the environment. He has authored approximately 20 peer-reviewed articles and serves as a Guest Editor for the Processes journal. His research contributions have been recognized with several awards, including the 2018 Hargreaves Research Project award.

Olusola Muritala Adeniyi is a Nigerian Inorganic Chemist, born and bred in Ibadan,Oyo State and he's known for his research in bioinorganic drugs, metal coordination, catalysis, computational chemistry, and environmental studies. He earned his Bachelor's degree, Master's degree and Doctor of Philosophy in Inorganic Chemistry from the prestigious Federal University of Agriculture, Abeokuta (FUNAAB) in Nigeria. He has collaborated with both local and international researchers.

Ahmed Olalekan Omoniyi is a materials scientist specializing in the design and functionalization of sustainable polymer materials to address global challenges in security, energy, and environmental sustainability. He earned his Ph.D. in Materials Science and Engineering from Changchun University of Science and Technology, China, and has authored more than 25 SCI articles

Elizabeth Omotola is an enthusiastic Chemistry teacher, a seasoned academic writer, and a dedicated researcher.

Ivan Oyege was a Fulbright Student from Uganda. He has a PhD in Earth Systems Science (Agroecology) from Florida International University, Miami, Florida, USA. He also has a Master of Science in Chemistry from Makerere University, Kampala, Uganda, and a Bachelor of Science and Education (Chemistry - major and Mathematics - minor) from Busitema University, Tororo, Uganda

Index

C

D

E

F

G

H

I

M

N

P

R

S

T

W

www.ingramcontent.com/pod-product-compliance
Lightning Source LLC
LaVergne TN
LVHW062340300926
849161LV00007B/1768